ELECTROMAGNETISM AND LINEAR CIRCUITS

THE MODERN UNIVERSITY PHYSICS SERIES

This series is intended for readers whose main interest is in physics, or who need the methods of physics in the study of science and technology. Some of the books will provide a sound treatment of topics essential in any physics training, while other, more advanced, volumes will be suitable as preliminary reading for research in the field covered. New titles will be added from time to time.

LITTLEFIELD and THORLEY: *Atomic and Nuclear Physics* (2nd Edn)
PERINA: *Coherence of Light*
BAGGULEY: *Electromagnetism and Linear Circuits*

ELECTROMAGNETISM AND LINEAR CIRCUITS

D. M. S. BAGGULEY

Clarendon Laboratory,
Department of Physics, University of Oxford
and
Brasenose College, Oxford

VAN NOSTRAND REINHOLD COMPANY
LONDON
NEW YORK CINCINNATI TORONTO MELBOURNE

VAN NOSTRAND REINHOLD COMPANY LTD.
25–28 Buckingham Gate, London, SW1E 6LQ

INTERNATIONAL OFFICES
New York Cincinnati Toronto Melbourne

First published 1973

Library of Congress Catalog Card No. 72-5918

ISBN 0 442 00492 3

Printed in Great Britain by
Butler and Tanner Ltd, Frome and London

Preface

This book has been written for use by students who are taking an honours degree course in physics or engineering (having in mind the 'experimental' rather than the 'theoretical' option) and for those who teach these courses. The material content of the book is mostly at advanced undergraduate level and is intended for reading rather than for reference. I have therefore tried to produce a text which is stimulating, imaginative and instructive. The difficulties inherent in such a project are well known. Stimulating and imaginative texts are not necessarily rigorous and accurate. I may have aimed at perfection in these respects, but it is unlikely that I shall have achieved so satisfying a result. My errors will be immediately obvious to my friends. Should they suddenly acquire an uncharacteristic reticence I shall depend upon other, less-inhibited, colleagues to bring to my notice the shortcomings of my endeavour. In any case I shall certainly welcome suggestions for improving this book from anyone who is seriously interested in increasing its usefulness.

Two significant and personal choices have been made in the writing of this book. The first and most important is the selection of topics. This is a choice for which I offer no apology. It is my own selection of those topics which I think are instructive, interesting and relevant. The second choice concerns units. It seems to be common form for an author to apologize for whatever system of units has been chosen in the writing of a book on electromagnetism. The reader will observe that, for the most part, I have used c.g.s. units. The decision to use this system was taken after careful consideration and against the advice of some of my colleagues whose judgement I respect. I accept full responsibility for this decision, but since I do not believe that there is an absolute 'better' or 'worse' in this matter I should explain that the decision was taken simply because, in my view, this system of units is the more convenient and improves the clarity of this particular textbook. If a reader would prefer an alternative set of units then by all means let him (or her) modify where necessary those passages which offend his (or her) taste. I regard units on roughly the same level as sign conventions and I am prepared to allow my colleagues considerable flexibility in selecting a system which is relevant to the project they are undertaking. My own decision was taken having in mind those other works, listed in the bibliographies, which a student may wish to consult.

I am particularly concerned that what I have written should be an encouragement to students and should stimulate their interest in the subject matter of this book. If I have succeeded to some small extent in this endeavour then I should acknowledge my indebtedness to all my teachers and colleagues who have so patiently encouraged my own devotion to physics. In particular I should mention Prof. B. Bleaney, Prof. R. Loudon, Prof. N.

Kurti, the late Prof. H. M. Last and Dr J. H. E. Griffiths. But I should also explain that the responsibility for what I have chosen to write is mine alone: for that I must be judged.

Oxford, May 1972 D. M. S. BAGGULEY

Contents

Useful Constants

	Symbol	S.I. unit	c.g.s. unit
1. Speed of light in vacuum	c	$2{\cdot}998 \times 10^{8}$ m s^{-1}	$2{\cdot}998 \times 10^{10}$ cm s^{-1}
2. Electron charge	$\lvert e\rvert$	$1{\cdot}602 \times 10^{-19}$ C	$4{\cdot}803 \times 10^{-10}$ e.s.u.
3. Planck's constant	h	$6{\cdot}626 \times 10^{-34}$ J s	$6{\cdot}626 \times 10^{-27}$ erg s
4. h-bar	$\hbar$	$1{\cdot}054 \times 10^{-34}$ J s	$1{\cdot}054 \times 10^{-27}$ erg s
5. Rest mass of electron	m_e	$9{\cdot}110 \times 10^{-31}$ kg	$0{\cdot}911 \times 10^{-27}$ g
6. Rest mass of proton	M_p	$1{\cdot}672 \times 10^{-27}$ kg	$1{\cdot}672 \times 10^{-24}$ g
7. $\lvert e\rvert/m$, electron	$\lvert e\rvert/m$	$1{\cdot}759 \times 10^{11}$ C kg^{-1}	$5{\cdot}272 \times 10^{7}$ e.s.u. g^{-1}
8. Avogadro Number	N_A	$6{\cdot}023 \times 10^{23}$ mol^{-1}	$6{\cdot}023 \times 10^{23}$ mol^{-1}
9. Boltzmann Constant	k	$1{\cdot}381 \times 10^{-23}$ J K^{-1}	$1{\cdot}381 \times 10^{-16}$ erg K^{-1}
10. Electronvolt	eV	$1{\cdot}602 \times 10^{-19}$ J	$1{\cdot}602 \times 10^{-12}$ erg
11. Wavelength corresponding to 1 eV		$12{\cdot}4 \times 10^{-3}$ nm	$1{\cdot}240 \times 10^{-4}$ cm
12. Frequency corresponding to 1 eV		$2{\cdot}418 \times 10^{14}$ s^{-1}	$2{\cdot}418 \times 10^{14}$ s^{-1}
13. Wavenumber corresponding to 1 eV		$8{\cdot}067 \times 10^{5}$ m^{-1}	8067 cm^{-1}
14. Bohr magneton	β	$9{\cdot}274 \times 10^{-24}$ A m^{2}(J T^{-1})	$0{\cdot}927 \times 10^{-20}$ erg G^{-1}

In S.I. units m = metre, kg = kilogramme, s = second, C = coulomb, A = ampere, J = joule, K = degrees Kelvin, T = tesla, H = henry, F = farad.

Permeability of a vacuum $\mu_0 = 4\pi \times 10^{-7}$ kg m s^{-2} A^{-2}(H m^{-1})

Permittivity of a vacuum $\varepsilon_0 = 8{\cdot}854 \times 10^{-12}$ kg^{-1} m^{-3} s^{4} A^{2}(F m^{-1})

Conversion Tables

Quantity	*c.g.s. Unit*	*Fundamental Relations*	*Practical Unit*	*Relation of Practical Unit to c.g.s. Unit*
1. Momentum, $\mathbf{p}$	g-cm/s	$\mathbf{p} = m\mathbf{v}$ ($\mathbf{v}$ = velocity)		
2. Force, $\mathbf{F}$,	dyne (= g-cm/s^2)	$\mathbf{F} = \frac{d}{dt}\mathbf{p}$		
3. Energy, W,	erg (= g-cm^2/s^2)	$W = \int \mathbf{F}.d\mathbf{s}$	joule	1 joule = 10^7 ergs
4. Power, P,	erg/s	$P = \frac{dW}{dt}$	watt	1 watt = 10^7 ergs/s
5. Charge, q,	e.s.u.	$\mathbf{F} = \left(\frac{q_1 q_2}{r^3}\right)\mathbf{r}$	coulomb	1 coulomb = $2{\cdot}998 \times 10^9$ e.s.u. = 10^{-1} e.m.u.
6. Charge density, ρ,	e.s.u./cm^3	$q = \int \rho\, dv$ (v = volume)		
7. Current density, $\mathbf{J}$,	e.m.u./cm^{2}s (1 e.m.u. = c e.s.u.)	$\mathbf{J} = \frac{1}{c}\rho\mathbf{v}$		
8. Current, I,	e.m.u./s (1 e.m.u. = c e.s.u.)	$I = \int_s \mathbf{J}.\mathbf{n}\, dS$ (S = area)	ampere	1 ampere = $2{\cdot}998 \times 10^9$ e.s.u./s = 10^{-1} e.m.u./s
9. Electric field, $\mathbf{E}$,	statvolt/cm (= dyne/e.s.u.)	$\mathbf{E} = \lim_{q \to 0} (\mathbf{F}_s/q)$	volt/cm	
10. Electric potential, ϕ,	statvolt (= erg/e.s.u.)	$\mathbf{E} = -\nabla\phi$, $\phi = -\int \mathbf{E}.d\mathbf{l}$ (static)	volt	1 volt = $\frac{1}{299{\cdot}8}$ statvolt = 10^8 e.m.u. of voltage
11. Magnetic induction, $\mathbf{B}$,	gauss (= dyne/e.s.u.)	$\mathbf{F} = \int \mathbf{J} \wedge \mathbf{B}\, dv$		
12. Magnetic field, $\mathbf{H}$,	oersted = gauss	$\nabla \wedge \mathbf{H} = 4\pi\mathbf{J}$ (static)		
13. Electrical conductivity, σ,	s^{-1}	$\mathbf{J} = \frac{1}{c}\sigma\mathbf{E}$	(ohm-cm)$^{-1}$	
14. Electrical resistance, R,	s/cm (ϕ, I, in e.s.u.)	$\phi = RI$	ohm	1 ohm = $1{\cdot}113 \times 10^{-12}$ e.s.u. = 10^9 e.m.u.
15. Capacitance, C,	cm (ϕ, I, in e.s.u.)	$\phi = C^{-1}D^{-1}I$	farad	1 farad = $0{\cdot}899 \times 10^{12}$ e.s.u. = 10^{-9} e.m.u.
16. Inductance, L,	s^2/cm (ϕ, I, in e.s.u.)	$\phi = LDI$	henry	1 henry = $1{\cdot}113 \times 10^{-12}$ e.s.u. = 10^9 e.m.u.

For a clear exposition of the different types of units see BLEANEY, B., and BLEANEY, B. I., *Electricity and Magnetism*, 2nd edn, O.U.P., 1965, or KIP, A. F., *Electricity and Magnetism*, McGraw-Hill, 1962.

Chapter 1

Electromagnetic Field Equations

1.1 Electromagnetic Field Equations

In this chapter we shall summarize the formal description of macroscopic electric and magnetic phenomena taking Maxwell's equations for an electromagnetic field as the basis for the discussion. We shall not question the validity of these equations but will proceed immediately to examine some of their implications. Our real interest is to indicate the ways in which the equations for the field, when taken with appropriate limiting conditions, can provide a description of experimental observations. For this purpose it is not necessary to concentrate too narrowly on a general and fully rigorous mathematical formulation of the theory, we may use physical arguments to provide ideal limiting conditions, to simplify the discussion, and also to 'bend' the mathematics so that we can see more easily how the theory tries to give an adequate description of particular electric and magnetic phenomena. On the other hand, we are trying to set out a reasonably unified description of these phenomena and this does require a somewhat abstract and formal presentation.

An electromagnetic field is formally described by the following vector equations

$$\nabla \wedge \mathbf{E} + \frac{1}{c}\frac{\partial \mathbf{B}}{\partial t} = 0 \tag{1.1.1}$$

$$\nabla \wedge \mathbf{H} - \frac{1}{c}\frac{\partial \mathbf{D}}{\partial t} - 4\pi \mathbf{J} = 0 \tag{1.1.2}$$

$$u = \frac{1}{4\pi}\int_0^{\mathbf{D}} \mathbf{E}.\mathrm{d}\mathbf{D} \tag{1.1.3}$$

$$w = \frac{1}{4\pi}\int_0^{\mathbf{B}} \mathbf{H}.\mathrm{d}\mathbf{B} \tag{1.1.4}$$

$$\mathbf{S} = \frac{c}{4\pi}(\mathbf{E} \wedge \mathbf{H}) \tag{1.1.5}$$

The vectors are given specific names, **E** is the electric field vector, **H** the magnetic field vector, **D** the electric displacement, **B** the magnetic induction and **J** the electric current density. The terms u and w represent the electric energy per unit volume and the magnetic energy per unit volume, and 4π and c are numerical constants which are determined from experimental observations when the measurements are carried out in c.g.s. units. The constant c has dimensions of velocity. The vector **S** (the Poynting Vector) represents the instantaneous rate at which electromagnetic energy is crossing a unit area normal to the direction $(\mathbf{E} \wedge \mathbf{H})$.

We generally assume that the vectors which describe the electromagnetic field are finite throughout the domain of interest and are continuous functions of position and time, with continuous derivatives. A particular region may be bounded by a surface across which there is a discontinuous (very rapid) change in the field vectors; this will indicate an abrupt change in the physical properties of the medium.

The electric and magnetic energy terms u, w, refer to work done (or energy supplied) by external sources in creating the field distributions **E**, **D**, and **H**, **B**. These terms are well defined for lossless media where the physical processes are reversible. For isothermal processes the increase in the Helmholtz Free Energy, F, for a region is given by

$$\Delta F_{\mathrm{e}} = \frac{1}{4\pi}\int_V \mathrm{d}v \int_0^{\mathbf{D}} \mathbf{E}.\mathrm{d}\mathbf{D} \tag{1.1.3a}$$

$$\Delta F_{\mathrm{m}} = \frac{1}{4\pi}\int_V \mathrm{d}v \int_0^{\mathbf{B}} \mathbf{H}.\mathrm{d}\mathbf{B} \tag{1.1.4a}$$

and we now regard the electromagnetic field as part of the complete thermodynamic system. When energy dissipation takes place, however, the processes are no longer reversible and thermodynamic relations must be applied with some caution. We may still identify the work done (or energy supplied) by external sources as

$$\Delta W_{\mathrm{e}} = \frac{1}{4\pi}\int_V \mathrm{d}v \int_0^{\mathbf{D}} \mathbf{E}.\mathrm{d}\mathbf{D} \tag{1.1.3b}$$

$$\Delta W_{\mathrm{m}} = \frac{1}{4\pi}\int_V \mathrm{d}v \int \mathbf{H}.\mathrm{d}\mathbf{B} \tag{1.1.4b}$$

but it is probably more satisfactory to discuss these relations in terms of the internal energy function U from the First Law of Thermodynamics. For example we may consider a medium in which energy dissipation occurs and carry through a cycle for which the field vectors **E**, **D**, are zero both at the beginning and the termination of the cycle. There will be no net energy stored in the electric field as a result of this process but energy will have been supplied by the sources and dissipated as heat. If the temperature of the medium is maintained constant the internal energy of the medium will be unchanged and so,

$$\Delta U = \Delta Q + \Delta W_{\mathrm{e}} = 0$$

To maintain the temperature constant we must therefore supply an amount of heat from another source (or sink)

$$\Delta Q = -\Delta W_{\mathrm{e}}$$

and so the energy dissipated in the medium as heat (the Joule heating) is

$$\Delta Q_J = -\Delta Q = \Delta W_{\mathrm{e}} = \frac{1}{4\pi}\int \mathrm{d}v \oint \mathbf{E}.\mathrm{d}\mathbf{D} \tag{1.1.3c}$$

An electric charge, q, may be recognized by the force it exerts on another charge. We assert that the law of force between two charges, q_1, q_2 separated by a distance, $\mathbf{r}$, in free space, is

$$\mathbf{F} = \left(\frac{q_1 q_2}{r^3}\right)\mathbf{r} \tag{1.1.6}$$

The presence of electric and magnetic fields may also be recognized by the force exerted on an electric charge.

$$\mathbf{F} = q\left(\mathbf{E} + \frac{1}{c}\mathbf{v} \wedge \mathbf{B}\right) \tag{1.1.7}$$

where $\mathbf{v}$ is the velocity of the charge q. Evidently the force on a stationary charge, $\mathbf{F}_s$, may be used to measure the intensity of an electric field by the relation

$$\mathbf{E} = \operatorname*{limit}_{q\to 0} (\mathbf{F}_s/q) \tag{1.1.8}$$

Here the limit $q \longrightarrow 0$ is chosen in order that the test charge should not itself modify the field which is being measured.

If a net charge Δq is contained within a volume element Δv, the charge density ρ at any point within is defined by the relation

$$\Delta q = \rho\, \Delta v \tag{1.1.9}$$

The charge density at a point is therefore the average charge per unit volume in the neighbourhood of that point. An ordered motion of this charge density is assumed to give rise to the current density $\mathbf{J}$. Such an ordered motion can be characterized by a vector field which specifies at each point the velocity $\mathbf{v}$ of the moving charges. The vector $\mathbf{v}$ defines a set of stream lines which are everywhere tangential to the flow of charge, and we define the current density by the relation

$$\mathbf{J} = \frac{1}{c}\rho\mathbf{v} \tag{1.1.10}$$

Evidently $\mathbf{J}$ is directed along the stream line at every point and is equal in magnitude to the charge (scaled by the numerical factor c) which in unit time crosses unit area normal to the stream line.

The presence of a magnetic induction field $\mathbf{B}$ may be recognized by the force exerted on a volume element of current. From equations (1.1.7) and (1.1.10) we have

$$\mathrm{d}\mathbf{F} = \mathbf{J} \wedge \mathbf{B}\, \mathrm{d}v \tag{1.1.11}$$

We note that whereas the force exerted by an electric field on an element of charge is directed along the vector $\mathbf{E}$ (equation (1.1.7)), the force on a volume element of current in a magnetic field is normal to the plane defined by the vectors $\mathbf{J}$ and $\mathbf{B}$.

The integral of $\mathbf{J}$ over a closed surface must measure the net rate of loss of charge from the region enclosed by the surface.

$$\int_S \mathbf{J}.\mathbf{n}\, \mathrm{d}S = -\frac{1}{c}\frac{\mathrm{d}}{\mathrm{d}t}\int_V \rho\, \mathrm{d}v \tag{1.1.12}$$

Here V is the volume enclosed by the surface S. If the surface is fixed in space, the charge density must be a function of the time coordinate and so

$$\int_V \left(\nabla.\mathbf{J} + \frac{1}{c}\frac{\partial\rho}{\partial t}\right) \mathrm{d}v = 0 \tag{1.1.13}$$

The integrand of this equation is, by assumption, a continuous function of the coordinates and so there must be some regions within which the integrand does not change sign. If therefore the integral is to vanish for *all* arbitrary regions specified by V then the integrand must be identically zero and we obtain the equation of continuity which expresses the conservation of charge

$$\nabla . \mathbf{J} + \frac{1}{c}\frac{\partial \rho}{\partial t} = 0 \tag{1.1.14}$$

The electric current I is the rate at which charge crosses a surface. If $\mathbf{n}$ is a vector drawn normal to a small element ΔS of a surface, as in Figs. 1.1(a) and (b),

$$\Delta I = \mathbf{J}.\mathbf{n}\,\Delta S \tag{1.1.15}$$

and

$$I = \int_S \mathbf{J}.\mathbf{n}\,\mathrm{d}S \tag{1.1.16}$$

Hence I is really a scalar quantity representing the *total* flow of charge through a surface, but in some applications it is convenient, when $\mathbf{J}$ has a well-defined direction over a surface element ΔS to visualize the current as itself a vector quantity. This vector is rigorously $\mathbf{J}\,\Delta S$. Again if the current is confined to a linear conductor of sufficiently small cross-section the current stream lines may be considered parallel to an element of the conductor of length $\mathrm{d}\mathbf{l}$. The force acting on the current element in a magnetic induction field $\mathbf{B}$ is then obtained from equation (1.1.11),

$$\mathrm{d}\mathbf{F} = I\,\mathrm{d}\mathbf{l} \wedge \mathbf{B} \tag{1.1.17}$$

There may be particular regions where the charge density is constant in time. For these regions, from equation (1.1.14)

$$\nabla . \mathbf{J} = 0 \tag{1.1.18}$$

and so

$$\oint \mathbf{J}.\mathbf{n}\,\mathrm{d}S = 0 \tag{1.1.19}$$

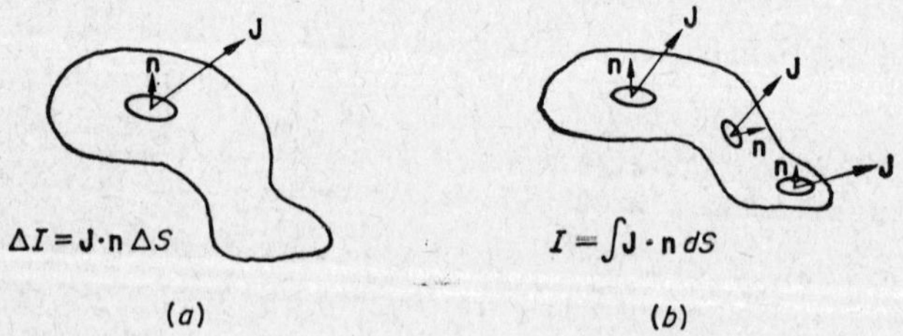

FIG. 1.1. The current density is a vector quantity, the current is a scalar quantity.

where the integral is taken over the surface enclosing the region of interest. In this case the current flowing into the region through the bounding surface must at all times be equal to the current flowing outward. This is the underlying basis of Kirchhoff's current law used in the analysis of electrical networks. This law states that the algebraic sum of all the currents flowing into a junction must be zero.

There are two further relations, satisfied by the vectors **B** and **D**, which are of fundamental importance. The divergence of the curl of any vector vanishes identically and so, from equation (1.1.1)

$$\nabla \cdot \frac{\partial \mathbf{B}}{\partial t} = \frac{\partial}{\partial t} \nabla \cdot \mathbf{B} = 0 \tag{1.1.20}$$

It follows from this equation that at every point in the field $\nabla \cdot \mathbf{B}$ is a constant which does not change with time, and so it is reasonable to take

$$\nabla \cdot \mathbf{B} = 0 \tag{1.1.21}$$

Similarly, from equation (1.1.2) we obtain

$$\frac{1}{c} \frac{\partial}{\partial t} \nabla \cdot \mathbf{D} + 4\pi \nabla \cdot \mathbf{J} = 0 \tag{1.1.22}$$

so that, using equation (1.1.14)

$$\frac{\partial}{\partial t}(\nabla \cdot \mathbf{D} - 4\pi\rho) = 0 \tag{1.1.23}$$

which again we may interpret as

$$\nabla \cdot \mathbf{D} = 4\pi\rho \tag{1.1.24}$$

The statements in equations (1.1.21) and (1.1.24) are equivalent to the assertion that the electromagnetic field is derived only from the fundamental sources, current density **J**, and charge density ρ.

Note that, if equation (1.1.24) is integrated throughout a volume V,

$$\int_V \nabla \cdot \mathbf{D} \, dv = 4\pi \int_V \rho \, dv \tag{1.1.25}$$

But by Green's transformation

$$\int_V \nabla \cdot \mathbf{D} \, dv = \oint_S \mathbf{D} \cdot \mathbf{n} \, dS \tag{1.1.26}$$

where S is the closed surface bounding V.
Hence

$$\oint_S \mathbf{D} \cdot \mathbf{n} \, dS = 4\pi \int_V \rho \, dv \tag{1.1.27}$$

This is Gauss' law.

1.2 General Properties of Material Media

The basic equations for the electromagnetic field set out in equations (1.1.1)–(1.1.5) involve five fundamental vectors **D**, **E**, **B**, **H** and **J**, but they do not provide definitive relations between these vectors. It is necessary therefore to introduce some additional relations. The justification for these relations will lie in their ability to account for a body of experimental phenomena.

We shall assume that **D** is a function of **E** only and that **B** is a function of **H** only. In 'free space' we shall assume that **D** reduces to **E** and that **B** reduces to **H**; here equations (1.1.21) and (1.1.24) may be written

$$\nabla \cdot \mathbf{H} = 0 \tag{1.2.1}$$

$$\nabla \cdot \mathbf{E} = 4\pi\rho \tag{1.2.2}$$

and the field equations (1.1.1)–(1.1.4) are modified in a corresponding manner. We are evidently also implying by these statements that other bounded regions of space exist for which the electromagnetic properties are recognizably different from 'free space'. These regions are called 'material media'.

Three limiting classes of material media may be identified. *Dielectric media* are regions in which **D** is recognizably different from **E**. *Magnetic media* are regions where **B** is recognizably different from **H**. *Conducting media* are regions in which **J** need not necessarily be zero. (The converse of this last statement identifies insulating media as regions in which **J** is always zero.) In practice, a particular material will possess, to some degree, a combination of dielectric magnetic and conducting properties. However, it is useful to simplify the discussion and consider the ideal cases where electric, magnetic and current phenomena can be treated separately. The justification for this simplification derives essentially from the time invariant form of the electromagnetic field equations but we may expect it to continue to be a reasonable assumption provided that the time variation is sufficiently slow.

It can be seen from equations (1.1.1), (1.1.2), (1.1.21) and (1.1.24) that the electrostatic and magnetostatic fields are solutions of independent pairs of differential equations

$$\nabla \wedge \mathbf{E} = 0 \qquad \nabla . \mathbf{D} = 4\pi\rho \tag{1.2.3}$$

$$\nabla \wedge \mathbf{H} = 4\pi\mathbf{J} \qquad \nabla . \mathbf{B} = 0 \tag{1.2.4}$$

For the time invariant case, therefore, the electric and magnetic vector fields may be superposed without interaction.

In the following discussion of the different classes of material media we shall be considering properties based essentially upon these time invariant equations. Phenomena which specifically involve the complete set of coupled differential equations will be discussed in Chapter 2.

1.2.1 The Equations of Poisson and Laplace

A further development of equations (1.2.3) is possible when **D** is linearly related to **E**. For a homogeneous linear, isotropic material we may write

$$\mathbf{D} = \varepsilon\mathbf{E} \tag{1.2.5}$$

where ε is a constant called the dielectric constant (compare equation (1.3.18)) which is equal to unity in free space.

Hence

$$\nabla . \mathbf{E} = \frac{4\pi\rho}{\varepsilon} \tag{1.2.6}$$

In addition, since $\nabla \wedge \mathbf{E} = 0$ everywhere, we may also write

$$\mathbf{E} = -\nabla\, \phi(x, y, z) \tag{1.2.7}$$

where $\phi(x, y, z)$ is a single valued scalar function called the electrostatic potential. Combining equations (1.2.6) and (1.2.7) we obtain

$$\nabla^2 \phi(x, y, z) = -\frac{4\pi}{\varepsilon}\rho(x, y, z) \tag{1.2.8}$$

This is Poisson's equation.

Whenever $\rho = 0$, that is for all regions with no free charge, equation (1.2.8) reduces to

$$\nabla^2 \phi(x, y, z) = 0 \tag{1.2.9}$$

This is Laplace's equation.

When there are no macroscopic currents present equations (1.2.4) give

$$\nabla \wedge \mathbf{H} = 0 \qquad \nabla . \mathbf{H} = 0 \tag{1.2.10}$$

for a homogeneous linear, isotropic material where **B** is linearly related to **H**. For this case we may define a magnetostatic potential $\phi^*(x, y, z)$ such that

$$\mathbf{H} = -\nabla \phi^*(x, y, z)$$
$$\nabla^2 \phi^*(x, y, z) = 0 \tag{1.2.11}$$

and the magnetostatic description is mathematically equivalent to the electrostatic description. However, this particular technique does not find such a wide application because in the majority of magnetic problems a source current is present.

When a source current is present the potential function $\phi^*(x, y, z)$ will still exist outside the current carrying regions but it will no longer be single valued. We can see how this arises by considering the integral of **H** around a closed contour which is itself everywhere in a region for which $\mathbf{J} = 0$ but encloses another region for which $\mathbf{J} \neq 0$. This situation is illustrated in Fig. 1.2.

$$\oint_c \mathbf{H}.\mathrm{d}\mathbf{l} = \int_{S_1+S_2} \nabla \wedge \mathbf{H}.\mathbf{n}\, \mathrm{d}S = 4\pi \int_{S_1} \mathbf{J}.\mathbf{n}\, \mathrm{d}S = 4\pi I = \Delta\phi^* \tag{1.2.12}$$

Hence the value of ϕ^* at a point may be increased by an amount $4\pi NI$ by performing N complete circuits about the current I.

When source currents are present in the magnetostatic problem it is frequently useful to define a magnetic vector potential **A**, such that

$$\mathbf{B} = \nabla \wedge \mathbf{A} \tag{1.2.13}$$

The vector equation $\nabla . \mathbf{B} = 0$ is then automatically satisfied. In a homogeneous isotropic medium, with **B** linearly related to **H** by the equation

$$\mathbf{B} = \mu \mathbf{H} \tag{1.2.14}$$

the equation for the magnetostatic vector potential in rectangular coordinates (see equation (2.11.10) page 99) may be written

$$\nabla^2 \mathbf{A} = -4\pi\mu \mathbf{J} \tag{1.2.15}$$

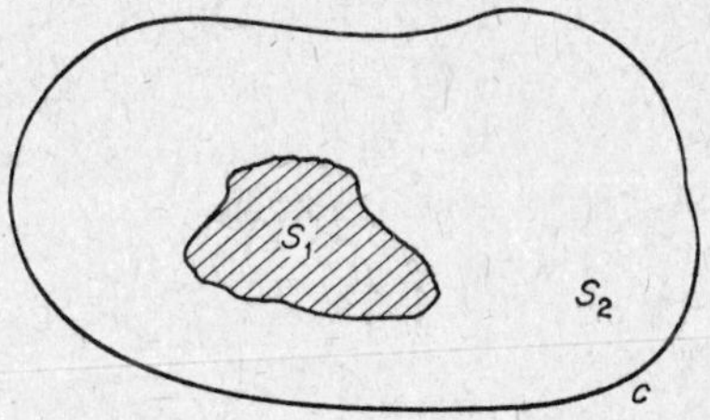

FIG. 1.2 Contour for evaluating $\oint_c H.\mathrm{d}\mathbf{l}$ in equation (1.2.12). The shaded area S_1 is a cross-section of the current-bearing conductor, S_2 is the region where no current is flowing.

In component form this equation is mathematically equivalent to Poisson's equation above. Note that in equation (1.2.14) the constant μ is the magnetostatic permeability (see page 38) and that we have not derived equation (1.2.15). The derivation of equation (1.2.15) is carried through in Chapter 2, Section 2.11.

1.2.2. Example Using the Equations of Poisson and Laplace

As an illustration of the use of equations (1.2.8) and (1.2.9) we calculate the potential function for a spherical charge distribution in free space, radius a, with a uniform charge density ρ_0.

Inside the charge distribution we have

$$\nabla^2 \phi_i = -4\pi\rho_0 \qquad r \leqslant a \tag{1.2.16}$$

whereas outside the charge distribution

$$\nabla^2 \phi_0 = 0 \qquad r \geqslant a \tag{1.2.17}$$

Since there is spherical symmetry, write $\nabla^2 \phi$ in spherical polar coordinates and retain only the term in r. From equation (1.2.16)

$$\frac{1}{r^2}\frac{\partial}{\partial r}\left(r^2 \frac{\partial \phi_i}{\partial r}\right) = -4\pi\rho_0 \qquad r \leqslant a \tag{1.2.18}$$

$$r^2 \frac{\partial \phi_i}{\partial r} = C - \frac{4\pi}{3}\rho_0 r^3 \tag{1.2.19}$$

and $C = 0$ since $(\partial\phi_i/\partial r) = 0$ at $r = 0$ from Gauss' law, equation (1.1.27).

Hence

$$\phi_i = B - \frac{2\pi}{3}\rho_0 r^2 \qquad r \leqslant a \tag{1.2.20}$$

Outside the spherical charge distribution

$$\frac{\partial}{\partial r}\left(r^2 \frac{\partial \phi_0}{\partial r}\right) = 0$$

$$\frac{\partial \phi_0}{\partial r} = \frac{C'}{r^2} = -E_r \qquad r \geqslant a \tag{1.2.21}$$

where E_r is the radial component of the electrostatic field.

Hence

$$\phi_0 = -\frac{C'}{r} \tag{1.2.22}$$

and the integration constant is zero because we take $\phi_0 = 0$ at $r = \infty$.

To evaluate the constant C' integrate $\mathbf{E}.\mathbf{n}$ over the surface of a sphere surrounding the charge distribution, that is, taking $r > a$

$$\oint \mathbf{E}.\mathbf{n}\,\mathrm{d}S = \int \nabla.\mathbf{E}\,\mathrm{d}v = 4\pi \int \rho_0\,\mathrm{d}v \tag{1.2.23}$$

Hence

$$E_r = \frac{\int \rho_0\,\mathrm{d}v}{r^2} \tag{1.2.24}$$

and from equation (1.2.21)

$$C' = -\frac{4\pi}{3}\rho_0 a^3 \tag{1.2.25}$$

The exterior potential function is, therefore,

$$\phi_0 = \frac{(4\pi/3)\rho_0 a^3}{r} \qquad r \geqslant a \tag{1.2.26}$$

We may now evaluate the constant B in equation (1.2.20). At the surface of the charge distribution the interior and exterior potential functions are equal.

$$\phi_i(a) = \phi_0(a) \qquad r = a \tag{1.2.27}$$

From equations (1.2.20) and (1.2.26)

$$B = 2\pi\rho_0 a^2 \tag{1.2.28}$$

and so the interior potential function is

$$\phi_i = \frac{2\pi}{3}\rho_0(3a^2 - r^2) \qquad r \leqslant a \tag{1.2.29}$$

1.2.3 Electrostatic Energy Relations

The potential energy of an arbitrary charge distribution with volume density ρ and surface density σ may be evaluated by observing that the work done in moving a charge q from a point where the electrostatic potential is zero to a point where its value is $\phi(x, y, z)$, is given by

$$\Delta U = \phi(x, y, z)q \tag{1.2.30}$$

This result derives from

$$\Delta U = -q\int \mathbf{E}.\mathbf{dl} = q\int \nabla\phi.\mathbf{dl} \tag{1.2.31}$$

Since the work done in assembling the charge distribution is independent of the particular order in which the charges are brought together we may discuss any particular process which can be conveniently calculated. Suppose that, at any instant, all the charge densities have the same fraction α of their final values. For final values $\rho(x, y, z)$ and $\sigma(x, y, z)$ the instantaneous values during assembly will be $\alpha\,\rho(x, y, z)$ and $\alpha\,\sigma(x, y, z)$. Since all the charges are at the same fraction of their final values the potential distribution will also be at that fraction of its final value $\phi(x, y, z)$. That is, the corresponding instantaneous distribution of potential is $\alpha\,\phi(x, y, z)$.†

Consider the work done in bringing up charges to provide an increment $\delta\rho = \rho(x, y, z)\,\delta\alpha$ and $\delta\sigma = \sigma(x, y, z)\,\delta\alpha$ in the charge densities. The work involved is

$$\Delta U = \int_V \alpha\,\phi(x, y, z)\,\rho(x, y, z)\,\delta\alpha\,\mathrm{d}v + \int_S \alpha\,\phi(x, y, z)\,\sigma(x, y, z)\,\delta\alpha\,\mathrm{d}S \tag{1.2.32}$$

† This is evident from equation (1.2.8), Poisson's equation.

where V denotes the volumes occupied by the charge distribution ρ and S represents the surfaces for which σ is non-zero. The total work done in assembling the final charge distribution is $\int \Delta U$ from $\alpha = 0$ to $\alpha = 1$. Hence

$$\int \Delta U = \int_0^1 \alpha\, \delta\alpha \int_V \rho(x, y, z)\, \phi(x, y, z)\, \mathrm{d}v + \int_0^1 \alpha\, \delta\alpha \int_S \sigma(x, y, z)\, \phi(x, y, z)\, \mathrm{d}S \quad (1.2.33)$$

that is

$$U = \frac{1}{2}\int_V \rho\phi\, \mathrm{d}v + \frac{1}{2}\int_S \sigma\phi\, \mathrm{d}S \quad (1.2.34)$$

This equation may be rewritten concisely in terms of the field vectors **D** and **E**, when we consider explicitly the properties of a free charge distribution only. The distinction between 'free' and 'bound' charges will be discussed in Section 1.3, page 18. For convenience assume that the surface density σ resides only on conductors.† From equation (1.1.24) we may write

$$\frac{1}{2}\int_V \rho\phi\, \mathrm{d}v = \frac{1}{8\pi}\int \phi\nabla . \mathbf{D}\, \mathrm{d}v \quad (1.2.35)$$

but

$$\nabla . (\phi\mathbf{D}) = \phi\nabla . \mathbf{D} + \mathbf{D} . \nabla\phi \quad (1.2.36)$$

Hence

$$\frac{1}{2}\int_V \rho\phi\, \mathrm{d}v = \frac{1}{8\pi}\int_V \mathbf{D}.\mathbf{E}\, \mathrm{d}v + \frac{1}{8\pi}\int_{S+S'} \phi\mathbf{D}.\mathbf{n}\, \mathrm{d}S \quad (1.2.37)$$

The surface integral is to be evaluated over the entire surface bounding V. It consists in part of the surface S which bounds the conductors and also of S' which surrounds the complete system and may for convenience be taken at infinity. The integral over the infinite surface is zero for any finite charge distribution and so the surface integral reduces to an integral over the conductors S. From Gauss' law (equation (1.1.27))

$$\frac{1}{4\pi}\int \mathbf{D}.\mathbf{n}\, \mathrm{d}S = -\int \sigma\, \mathrm{d}S \quad (1.2.38)$$

where the negative sign arises because the normal vector, **n**, is now directed into the conductor. This equation is true for all elements of surface (as can be seen by considering a small cylinder with its axis normal to the surface) and so

$$\frac{1}{4\pi}\mathbf{D}.\mathbf{n} = -\sigma \quad (1.2.39)$$

and

$$\frac{1}{8\pi}\int_{S+S'} \phi\mathbf{D}.\mathbf{n}\, \mathrm{d}S = \frac{1}{8\pi}\int_S \phi\mathbf{D}.\mathbf{n}\, \mathrm{d}S = -\frac{1}{2}\int_S \sigma\phi\, \mathrm{d}S \quad (1.2.40)$$

† This is not a restriction since it is possible to imagine any free charge surface density on dielectrics as spread out slightly and then included in the volume density.

Hence, from equations (1.2.37) and (1.2.40)

$$\frac{1}{2}\int_V \rho\phi \, dv + \frac{1}{2}\int_S \sigma\phi \, dS = \frac{1}{8\pi}\int_V \mathbf{E}.\mathbf{D} \, dv \tag{1.2.41}$$

or, from equation (1.2.34)

$$U_{\text{free}} = \frac{1}{8\pi}\int \mathbf{E}.\mathbf{D} \, dv \tag{1.2.42}$$

the integral being taken over all space except that occupied by conductors.

In equation (1.2.42) the free charges represent the sources which give rise to the field distribution **E**, **D**. The energy required to produce this field distribution may therefore also be derived from equation (1.1.3)

$$U_{\text{free}} = \frac{1}{4\pi}\int dv \int \mathbf{E}.d\mathbf{D} \tag{1.2.43}$$

and so

$$U_{\text{free}} = \frac{1}{8\pi}\int \mathbf{E}.\mathbf{D} \, dv$$

when **D** is linearly related to **E**. This assumption was implicit in the linear relation between ρ and ϕ which was used in writing down equation (1.2.32).

We may also consider separately the energy associated with the bound charge distribution which characterizes a dielectric material. This energy will represent that part of the total energy, supplied by the external sources, which may be regarded as being localized within the dielectric itself. It is equal to the work done in assembling the system of bound charges to form the polarized dielectric.

The bound charges are distributed continuously with charge density ρ' given by equation (1.3.5),

$$\rho' = -\nabla.\mathbf{P} \tag{1.2.44}$$

Using the same method of calculation as for the free charges, we obtain

$$U_{\text{bound}} = \frac{1}{2}\int \rho'\phi \, dv = -\frac{1}{2}\int \phi\nabla.\mathbf{P} \, dv$$

or

$$U_{\text{bound}} = -\frac{1}{2}\int_{\text{dielectric}} \mathbf{E}.\mathbf{P} \, dv \tag{1.2.45}$$

This integral need only be evaluated throughout the dielectric because **P** is zero outside. It represents the energy localized in the dielectric as distinct from the field. It is the variation of this energy with respect to a small displacement that leads to mechanical forces on polarized dielectrics.

Equation (1.2.45) does not describe the energy of a permanent dipole in an external field. For the permanent dipole the charge configuration is independent of the external field and so the linear relation between ρ and ϕ is no longer valid. For this case we obtain directly from equation (1.2.31)

$$U_{\text{dipole}} = q\frac{\partial\phi}{\partial l}\,\delta l = -\mathbf{p}.\mathbf{E} \tag{1.2.46}$$

where **p** is the dipole moment.

1.2.4 Magnetostatic Energy Relations

The energy relations for magnetostatic systems are similar in form to those for the electrostatic systems discussed in the previous sub-section. However, the derivation of these relations must start from equations involving the source currents, **J**, since the field equations do not require the existence of magnetic 'charges' or poles.

Consider a small change in a given current distribution, **J**, which produces a corresponding change in the magnetic field distribution. From equation (1.1.1), the change in magnetic field will give rise to an electric field such that

$$\nabla \wedge \mathbf{E} = -\frac{1}{c}\frac{\partial \mathbf{B}}{\partial t} \tag{1.2.47}$$

and this field **E** will of itself tend to alter **J**. In a time Δt, these electric fields will do work on the currents **J**, to the amount

$$\Delta W' = \Delta t \int c\mathbf{J}.\mathbf{E}\,\mathrm{d}v \tag{1.2.48}$$

If the source currents are to be maintained unaffected by the magnetic field the source e.m.f.s must provide an additional electric field configuration which exactly compensates **E**. Hence the sources must provide an additional field $-\mathbf{E}$ and supply an additional amount of work in time Δt,

$$\begin{aligned}\Delta W_{\mathrm{m}} &= -\Delta t \int c\mathbf{J}.\mathbf{E}\,\mathrm{d}v \\ &= -\frac{\Delta t}{4\pi}c \int (\nabla \wedge \mathbf{H}).\mathbf{E}\,\mathrm{d}v \end{aligned} \tag{1.2.49}$$

when the displacement current in equation (1.1.2) is neglected. This work is evidently work done against the 'back e.m.f.s' and is specifically associated with the establishment of the magnetic field configuration. It is therefore regarded as giving rise to an amount of energy ΔW_{m} stored in the magnetic field.

Using the vector relation

$$\nabla.(\mathbf{a} \wedge \mathbf{b}) = \mathbf{b}.\nabla \wedge \mathbf{a} - \mathbf{a}.\nabla \wedge \mathbf{b} \tag{1.2.50}$$

we obtain

$$\Delta W_{\mathrm{m}} = \frac{\Delta t}{4\pi}c \int \{\nabla.(\mathbf{E} \wedge \mathbf{H}) - \mathbf{H}.\nabla \wedge \mathbf{E}\}\,\mathrm{d}v \tag{1.2.51}$$

This integral is to be taken over all space. The first term in the integrand may be transformed to an integral over an infinitely distant surface and is zero; the second term involves

$$\Delta t(\nabla \wedge \mathbf{E}) = -\frac{1}{c}\frac{\partial \mathbf{B}}{dt}\Delta t = -\frac{1}{c}\Delta \mathbf{B}$$

so that equation (1.2.51) becomes

$$\Delta W_{\mathrm{m}} = \frac{1}{4\pi}\int (\mathbf{H}.\mathrm{d}\mathbf{B})\,\mathrm{d}v \tag{1.2.52}$$

integrated over all space. This equation may be compared with equation (1.1.4). It should be noted that equation (1.2.52) is explicitly the field energy associated with establishing a magnetic field distribution by source currents, **J**; it does not include the self-energy of permanent magnets.

When there is a linear relation between **B** and **H**, equation (1.2.52) gives

$$W_{\mathrm{m}} = \frac{1}{8\pi}\int \mathbf{H}.\mathbf{B}\,\mathrm{d}v \tag{1.2.53}$$

which is the analogue of the electrostatic energy relation given in equation (1.2.42).

Equation (1.2.53) may be written in an alternative form making use of the magnetic vector potential **A**. From equation (1.2.13)

$$\mathbf{B} = \nabla \wedge \mathbf{A}$$

and so

$$W_{\mathrm{m}} = \frac{1}{8\pi}\int \mathbf{H}.(\nabla \wedge \mathbf{A})\,\mathrm{d}v \tag{1.2.54}$$

Using the vector relation from equation (1.2.50) and dropping the surface integral as before, we obtain

$$W_{\mathrm{m}} = \frac{1}{2}\int \mathbf{J}.\mathbf{A}\,\mathrm{d}v \tag{1.2.55}$$

which is now a volume integral over the source currents.

Suppose that the field distribution arises from a set of source current filaments $\mathbf{J}_k$. Since the field equations are linear we may regard the resultant fields **H**, **B**, as the superposition of fields $\mathbf{H}_k$, $\mathbf{B}_k$ arising from the individual current filaments. We may imagine a process for creating the final fields **H**, **B**, in which the current filaments are initially open circuited with no currents flowing and are then switched on in succession. Switching on the first filament will give a contribution to the magnetic energy

$$W_{\mathrm{m},1} = \frac{1}{8\pi}\int \mathbf{H}_1.\mathbf{B}_1\,\mathrm{d}v = \frac{1}{2}\int \mathbf{J}_1.\mathbf{A}_1\,\mathrm{d}v \tag{1.2.56}$$

Switching on the second current filament gives an additional contribution to the energy

$$W_{\mathrm{m},2} = \frac{1}{8\pi}\int \mathbf{H}_2.\mathbf{B}_2\,\mathrm{d}v + \frac{1}{4\pi}\int \mathbf{H}_1.\mathbf{B}_2\,\mathrm{d}v \tag{1.2.57}$$

$$= \frac{1}{2}\int \mathbf{J}_2.\mathbf{A}_2\,\mathrm{d}v + \int \mathbf{J}_1.\mathbf{A}_2\,\mathrm{d}v \tag{1.2.58}$$

In these equations the second term on the right-hand side arises from the e.m.f. induced in circuit 1 by the changing field distribution due to circuit 2. There is no longer the factor $\frac{1}{2}$ in this term because $\mathbf{B}_2$ does not depend upon $\mathbf{H}_1$, and so equation (1.2.52) can be integrated directly.

Continuing the process of switching on the current filaments, we see that the total magnetic energy is given by

$$W_{\mathrm{m}} = W_{\mathrm{m},1} + W_{\mathrm{m},2} + \cdots W_{\mathrm{m},k} + W_{\mathrm{m},l}$$

$$= \frac{1}{8\pi}\sum_{k}\int \mathbf{H}_k.\mathbf{B}_k\, dv + \frac{1}{4\pi}\sum_{\substack{k,l\\ l>k}}\int \mathbf{H}_k.\mathbf{B}_l\, dv \qquad (1.2.59)$$

$$= \frac{1}{2}\sum_{k}\int \mathbf{J}_k.\mathbf{A}_k\, dv + \sum_{\substack{k,l\\ l>k}}\int \mathbf{J}_k.\mathbf{A}_l\, dv \qquad (1.2.60)$$

Note that equation (1.2.60) may also be written as

$$W_{\mathrm{m}} = \frac{1}{2}\sum_{k}\sum_{l}\int \mathbf{J}_k.\mathbf{A}_l\, dv$$

When the current distribution over the filament is given, equation (1.2.60) can be written in terms of the currents I_k, I_l, rather than the current densities **J**. Making use of the fact that the vector potential $\mathbf{A}_k$ is linearly related to the source current I_k, the integrations can be carried through formally to give,

$$W_{\mathrm{m}} = \frac{1}{2}\sum_{k} L_{k,k}I_k^2 + \sum_{\substack{k,l\\ l>k}} L_{k,l}I_kI_l \qquad (1.2.61)$$

The coefficient $L_{k,k}$ is called the self-inductance of kth circuit and the coefficient $L_{k,l}$ is called the mutual inductance of the circuits k and l. Evidently we may regard the terms in $L_{k,k}$ as representing the self-energies of the different circuits whilst the terms in $L_{k,l}$ represent mutual energies and these will be related to the mutual forces between the circuits.

The condition that $W_{\mathrm{m}} \geqslant 0$ for all values of I_k, I_l, requires†

$$L_{k,k} \geqslant 0$$

and

$$L_{k,k}L_{l,l} \geqslant L_{k,l}^2 \qquad (1.2.62)$$

1.2.5 Energy Associated with a Magnetic Material in a Magnetostatic Field

In order to determine the energy of a magnetized material in a magnetic field we calculate the difference in field energy with and without the material. When there is no magnetic material present, let a field distribution $\mathbf{H}_0$ be established by source currents **J**. The energy of the field is obtained from equation (1.2.53),

$$W_0 = \frac{1}{8\pi}\int_{\text{all space}} \mathbf{H}_0.\mathbf{H}_0\, dv \qquad (1.2.63)$$

Now reduce all the source currents to zero, introduce the magnetic material (assumed to be unmagnetized) and then return the source currents to their original value **J**. The material will now become magnetized and there will be

† See for example the discussion of equation (6.6.6) in Chapter 6.

a new field distribution **H**. Assuming that the relation between **B** and **H** for the material is linear, the new magnetostatic energy is

$$W = \frac{1}{8\pi} \int_{\text{all space}} \mathbf{H}.\mathbf{B}\, dv \tag{1.2.64}$$

The difference in energy $W - W_0$ arises from the presence of the magnetic material

$$W - W_0 = \frac{1}{8\pi} \int_{\text{all space}} (\mathbf{H}.\mathbf{B} - \mathbf{H}_0.\mathbf{H}_0)\, dv \tag{1.2.65}$$

which may be written in the form

$$W - W_0 = \frac{1}{8\pi} \int_{\text{all space}} \{(\mathbf{H} - \mathbf{H}_0).\mathbf{B} + (\mathbf{H} - \mathbf{H}_0).\mathbf{H}_0 + (\mathbf{B} - \mathbf{H}).\mathbf{H}_0\}\, dv \tag{1.2.66}$$

Introduce the vector potential $\mathbf{A}_0$ for the initial field distribution $\mathbf{H}_0$, and the vector potential **A** for the final induction field **B**

$$\begin{aligned} \mathbf{H}_0 &= \nabla \wedge \mathbf{A}_0 \\ \mathbf{B} &= \nabla \wedge \mathbf{A} \end{aligned} \tag{1.2.67}$$

In addition note that since the initial and final source current distributions, **J**, are identical

$$\nabla \wedge (\mathbf{H} - \mathbf{H}_0) = 0 \tag{1.2.68}$$

Using equations (1.2.67), we obtain from equation (1.2.66)

$$W - W_0 = \frac{1}{8\pi} \int_{\text{all space}} \{(\mathbf{H} - \mathbf{H}_0).\nabla \wedge \mathbf{A} + (\mathbf{H} - \mathbf{H}_0).\nabla \wedge \mathbf{A}_0 + (\mathbf{B} - \mathbf{H}).\mathbf{H}_0\}\, dv \tag{1.2.69}$$

The vector relation given in equation (1.2.50), when applied to the first two terms in this integral gives rise to terms involving $\nabla.\{(\mathbf{H} - \mathbf{H}_0) \wedge \mathbf{A}\}$ which are seen to be zero when transformed to surface integrals over an infinitely distant surface and to terms involving $\nabla \wedge (\mathbf{H} - \mathbf{H}_0)$ which are zero because of equation (1.2.68). Hence the first two terms in equation (1.2.69) are zero and so

$$W - W_0 = \frac{1}{8\pi} \int_{\text{all space}} (\mathbf{B} - \mathbf{H}).\mathbf{H}_0\, dv \tag{1.2.70}$$

But $\mathbf{B} = \mathbf{H} + 4\pi\mathbf{M}$ and the vector field **M** exists only within the magnetized material (see Section 1.4).
Hence

$$W - W_0 = \frac{1}{2} \int_{\substack{\text{magnetized}\\ \text{material}}} \mathbf{M}.\mathbf{H}_0\, dv \tag{1.2.71}$$

This energy is therefore regarded as localized within the magnetic material and it is the variation of this energy with respect to small displacements which leads to the mechanical forces. Equation (1.2.71) is the magnetic analogue of equation (1.2.45) for a dielectric material. The difference in sign for the localized energy should be noted.

1.2.6 Current Elements and Permanent Magnets

We shall now seek an energy function W from which the forces between current elements may be derived using the equation

$$\mathbf{F} = -\nabla W \tag{1.2.72}$$

This discussion will also lead to an energy function for a permanent magnet and a magnetic dipole.

Consider two small current filaments, labelled I and II which are in equilibrium under the action of their mutual forces (these arise from the magnetic field due to one filament acting on the current in the other filament), balanced, for example, by two mechanical springs. Suppose that filament I is allowed to undergo a virtual displacement Δx in time Δt, while the current distribution is maintained constant. Since there is no net force acting on a current filament due to its own magnetic field, the energy change of filament I will arise from the mechanical work done in its displacement in the magnetic field of filament II and from the electrical energy which must be provided by the source e.m.f. of filament I in order to maintain the current distribution unchanged despite the induced e.m.f. generated by its motion. Since filament I is initially in equilibrium the Principle of Virtual Work gives

$$\delta W_{\mathrm{I}} = \delta W^{\mathrm{I}}_{\mathrm{mech}} + \delta W^{\mathrm{I}}_{\mathrm{e.m.f.}} = 0 \tag{1.2.73}$$

This, however, is not the total energy change for the system of two current elements. Additional energy will have been provided by the source e.m.f. of filament II in order to maintain its current distribution constant in spite of the induced e.m.f. generated by the motion of filament I. The total energy change will be

$$\delta W = \delta W^{\mathrm{I}}_{\mathrm{mech}} + \delta W^{\mathrm{I}}_{\mathrm{e.m.f.}} + \delta W^{\mathrm{II}}_{\mathrm{e.m.f.}} \tag{1.2.74}$$

We may evaluate $\delta W^{\mathrm{II}}_{\mathrm{e.m.f.}}$ by fixing filament I and allowing filament II to undergo an equal and opposite displacement, $-\Delta x$, in an equal time Δt, again maintaining the currents constant. The mechanical work done will be the same as that done previously since it arises from the mutual forces between the two current filaments. The energy provided by the source e.m.f. for filament II will also be unchanged since it depends on the relative motion of the filaments. We have, corresponding to equation (1.2.73),

$$\delta W^{\mathrm{II}}_{\mathrm{mech}} + \delta W^{\mathrm{II}}_{\mathrm{e.m.f.}} = 0 \tag{1.2.75}$$

so that,

$$\delta W^{\mathrm{I}}_{\mathrm{mech}} = \delta W^{\mathrm{II}}_{\mathrm{mech}} = -\delta W^{\mathrm{I}}_{\mathrm{e.m.f.}} = -\delta W^{\mathrm{II}}_{\mathrm{e.m.f.}} \tag{1.2.76}$$

and hence, from equation (1.2.74)

$$\delta W = -\delta W_{\mathrm{mech}} = \delta W_{\mathrm{e.m.f.}} \tag{1.2.77}$$

The energy provided by a source e.m.f. will be

$$\delta W_{\mathrm{e.m.f.}} = \Delta t \int c\mathbf{J}.\mathbf{E}\, dv \tag{1.2.78}$$

But $\nabla \wedge \mathbf{E} = -1/c(\partial \mathbf{B}/\partial t)$ where $\mathbf{B}$ is the induction field at a current filament arising from the other source. Writing

$$\mathbf{B} = \nabla \wedge \mathbf{A}$$

we may take (see equation (2.11.27) p. 101)

$$\mathbf{E} = -\frac{1}{c}\frac{\partial \mathbf{A}}{\partial t} \tag{1.2.79}$$

and so

$$\delta W = \delta W_{\text{e.m.f.}} = -\int (\mathbf{J}.\Delta\mathbf{A})\, \mathrm{d}v \tag{1.2.80}$$

Thus, for a change in which the current distribution is maintained constant, we may define an energy function

$$W = -\int \mathbf{J}.\mathbf{A}\, \mathrm{d}v \tag{1.2.81}$$

from which the mechanical forces may be derived using equation (1.2.72).

It now follows that the potential energy function for a permanent magnet of volume V, in the field of other permanent magnets is given by

$$W_{\mathrm{p}} = -\int_V \mathbf{j}_{\mathrm{m}}.\mathbf{A}\, \mathrm{d}v \tag{1.2.82}$$

where $\mathbf{j}_{\mathrm{m}} = \nabla \wedge \mathbf{M}$ is the bound current density distribution corresponding to the magnetization $\mathbf{M}$. (This relation is discussed in Section 1.4.) From equation (1.2.50)

$$\mathbf{j}_{\mathrm{m}}.\mathbf{A} = (\nabla \wedge \mathbf{M}).\mathbf{A} = \nabla.(\mathbf{M} \wedge \mathbf{A}) + \mathbf{M}.(\nabla \wedge \mathbf{A})$$

so that upon transforming the first term to a surface integral, which is zero since $\mathbf{M} = 0$ everywhere outside the magnet, and writing $\nabla \wedge \mathbf{A} = \mathbf{B}$ in the second term, we obtain

$$W_{\mathrm{p}} = -\int_V \mathbf{M}.\mathbf{B}\, \mathrm{d}v \tag{1.2.83}$$

The field $\mathbf{B}$ in this integral is the field at a small elemental magnet with moment $\mathbf{M}\, \mathrm{d}v$. This field will arise from all external sources (say, $\mathbf{B}_1$) together with that from all other elements of the same total magnet occupying the volume V (say, $\mathbf{B}_2$). Hence the work required to introduce the magnet as a whole into a given external field will be

$$\Delta W_{\mathrm{p}} = -\int_V \mathbf{M}.\mathbf{B}_1\, \mathrm{d}v \tag{1.2.84}$$

Evidently the energy function for a magnetic dipole with moment $\mathbf{m} = \mathbf{M}\, \delta v$ situated in an external field $\mathbf{B}$ is

$$W_{\text{dipole}} = -\mathbf{m}.\mathbf{B} \tag{1.2.85}$$

The work required to construct a permanent magnet in the absence of an external field will be derived from

$$\Delta W'_{\mathrm{p}} = -\int \mathbf{M}.\mathbf{B}_2\, \mathrm{d}v \tag{1.2.86}$$

Evidently the mutual potential energy of two dipoles is given from this equation or from equation (1.2.85), as

$$W_{dd} = -\mathbf{m}_1.\mathbf{H}_2 = -\mathbf{m}_1.\left\{-\frac{\mathbf{m}_2}{r^3} + \frac{3(\mathbf{m}_2.\mathbf{r})}{r^5}\mathbf{r}\right\} \tag{1.2.87}$$

$$= \left\{\frac{\mathbf{m}_1.\mathbf{m}_2}{r^3} - \frac{3(\mathbf{m}_1.\mathbf{r})(\mathbf{m}_2.\mathbf{r})}{r^5}\right\} = -\mathbf{m}_2.\mathbf{H}_1 \tag{1.2.88}$$

If a permanent magnet is in the form of a uniformly magnetized ellipsoid, the internal self-field $\mathbf{H}_i$ may be written in terms of demagnetizing factors. For a specimen magnetized along one of the principal axes of the ellipsoid

$$\mathbf{H}_i = N\mathbf{M} \tag{1.2.89}$$

The self-energy of such a permanent magnet is

$$W_{\text{perm}} = \tfrac{1}{2}NM^2 \quad \text{per unit volume} \tag{1.2.90}$$

Here the factor $\frac{1}{2}$ arises because if it were omitted we should be counting the mutual interactions twice. We can see this by reference to equations (1.2.87) and (1.2.88) which show that the mutual energy of two dipoles is given by the product of one dipole moment and the field due to the other dipole. This is just one half of the same quantity summed over both dipoles.

From equations (1.2.84) and (1.2.90), the energy function per unit volume for an ellipsoidal permanent magnet (or a saturated ferromagnet) in a uniform external field $\mathbf{H}$ is

$$W = -\mathbf{M}.\mathbf{H} + \tfrac{1}{2}[N_x M_x^2 + N_y M_y^2 + N_z M_z^2] \tag{1.2.91}$$

Here N_x, N_y, N_z are demagnetizing factors for the principal axes of the ellipsoid which are also the axes of Cartesian coordinates to which $\mathbf{M}$ and $\mathbf{H}$ are referred.

1.3 Dielectric Insulating Media

In a dielectric medium $\mathbf{D}$ differs from $\mathbf{E}$ and so there must exist within the material an additional vector field which is not present in free space. In view of equation (1.1.24) it is reasonable to suppose that the sources for this additional field are separated charges within the material. These charges, however, give rise to zero net charge over any macroscopic volume elements.

If ρ' is the density of these separated charges, then

$$\int_{\Delta v} \rho' \, dv = 0 \tag{1.3.1}$$

where Δv is a small volume element within the material whose linear dimensions are large compared with the microscopic separation of the displaced charges. The displacements may reasonably be expected to be of such a microscopic nature if a material medium is bonded together by forces, electrical in origin, which are large compared with those arising from the field intensities we consider here. The charges ρ' are then essentially inaccessible to experiment on the scale of observation defined by Δv and so they are not included in the charge density ρ of equation (1.1.24), which is concerned with macroscopic properties of the material.

The macroscopic charge density is concerned with charges which in principle can be taken away from or added to the volume element. It is convenient to distinguish these charges by the name 'free charges' and they will obey the relation

$$q = \int_{\Delta v} \rho \, \mathrm{d}v \tag{1.3.2}$$

which is non-zero. Equation (1.3.1) may be regarded as referring to a set of 'bound' charges.

1.3.1 Electric Polarization

Since **D** differs from **E** only within the material medium, the additional vector field arising from the separated charges ρ' will also exist only within the material and will be zero elsewhere. Denote this additional field by a vector **P**, it is called the electric polarization vector. Integrate the normal component of **P** over a surface S which encloses the volume V of the dielectric material but which nowhere enters it.

Then

$$\oint_S \mathbf{P}.\mathbf{n} \, \mathrm{d}S = 0 \tag{1.3.3}$$

and so†

$$\int_V \nabla . \mathbf{P} \, \mathrm{d}v = \int_{\text{dielectric}} \nabla . \mathbf{P} \, \mathrm{d}v = 0 \tag{1.3.4}$$

In view of equation (1.3.1), the vector field **P** may be related to the separated charge density ρ' by the equation

$$\nabla . \mathbf{P} = -\rho' \tag{1.3.5}$$

in which the choice of the negative sign is arbitrary. This is not a unique definition of **P** since equation (1.3.5) is still satisfied if any further vector field, **A**, which satisfies the relation $\mathbf{A} = \nabla \wedge \mathbf{a}$, is superimposed on **P**. This arises because the divergence of the curl of any vector is identically zero. However, the vector field **P** is a convenient choice because it can be directly related to the density of charge σ'_s on the boundary of the dielectric (see equation (1.3.7)). Moreover there is an alternative unique definition of **P** in terms of the dipole moment per unit volume of the material which is entirely consistent with equation (1.3.5).

If equation (1.3.5) is integrated throughout the volume of a small right circular cylinder, having its axis normal to the boundary of the medium, as in Fig. 1.3, we obtain

$$\int_{\text{cylinder}} \nabla . \mathbf{P} \, \mathrm{d}v = \oint_{\text{cylinder}} \mathbf{P}.\mathbf{n} \, \mathrm{d}S = -\int_{\text{cylinder}} \rho' \, \mathrm{d}v \tag{1.3.6}$$

† In order that the mathematical requirements of continuity and differentiability be satisfied when applying Green's transformation in equation (1.3.3) it is necessary to suppose that **P** falls smoothly to zero through a very thin transition layer at the boundary surface.

As the axial length of the cylinder shrinks to zero, $\int \rho'\, dv$ becomes $\int \sigma_s'\, dS$ and since **P** is zero outside the boundary

$$\sigma_s' = -(\mathbf{P}.\mathbf{n}) \equiv P_s \tag{1.3.7}$$

The surface density of charge is therefore equal to the component, P_s, of the vector **P** along the outward normal at the surface.

The relation between **P** and the dipole moment per unit volume of the material may be derived from the definition of dipole moment for the whole body,

$$\mathscr{M} = \int \rho' \mathbf{r}\, dv \tag{1.3.8}$$

$$= -\int \mathbf{r} \nabla . \mathbf{P}\, dv \tag{1.3.9}$$

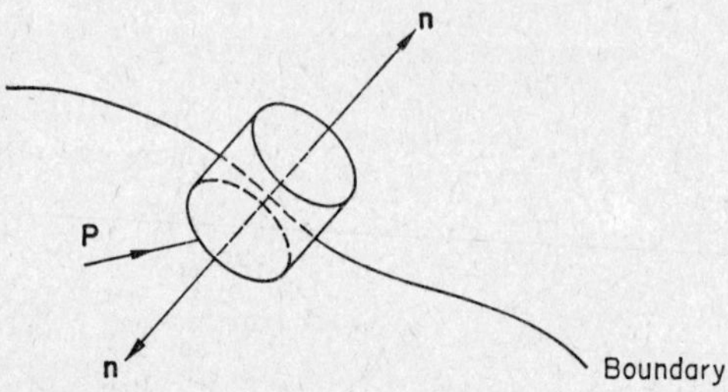

FIG. 1.3 Showing the relationship between the electric polarization **P** and the surface charge σ_s^1

Writing this equation out in component form, and using the vector identity

$$\nabla . (x\mathbf{P}) = x\nabla . \mathbf{P} + \mathbf{P} . \nabla x = x\nabla . \mathbf{P} + P_x \tag{1.3.10}$$

we have

$$\mathscr{M} = -\oint \mathbf{r}(\mathbf{P}.\mathbf{n})\, dS + \int \mathbf{P}\, dv \tag{1.3.11}$$

The surface integral in this equation is taken over a surface which encloses the material but which nowhere penetrates within and so **P** is everywhere zero. Hence

$$\mathscr{M} = \int \mathbf{P}\, dv \tag{1.3.12}$$

or

$$\int \rho' \mathbf{r}\, dv = \int \mathbf{P}\, dv \tag{1.3.13}$$

The vector **P** may therefore be identified with the dipole moment per unit volume of the dielectric material.†

† Note that the dimensions of **P** are charge-cm/cm³ or charge/cm², which are also the dimensions of the electric field.

The vector field **D** in a material medium may now be defined by superimposing the field from **P** on that from **E**, by the relation

$$\mathbf{D} = \mathbf{E} + 4\pi\mathbf{P} \tag{1.3.14}$$

The choice of the constant 4π in this equation implies that in an uncharged dielectric, for which $\nabla . \mathbf{D} = 0$ (equation (1.1.24))

$$\nabla . \mathbf{E} = -4\pi \nabla . \mathbf{P} = 4\pi\rho' \tag{1.3.15}$$

whilst within a charged dielectric, for which $\nabla . \mathbf{D} = 4\pi\rho$

$$\nabla . \mathbf{E} = 4\pi(\rho - \nabla . \mathbf{P}) = 4\pi(\rho + \rho') \tag{1.3.16}$$

In these equations ρ' includes not only the volume distribution of bound charges within the dielectric but also those charges in the surface transition layer which are described by equation (1.3.7). So far as the electric field is concerned therefore, a dielectric medium may be replaced by an appropriate distribution of charge in free space. This description is only strictly true for a rigid dielectric but it does not require the material to be either isotropic or homogeneous.

1.3.2 Polarization, the Electric Displacement and the Electric Field

The solution of a particular physical problem will be dependent upon the functional relation between the polarization and the electric field. For a large class of materials it is a good approximation to assume that there is a linear relation between **P** and **E**. The exclusion of higher powers of **E** is not unreasonable in view of the comment following equation (1.3.1) where we noted that the field intensities considered in this discussion are small compared with the internal binding fields in dielectrics. Non-linear behaviour will of course be anticipated for intense electric field devices. We are not, however, implying that **P** (and likewise **D**) depends only upon the value of **E** at the particular instant of time under consideration. We shall include the influence of fields which have been experienced at times in the past.

The linear relations for an isotropic dielectric will be written in the form

$$\mathbf{P}(t) = \chi^{(e)}(\omega)\,\mathbf{E}(j\omega t) \tag{1.3.17}$$

and

$$\mathbf{D}(t) = \varepsilon(\omega)\mathbf{E}(j\omega t) \tag{1.3.18}$$

for fields which have a harmonic time variation (and including static fields). Evidently we obtain from equation (1.3.14)

$$\varepsilon(\omega) = 1 + 4\pi\chi^{(e)}(\omega) \tag{1.3.19}$$

In these equations $\chi^{(e)}(\omega)$ and $\varepsilon(\omega)$ are dimensionless parameters which have characteristic values for each particular dielectric material measured under specific experimental conditions (e.g., temperature and frequency). $\chi^{(e)}(\omega)$ is called the electric susceptibility, $\varepsilon(\omega)$ is the dielectric constant. In free space, $\chi^{(e)}(\omega) = 0$ and $\varepsilon(\omega) = 1$.

For anisotropic dielectrics **D** and **P** are not necessarily in the same direction as the vector **E** and so there is a tensor relation between these quantities. We may for example write

$$\mathbf{D} = \overleftrightarrow{\varepsilon}.\mathbf{E} \tag{1.3.20}$$

and a particular component of $\mathbf{D}(i = x, y, z)$

$$D_i = \varepsilon_{ii}E_i + \varepsilon_{ij}E_j + \varepsilon_{ik}E_k \tag{1.3.21}$$

Energy considerations show that $\overleftrightarrow{\varepsilon}$ is a symmetric tensor. We obtain from equation (1.1.3) the expression for an increment in energy per unit volume of the material

$$\mathrm{d}u = \frac{1}{4\pi}\mathbf{E}.\mathrm{d}\mathbf{D} \tag{1.3.22}$$

Hence

$$\frac{\partial}{\partial E_i}\left(\frac{\partial u}{\partial E_j}\right) = \frac{\partial^2 u}{\partial E_i\,\partial E_j} = \varepsilon_{ij}; \qquad \frac{\partial}{\partial E_j}\left(\frac{\partial u}{\partial E_i}\right) = \frac{\partial^2 u}{\partial E_j\,\partial E_i} = \varepsilon_{ji}$$

but the energy function u and its derivatives are single valued continuous functions of the independent variables E_i when there is no power loss in the medium, and so

$$\frac{\partial^2 u}{\partial E_i\,\partial E_j} = \frac{\partial^2 u}{\partial E_j\,\partial E_i}$$

so that

$$\varepsilon_{ji} = \varepsilon_{ij} \tag{1.3.23}$$

It is therefore possible to choose three mutually perpendicular coordinate axes in the material such that $\overleftrightarrow{\varepsilon}$ is a diagonal tensor. In this coordinate system of 'principal dielectric axes' the material equations take the form

$$D_x = \varepsilon_x E_x; \qquad D_y = \varepsilon_y E_y; \qquad D_z = \varepsilon_z E_z \tag{1.3.24}$$

and ε_x, ε_y, ε_z are called the 'principal dielectric constants'. It is evident that, unless **E** coincides with one of the principal axes, **D** and **E** will have different directions.

There is a similar set of tensor relations between **P** and **E** for anisotropic media.

The polarization existing at any particular time, t, will depend upon the electric fields which have been experienced at previous times. We have argued at the beginning of this section (1.3) that the polarization **P** arises from the microscopic separation of bound charges which takes place when the dielectric material is subject to an electric field **E**. In general, $\mathbf{P}(t)$ will not vanish instantaneously when the electric field is removed but will decay towards zero as the displaced charges move back to their undisturbed positions. The polarization, $\mathbf{P}(t)$, measured at time, t, will therefore depend on the electric fields, $\mathbf{E}(t - \lambda)$, which have been present at previous times, $(t - \lambda)$.

Consider the polarization $\mathrm{d}\mathbf{P}(t)$ which remains at time, t, in an isotropic dielectric, and arises from an impulsive field $\mathbf{E}(t - \lambda_i)$ which was applied at the time $(t - \lambda_i)$. The polarization will decay away towards zero after the impulse excitation in a manner which is characteristic of the particular dielectric material. This is illustrated in Fig. 1.4, where the particular form of the decay curve for $d\mathbf{P}(t)$ is arbitrary.

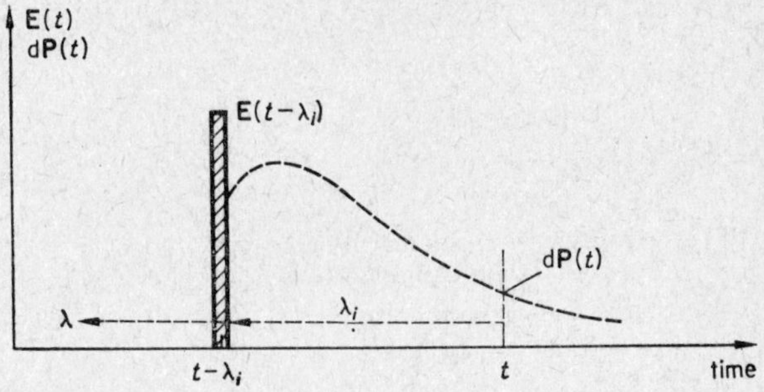

FIG. 1.4 The decay of the polarization following impulse excitation.

We write the polarization $d\mathbf{P}(t)$ remaining at the time of observation as

$$d\mathbf{P}(t) = h(\lambda_i)\,\mathbf{E}(t - \lambda_i) \qquad (1.3.25)$$

where $h(\lambda_i)$ is a scalar weighting factor which depends upon the time interval λ_i which has elapsed between the application of the impulsive field and the moment of observation. Evidently $h(t)$ is a function which describes the way in which the polarization decays to zero after an impulse excitation.

If a succession of impulses is applied, the total polarization at time t, will be the sum of the responses remaining from all the individual impulses,

$$\mathbf{P}(t) = \sum d\mathbf{P}(t) = \sum_{\lambda_i = 0^-}^{\infty} h(\lambda_i)\,\mathbf{E}(t - \lambda_i) \qquad (1.3.26)$$

For continuous excitation, the summation becomes an integration

$$\mathbf{P}(t) = \int_{0^-}^{\infty} h(\lambda)\,\mathbf{E}(t - \lambda)\,d\lambda \qquad (1.3.27)$$

Note that in equations (1.3.25)–(1.3.27), the positive values of λ are measured backwards in time from the moment of observation. For a causal system there can be no contribution to $\mathbf{P}(t)$ in equations (1.3.26), (1.3.27), from electric fields which have not yet been experienced. Hence the weighting functions $h(\lambda_i)$, $h(\lambda)$, must be zero for all $\lambda_i < 0^-$ and $\lambda < 0^-$. This requirement sets the lower limit of the summation and the integral in these equations.†

If the electric field excitation has the form of a unit impulse function (a Dirac δ-function) we may write

$$\mathbf{E}(t) = \mathbf{n}\,E_0\,\delta(t)$$
$$\mathbf{E}(t - \lambda) = \mathbf{n}\,E_0\,\delta(t - \lambda) \qquad (1.3.28)$$

† We take account of impulse excitation and response at the moment of observation by writing the limit as 0^- rather than 0.

where **n** is a unit vector and we have for convenience taken the origin of time at the impulse.

Then, from equation (1.3.27)

$$\mathbf{P}_\delta(t) = \mathbf{n}\, E_0 \int_{0^-}^{\infty} h(\lambda)\, \delta(t - \lambda)\, \mathrm{d}\lambda \tag{1.3.29}$$

and so

$$P_\delta(t) = \mathbf{n}\, E_0\, h(t) \qquad \text{with } h(t) = 0 \text{ for } t < 0^- \tag{1.3.30a}$$

which may be written quite generally as

$$\mathbf{P}_\delta(\xi) = \mathbf{n} E_0\, h(\xi) \qquad \text{with } h(\xi) = 0 \text{ for } \xi < 0^- \tag{1.3.30b}$$

where ξ is the time interval measured after the impulse excitation.

These relations again emphasize that $h(t)$ describes the response of the dielectric to an impulsive electric field. The function $h(\lambda)$, or $h(t)$, is therefore called either the weighting function (as in equation 1.3.25) or the impulse response for the material.

The expression for the electric displacement is obtained from equations (1.3.14) and (1.3.27).

$$\mathbf{D}(t) = \mathbf{E}(t) + 4\pi \int_{0^-}^{\infty} h(\lambda)\, \mathbf{E}(t - \lambda)\, \mathrm{d}\lambda \tag{1.3.31}$$

for an isotropic material.

1.3.3 Harmonic Time Functions

In many physical problems the dependence of **E** upon t is harmonic in form. In these cases the actual field, which is correctly described by real vector quantities, may be written in terms of the real part or the imaginary part of a complex vector function.

We can see this by the following argument. Let $\mathbf{A}(\mathbf{r}, t)$ represent the real vector function with harmonic time dependence. The components of **A** can be written

$$A_i(\mathbf{r}, t) = u_i(\mathbf{r}) \cos \omega t + v_i(\mathbf{r}) \sin \omega t \tag{1.3.32}$$

where $i \equiv x, y, z$, and ω is the angular frequency of the harmonic time function.

The space functions u_x, u_y, u_z, and v_x, v_y, v_z, may be regarded as components of two real vectors **u** and **v** so that we can write

$$\mathbf{A}(\mathbf{r}, t) = \mathbf{u}(\mathbf{r}) \cos \omega t + \mathbf{v}(\mathbf{r}) \sin \omega t \tag{1.3.33}$$

$$\mathbf{A}(\mathbf{r}, t) = \text{Real part } [\mathbf{G}(\mathbf{r})\, \mathrm{e}^{\mathrm{j}\omega t}] \tag{1.3.34}$$

where **G** is the complex vector,

$$\mathbf{G}(\mathbf{r}) = \mathbf{u}(\mathbf{r}) - \mathrm{j}\mathbf{v}(\mathbf{r}) \tag{1.3.35}$$

When the operations on $\mathbf{A}(\mathbf{r}, t)$ are linear,† it is possible to interchange the order of operation as between the Real part operator in equation (1.3.34) and

† A linear operator O is defined by the relations

$$O(f + g) = Of + Og \quad \text{and} \quad O(cf) = cOf$$

where c is any constant.

the linear operator. The real part of the resulting expression then represents the actual physical quantity which is being derived. Suppose O is such a linear operator, then

$$\phi \equiv O\mathbf{A}(\mathbf{r}, t) = O\{\text{Real part } [\mathbf{G}(\mathbf{r})\, e^{j\omega t}]\} \tag{1.3.36}$$

$$\phi = \text{Real part } [O\, \mathbf{G}(\mathbf{r})\, e^{j\omega t}] \tag{1.3.37}$$

A similar argument may be used to write $\mathbf{A}(\mathbf{r}, t)$ as the imaginary part of a complex vector function.

1.3.4 Complex Electric Susceptibility and Dielectric Constant

We may write $E(t)$ as $\mathbf{E}_0{}^{j\omega t}$ in the harmonic representation. We omit the explicit designation of the spatial dependence in order to simplify the notation, and obtain from equation (1.3.27),

$$\begin{aligned}\mathbf{P}(j\omega t) &= \int_{0^-}^{\infty} h(\lambda)\, \mathbf{E}_0\, e^{j\omega(t-\lambda)}\, d\lambda \\ &= \left\{\int_{0^-}^{\infty} h(\lambda)\, e^{-j\omega\lambda}\, d\lambda\right\} E_0\, e^{j\omega t}\end{aligned} \tag{1.3.38}$$

or

$$\mathbf{P}(j\omega t) = \chi^{(e)}(\omega)\, \mathbf{E}(j\omega t)$$

with the complex electric susceptibility defined by the relation

$$\chi^{(e)}(\omega) = \int_{0^-}^{\infty} h(\lambda)\, e^{-j\omega\lambda}\, d\lambda \tag{1.3.39}$$

The complex dielectric constant is derived from equation (1.3.31), or equation (1.3.19),

$$\varepsilon(\omega) = 1 + 4\pi \int_{0^-}^{\infty} h(\lambda)\, e^{-j\omega\lambda}\, d\lambda \tag{1.3.40}$$

Let us denote the real and imaginary parts of $\varepsilon(\omega)$ by $\varepsilon_1(\omega)$ and $\varepsilon_2(\omega)$. We write

$$\varepsilon(\omega) = \varepsilon_1(\omega) - j\varepsilon_2(\omega) \tag{1.3.41}$$

From equation (1.3.40), since $h(\lambda)$ is a real function, we can see that

$$\varepsilon(-\omega) = \varepsilon^*(\omega) \tag{1.3.42}$$

where the asterisk denotes the complex conjugate of $\varepsilon(\omega)$,

$$\varepsilon^*(\omega) = \varepsilon_1(\omega) + j\varepsilon_2(\omega)$$

separating the real and imaginary parts of equation (1.3.42), we have

$$\begin{aligned}\varepsilon_1(-\omega) &= \varepsilon_1(\omega) \\ \varepsilon_2(-\omega) &= -\varepsilon_2(\omega)\end{aligned} \tag{1.3.43}$$

Thus $\varepsilon_1(\omega)$ is an even function of the angular frequency ω, whereas $\varepsilon_2(\omega)$ is an odd function of ω. There are similar relations for the complex electric susceptibility $\chi^{(e)}(\omega) = \chi_1^{(e)}(\omega) - j\chi_2^{(e)}(\omega)$. Evidently $\chi_1^{(e)}(-\omega) = \chi_1^{(e)}(\omega)$ and $\chi_2^{(e)}(-\omega) = -\chi_2^{(e)}(\omega)$.

At low frequencies, the expansion of $\varepsilon_1(\omega)$ as a power series in ω will be of the form

$$\varepsilon_1(\omega) = \varepsilon(0) + a_2\omega^2 + a_4\omega^4 + \cdots \quad \text{even powers} \tag{1.3.44}$$

whilst the corresponding expression for $\varepsilon_2(\omega)$ will be

$$\varepsilon_2(\omega) = b_1\omega + b_3\omega^3 + b_5\omega^5 + \cdots \quad \text{odd powers} \tag{1.3.45}$$

The imaginary part of the dielectric constant is therefore a linear function of ω at sufficiently low frequencies.

Note that terms of the form ω^{-n} have not been included in these expansions because we are thinking of insulating dielectric media. A dielectric constant for a conducting material may also be defined (see Section 2.1, Chapter 2) which has a singularity at zero frequency arising from a term in $(\mathrm{j}\omega)^{-1}$.

The dielectric constant for electrostatic conditions is given by the limit $\omega \to 0$. Equation (1.3.40) reduces correctly to a real function $\varepsilon(0)$ giving

$$\varepsilon(0) = 1 + 4\pi \int_{0^-}^{\infty} h(\lambda)\, \mathrm{d}\lambda \tag{1.3.46}$$

Since $\varepsilon(0)$ is finite for a physical material, $\chi^{(e)}(0)$ is also finite so that

$$\left|\int_{0^-}^{\infty} h(\lambda)\, \mathrm{d}\lambda\right| = |\chi^{(e)}(0)| \leqslant |M|$$

where M is finite.

The behaviour of the dielectric constant in the high frequency limit may be derived by observing that equation (1.3.39) shows that $\chi^{(e)}(\omega)$ and $h(\lambda)$ constitute a Fourier Transform pair. Since $h(\lambda)$ is zero for $\lambda < 0^-$, we may extend the lower limit to $-\infty$ for a valid function $h(\lambda)$, and so we may write

$$\begin{aligned} \chi^{(e)}(\omega) &= \int_{-\infty}^{\infty} h(\lambda)\, \mathrm{e}^{-\mathrm{j}\omega\lambda}\, \mathrm{d}\lambda \\ h(\lambda) &= \frac{1}{2\pi}\int_{-\infty}^{\infty} \chi^{(e)}(\omega)\, \mathrm{e}^{\mathrm{j}\omega\lambda}\, \mathrm{d}\omega \end{aligned} \tag{1.3.47}$$

provided that these relations define a function $h(\lambda)$ which is zero for $\lambda < 0^-$. We expect $h(\lambda)$ to be a good mathematical function which is bounded and well behaved at $\omega = \infty$, with finite derivatives at $\lambda \geqslant 0$, when it properly describes a real physical material. For such a function, the high frequency limit of $\chi^{(e)}(\omega)$ is given by†

$$\begin{aligned} \chi^{(e)}(\omega) &= \operatorname*{Limit}_{\omega\to\infty} \int_{-\infty}^{\infty} \mathrm{e}^{-\mathrm{j}\omega\lambda}\, h(\lambda)\, \mathrm{d}\lambda \\ &= \left\{-\frac{h^{\mathrm{I}}(0)}{\omega^2} + \frac{h^{\mathrm{III}}(0)}{\omega^4} - \frac{h^{\mathrm{V}}(0)}{\omega^6} + \cdots\right\} \\ &\quad -\mathrm{j}\left\{\frac{h(0)}{\omega} - \frac{h^{\mathrm{II}}(0)}{\omega^3} + \frac{h^{\mathrm{IV}}(0)}{\omega^5} - \cdots\right\} \end{aligned} \tag{1.3.48}$$

In this equation $h(0)$, $h^{\mathrm{I}}(0)$, $h^{\mathrm{II}}(0)$ . . . are the values of $h(\lambda)$ and its derivatives at $\lambda = 0$. The high frequency behaviour of $\chi^{(e)}(\omega)$ is therefore determined by the properties of $h(\lambda)$ near $\lambda = 0$.

† LIGHTHILL, M. J., *Fourier Analysis and Generalised Functions*, Cambridge University Press, 1959.

1.3.5 Examples

We illustrate this discussion by deriving expressions for the electric susceptibility on two different models—the Debye model and the Lorentz simple harmonic oscillator model.

Consider a material for which the polarization decays exponentially towards zero after excitation by an electric field impulse.

$$\mathbf{P}(\xi) \sim \mathrm{e}^{-\xi/\beta} \qquad \text{for } \xi \geqslant 0 \tag{1.3.49}$$

and so, from equation (1.3.30), we may write,

$$h(\lambda) = \frac{\chi_0}{\beta}\,\mathrm{e}^{-\lambda/\beta} \qquad \text{for } \lambda \geqslant 0^- \tag{1.3.50}$$

$$h(\lambda) = 0 \qquad \text{for } \lambda < 0^-$$

where χ_0, β are positive constants.

From equation (1.3.39)

$$\chi^{(e)}(\omega) = \frac{\chi_0}{\beta}\int_{0^-}^{\infty} \mathrm{e}^{-\lambda/\beta}\,\mathrm{e}^{-\mathrm{j}\omega\lambda}\,\mathrm{d}\lambda \tag{1.3.51}$$

that is

$$\chi^{(e)}(\omega) = \frac{\chi_0}{1 + \mathrm{j}\omega\beta} \tag{1.3.52}$$

The static electric susceptibility is obtained by putting $\omega = 0$ in equation (1.3.52), and agrees with the value derived from equations (1.3.46) and (1.3.50).†

$$\chi^{(e)}(0) = \chi_0$$

The low frequency expansion is

$$\chi^{(e)}(\omega) = \chi_0(1 - \omega^2\beta^2 + \omega^4\beta^4 - \cdots) - \mathrm{j}\chi_0(\omega\beta - \omega^3\beta^3 + \omega^5\beta^5 - \cdots)$$

in agreement with equations (1.3.44) and (1.3.45).

The high frequency expansion may be obtained from equations (1.3.52) or equation (1.3.48),

$$\chi^{(e)}(\omega) = \frac{\chi_0}{\beta^2}\omega^{-2} - \mathrm{j}\frac{\chi_0}{\beta}\omega^{-1}$$

and so $\chi_1^{(e)}(\omega) \to 0$ from positive values. This contrasts with the high frequency behaviour of the next example for the Lorentz simple harmonic oscillator model.

Consider a material for which

$$h(\lambda) = \frac{(Ne^2/m)}{\omega_1}\,\mathrm{e}^{-b\lambda}\sin\omega_1\lambda \qquad \text{for } \lambda \geqslant 0^-$$

$$h(\lambda) = 0 \qquad \text{for } \lambda < 0^- \tag{1.3.53}$$

where (Ne^2/m) and b are positive quantities. Then

$$\chi^{(e)}(\omega) = \frac{(Ne^2/m)}{\omega_0^2 - \omega^2 + 2\mathrm{j}b\omega} \tag{1.3.54}$$

† This is the reason for taking χ_0 positive; see page 33.

At high frequencies therefore

$$\chi^{(e)}(\omega) \rightarrow -\frac{(Ne^2/m)}{\omega^2} - j\frac{2b(Ne^2/m)}{\omega^3}$$

and so $\chi^{(e)}(\omega) \rightarrow 0$ from negative values.

Thus on the Debye model for a dielectric material the real part of the dielectric constant tends to unity at high frequencies from a value greater than one, whereas on the Lorentz model the real part of the dielectric constant tends to unity at high frequencies from a value less than one.

1.3.6 Kramers–Kronig Relations

Equations (1.3.39) and (1.3.40) show that the complex electric susceptibility $\chi^{(e)}(\omega)$ and dielectric constant $\varepsilon(\omega)$ are strictly functions of $j\omega$. We should write explicitly $\chi^{(e)}(j\omega)$ and $\varepsilon(j\omega)$. It is possible therefore to regard these functions as limiting forms of more general functions of a complex variable $\chi^{(e)}(s)$, $\varepsilon(s)$ where $s = q + j\omega$ with q and ω real variables. The functions $\chi^{(e)}(j\omega)$, $\varepsilon(j\omega)$ are simply the particular forms which $\chi^{(e)}(s)$, $\varepsilon(s)$ take on the imaginary axis of coordinates. It is now possible to derive relations between the real and imaginary parts of $\chi^{(e)}(j\omega)$ and of $\varepsilon(j\omega)$ from consideration of more general properties of a function of a complex variable.

In carrying through this analysis we shall have in mind the particular properties of the electric susceptibility and the dielectric constant but the

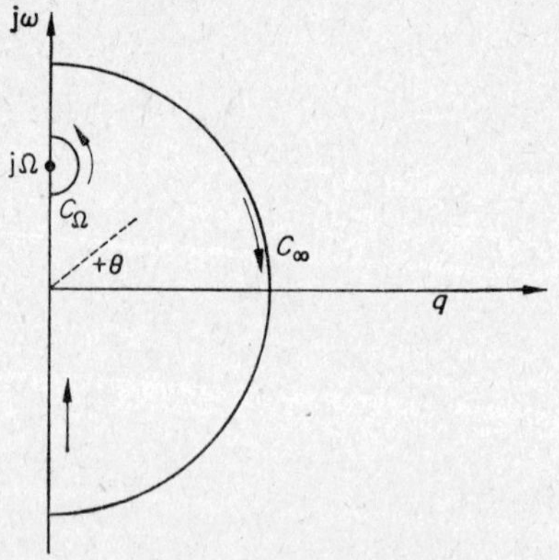

Fig. 1.5 Contour for evaluating $\oint \frac{\phi(s)}{s - j\,\Omega}\,ds$.

results we obtain will express general properties of a function of a complex variable which have wider application. It will not be surprising therefore to find them useful in general network theory also. It is, however, important to note that the application of these relations will involve the behaviour of physical systems as ω tends to infinity and so they can only be used for functions which are satisfactory in this limit.

Suppose that the function $\phi(s)$ is regular in the right half plane and has no poles on the entire $j\omega$ axis. It is necessary to assume in addition that $\phi(s)$

converges uniformly to a finite value as $s \longrightarrow \infty$. Consider integrating the function

$$\left\{\frac{\phi(s)}{s - \mathrm{j}\Omega}\right\}$$

around the contour shown in Fig. 1.5, which consists of the infinite semicircular arc C_∞ to the right, and the entire jω axis, but bypassing the point jΩ by a small semicircular arc C_Ω. From Cauchy's integral theorem,† we obtain for the complete contour,

$$\oint \frac{\phi(s)}{s - \mathrm{j}\Omega}\,\mathrm{d}s = 0 \tag{1.3.55}$$

The contributions from the three distinct parts of the contour may also be separately evaluated. That from the infinite semicircular arc C_∞ may be derived by allowing the radius of the semicircle, R_0, to tend to infinity. Since $s = R_0\,\mathrm{e}^{\mathrm{j}\theta}$ along such a contour

$$\int_{C_\infty} \frac{\phi(s)}{s - \mathrm{j}\Omega}\,\mathrm{d}s = \operatorname*{Limit}_{R_0 \to \infty} \int \frac{\phi(R_0\,\mathrm{e}^{\mathrm{j}\theta})\,R_0\,\mathrm{e}^{\mathrm{j}\theta}}{R_0\,\mathrm{e}^{\mathrm{j}\theta} - \mathrm{j}\Omega}\,\mathrm{j}\,\mathrm{d}\theta = \mathrm{j}\,\phi(\infty)\int_{+\pi/2}^{-\pi/2} \mathrm{d}\theta \tag{1.3.56}$$

$$= -\mathrm{j}\pi\,\phi(\infty) \tag{1.3.57}$$

The contribution from the arc which bypasses the pole at jΩ is given by Cauchy's integral formula

$$\int_{C_\Omega} \frac{\phi(s)}{s - \mathrm{j}\Omega}\,\mathrm{d}s = \mathrm{j}\pi\,\phi(\Omega) \tag{1.3.58}$$

and for the remainder of the jω axis, we shall have

$$\operatorname*{Limit}_{\substack{R_0 \to \infty \\ \rho \to 0}} \left\{\int_{-R_0}^{\Omega-\rho} \frac{\phi(\mathrm{j}\omega)}{\mathrm{j}(\omega - \Omega)}\,\mathrm{j}\,\mathrm{d}\omega + \int_{\Omega+\rho}^{R_0} \frac{\phi(\mathrm{j}\omega)}{\mathrm{j}(\omega - \Omega)}\,\mathrm{j}\,\mathrm{d}\omega\right\}$$

$$\equiv \mathrm{P}\int_{-\infty}^{\infty} \frac{\phi(\mathrm{j}\omega)}{\omega - \Omega}\,\mathrm{d}\omega \tag{1.3.59}$$

Here ρ is the radius of the semicircle used for bypassing the point jΩ and $\mathrm{P}\int_{-\infty}^{\infty}$ stands for the principal value of the integral which is defined by equation (1.3.59). Substituting the results from equations (1.3.57)–(1.3.59) into equation (1.3.55), we obtain

$$-\mathrm{j}\pi\,\phi(\infty) + \mathrm{j}\pi\,\phi(\Omega) + \mathrm{P}\int_{-\infty}^{\infty} \frac{\phi(\mathrm{j}\omega)}{\omega - \Omega}\,\mathrm{d}\omega = 0 \tag{1.3.60}$$

The function $\phi(s)$ may be written in terms of its real and imaginary parts as

$$\phi(s) = U(q, \omega) + \mathrm{j}V(q, \omega) \tag{1.3.61}$$

† See for example, Wylie, C. R., *Advanced Engineering Mathematics*, McGraw-Hill, 1960; and Sokolnikoff, I. S. and Redheffer, R. M., *Mathematics of Physics and Modern Engineering*, McGraw-Hill, 1958.

and so we obtain from equation (1.3.60)

$$U(\Omega) = U(\infty) - \frac{1}{\pi}\mathrm{P}\int_{-\infty}^{\infty} \frac{V(\omega)}{\omega - \Omega}\,\mathrm{d}\omega \tag{1.3.62}$$

$$V(\Omega) = V(\infty) + \frac{1}{\pi}\mathrm{P}\int_{-\infty}^{\infty} \frac{U(\omega)}{\omega - \Omega}\,\mathrm{d}\omega \tag{1.3.63}$$

These relations are called 'dispersion relations' or Kramers–Kronig relations. They have immediate application to the electric susceptibility. Evaluating $\chi^{(e)}(\omega)$ at some particular frequency Ω we have

$$\chi_1^{(e)}(\Omega) = \frac{1}{\pi}\mathrm{P}\int_{-\infty}^{\infty} \frac{\chi_2^{(e)}(\omega)}{\omega - \Omega}\,\mathrm{d}\omega \tag{1.3.64}$$

$$\chi_2^{(e)}(\Omega) = -\frac{1}{\pi}\mathrm{P}\int_{-\infty}^{\infty} \frac{\chi_1^{(e)}(\omega)}{\omega - \Omega}\,\mathrm{d}\omega \tag{1.3.65}$$

since $\chi^{(e)}(\infty) = 0$.

In many physical applications of equations (1.3.62) and (1.3.63) the quantity represented by $U(\omega)$ is an even function of frequency whereas that represented by $V(\omega)$ is an odd function of frequency. We have already shown that this is the case for the electric susceptibility since $\chi_1^{(e)}(\omega)$ is even whilst $\chi_2^{(e)}(\omega)$ is odd. In such cases (1.3.62) may be rewritten

$$U(\Omega) = U(\infty) - \frac{1}{\pi}\mathrm{P}\int_{-\infty}^{0} \frac{V(\omega)}{\omega - \Omega}\,\mathrm{d}\omega - \frac{1}{\pi}\mathrm{P}\int_{0}^{\infty} \frac{V(\omega)}{\omega - \Omega}\,\mathrm{d}\omega \tag{1.3.66}$$

and substituting $V(-\omega) = -V(\omega)$, we obtain

$$U(\Omega) = U(\infty) - \frac{1}{\pi}\mathrm{P}\int_{0}^{\infty} V(\omega)\left(\frac{1}{\omega + \Omega} + \frac{1}{\omega - \Omega}\right)\mathrm{d}\omega \tag{1.3.67}$$

so that

$$U(\Omega) = U(\infty) - \frac{2}{\pi}\mathrm{P}\int_{0}^{\infty} \frac{\omega V(\omega)}{\omega^2 - \Omega^2}\,\mathrm{d}\omega \tag{1.3.68}$$

Similarly,

$$V(\Omega) = V(\infty) + \frac{2\Omega}{\pi}\mathrm{P}\int_{0}^{\infty} \frac{U(\omega)}{\omega^2 - \Omega^2}\,\mathrm{d}\omega \tag{1.3.69}$$

Moreover, direct integration shows that

$$\mathrm{P}\int_{0}^{\infty} \frac{\mathrm{d}\omega}{\omega^2 - \Omega^2} = 0 \tag{1.3.70}$$

and so we can subtract from equations (1.3.68) and (1.3.69) any constant multiple of (1.3.70) to obtain more symmetric formulations. For example, we may write

$$U(\Omega) = U(\infty) - \frac{2}{\pi}\mathrm{P}\int_{0}^{\infty} \frac{\omega V(\omega) - \Omega V(\Omega)}{\omega^2 - \Omega^2}\,\mathrm{d}\omega \tag{1.3.71}$$

$$V(\Omega) = V(\infty) + \frac{2\Omega}{\pi}\mathrm{P}\int_{0}^{\infty} \frac{U(\omega) - U(\Omega)}{\omega^2 - \Omega^2}\,\mathrm{d}\omega \tag{1.3.72}$$

Writing down the expressions for the electric susceptibility, we have

$$\chi_1^{(e)}(\Omega) = \frac{2}{\pi}\mathrm{P}\int_0^\infty \frac{\omega\chi_2^{(e)}(\omega)}{\omega^2 - \Omega^2}\,\mathrm{d}\omega \tag{1.3.73}$$

$$\chi_2^{(e)}(\Omega) = -\frac{2\Omega}{\pi}\mathrm{P}\int_0^\infty \frac{\chi_1^{(e)}(\omega)}{\omega^2 - \Omega^2}\,\mathrm{d}\omega \tag{1.3.74}$$

These equations have a practical significance. It is not necessary to devise experiments which measure both $\chi_1^{(e)}(\omega)$ and $\chi_2^{(e)}(\omega)$. It is sufficient to measure one of these quantities over a range of frequencies, the other quantity can then be derived using the expressions given above.

There will of course be analogous relations for the real and imaginary parts of the dielectric constant, since

$$\varepsilon(\omega) = \varepsilon_1(\omega) - \mathrm{j}\,\varepsilon_2(\omega) = 1 + 4\pi\chi_1^{(e)}(\omega) - \mathrm{j}4\pi\chi_2^{(e)}(\omega) \tag{1.3.75}$$

we obtain

$$\varepsilon_1(\Omega) = 1 + \frac{2}{\pi}\mathrm{P}\int_0^\infty \frac{\omega\varepsilon_2(\omega)}{\omega^2 - \Omega^2}\,\mathrm{d}\omega \tag{1.3.76}$$

$$\varepsilon_2(\Omega) = -\frac{2\Omega}{\pi}\mathrm{P}\int_0^\infty \frac{\varepsilon_1(\omega)}{\omega^2 - \Omega^2}\,\mathrm{d}\omega \tag{1.3.77}$$

1.3.7 Power Loss in a Dielectric

A dielectric medium will be indistinguishable from free space unless $\varepsilon(\omega)$ differs from unity over some frequency interval. Moreover, since we have argued that $\varepsilon(\omega)$ tends to unity as ω becomes infinite, there must be a frequency range in which, say, $\varepsilon_1(\omega)$ is varying. The Kramers–Kronig integral equation (1.3.77) now indicates that, in general, $\varepsilon_2(\omega)$ will be different from zero over some frequency range and this means that there will be power dissipation within the material.

We may evaluate the power dissipation for the harmonic case by writing out the expressions for the actual physical vector fields in terms of real quantities.

$$\mathbf{E} = \text{Real part } [\mathbf{E}_0\,\mathrm{e}^{\mathrm{j}\omega t}] = \mathbf{E}_0 \cos \omega t \tag{1.3.78}$$

$$\mathbf{D} = \text{Real part } [\varepsilon(\omega)\,\mathbf{E}_0\,\mathrm{e}^{\mathrm{j}\omega t}] = \varepsilon_1(\omega)\,\mathbf{E}_0 \cos \omega t + \varepsilon_2(\omega)\,\mathbf{E}_0 \sin \omega t \tag{1.3.79}$$

We note that we have written the time and phase relations in a particular form in these equations. These relations are $\mathbf{E} = \mathbf{E}_0\,\mathrm{e}^{\mathrm{j}\omega t}$, $\mathbf{D} = \mathbf{D}_0\,\mathrm{e}^{\mathrm{j}(\omega t-\delta)}$ where $\tan\delta = \varepsilon_2(\omega)/\varepsilon_1(\omega)$; $\mathbf{E}_0$ and $\mathbf{D}_0$ are real vectors. The choice of the sign of δ is significant and expresses the causal relation between the electric field and the polarization. We have argued previously that the polarization must always lag behind the exciting field—it is determined by $E(t - \lambda)$ in equation (1.3.27)—and so the induction field must also lag by the phase angle δ. If therefore we had chosen $\mathbf{E}$ as the real part of $\mathbf{E}_0\,\mathrm{e}^{-\mathrm{j}\omega t}$ in equation (1.3.78) we should eventually have to change the sign of δ to put $\mathbf{D}$ in the new form $D_0{}^{-\mathrm{j}(\omega t-\delta')}$.

The work done per unit volume of material when the electric displacement changes from $\mathbf{D}_1$ to $\mathbf{D}_2$ is obtained from equation (1.1.3)

$$u = \frac{1}{4\pi}\int_{\mathbf{D}_1}^{\mathbf{D}_2} \mathbf{E}.\mathrm{d}\mathbf{D} \tag{1.3.80}$$

that is

$$u = \frac{1}{4\pi}\int \mathbf{E}_0 \cos \omega t.\mathrm{d}[\varepsilon_1(\omega)\, \mathbf{E}_0 \cos \omega t + \varepsilon_2(\omega)\, \mathbf{E}_0 \sin \omega t] \tag{1.3.81}$$

The energy dissipated in one cycle per unit volume of material U' is the net work done on the material in one cycle according to equation (1.1.3c) and is obtained by evaluating this integral over the time interval $2\pi/\omega$

$$U' = \frac{\varepsilon_2(\omega)}{4\pi} E_0^2 \int_0^{2\pi/\omega} \cos \omega t \ \mathrm{d}[\sin \omega t]$$
$$= \tfrac{1}{4}\, \varepsilon_2(\omega)\, E_0^2 \tag{1.3.82}$$

The energy dissipated per second per unit volume of material, is therefore given by

$$\bar{U} = \frac{\omega}{8\pi}\, \varepsilon_2(\omega)\, E_0^2 \tag{1.3.83}$$

or equivalently

$$\bar{U} = \frac{\omega}{2}\chi^{(e)}(\omega)\, E_0^2 \tag{1.3.83a}$$

The energy dissipated per second may also be written in terms of the complex vector notation (see page 51) when the time variation is harmonic. From equation (1.3.81) we obtain

$$\bar{U} = \frac{\omega}{2\pi}\cdot\frac{1}{4\pi}\oint \frac{1}{4}(\mathbf{E} + \mathbf{E}^*).(\dot{\mathbf{D}} + \dot{\mathbf{D}}^*)\, \mathrm{d}t$$
$$= \frac{\omega}{2\pi}\cdot\frac{1}{4\pi}\,\tfrac{1}{2}\, \mathrm{Re}\,[\mathbf{E}.\dot{\mathbf{D}}^*] \oint \mathrm{d}t$$

that is,

$$\bar{U} = \tfrac{1}{2}\, \mathrm{Re}\left[\frac{1}{4\pi}\mathbf{E}.\dot{\mathbf{D}}^*\right] \tag{1.3.84}$$

Here $\mathbf{E}$ and $\mathbf{D}$ are the complex vectors, $\mathbf{E} = \mathbf{E}_0\, \mathrm{e}^{\mathrm{j}\omega t}$, and $\mathbf{D} = \mathbf{D}_0\, \mathrm{e}^{\mathrm{j}(\omega t - \delta)}$. The time average of the energy stored per unit volume may be identified as

$$\langle U_s \rangle = \frac{1}{16\pi}\, \varepsilon_1(\omega)\, E_0^2 = \tfrac{1}{2}\, \mathrm{Re}\left[\frac{1}{8\pi}\mathbf{E}.\mathbf{D}^*\right] \tag{1.3.85}$$

This formulation corresponds to choosing the reference value for the internal energy U at the value when $\mathbf{E}(t) = 0$. With this choice $\langle U_s \rangle$ reduces to the loss free value when $\varepsilon_2(\omega)$ is zero.

Since the power dissipated is essentially a positive quantity or zero equations (1.3.83), (1.3.83a) seem to imply that $\varepsilon_2(\omega)$, $\chi_2^{(e)}(\omega)$ must also be positive or zero for positive angular frequencies. This argument is, however, only rigorous at low frequencies where the electric and magnetic fields can be treated independently. It is not necessarily valid at high frequencies where the influence of the fields cannot be separated and the total power loss is derived from the combined influence of these fields. Nevertheless, even with this qualification, we may guess that since the electric field usually produces first order effects while the magnetic field usually gives rise to second order effects in a dielectric material the dominant term in the expression for the total power loss should usually arise from $\chi_2^{(e)}(\omega)$. We therefore expect $\chi_2^{(e)}(\omega)$ and $\varepsilon_2(\omega)$ to be predominantly positive quantities. We may use this information to discuss the properties of the electrostatic dielectric constant.

Equations (1.3.76) and (1.3.77) provide expressions for the static (zero frequency) dielectric constant

$$\varepsilon_1(0) = 1 + \frac{2}{\pi}\int_0^\infty \frac{\omega\, \varepsilon_2(\omega)}{\omega^2}\, d\omega \tag{1.3.86}$$

$$\varepsilon_2(0) = 0 \tag{1.3.87}$$

We note that these expressions correctly show $\varepsilon(0)$ to be a real quantity. Equation (1.3.86) also shows that the value of the electrostatic dielectric constant reflects the sum total of all dissipative processes which occur throughout the entire positive frequency range. We have argued above that $\varepsilon_2(\omega)$ is predominantly a positive quantity and so equation (1.3.86) also leads us to expect that in general $\varepsilon_1(0)$ will be greater than unity, and correspondingly $\chi^{(e)}(0)$ is positive.

Within the limits of these assumptions we may now determine in principle whether significant dissipative processes occur above any particular frequency ω_0. We need only to compare the magnitude of the integral in equation (1.3.86) evaluated over the range 0 to ω_0 with the quantity $\{\varepsilon_1(0) - 1\}$ derived from the measured static dielectric constant.

It does not follow from our guess that $\varepsilon_1(0)$ is greater than unity that $\varepsilon_1(\omega)$ continues to be greater than unity over the entire frequency range. Particular dissipative mechanisms, such as Debye relaxation processes (see page 27), may give rise to a contribution to $\varepsilon_1(\omega)$ which is always positive, but this is by no means general. We saw, when considering a damped simple harmonic oscillator model (page 28), that $\chi_1(\omega)$ was negative at high frequencies and so $\varepsilon_1(\omega)$ was less than unity in this case. A more general example of this phenomenon arises when all the absorption processes are confined to the frequency range bounded by some upper limit ω_0. We may rewrite equation (1.3.76) in the form

$$\varepsilon_1(\Omega) = 1 + \frac{2}{\pi}\mathrm{P}\int_0^{\omega_0} \frac{\omega\, \varepsilon_2(\omega)}{\omega^2 - \Omega^2}\, d\omega \tag{1.3.88}$$

At sufficiently high frequencies, $\Omega \gg \omega_0, \omega$, we may neglect ω as compared with Ω in the integrand, so that

$$\varepsilon_1(\Omega) = 1 - \frac{2}{\pi}\frac{1}{\Omega^2}\int_0^{\omega_0} \omega\, \varepsilon_2(\omega)\, d\omega \tag{1.3.89}$$

$$= 1 - \frac{a}{\Omega^2} \tag{1.3.90}$$

The coefficient a is essentially a positive quantity and so $\varepsilon_1(\omega)$ tends to one from values less than unity.

1.4 Magnetic Media

Magnetic media are bounded regions of space in which **B** is recognizably different from **H**. In such regions therefore, just as in the previous case of dielectric media, there must exist an additional vector field which is not present in free space. It is possible to take over in formal language much of the discussion presented in Section 1.3. However, there appear to be fundamental differences between magnetic and dielectric media which must be included within the formal description. For example, equation (1.1.24) indicates that the fundamental source for the electric vector field is the charge density. On the other hand, equation (1.1.21) does not provide any evidence for an analogous distribution of magnetic charges (or poles). Moreover, equation (1.1.2) strongly suggests that the fundamental source of the magnetic vector field is more likely to be associated with the current density.

1.4.1 Magnetization Field

We shall suppose that the sources of the additional vector field, which exists within a magnetic material, are microscopic currents. These currents are essentially inaccessible to experiment on the scale considered here. They will be described in terms of a current density vector $\mathbf{J}_m$, and will play a role in magnetic phenomena similar to that played by the bound charges in the case of dielectric media. Since they do not give rise to any macroscopic current flow, we must have, from equation (1.1.16)

$$\int \mathbf{j}_m \cdot \mathbf{n}\, \mathrm{d}S = 0 \tag{1.4.1}$$

This equation includes surfaces which intersect the magnetic material in only one cross-sectional area A as shown in Fig. 1.6. For such an area we shall always have

$$\int_A \mathbf{j}_m \cdot \mathbf{n}\, \mathrm{d}S = 0 \tag{1.4.2}$$

whereas, if macroscopic currents are flowing in the material,

$$\int_A \mathbf{J} \cdot \mathbf{n}\, \mathrm{d}S = I \tag{1.4.3}$$

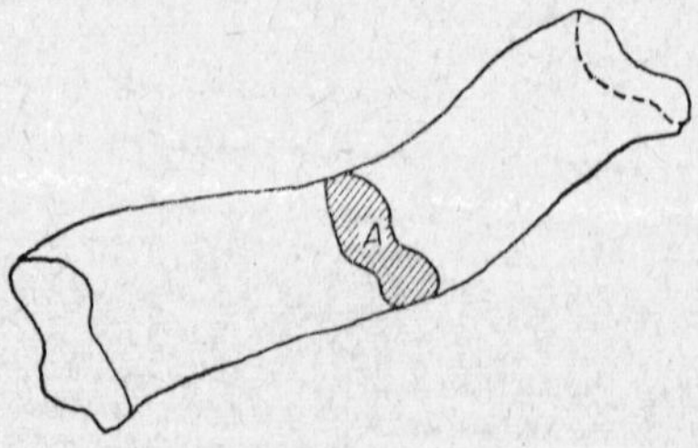

FIG. 1.6 A cross-sectional area in a magnetic material $\int \mathbf{j}_m \cdot \mathbf{n}\, \mathrm{d}S = 0$.

For the special cases of insulating materials or conductors in which no current is flowing I will, of course, vanish.

Equation (1.4.2) is satisfied for every cross-sectional area and so

$$\nabla . \mathbf{j}_m = 0 \tag{1.4.4}$$

We may therefore identify $\mathbf{j}_m$ with the curl of another vector $\mathbf{M}$

$$\mathbf{j}_m = \nabla \wedge \mathbf{M} \tag{1.4.5}$$

and $\mathbf{M}$, like the vector $\mathbf{P}$ in the previous case of dielectric materials, will vanish at all points outside the magnetic medium.

At the boundary of a magnetic material there will be surface currents derived from the current density $\mathbf{j}_m$ flowing within a narrow transition layer. These currents may be described by a surface current density $\mathbf{k}'$, which is the current crossing unit length of a line drawn in the boundary surface. Consider

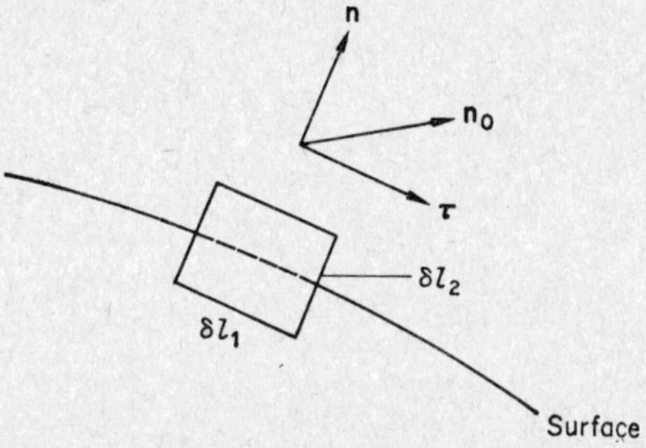

FIG. 1.7 Diagram illustrating equation (1.4.6).

a small rectangle with sides tangential and normal to this surface. If δl_1 is the length of a side in the surface and δl_2 is the length of the edge normal to the surface, $\mathbf{k}'$ is defined by the relation

$$\underset{\delta l_2 \to 0}{\text{Limit}}\, \mathbf{j}_m . \mathbf{n}_0\, \delta l_1\, \delta l_2 = \mathbf{k}' . \mathbf{n}_0\, \delta l_1 \tag{1.4.6}$$

or equivalently

$$\underset{\delta l_2 \to 0}{\text{Limit}}\, \mathbf{j}_m\, \delta l_2 = \mathbf{k}'$$

Here $\mathbf{n}_0$ is unit vector drawn normal to the element of area $\Delta S = \delta l_1\, \delta l_2$ and the limit $\delta l_2 \longrightarrow 0$ is taken to give a true discontinuity.

Evidently, from equation (1.4.5)

$$\mathbf{k}' . \mathbf{n}_0\, \delta l_1 = \underset{\delta l_2 \to 0}{\text{Limit}} \int_{\Delta S} \mathbf{j}_m . \mathbf{n}_0\, \mathrm{d}S = \underset{\delta l_2 \to 0}{\text{Limit}} \int_{\Delta S} (\nabla \wedge \mathbf{M}) . \mathbf{n}_0\, \mathrm{d}S \tag{1.4.7}$$

and so

$$\mathbf{k}' . \mathbf{n}_0 = \underset{\delta l_2 \to 0}{\text{Limit}} \oint_{\text{rectangle}} \mathbf{M} . \mathbf{ds} \tag{1.4.8}$$

Since $\mathbf{M}$ is zero outside the surface and the contribution from the two edges length δl_2 vanishes

$$\mathbf{k}' . \mathbf{n}_0 = -\mathbf{M} . \boldsymbol{\tau} \tag{1.4.9}$$

where $\boldsymbol{\tau}$ is unit vector along the side δl_1. The surface current density is therefore equal to the tangential component of the vector **M** evaluated just inside the surface of the body. The equation may be written in an alternative form since $\boldsymbol{\tau} = \mathbf{n}_0 \wedge \mathbf{n}$

$$\mathbf{k}' = \mathbf{M} \wedge \mathbf{n} \tag{1.4.10}$$

1.4.2 Magnetization and the Moment per Unit Volume

The vector **M** may be identified with the magnetic moment per unit volume of the material. The magnetic moment arising from a current distribution is defined to be

$$\mathscr{M}' = \tfrac{1}{2} \int \mathbf{r} \wedge (\mathbf{j}_m + \mathbf{J})\, \mathrm{d}v \tag{1.4.11}$$

At present we are interested in the properties arising from the inaccessible current density $\mathbf{j}_m$ and so we will assume that no macroscopic currents are flowing, that is $\mathbf{J} = 0$. Moreover, $\mathbf{j}_m = 0$ outside the magnetic material and so the integral in equation (1.4.11) may be taken over any volume which includes the magnetic material. We now have

$$\mathscr{M}' = \tfrac{1}{2} \int_{\substack{\text{volume}>\text{magnetic}\\ \text{material}}} \mathbf{r} \wedge \mathbf{j}_m\, \mathrm{d}v = \tfrac{1}{2} \int \mathbf{r} \wedge (\nabla \wedge \mathbf{M})\, \mathrm{d}v \tag{1.4.12}$$

$$= \tfrac{1}{2} \int \{\nabla(\mathbf{r}.\mathbf{M}) - (\mathbf{r}.\nabla)\mathbf{M} - (\mathbf{M}.\nabla)\mathbf{r}\}\, \mathrm{d}v$$

$$= \tfrac{1}{2} \oint (\mathbf{r}.\mathbf{M})\mathbf{n}\, \mathrm{d}S - \tfrac{1}{2} \int \{(\mathbf{r}.\nabla)\mathbf{M} + \mathbf{M}\}\, \mathrm{d}v \tag{1.4.13}$$

But

$$\nabla \wedge (\mathbf{r} \wedge \mathbf{M}) = \mathbf{r}(\nabla.\mathbf{M}) - \mathbf{M}(\nabla.\mathbf{r}) + (\mathbf{M}.\nabla)\mathbf{r} - (\mathbf{r}.\nabla)\mathbf{M}$$

and so

$$(\mathbf{r}.\nabla)\mathbf{M} = \mathbf{r}(\nabla.\mathbf{M}) - 2\mathbf{M} - \nabla \wedge (\mathbf{r} \wedge \mathbf{M}) \tag{1.4.14}$$

Using the vector relation derived for equation (1.3.10) to transform

$$\int \mathbf{r}(\nabla.\mathbf{M})\, \mathrm{d}v$$

we obtain,

$$\mathscr{M}' = \tfrac{1}{2} \oint \{(\mathbf{r}.\mathbf{M})\mathbf{n} - \mathbf{r}(\mathbf{M}.\mathbf{n}) - \mathbf{n} \wedge (\mathbf{r} \wedge \mathbf{M})\}\, \mathrm{d}S + \int \mathbf{M}\, \mathrm{d}v \tag{1.4.15}$$

$$\mathscr{M}' = -\tfrac{1}{2} \oint \mathbf{r} \wedge (\mathbf{M} \wedge \mathbf{n})\, \mathrm{d}S + \int \mathbf{M}\, \mathrm{d}v \tag{1.4.16}$$

The surface integral is evaluated over a surface which encloses the magnetic material but which nowhere penetrates within and so **M** is everywhere zero for this integral. Hence

$$\mathscr{M}' = \int_{\substack{\text{magnetic}\\ \text{material}}} \mathbf{M}\, \mathrm{d}v \tag{1.4.17}$$

and the vector **M** may be identified with the moment per unit volume of the material. It is called the magnetization.

We note here that the identification implied by equation (1.4.17), namely

$$\int \mathbf{M}\, dv = \tfrac{1}{2} \int (\mathbf{r} \wedge \mathbf{j}_{\mathrm{m}})\, dv \tag{1.4.18}$$

also relates **M** directly with the angular momentum of the material. The current density $\mathbf{j}_{\mathrm{m}}$ may be written in terms of a microscopic movement of charge, from equation (1.1.10)

$$\mathbf{j}_{\mathrm{m}} = \frac{1}{c}\rho'\mathbf{v}$$

and so

$$\int \mathbf{M}\, dv = \frac{1}{2c} \int \rho'(\mathbf{r} \wedge \mathbf{v})\, dv \tag{1.4.19}$$

If the mass associated with the charge ρ' is m,

$$\int \mathbf{M}\, dv = \frac{1}{2mc} \int \rho'(\mathbf{r} \wedge m\mathbf{v})\, dv \tag{1.4.20}$$

The magnetization of a body is therefore directly given by the angular momentum per unit volume multiplied by the factor $\rho'/2mc$. This relation is the underlying basis for the relationship between magnetic moment and orbital angular momentum in the case of atoms and molecules. In the case of an atom, $\int (\rho'/2mc)(\mathbf{r} \wedge m\mathbf{v})\, dv$ becomes $-(|e|/2mc)\langle \mathbf{L} \rangle$ where $|e|/m$ is just the ratio of charge to mass for an electron. Our discussion does not, however, provide any relation between magnetic moment and the intrinsic spin angular momentum of an electron or other fundamental particle.

1.4.3 Magnetic Induction Field

The vector field **B** in the medium is defined by superimposing the field **M** and the field **H** in the form

$$\mathbf{B} = \mathbf{H} + 4\pi\mathbf{M} \tag{1.4.21}$$

From equations (1.1.2) and (1.4.5) we derive

$$\nabla \wedge \mathbf{B} = \nabla \wedge (\mathbf{H} + 4\pi\mathbf{M}) = 4\pi(\mathbf{J} + \mathbf{j}_{\mathrm{m}}) + \frac{1}{c}\frac{\partial \mathbf{D}}{\partial t} \tag{1.4.22}$$

which implies that the microscopic current density $\mathbf{j}_{\mathrm{m}}$ is to be regarded as a source for the vector field **B** and not as a source for the field **H**. We should therefore take **B** rather than **H** as the fundamental magnetic field vector. This is not analogous to the dielectric case. In equation (1.3.16) we showed that the bound charges were also sources for the vector field **E**. We conclude therefore that the fundamental vectors for the electromagnetic field are **E** and **B** while **D** and **H** are subsidiary vectors.

1.4.4 Digression on Magnetic Poles

An alternative approach to magnetic phenomena is possible in which the mathematical equations are written in a form closely parallel to that used for describing dielectrics. To do this we make use of the idea of a magnetic charge (or pole) together with the experimental observation that materials exist for which **M** does not vanish even when there are no true currents available to produce magnetizing fields. These materials are 'permanent magnets'. For these materials we may have

$$\nabla \wedge \mathbf{H} = 0 \tag{1.4.23}$$

so that the vector **H** behaves formally like an electrostatic field (compare equations (1.2.10) and (1.2.11).

Moreover from equation (1.1.21)

$$\nabla . \mathbf{H} = -4\pi \nabla . \mathbf{M}$$

which we may write in terms of a formal magnetic source density

$$\nabla . \mathbf{H} = 4\pi \rho_{\mathrm{m}}$$

with

$$\rho_{\mathrm{m}} = -\nabla . \mathbf{M} \tag{1.4.24}$$

The field arising from permanent magnets may therefore be derived from a magnetic source density equal to the negative divergence of the magnetization. This description is equivalent to that used for dielectric materials (see equation (1.3.5)) but in the present case the magnetic charge density (and its associated unit magnetic charge or pole) are formal concepts without any immediate justification in the equations for the electromagnetic field as set out in Section 1.1. On the other hand this formulation may have some particular practical advantages. For example, when considering demagnetizing effects in bounded media, it is often more convenient to visualize the physical situation in terms of a distribution of poles on the boundary surface than as a surface current density.

1.4.5 Magnetic Permeability and Magnetic Susceptibility

For many materials the relation between **B** and **H** set out in equation (1.4.21) may be written in the linear form

$$\mathbf{B} = \mu \mathbf{H} \tag{1.4.25}$$

Here μ is a constant known as the magnetic permeability. For such materials, equation (1.4.21) implies also that

$$\mathbf{M} = \chi^{(\mathrm{m})} \mathbf{H} \tag{1.4.26}$$

and

$$\mu = 1 + 4\pi \chi^{(\mathrm{m})} \tag{1.4.27}$$

The constant $\chi^{(\mathrm{m})}$ is known as the magnetic susceptibility. Evidently μ and $\chi^{(\mathrm{m})}$ are analogous to the corresponding dielectric quantities ε and $\chi^{(\mathrm{e})}$. They will characterize the magnetic properties of a particular material under specific experimental conditions (for example the particular temperature and frequency at which measurements are being made). For the system of c.g.s. units used here, μ and $\chi^{(\mathrm{m})}$ are dimensionless quantities, and in free space $\chi^{(\mathrm{m})} = 0$ and $\mu = 1$.

When the linear relation between **B** and **H** is valid for an anisotropic material, equation (1.4.25) should be written in tensor form

$$\mathbf{B} = \overleftrightarrow{\mu}.\mathbf{H} \tag{1.4.28}$$

with components

$$B_i = \mu_{ii}H_i + \mu_{ij}H_j + \mu_{ik}H_k \tag{1.4.29}$$

Considerations, which are formally the same as those leading to equation (1.3.23) in the case of dielectrics, show that $\overleftrightarrow{\mu}$ is a symmetric tensor. It is therefore possible to choose a system of three mutually perpendicular co-ordinate axes, with corresponding 'principal magnetic permeabilities', such that in these directions **B** and **H** are parallel vectors.

The time and frequency dependence of μ and $\chi^{(m)}$ follows immediately from the discussion of dielectrics given on pages 23–31. We will summarize the results for harmonic conditions. The complex magnetic permeability is defined by

$$\mathbf{B} = \mu(\omega)\,\mathbf{H} \tag{1.4.30}$$

with

$$\mu(\omega) = 1 + 4\pi \int_{0^-}^{\infty} h(\lambda)\, e^{-j\omega\lambda}\, d\lambda \tag{1.4.31}$$

Writing

$$\mu(\omega) = \mu_1(\omega) - j\,\mu_2(\omega) \tag{1.4.32}$$

then $\mu_1(\omega)$ is an even function of frequency whilst $\mu_2(\omega)$ is an odd function of frequency. From the dispersion relations given in equations (1.3.62) and (1.3.63), we obtain

$$\mu_1(\Omega) = 1 + \frac{1}{\pi}\mathrm{P}\int_{-\infty}^{\infty} \frac{\mu_2(\omega)}{\omega - \Omega}\, d\omega \tag{1.4.33}$$

$$= 1 + \frac{2}{\pi}\mathrm{P}\int_{0}^{\infty} \frac{\omega\,\mu_2(\omega)}{\omega^2 - \Omega^2}\, d\omega \tag{1.4.34}$$

and

$$\mu_2(\Omega) = -\frac{1}{\pi}\mathrm{P}\int_{-\infty}^{\infty} \frac{\mu_1(\omega)}{\omega - \Omega}\, d\omega \tag{1.4.35}$$

$$= -\frac{2\Omega}{\pi}\mathrm{P}\int_{0}^{\infty} \frac{\mu_1(\omega)}{\omega^2 - \Omega^2}\, d\omega \tag{1.4.36}$$

There will be corresponding relations for the complex magnetic susceptibility

$$\chi^{(m)}(\omega) = \chi_1^{(m)}(\omega) - j\,\chi_2^{(m)}(\omega) = \frac{\mu(\omega) - 1}{4\pi}$$

$$\chi_1^{(m)}(\Omega) = \frac{2}{\pi}\mathrm{P}\int_{0}^{\infty} \frac{\omega\,\chi_2^{(m)}(\omega)}{\omega^2 - \Omega^2}\, d\omega \tag{1.4.37}$$

$$\chi_2^{(m)}(\Omega) = -\frac{2\Omega}{\pi}\mathrm{P}\int_{0}^{\infty} \frac{\chi_1^{(m)}(\omega)}{\omega^2 - \Omega^2}\, d\omega \tag{1.4.38}$$

1.4.6 Power Loss in a Magnetic Material

The energy dissipated per second per unit volume of material $\bar{W}$ is analogous to $\bar{U}$ in equation (1.3.83). It is given by

$$\bar{W} = \frac{\omega}{8\pi}\mu_2(\omega)\, H_0^2 \tag{1.4.39}$$

or equivalently

$$\bar{W} = \frac{\omega}{2}\chi_2^{(m)}(\omega)\, H_0^2 \tag{1.4.40}$$

and by analogy with equations (1.3.86) and (1.3.87), the static permeability is

$$\mu_1(0) = 1 + \frac{2}{\pi}\int_0^\infty \frac{\omega\, \mu_2(\omega)}{\omega^2}\, d\omega \tag{1.4.41}$$

$$\mu_2(0) = 0 \tag{1.4.42}$$

The energy relations corresponding to equations (1.3.84) and (1.3.85), using the complex vector notation, are

Average power dissipated

$$\bar{W} = \tfrac{1}{2}\,\mathrm{Re}\left[\frac{1}{4\pi}\mathbf{H}.\dot{\mathbf{B}}^*\right] \text{ per unit volume}$$

Average energy stored

$$\langle W_s\rangle = \frac{1}{16\pi}\mu_1(\omega)\, H_0^2 = \tfrac{1}{2}\,\mathrm{Re}\left[\frac{1}{8\pi}\mathbf{H}.\mathbf{B}^*\right] \text{ per unit volume}$$

These equations lead to an additional distinction between dielectric materials and magnetic materials. In the discussion of equations (1.3.84) and (1.3.85) we commented that the total power dissipation at any frequency was essentially a positive quantity. This led to the supposition that the electrostatic dielectric constant would be greater than unity. We must draw a different conclusion for magnetic materials. If it is true that the electrical terms by and large dominate the high frequency losses for a material media where the electric and magnetic fields cannot be treated separately, then the smaller magnetic terms may be either positive or negative and yet leave the total power dissipated at any frequency a positive quantity. The integral in equation (1.4.41) may therefore have a positive or negative value and the corresponding static permeability will be greater than or less than unity. Materials with static permeabilities greater than unity will have positive static susceptibilities $\chi_1^{(m)}(0) > 0$ and are known as paramagnetic materials. Materials with static permeabilities less than unity, and so having negative static susceptibilities, are known as diamagnetic materials.

The imaginary part of the low frequency susceptibility must nevertheless be positive since in this frequency range the magnetic and electric power losses can be separated. We may therefore anticipate that paramagnetic materials, which need a net positive contribution from the integral in equation (1.4.41), will be associated with power loss at relatively low frequencies.

The positive contribution from $\mu_2(\omega) = 4\pi\,\chi_2^{(m)}(\omega)$ in this range where ω is small may then be expected to outweigh any negative contribution which may arise at high frequencies. Such losses are found in many paramagnetic materials at frequencies of about 1 Mc/s; they give rise to the phenomenon known as 'paramagnetic relaxation'.

1.4.7 Magnetic Hysteresis

In our discussion so far we have seen that a wide range of physical phenomena in dielectric and magnetic materials may be described by taking a linear relation between the field vectors **D** and **E**, and between **B** and **H**. For magnetic media, however, there is, in addition, an important class of materials for which **M**, and consequently **B**, is not a linear function of **H**. Furthermore there is not necessarily a single valued, nor even a unique, functional relation between the vector fields. In these materials a magnetic moment per unit volume and vector field **M** may also exist even in the absence of macroscopic currents and exciting fields **H**. Examples of such materials are the ferromagnetic metals, iron cobalt and nickel.

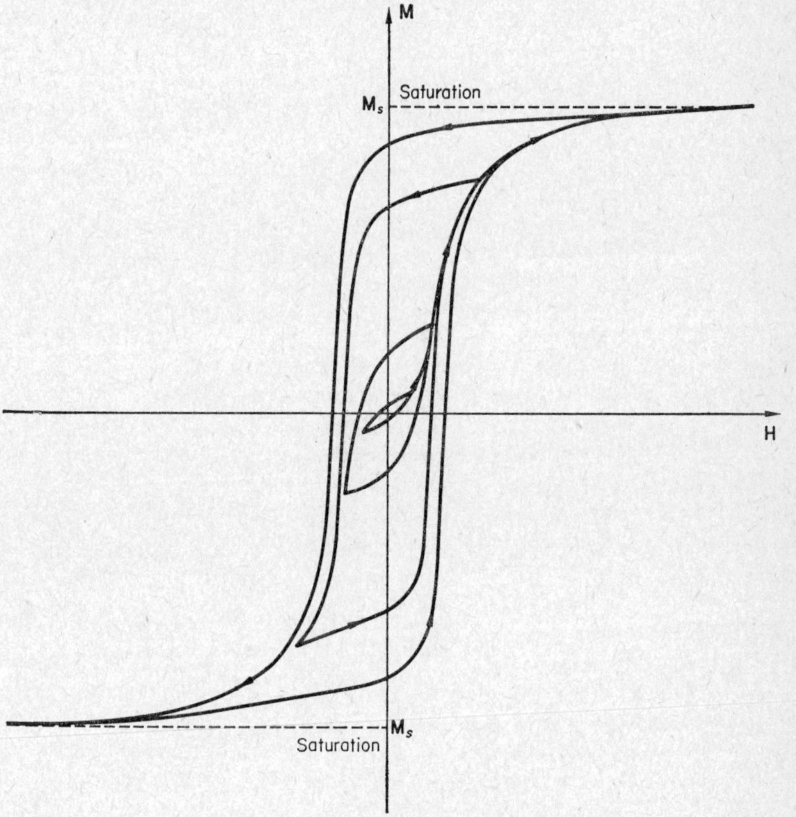

Fig. 1.8 Hysteresis loops for a ferromagnetic material.

A set of magnetization curves for such a non-linear material is shown in Fig. 1.8. Each loop is derived from a single cycle in **H** and it can be seen that in general there are two values of **M** for a given value of **H** in each cycle. The different individual curves correspond to the different extreme values of **H** which characterize a cycle. They are known as hysteresis loops. When the maximum field strength is small the loop is generally lens shaped with parabolic sides. As the maximum field strength is increased the shape becomes more nearly rectangular.

The phenomenon described by these hysteresis loops is non-conservative and has essential frictional aspects which lead to power dissipation within the material. The energy dissipated in one cycle is given by the area of the hysteresis loop. We can show this as follows.

The work done per unit volume of material in changing the state from $\mathbf{B}_1$ to $\mathbf{B}_2$ is given by

$$\Delta W = \frac{1}{4\pi}\int_{\mathbf{B}_1}^{\mathbf{B}_2} \mathbf{H}.\mathrm{d}\mathbf{B} = \frac{1}{4\pi}[\mathbf{H}.\mathbf{B}]_{\mathbf{H}_1,\mathbf{B}_1}^{\mathbf{H}_2,\mathbf{B}_2} - \frac{1}{4\pi}\int_{\mathbf{H}_1}^{\mathbf{H}_2} \mathbf{B}.\mathrm{d}\mathbf{H} \tag{1.4.43}$$

where $\mathbf{H}_1$ and $\mathbf{H}_2$ are the fields corresponding to $\mathbf{B}_1$ and $\mathbf{B}_2$. If the field variation carries the material through a complete cycle around a hysteresis loop, the net work done is

$$W = -\frac{1}{4\pi}\oint \mathbf{B}.\mathrm{d}\mathbf{H} \tag{1.4.44}$$

$$= -\oint \mathbf{M}.\mathrm{d}\mathbf{H} \tag{1.4.45}$$

The quantity W is evidently just the area of the hysteresis loop, it is called the hysteresis loss per unit volume of material and represents energy which is dissipated in the form of heat.

1.5 Conducting Media

In Section 1.2 we have defined a conducting medium as a region where **J** is not necessarily zero. This definition will effectively identify all materials as conductors in some degree. It will include not only the conventional conductors such as metals and alloys but also semiconductors (for example germanium and silicon), liquids and gases.

In practice it is usual to characterize the conducting properties of a material by a relation between the current density **J** and the electric field vector **E**. We write

$$\mathbf{J} = f(\mathbf{E}) \tag{1.5.1}$$

This relation is not to be regarded as completely general since equations (1.1.7) and (1.1.10) indicate that **J** should also depend upon the magnetic vector **B**. In fact the dependence of **J** upon **B** must be taken into account to explain, for example, the Hall effect. However, we shall ignore these magnetic effects for the moment and take equation (1.5.1) as the basis for discussion.

For the metals and alloys it is usually found that **J** is large even when **E** is small. Such materials are 'good' conductors. At the other extreme materials like quartz and sulphur pass negligible currents even in relatively strong electric fields. These materials are sufficiently near to the ideal limit of an insulating material (for which **J** is always zero) to be adequately represented as such in most instances. We are usually thinking of reasonably good conductors in the present section.

For reasonably good conductors it is adequate for most purposes to regard **J** as a linear function of **E** in equation (1.5.1). We write

$$\mathbf{J} = \frac{\sigma}{c}\mathbf{E} \tag{1.5.2}$$

for a homogeneous isotropic material. In this equation all quantities are real and σ is a constant of proportionality called the electrical conductivity. The constant c has occurred previously in equations (1.1.1) to (1.1.5) and is only written separately from σ for convenience. It is clear that σ will be large for a 'good' conductor ($\sigma = \infty$ for a perfect conductor) and small for a 'poor' conductor ($\sigma = 0$ for an ideal insulator).†

1.5.1 Ohm's Law

Equation (1.5.2) is one formulation of Ohm's law. It implies that the carriers of the current in a conductor acquire a constant velocity (a drift velocity) under the influence of a constant electric field (equation (1.1.10) states that $\mathbf{J} = (1/c)\rho\mathbf{v}$). This type of motion is not predicted by the fundamental force equation (equation (1.1.7)) which would lead to a constant acceleration for the carriers in a constant electric field. Evidently the motion of the carriers in a conducting material is determined not only by the applied electric field but also by some internal dissipative mechanism within the medium. This mechanism can remove the energy acquired by the carriers from the applied electric field at some definite rate. This dissipated energy gives rise to Joule heating. It is an irreversible process.

We may describe this loss process on a simple model where a 'collision' represents an event in which a charge carrier loses all sense of its previous velocity direction and starts off again in some random direction. Denote by $\mathbf{v}_c$ the randomly directed velocity for a charge carrier immediately after a 'collision'. The electric field **E** will impose a force field on the charge carrier always in the direction of **E**, and so at time t after a 'collision' the momentum of the charge carrier will be

$$\mathbf{p} = m\mathbf{v}_c + q\mathbf{E}t \tag{1.5.3}$$

and the average velocity for the complete set of charge carriers, which is also the drift velocity, is

$$\langle \mathbf{v}_d \rangle = \langle \mathbf{v}_c \rangle + \frac{q}{m}\mathbf{E}\langle t \rangle \tag{1.5.4}$$

In this equation $\langle t \rangle$ is the average time since the last collision. This is also the average time between collisions.

† In c.g.s. units σ has the dimensions of reciprocal time and the unit has no special name. In practical units with **J** measured in amperes per square centimetre and **E** in volts per centimetre, the unit of σ is the reciprocal ohm-cm, written $(\text{ohm-cm})^{-1}$.

The velocities $\mathbf{v}_c$ are completely random in direction and magnitude and so make no contribution to the average in equation (1.5.4). Hence,

$$\langle \mathbf{v}_d \rangle = \frac{q}{m}\mathbf{E}\langle t \rangle \tag{1.5.5}$$

and writing

$$\mathbf{J} = \frac{Nq}{c}\langle \mathbf{v}_d \rangle \tag{1.5.6}$$

from equation (1.1.10), with N the number of charge carriers per unit volume, we obtain from equation (1.5.2),†

$$\sigma = \frac{Nq^2}{m}\langle t \rangle \tag{1.5.7}$$

On this model a 'collision' process is chiefly responsible for producing a random distribution of velocities. However, since this process is unlikely to be completely elastic, there will also be an associated energy exchange. This energy exchange may be quite small so that the effect of the electric field is first of all to increase the total kinetic energy of the charge carriers to a point at which they eventually lose as much energy in a 'collision' as they gain from the electric field between 'collisions'. In this way the energy gained from the electric field is passed on to the material medium in which the charge carriers are moving.

When equilibrium has been achieved W_J, the energy which is dissipated per unit time within the conducting medium must equal the rate at which energy is acquired from the electric field. Equations (1.1.7) and (1.5.6) show that, for unit volume, this is

$$\begin{aligned} W_J &= \mathbf{F}.\langle \mathbf{v}_d \rangle \\ &= c\mathbf{E}.\mathbf{J} \end{aligned} \tag{1.5.8}$$

since

$$\langle \mathbf{v}_d \rangle.(\langle \mathbf{v}_d \rangle \wedge \mathbf{B}) = 0$$

For a linear homogeneous isotropic material this equation may also be written as

$$W_J = \sigma\mathbf{E}.\mathbf{E} = \frac{c^2}{\sigma}\mathbf{J}.\mathbf{J} \tag{1.5.9}$$

If the material is anisotropic, but still linear, equation (1.5.2) is written

$$\mathbf{J} = \frac{\overleftrightarrow{\sigma}.\mathbf{E}}{c} \tag{1.5.10}$$

where $\overleftrightarrow{\sigma}$ is a tensor and a particular component of $\mathbf{J}$ is given by

$$J_i = \frac{1}{c}(\sigma_{ii}E_i + \sigma_{ij}E_j + \sigma_{ik}E_k) \tag{1.5.11}$$

with $i, j, k \equiv x, y, z$.

† Taking $\sigma = 0{\cdot}5.10^6$ (ohm-cm)$^{-1}$, we calculate $\langle t \rangle \sim 10^{-14}$s.

This equation is mathematically equivalent to equation (1.3.21) for the anisotropic dielectric constant and to equation (1.4.29) for the anisotropic magnetic permeability. However, it is not possible to make use of the properties of the thermodynamic energy functions to determine the symmetry properties of the conductivity tensor since we are now discussing essentially irreversible phenomena. The symmetry properties are usually derived from Onsager's Principle and it is shown† that if no magnetic fields are present

$$\sigma_{ij} = \sigma_{ji} \tag{1.5.12}$$

whereas in the presence of a magnetic field

$$\sigma_{ij}(\mathbf{H}) = \sigma_{ji}(-\mathbf{H}) \tag{1.5.13}$$

It is again possible to choose a set of three mutually perpendicular coordinate directions for which $\overleftrightarrow{\sigma}$ is a diagonal tensor. In this system the principal conductivities are given by

$$J_x = \frac{\sigma_x}{c}\mathbf{E}_x, \qquad J_y = \frac{\sigma_y}{c}\mathbf{E}_y, \qquad J_z = \frac{\sigma_z}{c}\mathbf{E}_z \tag{1.5.14}$$

J and **E** will only have the same direction when the field **E** is along one of these coordinate axes.

1.5.2 Electromotive Force

The power W_J which is continuously being dissipated as Joule heat when a current flows in a conductor must be supplied by an appropriate source. It cannot be derived from a straightforward electrostatic field distribution described by equation (1.1.1) with $\partial\mathbf{B}/\partial t = 0$ since, for this case $\nabla \wedge \mathbf{E} = 0$ and the corresponding vector field **E** is conservative. If, everywhere

$$\nabla \wedge \mathbf{E} = 0 \tag{1.5.15}$$

then

$$\oint \mathbf{E}.d\mathbf{l} = 0$$

for a closed path, and so no net energy is delivered up to charges going completely around a circuit. At the most a completely conservative electrostatic field will give rise to transient currents which are the result of charge migration within the conductors.

When there is no continued motion of charge within a conductor there can be no steady state electric field in the material. The initial transient motion of the charges must have given rise to a secondary field distribution which exactly compensates that field which was applied. Thus, if the applied field is $\mathbf{E}_0$, the secondary field $\mathbf{E}_1$, then the resultant field **E** is

$$\mathbf{E} = \mathbf{E}_0 + \mathbf{E}_1 = 0 \tag{1.5.16}$$

everywhere within the conducting material. This secondary field arises from a surface distribution on the boundary of the conductor. It is evident that

† See for example Landau, L. D. and Lifshitz, E. M., *Electrodynamics of Continuous Media*, Pergamon, 1960; and Nye, J. F., *Physical Properties of Crystals*, O.U.P., 1957.

there can be no steady state charge distribution within the conductor since on taking an isotropic, linear and homogeneous material characterized by a dielectric constant ε we have, from equation (1.1.24),

$$4\pi\rho = \varepsilon \, \nabla . \mathbf{E}$$

and $\mathbf{E}$ is zero everywhere within the conductor.

The order of magnitude of time required to establish electrostatic equilibrium may be estimated from equations (1.1.14) and (1.1.24). We have

$$\nabla . \mathbf{J} + \frac{1}{c}\frac{\partial \rho}{\partial t} = 0$$

which may be written

$$\nabla . \frac{\sigma \mathbf{E}}{c} + \frac{1}{c}\frac{\partial \rho}{\partial t} = 0 \tag{1.5.17}$$

where σ is the electrical conductivity of the material. Also

$$\varepsilon \, \nabla . \mathbf{E} = 4\pi\rho \tag{1.5.18}$$

so that, combining equations (1.5.17) and (1.5.18)

$$\frac{\partial \rho}{\partial t} + \frac{4\pi\sigma}{\varepsilon}\rho = 0 \tag{1.5.19}$$

$$\rho = \rho_0 \mathrm{e}^{-t/\tau} \tag{1.5.20}$$

with $\tau = \varepsilon/4\pi\sigma$. For a good conductor we take $\sigma = 2.10^{17}$ c.g.s. units and $\varepsilon = 1$ and derive $\tau \sim 0{\cdot}4.10^{-18}$ seconds. We therefore expect the transient effects to be completed in a negligibly short time.

A steady state current requires a source which in principle can give rise to a non-conservative electric field. One possible source of such a field is a changing magnetic flux. Equation (1.1.1) shows that in this case $\nabla \wedge \mathbf{E}$ is not zero. Evidently

$$\oint_C \mathbf{E} . \mathrm{d}\mathbf{l} = -\frac{1}{c}\frac{\partial}{\partial t}\int_S \mathbf{B} . \mathbf{n} \, \mathrm{d}S \tag{1.5.21}$$

$$= -\frac{1}{c}\frac{\partial \Phi}{\partial t} \neq 0 \tag{1.5.22}$$

where Φ is the magnetic flux through any surface S whose perimeter is the circuit C. A continuous current can now be maintained in this circuit and we speak of an 'electromotive force' given in this case by

$$\mathscr{E} = \oint \mathbf{E} . \mathrm{d}\mathbf{l} = -\frac{1}{c}\frac{\partial \Phi}{\partial t} \tag{1.5.23}$$

which is responsible for the current flow in the conductors. In this particular example the energy required to maintain the current may be obtained either from the mechanical work done by the rotating machinery of a generator or from another electrical source which is capable of producing a changing magnetic flux at the circuit we are considering. The changing magnetic field gives rise to vector field $\mathbf{E}$ which is non-conservative and so can provide the driving force for the current vector $\mathbf{J}$.

Other more subtle sources of electromotive force derive their energy from chemical transformations or from thermo-electric effects. These sources are not explicitly included in the field equations set out in equations (1.1.1)–(1.1.5), and it is not so useful to try to visualize the distributed electric field **E** as to consider the electromotive force itself. The electromotive force is just the work done in taking unit charge reversibly round the closed circuit. We now therefore write, formally

$$\mathscr{E} = \oint \mathbf{E}.\mathrm{d}\mathbf{l}$$

the integral being taken round a closed circuit, with **E** no longer to be regarded as a conservative vector field. When this is carried out at constant temperature and pressure the e.m.f. is equal to the change in the Gibbs thermodynamic potential for the system.

To illustrate this point, consider the particular case of a Daniell cell; we may write the chemical transformation as

$$\mathrm{Zn}^{++} + \mathrm{CuSO}_4 \rightarrow \mathrm{Cu}^{++} + \mathrm{ZnSO}_4$$

which represents a process in which a positive zinc ion passes into solution from the zinc electrode and a copper ion is deposited at the copper electrode. At the same time an electron is transferred from the zinc electrode to the copper electrode by the external conducting circuit. If the thermodynamic potentials for the constituent parts of the system are written G(Zn), G($ZnSO_4$), G(Cu) and G($CuSO_4$) per particle, we have

$$e\mathscr{E} = \{\mathrm{G(Cu)} + \mathrm{G(ZnSO_4)}\} - \{\mathrm{G(Zn)} + \mathrm{G(CuSO_4)}\}.$$

This equation expresses the electromotive force of the cell in terms of the thermodynamic properties of the electrodes and the electrolyte solution.

If the source of non-conservative electric field is confined within a bounded region, as for example is the case for a battery or a screened generator from which a pair of accessible terminals is brought out, the field $\mathbf{E}_0$ external to the generating element will be irrotational and everywhere $\nabla \wedge \mathbf{E}_0 = 0$. We can therefore define a potential function outside the generator such that $\mathbf{E}_0 = -\mathrm{grad}\,\phi$ and the accessible terminals, say A, B, will have well-defined potentials ϕ_A and ϕ_B given by the relation

$$\int_A^B \mathbf{E}_0.\mathrm{d}\mathbf{l} = \phi_A - \phi_B \tag{1.5.24}$$

The integration can be performed along any path between A and B which is external to the generating element. If there is no electrical connection between the terminals there will be no current flowing in the circuit and so there can be no resultant force acting on the charges within the generating element itself. The work done in taking a unit charge round the circuit is therefore just $\int_A^B \mathbf{E}_0.\mathrm{d}\mathbf{l}$ externally and so the open circuit voltage for a generating element is equal to its electromotive force.

If the terminals A, B are now joined together by a conducting system a current will flow, and from equation (1.5.2)

$$\mathbf{J} = \frac{\sigma}{c}\mathbf{E}$$

where $\mathbf{E}$ is the electric field at any point in the circuit. Outside the generating element $\mathbf{E}$ remains an irrotational field whereas within the element $\mathbf{E}$ need not necessarily be irrotational and certainly will not match onto the field outside to give a vanishing line integral for any path which crosses the boundary of the element. For such a path, the line integral will give the electromotive force

$$\mathscr{E} = \oint \mathbf{E}.\mathrm{d}\mathbf{l} = c\oint \frac{\mathbf{J}}{\sigma}\cdot \mathrm{d}\mathbf{l} \tag{1.5.25}$$

The current flowing across any cross-section of the conducting circuit is constant and given by

$$I = \int_A \mathbf{J}.\mathbf{n}\,\mathrm{d}S$$

where A is the cross-section of the conductor at the point considered. Hence, if $\mathbf{J}$ and σ are constant across a cross-sectional area A_n drawn normal to the direction $\mathrm{d}\mathbf{l}$, which is also the direction of current flow, we may follow this current filament now labelled λ, around the circuit to obtain

$$I_\lambda = |\mathbf{J}_\lambda|\ A_n^\lambda; \quad \mathbf{J}_\lambda\cdot \mathrm{d}\mathbf{l}_\lambda = |\mathbf{J}_\lambda|\ \mathrm{d}l_\lambda$$

and

$$\mathscr{E} = I_\lambda \oint \frac{c\ \mathrm{d}l_\lambda}{\sigma_\lambda A_n^\lambda} \equiv I_\lambda \sum R_{ij}^{(\lambda)} \tag{1.5.26}$$

where $R_{ij}^{(\lambda)}$ is called the resistance of an element of the circuit connected between points i and j,† that is,

$$R_{ij}^{(\lambda)} = \int_i^j \frac{c\ \mathrm{d}l_\lambda}{\sigma_\lambda A_n^\lambda} \tag{1.5.27}$$

Thus the electromotive force in a circuit is equal to the product of the current and the sum of the resistances for the elements connected in series round the complete circuit. This is the basis for Kirchoff's voltage law for electrical networks. We emphasize that the summation of the resistances is to be carried out round the complete circuit.

Separating those contributions which arise from within the generator element, we have from equation (1.5.26)

$$\mathscr{E} = I(R_{\text{generator}} + R_{\text{external}}) \tag{1.5.28}$$

From this equation it follows that, when a current is drawn from the system, the potential difference between the accessible terminals of the generator is no longer equal to the electromotive force. This potential difference is given by an integration external to the generating element,

† This argument may be extended further by considering current filaments in parallel. Evidently $\phi_i - \phi_j = I_1 R_{ij}^{(1)} = I_2 R_{ij}^{(2)} = I_\lambda R_{ij}^{(\lambda)}$, etc. and so the total current

$$I = \sum I_\lambda = (\phi_i - \phi_j) \sum_\lambda \frac{1}{R_{ij}^{(\lambda)}} \quad \text{or} \quad \frac{1}{R_{ij}} = \sum_\lambda \frac{1}{R_{ij}^\lambda}$$

$$\phi_{AB} = \int_A^B \mathbf{E}_0 \cdot \mathrm{d}\mathbf{l} = c \int_A^B \frac{\mathbf{J}}{\sigma} \cdot \mathrm{d}\mathbf{l} = IR_{\text{external}} \tag{1.5.29}$$

that is,

$$\phi_{AB} = \mathscr{E} - IR_{\text{generator}} \tag{1.5.30}$$

Equation (1.5.28) is an example of Thévenin's theorem. This states that a general circuit element having two accessible terminals and containing sources of electromotive force may be replaced by an ideal electromotive force in series with an impedance. The magnitude of the electromotive force is equal to the potential difference between the terminals when they are open circuit and the series impedance is equal to that measured between the terminals when all the generators within the network are replaced by their internal impedances.

Note that, since the e.m.f. in the circuit is the work done in taking unit charge round the circuit, the rate at which energy has to be supplied when a current I is flowing is given by

$$\mathscr{E} \frac{\mathrm{d}q}{\mathrm{d}t} = c\,\mathscr{E} I$$

The total power dissipated in the circuit is, therefore, from equation (1.5.28)

$$c\,\mathscr{E} I = c(I^2 R_{\text{generator}} + I^2 R_{\text{external}}) \tag{1.5.31}$$

1.5.3 Explicitly Time-Dependent Electromotive Force

When the electromotive force in a circuit is a function of the time, the current will likewise be time dependent. For reasonably slowly varying functions of time it is still possible to visualize $\mathscr{E}(t)$ and $I(t)$ closely similar to the time invariant functions discussed previously and it is to be expected that they will be linearly related. We should not, however, restrict the relation between $\mathscr{E}(t)$ and $I(t)$ to one in which the coefficients are constant multipliers. Thus $\mathscr{E}(t)$ and $I(t)$ may also be related through differential operators. This type of linear relation is, in fact, suggested by the electromagnetic field equations (1.1.1) and (1.1.2) from which we might expect $\mathscr{E}(t)$ to be connected through $\nabla \wedge \mathbf{E}$ with $\partial\mathbf{B}/\partial t$ and so through $(\partial/\partial t)(\nabla \wedge \mathbf{H})$ to $\partial\mathbf{J}/\partial t$. Again, equation (1.1.24) indicates a dependence of $\mathscr{E}(t)$ through $\nabla . \mathbf{E}$ upon ρ and this is essentially an integral relation with the current, $\int I(t)\,\mathrm{d}t$, from equation (1.1.12).

We shall discuss the properties of these elements in detail in a later section, when we treat general circuit theory; for the moment we only wish to draw the conclusion that the electromotive force of a generator can be related to the current drawn from it quite generally through a set of differential operators. The general relation may be written

$$\mathscr{E}(t) = Z(D)\, I(t) \tag{1.5.32}$$

where $Z(D)$ is some function of the operator $D = \mathrm{d}/\mathrm{d}t$. It is usually called the impedance operator (more specifically the driving point impedance operator).

If the functions $\mathscr{E}(t)$, $I(t)$ have a time dependence which is harmonic in form, the discussion leading to equations (1.3.34) and (1.3.36) shows that we can write

$$I(t) = \mathrm{Re}\,[I_0(\omega)\,\mathrm{e}^{\mathrm{j}\omega t}] \equiv \mathrm{Re}\,[I] \tag{1.5.33}$$

and

$$\mathscr{E}(t) = \mathrm{Re}\,[Z(D)\,I_0(\omega)\,\mathrm{e}^{\mathrm{j}\omega t}] \tag{1.5.34}$$

Moreover, since the function $Z(D)$ is operating on an exponential term, we can carry out the operation formally and obtain

$$\mathscr{E}(t) = \mathrm{Re}\,[Z(\mathrm{j}\omega)\,I_0(\omega)\,\mathrm{e}^{\mathrm{j}\omega t}] \equiv \mathrm{Re}\,[ZI] \tag{1.5.35}$$

The function $Z(\mathrm{j}\omega)$ is called the impedance function, or simply the impedance of the circuit (more correctly the driving point impedance). It has symmetry properties which are formally analogous to those derived in Sections 1.3 and 1.4 where we discussed the linear relations leading to a complex dielectric constant and a complex magnetic permeability. Evidently

$$Z^*(\mathrm{j}\omega) = Z(-\mathrm{j}\omega) \tag{1.5.36}$$

so that, if we write

$$Z(\mathrm{j}\omega) = R(\omega) + \mathrm{j}\,X(\omega) \tag{1.5.37}$$

then

$$R(-\omega) = R(\omega) \quad \text{and} \quad X(-\omega) = -X(\omega) \tag{1.5.38}$$

Hence R is an even function of frequency, and X is an odd function of frequency. In circuit theory $R(\omega)$, the real part of Z, is called the resistive part of the impedance, whilst $X(\omega)$ is known as the reactive part of the impedance.

The constant current resistance, defined previously in equation (1.5.27), is evidently the zero order term in the expansion of $Z(\mathrm{j}\omega)$ as a power series in ω. The frequency-dependent terms are related by the Kramers–Kronig formulae, which were derived in sub-section 1.3.6, provided that the impedance function $Z(\mathrm{j}\omega)$ satisfies the analytical conditions for these relations to be valid. In general, physically realizable, stable (non-oscillatory) circuits, constructed from passive elements, do satisfy these conditions but we must ignore the difficulties which arise in principle in carrying out the integration to infinite frequencies. At very high frequencies the idea of a 'lumped' circuit element breaks down and the problem should really be treated in terms of wave propagation (see Chapter 2). For the practical purpose of computation, however, it is usually sufficient for $Z(\mathrm{j}\omega)$ to converge uniformly to a finite value as $\omega \to \infty$ when $Z(\mathrm{j}\omega)$ is formulated in terms of the low frequency approximation. We may then use the various relationships derived from equations (1.3.62) and (1.3.63). For example,

$$R(\Omega) = R(\infty) - \frac{2}{\pi}\mathrm{P}\int_0^\infty \frac{\omega\,X(\omega)}{\omega^2 - \Omega^2}\,\mathrm{d}\omega \tag{1.5.39}$$

$$X(\Omega) = \frac{2\Omega}{\pi}\mathrm{P}\int_0^\infty \frac{R(\omega)}{\omega^2 - \Omega^2}\,\mathrm{d}\omega \tag{1.5.40}$$

1.5.4 Power Loss for Harmonic Time Functions

The time average of the power loss per unit volume in a conducting material is obtained from equation (1.5.8)

$$\langle W_J(t)\rangle = c\langle \mathbf{E}(t)\,.\,\mathbf{J}(t)\rangle \tag{1.5.41}$$

Here $\mathbf{E}(t)$, $\mathbf{J}(t)$ are real vectors. Using the complex vector representation we may write

$$\mathbf{E}(t) = \tfrac{1}{2}(\mathbf{E} + \mathbf{E}^*)$$
$$\mathbf{J}(t) = \tfrac{1}{2}(\mathbf{J} + \mathbf{J}^*) \qquad (1.5.42)$$

and derive

$$W_J(t) = \tfrac{1}{2}\,\mathrm{Re}\,[c\mathbf{E}.\mathbf{J}] + \tfrac{1}{2}\,\mathrm{Re}\,[c\mathbf{E}.\mathbf{J}^*] \qquad (1.5.43)$$

The term $\mathbf{E}.\mathbf{J}$ oscillates at twice the fundamental frequency about zero mean value and hence the time average is zero. The term $\mathbf{E}.\mathbf{J}^*$ is independent of the time and so

$$\langle W_J(t)\rangle = \tfrac{1}{2}\,\mathrm{Re}\,[c\mathbf{E}.\mathbf{J}^*] = \tfrac{1}{2}\,\mathrm{Re}\,[c\mathbf{J}.\mathbf{E}^*] \qquad (1.5.44)$$

The alternative forms of this equation corresponding to equation (1.5.9) are

$$\langle W_J(t)\rangle = \tfrac{1}{2}\,\mathrm{Re}\,\sigma^*\,|\mathbf{E}|^2 = \tfrac{1}{2}\,\mathrm{Re}\,\sigma\,|\mathbf{E}|^2 \qquad (1.5.45)$$

and

$$\langle W_J(t)\rangle = \tfrac{1}{2}\,\mathrm{Re}\left[\frac{1}{\sigma}\right] c^2\,|\mathbf{J}|^2 \qquad (1.5.46)$$

Bibliography

ABRAHAM, M. and BECKER, R., *The Classical Theory of Electricity and Magnetism*, Blackie and Son, 1946

BLEANEY, B. I. and BLEANEY, B., *Electricity and Magnetism*, O.U.P., 1965

FEYNMAN, R. P., LEIGHTON, R. B. and SANDS, M., *The Feynman Lectures on Physics*, Addison-Wesley, 1964

KIP, A. F., *Electricity and Magnetism*, McGraw-Hill, 1962

LANDAU, L. D. and LIFSHITZ, E. M., *Electrodynamics of Continuous Media*, Pergamon, 1960

PURCELL, E. M., *Electricity and Magnetism*, McGraw-Hill, 1965

STRATTON, J. A., *Electromagnetic Theory*, McGraw-Hill, 1941

Table of Useful Mathematical Relations

1. $\mathbf{a}.\mathbf{b} \wedge \mathbf{c} = \mathbf{b}.\mathbf{c} \wedge \mathbf{a} = \mathbf{c}.\mathbf{a} \wedge \mathbf{b}$
2. $\mathbf{a} \wedge (\mathbf{b} \wedge \mathbf{c}) = (\mathbf{a}.\mathbf{c})\mathbf{b} - (\mathbf{a}.\mathbf{b})\mathbf{c}$
3. $\nabla(uw) = u\,\nabla w + w\,\nabla u$
4. $\nabla.(u\mathbf{a}) = \mathbf{a}.\nabla u + u\,\nabla.\mathbf{a}$
5. $\nabla \wedge (u\mathbf{a}) = \nabla u \wedge \mathbf{a} + u\,\nabla \wedge \mathbf{a}$
6. $\nabla(\mathbf{a}.\mathbf{b}) = (\mathbf{a}.\nabla)\mathbf{b} + (\mathbf{b}.\nabla)\mathbf{a} + \mathbf{a} \wedge (\nabla \wedge \mathbf{b}) + \mathbf{b} \wedge (\nabla \wedge \mathbf{a})$
7. $\nabla.(\mathbf{a} \wedge \mathbf{b}) = \mathbf{b}.\nabla \wedge \mathbf{a} - \mathbf{a}.\nabla \wedge \mathbf{b}$
8. $\nabla \wedge (\mathbf{a} \wedge \mathbf{b}) = \mathbf{a}\,\nabla.\mathbf{b} - \mathbf{b}\,\nabla.\mathbf{a} + (\mathbf{b}.\nabla)\mathbf{a} - (\mathbf{a}.\nabla)\mathbf{b}$
9. $\nabla \wedge \nabla \wedge \mathbf{a} = \nabla\nabla.\mathbf{a} - \nabla^2\mathbf{a}$
10. $\nabla \wedge \nabla u = 0$
11. $\nabla.\nabla \wedge \mathbf{a} = 0$
12. $\nabla.\mathbf{r} = 3, \quad \nabla \wedge \mathbf{r} = 0, \quad \mathbf{r} = \hat{\mathbf{i}}x + \hat{\mathbf{j}}y + \hat{\mathbf{k}}z$
13. $\int_V \nabla u\,\mathrm{d}v = \int_S u\,\mathbf{n}\,\mathrm{d}S$

14. $\int_V \nabla . \mathbf{a}\, dv = \int_S \mathbf{a}.\mathbf{n}\, dS$

15. $\int_V \nabla \wedge \mathbf{a}\, dv = \int_S \mathbf{n} \wedge \mathbf{a}\, dS$

16. $\int_S \mathbf{n} \wedge \nabla u\, dS = \int_C u\, d\mathbf{l}$

17. $\int_S \nabla \wedge \mathbf{a}.\mathbf{n}\, dS = \int_C \mathbf{a}.d\mathbf{l}$

18. $\int_V \nabla u.(\nabla \wedge \mathbf{a})\, dv = \int_S u\,(\nabla \wedge \mathbf{a}).\mathbf{n}\, dS = \int_S (\mathbf{a} \wedge \nabla u).\mathbf{n}\, dS$

Cartesian Coordinates

19. $\mathbf{a} \wedge \mathbf{b} = \begin{vmatrix} \hat{\mathbf{i}} & \hat{\mathbf{j}} & \hat{\mathbf{k}} \\ a_x & a_y & a_z \\ b_x & b_y & b_z \end{vmatrix}$

20. $\nabla . \mathbf{a} = \frac{\partial a_x}{\partial x} + \frac{\partial a_y}{\partial y} + \frac{\partial a_z}{\partial z}$

21. $\nabla \wedge \mathbf{a} = \begin{vmatrix} \hat{\mathbf{i}} & \hat{\mathbf{j}} & \hat{\mathbf{k}} \\ \frac{\partial}{\partial x} & \frac{\partial}{\partial x} & \frac{\partial}{\partial x} \\ a_x & a_y & a_z \end{vmatrix}$

22. $\nabla u = \hat{\mathbf{i}} \frac{\partial u}{\partial x} + \hat{\mathbf{j}} \frac{\partial u}{\partial y} + \hat{\mathbf{k}} \frac{\partial u}{\partial z}$

23. $\nabla.(\hat{\mathbf{k}} \wedge \mathbf{a}) = -\hat{\mathbf{k}}.\nabla \wedge \mathbf{a}$

24. $\nabla \wedge (\hat{\mathbf{k}} \wedge \mathbf{a}) = \hat{\mathbf{k}}\, \nabla.\mathbf{a} - \frac{\partial \mathbf{a}}{\partial z}$

25. $\nabla \wedge (u\hat{\mathbf{k}}) = \nabla u \wedge \hat{\mathbf{k}}$

26. $\nabla^2 u = \frac{\partial^2 u}{\partial x^2} + \frac{\partial^2 u}{\partial y^2} + \frac{\partial^2 u}{\partial z^2}$

27. $\nabla^2 \mathbf{a} = \hat{\mathbf{i}}\, \nabla^2 a_x + \hat{\mathbf{j}}\, \nabla^2 a_y + \hat{\mathbf{k}}\, \nabla^2 a_z$

28. $\nabla^2 (u\hat{\mathbf{k}}) = \hat{\mathbf{k}}\, \nabla^2 u$

29. $\nabla^2 (\hat{\mathbf{k}} \wedge \mathbf{a}) = \hat{\mathbf{k}} \wedge \nabla^2 \mathbf{a}$

Cylindrical Coordinates

30. Orthogonal line elements: dr, $r\, d\phi$, dz

31. $\mathbf{a}_r = \hat{\mathbf{i}} \cos\phi + \hat{\mathbf{j}} \sin\phi$
$\mathbf{a}\phi = -\hat{\mathbf{i}} \sin\phi + \hat{\mathbf{j}} \cos\phi$

32. $\hat{\mathbf{i}} = \mathbf{a}_r \cos\phi - \mathbf{a}_\phi \sin\phi$
$\hat{\mathbf{j}} = \mathbf{a}_r \sin\phi + \mathbf{a}_\phi \cos\phi$

33. $\mathbf{A}.\mathbf{B} = A_r B_r + A_\phi B_\phi + A_z B_z$

34. $\mathbf{A} \wedge \mathbf{B} = \begin{vmatrix} \mathbf{a}_r & \mathbf{a}_\phi & \mathbf{a}_z \\ A_r & A_\phi & A_z \\ B_r & B_\phi & B_z \end{vmatrix}$

35. $\nabla u = \frac{\partial u}{\partial r}\mathbf{a}_r + \frac{1}{r}\frac{\partial u}{\partial \phi}\mathbf{a}_\phi + \frac{\partial u}{\partial z}\mathbf{a}_z$

36. $\nabla . \mathbf{A} = \frac{1}{r}\frac{\partial}{\partial r}(rA_r) + \frac{1}{r}\frac{\partial A_\phi}{\partial \phi} + \frac{\partial A_z}{\partial z}$

37. $\nabla \wedge \mathbf{A} = \left(\frac{1}{r}\frac{\partial A_z}{\partial \phi} - \frac{\partial A_\phi}{\partial z}\right)\mathbf{a}_r + \left(\frac{\partial A_r}{\partial z} - \frac{\partial A_z}{\partial r}\right)\mathbf{a}_\phi + \left(\frac{1}{r}\frac{\partial}{\partial r}(rA_\phi) - \frac{1}{r}\frac{\partial A_r}{\partial \phi}\right)\mathbf{a}_z$

38. $\nabla . \mathbf{a}_r = \frac{1}{r}; \quad \nabla . \mathbf{a}_\phi = \nabla . \mathbf{a}_z = 0$

39. $\nabla \wedge \mathbf{a}_\phi = \frac{\mathbf{a}_z}{r}; \quad \nabla \wedge \mathbf{a}_r = \nabla \wedge \mathbf{a}_z = 0$

40. $\nabla^2 u = \frac{1}{r}\frac{\partial}{\partial r}\left(r\frac{\partial u}{\partial r}\right) + \frac{1}{r^2}\frac{\partial^2 u}{\partial \phi^2} + \frac{\partial^2 u}{\partial z^2}$

41. $\nabla^2 \mathbf{A} = \left[\nabla^2 A_r - \frac{A_r}{r^2} - \frac{2}{r^2}\frac{\partial A_\phi}{\partial \phi}\right]\mathbf{a}_r + \left[\nabla^2 A_\phi - \frac{A_\phi}{r^2} + \frac{2}{r^2}\frac{\partial A_r}{\partial \phi}\right]\mathbf{a}_\phi + (\nabla^2 A_z)\mathbf{a}_z$

Spherical Coordinates

42. Orthogonal line elements: $\mathrm{d}r$, $r\,\mathrm{d}\theta$, $r \sin\theta\, \mathrm{d}\phi$

43. $\mathbf{a}_r = \hat{\mathbf{i}} \sin\theta\cos\phi + \hat{\mathbf{j}}\sin\theta\sin\phi + \hat{\mathbf{k}}\cos\theta$
$\mathbf{a}_\theta = \hat{\mathbf{i}}\cos\theta\cos\phi + \hat{\mathbf{j}}\cos\theta\sin\phi - \hat{\mathbf{k}}\sin\theta$
$\mathbf{a}_\phi = -\hat{\mathbf{i}}\sin\phi + \hat{\mathbf{j}}\cos\phi$

44. $\hat{\mathbf{i}} = \mathbf{a}_r\sin\theta\cos\phi + \mathbf{a}_\theta\cos\theta\cos\phi - \mathbf{a}_\phi\sin\phi$
$\hat{\mathbf{j}} = \mathbf{a}_r\sin\theta\sin\phi + \mathbf{a}_\theta\cos\theta\sin\phi + \mathbf{a}_\phi\cos\phi$
$\hat{\mathbf{k}} = \mathbf{a}_r\cos\theta - \mathbf{a}_\theta\sin\theta$

45. $\mathbf{A}.\mathbf{B} = \mathbf{A}_r\mathbf{B}_r + \mathbf{A}_\theta\mathbf{B}_\theta + \mathbf{A}_\phi\mathbf{B}_\phi$

46. $\mathbf{A} \wedge \mathbf{B} = \begin{vmatrix} \mathbf{a}_r & \mathbf{a}_\theta & \mathbf{a}_\phi \\ A_r & A_\theta & A_\phi \\ B_r & B_\theta & B_\phi \end{vmatrix}$

47. $\nabla u = \frac{\partial u}{\partial r}\mathbf{a}_r + \frac{1}{r}\frac{\partial u}{\partial \theta}\mathbf{a}_\theta + \frac{1}{r\sin\theta}\frac{\partial u}{\partial \phi}\mathbf{a}_\phi$

48. $\nabla . \mathbf{A} = \frac{1}{r^2}\frac{\partial}{\partial r}(r^2 A_r) + \frac{1}{r\sin\theta}\frac{\partial}{\partial \theta}(\sin\theta A_\theta) + \frac{1}{r\sin\theta}\frac{\partial A_\theta}{\partial \phi}$

49. $\nabla \wedge \mathbf{A} = \frac{1}{r\sin\theta}\left\{\frac{\partial}{\partial \theta}(\sin\theta A_\phi) - \frac{\partial A_\theta}{\partial \phi}\right\}\mathbf{a}_r + \frac{1}{r}\left\{\frac{1}{\sin\theta}\frac{\partial A_r}{\partial \phi} - \frac{\partial}{\partial r}(rA_\phi)\right\}\mathbf{a}_\theta + \frac{1}{r}\left\{\frac{\partial}{\partial r}(rA_\theta) - \frac{\partial A_r}{\partial \theta}\right\}\mathbf{a}_\phi$

50. $\nabla . \mathbf{a}_r = \dfrac{2}{r}; \quad \nabla . \mathbf{a}_\theta = \dfrac{1}{r \tan \theta}; \quad \nabla . \mathbf{a}_\phi = 0$

51. $\nabla \wedge \mathbf{a}_r = 0; \quad \nabla \wedge \mathbf{a}_\theta = \dfrac{\mathbf{a}_\phi}{r}; \quad \nabla \wedge \mathbf{a}_\phi = \dfrac{1}{r \tan \theta} \mathbf{a}_r - \dfrac{1}{r} \mathbf{a}_\theta$

52. $$\nabla^2 u = \frac{1}{r^2} \frac{\partial}{\partial r}\left(r^2 \frac{\partial u}{\partial r}\right) + \frac{1}{r^2 \sin \theta} \frac{\partial}{\partial \theta}\left(\sin \theta \frac{\partial u}{\partial \theta}\right) + \frac{1}{r^2 \sin^2 \theta} \frac{\partial^2 u}{\partial \phi^2}$$

53. $$\begin{aligned} \nabla^2 \mathbf{A} = {} & \left(\nabla^2 A_r - \frac{2A_r}{r^2} - \frac{2A_\theta \cot \theta}{r^2} - \frac{2}{r^2} \frac{\partial A_\theta}{\partial \theta} - \frac{2}{r^2 \sin \theta} \frac{\partial A_\phi}{\partial \phi}\right) \mathbf{a}_r \\ & + \left(\nabla^2 A_\theta + \frac{2}{r^2} \frac{\partial A_r}{\partial \theta} - \frac{A_\theta}{r^2 \sin^2 \theta} - \frac{2 \cos \theta}{r^2 \sin^2 \theta} \frac{\partial A_\phi}{\partial \phi}\right) \mathbf{a}_\theta \\ & + \left(\nabla^2 A_\phi + \frac{2}{r^2 \sin \theta} \frac{\partial A_r}{\partial \phi} - \frac{1}{r^2 \sin^2 \theta} A_\phi + \frac{2 \cos \theta}{r^2 \sin^2 \theta} \frac{\partial A_\theta}{\partial \phi}\right) \mathbf{a}_\phi \end{aligned}$$

Problems

1. Given the vector function

$$\mathbf{F} = \hat{\mathbf{i}}(6xy) + \hat{\mathbf{j}}(3x^2 - y^2) + \hat{\mathbf{k}}.0$$

show that $\mathbf{F}$ represents a possible electrostatic field. Determine the electrostatic potential function $\phi(x, y, z)$ and a charge density $\rho(x, y, z)$ corresponding to this field.

2. Show that the electrostatic potential at a point with coordinates (x, y, z) due to a charge distribution $\rho(\xi, \eta, \zeta)$ with coordinates (ξ, η, ζ) is given by

$$\phi(x, y, z) = \int_{\xi, \eta, \zeta} \frac{\rho(\xi, \eta, \zeta) \, d\xi \, d\eta \, d\zeta}{\{(x - \xi)^2 + (y - \eta)^2 + (z - \zeta)^2\}^{1/2}}$$

Find the electrostatic potential (a) at a point along the axis of a uniformly charged disc of radius R, (b) at a point on the rim of the disc.

Show that the uniformly charged disc is not an equipotential surface.

3. An uncharged dielectric sphere of radius, a, is placed in a uniform electrostatic field. Assume that the dielectric is linear, isotropic and homogeneous and that at large distances from the sphere the electric field is a constant vector $\mathbf{E}_0$. Derive the potential functions inside and outside the sphere taking

$$\phi_i = Ar \cos \theta \qquad \text{for } r \leqslant a$$

$$\phi_0 = \left(Br + \frac{b}{r^2}\right) \cos \theta \qquad \text{for } r \geqslant a$$

with θ the angle between $\mathbf{r}$ and $\mathbf{E}_0$.

Show that the sphere is uniformly polarized and that

$$\mathbf{P} = \frac{3}{4\pi}\left(\frac{\varepsilon - 1}{\varepsilon + 2}\right) \mathbf{E}_0$$

4. A small circular disc of radius a is removed from a spherical shell of radius

$R(a \ll R)$. If the shell (R) is uniformly charged with a density σ per unit area calculate the magnitude and direction of the field at the centre of the aperture.

5. A spherical distribution of charge, radius a, has a uniform charge density ρ_0. Determine the self-energy of the distribution (a) using equation (1.2.34), (b) using equation (1.2.42).

6. Show that the self-energy of a small uniformly magnetized sphere of radius a is $\frac{8}{9}\pi^2 a_0^3 M_0$ where M_0 is the moment per unit volume.

7. Show that the mutual potential energy of two dipoles may be written

$$W_m = \left\{\frac{\mathbf{m}_1 . \mathbf{m}_2}{r^3} - \frac{3(\mathbf{m}_1 . \mathbf{r})(\mathbf{m}_2 . \mathbf{r})}{r^5}\right\}$$

Use this potential function to evaluate the forces and torques acting on the two dipoles when they lie in the same plane.

Note that the torque acting on $\mathbf{m}_1$ is not equal and opposite to that acting on $\mathbf{m}_2$. Explain this paradox.

8. A single circular loop of wire, radius a, is placed coplanar and at the centre of a larger circular loop of radius b $(b \gg a)$. The loops carry constant currents I_a and I_b, respectively.

Determine the torque acting on the smaller loop when it has been rotated from the coplanar position through an angle θ, the axis of rotation being a diameter located in the plane of the larger loop.

9. Derive the time invariant relation

$$\int \mathbf{H} . d\mathbf{l} = 4\pi \int \mathbf{J} . \mathbf{n} \, dS$$

Show that the magnetic field inside an ideal straight solenoid is uniform across a cross-section, with magnitude $H = 4\pi n I$, where n is the number of turns per unit length and I is the current in the wire.

10. Show that the magnetic field inside an ideal torus is everywhere tangential to circles having the axis of the torus as centre. Find the magnitude of the field inside the torus and show that the field outside (including the central hole) is zero.

11. Show that a magnetic scalar potential at a point on the axis (the z-axis) of a single circular loop of radius a is given by

$$\phi^*(z) = \tfrac{1}{2} I \left\{1 - \frac{z}{\sqrt{(a^2 + z^2)}}\right\}$$

12. Show that the magnetic field due to a dipole situated at the origin of coordinates is given by

$$\mathbf{H} = -\frac{\mathbf{m}}{r^3} + \frac{3(\mathbf{m} . \mathbf{r})}{r^5}\mathbf{r}$$

Verify that this field may be derived from either

(a) the vector potential $\mathbf{A} = \dfrac{\mathbf{m} \wedge \mathbf{r}}{r^3}$ or

(b) the scalar potential $\phi^* = \dfrac{\mathbf{m} \cdot \mathbf{r}}{r^3}$

13. Two parallel conducting planes are maintained at electric potentials ϕ_1, ϕ_2 by means of a battery. The space between the planes is filled with two linear, isotropic and homogeneous dielectric materials, the interface between the materials being parallel to the conducting planes. If the thickness of one dielectric is d_1 the dielectric constant ε_1, the conductivity σ_1, and the corresponding quantities for the second dielectric are d_2, ε_2 and σ_2, determine the electric potential and charge distribution at the interface.

14. The surface of a homogeneous isotropic conducting sphere has a potential distribution $\phi = \phi_0 \cos \theta$ maintained by means of an external source of e.m.f. Here θ is the angle with respect to a diameter of the sphere. Determine the current density $\mathbf{J}$ at any point inside the sphere and the total Joule losses for the sphere.

15. The electrical conductivity of a copper wire is $0{\cdot}5 . 10^6$ (ohm-cm)$^{-1}$. Determine the average collision time for an electron assuming that each copper atom contributes one conduction electron. [Take the atomic weight of copper as 63·6, the density $= 8{\cdot}93$ gram cm^{-3}, Avogadro's number $= 6{\cdot}02 . 10^{23}$ atoms mole^{-1}, $m = 9 \times 10^{-28}$ gram and $|e| = 4{\cdot}8 . 10^{-10}$ e.s.u.]

Calculate the drift velocity when the wire carries a current density of 100 A cm^{-2}.

16. A linear, isotropic and homogeneous dielectric cylinder rotates with angular velocity ω about its axis. Determine the induced polarization charge when a steady magnetic field $\mathbf{H}$ is maintained along the axis of rotation.

17. The complex amplitude of reflection for an electromagnetic wave at a surface may be written

$$R = |R| \, e^{-j\theta}$$

Assuming that the Kramers–Kronig relations may be applied to this function, discuss the validity of the relation

$$\theta(\Omega_1) - \theta(\Omega_2) = -\frac{2}{\pi} \mathrm{P} \int_0^\infty \log_e |R| \left(\frac{\Omega_1}{\omega^2 - \Omega_1^2} - \frac{\Omega_2}{\omega^2 - \Omega_2^2} \right) d\omega$$

18. An insulating dielectric material has a single absorption line at frequency ω_0. Writing $\varepsilon_2(\omega) = \alpha\{\delta(\omega - \omega_0) - \delta(\omega + \omega_0)\}$ with α a positive constant, derive the zero frequency dielectric constant $\varepsilon(0)$. If ω_1 is the frequency at which $\varepsilon_1(\omega) = 0$, show that

$$\varepsilon_1(0) = \left(\frac{\omega_1}{\omega_0} \right)^2$$

Chapter 2

Electromagnetic Waves

2.1 Introduction

In the previous chapter we have given a simplified description of electromagnetic phenomena based on the identification of three ideal categories of material media: dielectric media, magnetic media, and conducting media. This approach was suggested by the time invariant form of the electromagnetic field equations together with the assumptions that **D** is a linear function of **E**, and **B** is a linear function of **H**. For the time invariant case equations (1.1.1)–(1.1.5) and (1.1.21), (1.1.24), give

$$\nabla \wedge \mathbf{E} = 0 \qquad \nabla . \varepsilon \mathbf{E} = 4\pi\rho \tag{2.1.1}$$

$$\nabla \wedge \mathbf{H} = 0 \qquad \nabla . \mu \mathbf{H} = 0 \tag{2.1.2}$$

in regions where no macroscopic currents are flowing. According to these equations, electric and magnetic phenomena are separable and the fields are solutions of independent pairs of differential equations. The electric and magnetic fields may be determined independently of each other and superposed without mutual interaction.

In Chapter 1 electric and magnetic phenomena were treated separately even when the field vectors were functions of time. This method may be a reasonable approximation for slowly varying time functions but strictly, whenever the time variation has to be included, the complete set of coupled differential equations should be solved simultaneously. The solutions of the field equations now describe truly electromagnetic phenomena. The concepts of dielectric constant, magnetic permeability and electrical conductivity developed in Chapter 1, may continue to be useful, but in practice a physical measurement will involve all three of these quantities. A description given in terms of one of them may not be unique.

We may illustrate the way in which the properties of a material medium are combined in electromagnetic phenomena by considering the relationship between the dielectric constant and the electrical conductivity. For harmonic time functions, taken in the form $e^{j\omega t}$, the field equations are

$$\nabla \wedge \mathbf{E} + \frac{j\omega\mu}{c}\mathbf{H} = 0 \tag{2.1.3}$$

$$\nabla \wedge \mathbf{H} - \frac{j\omega\varepsilon}{c}\mathbf{E} - \frac{4\pi\sigma}{c}\mathbf{E} = \nabla \wedge \mathbf{H} - \frac{j\omega}{c}\left(\varepsilon - j\frac{4\pi\sigma}{\omega}\right)\mathbf{E} = 0 \tag{2.1.4}$$

Here it has been assumed that there are no current sources in the medium and that the current density **J** is derived from the action of the electric field vector according to the relation

$$\mathbf{J} = \frac{\sigma}{c}\mathbf{E} \tag{2.1.5}$$

Equation (2.1.4) has the same mathematical form as that which describes an insulating dielectric medium, namely,

$$\nabla \wedge \mathbf{H} - \frac{j\omega\varepsilon}{c}\mathbf{E} = 0 \tag{2.1.6}$$

We may therefore introduce a composite dielectric constant for a conducting medium by the relation

$$\varepsilon_c(\omega) = \varepsilon(\omega) - j\,\frac{4\pi\,\sigma(\omega)}{\omega} \tag{2.1.7}$$

and the field equations take the symmetrical form

$$\nabla \wedge \mathbf{E} + \frac{j\omega\mu}{c}\mathbf{H} = 0 \tag{2.1.8}$$

$$\nabla \wedge \mathbf{H} - \frac{j\omega\varepsilon_c}{c}\mathbf{E} = 0 \tag{2.1.9}$$

In practice, an experiment for measuring the dielectric constant of a material will usually determine $\varepsilon_c(\omega)$. Since $\varepsilon(\omega)$ and $\sigma(\omega)$ are both functions of frequency and complex, it is possible to describe a lossy dielectric material either in terms of a polarization effect, or a conductivity effect, or a combination of these effects. When describing the electromagnetic properties of a material, therefore, it is not always useful to identify separately the dielectric, magnetic and conductivity contributions. Instead a new parameter, such as the refractive index, may be defined which combines ε, μ and σ in a relation for the velocity of an electromagnetic wave, or a 'dispersion relation' may be used which characterizes a medium by an equation relating the angular frequency, ω, and the wave vector $\mathbf{k}$.

2.2 Coupled Field Equations

A solution of the coupled differential equations set out below, which also satisfies the prescribed boundary conditions, is a valid electromagnetic field distribution in the linear approximation for **D** and **B**.

$$\nabla \wedge \mathbf{E} + \frac{\mu}{c}\frac{\partial \mathbf{H}}{\partial t} = 0 \tag{2.2.1}$$

$$\nabla \wedge \mathbf{H} - \frac{\varepsilon}{c}\frac{\partial \mathbf{E}}{\partial t} - 4\pi\mathbf{J} = 0 \tag{2.2.2}$$

$$\nabla . \mu\mathbf{H} = 0 \qquad \nabla . \varepsilon\mathbf{E} = 4\pi\rho \tag{2.2.3}$$

In practice it does not prove to be useful to seek a completely general set of solutions for these equations, the mathematical difficulties are too formidable. Instead, solutions of a desired form (or type) are sought in the expectation that these will be of particular significance for the problem under investigation.

When it is convenient to seek solutions in rectangular Cartesian coordinates, it is usual to rearrange equations (2.2.1)–(2.2.3) at the outset so that an equation involving only **E** or **H** can be used. Thus, if ε, μ, σ are spatially constant and non-zero, and there are no sources present so that **J** is derived from the action of the electromagnetic field alone, $\mathbf{J} = (\sigma/c)\mathbf{E}$, whilst $\rho = 0$, we have from equation (2.2.1)

$$\nabla \wedge \nabla \wedge \mathbf{E} + \frac{\mu}{c}\frac{\partial}{\partial t}(\nabla \wedge \mathbf{H}) = 0 \tag{2.2.4}$$

Substituting from equation (2.2.2)

$$\nabla \wedge \nabla \wedge \mathbf{E} + \frac{\varepsilon\mu}{c^2}\frac{\partial^2 \mathbf{E}}{\partial t^2} + \frac{4\pi\,\sigma\mu}{c^2}\frac{\partial \mathbf{E}}{\partial t} = 0 \tag{2.2.5}$$

For rectangular Cartesian coordinates

$$\nabla \wedge \nabla \wedge \mathbf{E} = \nabla\,\nabla . \mathbf{E} - \nabla^2 \mathbf{E} \tag{2.2.6}$$

with $\nabla^2 \mathbf{E}$ given by

$$\nabla^2 \mathbf{E} = \hat{\mathbf{i}}\left(\frac{\partial^2 E_x}{\partial x^2} + \frac{\partial^2 E_x}{\partial y^2} + \frac{\partial^2 E_x}{\partial z^2}\right) + \hat{\mathbf{j}}\left(\frac{\partial^2 E_y}{\partial x^2} + \frac{\partial^2 E_y}{\partial y^2} + \frac{\partial^2 E_y}{\partial z^2}\right) + \hat{\mathbf{k}}\left(\frac{\partial^2 E_z}{\partial x^2} + \frac{\partial^2 E_z}{\partial y^2} + \frac{\partial^2 E_z}{\partial z^2}\right) \tag{2.2.7}$$

which may be written concisely as

$$\nabla^2 \mathbf{E} = \hat{\mathbf{i}}\,\nabla^2 E_x + \hat{\mathbf{j}}\,\nabla^2 E_y + \hat{\mathbf{k}}\,\nabla^2 E_z \tag{2.2.8}$$

Since ε, μ, $\neq 0$ by assumption, equation (2.2.3) gives

$$\nabla . \mathbf{H} = 0 \qquad \nabla . \mathbf{E} = 0 \tag{2.2.9}$$

and so equations (2.2.5), (2.2.6) now give

$$\nabla^2 \mathbf{E} - \frac{\varepsilon\mu}{c^2}\frac{\partial^2 \mathbf{E}}{\partial t^2} - \frac{4\pi\,\sigma\mu}{c^2}\frac{\partial \mathbf{E}}{\partial t} = 0 \tag{2.2.10}$$

Using a similar technique with equation (2.2.2) we derive

$$\nabla^2 \mathbf{H} - \frac{\varepsilon\mu}{c^2}\frac{\partial^2 \mathbf{H}}{\partial t^2} - \frac{4\pi\,\sigma\mu}{c^2}\frac{\partial \mathbf{H}}{\partial t} = 0 \tag{2.2.11}$$

Equations (2.2.10) and (2.2.11) may be written out in terms of the Cartesian vector components to provide three separate scalar differential equations for each of the field vectors **E** and **H**. These equations all have the same mathematical form, which may be written,

$$\nabla^2\psi(x, y, z, t) - \frac{\varepsilon\mu}{c^2}\frac{\partial^2}{\partial t^2}\psi(x, y, z, t) - \frac{4\pi\,\sigma\mu}{c^2}\frac{\partial\psi}{\partial t}(x, y, z, t) = 0 \tag{2.2.12}$$

It is not permissible, however, to select independent solutions of this differential equation arbitrarily and then to regard the resulting field components as a valid electromagnetic field. Even when solutions are found which satisfy the boundary conditions the vectors **E** and **H** must be further restricted to functions which are related by equations (2.2.1), (2.2.2) and (2.2.9).

A useful technique, when the field vectors are harmonic functions of the time, is to solve equation (2.2.12) together with equation (2.2.9) for one field vector. The other field vector is then derived by means of the relation

$$\mathbf{H} = -\frac{c}{j\omega\mu}\nabla \wedge \mathbf{E} \tag{2.2.13}$$

or

$$\mathbf{E} = \frac{c}{j\omega\varepsilon_c}\nabla \wedge \mathbf{H} \tag{2.2.14}$$

where the time dependence is $e^{j\omega t}$ and ε_c is given by equation (2.1.7). For more general time functions, however, equations (2.2.1) and (2.2.2) provide only differential equations relating **E** and **H**. In these cases it may be possible to use alternative mathematical techniques involving 'potential functions', or Laplace Transforms.

If it is not convenient to seek solutions for the field equations in rectangular Cartesian coordinates (for example when cylindrical or spherical polar coordinates may be more appropriate for the boundary conditions) the most direct method of solution is often to be preferred. In this case equations (2.2.1), (2.2.2) and (2.2.3) have to be written out separately in terms of the individual field components. General solutions are usually complicated mathematical functions involving cylindrical harmonics, spherical harmonics and Bessel functions. In these circumstances it may be advantageous to solve the equations indirectly by using 'potential functions' from which the field vectors can be derived at a later stage in the calculation.

When sources of the electromagnetic field, $\mathbf{J}_s$, ρ, have to be included in the calculation the standard technique for solving the field equations makes use of vector and scalar potential functions. Formal solutions can be found for a vector potential $\mathbf{A}(\mathbf{r}, t)$ and a scalar potential $\phi(\mathbf{r}, t)$ in terms of integrals over the current and charge distributions. The field vectors themselves are then derived from the relations

$$\mathbf{B} = \nabla \wedge \mathbf{A} \tag{2.2.15}$$

and

$$\mathbf{E} = -\frac{1}{c}\frac{\partial \mathbf{A}}{\partial t} - \nabla\phi \tag{2.2.16}$$

This technique will be discussed in more detail in Section 2.11.

2.3 Plane Waves

Consider the scalar differential equation for the field components, using Cartesian coordinates. Taking ε, μ, σ, to be constants independent of x, y, z, t, and each non-zero, we shall seek solutions of equation (2.2.12) based on the method of separation of variables. We have

$$\nabla^2\psi(x, y, z, t) - \frac{\varepsilon\mu}{c^2}\frac{\partial^2}{\partial t^2}\psi(x, y, z, t) - \frac{4\pi\,\sigma\mu}{c^2}\frac{\partial}{\partial t}\psi(x, y, z, t) = 0 \tag{2.3.1}$$

Write

$$\psi(x, y, z, t) = F(x, y, z)\,T(t) \tag{2.3.2}$$

so that

$$\frac{1}{F}\nabla^2 F = \frac{\varepsilon\mu}{c^2}\frac{1}{T}\frac{d^2T}{dt^2} + \frac{4\pi\,\sigma\mu}{c^2}\frac{1}{T}\frac{dT}{dt} \tag{2.3.3}$$

The left-hand side of this equation is a function of the space coordinates only whereas the right-hand side is a function of the time coordinates only. In order to have a non-trivial solution each side of the equation must be equal to the same constant since a change in space coordinates will not affect the time function and conversely a change in the time coordinate will not modify the space function.

Taking the separation constant equal to $-k^2$, where k may be a complex quantity, we obtain from the right-hand side of equation (2.3.3),

$$\frac{\varepsilon\mu}{c^2}\frac{\mathrm{d}^2T}{\mathrm{d}t^2} + \frac{4\pi\,\sigma\mu}{c^2}\frac{\mathrm{d}T}{\mathrm{d}t} + k^2T = 0 \tag{2.3.4}$$

which has a solution of the form

$$T(t) = A\,\mathrm{e}^{st} \tag{2.3.5}$$

with s related to k through the equation

$$\frac{\varepsilon\mu}{c^2}s^2 + \frac{4\pi\,\sigma\mu}{c^2}s + k^2 = 0 \tag{2.3.6}$$

This equation is the 'dispersion relation' between s and k. It characterizes the properties of the medium. Note that this dispersion relation is entirely distinct from the Kramers–Kronig relations discussed previously (Chapter 1, pages 28–31) and should not be confused with them.

Writing $s = q_k + \mathrm{j}\omega_k$, the steady state solution will have $q_k = 0$, $s = \mathrm{j}\omega_k$, and will correspond to a harmonic time function with real angular frequency ω_k related to k, by

$$\frac{\varepsilon\mu}{c^2}\omega_k^2 - \mathrm{j}\frac{4\pi\,\sigma\mu}{c^2}\omega_k - k^2 = 0 \tag{2.3.7}$$

or, in view of equation (2.1.7),

$$\frac{\varepsilon_c\mu}{c^2}\omega_k^2 - k^2 = 0 \tag{2.3.8}$$

Writing $k = k_1 - \mathrm{j}k_2$ and taking ε, μ, σ to be real quantities we may derive k_1 and k_2 explicitly by equating the real and imaginary parts of equation (2.3.7).

$$k_1^2 - k_2^2 = \frac{\varepsilon\mu}{c^2}\omega_k^2, \qquad 2k_1k_2 = -\frac{4\pi\,\sigma\mu}{c^2}\omega_k \tag{2.3.9}$$

$$k_1 = \pm\frac{\omega_k}{c}\left(\frac{\varepsilon\mu}{2}\right)^{\frac{1}{2}}\left[1 + \sqrt{\left\{1 + \left(\frac{4\pi\sigma}{\varepsilon\omega_k}\right)^2\right\}}\right]^{1/2} \tag{2.3.10}$$

$$k_2 = \pm\frac{4\pi\sigma}{c}\left(\frac{\mu}{2\varepsilon}\right)^{\frac{1}{2}}\left[1 + \sqrt{\left\{1 + \left(\frac{4\pi\sigma}{\varepsilon\omega_k}\right)^2\right\}}\right]^{-1/2} \tag{2.3.11}$$

For a loss-free medium, ε, μ are real and $\sigma = 0$.
Equations (2.3.10) and (2.3.11) give

$$\omega_k = \frac{c}{\sqrt{(\varepsilon\mu)}}k \tag{2.3.12}$$

$$k_2 = 0$$

and so ω_k is linearly related to k which is a real quantity.

Consider now the spatial function $F(x, y, z)$. From the left-hand side of equation (2.3.3) we obtain

$$\nabla^2 F + k^2 F = 0 \tag{2.3.13}$$

Again seek a solution based upon separation of the variables by writing,

$$F = X(x)\, Y(y)\, Z(z) \tag{2.3.14}$$

We obtain

$$\frac{d^2X}{dx^2} + k_x^2 X = 0, \quad \frac{d^2Y}{dy^2} + k_y^2 Y = 0, \quad \frac{d^2Z}{dz^2} + k_z^2 Z = 0 \tag{2.3.15}$$

with

$$k_x^2 + k_y^2 + k_z^2 = k^2 \tag{2.3.16}$$

The function $F(x, y, z)$ may therefore be written in the form

$$F(x, y, z) = F_{\mathbf{k}} \exp \{j(\pm k_x x \pm k_y y \pm k_z z)\} = F_{\mathbf{k}} \exp (j\mathbf{k}.\mathbf{r}) \tag{2.3.17}$$

$$k_x^2 + k_y^2 + k_z^2 = |\mathbf{k}|^2 = k^2 \tag{2.3.18}$$

Combining equations (2.3.5) and (2.3.17), particular steady state solutions of the scalar differential equation, equation (2.3.1), may be written with $s = j\omega_k$, as

$$\psi_{\mathbf{k}} = a(\mathbf{k}) \exp \{j(\omega_k t \pm \mathbf{k}.\mathbf{r})\} \tag{2.3.19}$$

and general solutions

$$\psi(x, y, z, t) = \sum_{\mathbf{k}} a(\mathbf{k}) \exp \{j(\omega_k t \pm \mathbf{k}.\mathbf{r})\} \tag{2.3.20}$$

or

$$\psi(x, y, z, t) = \int a(\mathbf{k}) \exp \{j(\omega_k t \pm \mathbf{k}.\mathbf{r})\}\, d^3\mathbf{k} \tag{2.3.21}$$

where $d^3\mathbf{k} = dk_x\, dk_y\, dk_z$, and ω_k is related to $\mathbf{k}$ through equations (2.3.7) or (2.3.8).

The functions $\psi_{\mathbf{k}}$ of equation (2.3.19) describe simple plane waves when $\mathbf{k}$ is real. For this case consider a particular $\psi_{\mathbf{k}}$. At any instant of time the value of $\psi_{\mathbf{k}}$ will be constant over a surface for which

$$\mathbf{k}.\mathbf{r} = \text{constant} \tag{2.3.22}$$

This equation defines a plane perpendicular to the vector $\mathbf{k}$. These planes also coincide with those over which the phase factor $(\omega_k t \pm \mathbf{k}.\mathbf{r})$ is constant at any instant in time.

These plane waves are periodic in space at a given instant of time. Thus $\psi_{\mathbf{k}}$ is multiplied by unity if $\mathbf{k}.\mathbf{r}$ is replaced by

$$\mathbf{k}.\left(\mathbf{r} + \frac{2\pi}{|\mathbf{k}|}\mathbf{n}\right)$$

in equation (2.3.19). Here $\mathbf{n}$ is a unit vector (or integral multiple of unit vector) in the direction of $\mathbf{k}$. A scalar wavelength λ, may therefore be defined by the relation

$$\lambda = \frac{2\pi}{|\mathbf{k}|} = \frac{2\pi}{\omega_k} \frac{c}{\sqrt{(\varepsilon\mu)}} \tag{2.3.23}$$

upon using equation (2.3.12) and remembering that we have assumed $\mathbf{k}$ to be real. The quantity $|\mathbf{k}| = (2\pi/\lambda)$ is often called the *wave number* for the plane wave, whilst $\mathbf{k}$ is called the *wave vector*.

A plane of constant phase advances with a definite velocity in the direction of $\mathbf{k}$ and so the function $\psi_{\mathbf{k}}$ describes a propagating wave. This *phase velocity* can be derived by evaluating the distance $\mathbf{n}\,\delta\xi$ a given plane of constant phase will have travelled in a time interval δt. We require the phase at the planes to be the same for the initial and final conditions. That is

$$\omega_k t - \mathbf{k}.\mathbf{r} = \omega_k(t + \delta t) - \mathbf{k}.(\mathbf{r} + \mathbf{n}\,\delta\xi) \tag{2.3.24}$$

or

$$\mathbf{k}.\mathbf{n}\,\delta\xi = \omega_k\,\delta t \tag{2.3.25}$$

The phase velocity V^{p} is defined to be

$$V^{\mathrm{p}} = \frac{\delta\xi}{\delta t} = \frac{\omega_k}{\mathbf{k}.\mathbf{n}} = \frac{\omega_k}{|\mathbf{k}|} = \frac{c}{\sqrt{(\varepsilon\mu)}} \tag{2.3.26}$$

The negative signs in equation (2.3.24) correspond to propagation in the direction of $(+)\mathbf{k}$; had positive signs been chosen propagation would have been in the direction $-\mathbf{k}$.

In the example we have just considered ω_k, $\mathbf{k}$ were real quantities and $\psi_{\mathbf{k}}$ described a wave motion which was undamped both in space and time. We now consider a more general plane wave structure which is damped in space such as occurs for example in an attenuating medium. For this case ω_k remains a real quantity but $\mathbf{k}$ is complex. We write

$$\psi_{\mathbf{k}} = a(\mathbf{k}) \exp[\mathrm{j}\{\omega_k t - (\mathbf{k}_1 - \mathrm{j}\mathbf{k}_2).\mathbf{r}\}] \tag{2.3.27}$$

$$= a(\mathbf{k}) \exp(-\mathbf{k}_2.\mathbf{r}) \exp\{\mathrm{j}(\omega_k t - \mathbf{k}_1.\mathbf{r})\} \tag{2.3.28}$$

For this wave the planes over which the phase is constant at any instant of time are given by

$$\mathbf{k}_1.\mathbf{r} = \text{constant} \tag{2.3.29}$$

The phase velocity is along the direction $+\mathbf{k}$, and is given by

$$V^{\mathrm{p}} = \frac{\omega_k}{|\mathbf{k}_1|} \tag{2.3.30}$$

When the losses in the medium arise only from the electrical conductivity, ε, μ, σ are real quantities. If one of the Cartesian coordinate axes is taken along $\mathbf{k}$ (say, the z-axis), then $|\mathbf{k}_1|$ is given directly by equation (2.3.10).† The phase velocity is, therefore,

$$V^{\mathrm{p}} = \frac{c}{\sqrt{(\varepsilon\mu/2)}(1 + \sqrt{[1 + (4\pi\sigma/\varepsilon\omega_k)^2]})^{1/2}} \qquad \varepsilon, \mu, \sigma \text{ real} \tag{2.3.31}$$

Note that the phase velocity has an explicit frequency dependence.

The planes over which the amplitude is constant are given by (from equation (2.3.28))

$$\mathbf{k}_2.\mathbf{r} = \text{constant} \tag{2.3.32}$$

† Alternatively equate real and imaginary parts of the complex vector relation $(\mathbf{k}_1 - \mathrm{j}\mathbf{k}_2) = \hat{\mathbf{i}}(k_{1x} - \mathrm{j}k_{2x}) + \hat{\mathbf{j}}(k_{1y} - \mathrm{j}k_{2y}) + \mathbf{k}(k_{1z} - \mathrm{j}k_{2z})$.

These are perpendicular to the vector $\mathbf{k}_2$. The planes of equal amplitude do not necessarily coincide with the planes of constant phase.

The components of the electromagnetic field vectors derived from $\psi_{\mathbf{k}}$ will now be constant over surfaces at a given instant of time. The equation for such a surface is obtained from Re $[\psi_{\mathbf{k}}]$, namely,

$$\exp(-\mathbf{k}_2 . \mathbf{r}) \cos \mathbf{k}_1 . \mathbf{r} = \text{constant} \tag{2.3.33}$$

This modified form of a plane wave in an attentuating medium is frequently referred to as an inhomogeneous plane wave. A wave of this type occurs in practice under the conditions where total internal reflection takes place. An inhomogeneous wave is then propagated along the boundary between the two dielectric media.

2.3.1 Field Vectors in a Harmonic Plane Wave

In a harmonic plane wave each field component will have the form of $\psi_{\mathbf{k}}$ in equation (2.3.19). For a particular frequency ω_k we may write

$$\mathbf{E} = \mathbf{E}_0 \exp\{\mathrm{j}(\omega_k t - \mathbf{k}.\mathbf{r})\} \qquad \mathbf{H} = \mathbf{H}_0 \exp\{\mathrm{j}(\omega_k t - \mathbf{k}.\mathbf{r})\} \tag{2.3.34}$$

Here we have dropped the alternative $\pm$ sign since this only indicates different directions of propagation. The wave vector $\mathbf{k}$, which we shall take to be real, now gives the direction of propagation explicitly. Although these fields satisfy the scalar differential equation, equation (2.3.1), they do not necessarily represent an electromagnetic field. For this to be the case $\mathbf{E}_0$ and $\mathbf{H}_0$ are not independent constants. Their relationship must be determined from equations (2.2.1) and (2.2.2).

Substituting for example into equations (2.2.13) and (2.2.14), and taking ε, μ real, $\sigma = 0$, so that $\mathbf{k}$ is also real, we have†

$$\nabla \wedge \mathbf{E} = -\mathrm{j}\mathbf{k} \wedge \mathbf{E} = -\mathrm{j}\frac{\omega_k \mu}{c}\mathbf{H} \tag{2.3.35}$$

$$\nabla \wedge \mathbf{H} = -\mathrm{j}\mathbf{k} \wedge \mathbf{H} = \mathrm{j}\frac{\omega_k \varepsilon}{c}\mathbf{E} \tag{2.3.36}$$

or

$$\mathbf{k} \wedge \mathbf{E}_0 = \frac{\omega_k \mu}{c}\mathbf{H}_0 \tag{2.3.37}$$

$$\mathbf{k} \wedge \mathbf{H}_0 = -\frac{\omega_k \varepsilon}{c}\mathbf{E}_0 \tag{2.3.38}$$

We also have

$$\nabla . \mathbf{D} = 0 \qquad \varepsilon \mathbf{k}.\mathbf{E}_0 = 0 \tag{2.3.39}$$

$$\nabla . \mathbf{B} = 0 \qquad \mu \mathbf{k}.\mathbf{H}_0 = 0 \tag{2.3.40}$$

From equation (2.3.37), if $\mathbf{E}_0 = 0$ and $\mu \neq 0$ then $\mathbf{H}_0 = 0$, and similarly from equation (2.3.38), if $\mathbf{H}_0 = 0$ and $\varepsilon \neq 0$ then $\mathbf{E}_0 = 0$. The solutions with both $\mathbf{E}_0$, $\mathbf{H}_0$ zero are trivial so far as wavelike processes are concerned.

† Equations (2.3.35) to (2.3.40) are also valid when $\mathbf{k}$ is complex provided that ε_c is written for ε in a conducting medium in equation (2.3.36).

In these cases the solutions of the general field equations correspond to electrostatic and magnetostatic field distributions.

Assume that ε, μ are non-zero. From equation (2.3.39)

$$\mathbf{k}.\mathbf{E}_0 = 0 \tag{2.3.41}$$

Since $\mathbf{E}_0$ is non-zero, $\mathbf{E}_0$ is perpendicular to $\mathbf{k}$. Moreover from equation (2.3.37), since μ is non-zero, $\mathbf{H}_0$ is perpendicular to both $\mathbf{k}$ and $\mathbf{E}_0$. Hence the vectors $\mathbf{E}_0$, $\mathbf{H}_0$, $\mathbf{k}$ form a right-handed set with both $\mathbf{E}_0$, $\mathbf{H}_0$ perpendicular to the direction of propagation $\mathbf{k}$. This is a transverse wave. From equations (2.3.37), (2.3.38) and (2.3.12)

$$\left|\frac{\mathbf{E}_0}{\mathbf{H}_0}\right| = \frac{\omega_k \mu}{ck} = \frac{ck}{\varepsilon \omega_k} = \sqrt{\left(\frac{\mu}{\varepsilon}\right)} \tag{2.3.42}$$

Consider the case $\varepsilon = 0$, but μ, $\mathbf{E}_0$ non-zero. We obtain from equations (2.3.38) and (2.3.40)

$$\mathbf{k} \wedge \mathbf{H}_0 = 0 \quad \text{and} \quad \mathbf{k}.\mathbf{H}_0 = 0 \tag{2.3.43}$$

Since $\mathbf{k} \neq 0$, these relations require

$$\mathbf{H}_0 = 0 \tag{2.3.44}$$

From equation (2.3.37), it follows now that

$$\mathbf{k} \wedge \mathbf{E}_0 = 0 \tag{2.3.45}$$

but $\mathbf{k}.\mathbf{E}_0 \neq 0$ and so $\mathbf{E}_0$ is parallel to the direction of propagation $\mathbf{k}$. Hence it is possible to propagate a longitudinal electric wave having no associated magnetic field if $\varepsilon = 0$. This situation can in principle arise on the high frequency side of an ideal dielectric dispersion line. Note that in this case the wave vector $\mathbf{k}$ is independent of the particular frequency ω_L for which $\varepsilon(\omega) = 0$. All $\mathbf{k}$ vectors for these waves correspond to the same value of ω. The phase velocity is continuously variable, the group velocity is zero and there is no energy flow as given by the Poynting Vector (compare equation (2.7.1)).

2.3.2 Plane Waves in a Conducting Medium

We may simplify the algebra for this case without losing any essential physical feature by taking the z-axis of Cartesian coordinates along the direction of propagation. We will also assume that the wave is attentuated in the z-direction only, so that we may write

$$\mathbf{k} = (k_1 - \mathrm{j}k_2)\mathbf{n}_z = k\mathbf{n}_z \tag{2.3.46}$$

where $\mathbf{n}_z$ is a unit vector along the z-axis.

For this case, with

$$\mathbf{E}, \mathbf{H} \sim \exp\{\mathrm{j}(\omega_k t - k\mathbf{n}_z.\mathbf{r})\} \tag{2.3.47}$$

we obtain from equation (2.3.35)

$$\nabla \wedge \mathbf{E} = -\mathrm{j}k\mathbf{n}_z \wedge \mathbf{E} = -\mathrm{j}\frac{\omega_k \mu}{c}\mathbf{H} \tag{2.3.48}$$

and so $\mathbf{E}$, $\mathbf{H}$, $\mathbf{n}_z$ form a right-handed set. We may therefore simplify the notation further by taking the x-axis along $\mathbf{E}_0$ and the y-axis along $\mathbf{H}_0$. Write the field vectors as,

$$\begin{aligned} \mathbf{E} &= \mathbf{n}_x E_{0x} \exp\{\mathrm{j}(\omega_k t - k\mathbf{n}_z.\mathbf{r})\} \\ \mathbf{H} &= \mathbf{n}_y H_{0y} \exp\{\mathrm{j}(\omega_k t - k\mathbf{n}_z.\mathbf{r})\} \end{aligned} \tag{2.3.49}$$

with $\mathbf{n}_x$, $\mathbf{n}_y$ unit vectors along the x- and y-axes, and derive from equations (2.3.37) and (2.3.38),

$$kE_{0x}(\mathbf{n}_z \wedge \mathbf{n}_x) = \frac{\omega_k \mu}{c} H_{0y}\mathbf{n}_y \tag{2.3.50a}$$

$$kH_{0y}(\mathbf{n}_z \wedge \mathbf{n}_y) = -\frac{\omega_k \varepsilon_c}{c} E_{0x}\mathbf{n}_x \tag{2.3.50b}$$

The ratio of the amplitudes of the field vectors in the attenuated wave may be obtained from these equations

$$(E_{0x})^2 = \frac{\mu}{\varepsilon_c}(H_{0y})^2 \tag{2.3.51}$$

$$\frac{E_{0x}}{H_{0y}} = \sqrt{\left(\frac{\mathrm{j}\omega_k \mu}{\mathrm{j}\omega_k \varepsilon + 4\pi\sigma}\right)} \tag{2.3.52}$$

Note that the dispersion relation given in equation (2.3.8) follows immediately from equations (2.3.50).

$$kE_{0x} = \frac{\omega_k \mu}{c} H_{0y} = \frac{\omega_k \mu}{c}\left(\frac{\omega_k \varepsilon_c}{kc}\right) E_{0x} \tag{2.3.53}$$

Hence

$$k^2 = \frac{\omega_k^2 \varepsilon_c \mu}{c^2} \tag{2.3.54}$$

This equation expresses in concise form a complicated relation for k as a function of ε, μ, σ. Even when ε, μ, σ are real quantities and $k = k_1 - \mathrm{j}k_2$ is given by equations (2.3.10) and (2.3.11), the relation is by no means straightforward. A case of great practical importance arises, however, when ε, μ, σ are real quantities and

$$\frac{4\pi\sigma}{\varepsilon\omega_k} \gg 1 \tag{2.3.55}$$

This inequality is valid for practically all metals at frequencies below the optical range. In this case we have

$$k_1 = +\sqrt{\left(\frac{2\pi\mu\sigma\omega_k}{c^2}\right)} \qquad k_2 = +\sqrt{\left(\frac{2\pi\mu\sigma\omega_k}{c^2}\right)} \tag{2.3.56}$$

and the signs before the radicals have been chosen to ensure that the wave is attenuated in the direction of propagation. We may now write

$$k = (1 - \mathrm{j})\frac{1}{\delta} \tag{2.3.57}$$

with

$$\delta = \sqrt{\left(\frac{c^2}{2\pi\mu\sigma\omega_k}\right)} \tag{2.3.58}$$

The parameter δ is called the *penetration depth* or *skin depth* since it is the distance a wave travels when the field vectors decrease by a factor $1/e$. In particular the field vectors will have decreased by $1/e$ of their values at the surface of a conductor when the wave has penetrated a distance δ. If σ is infinite, $\delta = 0$ and so both **E** and **H** are zero inside a perfect conductor. Taking σ finite and equal to $5{\cdot}8 \times 10^5$ (ohm cm)$^{-1}$, a typical value for copper, we calculate†

$$\begin{aligned} \delta &= 0{\cdot}95 \text{ cm at } 50 \text{ c/s} \\ \delta &= 6{\cdot}7 \times 10^{-5} \text{ cm at } 10^{10} \text{ c/s} \end{aligned} \qquad (2.3.59)$$

Thus at microwave frequencies ($\sim 10^{10}$ c/s) the penetration depth is already very small.

2.4 Spherical Waves

A useful technique for finding solutions of the electromagnetic field equations in spherical polar coordinates (with a harmonic time dependence) is based on a potential function $\Psi(r, \theta, \phi)$.‡ The field vectors **E**, **H** are derived by operating on the function $(\mathbf{r} \wedge \nabla\Psi)$, and $\Psi(r, \theta, \phi)$ is itself a solution of the scalar wave equation. The need for this indirect method of solution arises because the functions $\nabla \wedge \nabla \wedge \mathbf{E}$, $\nabla \wedge \nabla \wedge \mathbf{H}$ only simplify to separate vector component relations when rectangular Cartesian coordinates are used for describing the field.

As an example of this method of solution consider a lossless medium with ε, μ real and non-zero, and $\sigma = 0$. Write the spatial part of **E** as

$$\mathbf{E}(r, \theta, \phi) = \mathbf{r} \wedge \nabla\Psi \qquad (2.4.1)$$

where the function $\Psi(r, \theta, \phi)$ is as yet unspecified. Equivalently

$$\mathbf{E}(r, \theta, \phi) = -\nabla \wedge (\mathbf{r}\Psi) \qquad (2.4.2)$$

since $\nabla \wedge \mathbf{r} = 0$.

When there is no free charge distribution, $\rho = 0$, and we require

$$\nabla . \mathbf{E} = 0 \qquad (2.4.3)$$

This equation is satisfied automatically since

$$\nabla . \mathbf{E}(r, \theta, \phi) = \nabla . \{-\nabla \wedge (\mathbf{r}\Psi)\} \equiv 0 \qquad (2.4.4)$$

Next consider the function $\nabla \wedge \nabla \wedge \mathbf{E}(r, \theta, \phi)$.
We have

$$\begin{aligned} \nabla \wedge \mathbf{E}(r, \theta, \phi) &= \nabla \wedge (\mathbf{r} \wedge \nabla\Psi) \\ &= \mathbf{r}\nabla^2\Psi - \nabla\Psi(\nabla . \mathbf{r}) + (\nabla\Psi . \nabla)\mathbf{r} - (\mathbf{r} . \nabla)\nabla\Psi \end{aligned} \qquad (2.4.5)$$

but $\nabla . \mathbf{r} = 3$ and $(\mathbf{A} . \nabla)\, \mathbf{r} = \mathbf{A}$ where **A** is any vector. Hence

$$\nabla \wedge \mathbf{E}(r, \theta, \phi) = \mathbf{r}\nabla^2\Psi - \nabla\Psi - \nabla(\mathbf{r} . \nabla\Psi) \qquad (2.4.6)$$

and so

$$\nabla \wedge \nabla \wedge \mathbf{E}(r, \theta, \phi) = \nabla \wedge (\mathbf{r}\, \nabla^2\Psi) \qquad (2.4.7)$$

If $\Psi(r, \theta, \phi)$ satisfies the scalar wave equation

$$\nabla^2\Psi + \frac{\varepsilon\mu\omega^2}{c^2}\Psi = 0 \qquad (2.4.8)$$

† σ(e.s.u.) = $(0{\cdot}899 \times 10^{12}) \times \sigma$ (ohm cm)$^{-1}$.
‡ See for example, Reitz, J. R. and Milford, F. J., *Foundations of Electromagnetic Theory*, Addison-Wesley, 1962; or Slater, J. C. and Frank, N. H., *Electromagnetism*, McGraw-Hill, 1947.

then

$$\nabla \wedge (\mathbf{r}\nabla^2\Psi) = \nabla \wedge \left(-\frac{\varepsilon\mu\omega^2}{c^2}\mathbf{r}\Psi\right) = \frac{\varepsilon\mu\omega^2}{c^2}\mathbf{r} \wedge \nabla\Psi \tag{2.4.9}$$

and we obtain from equations (2.4.7) and (2.4.1)

$$\nabla \wedge \nabla \wedge \mathbf{E} - \frac{\varepsilon\mu\omega^2}{c^2}\mathbf{E} = 0 \tag{2.4.10}$$

This is just the vector differential equation for the electromagnetic field (compare equation (2.2.5)), with no sources present. Hence we may derive $\mathbf{E}(r, \theta, \phi)$ from

$$\mathbf{E}(r, \theta, \phi) = \mathbf{r} \wedge \nabla\Psi \tag{2.4.11}$$

when Ψ satisfies the scalar wave equation, equation (2.4.8). The magnetic vector field corresponding to this set of solutions is obtained from equation (2.2.13),

$$\mathbf{H}(r, \theta, \phi) = -\frac{c}{j\omega\mu}\nabla \wedge (\mathbf{r} \wedge \nabla\Psi) \tag{2.4.12}$$

and again $\nabla . \mathbf{H} = 0$ is automatically satisfied.

Evidently there is an alternative set of solutions which may be derived from the potential $\Psi(r, \theta, \phi)$ by defining

$$\mathbf{H}'(r, \theta, \phi) = \mathbf{r} \wedge \nabla\Psi \tag{2.4.13}$$

For this case, we obtain from equation (2.2.14)

$$\mathbf{E}'(r, \theta, \phi) = \frac{c}{j\omega\varepsilon}\nabla \wedge (\mathbf{r} \wedge \nabla\Psi) \tag{2.4.14}$$

The two sets of solutions given by equations (2.4.11)–(2.4.14), differ in the spatial orientations of the field vectors. Equation (2.4.11) shows that for this solution the vector $\mathbf{E}(r, \theta, \phi)$ is perpendicular to the radius vector $\mathbf{r}$. Hence $\mathbf{E}(r, \theta, \phi)$ is tangential to a spherical surface centred on the origin of coordinates. Such a wave is called a Transverse Electric (TE) wave. Similarly equation (2.4.13) describes a Transverse Magnetic (TM) wave in spherical polar coordinates.†

We now seek solutions of equation (2.4.8) written in spherical polar coordinates

$$\frac{1}{r^2}\frac{\partial}{\partial r}\left(r^2\frac{\partial\Psi}{\partial r}\right) + \frac{1}{r^2\sin\theta}\frac{\partial}{\partial\theta}\left(\sin\theta\frac{\partial\Psi}{\partial\theta}\right) + \frac{1}{r^2\sin^2\theta}\frac{\partial^2\Psi}{d\phi^2} + \frac{\varepsilon\mu\omega^2}{c^2}\Psi = 0 \tag{2.4.15}$$

Assume that Ψ is of the form

$$\Psi(r, \theta, \phi) = R(r)\,\Theta(\theta)\,\Phi(\phi) \tag{2.4.16}$$

† It is for this reason that the technique described above is most relevant to spherical polar coordinates. Note, however, that although the derivation of equations (2.4.11)–(2.4.14) has indicated coordinates (r, θ, ϕ) for clarity, these solutions have made no use of any specific property of spherical polar coordinates and are applicable to other coordinate systems.

then

$$\left\{\frac{1}{R}\sin^2\theta\frac{\mathrm{d}}{\mathrm{d}r}\left(r^2\frac{\mathrm{d}R}{\mathrm{d}r}\right)+\frac{1}{\Theta}\sin\theta\frac{\mathrm{d}}{\mathrm{d}\theta}\left(\sin\theta\frac{\mathrm{d}\Theta}{\mathrm{d}\theta}\right)-\frac{\varepsilon\mu\omega^2}{c^2}r^2\sin^2\theta\right\}$$
$$=-\frac{1}{\Phi}\frac{\mathrm{d}^2\Phi}{\mathrm{d}\phi^2} \tag{2.4.17}$$

The right-hand side of this equation depends only upon the variable ϕ, the left-hand side is independent of ϕ. Taking the separation constant to be $-m^2$,

$$\frac{\mathrm{d}^2\Phi}{\mathrm{d}\phi^2}+m^2\Phi=0 \tag{2.4.18}$$

$$\Phi_m=A_m\,\mathrm{e}^{\pm \mathrm{j}m\phi} \tag{2.4.19}$$

where the subscript m indicates explicitly that Φ depends upon m.

The left-hand side of equation (2.4.17) may now be written

$$\left\{\frac{1}{R}\frac{\mathrm{d}}{\mathrm{d}r}\left(r^2\frac{\mathrm{d}R}{\mathrm{d}r}\right)+\frac{\varepsilon\mu\omega^2}{c^2}r^2\right\}+\left\{\frac{1}{\Theta}\cdot\frac{1}{\sin\theta}\cdot\frac{\mathrm{d}}{\mathrm{d}\theta}\left(\sin\theta\frac{\mathrm{d}\Theta}{\mathrm{d}\theta}\right)-\frac{m^2}{\sin^2\theta}\right\}=0 \tag{2.4.20}$$

Introducing an additional separation constant in the form $l(l+1)$ we obtain†

$$\frac{1}{r^2}\frac{\mathrm{d}}{\mathrm{d}r}\left(r^2\frac{\mathrm{d}R}{\mathrm{d}r}\right)+\left\{\frac{\varepsilon\mu\omega^2}{c^2}-\frac{l(l+1)}{r}\right\}R=0 \tag{2.4.21}$$

$$\frac{1}{\sin\theta}\frac{\mathrm{d}}{\mathrm{d}\theta}\left(\sin\theta\frac{\mathrm{d}\Theta}{\mathrm{d}\theta}\right)+\left\{l(l+1)-\frac{m^2}{\sin^2\theta}\right\}\Theta=0 \tag{2.4.22}$$

Equation (2.4.21) may be transformed into Bessel's differential equation by substituting

$$\rho=\left(\frac{\varepsilon\mu\omega^2}{c^2}\right)^{1/2}r$$

and $Q=R\sqrt{\rho}$

$$\rho^2\frac{\mathrm{d}^2Q}{\mathrm{d}\rho^2}+\rho\frac{\mathrm{d}Q}{\mathrm{d}\rho}+\{\rho^2-(l+\tfrac{1}{2})^2\}Q=0 \tag{2.4.23}$$

The solutions for $R(r)$ are therefore related to Bessel functions or Neumann functions of order $(l+\frac{1}{2})$

$$R(r)=\frac{1}{(\varepsilon\mu)^{1/4}}\left(\frac{c}{\omega r}\right)^{\frac{1}{2}}J_{(l+\frac{1}{2})}\left\{\sqrt{(\varepsilon\mu)}\frac{\omega r}{c}\right\} \tag{2.4.24}$$

or

$$R(r)=\frac{1}{(\varepsilon\mu)^{1/4}}\left(\frac{c}{\omega r}\right)^{\frac{1}{2}}N_{(l+\frac{1}{2})}\left\{\sqrt{(\varepsilon\mu)}\frac{\omega r}{c}\right\} \tag{2.4.25}$$

If l is an integer (which is the case for our problem) functions $j_l(\xi)$, $n_l(\xi)$ may be defined which have simple analytical forms.

$$j_l(\xi)=\left(\frac{\pi}{2\xi}\right)^{\frac{1}{2}}J_{(l+\frac{1}{2})}(\xi),\qquad n_l(\xi)=\left(\frac{\pi}{2\xi}\right)^{\frac{1}{2}}N_{(l+\frac{1}{2})}(\xi) \tag{2.4.26}$$

† The constants l, m are integers: m to ensure that Φ is single valued, and l in order that Θ does not diverge for $\cos\theta=\pm 1$.

The first few functions are

$$j_0(\xi) = \frac{\sin \xi}{\xi} \qquad n_0(\xi) = -\frac{\cos \xi}{\xi}$$

$$j_1(\xi) = \frac{\sin \xi}{\xi^2} - \frac{\cos \xi}{\xi} \qquad n_1(\xi) = -\frac{\sin \xi}{\xi} - \frac{\cos \xi}{\xi^2} \tag{2.4.27}$$

whereas generally, for large values of ξ,

$$j_l(\xi) \rightarrow \frac{1}{\xi}\cos\left(\xi - \frac{l+1}{2}\pi\right)$$

$$n_l(\xi) \rightarrow \frac{1}{\xi}\sin\left(\xi - \frac{l+1}{2}\pi\right) \tag{2.4.28}$$

Equation (2.4.22) is the differential equation for associated Legendre functions denoted by $P_l^m(\cos\theta)$.† A few of these functions are listed below

$$P_0^0(\cos\theta) = 1$$

$$P_1^0(\cos\theta) = \cos\theta \qquad P_1^1(\cos\theta) = (1-\cos^2\theta)^{1/2} = \sin\theta$$

$$P_2^0(\cos\theta) = \tfrac{1}{2}(3\cos^2\theta - 1) \qquad P_2^1(\cos\theta) = \tfrac{3}{2}\sin 2\theta \qquad P_2^2(\cos\theta) = 3\sin^2\theta$$

$$P_3^0(\cos\theta) = \tfrac{1}{2}(5\cos^3\theta - 3\cos\theta) \qquad P_3^1(\cos\theta) = \tfrac{3}{2}\sin\theta(5\cos^2\theta - 1)$$

$$P_3^2 = 15\sin^2\theta\cos\theta \qquad P_3^3(\cos\theta) = 15\sin^3\theta \tag{2.4.29}$$

The function $\Psi(r, \theta, \phi)$ may therefore be written generally in the form of equation (2.4.16) as

$$\Psi_{l,m}(r, \theta, \phi) \sim \left[j_l\left\{\sqrt{(\varepsilon\mu)}\frac{\omega r}{c}\right\}, n_l\left\{\sqrt{(\varepsilon\mu)}\frac{\omega r}{c}\right\}\right] P_l^m(\cos\theta)\, e^{\pm jm\phi} \tag{2.4.30}$$

The vector fields are derived using equations (2.4.11)–(2.4.14). Consider a simple choice of $\Psi(r, \theta, \phi)$ with $l = 1$, $m = 0$, taken as

$$\Psi_{10} = \left(\frac{\mu}{\varepsilon}\right)^{\frac{1}{2}} \frac{k^2}{r} m_0\, e^{-jkr}\left(1 - \frac{j}{kr}\right)\cos\theta \tag{2.4.31}$$

with $k = (\omega/c)\sqrt{(\varepsilon\mu)}$ and m_0 a constant. From equation (2.4.11) and equation (2.4.7)

$$\mathbf{E}(r, \theta, \phi) = -\mathbf{a}_\phi k^2\left(\frac{\mu}{\varepsilon}\right)^{\frac{1}{2}}\left(\frac{1}{r} - \frac{j}{kr^2}\right) e^{-jkr}\, m_0 \sin\theta \tag{2.4.32}$$

and from equation (2.4.12)

$$\mathbf{H}(r, \theta, \phi) = -\frac{1}{jk}\left(\frac{\varepsilon}{\mu}\right)^{\frac{1}{2}} \nabla \wedge \mathbf{E}$$

so that

$$\mathbf{H}(r, \theta, \phi) = \mathbf{a}_r 2\left(\frac{1}{r^3} + \frac{jk}{r^2}\right) e^{-jkr}\, m_0 \cos\theta$$

$$-\mathbf{a}_\theta\left(\frac{1}{r^3} + \frac{jk}{r^2} - \frac{k^2}{r^2}\right) e^{-jkr}\, m_0 \sin\theta \tag{2.4.33}$$

† See for example, Margenau, H. and Murphy, G. M., *The Mathematics of Physics and Chemistry*, Van Nostrand, 1944.

These fields are in fact the fields due to an oscillating magnetic dipole of moment m_0.

2.5 Boundary Conditions for the Field Vectors

Rapid changes occur in the field vectors across any surface which bounds a material medium or separates two different media. These changes generally appear to be discontinuities on the macroscopic scale which is relevant to this discussion and so may be described in terms of boundary conditions. Rigorously, however, the surface of discontinuity should be regarded as the limit of a transition region within which the field vectors and their derivatives are continuous bounded functions of position and time.

Imagine such a transition region separating a medium labelled (1) from another medium labelled (2). The characteristic parameters describing medium (1), for example ε_1, μ_1, σ_1, will change rapidly in the transition region to new values, for example ε_2, μ_2, σ_2, describing medium (2). By regarding the transition region as a true discontinuity we shall be able to derive from the electromagnetic field equations relationships between the field vectors themselves which do not explicitly involve any specific set of parameters. Erect a small right circular cylinder with its axis normal to the boundary between the two media as in Fig. 2.1, and consider first the field vector **B**.

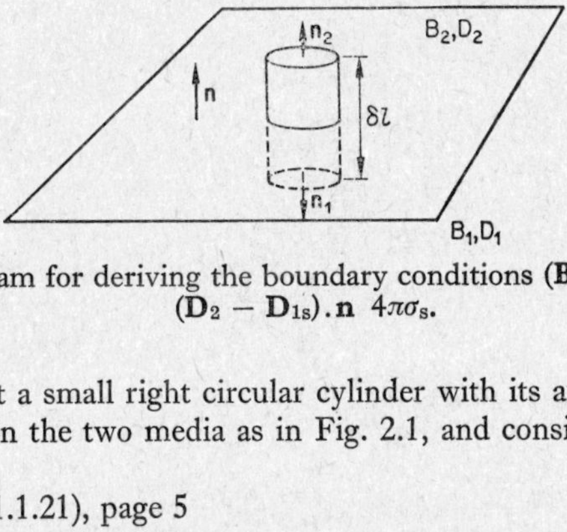

FIG. 2.1 Diagram for deriving the boundary conditions $(\mathbf{B}_2 - \mathbf{B}_1).\mathbf{n} = 0$, $(\mathbf{D}_2 - \mathbf{D}_{1s}).\mathbf{n}\ 4\pi\sigma_s$.

From equation (1.1.21), page 5

$$\nabla . \mathbf{B} = 0$$

and, integrating throughout the volume of the cylinder, we find that

$$\int_{\text{cylinder}} \nabla . \mathbf{B}\, dv = \oint_{\substack{\text{surface of}\\ \text{cylinder}}} \mathbf{B} . d\mathbf{S} = 0 \tag{2.5.1}$$

As the transition region shrinks into an ideal boundary surface the axial length of the cylinder, δl, can approach zero so that **B** in equation (2.5.1) will refer to $\mathbf{B}_1$ and $\mathbf{B}_2$ evaluated immediately at each side of the surface. Moreover as $\delta l \to 0$ the contribution to the surface integral from the walls of the cylinder will vanish. Taking the cross-sectional area of the cylinder as ΔS, equation (2.5.1) gives,

$$(\mathbf{B}_2 . \mathbf{n}_2 + \mathbf{B}_1 . \mathbf{n}_1)\, \Delta S = 0 \tag{2.5.2}$$

or

$$(\mathbf{B}_2 - \mathbf{B}_1).\mathbf{n} = 0 \tag{2.5.3}$$

where **n** is a unit vector normal to the boundary surface. The normal component of **B** is therefore continuous across a boundary.

For the field vector **D**, we obtain from equation (1.1.24), page 5, and Fig. 2.1.

$$\nabla.\mathbf{D} = 4\pi\rho$$

$$\oint_{\substack{\text{surface of}\\ \text{clinder}}} \mathbf{D}.\mathrm{d}\mathbf{S} = 4\pi \int_{\text{cylinder}} \rho\, \mathrm{d}v \tag{2.5.4}$$

As the axial length of the cylinder shrinks to zero the integral $\int\rho\, dv$ tends to $\int\sigma_s\, dS$, where $\sigma_s \equiv \lim_{\delta l \to 0} \rho\, \delta l$ is the surface density of free charge.

Hence, following the argument given above for **B**, we have

$$(\mathbf{D}_2 - \mathbf{D}_1).\mathbf{n} = 4\pi\sigma_s \tag{2.5.5}$$

The change in the normal component of **D** is equal to $4\pi \times$ surface density of free charge. The normal component of **D** is therefore continuous if there is no free charge on the surface of discontinuity.

Information about tangential components of **E** and **H**, may be derived from the field equations by constructing a small rectangular contour with sides normal and tangential to the boundary surface as shown in Fig. 2.2.

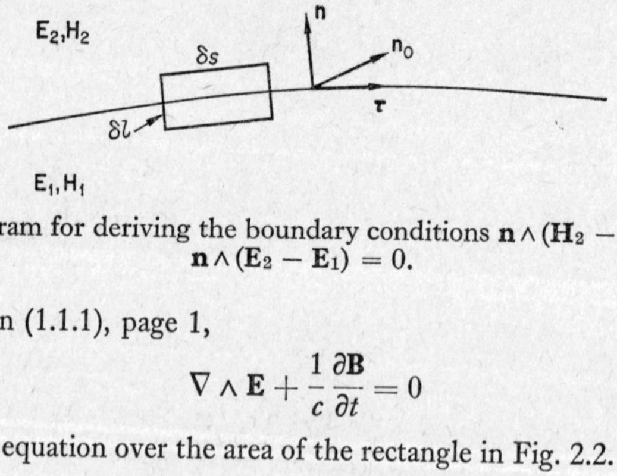

FIG. 2.2 Diagram for deriving the boundary conditions $\mathbf{n} \wedge (\mathbf{H}_2 - \mathbf{H}_1) = 4\pi\mathbf{K}_s$, $\mathbf{n} \wedge (\mathbf{E}_2 - \mathbf{E}_1) = 0$.

From equation (1.1.1), page 1,

$$\nabla \wedge \mathbf{E} + \frac{1}{c}\frac{\partial \mathbf{B}}{\partial t} = 0$$

Integrate this equation over the area of the rectangle in Fig. 2.2.

$$\int_{\delta s\, \delta l} \left(\nabla \wedge \mathbf{E} + \frac{1}{c}\frac{\partial \mathbf{B}}{\partial t}\right).\mathbf{n}_0\, ds\, dl = 0 \tag{2.5.6}$$

from which

$$\oint_{\text{perimeter}} \mathbf{E}.\mathrm{d}\mathbf{s} + \frac{1}{c}\int_{\delta s\, \delta l} \frac{\partial \mathbf{B}}{\partial t}.\mathbf{n}_0\, ds\, dl = 0 \tag{2.5.7}$$

In the limit $\delta l \to 0$,

$$\left\{(\mathbf{E}_2 - \mathbf{E}_1).\boldsymbol{\tau} + \lim_{\delta l \to 0} \frac{1}{c}\frac{\partial \mathbf{B}}{\partial t}.\mathbf{n}_0\, \delta l\right\} \delta s = 0 \tag{2.5.8}$$

where τ is a unit vector tangential to the boundary. The second term in equation (2.5.8) tends to zero as $\delta l \rightarrow 0$ since the field vectors and their derivatives are assumed to be finite within the transition region. Hence

$$(\mathbf{E}_2 - \mathbf{E}_1).\boldsymbol{\tau} = 0 \tag{2.5.9}$$

Hence the component of **E** along $\boldsymbol{\tau}$ is continuous. However, the orientation of $\boldsymbol{\tau}$ in the boundary surface is completely arbitrary and equation (2.5.9) must hold for all orientations of $\boldsymbol{\tau}$. Hence the component of **E** tangential to the surface of discontinuity must be continuous, and this may be written

$$\mathbf{n} \wedge (\mathbf{E}_2 - \mathbf{E}_1) = 0 \tag{2.5.10}$$

where **n** is unit vector normal to the surface of discontinuity.

From equation (1.1.2), page 1,

$$\nabla \wedge \mathbf{H} - \frac{1}{c}\frac{\partial \mathbf{D}}{\partial t} - 4\pi \mathbf{J} = 0$$

Integrate this equation over the area of the rectangle shown in Fig. 2.2.

$$\oint_{\text{perimeter}} \mathbf{H}.\mathrm{d}\mathbf{s} - \int_{\delta s\, \delta l} \left(\frac{1}{c}\frac{\partial \mathbf{D}}{\partial t} + 4\pi \mathbf{J}\right).\mathbf{n}_0 \, \mathrm{d}s \, \mathrm{d}l = 0 \tag{2.5.11}$$

The term $(\partial \mathbf{D}/\partial t)$ is finite and so this term will not contribute to equation (2.5.11) as $\delta l \rightarrow 0$. In this limit

$$(\mathbf{H}_2 - \mathbf{H}_1).\boldsymbol{\tau}\, \delta s = \lim_{\delta l \rightarrow 0} \int 4\pi \mathbf{J}.\mathbf{n} \, \mathrm{d}s \, \mathrm{d}l \tag{2.5.12}$$

It is usually convenient to imagine that the currents which are actually flowing in a finite region of the material (for example within the skin depth) may for the purpose of analysis be entirely confined to the boundary surface. In this case we write

$$\lim_{\delta l \rightarrow 0} 4\pi \int \mathbf{J}.\mathbf{n}_0 \, \mathrm{d}s \, \mathrm{d}l \longrightarrow 4\pi \mathbf{K}_s.\mathbf{n}_0 \; \delta s \tag{2.5.13}$$

where $\mathbf{K}_s$ is called the surface current density. Equation (2.5.13) states formally that the current which originally flowed through the total cross-section of the rectangle is now confined to the surface of discontinuity. From equation (2.5.12) and (2.5.13), therefore

$$(\mathbf{H}_2 - \mathbf{H}_1).\boldsymbol{\tau} = 4\pi \mathbf{K}_s.\mathbf{n}_0 \tag{2.5.14}$$

But, from Fig. 2.2,

$$\boldsymbol{\tau} = \mathbf{n}_0 \wedge \mathbf{n} \tag{2.5.15}$$

and so using the vector relation $\mathbf{a}.\mathbf{b} \wedge \mathbf{c} = \mathbf{b}.\mathbf{c} \wedge \mathbf{a}$

$$\mathbf{n} \wedge (\mathbf{H}_2 - \mathbf{H}_1) = 4\pi \mathbf{K}_s \tag{2.5.16}$$

The tangential component of **H** is therefore discontinuous across a boundary surface in which macroscopic surface currents are flowing. Note, however, that for the significant case when $\mathbf{K}_s = 0$

$$\mathbf{n} \wedge (\mathbf{H}_2 - \mathbf{H}_1) = 0 \qquad \mathbf{K}_s = 0 \tag{2.5.17}$$

We may summarize the boundary conditions for the field vectors in the form

$$\mathbf{n}.(\mathbf{B}_2 - \mathbf{B}_1) = 0 \qquad \mathbf{n}.(\mathbf{D}_2 - \mathbf{D}_1) = 4\pi\sigma_s$$
$$\mathbf{n} \wedge (\mathbf{H}_2 - \mathbf{H}_1) = 4\pi\mathbf{K}_s \qquad \mathbf{n} \wedge (\mathbf{E}_2 - \mathbf{E}_1) = 0 \tag{2.5.18}$$

We observed on page 67 that the penetration depth for a wave in a perfect conductor is zero. The field vectors **E, H** and correspondingly **D, B** are zero inside such a material. Hence the boundary conditions at the surface of a perfect conductor are

$$\mathbf{n}.\mathbf{B} = 0 \qquad \mathbf{n}.\mathbf{D} = 4\pi\sigma_s$$
$$\mathbf{n} \wedge \mathbf{H} = 4\pi\mathbf{K}_s \qquad \mathbf{n} \wedge \mathbf{E} = 0 \tag{2.5.19}$$

2.6 Reflection and Refraction of Plane Waves

As an example of the use of the boundary conditions derived in Section 2.5 we shall investigate the reflection and refraction of a plane wave (of single frequency) at a plane surface of discontinuity between two media. Take this boundary to be the xy plane ($z = 0$) and characterize the medium below the plane by the parameters ε_1, μ_1, σ_1. The medium above the plane is characterized by parameters ε_2, μ_2, σ_2. Let ε_1, μ_1 be real and $\sigma_1 = 0$ so that the first medium is non-absorbing. In addition suppose that any attenuation in the second medium occurs only along the z-axis. This appears to be an artificial restriction but it provides a considerable simplification in the mathematics and is at the same time a valid representation for the significant case of a wave which is incident normal to the plane of discontinuity.

Assume that a wave is incident upon the boundary from below the xy plane, as shown in Fig. 2.3, and choose the orientation of the x-axis so that the direction of propagation lies in the xz plane. We shall attempt to satisfy the boundary conditions by means of three waves: the incident wave together with a refracted and a reflected wave. Write the electric fields of these three waves in the forms

$$\mathbf{E} = \mathbf{E}_0 \, e^{j(\omega t - \mathbf{k}.\mathbf{r})} \tag{2.6.1}$$

$$\mathbf{E}' = \mathbf{E}'_0 \, e^{j(\omega' t - \mathbf{k}'.\mathbf{r})} \tag{2.6.2}$$

$$\mathbf{E}'' = \mathbf{E}''_0 \, e^{j(\omega'' t - \mathbf{k}''.\mathbf{r})} \tag{2.6.3}$$

with $\mathbf{E}_0$, $\mathbf{E}'_0$ and $\mathbf{E}''_0$ independent of the coordinates.

In these equations $\mathbf{k}$ and $\mathbf{k}''$ refer to a non-absorbing medium and so are real vectors. The vector $\mathbf{k}'$ will in general have components k'_x, k'_y, k'_z, with k'_z a complex quantity in our case.

From equation (2.5.18) we require the tangential component of **E** to be continuous across the boundary.

That is

$$E_x + E''_x = E'_x \qquad E_y + E''_y = E'_y \qquad (z = 0) \tag{2.6.4}$$

for all values of x, y and t. For these relations to be satisfied the exponents of e in equations (2.6.1)–(2.6.3) must all be equal, at $z=0$. We can see that this is a necessary requirement by putting two of the variables x, y, t equal to zero

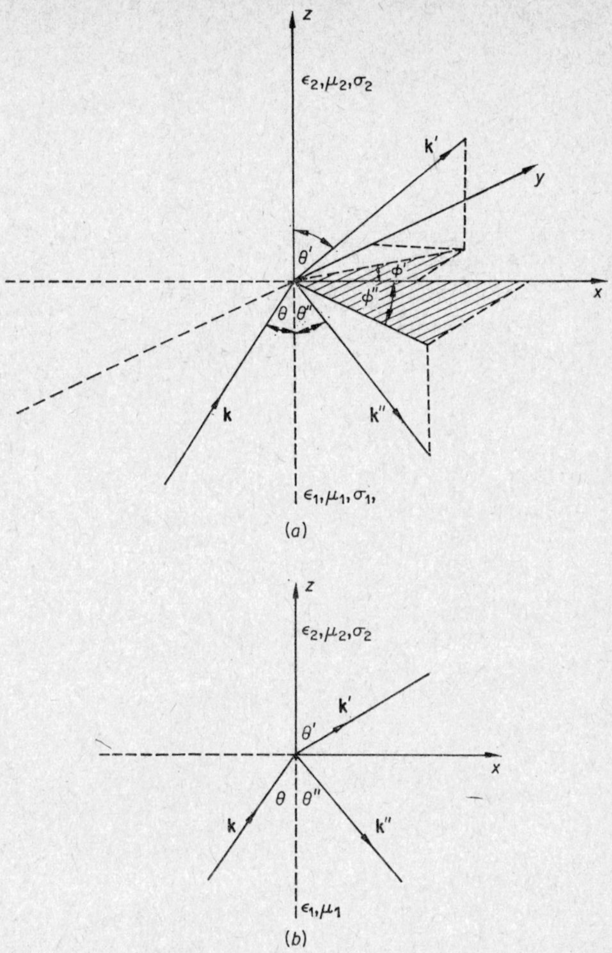

FIG. 2.3 The wave vectors for refraction at a plane interface. In (*b*) the diagram is simplified by knowing that all the wave vectors lie in the xz plane.

and considering the implications of equation (2.6.4) in terms of the remaining variable. Hence, taking $x = y = 0$, $t \neq 0$ we find that

$$\omega = \omega' = \omega'' \tag{2.6.5}$$

There is no change of frequency when the surface of discontinuity is stationary.

Since the incident wave is in the xz plane

$$k_y = 0 \quad \text{and so} \quad k'_y = k''_y = 0 \tag{2.6.6}$$

upon considering $x = t = 0$, $y \neq 0$.

The directions of propagation of all three waves lie in the same plane (in our case it is the xz plane). Also

$$k_x = k'_x = k''_x \tag{2.6.7}$$

The waves characterized by $\mathbf{k}$, $\mathbf{k}''$ are travelling in the same medium and have the same frequency. Hence, from equations (2.3.12), (2.3.18),

$$\frac{\omega}{c}\sqrt{(\varepsilon_1\mu_1)} = |\mathbf{k}| = |\mathbf{k}''| \tag{2.6.8}$$

and so, using equation (2.6.7) we obtain

$$k_z = -k_z'' \tag{2.6.9}$$

These equations require that $\theta'' = \theta$ in Fig. 2.3. The angle of reflection is equal to the angle of incidence.

From equations (2.3.8) and (2.3.16) we may write

$$\omega\sqrt{\left(\frac{\varepsilon_{c2}\mu_2}{c^2}\right)} \equiv k' = \sqrt{\{k_x'^2 + (k_{1z}' - jk_{2z}')^2\}} \tag{2.6.10}$$

with $\varepsilon_{c2} = \varepsilon_2 - j\left(\dfrac{4\pi\sigma_2}{\omega}\right)$. Here k' is complex because we have allowed k_z' to be complex. Hence

$$k_z' = \sqrt{\{(k')^2 - (k_x')^2\}} = \frac{\omega}{c}\sqrt{\{\varepsilon_{c2}\mu_2 - \varepsilon_1\mu_1 \sin\theta\}} \tag{2.6.11}$$

Here the sign of the square root factor must be so chosen that the wave is attenuated in the direction of $+z$.

When the second medium is lossless, ε_2, μ_2 are real and $\sigma_2 = 0$. For this case, we obtain from equations (2.6.7), (2.6.8), (2.6.10),

$$\frac{\omega}{c}\sqrt{(\varepsilon_1\mu_1)}\sin\theta = \frac{\omega}{c}\sqrt{(\varepsilon_2\mu_2)}\sin\theta' \tag{2.6.12}$$

that is

$$\frac{\sin\theta'}{\sin\theta} = \sqrt{\left(\frac{\varepsilon_1\mu_1}{\varepsilon_2\mu_2}\right)} \qquad (\sigma_2 = 0) \tag{2.6.13}$$

This is Snell's law of refraction.

Note also that from equations (2.6.8) and (2.6.10), since k' is now real,

$$\frac{k'}{|\mathbf{k}|} \rightarrow \frac{|\mathbf{k}'|}{|\mathbf{k}|} = \sqrt{\left(\frac{\varepsilon_2\mu_2}{\varepsilon_1\mu_1}\right)} \qquad (\sigma_2 = 0) \tag{2.6.14a}$$

and

$$\frac{k_z'}{k_z} = \frac{\cos\theta'}{\cos\theta}\sqrt{\left(\frac{\varepsilon_2\mu_2}{\varepsilon_1\mu_1}\right)} = \frac{\sin\theta\cos\theta'}{\sin\theta'\cos\theta} \qquad (\sigma_2 = 0) \tag{2.6.14b}$$

When $\varepsilon_2\mu_2 < \varepsilon_1\mu_1$ equation (2.6.13) indicates that $\theta' = \pi/2$ for a particular of incidence θ_c which is given by

$$\sin\theta_c = \sqrt{\left(\frac{\varepsilon_2\mu_2}{\varepsilon_1\mu_1}\right)} \tag{2.6.15a}$$

This is the angle at which total internal reflection first occurs. Angles of incidence greater than θ_c require special consideration. Again taking ε_2, μ_2 real and $\sigma_2 = 0$ in equation (2.6.11), we obtain

$$k_z' = \sqrt{\{(k')^2 - (k_x')^2\}} = \frac{\omega j}{c}\sqrt{(\varepsilon_1\mu_1\sin\theta - \varepsilon_2\mu_2)} \qquad (\theta > \theta_c) \tag{2.6.15b}$$

Evidently k'_z is now an imaginary quantity and the electromagnetic field is attenuated in the z-direction. On the other hand, k'_x is still a real quantity and is given by $k'_x = k_x$ (from equation (2.6.7)), describing, therefore, an excitation which propagates along the boundary in the x-direction in direct correspondence with the field variation along the boundary arising from the particular angle of incidence for the incident and reflected waves. For this excitation, the planes of constant amplitude are perpendicular to the z-axis whereas the planes of constant phase are perpendicular to the x-axis. This is an example of a non-homogeneous wave of the type referred to previously in connection with equations (2.3.28) and (2.3.32). Although an electromagnetic field is present in the second medium, the time average of the component of the Poynting vector normal to the interface is zero for this field and there is no net energy flow across the boundary.

2.6.1 Fresnel's Equations

In order to determine the amplitudes of the reflected and refracted waves we consider separately two cases (a) $\mathbf{E}_0$ perpendicular to the plane of incidence (the xz plane) and (b) $\mathbf{E}_0$ parallel to the plane of incidence. The solution for a general case is obtained by resolving the electric vector into these two components.

For case (a) $\mathbf{E}_0$ is directed along the y-axis. We expect that this will also be the case for $\mathbf{E}'_0$ and $\mathbf{E}''_0$. The boundary conditions are, from equations (2.5.18)

$$E_y + E''_y = E'_y \qquad (z = 0) \qquad (2.6.16)$$

for the tangential components of $\mathbf{E}$,

$$H_x + H''_x = H'_x \qquad (z = 0) \qquad (2.6.17)$$

for the tangential components of $\mathbf{H}$. Equation (2.6.17) may be rewritten, using equation (2.3.37), as

$$\frac{k_z}{\mu_1}(E_y - E''_y) = \frac{k'_z}{\mu_2}E'_y \qquad (z = 0) \qquad (2.6.18)$$

Since the exponential factors are all equal, equations (2.6.16), (2.6.18) provide relationships between the amplitudes of the electric vectors.

$$E''_{0y} = \frac{\mu_2 k_z - \mu_1 k'_z}{\mu_2 k_z + \mu_1 k'_z} E_{0y} \qquad (2.6.19)$$

$$E'_{0y} = \frac{2\mu_2 k_z}{\mu_2 k_z + \mu_1 k'_z} E_{0y} \qquad (2.6.20)$$

When both media are lossless dielectrics, for which $\sigma_1 = \sigma_2 = 0$, $\mu_1 = \mu_2 = 1$, we obtain from equation (2.6.14b)

$$E''_{0y} = \frac{\sin(\theta' - \theta)}{\sin(\theta' + \theta)} E_{0y} \qquad (2.6.21)$$

$$E'_{0y} = \frac{2 \sin\theta' \cos\theta}{\sin(\theta' + \theta)} \qquad (2.6.22)$$

For case (b) where the electric vector lies in the plane of incidence note that the magnetic vector is along the y-axis. For the tangential components of **H** at the boundary,

$$H_y + H_y'' = H_y' \qquad (z = 0) \qquad (2.6.23)$$

Using equation (2.3.38), the continuity of the tangential components of **E** gives,

$$\frac{k_z}{\varepsilon_1}(H_y - H_y'') = \frac{k_z'}{\varepsilon_{c2}} H_y' \qquad (z = 0) \qquad (2.6.24)$$

Hence

$$H_{0y}'' = \frac{\varepsilon_{c2} k_z - \varepsilon_1 k_z'}{\varepsilon_{c2} k_z + \varepsilon_1 k_z'} H_{0y} \qquad (2.6.25)$$

$$H_{0y}' = \frac{2\varepsilon_{c2} k_z}{\varepsilon_{c2} k_z + \varepsilon_1 k_z'} H_{0y} \qquad (2.6.26)$$

When both media are lossless dielectrics (that is $\sigma = 0$, $\mu = 1$, again) we obtain from equations (2.6.13) and (2.6.14)

$$H_{0y}'' = \frac{\sin\theta\cos\theta - \sin\theta'\cos\theta'}{\sin\theta\cos\theta + \sin\theta'\cos\theta'} H_{0y} = \frac{\tan(\theta - \theta')}{\tan(\theta + \theta')} H_{0y} \qquad (2.6.27)$$

$$H_{0y}' = \frac{\sin 2\theta}{\sin(\theta + \theta')\cos(\theta - \theta')} H_{0y} \qquad (2.6.28)$$

Equations (2.6.21), (2.6.22), (2.6.27) and (2.6.28) are usually called Fresnel's formulae.

There is an interesting application of equation (2.6.27). When $\sin 2\theta = \sin 2\theta'$ there is no reflected wave. This condition is satisfied for the trivial case $\theta = \theta'$, $\varepsilon_1 = \varepsilon_2$, and also for the significant case $\theta = (\pi/2 - \theta')$. Let this second condition be satisfied when the wave is incident at an angle $\theta = \theta_p$. Then, from equation (2.6.13)

$$\tan\theta_p = \sqrt{\left(\frac{\varepsilon_2}{\varepsilon_1}\right)} \qquad (2.6.29)$$

This is Brewster's law. If an electromagnetic wave of arbitrary polarization is incident on the boundary surface at the angle θ_p, the reflected wave will be polarized so that its electric field is perpendicular to the plane of incidence. We have therefore a method for producing polarized light.

2.7 Poynting Vector

In Section 1.1 we set out the fundamental electromagnetic field equations and defined the energy flow to be the Poynting Vector **S**,

$$\mathbf{S} = \frac{c}{4\pi}(\mathbf{E} \wedge \mathbf{H}) \qquad (2.7.1)$$

In this equation **S** is the instantaneous rate at which electromagnetic energy flows across a unit area having its normal in the direction of the vector $(\mathbf{E} \wedge \mathbf{H})$. The vectors **S**, **E**, **H** are all real quantities in equation (2.7.1). This definition of energy flow is to be regarded as a fundamental postulate of electromagnetism but we can show that it is a reasonable assumption and that it is consistent with the other field equations.

Write the total current density $\mathbf{J}$ in an arbitrary region of space as the sum of two components. These will be a current density $\mathbf{J}_F$ arising from the action of the electromagnetic field itself (for example $\mathbf{J}_F = (\sigma/c)\mathbf{E}$ in a linear conducting medium) together with an additional current density $\mathbf{J}_S$ which represents current sources. The current sources $\mathbf{J}_S$ will themselves give rise to an electromagnetic field. Thus, writing,

$$\mathbf{J} = \mathbf{J}_F + \mathbf{J}_S \tag{2.7.2}$$

the source currents will deliver energy to the electromagnetic field in the region of space under consideration at the rate

$$-c\mathbf{E}.\mathbf{J}_S = c\mathbf{E}.\mathbf{J}_F - c\mathbf{E}.\mathbf{J} \tag{2.7.3}$$

This relation is derived from the rate at which work is done when the source currents interact with the electromagnetic field (compare equations (1.1.7) (1.5.8)). We have

$$-\mathbf{F}.\mathbf{v} = -\rho_S\left(\mathbf{E} + \frac{\mathbf{v} \wedge \mathbf{B}}{c}\right).\mathbf{v} = -c\mathbf{E}.\mathbf{J}_S \tag{2.7.4}$$

The term $c\mathbf{E}.\mathbf{J}_F$ in equation (2.7.3) represents the rate of loss of energy per unit volume due to Joule heating.

Form the scalar products

$$\mathbf{H}.\nabla \wedge \mathbf{E} = -\frac{1}{c}\mathbf{H}.\frac{\partial \mathbf{B}}{\partial t} \tag{2.7.5}$$

$$\mathbf{E}.\nabla \wedge \mathbf{H} = \frac{1}{c}\mathbf{E}.\frac{\partial \mathbf{D}}{\partial t} + 4\pi\mathbf{E}.(\mathbf{J}_F + \mathbf{J}_S) \tag{2.7.6}$$

Using the vector identity

$$\nabla.(\mathbf{E} \wedge \mathbf{H}) = \mathbf{H}.\nabla \wedge \mathbf{E} - \mathbf{E}.\nabla \wedge \mathbf{H} \tag{2.7.7}$$

we obtain, upon integrating throughout a volume V

$$-c\int_V \mathbf{E}.\mathbf{J}_S\,dv - \frac{c}{4\pi}\oint_A (\mathbf{E} \wedge \mathbf{H}).\mathbf{n}\,dA = c\int_V \mathbf{E}.\mathbf{J}_F\,dv + \frac{1}{4\pi}\int_V\left(\mathbf{E}.\frac{\partial \mathbf{D}}{\partial t} + \mathbf{H}.\frac{\partial \mathbf{B}}{\partial t}\right)dv \tag{2.7.8}$$

where A is the surface bounding V.

The right-hand side of this equation contains first a term which is equal to the Joule heating within the volume V and second a term which is equal to the rate of increase of energy stored in the electromagnetic field within V. We may therefore regard this side of the equation as evaluating the total rate at which energy is being accumulated within V. This energy is derived from the source currents $\mathbf{J}_S$ within V and from any net energy flow across the surface A bounding V. Since the first term on the left-hand side of equation (2.7.8) gives the rate at which energy is provided by the source currents it is reasonable to regard the second term as evaluating the energy flow across A. We shall therefore define the Poynting Vector $\mathbf{S}$,

$$\mathbf{S} = \frac{c}{4\pi}(\mathbf{E} \wedge \mathbf{H}) \tag{2.7.9}$$

to be the instantaneous rate at which energy is flowing across a unit surface area having its normal in the direction of $(\mathbf{E} \wedge \mathbf{H})$. This is not a rigorous derivation of the properties of $\mathbf{S}$ because in equation (2.7.8) the significant quantity is the integral over a closed surface, $\oint \mathbf{S}.\mathbf{n}\, \mathrm{d}A$. We can therefore add to $\mathbf{S}$ another vector $\mathbf{s} = \nabla \wedge \mathbf{b}$, where $\mathbf{b}$ is arbitrary without affecting equation (2.7.8). The statement about the Poynting Vector must therefore be regarded as a postulate of the theory and it was for this reason that it was written separately into the field equations set out in Section 1.1.

2.7.1 Time Average of the Poynting Vector and the Wave Impedance

When the time functions are harmonic, the time average of the Poynting Vector may be written in a simple form by using complex notation. The Poynting Vector is defined by equations (2.7.1) and (2.7.9) in terms of real quantities. For harmonic time functions, using complex notation, we should write these equations as,

$$\mathbf{S} = \frac{c}{4\pi}(\mathrm{Re}\,\mathbf{E}) \wedge (\mathrm{Re}\,\mathbf{H}) \tag{2.7.10}$$

We shall show that the time average of $\mathbf{S}$ is given by

$$\langle \mathbf{S} \rangle = \frac{c}{4\pi}\, \tfrac{1}{2}\, \mathrm{Re}\,[\mathbf{E} \wedge \mathbf{H}^*]$$

Recall that, for a complex quantity $\mathbf{A}$,

$$\mathrm{Re}\,\mathbf{A} = \tfrac{1}{2}(\mathbf{A} + \mathbf{A}^*) \tag{2.7.11}$$

Hence, from equation (2.7.10)

$$\langle \mathbf{S} \rangle = \frac{c}{4\pi}\langle \tfrac{1}{4}(\mathbf{E} \wedge \mathbf{H} + \mathbf{E}^* \wedge \mathbf{H}^* + \mathbf{E} \wedge \mathbf{H}^* + \mathbf{E}^* \wedge \mathbf{H})\rangle \tag{2.7.12}$$

which may be written, using equation (2.7.11) again, as

$$\langle \mathbf{S} \rangle = \frac{c}{4\pi}\langle \tfrac{1}{2}\, \mathrm{Re}\,[\mathbf{E} \wedge \mathbf{H}] + \tfrac{1}{2}\, \mathrm{Re}\,[\mathbf{E} \wedge \mathbf{H}^*]\rangle \tag{2.7.13}$$

In this equation, the term involving $\mathbf{E} \wedge \mathbf{H}$ varies at twice the harmonic frequency and has a mean value zero. The term involving $\mathbf{E} \wedge \mathbf{H}^*$ contains the factor $\mathrm{e}^{\mathrm{j}\omega t}.\mathrm{e}^{-\mathrm{j}\omega t}$ and so is independent of the time. Hence the time average of the Poynting Vector is

$$\langle \mathbf{S} \rangle = \frac{c}{4\pi}\, \tfrac{1}{2}\, \mathrm{Re}\,[\mathbf{E} \wedge \mathbf{H}^*] \tag{2.7.14}$$

2.7.2 Complex Poynting Vector

Equation (2.7.14) leads us to consider the properties of the complex Poynting Vector $\langle \tilde{\mathbf{S}} \rangle$, which is defined for harmonic time functions by the relation

$$\langle \tilde{\mathbf{S}} \rangle = \frac{c}{4\pi}\,\tfrac{1}{2}(\mathbf{E} \wedge \mathbf{H}^*) \tag{2.7.15}$$

obviously

$$\langle \mathbf{S} \rangle = \operatorname{Re} \langle \tilde{\mathbf{S}} \rangle \tag{2.7.16}$$

From equation (2.7.7)

$$\nabla . (\mathbf{E} \wedge \mathbf{H}^*) = \mathbf{H}^* . \nabla \wedge \mathbf{E} - \mathbf{E} . \nabla \wedge \mathbf{H}^* \tag{2.7.17}$$

For a linear isotropic medium equations (2.7.5) and (2.7.6) give

$$\nabla . (\mathbf{E} \wedge \mathbf{H}^*) = -4\pi \mathbf{E} . (\mathbf{J}_{\mathrm{F}}^* + \mathbf{J}_{\mathrm{s}}^*) + \frac{j\omega\varepsilon^*}{c}|\mathbf{E}|^2 - \frac{j\omega\mu}{c}|\mathbf{H}|^2 \tag{2.7.18}$$

From which

$$\begin{aligned} -\tfrac{1}{2}c \int_V \mathbf{E} . \mathbf{J}_{\mathrm{s}}^* \, \mathrm{d}v - \oint_A \langle \tilde{\mathbf{S}} \rangle . \mathbf{n} \, \mathrm{d}A \\ = \int_V \left(\tfrac{1}{2}c\mathbf{E} . \mathbf{J}_{\mathrm{F}}^* + \frac{\omega}{8\pi}\varepsilon_2 |\mathbf{E}|^2 + \frac{\omega}{8\pi}\mu_2 |\mathbf{H}|^2 \right) \mathrm{d}v \\ - \frac{j\omega}{4\pi} \int_V \tfrac{1}{2}(\varepsilon_1 |\mathbf{E}|^2 - \mu_1 |\mathbf{H}|^2) \, \mathrm{d}v \end{aligned} \tag{2.7.19}$$

where we have written explicitly $\varepsilon = \varepsilon_1 - j\varepsilon_2$ and $\mu = \mu_1 - j\mu_2$. We may write this equation formally as

$$\tilde{W} = W_J - 2j\omega(U - V) \tag{2.7.20}$$

where $\tilde{W}$ represents the complex power delivered into the volume under consideration both from current sources and from energy flow in the electromagnetic field, W_J is the time average power loss arising from conductivity, dielectric, and magnetic losses taken together and U, V are the time averages of the stored electric and magnetic energies.

In low frequency circuits it is customary to define a driving point impedance function $Z_{\mathrm{d}}(j\omega)$ by means of the relation

$$V(j\omega) = Z_{\mathrm{d}}(j\omega) I(j\omega) \tag{2.7.21}$$

where $V(j\omega)$ is the harmonic voltage excitation and $I(j\omega)$ is the current response at the same terminals (see for example page 50). A complex power relation may be derived from equation (2.7.21), namely,

$$S = \tfrac{1}{2}V(j\omega)\, I^*(j\omega) = \tfrac{1}{2}Z_{\mathrm{d}}(j\omega)|I(j\omega)|^2 \tag{2.7.22}$$

and this relation can also be regarded as defining the driving point impedance function. We may define an impedance function for an electromagnetic wave by analogy with equation (2.7.22).

Write, if possible,

$$\langle \tilde{\mathbf{S}} \rangle = \frac{c}{4\pi}\tfrac{1}{2}(\mathbf{E} \wedge \mathbf{H}^*) = \frac{c}{4\pi}\,\tfrac{1}{2}\, Z_{\mathrm{W}}(j\omega)(\mathbf{H} . \mathbf{H}^*)\mathbf{n} \tag{2.7.23}$$

where $\mathbf{n}$ is a unit vector in the direction of $\langle \mathbf{S} \rangle$.

Hence

$$Z_W(j\omega) = \frac{2\langle\tilde{\mathbf{S}}\rangle}{\frac{c}{4\pi}|\mathbf{H}|^2\mathbf{n}} \tag{2.7.24}$$

For a plane wave we have, from equations (2.3.38) and (2.3.40),

$$\mathbf{k} \wedge \mathbf{H} = -\frac{\omega\varepsilon_c}{c}\mathbf{E} \tag{2.7.25}$$

$$\mathbf{k}.\mathbf{H} = 0 \tag{2.7.26}$$

so that

$$\langle\tilde{\mathbf{S}}\rangle = \frac{c}{4\pi}\,\tfrac{1}{2}(\mathbf{E} \wedge \mathbf{H}^*) = \frac{c}{4\pi}\,\tfrac{1}{2}\left(\frac{c}{\omega\varepsilon_c}\right)(\mathbf{H}.\mathbf{H}^*)\mathbf{k} \tag{2.7.27}$$

upon using the vector identity

$$\mathbf{a} \wedge (\mathbf{b} \wedge \mathbf{c}) = (\mathbf{a}.\mathbf{c})\mathbf{b} - (\mathbf{a}.\mathbf{b})\mathbf{c} \tag{2.7.28}$$

The wave impedance $Z_W(j\omega)$ is obtained from equation (2.7.24) and (2.3.8)

$$Z_W(j\omega) = \frac{c}{\omega\varepsilon_c}\frac{\mathbf{k}}{\mathbf{n}} = \sqrt{\left(\frac{\mu}{\varepsilon_c}\right)} \tag{2.7.29}$$

$$= \sqrt{\left(\frac{j\omega\mu}{j\omega\varepsilon + 4\pi\sigma}\right)} \tag{2.7.30}$$

Thus $Z_W(j\omega)$ is characteristic of the medium through which the wave is propagating.

In free space $Z_W(j\omega) \equiv Z_0$ is equal to unity for our system of units where the electric field **E** is measured in electrostatic units (e.s.u.) and the magnetic field **H** is measured in electromagnetic units (e.m.u.). We shall use the symbol Z_0 to represent the impedance of free space, however, because in other systems of units Z_0 takes on different values. Thus, if rationalized m.k.s. units are used $Z_0 = 376{\cdot}6$ ohms.

Note that for this example of a plane wave, we also have the following relation from equations (2.7.30) and (2.3.53)

$$Z_W(j\omega) = \sqrt{\left(\frac{j\omega\mu}{j\omega\varepsilon + 4\pi\sigma}\right)} = \frac{E_{0x}}{H_{0y}} \tag{2.7.31}$$

Equation (2.7.24) implicitly defines a unit area impedance function since the complex Poynting Vector $\langle\tilde{\mathbf{S}}\rangle$ is related to $\langle\mathbf{S}\rangle$ which refers to unit area having its normal in the direction of $(\mathbf{E} \wedge \mathbf{H})$. We may also define an impedance function for a finite area of cross-section by the equation

$$\int_A \langle\tilde{\mathbf{S}}\rangle.\mathbf{n}\,\mathrm{d}A = \frac{c}{4\pi}\,\tfrac{1}{2}Z_A(j\omega)\int_A (\mathbf{H}.\mathbf{H}^*)\,\mathrm{d}A \tag{2.7.32}$$

$$Z_A(j\omega) = \frac{2\int_A \langle\tilde{\mathbf{S}}\rangle.\mathbf{n}\,\mathrm{d}A}{(c/4\pi)\int_A (\mathbf{H}.\mathbf{H}^*)\,\mathrm{d}A} \tag{2.7.33}$$

This impedance function is useful when considering wave propagation within bounded regions, for example in waveguides and transmission lines.

2.8 Group Velocity

In the previous sections of this chapter we have described the properties of the electromagnetic field vectors by considering harmonic waves characterized by a single frequency ω and a corresponding single wave vector **k**. A pure monochromatic wave of this type is an abstraction which is never fully realized in practice and so we shall now consider the propagation of a slightly more general wave form.

Consider the propagation of an electromagnetic field distribution which takes the form of a pulse travelling in a particular direction which we identify as the z-axis of coordinates. Represent the actual field distribution by a system of plane waves in accordance with equation (2.3.21). A particular component of the field, say E_x, will have the form

$$E_x(z, t) = \int_{-\infty}^{\infty} a_x(k) \exp\{\mathrm{j}(\omega_k t - kz)\}\, \mathrm{d}k \tag{2.8.1}$$

If we take the distribution function $a_x(k)$ to be real and positive, for mathematical convenience, the pulse can be regarded as centred on $z = 0$ at $t = 0$. Thus, at time $t = 0$, the exponential factor in equation (2.8.1) is unity over the plane $z = 0$ for all values of k, and E_x has the maximum value $E_x(0, 0)$, where

$$E_x(0, 0) = \int_{-\infty}^{\infty} a_x(k)\, \mathrm{d}k \tag{2.8.2}$$

If the pulse is propagating in a non-dispersive medium, for which ω_k is a linear function of k, we have

$$\omega_k = V_0^{\mathrm{p}} k \tag{2.8.3}$$

with the phase velocity V_0^{p} a constant. In these circumstances equation (2.8.1) may be written as

$$E_x(z, t) = \int_{-\infty}^{\infty} a_x(k) \exp\{\mathrm{j}k(V_0^{\mathrm{p}} t - z)\}\, \mathrm{d}k \tag{2.8.4}$$

and at time t the maximum value of $E_x(z, t)$ will occur at a unique coordinate z, given by

$$V_0^{\mathrm{p}} t - z = 0 \tag{2.8.5}$$

which is independent of k. Hence we may write

$$E_x(z = tV_0^{\mathrm{p}}, t) = E_x(0, 0) \tag{2.8.6}$$

showing that the maximum value of E_x is propagated at the constant velocity V_0^{p}. A similar relation will hold for all the other field components and so the pulse will propagate through a non-dispersive medium at the phase velocity of the component harmonic waves.

In general, however, ω_k is not a simple linear function of k. The medium is dispersive and we should write

$$\omega_k = f(k) \tag{2.8.7}$$

If the medium is very strongly dispersive, for example in the vicinity of an absorption line, ω_k will vary as a high power of k and the pulse shape may change so significantly as the field propagates that it is no longer possible to identify a constant feature from which to define a characteristic velocity. We shall consider therefore a less extreme situation where ω_k can be approximated by a Taylor expansion about the value for $k = k_0$, and the coefficient function $a(k)$ differs appreciably from zero only over a narrow range of k centred on $k = k_0$. The component waves in equation (2.8.1) are now restricted to a narrow range of k in the vicinity of k_0. Write,

$$\omega_k = \omega(k_0) + \left(\frac{\partial\omega_k}{\partial k}\right)_{k_0}(k - k_0) + \tfrac{1}{2}\left(\frac{\partial^2\omega_k}{\partial k^2}\right)_{k_0}(k - k_0)^2 + \cdots \tag{2.8.8}$$

and assume that it is only necessary to retain the first two terms on the right-hand side of this equation. From equation (2.8.1), the field component E_x has the form,

$$E_x(z, t) = \int_{\Delta k} a_x(k) \exp\left[\left[\mathrm{j}\left[\left\{\omega(k_0) + \left(\frac{\partial\omega_k}{\partial k}\right)_{k_0}(k - k_0)\right\}t - kz\right]\right]\right] \mathrm{d}k$$

$$= \exp\left[\mathrm{j}\{\omega(k_0)t - k_0z\}\right]\int_{\Delta k} a_x(k) \exp\left[\mathrm{j}(k - k_0)\left\{\left(\frac{\partial\omega_k}{\partial k}\right)_{k_0} t - z\right\}\right] \mathrm{d}(k - k_0) \tag{2.8.9}$$

where Δk indicates the small range over which $a_x(k)$ differs appreciably from zero. Similar relations will hold for the other field components. The amplitude function

$$A_{k_0} \equiv \int_{\Delta k} a_x(k) \exp\left[\mathrm{j}(k - k_0)\left\{\left(\frac{\partial\omega_k}{\partial k}\right)_{k_0} t - z\right\}\right] \mathrm{d}(k - k_0) \tag{2.8.10}$$

is a superposition of slowly varying components with frequencies

$$\omega = \left(\frac{\partial\omega_k}{\partial k}\right)_{k_0}(k - k_0)$$

This function may therefore be regarded as modulating the high frequency wave $\omega(k_0)$. At time t, the function A_{k_0} has its maximum value at the coordinate z given by

$$\left(\frac{\partial\omega_k}{\partial k}\right)_{k_0} t - z = 0 \tag{2.8.11}$$

which defines a set of planes. The velocity at which the maximum value of A_{k_0} is advancing is called the group velocity V^{g}. Evidently,

$$V^{\mathrm{g}}(k_0) = \frac{\mathrm{d}z}{\mathrm{d}t} = \left(\frac{\partial\omega_k}{\partial k}\right)_{k_0} \tag{2.8.12}$$

The group velocity is frequently written simply as a function of k, leaving the particular value $k = k_0$ to be inserted when V^{g} needs to be evaluated. For example

$$V^g = V^g(k) = \left(\frac{\partial \omega_k}{\partial k}\right) \tag{2.8.13}$$

The group velocity will, in general, differ from the phase velocity. We have

$$V^g = \frac{\partial}{\partial k}(V^p k) = V^p + k\,\frac{\partial V^p}{\partial k} \tag{2.8.14}$$

Hence $V^g = V^p \equiv V_0^p$ only for the case of the non-dispersive medium which was discussed previously in connection with equation (2.8.6).

The present discussion of group velocity may be extended to describe a more general three-dimensional wave representation. For example, we may write

$$E_x(\mathbf{r}, t) = \int a_x(\mathbf{k}) \exp \mathrm{j}\{\omega(\mathbf{k})t - \mathbf{k}.\mathbf{r}\}\, \mathrm{d}^3\mathbf{k} \tag{2.8.15}$$

and expand $\omega(\mathbf{k})$ to first order

$$\omega(\mathbf{k}) = \omega(\mathbf{k}_0) + \{\nabla_{\mathbf{k}}\omega(\mathbf{k})\}_{\mathbf{k}_0}.(\mathbf{k} - \mathbf{k}_0) \tag{2.8.16}$$

The corresponding group velocity is now

$$\mathbf{V}^g(\mathbf{k}) = \nabla_{\mathbf{k}}\omega(\mathbf{k}) \tag{2.8.17}$$

The group velocity is a well-defined quantity when the assumptions underlying our present discussion are satisfied. In particular we have assumed that ω_k is a sufficiently slowly varying function of k for terms higher than the first power of $(k - k_0)$ to be neglected in equation (2.8.8). This restriction has provided an amplitude function in equation (2.8.10) which is a function of $\{V^g(k_0)t - z\}$. Hence planes over which A_{k_0} has constant values are defined by

$$V^g(k_0)t - z = \text{constant} \tag{2.8.18}$$

and all these planes propagate with the same group velocity. In these circumstances the pulse is propagated without change of shape. Suppose, however, that it is necessary to retain the term in $(k - k_0)^2$ from equation (2.8.8). Writing $(\partial^2\omega_k/\partial k^2)_{k_0} = \alpha$, we obtain an amplitude function A'_{k_0} given by

$$A'_{k_0} = \int a_x(k) \exp \left[\!\left[\mathrm{j}(k - k_0)\left[\left\{V^g(k_0) + \frac{\alpha}{2}(k - k_0)\right\}t - z\right]\right]\!\right] \mathrm{d}(k - k_0) \tag{2.8.19}$$

The equation replacing equation (2.8.18) is now

$$\left\{V^g(k_0) + \frac{\alpha}{2}(k - k_0)\right\}t - z = \text{constant} \tag{2.8.20}$$

and the planes corresponding to different values of k propagate with different velocities. The pulse will now change shape as it propagates through the medium. In an extreme case it will spread out and become so diffuse that it is no longer possible to identify a useful feature from which to define a group velocity.

2.9 Rectangular Waveguide and the Coaxial Transmission Line

As examples of the general techniques discussed previously we shall now consider two aspects of wave propagation within hollow conductors of specified shape. The particular examples chosen are of considerable interest both because the solutions are widely applied in practice and also because they draw attention to the conditions imposed by the boundary surfaces. The field vectors are limited to the regions within the boundaries. We do not therefore expect the solutions to be simple plane waves having constant values over an infinite plane because now the field is confined within a given cross-section.

2.9.1 Rectangular Waveguide

A waveguide is simply a long tube with a conducting wall. Consider the propagation of an electromagnetic wave down such a tube having a rectangular cross-section. For simplicity, take the region within the waveguide to be free space ($\varepsilon = 1$, $\mu = 1$, $\sigma = 0$) and assume that all bounding surfaces are perfect conductors. Let the inside dimensions be a in the x-direction, and b in the y-direction, as shown in Fig. 2.4. Suppose the waveguide to

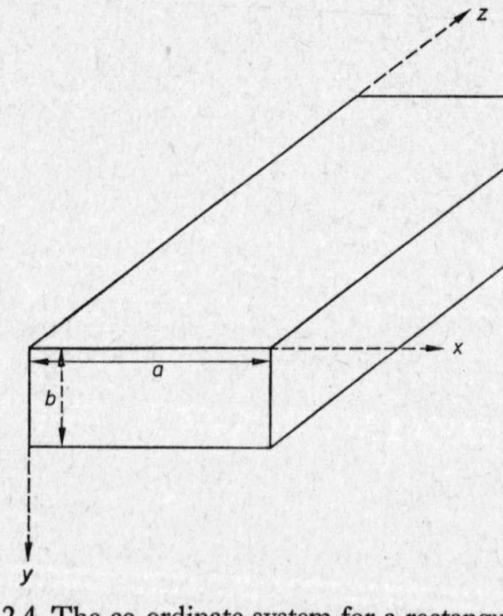

FIG. 2.4 The co-ordinate system for a rectangular waveguide.

extend indefinitely in the z-direction so that end effects need not be taken into account. We shall seek solutions for waves propagating in the z-direction having a harmonic time dependence, and guess that there will be two fundamental types. Type (a) will have a field configuration in which **E** lies entirely in the xy plane. This is a *transverse electric* or TE mode which may have a component of **H** in the direction of propagation. Type (b) will have a field configuration in which **H** lies entirely in the xy plane. This is a *transverse magnetic* or TM mode and this mode may have a component of **E** in the direction of propagation. A general solution of the electromagnetic field equations may require the superposition of a set of these fundamental modes.

2.9.2 Transverse Electric Modes

Consider first the transverse electric (TE) mode solutions and take the electric field in the form

$$\begin{aligned} \mathbf{E} &= \mathbf{E}_0(x, y)\, e^{(j\omega t - \Gamma z)} \\ E_z &= 0 \end{aligned} \tag{2.9.1}$$

Since we are using a Cartesian coordinate system we obtain from equation (2.2.10)

$$\nabla^2\mathbf{E} + \frac{\omega^2}{c^2}\mathbf{E} = 0 \tag{2.9.2}$$

For the x component of this equation

$$\frac{\partial^2 E_{0x}}{\partial x^2} + \frac{\partial^2 E_{0x}}{\partial y^2} + \left(\Gamma^2 + \frac{\omega^2}{c^2}\right)E_{0x} = 0 \tag{2.9.3}$$

Separate the variables by writing

$$E_{0x}(x, y) = X(x)\, Y(y) \tag{2.9.4}$$

so that

$$\begin{aligned} X(x) &= A_{1x} \cos k_x x + B_{1x} \sin k_x x \\ Y(y) &= A_{2x} \cos k_y y + B_{2x} \sin k_y y \end{aligned} \tag{2.9.5}$$

with

$$\Gamma^2 + \frac{\omega^2}{c^2} = k_x^2 + k_y^2 \equiv k^2 \tag{2.9.6}$$

Hence $E_{0x}(x, y)$ takes the form

$$E_{0x}(x, y) = (A_{1x} \cos k_x x + B_{1x} \sin k_x x)(A_{2x} \cos k_y y + B_{2x} \sin k_y y) \tag{2.9.7}$$

Since $E_{0y}(x, y)$ is a solution of

$$\frac{\partial^2 E_{0y}}{\partial x^2} + \frac{\partial^2 E_{0y}}{\partial y^2} + \left(\Gamma^2 + \frac{\omega^2}{c^2}\right)E_{0y} = 0 \tag{2.9.8}$$

which is of the same form as equation (2.9.3), we shall have

$$E_{0y}(x, y) = (A_{1y} \cos k_x' x + B_{1y} \sin k_x' x)(A_{2y} \cos k_y' y + B_{2y} \sin k_y' y) \tag{2.9.9}$$

We also require

$$\nabla . \mathbf{E} = \frac{\partial E_{0x}}{\partial x} + \frac{\partial E_{0y}}{\partial y} = 0 \tag{2.9.10}$$

for all values of x, y. The implication of this equation which we require immediately is that $k_x = k_x'$ and $k_y = k_y'$. We can see how this condition arises by evaluating equation (2.9.10) at a particular value of y. In this case $(\partial E_{0x}/\partial x)$ is a function of x only which we write as $A_x' \sin (k_x x + \phi)$. This function vanishes for values of x which are separated by the distance

$(x_1 - x_2) = (\pi/k_x)$. From equation (2.9.10), $-(\partial E_{0y}/\partial y)$ must be same function of x, vanishing at points separated by

$$(x_1 - x_2) = \frac{\pi}{k'_x} = \frac{\pi}{k_x}.$$

Hence $k_x = k'_x$. A similar argument for a fixed value of x, gives $k_y = k'_y$.

For perfectly conducting walls, the boundary conditions (equations (2.5.19)) require

$$\mathbf{n} \wedge \mathbf{E} = 0 \qquad \text{(at the walls)} \qquad (2.9.11)$$

at each surface. Hence, from Fig. 2.4

$$\begin{aligned} E_{0x} &= 0 \quad \text{at} \quad y = 0 \quad \text{and} \quad y = b \\ E_{0y} &= 0 \quad \text{at} \quad x = 0 \quad \text{and} \quad x = a \end{aligned} \qquad (2.9.12)$$

so that

$$k_y = \frac{n\pi}{b}, \quad k_x = \frac{m\pi}{a}, \quad A_{2x} = A_{1y} = 0 \qquad (2.9.13)$$

and n, m are integers. We may now write

$$\begin{aligned} E_{0x}(x, y) &= B_{2x} \sin k_y y (A_{1x} \cos k_x x + B_{1x} \sin k_x x) \\ E_{0y}(x, y) &= B_{1y} \sin k_x x (A_{2y} \cos k_y y + B_{2y} \sin k_y y) \end{aligned} \qquad (2.9.14)$$

Equation (2.9.10) now requires

$$\begin{aligned} k_x B_{2x} \sin k_y y (-A_{1x} \sin k_x x + B_{1x} \cos k_x x) \\ = -k_y B_{1y} \sin k_x x (-A_{2y} \sin k_y y + B_{2y} \cos k_y y) \end{aligned} \qquad (2.9.15)$$

for all x, y. This equation is satisfied if

$$\begin{aligned} B_{1x} &= B_{2y} = 0 \\ k_x B_{2x} A_{1x} &= -k_y B_{1y} A_{2y} \end{aligned} \qquad (2.9.16)$$

The solution for the electric vector field, therefore, simplifies to

$$\mathbf{E} = A_0 (\hat{\mathbf{i}} k_y \cos k_x x \sin k_y y - \hat{\mathbf{j}} k_x \sin k_x x \cos k_y y)\, e^{(j\omega t - \Gamma z)}$$

$$k_{mn}^2 = k_x^2 + k_y^2 = \left(\frac{m\pi}{a}\right)^2 + \left(\frac{n\pi}{b}\right)^2 = \Gamma^2 + \frac{\omega^2}{c^2} \qquad (2.9.17)$$

m, n, integers.

The magnetic field is derived using equation (2.2.13)

$$\mathbf{H} = -\frac{c}{j\omega} \nabla \wedge \mathbf{E}$$

we obtain

$$\begin{aligned} \mathbf{H} = \frac{c}{j\omega} A_0 \Big\{ & \hat{\mathbf{i}} \Gamma k_x \sin k_x x \cos k_y y + \hat{\mathbf{j}} \Gamma k_y \cos k_x x \sin k_y y \\ & + \hat{\mathbf{k}} \left(\Gamma^2 + \frac{\omega^2}{c^2}\right) \cos k_x x \cos k_y y \Big\}\, e^{(j\omega t - \Gamma z)} \end{aligned} \qquad (2.9.18)$$

Note that **H** has a component in the direction of propagation.

If the wave described by equations (2.9.17) and (2.9.18) is to be propagated without attenuation, then Γ must be pure imaginary and so Γ^2 is negative. From equation (2.9.17), it can be seen that for this to be true,

$$\frac{\omega^2}{c^2} > \left\{\left(\frac{m\pi}{a}\right)^2 + \left(\frac{n\pi}{b}\right)^2\right\} \tag{2.9.19}$$

Each wave mode has a cut-off frequency, denoted by ω_{mn}, below which the wave is attenuated. Evidently

$$\frac{\omega_{mn}^2}{c^2} = \left\{\left(\frac{m\pi}{a}\right)^2 + \left(\frac{n\pi}{b}\right)^2\right\} \tag{2.9.20}$$

A corresponding critical wavenumber is obtained from equation (2.9.17),

$$k_{mn}^2 = \frac{\omega_{mn}^2}{c^2} = \left\{\left(\frac{m\pi}{a}\right)^2 + \left(\frac{n\pi}{b}\right)^2\right\} \tag{2.9.21}$$

When Γ is a pure imaginary quantity, we may write $\Gamma = jk_g$. The wave vector describing the propagation in the waveguide is directed along the z-axis and has magnitude k_g. From equation (2.9.17)

$$k_g^2 = \frac{\omega^2}{c^2} - k_{mn}^2 \tag{2.9.22}$$

which may be written as

$$k_g^2 = k_f^2 - k_{mn}^2 \tag{2.9.23}$$

where $k_f = \omega/c$ is the wavenumber for a plane wave in free space having the frequency ω.

The phase velocity of the guided wave is

$$V^{\text{p}} = \omega/k_g \tag{2.9.24}$$

The corresponding group velocity is obtained from equations (2.8.13) and (2.9.22)

$$V^{\text{g}}(k_g) = \left(\frac{\partial\omega}{\partial k_g}\right) = \frac{c^2 k_g}{\omega} \tag{2.9.25}$$

Evidently

$$V^{\text{p}}V^{\text{g}} = c^2 \tag{2.9.26}$$

The time average of the Poynting Vector is derived using equations (2.7.14), (2.9.17) and (2.9.18)

$$\langle \mathbf{S} \rangle = \frac{c}{4\pi}\,\tfrac{1}{2}\,\mathrm{Re}\,[\mathbf{E} \wedge \mathbf{H}^*]$$

Remembering that we only require the real terms,

$$\langle \mathbf{S} \rangle = \frac{c}{8\pi}\left(\frac{ck_g}{\omega}\right)(k_x^2 \sin^2 k_x x \cos^2 k_y y + k_y^2 \cos^2 k_x x \sin^2 k_y y)A_0^2\hat{\mathbf{k}} \tag{2.9.27}$$

and is directed along the waveguide. This equation may also be written

$$\langle \mathbf{S} \rangle = \frac{c}{8\pi}\,\frac{\omega}{ck_g}(\mathbf{H}_t . \mathbf{H}_t^*)\hat{\mathbf{k}} \tag{2.9.28}$$

where $\mathbf{H}_t$ is the component of the magnetic vector field transverse to the direction of propagation (that is, the component in the xy plane).

The complex Poynting Vector is

$$\langle\tilde{\mathbf{S}}\rangle = \frac{c}{8\pi}(\hat{\mathbf{i}}k_x \sin 2k_x x \cos^2 k_y y + \hat{\mathbf{j}}k_y \sin 2k_y y \cos^2 k_x x)A_0^2\,\frac{1}{2}\frac{c}{\mathrm{j}\omega}\left(k_g^2 - \frac{\omega^2}{c^2}\right) + \langle\mathbf{S}\rangle \tag{2.9.29}$$

An impedance function for the waveguide is obtained from equation (2.7.33)

$$Z_{\mathrm{G}}(\mathrm{j}\omega) = \frac{2\int_A \langle\tilde{\mathbf{S}}\rangle.\hat{\mathbf{k}}\,\mathrm{d}A}{(c/4\pi)\int_A (\mathbf{H}.\mathbf{H}^*)\,\mathrm{d}A} \tag{2.9.30}$$

where A is the cross-section of the waveguide in the xy plane.

From equations (2.9.18), (2.9.28) and (2.9.29)

$$Z_{\mathrm{G}}(\mathrm{j}\omega) = \frac{(c/4\pi).(\omega/ck_g)\int_A (\mathbf{H}_{\mathrm{t}}.\mathbf{H}_{\mathrm{t}}^*)\,\mathrm{d}x\,\mathrm{d}y}{(c/4\pi)\int_A (\mathbf{H}_{\mathrm{t}}.\mathbf{H}_{\mathrm{t}}^*)\,\mathrm{d}x\,\mathrm{d}y} = \frac{\omega}{ck_g} \tag{2.9.31}$$

The $\hat{\mathbf{k}}$ component of $\mathbf{H}$ in equation (2.9.18) does not contribute to the integral in the denominator because of the boundary conditions.

This impedance function closely resembles that previously defined for a plane wave in free space. From equations (2.9.17) and (2.9.18) we can see that

$$Z_{\mathrm{G}}(\mathrm{j}\omega) = \frac{\omega}{ck_g} = -\frac{E_{0y}}{H_{0x}} = \frac{E_{0x}}{H_{0y}} \tag{2.9.32}$$

These relations should be compared with equation (2.7.31). Note also that $Z_{\mathrm{G}}(\mathrm{j}\omega)$ tends to unity (which is Z_0 the value previously obtained for a plane wave in free space) as the dimensions a, b of the waveguide tend to infinity.

2.9.3 Transverse Magnetic Modes

We shall now consider the transverse magnetic modes (TM modes), for which we take the magnetic field to be

$$\mathbf{H} = \mathbf{H}_0(x, y)\,\mathrm{e}^{(j\omega t - \Gamma z)} \tag{2.9.33}$$

$$H_z = 0$$

Solutions of the field equations can be found by following the method we have already used for the TE modes. The boundary conditions given in equations (2.5.19), however, now require

$$\mathbf{n}.\mathbf{H} = 0 \qquad \text{(at the walls)} \tag{2.9.34}$$

and we also make use of

$$\nabla.\mathbf{H} = 0 \tag{2.9.35}$$

The magnetic vector field is given by

$$\mathbf{H} = A_0(\hat{\mathbf{i}}k_y \sin k_x x \cos k_y y - \hat{\mathbf{j}}k_x \cos k_x x \sin k_y y)\,\mathrm{e}^{(\mathrm{j}\omega t - \Gamma z)} \tag{2.9.36}$$

with $k_x = (m\pi/a)$, $k_y = (n\pi/b)$, m, n integers and

$$k_{mn}^2 = k_x^2 + k_y^2 = \left(\frac{m\pi}{a}\right)^2 + \left(\frac{n\pi}{b}\right)^2 = \Gamma^2 + \frac{\omega^2}{c^2}.$$

The electric vector field is derived using equation (2.2.14)

$$\mathbf{E} = \frac{c}{\mathrm{j}\omega} \nabla \wedge \mathbf{H}$$

$$\mathbf{E} = -\frac{c}{\mathrm{j}\omega} A_0 \left\{ \hat{\mathbf{i}} \Gamma k_x \cos k_x x \sin k_y y + \hat{\mathbf{j}} \Gamma k_y \sin k_x x \cos k_y y + \hat{\mathbf{k}} \left(\Gamma^2 + \frac{\omega^2}{c^2} \right) \sin k_x x \sin k_y y \right\} \mathrm{e}^{(\mathrm{j}\omega t - \Gamma z)} \quad (2.9.37)$$

For these waves **E** has a component in the direction of propagation. Note that equations (2.9.19)–(2.9.26) apply both to TE and to TM modes.

The time average of the Poynting Vector for a propagating TM mode is,

$$\langle \mathbf{S} \rangle = \frac{c}{8\pi} \left(\frac{c k_g}{\omega} \right) (k_x^2 \cos^2 k_x x \sin^2 k_y y + k_y \sin^2 k_x x \cos^2 k_y y) A_0^2 \hat{\mathbf{k}} \quad (2.9.38)$$

$$= \frac{c}{8\pi} \left(\frac{c k_g}{\omega} \right) (\mathbf{H}.\mathbf{H}^*) \hat{\mathbf{k}} \quad (2.9.39)$$

The waveguide impedance function for the TM mode is,

$$Z_{\mathrm{G}}(\mathrm{j}\omega) = \frac{c k_g}{\omega} \quad (2.9.40)$$

Equation (2.9.40) indicates that the impedance function for a TM mode is just the reciprocal of that for a TE mode given in equation (2.9.32). This result, however, refers to the particular case where the interior of the waveguide is taken to be free space, that is, $\varepsilon = \mu = 1$. If the waveguide is filled with a material medium, equations (2.9.32) and (2.9.40) become

$$Z_{\mathrm{G}}(\mathrm{j}\omega) = \left(\frac{\mu}{\varepsilon} \right)^{\frac{1}{2}} \frac{\omega}{c k_g} \quad \text{for a TE mode} \quad (2.9.41)$$

$$Z_{\mathrm{G}}(\mathrm{j}\omega) = \left(\frac{\mu}{\varepsilon} \right)^{\frac{1}{2}} \frac{c k_g}{\omega} \quad \text{for a TM mode} \quad (2.9.42)$$

The integers m, n may be used to identify the different forms of wave propagation. Thus a particular TE mode will be designated as a $\mathrm{TE}_{m,n}$ mode whereas a TM mode will be designated as a $\mathrm{TM}_{m,n}$ mode. The lowest permissible TE mode has the designation TE_{10} or TE_{01}. All components of the mode TE_{00} are zero (this is evident from equations (2.9.17) and (2.9.18)). The lowest TM mode is TM_{11}. The TE_{10} mode is of considerable practical importance since the dimensions of the waveguide may be so chosen that for a given frequency only this mode is propagated.

2.9.4 Coaxial Transmission Line

Consider the propagation of an electromagnetic wave in a region bounded by two coaxial and perfectly conducting cylinders. Let the annular space between the cylinders be characterized by the parameters ε, μ, σ, and choose cylindrical polar coordinates as shown in Fig. 2.5.

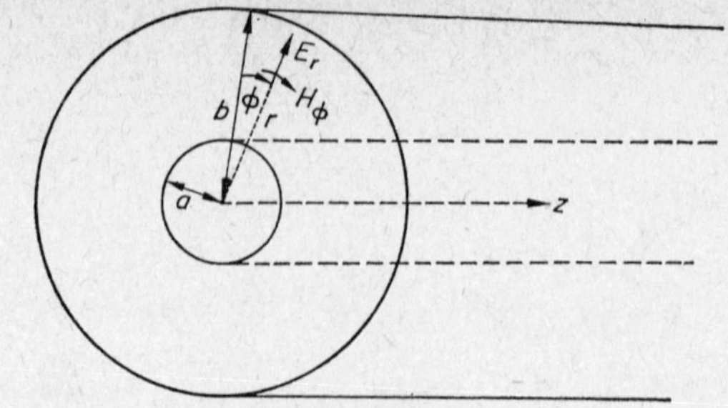

FIG. 2.5 The co-ordinate system for a coaxial transmission line.

In the present example we shall seek only the lowest mode solution of the field equations since this solution is particularly important in practical applications. Attempt to find a solution using a purely transverse waveform but allow the field vectors to be functions of position in the transverse plane. The presence of the inner conductor enables a transverse wave to satisfy the boundary conditions and distinguishes the coaxial line from the waveguide. Recall that, from equations (2.5.19), we require $\mathbf{n} \wedge \mathbf{E} = 0$ and $\mathbf{n} . \mathbf{H} = 0$ at the bounding surfaces. We may therefore take $E_\phi = E_z = 0$ and $H_r = H_z = 0$ to satisfy these conditions immediately for a transverse wave. We now seek solutions in the form

$$\mathbf{E} = \mathbf{a}_r E_r = \mathbf{a}_r E_{0r}(r)\, e^{(j\omega t - \Gamma z)} \qquad E_\phi = E_z = 0 \tag{2.9.43}$$

$$\mathbf{H} = \mathbf{a}_\phi H_\phi = \mathbf{a}_\phi H_{0\phi}(r)\, e^{(j\omega t - \Gamma z)} \qquad H_r = H_z = 0 \tag{2.9.44}$$

with $\mathbf{a}_r$, $\mathbf{a}_\phi$ unit vectors along the coordinate directions of r and ϕ.

The field equations, written in cylindrical polar coordinates, become

$$\nabla \wedge \mathbf{E} + \frac{j\omega\mu}{c}\mathbf{H} = 0: \qquad \frac{\partial E_r}{\partial z} + \frac{j\omega\mu}{c} H_\phi = 0, \quad \frac{\partial E_r}{\partial \phi} = 0 \tag{2.9.45}$$

$$\nabla \wedge \mathbf{H} - \frac{j\omega\varepsilon_c}{c}\mathbf{E} = 0: \qquad -\frac{\partial H_\phi}{\partial z} - \frac{j\omega\varepsilon_c}{c} E_r = 0, \quad \frac{\partial}{\partial r}(r H_\phi) = 0 \tag{2.9.46}$$

with

$$\varepsilon_c = \varepsilon - j\frac{4\pi\sigma}{\omega}$$

Evidently,

$$\begin{aligned} \frac{\partial^2 H_\phi}{\partial z^2} + \frac{\omega^2 \mu \varepsilon_c}{c^2} H_\phi &= 0 \\ \frac{\partial^2 E_r}{\partial z^2} + \frac{\omega^2 \mu \varepsilon_c}{c^2} E_r &= 0 \end{aligned} \tag{2.9.47}$$

and so

$$\Gamma = \pm j \frac{\omega}{c} \sqrt{(\mu \varepsilon_c)} \tag{2.9.48}$$

The choice of sign in this equation indicates that the wave may propagate in either the positive or the negative direction of z. For convenience we shall take only the positive sign in the following analysis and so refer to a wave travelling in the positive direction of z when the material filling the coaxial line is loss free. From equations (2.9.44), (2.9.46) and (2.9.48)

$$E_r(r) = \sqrt{\left(\frac{\mu}{\varepsilon_c}\right)} H_\phi(r) \tag{2.9.49}$$

$$H_\phi(r) = \frac{A_0}{r} e^{(j\omega t - \Gamma z)} \tag{2.9.50}$$

The constant A_0 in equation (2.9.50) may be evaluated in terms of a current amplitude, I_0, by using the boundary conditions at the surface of the inner conductor. From equations (2.5.19),

$$\mathbf{n} \wedge \mathbf{H} = 4\pi \mathbf{K}_s \tag{2.9.51}$$

and so, for $r = a$

$$H_\phi(a) = 4\pi K_s^{(z)} = \frac{A_0}{a} e^{(j\omega t - \Gamma z)} \tag{2.9.52}$$

where $K_s^{(z)}$ is the surface current density at the inner conductor and is directed along the z-axis. The total current, I, flowing along the surface of the inner conductor is

$$\begin{aligned} I &= 2\pi a K_s^{(z)} \\ &= \frac{A_0}{2} e^{(j\omega t - \Gamma z)} \qquad (2.9.53) \\ &= I_0 e^{(j\omega t - \Gamma z)} \qquad (2.9.54) \end{aligned}$$

with $A_0 = 2I_0$, where I_0 is the amplitude of the surface current. We may now write,

$$H_\phi = \frac{2I_0}{r} e^{(j\omega t - \Gamma z)} \tag{2.9.55}$$

$$E_r = \frac{2I_0}{r} \sqrt{\left(\frac{\mu}{\varepsilon_c}\right)} e^{(j\omega t - \Gamma z)} \tag{2.9.56}$$

Note that, since

$$aH_\phi(a) = bH_\phi(b)$$

there will be currents of equal magnitude flowing on the surface of each conducting cylinder. These currents will differ in phase by 180° because of the sign of the vector $\mathbf{n}$ which has to be included in equation (2.9.51).

If the space between the coaxial cylinders is filled with a lossless material, then ε, μ are real and $\sigma = 0$. For this case Γ is a pure imaginary quantity,

$$\Gamma = jk_L = j\frac{\omega}{c}\sqrt{(\mu\varepsilon)} \tag{2.9.57}$$

and the phase velocity is equal to the group velocity,

$$V^p = \frac{\omega}{k_L} = V^g = \frac{c}{\sqrt{(\mu\varepsilon)}} \tag{2.9.58}$$

These velocities are equal to the phase and group velocities for a plane wave propagating in the infinite dielectric medium. Evidently, for this case, all frequencies are propagated without attenuation. There is no cut-off phenomenon such as occurs in a waveguide. It can be seen from equation (2.9.48) that such dispersion and attenuation as does in fact arise will be due to the properties of the medium filling the coaxial line and will not depend upon the dimensions of the bounding surfaces.†

The time average of the Poynting Vector is,

$$\langle \mathbf{S} \rangle = \frac{c}{4\pi}\tfrac{1}{2}\,\mathrm{Re}\,[\mathbf{E}\wedge\mathbf{H}^*] = \frac{c}{8\pi}\,\mathrm{Re}\,[E_r H_\phi^*]\hat{\mathbf{k}} \tag{2.9.59}$$

$$= \frac{c}{8\pi}\,\mathrm{Re}\left[\sqrt{\left(\frac{\mu}{\varepsilon_c}\right)}\right]|H_\phi|^2\hat{\mathbf{k}} \tag{2.9.60}$$

A wave impedance function is obtained from the complex Poynting Vector using equation (2.7.33)

$$Z_A(\mathrm{j}\omega) = \frac{2\int_A \langle \tilde{\mathbf{S}}\rangle .\hat{\mathbf{k}}\,\mathrm{d}A}{(c/4\pi)\int_A (\mathbf{H}.\mathbf{H}^*)\,\mathrm{d}A} = \frac{(c/4\pi)\int_A \sqrt{(\mu/\varepsilon_c)}\,|H_\phi|^2\,\mathrm{d}A}{(c/4\pi)\int_A |H_\phi|^2\,\mathrm{d}A} \tag{2.9.61}$$

or

$$Z_A(\mathrm{j}\omega) = \sqrt{\left(\frac{\mu}{\varepsilon_c}\right)} = \sqrt{\left(\frac{\mathrm{j}\omega\mu}{\mathrm{j}\omega\varepsilon + 4\pi\sigma}\right)} \tag{2.9.62}$$

This impedance function refers to a single wave propagating along the z-direction since only the positive sign was retained from equation (2.9.48). Had the choice of signs been retained there would also be a choice of signs in equation (2.9.62) and the wave impedance would change sign for a wave which propagates in the opposite direction.

Another impedance function may be defined by observing that the coaxial line is very similar to a low frequency circuit element when it is operating in the mode that we have discussed here. The currents along the inner and outer cylinders are of equal magnitude but in opposing directions and so these conductors may be regarded as a terminal pair. A voltage function between the cylinders is defined by

$$V_b - V_a = -\int_a^b E_r(r)\,\mathrm{d}r \tag{2.9.63}$$

From equation (2.9.56)

$$V_a - V_b \equiv V = 2\sqrt{\left(\frac{\mu}{\varepsilon_c}\right)}\log_e(b/a).I \tag{2.9.64}$$

The ratio V/I defines an impedance function Z_c

$$Z_c = 2\sqrt{\left(\frac{\mu}{\varepsilon_c}\right)}\log_e(b/a) \tag{2.9.65}$$

† We assume the bounding surfaces to be perfectly conducting.

where Z_c is called the *characteristic impedance* of the coaxial line. This impedance function also refers to a single wave propagating along the line. From equation (2.9.64), we may describe a wave in terms of a 'low frequency' representation using V, I, as significant quantities rather than **E, H.**

$$V(z) = Z_c I_0 \, e^{(j\omega t - \Gamma z)} \tag{2.9.66}$$

$$\frac{\partial V}{\partial z} = -\Gamma Z_c I \tag{2.9.67}$$

$$\frac{\partial I}{\partial z} = -\frac{\Gamma}{Z_c} V \tag{2.9.68}$$

If the coaxial line is terminated at some point we may expect that the simplest representation of the complete wave motion will be the sum of two waves propagating in opposite directions.† We therefore write

$$V(z) = A \, e^{(j\omega t - \Gamma z)} + B \, e^{(j\omega t + \Gamma z)} \tag{2.9.69}$$

that is

$$V(z) = (A \, e^{-\Gamma z} + B \, e^{\Gamma z}) \, e^{j\omega t} \tag{2.9.70}$$

and from equation (2.9.67)

$$I(z) = \left(\frac{A}{Z_c} e^{-\Gamma z} - \frac{B}{Z_c} e^{\Gamma z}\right) e^{j\omega t} \tag{2.9.71}$$

These equations define a 'driving point' impedance function at a particular plane for the coaxial line,

$$\frac{V(z)}{I(z)} = Z(z) = Z_c \frac{A \, e^{-\Gamma z} + B \, e^{\Gamma z}}{A \, e^{-\Gamma z} - B \, e^{\Gamma z}} \tag{2.9.72}$$

Suppose the coaxial line is terminated at the plane $(z + l)$ by an impedance $Z(l)$, as shown in Fig. 2.6

$$Z(l) = Z_c \frac{A \, e^{-\Gamma(z+l)} + B \, e^{\Gamma(z+l)}}{A \, e^{-\Gamma(z+l)} - B \, e^{\Gamma(z+l)}} \tag{2.9.73}$$

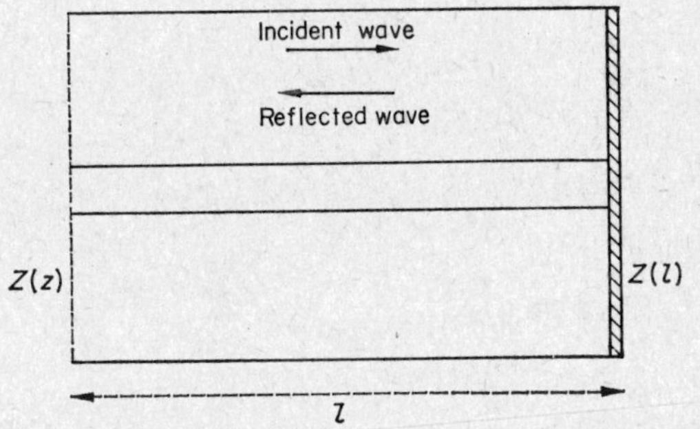

FIG. 2.6 Coaxial line terminated by an impedance $Z(l)$.

† This will not necessarily be a complete solution since the termination may give rise to higher mode forms which we have not considered.

From equation (2.9.73)

$$Z(l) = Z_c \frac{1 + (B/A)\, e^{2\Gamma(z+l)}}{1 - (B/A)\, e^{2\Gamma(z+l)}} \tag{2.9.74}$$

$$\frac{B}{A} e^{2\Gamma z} = \frac{Z(l) - Z_c}{Z(l) + Z_c} e^{-2\Gamma l} \tag{2.9.75}$$

The input impedance of the terminated section of coaxial line is given by equation (2.9.72),

$$Z_{in} = Z(z) = Z_c \frac{1 + (B/A)\, e^{2\Gamma z}}{1 - (B/A)\, e^{2\Gamma z}} \tag{2.9.76}$$

substituting in this equation for $(B/A)\, e^{2\Gamma z}$ from equation (2.9.75), we obtain

$$Z_{in} = Z_c \frac{Z(l) + Z_c \tanh \Gamma l}{Z_c + Z(l) \tanh \Gamma l} \tag{2.9.77}$$

In particular when the line is lossless, $\Gamma = jk_L$, and

$$Z_{in} = Z_c \frac{Z(l) + jZ_c \tan k_L l}{Z_c + jZ(l) \tan k_L l} \tag{2.9.78}$$

There are important applications of this equation.

(*a*) If the line is terminated by a short circuit,

$$Z_{in,s} = jZ_c \tan k_L l \tag{2.9.79}$$

(*b*) If the line is terminated by an open circuit,

$$Z_{in,o} = -jZ_c \cot k_L l \tag{2.9.80}$$

(*c*) If the length of the section of line is chosen so that $k_L l = \pi/2$ (that is, $l = (\lambda_L/4)$)

$$Z_{in,\pi/2} = \frac{Z_c^{\,2}}{Z(l)} \tag{2.9.81}$$

(*d*) If the length of the section of line is chosen so that $k_L l = \pi$, that is, $l = (\lambda_L/2)$

$$Z_{in,\pi} = Z(l) \tag{2.9.82}$$

The coaxial line may therefore be used as an impedance transformer to match a load to a generator.

2.10 Reflection Coefficient at Normal Incidence

The reflection coefficient, R, is defined to be the ratio of the magnitude of the time averaged Poynting Vector for the reflected wave to the same quantity for the incident wave.

$$R = \frac{\text{Magnitude } \{\text{Re } [\mathbf{E} \wedge \mathbf{H}^*]\} \text{ reflected}}{\text{Magnitude } \{\text{Re } [\mathbf{E} \wedge \mathbf{H}^*]\} \text{ incident}} \tag{2.10.1}$$

Consider a plane wave which is reflected at a plane surface of discontinuity between an absorbing medium and a loss free medium. Assume that the wave is propagating in the loss free medium and is incident normally upon the surface of discontinuity. For this case of normal incidence the two modes of polarization which were treated separately in Section 2.6 become equivalent, and so we only need to consider one wave type. For a plane wave, we know from equation (2.3.52) that

$$\frac{E_0}{H_0} = \sqrt{\left(\frac{\mu}{\varepsilon_c}\right)} \tag{2.10.2}$$

and so, in the notation of Section 2.6,

$$R = \frac{\sqrt{(\varepsilon_1/\mu_1)}\,|E_0''|^2}{\sqrt{(\varepsilon_1/\mu_1)}|E_0|^2} = \frac{|E_0''|^2}{|E_0|^2} \tag{2.10.3}$$

since both the incident and the reflected waves are propagating in the loss free medium. From equation (2.6.19)

$$R = \left|\frac{\mu_2 k_z - \mu_1 k_z'}{\mu_2 k_z + \mu_1 k_z'}\right|^2 \tag{2.10.4}$$

and using equations (2.6.8) and (2.6.11) with $\theta = 0$ for normal incidence

$$R = \left|\frac{\sqrt{(\varepsilon_1/\mu_1)} - \sqrt{(\varepsilon_{c2}/\mu_2)}}{\sqrt{(\varepsilon_1/\mu_1)} + \sqrt{(\varepsilon_{c2}/\mu_2)}}\right|^2 \tag{2.10.5}$$

For non-magnetic media, $\mu_1 = \mu_2 = 1$, and then

$$R = \left|\frac{\sqrt{\varepsilon_1} - \sqrt{\varepsilon_{c2}}}{\sqrt{\varepsilon_1} + \sqrt{\varepsilon_{c2}}}\right|^2 \tag{2.10.6}$$

This equation may be conveniently written in terms of a refractive index defined by

$$\eta - \mathrm{j}\varkappa = \sqrt{\varepsilon} \tag{2.10.7}$$

Remembering that the first medium is assumed to be loss free, so that ε_1 is real, equation (2.10.6) becomes

$$R = \frac{(\eta_1 - \eta_2)^2 + \varkappa_2^2}{(\eta_1 + \eta_2)^2 + \varkappa_2^2} \tag{2.10.8}$$

For practically all (non-magnetic) metals we shall have $(4\pi\sigma/\omega) \gg 1$, and so equations (2.3.54) and (2.3.56) may be used to evaluate η_2, $\varkappa_2$, for these materials. For a metal with $\mu = 1$,

$$\eta_2 \sim \varkappa_2 \sim \sqrt{\left(\frac{2\pi\sigma}{\omega}\right)} \gg 1 \tag{2.10.9}$$

and so

$$R_{\text{metal}} \sim \frac{\{(\eta_1/\varkappa_2) - 1\}^2 + 1}{\{(\eta_1/\varkappa_2) + 1\}^2 + 1} \sim \frac{1 - (\eta_1/\varkappa_2)}{1 + (\eta_1/\varkappa_2)} \sim 1 - 2\frac{\eta_1}{\varkappa_2} \tag{2.10.10}$$

$$R_{\text{metal}} \sim 1 - 2\eta_1 \sqrt{\left(\frac{\omega}{2\pi\sigma}\right)} \tag{2.10.11}$$

Hence the reflection coefficient is close to unity for most metals. For silver the value of R is about 0·98 and so essentially all the energy in the incident wave is reflected.

2.11 Scalar and Vector Potentials

In the preceding sections of this chapter we have been able to discuss a number of important electromagnetic phenomena by transforming the coupled differential equations (equations (2.2.1)–(2.2.3) page 58) into separate 'wave' equations for the vectors **E** and **H**. Solutions of these separated second order differential equations which satisfied the boundary conditions and were also consistent with the coupled equations provided an acceptable description of the electromagnetic field. This simplification of the fundamental equations arose essentially from restricting the discussion to linear, homogeneous media, with $\rho = 0$ and **J** derived the action of the electromagnetic field itself, so that $\mathbf{J} = (\sigma/c)\mathbf{E}$. However, even with these limitations the spherical wave solutions discussed in Section 2.4 have already provided an example of a case where it was more practicable to solve the equations less directly by making use of a potential function. If sources of the electromagnetic field are present it is usually most convenient to derive the field vectors from potential functions even when a rectangular Cartesian coordinate system is used.

The relation $\nabla . \mathbf{B} = 0$ is satisfied automatically if **B** is derived from another vector function of position and time, $\mathbf{A}(\mathbf{r}, t)$, by the equation

$$\mathbf{B} = \nabla \wedge \mathbf{A}(\mathbf{r}, t) \tag{2.11.1}$$

The function $\mathbf{A}(\mathbf{r}, t)$ is called the *vector potential function*, it must of course be chosen to be consistent with the remaining field equations. Thus equation (1.1.1) gives

$$\nabla \wedge \left(\mathbf{E} + \frac{1}{\partial} \frac{\partial \mathbf{A}}{ct}\right) = 0 \tag{2.11.2}$$

which is satisfied if

$$\mathbf{E} = -\frac{1}{c}\frac{\partial \mathbf{A}}{\partial t} - \nabla\phi \tag{2.11.3}$$

Here $\phi(\mathbf{r}, t)$ is a scalar function of the space and time coordinates. The function $\phi(\mathbf{r}, t)$ is called the *scalar potential function* for the electromagnetic field. The functions $\mathbf{A}(\mathbf{r}, t)$, $\phi(\mathbf{r}, t)$, must also satisfy equations (1.1.2) and (1.1.24); these are

$$\nabla \wedge \mathbf{H} - \frac{1}{c}\frac{\partial \mathbf{D}}{\partial t} - 4\pi\mathbf{J} = 0 \tag{2.11.4}$$

$$\nabla . \mathbf{D} = 4\pi\rho \tag{2.11.5}$$

For a linear, homogeneous and isotropic medium, for which $\mathbf{D} = \varepsilon\mathbf{E}$ $\mathbf{B} = \mu\mathbf{H}$, equation (2.11.4) gives

$$\frac{1}{\mu}\nabla \wedge \nabla \wedge \mathbf{A} + \frac{\varepsilon}{c}\left(\frac{1}{c}\frac{\partial^2 \mathbf{A}}{\partial t^2} + \nabla\frac{\partial\phi}{\partial t}\right) = 4\pi\mathbf{J} \tag{2.11.6}$$

For rectangular Cartesian coordinates,

$$\nabla^2\mathbf{A} - \frac{\varepsilon\mu}{c^2}\frac{\partial^2 \mathbf{A}}{\partial t^2} - \nabla\left(\nabla . \mathbf{A} + \frac{\varepsilon\mu}{c}\frac{\partial\phi}{\partial t}\right) = -4\pi\mu\mathbf{J} \tag{2.11.7}$$

with $\nabla^2\mathbf{A}$ given by equation (2.2.8).

Adding and subtracting the term

$$\left(\frac{\varepsilon\mu}{c^2}\frac{\partial^2\phi}{\partial t^2}\right)$$

to equation (2.11.5), we obtain

$$\nabla^2\phi - \frac{\varepsilon\mu}{c^2}\frac{\partial^2\phi}{\partial t^2} + \frac{1}{c}\frac{\partial}{\partial t}\left(\nabla.\mathbf{A} + \frac{\varepsilon\mu}{c}\frac{\partial\phi}{\partial t}\right) = -\frac{4\pi}{\varepsilon}\rho \tag{2.11.8}$$

If it is possible to choose the functions $\mathbf{A}(xyz, t)$, $\phi(xyz, t)$ so that

$$\nabla.\mathbf{A} + \frac{\varepsilon\mu}{c}\frac{\partial\phi}{\partial t} = 0 \tag{2.11.9}$$

equations (2.11.7) and (2.11.8) are considerably simplified. This requirement is called the Lorentz condition and when it is satisfied,

$$\nabla^2\mathbf{A} - \frac{\varepsilon\mu}{c^2}\frac{\partial^2\mathbf{A}}{\partial t^2} = -4\pi\mu\mathbf{J} \tag{2.11.10}$$

$$\nabla^2\phi - \frac{\varepsilon\mu}{c^2}\frac{\partial^2\phi}{\partial t^2} = -\frac{4\pi}{\varepsilon}\rho \tag{2.11.11}$$

$$\nabla.\mathbf{A} + \frac{\varepsilon\mu}{c}\frac{\partial\phi}{\partial t} = 0 \tag{2.11.12}$$

are the fundamental equations describing the electromagnetic field in the linear approximation for **D** and **B**.

Note that if we write the total current density **J** in terms of a component arising from the action of the electromagnetic field, $\mathbf{J}_F = (\sigma\mathbf{E}/c)$, together with an additional component $\mathbf{J}_S$ arising from external sources, we have

$$\mathbf{J} = \mathbf{J}_F + \mathbf{J}_S = \frac{\sigma}{c}\mathbf{E} + \mathbf{J}_S \tag{2.11.13}$$

The field equations (2.11.10)–(2.11.12), now become

$$\nabla^2\mathbf{A} - \frac{\varepsilon\mu}{c^2}\frac{\partial^2\mathbf{A}}{\partial t^2} - \frac{4\pi\sigma\mu}{c^2}\frac{\partial\mathbf{A}}{\partial t} = -4\pi\mu\mathbf{J}_S \tag{2.11.14}$$

$$\nabla^2\phi - \frac{\varepsilon\mu}{c^2}\frac{\partial^2\phi}{\partial t^2} - \frac{4\pi\sigma\mu}{c^2}\frac{\partial\phi}{\partial t} = -\frac{4\pi}{\varepsilon}\rho \tag{2.11.15}$$

$$\nabla.\mathbf{A} + \frac{\varepsilon\mu}{c}\frac{\partial\phi}{\partial t} + \frac{4\pi\mu\sigma}{c}\phi = 0 \tag{2.11.16}$$

We shall usually be concerned with situations for which $\sigma = 0$. For these cases equations (2.11.14)–(2.11.16) simplify to have the same form as equations (2.11.10)–(2.11.12) when we write $\mathbf{J} = \mathbf{J}_S$.

The Lorentz condition (equations (2.11.12) and (2.11.16)) does not impose a significant restriction on the choice of potential functions because these are not uniquely related to **E** and **H**. Thus **H** is derived from $\nabla \wedge \mathbf{A}$, but exactly the same field distribution will be derived from the vector function

$$\mathbf{a} = \mathbf{A} + \nabla\psi,$$

where ψ is an arbitrary scalar function of position and time, since $\nabla \wedge \nabla \psi = 0$. Suppose that $\mathbf{a}(\mathbf{r}, t)$ and $\phi'(\mathbf{r}, t)$ are potential functions which satisfy equations (2.11.7) and (2.11.8) but do not necessarily obey the Lorentz condition. Write another vector function, $\mathbf{A}(\mathbf{r}, t)$, according to the relation

$$\mathbf{A}(\mathbf{r}, t) = \mathbf{a}(\mathbf{r}, t) - \nabla \psi(\mathbf{r}, t) \tag{2.11.17}$$

The same magnetic field $\mathbf{H}$ will be derived from both $\mathbf{A}(\mathbf{r}, t)$ and from $\mathbf{a}(\mathbf{r}, t)$. The electric vector field, $\mathbf{E}$, will also be unchanged by the use of $\mathbf{A}$, if

$$\mathbf{E} = -\frac{1}{c}\frac{\partial \mathbf{a}}{\partial t} - \nabla \phi' = -\frac{1}{c}\frac{\partial \mathbf{A}}{\partial t} - \nabla \phi \tag{2.11.18}$$

which is satisfied if we use a scalar potential function given by the relation

$$\phi = \phi' + \frac{1}{c}\frac{\partial \psi}{\partial t} \tag{2.11.19}$$

We may now require the functions $\mathbf{A}(\mathbf{r}, t)$, $\phi(\mathbf{r}, t)$, to obey the Lorentz condition. This will restrict the choice of $\psi(\mathbf{r}, t)$ to functions which are solutions of the differential equation

$$\nabla^2 \psi - \frac{\varepsilon\mu}{c^2}\frac{\partial^2 \psi}{\partial t^2} = \nabla . \mathbf{a} + \frac{\varepsilon\mu}{c}\frac{\partial \phi'}{\partial t} \tag{2.11.20}$$

Hence, starting from entirely general solutions of equations (2.11.7) and (2.11.8), in the form $\mathbf{a}(\mathbf{r}, t)$, $\phi'(\mathbf{r}, t)$, it is possible to construct alternative functions which satisfy the Lorentz conditions and yet leave the field vectors themselves unchanged. Moreover, having found one pair of Lorentz potentials $\mathbf{A}(\mathbf{r}, t)$, $\phi(\mathbf{r}, t)$ alternative Lorentz potentials $\mathbf{A}_1(\mathbf{r}, t)$, $\phi_1(\mathbf{r}, t)$ can be formed (which may be more convenient for the particular problem under investigation) by the transformation

$$\mathbf{A}_1 = \mathbf{A} - \nabla \psi_1 \tag{2.11.21}$$

$$\phi_1 = \phi + \frac{1}{c}\frac{\partial \psi_1}{\partial t} \tag{2.11.22}$$

with

$$\nabla^2 \psi_1 - \frac{\varepsilon\mu}{c^2}\frac{\partial^2 \psi_1}{\partial t^2} = 0 \tag{2.11.23}$$

In this case $\psi_1(\mathbf{r}, t)$ can be any solution of the homogeneous wave equation. There is therefore a very large class of potential functions available which will obey the Lorentz condition and so no loss of generality is incurred by choosing this form of solution.

A solution of particular interest arises within a region where the charge density, ρ, is zero. In this case it is possible to derive the field distribution from the vector potential function alone. Evidently $\phi = 0$ is a solution of equation (2.11.11) when $\rho = 0$, but we must show that this does not necessarily lead to trivial solutions for the field vectors. Consider the transformation of the general vector and scalar potential functions $\mathbf{A}(\mathbf{r}, t)$, $\phi(\mathbf{r}, t)$, which already satisfy the Lorentz condition, equation (2.11.9), into new potential functions $\mathbf{A}_1(\mathbf{r}, t)$, $\phi_1(\mathbf{r}, t) = 0$, by choosing a suitable function $\psi_1(\mathbf{r}, t)$

from solutions of equation (2.11.23). We may show that this is always possible when $\rho = 0$, by observing that $\phi(\mathbf{r}, t) \neq 0$ is a solution of

$$\nabla^2\phi - \frac{\varepsilon\mu}{c^2}\frac{\partial^2\phi}{\partial t^2} = 0 \tag{2.11.24}$$

from equation (2.11.11). Hence, if we can choose

$$\psi_1 = -c\int \phi \, \mathrm{d}t \tag{2.11.25}$$

we shall have $\phi_1(\mathbf{r}, t) = 0$ which is our required solution of equation (2.11.11) with $\rho = 0$. The function $\psi_1(\mathbf{r}, t)$ must also be a solution of equation (2.11.23). Evidently this will be the case, because,

$$\nabla^2\psi_1 - \frac{\varepsilon\mu}{c^2}\frac{\partial^2\psi_1}{\partial t^2} = -c\int\left(\nabla^2\phi - \frac{\varepsilon\mu}{c^2}\frac{\partial^2\phi}{\partial t^2}\right)\mathrm{d}t = 0 \tag{2.11.26}$$

In a region where $\rho = 0$ therefore we may choose a vector potential $\mathbf{A}_1(\mathbf{r}, t)$ such that

$$\begin{aligned} \mathbf{B} &= \nabla \wedge \mathbf{A}_1 \\ \mathbf{E} &= -\frac{1}{c}\frac{\partial \mathbf{A}_1}{\partial t} \\ \nabla . \mathbf{A}_1 &= 0 \end{aligned} \tag{2.11.27}$$

This form of solution has applications in magnetism but its usefulness there is restricted because $\mathbf{A}_1(\mathbf{r}, t)$ is a simple function of the space coordinates only for the case of the elementary magnetostatic dipole (see page 111, Problem 12). Another important application arises in the treatment of optical phenomena using Quantum Mechanics. This aspect is discussed in Chapter 7, Section 7.4.

In a region where $\mathbf{J} = 0$, $\rho = 0$ and $\varepsilon = 1$, $\mu = 1$, $\sigma = 0$ (that is free space with no sources present), it is sometimes convenient to make use of another vector potential function which enables the Lorentz condition to be automatically satisfied. This potential function is called the Hertz Vector and is denoted by $\mathbf{\Pi}$. We are considering a region where $\mathbf{A}$ and ϕ are, in general, solutions of the equations

$$\nabla^2\mathbf{A} - \frac{1}{c^2}\frac{\partial^2\mathbf{A}}{\partial t^2} = 0 \tag{2.11.28}$$

$$\nabla^2\phi - \frac{1}{c^2}\frac{\partial^2\phi}{\partial t^2} = 0 \tag{2.11.29}$$

$$\nabla . \mathbf{A} + \frac{1}{c}\frac{\partial\phi}{\partial t} = 0 \tag{2.11.30}$$

Substitute for $\mathbf{A}$ and ϕ according to the relations

$$\mathbf{A} = \frac{1}{c}\frac{\partial\mathbf{\Pi}}{\partial t} \qquad \phi = -\nabla . \mathbf{\Pi} \tag{2.11.31}$$

Then equation (2.11.30) is satisfied and so also will be equations (2.11.28) and (2.11.29) if $\mathbf{\Pi}$ is a solution of the vector wave equation

$$\nabla^2\mathbf{\Pi} - \frac{1}{c^2}\frac{\partial^2\mathbf{\Pi}}{\partial t^2} = 0 \tag{2.11.32}$$

The field vectors $\mathbf{E}$, $\mathbf{H}$, are derived from $\mathbf{\Pi}$ by using the relations

$$\mathbf{H} = \frac{1}{c}\nabla \wedge \left(\frac{\partial\mathbf{\Pi}}{\partial t}\right) \tag{2.11.33}$$

$$\mathbf{E} = -\frac{1}{c^2}\frac{\partial^2\mathbf{\Pi}}{\partial t^2} + \nabla\nabla.\mathbf{\Pi} = \nabla \wedge \nabla \wedge \mathbf{\Pi} \tag{2.11.34}$$

2.11.1 Formal Solutions for the Vector and Scalar Potentials

Formal solutions of the field equations

$$\nabla^2\mathbf{A} - \frac{\varepsilon\mu}{c^2}\frac{\partial^2\mathbf{A}}{\partial t^2} = -4\pi\mu\mathbf{J}_s \tag{2.11.35}$$

$$\nabla^2\phi - \frac{\varepsilon\mu}{c^2}\frac{\partial^2\phi}{\partial t^2} = -\frac{4\pi}{\varepsilon}\rho \tag{2.11.36}$$

for a linear isotropic and homogeneous medium with $\sigma = 0$ are given by

$$\mathbf{A}(\mathbf{r}, t) = \mu\int\frac{\mathbf{J}_s(\mathbf{r}', t - R/\beta)}{R}\,\mathrm{d}^3\mathbf{r}' \tag{2.11.37}$$

$$\phi(\mathbf{r}, t) = \frac{1}{\varepsilon}\int\frac{\rho(\mathbf{r}', t - R/\beta)}{R}\,\mathrm{d}^3\mathbf{r}' \tag{2.11.38}$$

where $R = |\mathbf{r} - \mathbf{r}'| = \{(x - \xi)^2 + (y - \eta)^2 + (z - \zeta)^2\}^{\frac{1}{2}}$ is the distance of the point $\mathbf{r}(x, y, z)$ at which the potential functions are to be evaluated from the volume element $\mathrm{d}^3\mathbf{r}' = \mathrm{d}\xi\,\mathrm{d}\eta\,\mathrm{d}\zeta$ located at the point $r'(\xi, \eta, \zeta)$ where the sources exist. The factor $\beta = \{c/\sqrt{(\varepsilon\mu)}\}$, and the integration is to be carried throughout all space.

Equations (2.11.37) and (2.11.38) imply that the field potentials at the point $\mathbf{r}(x, y, z)$ are not derived from the instantaneous values of $\mathbf{J}_s$ and ρ existing at the point $\mathbf{r}'(\xi, \eta, \zeta)$ at the time instant t when the observation is made. Instead they are derived from values of $\mathbf{J}_s$ and ρ which refer to an earlier time $t' = t - R/\beta$. The quantity R/β is just the time taken by an electromagnetic wave in travelling from $\mathbf{r}'$ to $\mathbf{r}$, and evidently this time interval must elapse before a source contributes to the potential functions at $\mathbf{r}$. The functions $\mathbf{A}(\mathbf{r}, t)$, $\phi(\mathbf{r}, t)$ in equations (2.11.37) and (2.11.38) are therefore called *retarded potentials*. It is possible, in principle, to construct *advanced potentials* using the time function $(t + R/\beta)$ but these do not have any applications in electromagnetism.

We will show that $\mathbf{A}(\mathbf{r}, t)$ given by equation (2.11.37) is a solution of the inhomogeneous differential equation, equation (2.11.35). The same form of proof may be used to show that $\phi(\mathbf{r}, t)$ from equation (2.11.38) is a solution of equation (2.11.36).

Imagine that the point $\mathbf{r}$, where the potential functions are to be evaluated, is surrounded by a small sphere of radius a. Consider the component A_x of the vector potential $\mathbf{A}$ and write A_x in terms of two contributions

$$A_x = A_x^{(1)} + A_x^{(2)} \tag{2.11.39}$$

Here $A_x^{(1)}$ represents the contribution to the integral in equation (2.11.37) from the interior of the spherical volume whilst $A_x^{(2)}$ is the contribution from the rest of space. From equation (2.11.37)

$$A_x^{(2)} = \mu \int_{R \geqslant a} \frac{J_x(\mathbf{r}', t - R/\beta)}{R} \, \mathrm{d}^3\mathbf{r}' \tag{2.11.40}$$

Since $R \neq 0$ in this integral it is permissible to differentiate under the integral sign. Moreover, J_x, is a function of R only, so far as the differentiation with respect to x, y, z in equation (2.11.35) is concerned. We have, therefore,

$$\frac{\partial}{\partial x}\left(\frac{1}{R}J_x\right) = \frac{\partial}{\partial R}\left(\frac{1}{R}J_x\right)\frac{\partial R}{\partial x} = \frac{x-\xi}{R}\frac{\partial}{\partial R}\left(\frac{1}{R}J_x\right) \tag{2.11.41}$$

$$\frac{\partial}{\partial R}(x - \xi) = \frac{R}{x - \xi}$$

$$\nabla^2\left(\frac{1}{R}J_x\right) = \frac{1}{R}\frac{\partial^2}{\partial R^2}(J_x) \tag{2.11.42}$$

and

$$\frac{\partial}{\partial t}\left(\frac{1}{R}J_x\right) = \frac{1}{R}\frac{\partial J_x}{\partial t} = -\frac{\beta}{R}\frac{\partial J_x}{\partial R} \tag{2.11.43}$$

Hence

$$\nabla^2 A_x^{(2)} - \frac{1}{\beta^2}\frac{\partial^2 A_x^{(2)}}{\partial t^2} = \mu \int_{R \geqslant a} \left\{\frac{1}{R}\frac{\partial^2 J_x}{\partial R^2} - \frac{1}{\beta^2}\left(\frac{\beta^2}{R}\frac{\partial^2 J_x}{\partial R^2}\right)\right\} \mathrm{d}^3\mathbf{r}' \tag{2.11.44}$$
$$= 0$$

To evaluate the contribution from $A_x^{(1)}$, note that the retardation R/β will become negligible provided the radius a of the spherical volume is taken to be sufficiently small. For this case we may write

$$A_x^{(1)} = \mu \int_{R < a} \frac{J_x(\mathbf{r}', t)}{R} \, \mathrm{d}^3\mathbf{r}' \tag{2.11.45}$$

and

$$\frac{\partial^2 A_x^{(1)}}{\partial t^2} \to \mu \frac{\partial^2 J_x}{\partial t^2} \int_{R < a} \frac{\mathrm{d}^3\mathbf{r}'}{R} = 4\pi\mu \frac{\partial^2 J_x}{\partial t^2} \int_0^a R \, \mathrm{d}R \tag{2.11.46}$$

Evidently

$$\frac{\partial^2 A_x^{(1)}}{\partial t^2} \to 0 \quad \text{as} \quad a \to 0 \tag{2.11.47}$$

To evaluate $\nabla^2 A_x^{(1)}$, form the integral

$$\int_{R < a} \nabla^2 A_x^{(1)} \, \mathrm{d}^3\mathbf{r} = \oint \nabla A_x^{(1)} . \mathbf{n} \, \mathrm{d}S \tag{2.11.48}$$

From equation (2.11.45)

$$\oint \nabla A_x^{(1)} . \mathbf{n} \, dS = \mu \int_{R<a} \int_0^{\pi} J_x(\mathbf{r}', t) \, d^3\mathbf{r}' \, \nabla\left(\frac{1}{R}\right) . \mathbf{n} 2\pi R^2 \sin\theta \, d\theta \quad (2.11.49)$$

and so

$$\int_{R<a} \nabla^2 A_x^{(1)} \, d^3\mathbf{r} = -4\pi\mu \int_{R<a} J_x(\mathbf{r}', t) \, d^3\mathbf{r}' \quad (2.11.50)$$

The integral of $\nabla^2 A_x^{(1)}$ throughout an arbitrary region is equal to the integral of $-4\pi\mu J_x(\mathbf{r}', t)$ throughout the same region. Hence, by taking a sufficiently small

$$\nabla^2 A_x^{(1)} = -4\pi\mu J_x(\mathbf{r}', t) \quad (2.11.51)$$

From equations (2.11.44), (2.11.47) and (2.11.50) we have

$$\begin{aligned} \nabla^2 A_x - \frac{1}{c^2}\frac{\partial^2 A_x}{\partial t^2} &= \nabla^2(A_x^{(1)} + A_x^{(2)}) - \frac{1}{c^2}\frac{\partial^2}{\partial t^2}(A_x^{(1)} + A_x^{(2)}) \\ &= -4\pi\mu J_x(\mathbf{r}', t) \end{aligned} \quad (2.11.52)$$

and so A_x, derived from equation (2.11.37), is a solution for the x component of the inhomogeneous wave equation, equation (2.11.35). Similarly, we can show that A_y and A_z are solutions also. Hence equation (2.11.37) provides a valid vector potential from which to derive the electromagnetic field.

Equations (2.11.37) and (2.11.38) give particular solutions of the inhomogeneous wave equations. To obtain general solutions we may add other functions which are solutions of the equations

$$\begin{aligned} \nabla^2\mathbf{A} - \frac{\varepsilon\mu}{c^2}\frac{\partial^2\mathbf{A}}{\partial t^2} &= 0 \\ \nabla^2\phi - \frac{\varepsilon\mu}{c^2}\frac{\partial^2\phi}{\partial t^2} &= 0 \\ \nabla . \mathbf{A} + \frac{\varepsilon\mu}{c}\frac{\partial\phi}{\partial t} &= 0 \end{aligned} \quad (2.11.53)$$

The complete solutions formed in this way will have sufficient constants to enable the boundary conditions in a given problem to be satisfied.

2.12 Radiation from an Oscillating Dipole

An important example of the electromagnetic field due to a prescribed charge-current distribution is provided by the oscillating electric dipole. The dipole will be assumed to be in the form of two spheres located symmetrically about the origin of coordinates at the points $z = \pm l/2$ and connected together by a short, straight wire of negligible capacitance. We are therefore taking the dipole to be oriented along the z-coordinate axis as shown in Fig. 2.7. A more general dipole oriented along an arbitrary direction in space is equivalent to three linear dipoles with moments along three mutually perpendicular axes. The field for such a dipole may be obtained by symmetry arguments from the present discussion. At time, t, the charge on the upper

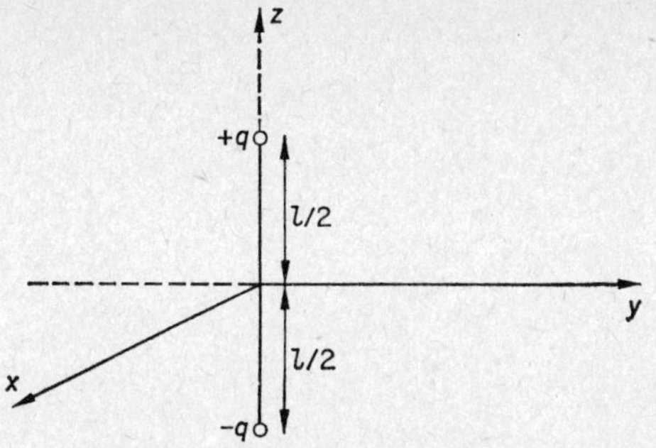

FIG. 2.7 Representation of an oscillating electric dipole.

sphere is $+q(t)$, that on the lower sphere is $-q(t)$. Consequently, the current in the connecting wire, $I(t)$, is given by

$$I(t) = +\frac{1}{c}\frac{\partial q}{\partial t} = \int_{\Delta S} \mathbf{J}.\mathbf{u}\,\mathrm{d}S \tag{2.12.1}$$

where ΔS is the cross-sectional area of the wire and $\mathbf{u}$ is a unit vector directed along the axis of the dipole, which in our case is the z-axis. The current, I, is positive in the positive z-direction, and it should be emphasized that the condition of negligible capacitance for the wire can only be satisfied if the length l is small compared with the wavelength of the resulting radiation. We also assume that l is much less than the distance to any point where the potential functions are to be evaluated. With this restriction, $\xi, \eta, \zeta, \ll x, y, z$ in equation (2.11.37) and so $R = r$. The vector potential is therefore

$$\mathbf{A}(\mathbf{r}, t) = \frac{\mu}{r}\int \mathbf{J}\left(\mathbf{r}', t - \frac{r}{\beta}\right)\mathrm{d}^3\mathbf{r}' \tag{2.12.2}$$

$$= \frac{\mu}{r}\int_{-l/2}^{l/2} \mathbf{u}\,\mathrm{d}\zeta \int_{\Delta S} \mathbf{J}\left(\mathbf{r}', t - \frac{r}{\beta}\right).\mathbf{u}\,\mathrm{d}S \tag{2.12.3}$$

$$= \frac{\mu}{rc}\frac{\partial}{\partial t}\int_{-l/2}^{l/2} q\left(l, t - \frac{r}{\beta}\right)\mathbf{u}\,\mathrm{d}\zeta \tag{2.12.4}$$

or

$$\mathbf{A}(\mathbf{r}, t) = \frac{\mu\mathbf{u}}{rc}\frac{\partial}{\partial t}m\left(t - \frac{r}{\beta}\right) = \frac{\mu\dot{m}}{rc}\mathbf{u} \tag{2.12.5}$$

where $m\mathbf{u}$ is the electric dipole moment of the oscillating charge system. Evidently the vector potential $\mathbf{A}(\mathbf{r}, t)$ is described by a simple mathematical expression of the form $(1/r)f\{t - (r/\beta)\}$ which represents a spherical wave travelling outwards from the dipole.

The scalar potential function $\phi(\mathbf{r}, t)$ is obtained by applying the Lorentz condition, equation (2.11.9).

$$\frac{\varepsilon\mu}{c}\frac{\partial\phi}{\partial t} = -\nabla.\mathbf{A} = -\frac{\partial A_z}{\partial z} = \frac{\mu}{c}\left(\frac{z\dot{m}}{r^3} + \frac{z\ddot{m}}{\beta r^2}\right) \tag{2.12.6}$$

From which

$$\phi(\mathbf{r}, t) = \frac{1}{\varepsilon}\left(\frac{m}{r^3} + \frac{\dot{m}}{\beta r^2}\right)(\mathbf{u}.\mathbf{r}) \tag{2.12.7}$$

where the integration constant has been taken to be zero.

The electromagnetic field is now obtained from equations (2.12.5) and (2.12.7) using the relations

$$\mathbf{H} = \frac{1}{\mu}\nabla \wedge \mathbf{A}$$

$$\mathbf{E} = -\frac{1}{c}\frac{\partial \mathbf{A}}{\partial t} - \nabla\phi \tag{2.12.8}$$

Simplify the mathematics by considering the case of free space with $\varepsilon = \mu = 1$, and obtain

$$\mathbf{H} = \left(\frac{\dot{m}}{cr^3} + \frac{\ddot{m}}{c^2r^2}\right)(\mathbf{u} \wedge \mathbf{r}) \tag{2.12.9}$$

$$\mathbf{E} = \left(\frac{3m}{r^5} + \frac{3\dot{m}}{cr^4} + \frac{\ddot{m}}{c^2r^3}\right)(\mathbf{u}.\mathbf{r})\mathbf{r} - \left(\frac{m}{r^3} + \frac{\dot{m}}{cr^2} + \frac{\ddot{m}}{c^2r}\right)\mathbf{u} \tag{2.12.10}$$

These equations may be written more neatly in terms of spherical polar co-ordinates. Using the relations

$$\mathbf{r} = r\mathbf{a}_r \qquad \mathbf{u} = \cos\theta\,\mathbf{a}_r - \sin\theta\mathbf{a}_\theta \tag{2.12.11}$$

$$\mathbf{u}.\mathbf{r} = r\cos\theta \qquad \mathbf{u}\wedge\mathbf{r} = r\sin\theta\,\mathbf{a}_\phi$$

Evidently

$$\mathbf{H} = \left(\frac{\dot{m}}{cr^2} + \frac{\ddot{m}}{c^2r}\right)\sin\theta\,\mathbf{a}_\phi \tag{2.12.12}$$

$$\mathbf{E} = \left(\frac{2m}{r^3} + \frac{2\dot{m}}{cr^2}\right)\cos\theta\,\mathbf{a}_r + \left(\frac{m}{r^3} + \frac{\dot{m}}{cr^2} + \frac{\ddot{m}}{c^2r}\right)\sin\theta\,\mathbf{a}_\theta \tag{2.12.13}$$

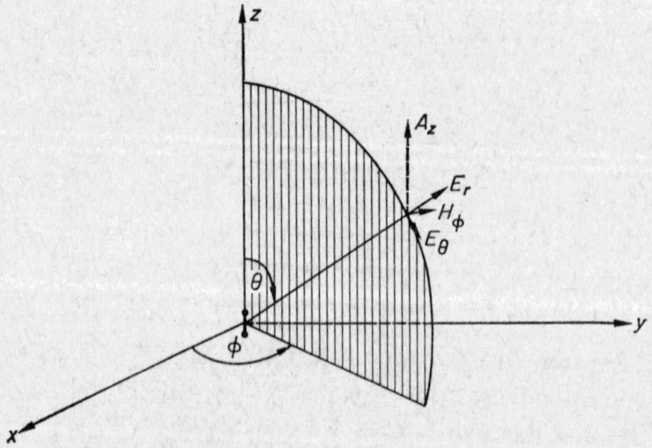

FIG. 2.8 The field vectors for the electric dipole.

Note that $H_r = H_\theta = 0$ and $E_\phi = 0$. The electric vector lies in the meridional plane through the axis of the dipole and the magnetic vector is perpendicular to this plane as shown in Fig. 2.8.

2.12.1 The Near Zone

Near the dipole, where the retardation term (r/c) is small, we should expect the field distribution to correspond to that for a simple electrostatic or magnetostatic source. In this region we expand by Taylor's theorem,

$$\begin{aligned} \mathcal{m}(t - r/c) &= \mathcal{m}(t) - \frac{r}{c}\dot{\mathcal{m}}(t) \\ \dot{\mathcal{m}}(t - r/c) &= \dot{\mathcal{m}}(t) - \frac{r}{c}\ddot{\mathcal{m}}(t) \end{aligned} \tag{2.12.14}$$

so that to terms in $(r/c)^2$

$$\mathbf{H} = \frac{\dot{\mathcal{m}}(t)}{cr^2} \sin\theta\, \mathbf{a}_\phi = \frac{\dot{\mathcal{m}}(t)}{cr^3}(\mathbf{u} \wedge \mathbf{r}) \tag{2.12.15}$$

$$\mathbf{E} = \frac{2\mathcal{m}(t)}{r^3} \cos\theta\, \mathbf{a}_r + \frac{\mathcal{m}(t)}{r^3} \sin\theta\, \mathbf{a}_\theta \tag{2.12.16}$$

Equation (2.12.16) reduces to that for the electrostatic dipole whilst equation (2.12.15) may be rewritten using equation (2.12.1) and the definition of $\mathcal{m}$, to give

$$\mathbf{H} = I\,\frac{\mathbf{u} \wedge \mathbf{r}}{r^3} l \tag{2.12.17}$$

or

$$\mathrm{d}\mathbf{H} = I\,\frac{\mathbf{u} \wedge \mathbf{r}}{r^3}\,\mathrm{d}l \tag{2.12.18}$$

This is the law of Biot and Savart for the magnetic field due to a current element.

2.12.2 The Radiation Zone

In a region sufficiently far from the dipole all terms excepting those in $(1/r)$ may be neglected in equations (2.12.12) and (2.12.13). This is the radiation zone where

$$H_\phi = E_\theta = \frac{\ddot{\mathcal{m}}\{t - (r/c)\}}{c^2 r} \sin\theta \tag{2.12.19}$$

and all other components are taken to be zero.

In this region the vectors **E**, **H** are perpendicular to each other and to the radius vector **r**. The structure of the field is therefore similar to that of a plane wave in that it is a transverse field, but on each 'wave surface' the field vectors decrease in magnitude away from the equatorial plane and become zero along the axis of the dipole. This means that no energy is radiated in the direction of the dipolar axis.

The transverse nature of the wave motion in the radiation zone is a characteristic common to all electromagnetic fields in unbounded media at large distances from the source. This is in contrast to the field distribution near to the source, where components in the direction of propagation will generally occur.

2.12.3 Energy Flow in the Radiation Zone

The instantaneous rate at which energy is crossing unit area in the radiation zone is obtained from equation (2.12.19), by forming the Poynting Vector.

$$\mathbf{S} = \frac{c}{4\pi}(\mathbf{E} \wedge \mathbf{H}) = \frac{c}{4\pi} E_\theta H_\phi \, \mathbf{a}_r \tag{2.12.20}$$

$$= \frac{c}{4\pi} \frac{[\ddot{m}\{t - (r/c)\}]^2}{c^4 r^2} \sin^2 \theta \, \mathbf{a}_r \tag{2.12.21}$$

The direction of the Poynting Vector is radially outwards from the dipole. In magnitude it varies as r^{-2} which is the familiar inverse square law for radiation intensity. The energy flux also varies with the angle as $\sin^2 \theta$. Thus, in the radiation zone, the energy flux is zero along the axis of the dipole and a maximum in the equatorial plane as shown in Fig. 2.9.

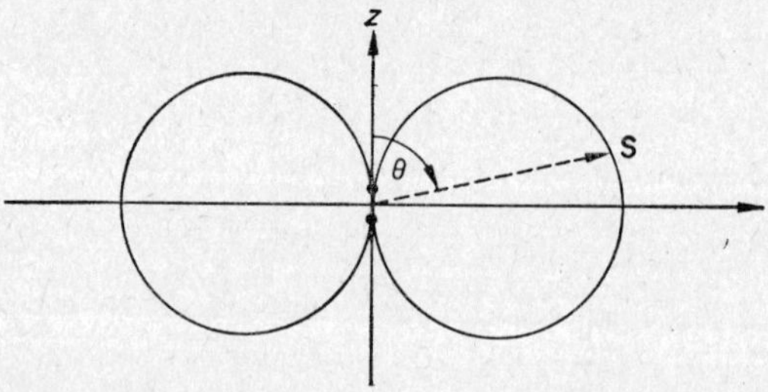

Fig. 2.9 The radiation pattern from an electric dipole.

The total energy flux at any instant is given by the integral of $\mathbf{S}.\mathbf{n}$ over a surface completely enclosing the dipole. Choose a spherical surface centred at the dipole.

$$S_t = \int_0^\pi \frac{c}{4\pi} \frac{[\ddot{m}\{t - (r/c)\}]^2}{c^4 r^2} \sin^2 \theta \, 2\pi r^2 \sin \theta \, d\theta \tag{2.12.22}$$

$$S_t = \frac{2}{3} \frac{[\ddot{m}\{t - (r/c)\}]^2}{c^3} \tag{2.12.23}$$

This is not necessarily the total power radiated at any instant by the dipole since some additional energy may be stored in the electromagnetic field. Consider the energy stored in the region between two concentric spherical surfaces having radii r and $r + \delta r$. The rate at which the stored energy is increasing in this region is

$$\frac{\partial}{\partial t}\int\left(\frac{E_\theta^2}{8\pi}+\frac{H_\phi^2}{8\pi}\right)\mathrm{d}v$$

$$=\frac{\partial}{\partial t}\int_0^\pi \frac{[\ddot{m}\{t-(r/c)\}]^2}{4\pi c^4 r^2}\sin^2\theta\, 2\pi r^2 \sin\theta\,\mathrm{d}\theta\,\delta r$$

$$=\frac{2}{3}\frac{\partial}{\partial t}\frac{[\ddot{m}\{t-(r/c)\}]^2}{c^4}\delta r \tag{2.12.24}$$

This is just equal to the difference between the total energy flux into the region and that flowing out of the region at any instant, which can be calculated directly from equation (2.12.23).

2.12.4 The Harmonic Electric Dipole

When the time variation of the electric dipole moment is harmonic we may write

$$m(t) = \mathrm{Re}\,[m_0\,\mathrm{e}^{\mathrm{j}\omega t}] \tag{2.12.25}$$

$$m\{t-(r/c)\} = \mathrm{Re}\,[m_0\,\mathrm{e}^{\mathrm{j}\omega\{t-(r/c)\}}] \tag{2.12.26}$$

The field in the radiation zone is now given by

$$E_\theta = H_\phi = -\frac{\omega^2}{c^2}\frac{m_0\,\mathrm{e}^{\mathrm{j}\omega\{t-(r/c)\}}}{r}\sin\theta \tag{2.12.27}$$

The time average of the Poynting Vector is

$$\langle \mathbf{S}\rangle = \frac{c}{4\pi}\,\tfrac{1}{2}\,\mathrm{Re}\,[\mathbf{E}\wedge\mathbf{H}^*]$$

$$=\frac{c}{8\pi}\frac{\omega^4\,|m_0|^2}{c^4 r^2}\sin^2\theta\,\mathbf{a}_r \tag{2.12.28}$$

The time average of the energy stored in the field is constant and so the mean power radiated by the harmonic dipole is given by the integral of $\langle \mathbf{S}\rangle$ over a surface enclosing the dipole. Choosing, again, a spherical surface

$$\langle \mathbf{S}\rangle_t = \oint \langle \mathbf{S}\rangle.\mathbf{n}\,\mathrm{d}S = \frac{c}{8\pi}\int_0^\pi \frac{\omega^4\,|m_0|^2}{c^4 r^2}\sin^2\theta\,2\pi r^2\sin\theta\,\mathrm{d}\theta$$

$$=\frac{1}{3}\frac{|m_0|^2}{c^3}\omega^4 \tag{2.12.29}$$

The mean power radiated is proportional to the fourth power of the frequency and to the square of the dipole moment.

Equation (2.12.29) may be written in terms of the source current flowing in the dipole to derive an expression for the *radiation resistance.* From equations (2.12.1), (2.12.3) and (2.12.5)

$$\frac{\dot{m}}{c} = Il \tag{2.12.30}$$

so that

$$m_0 = \frac{lc}{j\omega} I_0 \tag{2.12.31}$$

and

$$\langle \mathbf{S} \rangle_t = \frac{1}{3} \frac{\omega^2 l^2}{c} |I_0|^2 = \frac{4\pi^2 c}{3} \left(\frac{l}{\lambda}\right)^2 |I_0|^2 \tag{2.12.32}$$

This gives the mean radiated power in ergs per second. It may be written in watts, with the current measured in amperes, as

$$\langle w \rangle_t = \frac{4\pi^2 c}{3 \cdot 10^9} \left(\frac{l}{\lambda}\right)^2 |i_0|^2 \quad \text{watts} \tag{2.12.33}$$

The generator driving the dipole will supply this power in addition to the other resistive losses occurring in the circuit. The mean power supplied by the generator may therefore be written

$$\langle w_g \rangle = \tfrac{1}{2}(R_0 + R_r) |i_0|^2 \tag{2.12.34}$$

where R_0 is the resistive loss in the circuit without radiation and R_r is the *radiation resistance* representing the power loss to the radiation field. From equation (2.12.33), R_r is evidently given by,

$$R_r = 80\pi^2 (l/\lambda)^2 \text{ ohms} \qquad (\text{free space and } l \ll \lambda) \tag{2.12.35}$$

Thus, if the length of the dipole is $\lambda/10$, the radiation resistance is about 10 ohms which is probably large compared with the resistance of the wire forming the dipole.

Bibliography

Born, M. and Wolf, E. *Principles of Optics*, Pergamon, 1959

Lamont, H. R. L. *Waveguides*, Methuen, 1946

Landau, L. D. and Lifshitz, E. M. *Electrodynamics of Continuous Media*, Pergamon, 1963

Montgomery, C. G., Dicke, R. H. and Purcell, E. M. *Principles of Microwave Circuits*, Radiation Laboratory Series, Vol. 8, McGraw-Hill, 1948

Panofsky, W. K. H. and Phillips, M. *Classical Electricity and Magnetism*, Addison-Wesley, 1956

Reitz, J. R. and Milford, F. J. *Foundations of Electromagnetic Theory*, Addison-Wesley, 1962

Slater, J. C. and Frank, N. H. *Electromagnetism*, McGraw-Hill, 1947

Stratton, J. A. *Electromagnetic Theory*, McGraw-Hill, 1941

Wangsness, R. K. *Introduction to Theoretical Physics*, John Wiley, 1963

Problems

1. A plane monochromatic wave travelling in vacuum is incident normally on a plane dielectric sheet. The thickness of the sheet is d. Calculate the reflection coefficient as a function of d and ε.

2. Calculate the reflection coefficient for Problem (1) when the dielectric is backed by a perfect conductor.

3. A plane monochromatic wave is incident on a plane interface separating two media. Show that total reflection can occur at an angle of incidence

$\theta \geqslant \theta_r$, if the dielectric constant ε_1 of the first medium is greater than that for the second medium ε_2, and that

$$\sin \theta_r = \sqrt{\left(\frac{\varepsilon_2}{\varepsilon_1}\right)}$$

Discuss the waves which occur when the angle of incidence θ is greater than θ_r.

4. A plane monochromatic wave is incident on a plane interface separating two media. Calculate the time average Poynting Vector for the various waves under conditions of total reflection.

5. Discuss the propagation of a plane wave in an ionized atmosphere. Assume that the mean free path of the electrons in the gas is very long and that these electrons are responsible for the electrical conductivity. Calculate the wave velocity and the group velocity.

6. A plane wave with wavelength 100 m and electric field amplitude 100 V cm^{-1} is incident normally on a copper sheet 0·01 cm thick. Estimate the electric field amplitude in the wave after traversing the copper sheet. Take $\sigma = 10^5\ \Omega^{-1}\ \text{cm}^{-1}$ and $\varepsilon = \mu = 1$ for copper.

7. Sketch the configuration of the electric and magnetic field at a given instant of time for a TE_{10} mode in a rectangular waveguide.

8. Calculate the critical frequencies for waveguides having the following dimensions

(a) 2·286 cm × 1·016 cm
(b) 1·067 cm × 0·4318 cm
(c) 0·7112 cm × 0·3556 cm

9. Discuss the electromagnetic wave configuration and energy flow in a waveguide operating beyond cut-off.

10. A waveguide with dimensions 2·286 cm × 1·016 cm is used to transmit 100 mW of power at a frequency of 10 GHz. Calculate the maximum amplitude of the electric field for the TE_{10} mode. Take $\varepsilon = 1$, $\mu = 1$ for air.

11. Assuming that breakdown will occur when the electric field amplitude exceeds $30 \cdot 10^3\ \text{V cm}^{-1}$, calculate the maximum power which can be transmitted down the waveguide of Problem (10), at a frequency of 10 GHz. Take $\varepsilon = 1$, $\mu = 1$ for air.

12. Discuss the propagation of an electromagnetic wave in the space between two parallel conducting planes.

13. Calculate the total power radiated by an isolated electric dipole of length 30 cm operating at a frequency of 1 MHz in free space when the current amplitude is 2 A.

What is the radiation resistance for this dipole?

14. An isolated electric dipole radiates 5 mW power at 10 GHz. Calculate

the magnitudes of the electromagnetic field vectors at a distance of 3 m from the source in free space.

15. A positive and a negative charge are bound together by elastic forces to form a harmonic oscillator. Discuss how the motion of the charges is damped by the radiation loss from the oscillator.

16. Derive boundary conditions for the vector potential $\mathbf{A}(\mathbf{r},t)$ and the scalar potential $\phi(\mathbf{r}, t)$.

The subject matter of many of these Problems is discussed in the texts of books listed in the Bibliography.
Thus:
for Problem (4), *see* Landau and Lifshitz, Born and Wolf, or Wangsness.
for Problems (5), (15), *see* Stratton.
for Problems (7), (8), (9), (12), *see* Montgomery, Dicke and Purcell, Lamont, Reitz and Milford.

Chapter 3

Introductory Network Analysis

In this chapter we shall consider some of the standard techniques which have been developed for the analysis of passive electrical networks. This discussion will have in mind excitation and response functions (in the present case voltages or currents) which are exponential functions of the time. Thus a function of the form $F_0 \, e^{j\omega t}$, with ω a real angular frequency, will correspond to a harmonic time function, whilst $F_0 \, e^{st}$ with $s = q + j\omega$ will correspond to a damped harmonic time function. For practical purposes these are still probably the most widely used and familiar forms of time variation. However, the discussion should also be regarded as an introduction to the more general techniques of linear systems analysis which will be developed in later chapters. It is useful, therefore, to emphasize that signals are by no means always harmonic in form and that electrical circuits are only particular examples of a general class of linear systems.

3.1 Circuit Elements

Consider a set of abstract circuit elements each of which interacts with its surroundings only via two accessible terminals. As in the case of the generator of electromotive force discussed previously (see Section 1.5, Chapter 1) assume that all non-irrotational electric fields have been confined to the interior of the element so that there is a well-defined potential difference between the terminals when a current $I(t)$ flows into one terminal and out from the other; see Fig. 3.1(a).

Particular types of circuit element are identified by the operator equations

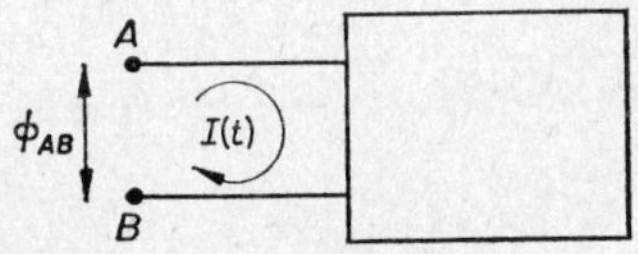

(*a*)

FIG 3.1

(a) An abstract circuit element with two accessible terminals.

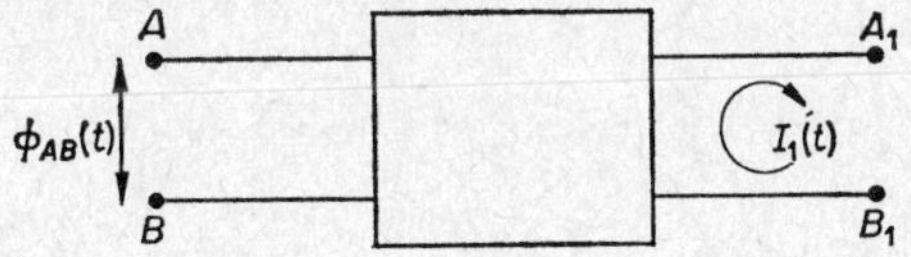

(*b*)

(b) Abstract representation of a mutual inductance element.

which relate the current to the potential difference between the terminals. We may identify a fundamental set of elements by means of the simplest formal equations which connect the current and the potential difference through the time differential operator multiplied by a constant coefficient. We use the notation $D \equiv d/dt$ to represent differentiation with respect to the time variable.

A resistance is identified by the relation

$$\phi_{AB}(t) = RD^0I(t) = RI(t) \tag{3.1.1}$$

An inductance is identified by the relation

$$\phi_{AB}(t) = LDI(t) \tag{3.1.2}$$

A capacitance is identified by the relation

$$\phi_{AB}(t) = C^{-1}D^{-1}I(t) \tag{3.1.3}$$

For completeness we introduce an additional circuit element which has two pairs of accessible terminals. The particular element of this class which we wish to include has the property that when a time varying current flows into and out of one pair of terminals a potential difference is produced at the other pair; see Fig. 3.1(b).

The relation

$$\phi_{AB}(t) = MDI_1(t) \tag{3.1.4}$$

identifies the element as a *mutual inductance.*

Care may be necessary in recognizing a terminal pair for this circuit element. A terminal pair is identified by requiring that the same current must flow in at one terminal of the pair as flows out at the other terminal of the pair. The formal circuit representations for these various elements are shown in Fig. 3.2.

Note that the elements representing both the inductance and the mutual

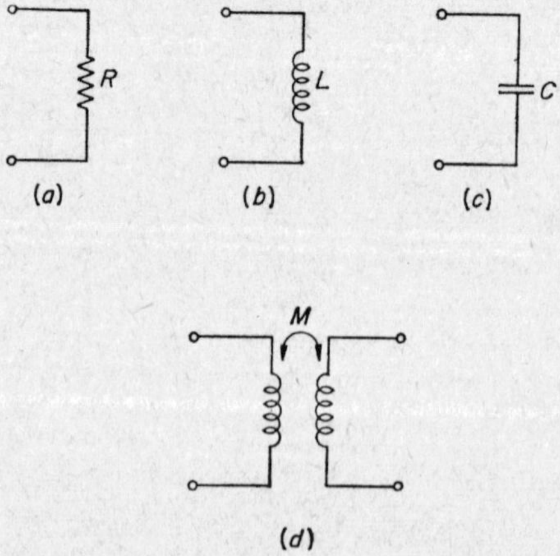

FIG. 3.2 Formal circuit representations for (a) a resistance, (b) an inductance, (c) a capacitance and (d) a mutual inductance.

inductance can be written in one formal operator relation. For the mutual inductance the potential difference between the terminal pair in a circuit labelled, *m*, is derived from a current in a different circuit labelled, *n*. On the other hand, for the inductance, the potential difference and the current refer to the same circuit (that labelled, *m*, say). Hence we can write

$$\phi_{AB}(t) = L_{m,n}DI_n(t) \tag{3.1.5}$$

and $L_{m,m}$ refers to an inductive element in circuit, *m*, whilst $L_{m,n}(m \neq n)$ refers to a mutual inductance coupling circuit, *m*, with circuit, *n*.

The constant coefficients *R*, *L*, *C*, are characteristics of the particular circuit elements and are called the resistance, the inductance and the capacitance, respectively. Equations (3.1.1)–(3.1.5) may, however, be summarized in one formal relation

$$\phi_{AB}(t) = Z(D)I(t) \tag{3.1.6}$$

Here $Z(D)$ is a factor which includes the time differential operator and is called the impedance operator for the circuit element. If circuit elements *R*, *L*, *C*, are connected together in series, the same current $I(t)$ will flow through each element and $Z(D)$ for the combination will be the sum of $Z(D)$ for each individual component. Similarly, if the circuit elements are connected in parallel, $\phi_{AB}(t)$ will be the same for all elements and now the reciprocal of $Z(D)$ for the combination will be the sum of the reciprocal of $Z(D)$ for each individual component.

In equations (3.1.1)–(3.1.3) the time differential operator has arbitrarily been assigned to operate on $I(t)$. An alternative set of equations can, however, be written down which will be summarized by a relation of the form

$$I(t) = Y(D)\phi_{AB}(t) \tag{3.1.7}$$

This equation defines a set of admittance operators for the circuit elements. In this case $Y(D)$ for the parallel combination of a set of elements is the sum of $Y(D)$ for each individual component and the reciprocal of $Y(D)$ for the series combination is the sum of the reciprocal of $Y(D)$ for each individual component. Note that the mutual inductance element is not included in this set of admittance operators since the definition of such an element specifically relates the potential difference at one pair of terminals to the differential of the current at the other pair of terminals.

3.2 Formal Network Theory

A set of elements described by equations (3.1.1)–(3.1.5) may be connected together by joining their terminals to form a network. A junction of two or more terminals is called a *node* of the network whereas the elements themselves constitute the *branches* of the network. Generators may be inserted in the branches to excite the network, and the work done at any instant in taking unit charge round a closed contour in the network, on which the terminals of a set of elements labelled by the running index *j* lie, will be equal to the electromotive force in that circuit. We may write

$$\sum_j \pm \phi_j(t) = \mathscr{E}_i(t) \tag{3.2.1}$$

for a contour which includes an electromotive force $\mathscr{E}_i(t)$. There will be many equations of this kind, each one corresponding to a particular contour in the network. The complete set may be written

$$\begin{aligned}
&\beta_{11}\phi_1(t) + \beta_{12}\phi_2(t) + \beta_{13}\phi_3(t) + \cdots \beta_{1n}\phi_n(t) = \mathscr{E}_1(t) \\
&\beta_{21}\phi_1(t) + \beta_{22}\phi_2(t) + \beta_{23}\phi_3(t) + \cdots \beta_{2n}\phi_n(t) = \mathscr{E}_2(t) \qquad (3.2.2) \\
&\cdot \\
&\cdot \\
&\cdot \\
&\beta_{n1}\phi_1(t) + \beta_{n2}\phi_2(t) + \beta_{n3}\phi_3(t) + \cdots \beta_{nn}\phi_n(t) = \mathscr{E}_n(t)
\end{aligned}$$

or, in matrix notation

$$[\beta].\phi] = \mathscr{E}] \qquad (3.2.3)$$

In these equations the terms β_{ij} are equal to 0, ± 1, and the $\mathscr{E}_i(t)$ will be zero when there is no generator lying on the particular contour chosen.

Electrical network problems usually involve the evaluation of the currents flowing in the individual circuit elements when the way in which these elements have been connected together with a set of generators is known. The currents $I_j(t)$ in the individual elements (called branch currents) are related to the potential differences $\phi_j(t)$ between the terminal pairs through equations (3.1.1)–(3.1.5). These equations may also be written in matrix form:

$$\begin{bmatrix} R_{11} & 0 & 0 & 0 \ldots & 0 \\ 0 & R_{22} & 0 & 0 \ldots & \\ 0 & 0 & R_{33} & 0 \ldots & \\ 0 & \ldots & \ldots & \ldots & R_{\lambda\lambda} \end{bmatrix} \begin{bmatrix} I_1^R \\ I_2^R \\ \vdots \\ I_\lambda^R \end{bmatrix} = \begin{bmatrix} \phi_1^R \\ \phi_2^R \\ \vdots \\ \phi_\lambda^R \end{bmatrix} \quad \text{or} \quad [R].I^R]_\lambda^1 = \phi^R]_\lambda^1 \quad (3.2.4)$$

$$D\begin{bmatrix} L_{11} & L_{12} & \ldots & L_{1\eta} \\ L_{21} & L_{22} & \ldots & L_{2\eta} \\ \vdots & & & \\ L_{\eta 1} & L_{\eta 2} & \ldots & L_{\eta\eta} \end{bmatrix} \begin{bmatrix} I_1^L \\ \cdot \\ \cdot \\ I_\eta^L \end{bmatrix} = \begin{bmatrix} \phi_1^L \\ \cdot \\ \cdot \\ \phi_\eta^L \end{bmatrix} \quad \text{or} \quad D[L].I^L]_\eta^1 = \phi^L]_\eta^1 \quad (3.2.5)$$

$$D^{-1}\begin{bmatrix} C_{11}^{-1} & 0 & 0 & 0 & 0 \\ 0 & C_{22}^{-1} & 0 & 0 & \cdot \\ 0 & 0 & C_{33}^{-1} & 0 & \cdot \\ \vdots & & & & \\ 0 & 0 & \ldots & \ldots & C_{\mu\mu}^{-1} \end{bmatrix} \begin{bmatrix} I_1^C \\ \cdot \\ \cdot \\ \cdot \\ I_\mu^C \end{bmatrix} = \begin{bmatrix} \phi_1^C \\ \cdot \\ \cdot \\ \cdot \\ \phi_\mu^C \end{bmatrix} \quad \text{or} \quad D^{-1}[C^{-1}].I^C]_\mu^1 = \phi^C]_\mu^1 \quad (3.2.6)$$

In these equations, the matrices $[R]$ and $[C^{-1}]$ are diagonal while the matrix $[L]$, in its general form which includes both inductance and mutual inductance elements, is a square symmetric matrix. The three sets of equations (3.2.4)–(3.2.6) may be combined into one matrix equation

$$[A].I] = \phi] \qquad (3.2.7)$$

where

$$[A] = \left[\begin{array}{c:c:c} [R] & 0 & 0 \\ \hdashline 0 & D[L] & 0 \\ \hdashline 0 & 0 & D^{-1}[C^{-1}] \end{array}\right] \qquad (3.2.8)$$

and $I]$, $\phi]$ are column matrices with elements equal in number to the total number of circuit elements. The equations which must be solved to obtain the branch currents in terms of the e.m.f.'s of the given set of generators may now be written formally by substituting for $\phi]$ from equation (3.2.3). We have

$$[\beta].[A].I] = \mathscr{E}] \tag{3.2.9}$$

These equations, however, are not in themselves sufficient to determine the currents $I(t)$ completely since these currents are subject to an additional set of constraints. Assuming that there are no current sources in the network, the total current flowing into any junction of the network must be zero. This condition may be written

$$[\mu].I] = [0] \tag{3.2.10}$$

where the μ_{ij} are equal to 0, ± 1. The complete solution of the network problem is therefore obtained by solving equation (3.2.9) subject to the constraint of equation (3.2.10).

3.3 Loop or Mesh Analysis

One technique of solving the complete set of network equations (3.2.9) and (3.2.10) involves choosing an alternative set of current variables which automatically satisfy the second of these equations. Consider a set of currents which circulate on closed contours of the network. Such currents, known as loop, or mesh currents automatically satisfy equation (3.2.10) since, for circulating currents, all currents entering a node or junction in the circuit must also leave it. By definition, a current circulating on a closed contour leaves all points which it enters. Let such a set of currents be denoted by $i_m(t)$. They will be related to the branch currents by a set of linear equations since $I_j(t)$, the branch current in a given circuit element labelled j, is just a particular linear combination of the new set of currents. We may, therefore, write

$$[\delta].i] = I] \tag{3.3.1}$$

with δ_{ij} equal to 0, ± 1. Moreover the matrix $[\delta]$ is just the transpose of the matrix $[\beta]$ occurring in equations (3.2.3) and (3.2.9) if the same contours are chosen to evaluate the voltage relations as are traversed by the loop currents $i_m(t)$.† Substituting from equation (3.3.1) into (3.2.9), we have

$$[\beta].[A].[\beta]^T.i] = \mathscr{E}] \tag{3.3.2}$$

or

$$[Z].i] = \mathscr{E}] \tag{3.3.3}$$

where

$$[Z] = [\beta].[A].[\beta]^T \tag{3.3.4}$$

is called the impedance operator matrix for the network. This matrix characterizes the particular way in which the circuit elements have been connected together to construct the network.

Note that $[Z]$ is a symmetric matrix. An element

$$Z_{ij} = \sum_{\lambda} \sum_{\mu} \beta_{i\lambda} A_{\lambda\mu} \beta_{j\mu} \tag{3.3.5}$$

† For proof see end of this chapter.

and

$$Z_{ji} = \sum_m \sum_l \beta_{jm} A_{ml} \beta_{il} \tag{3.3.6}$$

but $[A]$ is itself a symmetric matrix and so the summation is over equal terms identified by $\lambda = l$ and $\mu = m$. Hence

$$Z_{ij} = Z_{ji} \tag{3.3.7}$$

In addition, each Z_{ij} is a linear combination of terms such as R_{ll}, $L_{mn}D$, $C_{rr}^{-1}D^{-1}$ since β_{ij} is a number 0, ± 1.

The practical technique for deriving the matrix elements Z_{ij} from a given network configuration may be obtained as follows. Consider the network loop labelled, j, with generator e.m.f. $\mathscr{E}_j$. From equation (3.3.3), we have

$$Z_{j1}i_1(t) + Z_{j2}i_2(t) + Z_{j3}i_3(t) + \cdots Z_{jj}i_j(t) + \cdots Z_{jn}i_n(t) = \mathscr{E}_j(t) \tag{3.3.8}$$

where the total number of loop currents is taken equal to n. If all loop currents save $i_j(t)$ are put equal to zero (equivalent to open circuiting all these loops), we have

$$Z_{jj}i_{j0}(t) = \mathscr{E}_j(t) \tag{3.3.9}$$

with Z_{jj} still the same linear combination of R, LD, $C^{-1}D^{-1}$, as occurred in equation (3.3.8). Hence, if we write Z_{jj} in the form

$$Z_{jj} = R_{jj} + L_{jj}D + C_{jj}^{-1}D^{-1} \tag{3.3.10}$$

then, R_{jj} is the total series resistance in the loop labelled j, L_{jj} is the total series self-inductance in the loop labelled j and C_{jj}^{-1} is the total series combination of capacitance in the loop labelled j. If now one other loop, labelled r, has a current flowing in it

$$Z_{jj}i_j(t) + Z_{jr}i_r(t) = \mathscr{E}_j(t) \tag{3.3.11}$$

with Z_{jj} as before in equation (3.3.10), and Z_{jr} may be written

$$Z_{jr} = R_{jr} + L_{jr}D + C_{jr}^{-1}D^{-1} \tag{3.3.12}$$

Adding up the potential differences around the circuit we have

R_{jr} = $\pm$ total series resistance common to loops j and r,
L_{jr} = $\pm$ total series inductance (self-inductance and mutual inductance) common to loops j and r and
C_{jr}^{-1} = $\pm$ total series combination of capacitance common to loops j and r.

The + sign is used if $i_j(t)$, $i_r(t)$ flow in the same direction through a common element, the − sign if the currents flow in opposite directions.

When the e.m.f.s and the currents have a harmonic time dependence of the form $e^{j\omega t}$, Z_{jj} will be the total series impedance of loop j, (the self-impedance of loop, j,) whilst Z_{jr} will be the impedance of all the elements common to loop j and loop i.

3.3.1 Examples Using Loop Analysis

As a specific example of this general discussion consider the network shown in Fig. 3.3 which consists of three elements joined together in series

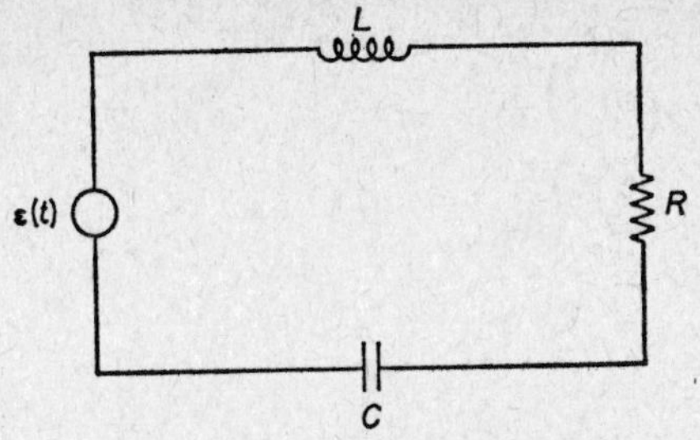

FIG. 3.3 A simple R, L, C, series combination.

to form a single closed loop. If $\mathscr{E}(t)$ is harmonic, of the form $\mathscr{E}_0\, e^{j\omega t}$, we assume $i = i_0\, e^{j\omega t}$, and the matrix $[Z]$ consists of one element only

$$[Z] = Z_{11} = R + LD + C^{-1}D^{-1} \tag{3.3.13}$$

so that

$$\left\{R + j\left(\omega L - \frac{1}{\omega C}\right)\right\} i_0 = \mathscr{E}_0 \tag{3.3.14}$$

If the amplitude of $\mathscr{E}$ is constant in time, ω_0 is real, and so the amplitude and phase of the loop current, $i(t)$, is determined directly from equation (3.3.14).

$$i(t) = i_{00}\, e^{j\delta}\, e^{j\omega t} \tag{3.3.15}$$

with

$$i_{00} = \frac{\mathscr{E}_0}{\sqrt{[R^2 + \{\omega L - (1/\omega C)\}^2]}} \tag{3.3.16}$$

$$\tan \delta = \frac{\omega L - (1/\omega C)}{R} \tag{3.3.17}$$

On the other hand, if the e.m.f. is, for example, a rectangular voltage pulse and we wish to describe the circulating current at subsequent times when the voltage has again fallen to zero, ignoring the initial conditions, assume a solution for $i(t)$ of the form $i(t) = i_{00}\, e^{st}$. From equation (3.3.13)

$$R + Ls + \frac{1}{sC} = 0 \tag{3.3.18}$$

If $R < 4L/C$

$$s = -\frac{R}{2L} \pm j\sqrt{\left(\frac{1}{LC} - \frac{R^2}{4L^2}\right)} \tag{3.3.19}$$

and we may write

$$i(t) = i_{00} \exp\{-(R/2L)t\} \exp(j\omega_\nu t) \tag{3.3.20}$$

where

$$\omega_\nu \equiv \sqrt{\left(\frac{1}{LC} - \frac{R^2}{4L^2}\right)}$$

is the 'natural' frequency of the circuit.

Equation (3.3.20) may also be written formally as

$$i(t) = i_{00} \exp\{-(\omega_v/2Q)t\}\exp(j\omega_v t) \tag{3.3.21}$$

which defines the quantity $Q = (\omega_v L/R)$. Q is called the quality factor of the circuit.

For a second example, consider the circuit shown in Fig. 3.4

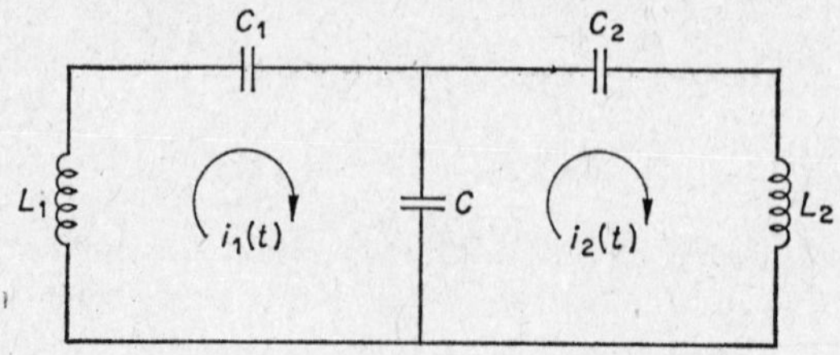

FIG. 3.4 A simple coupled circuit.

Taking the loop currents $i_1(t)$ and $i_2(t)$ as shown, the impedance operator matrix $[Z]$ is given by

$$[Z] = \begin{bmatrix} L_1D + (C_1^{-1} + C^{-1})D^{-1} & -C^{-1}D^{-1} \\ -C^{-1}D^{-1} & L_2D + (C_2^{-1} + C^{-1})D^{-1} \end{bmatrix} \tag{3.3.22}$$

By analogy with the previous example, the response to a rectangular voltage pulse may be taken in the form $i_1(t) = i_{01}\, e^{st}$, $i_2(t) = i_{02}\, e^{st}$, so that

$$\begin{bmatrix} L_1s + (C_1^{-1} + C^{-1})s^{-1} & -C^{-1}s^{-1} \\ -C^{-1}s^{-1} & L_2s + (C_2^{-1} + C^{-1})s^{-1} \end{bmatrix} \cdot \begin{bmatrix} i_{01} \\ i_{02} \end{bmatrix} = 0 \tag{3.3.23}$$

For this pair of homogeneous algebraic equations to have a solution with $i_{01}, i_{02} \neq 0$, the determinantal equation

$$\begin{vmatrix} L_1s + (C_1^{-1} + C^{-1})s^{-1} & -C^{-1}s^{-1} \\ -C^{-1}s^{-1} & L_2s + (C_2^{-1} + C^{-1})s^{-1} \end{vmatrix} = 0 \tag{3.3.24}$$

must be satisfied. We obtain

$$(s^2L_1C_1C + C_1 + C)(s^2L_2C_2C + C_2 + C) - C_1C_2 = 0 \tag{3.3.25}$$

This equation determines the natural frequencies of the circuit shown in Fig. 3.4. In particular, if $L_1 = L_2 = L_0$ and $C_1 = C_2 = C_0$,

$$s^2 = -\frac{2C_0 + C}{L_0C_0C} \quad \text{or} \quad s^2 = -\frac{1}{L_0C_0} \tag{3.3.26}$$

Note that there is no resistance in the network as drawn and so the variable s, is purely imaginary.

3.3.2 NATURAL FREQUENCIES AND RESONANCE FREQUENCIES

We may generalize the examples discussed above. Consider a network excited by a rectangular voltage pulse and write the loop currents at subsequent times, ignoring the initial transients, as

$$i_j(t) = i_{0j}\, e^{st} \tag{3.3.27}$$

Then

$$[Z(s)].i_0] = 0 \qquad (3.3.28)$$

upon carrying through the differential operations in equation (3.3.3). This set of homogeneous equations will have a non-trivial solution for the i_{0j}, only if the determinant of $[Z(s)]$ is zero. This determinant is a function of the complex frequency variable s, and the roots of the equation

$$\det [Z(s)] = 0 \qquad (3.3.29)$$

will give the natural frequencies of the network. If there are n rows and columns in $[Z(s)]$, since equation (3.3.29) contains terms in both s and s^{-1}, the resulting algebraic equation will be of degree $2n$ in the variable s and will in general have $2n$ roots. The coefficients of s in this equation are real, however, and so the roots will either be real or occur as conjugate complex pairs.† A complex root $s_\nu = q_\nu + j\omega_\nu$ will define a real frequency (from ω_ν) and an exponential factor (from q_ν). The term q_ν can never be positive for a realizable passive network since this would give rise to currents which continuously increase without limit and the circuit would eventually dissipate infinite energy in the resistive elements of the network without any generator supply being present. The roots of $\det [Z(s)]$ must therefore all lie in the left half of the complex plane and occur in conjugate pairs, that is, as image pairs in the real axis. The individual roots will give rise to currents $i_j^{(\nu)}(t) = i_{0j}^{(\nu)} \exp (s_\nu t)$ and the image pairs will give rise to currents $i_j^{(\nu)}(t) = i_{0j}^{(\nu)} \exp q_\nu t \cos \omega_\nu t$.

When the roots s_ν are all distinct, the ratios of the amplitudes $i_{0j}^{(\nu)}$ are given by the ratios of the cofactors in any row of $\det [Z(s)]$.

$$i_{01}^{(\nu)}: i_{02}^{(\nu)}: i_{03}^{(\nu)} \dots i_{0n}^{(\nu)} = \tilde{Z}_{11}(s_\nu): \tilde{Z}_{12}(s_\nu) \dots \tilde{Z}_{1n}(s_\nu) \qquad (3.3.30)$$

and also

$$= \tilde{Z}_{i1}(s_\nu): \tilde{Z}_{i2}(s_\nu): \dots \tilde{Z}_{in}(s_\nu)$$

Note that the distribution of current amplitudes in the network for these natural response conditions is independent of the point at which the excitation was applied.

If the network is continuously excited by a single generator with a harmonic, e.m.f., $\mathscr{E}_r(t) = \mathscr{E}_{0r}\, e^{j\omega t}$ in loop r, say, we may assume a steady state solution for the current of the form $i_j(t) = i_{0j}\, e^{j\omega t}$. From equation (3.3.3), after carrying through the differential operations

$$i_{0j} = \frac{\tilde{Z}_{rj}(j\omega)}{\det [Z(j\omega)]}\mathscr{E}_{0r} \qquad (3.3.31)$$

Here $Z_{rj}(j\omega)$ is the appropriate cofactor from $\det [Z(j\omega)]$. The $\det [Z(j\omega)]$ is not zero because, in general, $j\omega$ will not be a root of equation (3.3.29). If,

† Proof. Write $\det [Z(s)] = f(s)$. Then, if $s = q+j\omega$ is a root, we may write, with A and B real quantities

$$f(q+j\omega) = A+jB = 0$$
$$f(q-j\omega) = A-jB$$

If $A+jB = 0$, since A and B are real, $A = 0$, $B = 0$ and so $(A-jB) = 0$. Hence $f(q-j\omega) = 0$ and so $(q-j\omega)$ is also a root of $f(s)$.

however, the network has a natural frequency $s_\nu = q_\nu + j\omega_\nu$ for which $q_\nu \ll \omega_\nu$ then the value of det $[Z(j\omega_\nu)]$ will be close to zero and the corresponding current amplitudes $i_{0j}^{(\nu)}$ will become large. The network is now said to be resonant to the driving frequency ω_ν.

Note that the ratio of the current amplitudes

$$i_{01} : i_{02} : i_{03} \ldots : i_{0n} = \tilde{Z}_{j1}(j\omega) : \tilde{Z}_{j2}(j\omega) : \tilde{Z}_{j3}(j\omega) \ldots : \tilde{Z}_{jn}(j\omega) \qquad (3.3.32)$$

is again independent of the location of the source.

3.4 Node Analysis

In the previous section the solution of the general network equations (3.2.9) and (3.2.10) has been illustrated by the use of the technique of loop or mesh analysis. This is only one of the possible methods for solving these equations and we shall emphasize this point by describing an important alternative technique called node analysis. Loop analysis is a particularly useful method when the generators are sources of electromotive force and the circuit elements lie all in one plane. Node analysis is particularly useful when we may regard the network as being excited by current sources and

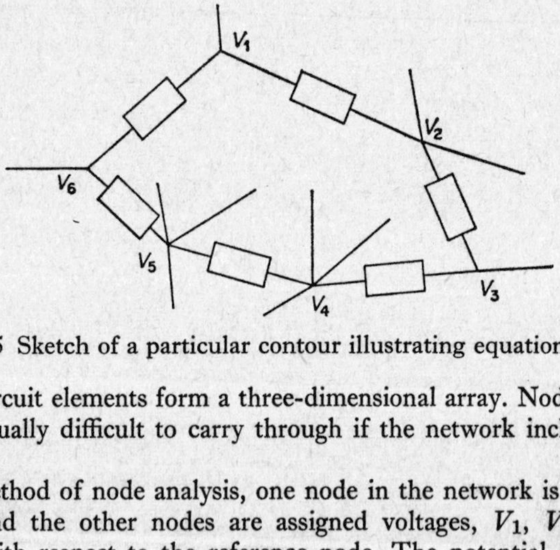

FIG. 3.5 Sketch of a particular contour illustrating equation (3.4.2).

when the circuit elements form a three-dimensional array. Node analysis is, however, usually difficult to carry through if the network includes mutual inductances.

In the method of node analysis, one node in the network is taken as the reference and the other nodes are assigned voltages, V_1, $V_2 \ldots V_j \ldots$ measured with respect to the reference node. The potential differences ϕ which occur in equation (3.2.3) are the voltage differences between adjacent nodes in the network. That is, $\phi_i = (V_{r+1} - V_r)$. When there are no electromotive forces in the network, equations (3.2.3) become

$$[\beta].\phi] = 0 \qquad (3.4.1)$$

These equations are automatically satisfied by the system of node voltage V_j, since upon traversing a closed contour in the network in a consistent manner (for example, clockwise), we shall have equations of the form (compare Fig. 3.5)

$$(V_2 - V_1) + (V_3 - V_2) + (V_4 - V_3) + \cdots (V_1 - V_n) \equiv 0 \qquad (3.4.2)$$

The individual node voltages always appear twice in equation (3.4.2), once with positive sign and again with negative sign so that the complete sum is zero.† We have therefore found a set of voltage variables which automatically satisfy equation (3.4.1). This is analogous to the case of loop analysis, where the loop currents automatically satisfied equation (3.2.10).

We may summarize the relationship between the ϕ's and the V's by the matrix equation

$$\phi] = [\alpha].V] \tag{3.4.3}$$

We know that in terms of the node voltages, V, equation (3.4.1) is automatically satisfied and it remains therefore to determine the V's in terms of the branch currents I.

Equation (3.2.7) which relates the branch currents, I, to the potential differences, ϕ, may be written as

$$I] = [B].\phi] \tag{3.4.4}$$

where

$$[B] = [A]^{-1}$$

Hence

$$I] = [B].[\alpha].V] \tag{3.4.5}$$

The branch currents are still subject to the constraint that the total current flowing into a junction is zero. If the source currents flowing to the different nodes are denoted by $I_j^{(s)}$, we shall have at each node of the network an equation of the form

$$\mu_{j1}I_1 + \mu_{j2}I_2 + \cdots \mu_{jj}I_j + \cdots \mu_{jn}I_n = I_j^{(s)}$$

which may be summarized as

$$[\mu].I] = I^{(s)}] \tag{3.4.6}$$

with μ_{ij} equal to 0, ± 1.

The matrix $[\mu]$ is the transpose of the matrix $[\alpha]$ in equation (3.4.3), and so, substituting from equation (3.4.5) into equation (3.4.6)

$$[\alpha]^T.[B].[\alpha].V] = I^{(s)}] \tag{3.4.7}$$

or

$$[Y].V] = I^{(s)}] \tag{3.4.8}$$

where $[Y] \equiv [\alpha]^T.[B].[\alpha]$ is called the admittance operator matrix for the network. It is a symmetric matrix.

The particular combination of network operators which occurs in an

† These positive and negative signs simply arise from the limits of integration for the integral $\int \mathbf{E}_0.\mathbf{dl}$ taken over the separate segments into which the complete contour is broken down by the branches:

$$\int \mathbf{E}_0.\mathbf{dl} = \int_1^2 \mathbf{E}_0.\mathbf{dl} + \int_2^3 \mathbf{E}_0.\mathbf{dl} + \cdots \int_n^1 \mathbf{E}_0.\mathbf{dl} = 0 = \sum(V_i - V_j)$$

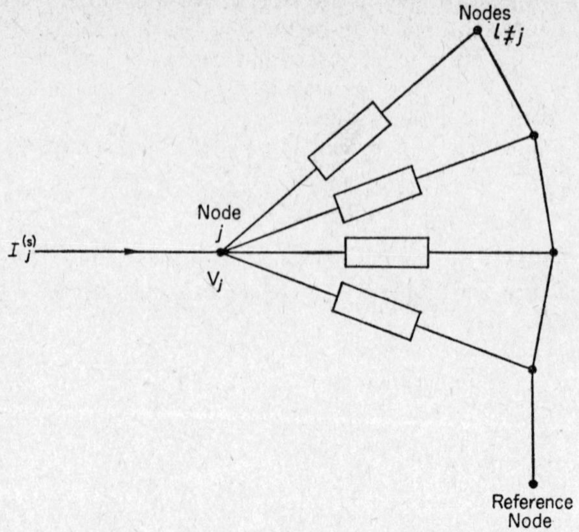

FIG. 3.6 Circuit configuration for the identification of Y_{jj}.

element Y_{ij} of the admittance matrix $[Y]$ may be derived as follows. Consider the node labelled j, then

$$Y_{j1}V_1 + Y_{j2}V_2 + \cdots Y_{jj}V_j + \cdots Y_{jn}V_n = I_j^{(s)} \tag{3.4.9}$$

where the total number of nodes in the network (not counting the reference node) is n. Suppose all nodes, excepting that labelled j, are short circuited to the reference node, so that $V_l = 0$ for all $l \neq j$.

From equation (3.4.9), and Fig. 3.6

$$Y_{jj}V_j = I_j^{(s)} \tag{3.4.10}$$

with Y_{jj} still the same combination of operators as that which occurred in equation (3.4.9). Hence, if there are no mutual inductances involved,

Y_{jj} is the sum of all admittance operators corresponding to the circuit elements which have node j as their common junction.

Next suppose that node i is disconnected from the reference node and node j is joined by short circuit to the reference node, so that $V_l = 0$ for all $l \neq i$. Disconnect the current source $I^{(s)}$.

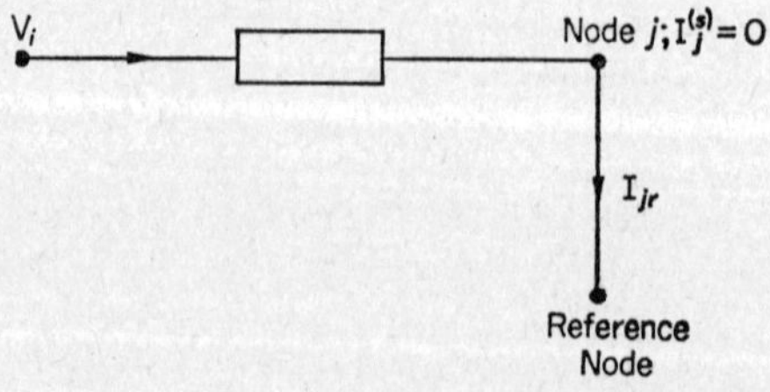

FIG. 3.7 Circuit configuration for the identification of Y_{ij}.

From equation (3.4.9) and Fig. 3.7

$Y_{ji}V_i$ = current flowing into node j due to the voltage V_i at node i.
= – (current flowing to the reference node out of node j)
$= - I_{jr}$

Hence

$Y_{ji} = Y_{ij} = -$ the sum of all admittance operators appropriate to the circuit elements which are in parallel and connect node i to node j.

When the time dependence of V and I is harmonic,

$Y_{jj}(j\omega)$ = sum of all admittances having node j as their common junction.
$Y_{ji}(j\omega) = Y_{ij}(j\omega) = -$ the sum of all admittances in parallel connecting node i to node j.

3.4.1 Example Using Node Analysis

As an example of the use of the admittance operator matrix consider again the circuit of Fig. 3.4. This circuit is redrawn in Fig. 3.8 from which

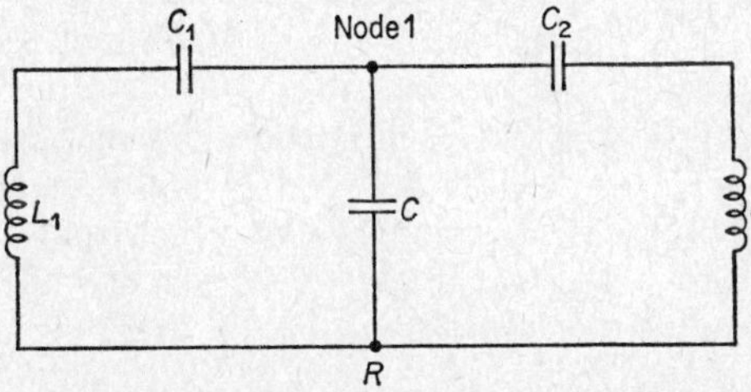

FIG. 3.8 A simple coupled circuit labelled for node analysis.

it can be seen that if the junction R is taken as the reference node, there is just one other significant node labelled 1. The admittance operator matrix therefore has only one term Y_{11}.

$$[Y] = [Y_{11}] = \left[\frac{1}{L_1D + C_1^{-1}D^{-1}} + CD + \frac{1}{L_2D + C_2^{-1}D^{-1}}\right] \quad (3.4.11)$$

The natural frequencies of the network may be obtained by considering the response to a rectangular current pulse. Assuming that the node voltage is $V_1 = V_{01}\, e^{st}$ we have

$$\frac{1}{L_1s + C_1^{-1}s^{-1}} + Cs + \frac{1}{L_2s + C_2^{-1}s^{-1}} = 0 \quad (3.4.12)$$

or

$$C + \frac{1}{1 + s^2L_1C_1} + \frac{1}{1 + s^2L_2C_2} = 0 \quad (3.4.13)$$

This equation is just a different algebraic form of equation (3.3.25) which was derived by loop analysis. The natural frequencies are of course the same whichever method of solution is used.

3.4.2 Natural Frequencies

The natural frequencies of a general network may be derived from node analysis by a similar argument to that which was used in the previous section for loop analysis. The node voltages are assumed to have the form $V_j = V_{0j}\, e^{st}$ following excitation by a rectangular current pulse. The natural frequencies, are given by the roots of the equation

$$\det [Y(s)] = 0 \tag{3.4.14}$$

The roots of det $[Y(s)]$ will all lie in the left half of the complex plane and occur as image points in the real axis (compare the discussion on page 121). The voltage amplitudes at the nodes will be given by the ratio of the cofactors in any row of this determinant

$$V_{01}^{(\nu)} : V_{02}^{(\nu)} : V_{03}^{(\nu)} : \ldots : V_{0n}^{(\nu)} = \tilde{Y}_{11}(s_\nu) : \tilde{Y}_{12}(s_\nu) : \tilde{Y}_{13}(s_\nu) \ldots \tilde{Y}_{1n}(s_\nu) \tag{3.4.15}$$

3.5 Comparison of Loop and Node Analysis

The choice of a particular technique for solving a given network problem is determined entirely by convenience. The aim is to achieve as simple a set of network equations as possible. A network having a total number of branches, b, will require b simultaneous equations to specify completely either the branch voltages ϕ or the branch currents I. However, these variables may be derived by first solving a smaller number of equations.

In the case of node analysis, a network having n significant junctions (where three or more branches join together) will require the solution of $(n - 1)$ equations to determine the complete set of node voltages. The branch voltages are then simply obtained from the voltage differences between adjacent nodes. Note that there are $(n - 1)$ equations for n significant nodes because one junction is taken as the reference node.

In the case of loop analysis, $(n - 1)$ equations, one for each significant junction are automatically satisfied by the choice of loop currents. This leaves $b - (n - 1)$ equations to be solved in order to determine these loop currents specifically. The individual branch currents are then derived by simple addition or subtraction.

When node analysis is used, it is usually easy to identify, by inspection, the significant junctions which give rise to the $(n - 1)$ independent voltage equations for the network. This is true even if the network is in the form of a three-dimensional array. It is not always so easy to find the $b - (n - 1)$ independent equations for loop analysis. If the network is planar, the independent loops may be found by first drawing out the network so that no branch crosses another. The independent loops then coincide with the 'windows' in the network, as shown in Fig. 3.9. For three-dimensional arrays it is necessary to ensure that each additional loop, when drawn, includes at least one branch of the network which is not covered by a previous loop. The set is complete when at least one loop current traverses each circuit element in the network.

Note that, although the formal description of node analysis given by equations (3.4.1)–(3.4.8) includes mutual inductance elements straightforwardly through the identification of the matrix $[B]$ with $[A]^{-1}$ in equation

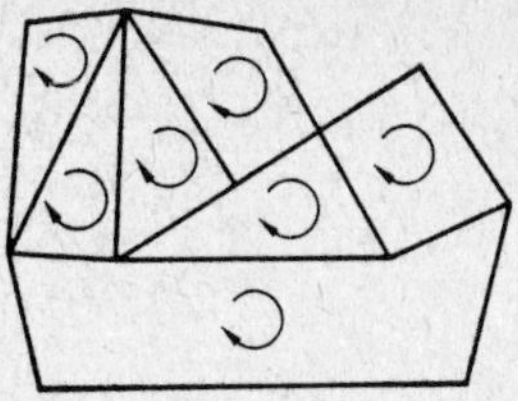

FIG. 3.9 Loop currents in the 'windows' of a planar network.

(3.4.4), it is in practice difficult to handle such elements when this method of analysis is used for an actual circuit. This arises, because, although the reciprocal of a diagonal matrix $[a_{ii}]$ is also a diagonal matrix, $[a_{ii}^{-1}]$, with diagonal elements equal to the reciprocal of the corresponding elements in the original matrix, the reciprocal of a general matrix $[a_{ij}]$ does not have its elements equal to the reciprocal of corresponding elements in the original matrix. Hence, when mutual inductances are present and the matrix is non-diagonal the Y_{jj} and Y_{ij} do not have the simple interpretation given on pages 124 and 125 and it is usually preferable to use loop analysis for such networks.

3.6 Signal Flow Graphs

In the previous sections we have discussed the solution of network problems in terms of loop and node analysis. These were set out as purely algebraic techniques. The signal flow graph provides a graphical method for solving the algebraic equations which arise in network problems. The graphical diagram representing the network again comprises branches and nodes but now these quantities are assigned a more general meaning. A node represents a signal variable, for example, the voltage or the current, and a branch between two nodes represents the functional relation (called the transmittance) which couples the variables at these two nodes. A branch has an associated direction (the direction being indicated on the signal flowgraph by an arrowhead) which indicates the way in which the signal is transmitted through the network. A node variable is multiplied by the transmittance of a branch as the signal traverses that branch in the direction of the arrowhead.

A node is also a summing junction. Signals *coming into* the node are added together to obtain the value of the node variable at that point. This summed value leaves the node as signal along each branch which *goes out* from the node point. A *source node* has only outgoing signals and represents an independent variable for the network. A *sink node* is a point at which all branches are directed inwards. All nodes except source nodes represent dependent variables.

Consider a typical signal flow graph such as that shown in Fig. 3.10. Here there are seven nodes and hence seven variables, $x_i = \{x_1, x_2 \ldots x_7\}$. Node x_1 is a source node while x_7 is a sink node. Hence x_1 is the independent variable in this network. Since there are six dependent variables the graph represents six simultaneous algebraic equations which can be written down by considering the incoming branches at each node. Remembering that node

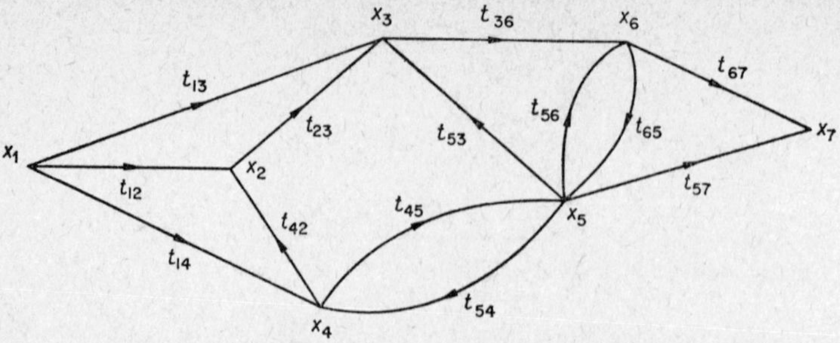

FIG. 3.10 A typical signal flow graph.

variable x_i is multiplied by the branch transmittance t_{ij} as it flows from node i to node j, we obtain on summing the incoming signals at each node,

$$\begin{aligned}
x_2 &= t_{12}x_1 + t_{42}x_4 \\
x_3 &= t_{13}x_1 + t_{23}x_2 + t_{53}x_5 \\
x_4 &= t_{14}x_1 + t_{54}x_5 \\
x_5 &= t_{45}x_4 + t_{65}x_6 \\
x_6 &= t_{36}x_3 + t_{56}x_5 \\
x_7 &= t_{57}x_5 + t_{67}x_6
\end{aligned} \tag{3.6.1}$$

In these equations each network variable appears once and in the labelled order on the left-hand side of the equations; the transmittance subscripts indicate the direction of signal flow along the branches. This is the usual convention for listing the signal flow equations. It can be seen that one advantage of the signal flow graph technique is the way in which features of the network such as the feedback loops $t_{54}t_{45}$, $t_{65}t_{56}$ are clearly illustrated by the graph although these are not immediately apparent in the algebraic equations.

Next consider the procedure for drawing out the signal flow graph to represent a set of algebraic equations. Suppose

$$\begin{aligned}
x_2 &= t_{12}x_1 + t_{22}x_2 + t_{32}x_3 \\
x_3 &= t_{13}x_1 + t_{23}x_2 + t_{43}x_4 \\
x_4 &= t_{24}x_2 + t_{34}x_3 + t_{44}x_4
\end{aligned} \tag{3.6.2}$$

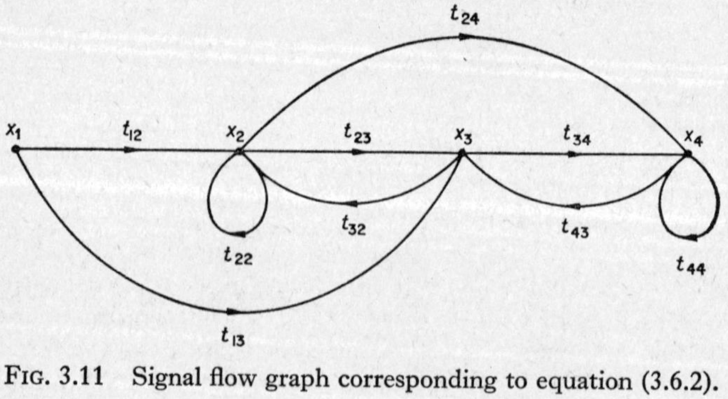

FIG. 3.11 Signal flow graph corresponding to equation (3.6.2).

First set out the separate nodes $x_i = (x_1, x_2, x_3, x_4)$ as a suitable array of points say

$$\begin{matrix} \cdot & \cdot & \cdot & \cdot \\ x_1 & x_2 & x_3 & x_4 \end{matrix}$$

Upon drawing in the directed branches, one branch for each term in the simultaneous equations (3.6.2), we obtain Fig. 3.11.

This signal flow graph representing the set of simultaneous algebraic equations (3.6.2) is not unique and any rearrangement of the variables will yield a different flow graph. For example, if the first two equations are interchanged, we obtain

$$x_2 = \frac{t_{13}}{t_{23}}x_1 - \frac{1}{t_{23}}x_3 + \frac{t_{43}}{t_{23}}x_4 = t'_{12}x_1 + t'_{32}x_3 + t'_{42}x_4$$

$$x_3 = \frac{t_{12}}{t_{32}}x_1 + \frac{(t_{22} - 1)}{t_{32}}x_2 = t'_{13}x_1 + t'_{23}x_2 \tag{3.6.3}$$

$$x_4 = t_{24}x_2 + t_{34}x_3 + t_{44}x_4$$

These algebraic equations are entirely equivalent to those set out in equation (3.6.2) and the new signal flow graph is shown in Fig. 3.12.

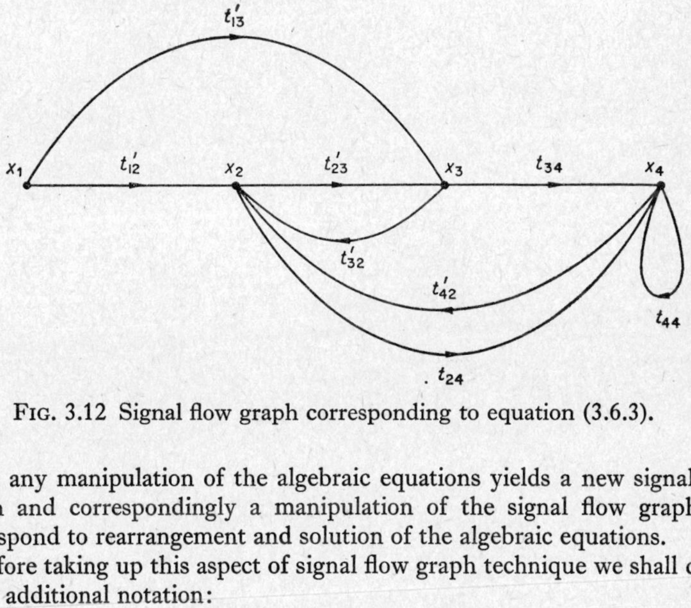

FIG. 3.12 Signal flow graph corresponding to equation (3.6.3).

Thus any manipulation of the algebraic equations yields a new signal flow graph and correspondingly a manipulation of the signal flow graph can correspond to rearrangement and solution of the algebraic equations.

Before taking up this aspect of signal flow graph technique we shall define some additional notation:

A *path* is the route taken by a signal in travelling from node i to node r, that is, a *path* is a continuous sequence of branches. For example, the route t'_{12}, t'_{23}, t_{34}, in Fig. 3.12.

A *feedback loop* or simply a *loop* is a path which begins and terminates on the same node but is restricted to paths such that no node is traversed more than once. For example, the path t'_{23}, t_{34}, t'_{42}, in Fig. 3.12.

A *self-loop* is a feedback which contains only a single node. For example, the path t_{44} in Fig. 3.12.

The *forward path gain* is the product of all transmittances along a path which involves no feedback loops.

The *loop gain* is the product of all transmittances associated with a loop. For example, the feedback loop described above has loop gain $t'_{23}t_{34}t'_{42}$.

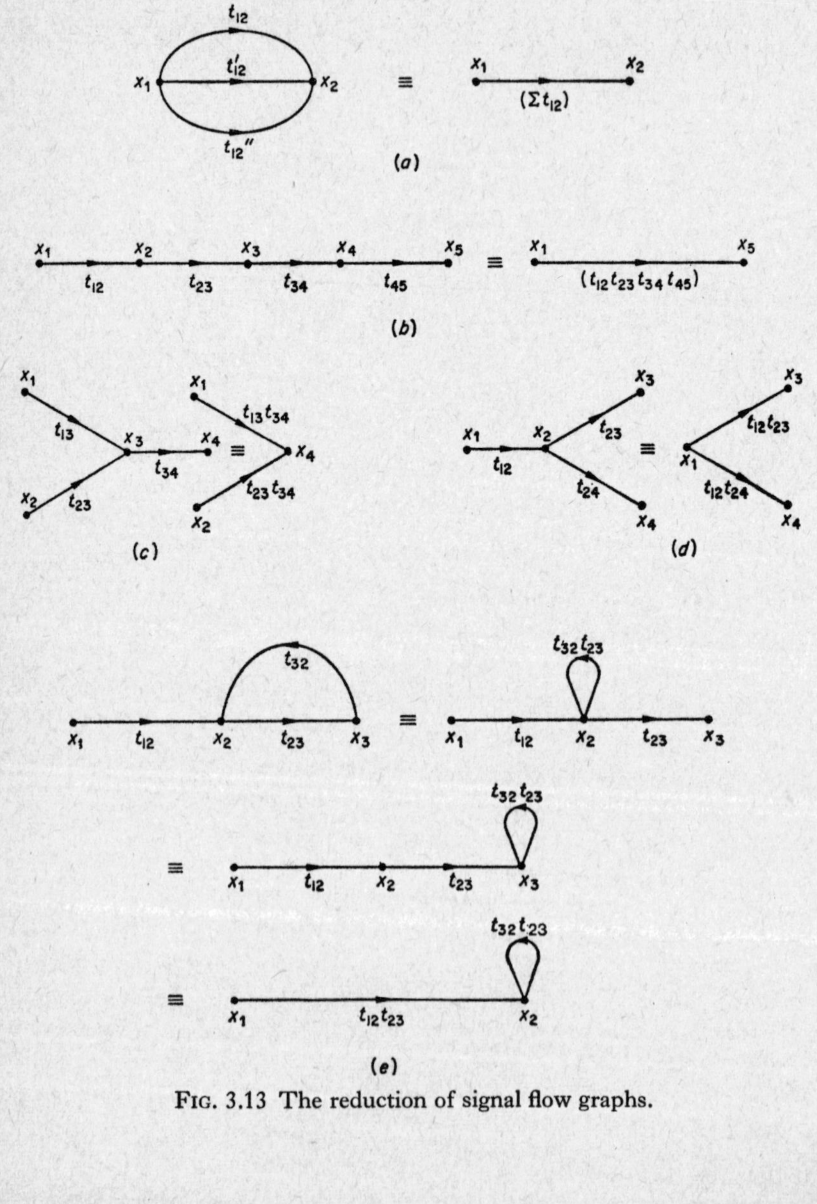

FIG. 3.13 The reduction of signal flow graphs.

The solution of a network problem requires a reduction of the signal flow graph by a systematic procedure. There are five fundamental operations.

1. The combination of parallel branches.
2. The combination of series branches and the elimination of internal nodes.
3. The reduction of a feedback loop to a self-loop.
4. The elimination of a self-loop.
5. Moving a node entrance.

1. Parallel branches between two nodes may be replaced by a single branch with transmittances equal to the sum of the transmittances of the individual branches. From Fig. 3.13(a)

$$x_2 = t_{12}x_1 + t'_{12}x + t''_{12}x_1 = (t_{12} + t'_{12} + t''_{12})x_1 \qquad (3.6.4)$$
$$= (\textstyle\sum t_{12})x_1$$

2. The transmittance between two nodes along an open path is the product of the transmittances for each branch of the path. This operation may also eliminate an internal node as shown in Fig. 3.13(b), (c) and (d).

3. A feedback loop can be reduced to a self-loop with the same loop gain as the original feedback loop.

From Fig. 3.13(e), we have

$$x_2 = t_{12}x_1 + t_{32}x_3$$

but $$x_3 = t_{23}x_2$$

therefore $$x_2 = t_{12}x_1 + t_{32}t_{23}x_2$$

also $$x_3 = t_{12}t_{23}x_1 + t_{23}t_{32}x_3 \qquad (3.6.5)$$

4. A self loop at node x_i with a loop gain t_{ii} is eliminated if the transmittance of each branch entering the node x_i is multiplied by $(1 - t_{ii})^{-1}$. This operation is illustrated by the rearrangement of node variables, for example

$$x_i = t_{1i}x_1 + t_{2i}x_2 + \cdots t_{ii}x_i + \cdots t_{ni}x_n$$

that is, $$x_i = \frac{t_{1i}}{1 - t_{ij}}x_1 + \frac{t_{2i}}{1 - t_{ii}}x_2 + \cdots 0x_i + \cdots \frac{t_{ni}}{1 - t_{ii}}x_n \qquad (3.6.6)$$

5. A node entrance may be moved by carrying through operation 2.

3.6.1 Examples

As an example of the reduction of a signal flow graph consider the graph shown below.

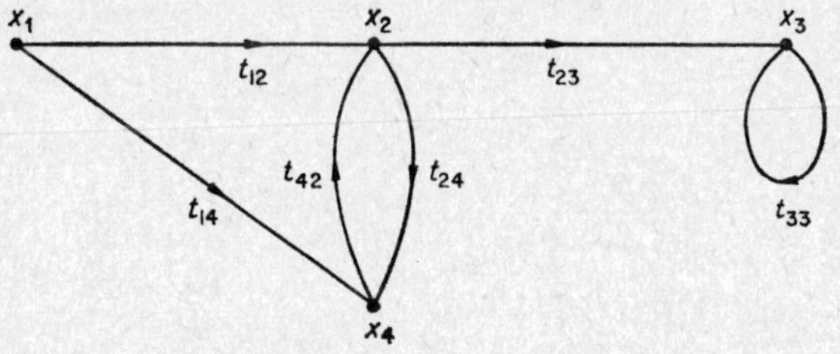

The following steps reduce the graph

(i)

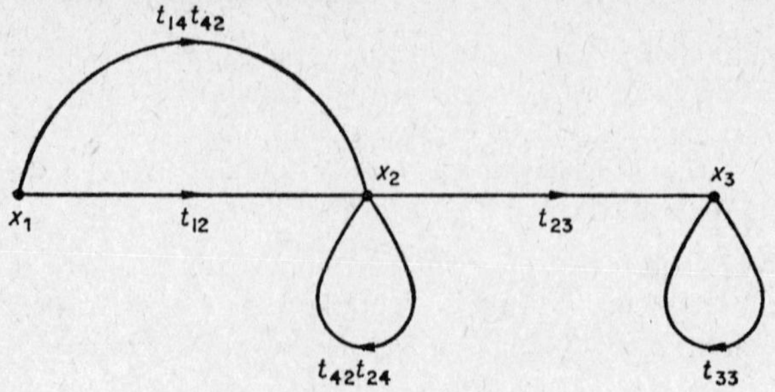

(ii)

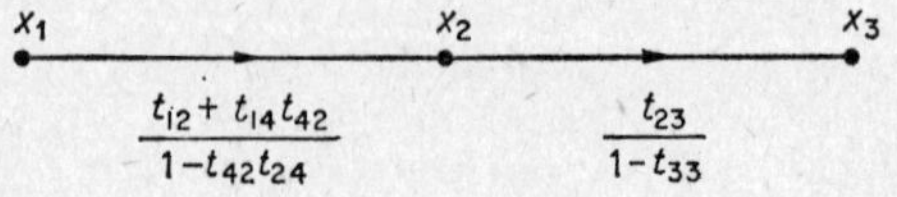

Note that step (i) follows from the algebraic relations

$$\begin{aligned} x_2 &= t_{12}x_1 + t_{42}x_4 \\ x_4 &= t_{14}x_1 + t_{24}x_2 \end{aligned} \qquad (3.6.7)$$

so that

$$x_2 = t_{12}x_1 + t_{14}t_{42}x_1 + t_{42}t_{24}x_4 \qquad (3.6.8)$$

This particular step indicates that care must be exercised when reducing a feedback loop to a self-loop. If this operation destroys an existing path then that path must also be drawn in together with the self-loop.

As a second example consider the construction of a signal flow diagram to represent the network shown in Fig. 3.14(a). From which it is required to determine the voltage V_4.

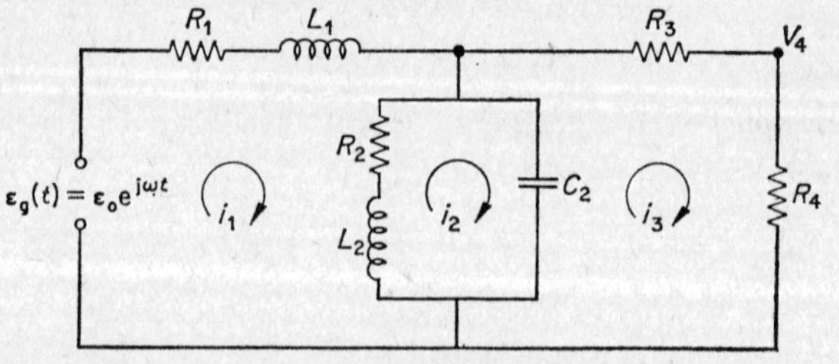

(a)

FIG. 3.14

(a) Circuit for the construction of the signal flow graph shown in Fig. 3.14 (b)

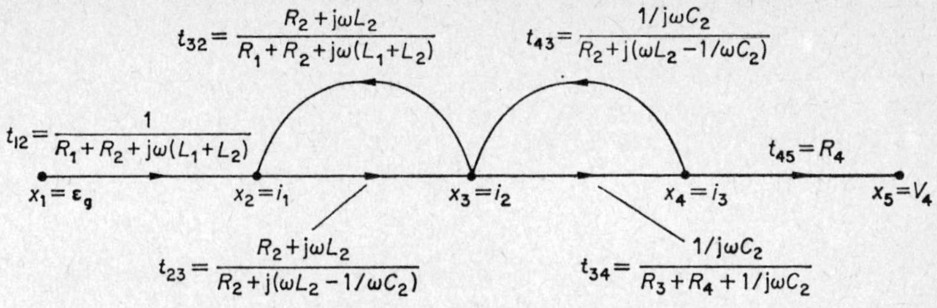

(b)

(b) A signal flow graph corresponding to the circuit shown in Fig. 3.14 (a).

From the loop equations we obtain

$$i_1 = \frac{\mathscr{E}g}{(R_1 + j\omega L_1 + R_2 + j\omega L_2)} + \frac{R_2 + j\omega L_2}{(R_1 + j\omega L_1 + R_2 + j\omega L_2)} i_2$$

$$i_2 = \frac{R_2 + j\omega L_2}{R_2 + j(\omega L_2 - 1/\omega C_2)} i_1 + \frac{1/j\omega C_2}{R_2 + j(\omega L_2 - 1/\omega C_2)} i_3$$

$$i_3 = \frac{1/j\omega C_2}{(R_3 + R_4 + 1/j\omega C_2)} i_2$$

$$V_4 = R_4\, i_3$$

and the corresponding signal flow graph is shown in Fig. 3.14(b).

The signal flow graph is obviously fundamentally distinct from a circuit diagram. It represents a useful graphic technique for representing a system of linear algebraic equations and for solving these equations. There is neither more nor less information contained in the signal flow graph than is present in the equations on which the graph is based but the graph may emphasize some features advantageously and the graphical technique for reducing the signal flow graph may introduce some simplifications. It should be noted, however, that although one can always draw one (or more) signal flow graphs to represent a circuit the reverse procedure is not necessarily possible with realizable circuit elements. Even when a realizable network does exist which corresponds to a given signal flow graph there is no general technique for finding it, except by trial and error.

3.7 Driving Point Impedance and Admittance Functions

Consider loop analysis of a network which is continuously excited in the loop labelled, r, by a generator with a harmonic e.m.f. $\mathscr{E}_r(t) = \mathscr{E}_{0r}\, e^{j\omega t}$. There

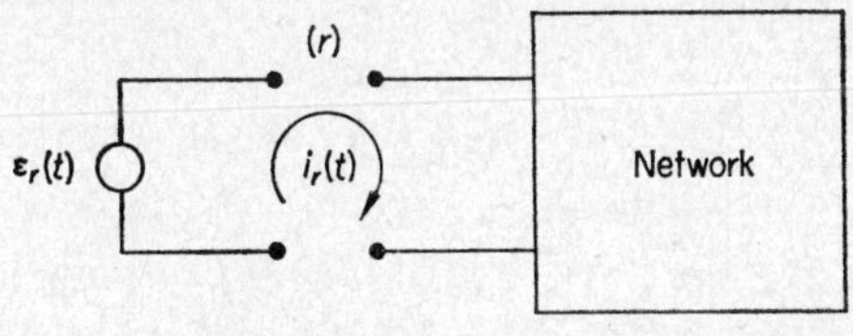

FIG. 3.15 The driving point configuration for a network.

are no other generators in the circuit. We may also label the terminal pair at the generator as r and imagine the leads to the generator brought outside the network as shown in Fig. 3.15.

The current supplied by the generator is now the loop current $i_r(t)$ and from equation (3.3.3.) after performing the time differential operations

$$i_{0r} = \frac{\tilde{Z}_{rr}(j\omega)}{\det [Z(j\omega)]} \mathscr{E}_{0r} \tag{3.7.1}$$

where $\tilde{Z}_{rr}(j\omega)$ is the cofactor of the element (r, r) in $\det [Z(j\omega)]$. Equation (3.7.1) defines a driving point impedance function $Z_{d,r}(j\omega)$ at the terminals, r, from the ratio $\mathscr{E}_{0r}/i_{0r}$. A driving point admittance function $Y_{d,r}(j\omega)$ may equally well be defined from the ratio $i_{0r}/\mathscr{E}_{0r}$.

Evidently

$$Z_{d,r}(j\omega) = \frac{\det [Z(j\omega)]}{\tilde{Z}_{rr}(j\omega)} = \frac{1}{Y_{d,r}(j\omega)} \tag{3.7.2}$$

and

$$\mathscr{E}_{0r} = Z_{d,r}(j\omega)\, i_{0r} \tag{3.7.3}$$

Equation (3.7.3) is a particular example of the more general relation between $\mathscr{E}(t)$ and $i(t)$ at a pair of accessible terminals for a linear passive network. The general relation may be written

$$\mathscr{E}(t) = Z_d(D)\, i(t) \tag{3.7.4}$$

where $Z_d(D)$ is a function of the time differential operator D and is characteristic of the network. $Z_d(D)$ is called the driving point impedance operator. A driving point admittance operator may also be defined by the relation

$$i(t) = Y_d(D)\, \mathscr{E}(t) \tag{3.7.5}$$

so that

$$Z_d(D) = \frac{1}{Y_d(D)} \tag{3.7.6}$$

3.7.1 Natural and Resonance Frequencies

The driving point impedance and admittance functions provide information about the characteristic frequencies of the network. From equation (3.7.2) we may write

$$Z_{d,r}(j\omega) = \frac{N(j\omega)}{\mathscr{D}(j\omega)} \tag{3.7.7}$$

where the numerator function $N(j\omega)$ and the denominator function $\mathscr{D}(j\omega)$ are each polynomials in $j\omega$ with real coefficients. If the roots of the numerator function when written as $N(s)$ are given by $s_{N1}, s_{N2}, \ldots s_{Nn}$, and the roots of the denominator function written $\mathscr{D}(s)$ are $s_{\mathscr{D}1}, s_{\mathscr{D}2} \ldots s_{\mathscr{D}m}$ we may write

$$Z_{d,r}(j\omega) = K\frac{(j\omega - s_{N1})(j\omega - s_{N2}) \cdots (j\omega - s_{Nn})}{(j\omega - s_{\mathscr{D}1})(j\omega - s_{\mathscr{D}2}) \cdots (j\omega - s_{\mathscr{D}m})} \tag{3.7.8}$$

The numerator function arises from $\det [Z(j\omega)]$ and is characteristic of the complete network; it does not depend upon the choice of the particular terminals, r, at which $Z_{d,r}(j\omega)$ is evaluated. The denominator function arises from the cofactor determinant $\tilde{Z}_{rr}(j\omega)$. This determinant is just the characteristic determinant of the network with loop, r, open circuit and so the denominator function depends upon the choice of terminals.

If the losses in the circuit are small (as indicated by $s_N \equiv q_N + j\omega_N$ with $q_N \ll \omega_N$) it can be seen from equations (3.7.1) and (3.7.8) that the current drawn from the generator will become large when the frequency of the generator, ω, is equal to the imaginary part of a root s_{Nj}. The frequencies ω_{Nj} are resonance frequencies of the network. The roots s_{Nj} which give rise to zeros of the generalized driving point impedance function are the natural frequencies of the network.

The natural frequencies obtained from the zeros of $Z_{d,r}(s)$ (or the infinities of $Y_{d,r}(s)$) refer to the network with the pair of accessible terminals short circuited. This arises because the constraint det $[Z(s)] = 0$ implies that all the current loops in the network, including the one which contains the terminal pair, are completed. These natural frequencies then correspond to those derived previously on pages 120, 125. There is, however, another set of natural frequencies for the network which refer to the situation with the accessible terminals open circuited. These frequencies are derived from the zeros of $\tilde{Z}_{rr}(s)$ since, as we have commented above, this determinant is just the characteristic determinant of the network with loop, r, open circuit. For these values of $s_{\mathscr{D}j}$, the generalized driving point impedance $Z_{d,r}(s)$ becomes infinite while $Y_{d,r}(s)$ becomes zero. The terminal current now falls to zero but there will still remain a voltage across the terminals.

There will be a set of resonance frequencies derived from the imaginary part of $s_{\mathscr{D}j}$. At these resonance frequencies the current drawn from the exciting generator will fall to a low value.

The natural frequencies derived from the zeros and infinites (poles) of the generalized driving point impedance and admittance functions will occur as complex conjugate pairs. The natural oscillations in a realizable passive network must eventually die away and so the real part of a natural frequency can never be positive and can only be zero for ideal loss-free circuit elements. Both the zeros and poles of a realizable driving point impedance or admittance function must therefore lie in the left half complex plane and occur as image pairs in the real axis.

3.7.2 Example

We may illustrate this discussion by considering again the circuit discussed previously in terms of impedance and admittance operator matrices. The circuit of Figs. 3.4 and 3.8 is redrawn in Fig. 3.16 with two terminals brought out from loop (1).

The generalized driving point impedance function at these terminals is obtained directly from equations (3.3.24) and (3.7.2).

$$Z_{d1}(s) = \frac{\det [Z(s)]}{\tilde{Z}_{11}(s)}$$
$$= \frac{(s^2L_1C_1C + C_1 + C)(s^2L_2C_2C + C_2 + C) - C_1C_2}{sC_1C(s^2L_2C_2C + C_2 + C)} \tag{3.7.9}$$

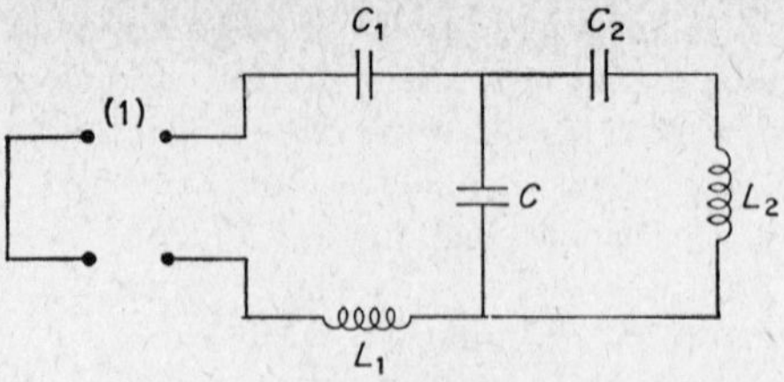

FIG. 3.16 A simple coupled circuit for driving point analysis.

The natural frequencies of oscillation when the terminals (1) are short circuited are given by the zeros of $Z_{d1}(s)$.

$$(s^2L_1C_1C + C_1 + C)(s^2L_2C_2C + C_2 + C) - C_1C_2 = 0 \quad (3.7.10)$$

This equation agrees with the previous derivations in equations (3.3.25) and (3.4.13). The natural frequencies of oscillation for the circuit when terminals (1) are open circuited are given by the poles of $Z_{d1}(s)$

$$sC_1C(s^2L_2C_2C + C_2 + C) = 0 \quad (3.7.11)$$

3.8 Power Dissipation in a Network

The rate at which energy is supplied to a network is, at any instant, equal to the rate at which work is being performed on the moving charges which constitute the electric currents in the circuit. The work done in taking unit charge round a closed circuit is, however, just the electromotive force in that circuit. The instantaneous power supplied to the circuit is therefore the product of the electromotive forces which are present and the currents flowing through these generators. If the current is measured in amperes and the electromotive force in volts, then the unit of power is the watt. The instantaneous power is given by

$$W_c(t) = \sum_{\nu} I_\nu(t)\mathscr{E}_\nu(t) \quad (3.8.1)$$

where $I_\nu(t)$ is the branch current through the generator with e.m.f. $\mathscr{E}_\nu$. When loop currents $i(t)$ are chosen so that one current flows through each generator

$$W_c(t) = \underline{i(t)}\,.\,\mathscr{E}(t)] \quad (3.8.2)$$

where $\underline{i(t)}$ represents the transpose of the column matrix $i(t)$. This equation may be written in terms of the impedance operator matrix, from equation (3.3.3).

$$W_c(t) = \underline{i(t)}\,.\,[Z]\,.\,i(t)] \quad (3.8.3)$$

or alternatively in terms of branch currents and potential differences from equations (3.2.3) and (3.3.1).

$$\begin{aligned} W_c(t) &= \underline{I(t)}\,.\,[\beta]^{-1}\,.\,[\beta]\,.\,\phi(t)] \\ &= \underline{I(t)}\,.\,\phi(t)] \end{aligned} \quad (3.8.4)$$

When the time functions are harmonic

$$I(t) = \text{Real part } [I_0\, e^{j\omega t}] = \text{Re}\,[I] = \tfrac{1}{2}(I + I^*) \qquad (3.8.5)$$

$$\phi(t) = \text{Real part } [\phi_0\, e^{j\omega t}] = \tfrac{1}{2}(\phi + \phi^*) \qquad (3.8.6)$$

The instantaneous power delivered to the circuit is therefore, from equation (3.8.4)

$$W_c(t) = \text{Re}\,\underline{I}\,.\,\text{Re}\,\phi] \qquad (3.8.7)$$

$$= \tfrac{1}{4}(\underline{I} + \underline{I^*}).(\phi] + \phi^*])$$

$$= \tfrac{1}{4}\{\underline{I}\,.\phi] + \underline{I^*}\,.\phi^*] + \underline{I}\,.\phi^*] + I^*\,.\phi]\}$$

that is,

$$W_c(t) = \tfrac{1}{2}\,\text{Re}\,\{\underline{I}\,.\phi]\} + \tfrac{1}{2}\,\text{Re}\,\{\underline{I^*}\,.\phi]\} \qquad (3.8.8)$$

The second term on the right-hand side of equation (3.8.8) contains terms of the form $I^*\phi_0\, e^{-j\omega t}\, e^{j\omega t}$ and so is independent of the time. The first term on the right-hand side involves the time factor $e^{2j\omega t}$ and varies harmonically at twice the driving frequency with a mean value zero. The time average power drawn from the source is therefore given by just the second term in equation (3.8.8)

$$\langle W_c(t)\rangle = \tfrac{1}{2}\,\text{Re}\,\{\underline{I^*}\,.\phi]\} \qquad (3.8.9)$$

If there is a phase angle, δ_j, between a branch current, I_j, and its corresponding potential difference, ϕ_j, this equation has the form

$$\langle W_c(t)\rangle = \tfrac{1}{2}\sum |I_{0j}|\,.\,|\phi_{0j}|\cos\delta_j \qquad (3.8.10)$$

and this represents the power which is irretrievably lost from the generator. The instantaneous power, $W_c(t)$, on the other hand varies in magnitude from

$$\Big\{\langle W_c(t)\rangle - \tfrac{1}{2}\sum_j |I_{0j}|\,.\,|\phi_{0j}|\Big\} \text{ to } \Big\{\langle W_c(t)\rangle + \tfrac{1}{2}\sum_j |I_{0j}|\,.\,|\phi_{0j}|\Big\}$$

and so may be negative during part of the cycle. Energy may therefore be returned to the source from the reactive elements in the network as is shown by the shaded areas shown in Fig. 3.17.

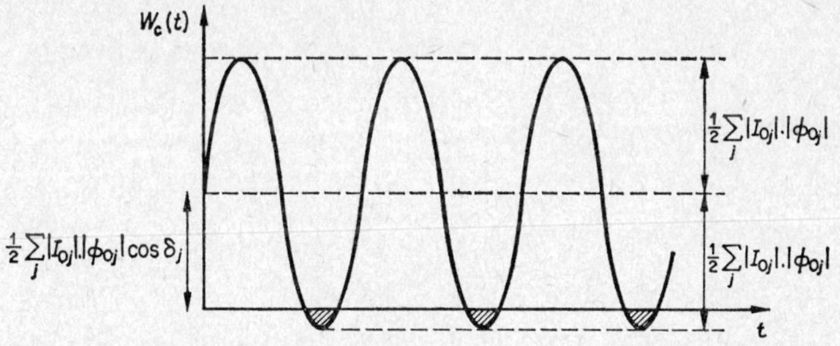

FIG. 3.17 The instantaneous power function $W_c(t)$ corresponding to equation (3.8.8).

3.8.1 Complex Power Function

The relation given in equation (3.8.9) may be extended to define a complex power function, $\tilde{S}$,

$$\tilde{S} \equiv \tfrac{1}{2}\underline{I^*}.\phi] \qquad (3.8.11)$$

since Re $[S]$ is $\langle W_c(t)\rangle$, we may also write

$$\tilde{S} = \langle W_c(t)\rangle + jQ \qquad (3.8.12)$$

The function $\tilde{S}$ may also be related to the driving point impedance function for a network. Consider a network driven by a single generator in loop, r,

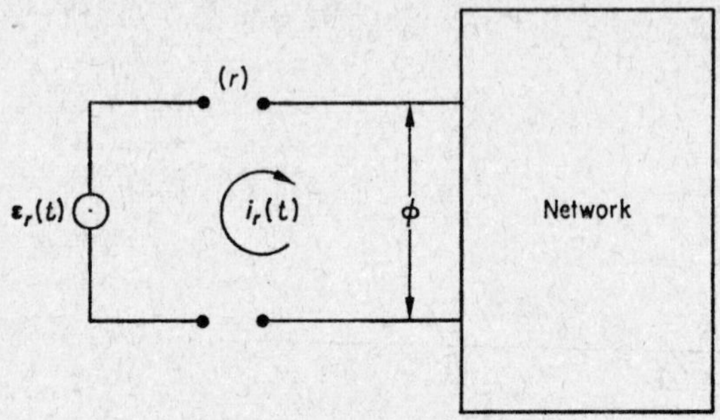

Fig. 3.18 The driving point configuration for the discussion of power loss in a network.

such as was discussed in the previous section. From Fig. 3.18 and equation (3.7.3) it can be seen that

$$\tilde{S} = \tfrac{1}{2}I^*\phi = \tfrac{1}{2}i_{0r}^*\mathscr{E}_{0r}$$

that is

$$\tilde{S} = \tfrac{1}{2}Z_{dr}(j\omega).|i_{0r}|^2 \qquad (3.8.13)$$

Hence, the mean power supplied to the network is given by

$$\langle W_c(t)\rangle = \text{Re}\,[\tilde{S}] = \tfrac{1}{2}\text{Re}\,[Z_{dr}(j\omega)].|i_{0r}|^2 \qquad (3.8.14)$$

There is a corresponding relation in terms of the driving point admittance function

$$\langle W_c(t)\rangle = \tfrac{1}{2}\,\text{Re}\,[Y_{dr}(j\omega)].|\mathscr{E}_{0r}|^2 \qquad (3.8.15)$$

Note that since $\langle W_c(t)\rangle$ is a positive quantity, or zero, the driving point functions must have real parts which are also positive or zero (i.e. can never be negative)

From equations (3.8.12) and (3.8.13)

$$Z_{dr}(j\omega) = \frac{2\{\langle W_c(t)\rangle + jQ\}}{|i_{0r}|^2} \qquad (3.8.16)$$

3.8.2 Resonance from Unity Power Factor

When the power dissipation in the network is small, the term $\langle W_c(t)\rangle$ in equation (3.8.16) is small, and the 'zeros' of $Z_{dr}(j\omega)$ correspond closely to the condition that $Q = 0$, or Im $[Z_{dr}(j\omega] = 0$. This defines a resonance condition for the network. If $\mathscr{E}_r(t) = |\mathscr{E}_{0r}|\, e^{j\omega t}$ and $i_r(t) = |i_{0r}|\, e^{j(\omega t-\delta)}$ then

$$\tan \delta = \frac{\text{Im}\,[Z_{dr}(j\omega)]}{\text{Re}\,[Z_{dr}(j\omega)]}$$

and this resonance condition requires the phase angle, δ, to be zero, which implies $\cos \delta = 1$. From equation (3.8.14)

$$\begin{aligned}\langle W_c(t)\rangle &= \tfrac{1}{2}\,\text{Re}\,[Z_{dr}(j\omega)]\,|i_{0r}|^2 \\ &= \tfrac{1}{2}|Z_{dr}(j\omega)|\,.\,|i_{0r}|^2 \cos \delta \\ &= \tfrac{1}{2}|i_{0r}|\,.\,|\mathscr{E}_{0r}| \cos \delta \qquad (3.8.17)\end{aligned}$$

The term $\cos \delta$ is called the power factor for the circuit and so this resonance condition is called the condition for unity power factor. The resonance frequencies derived from this relation will only agree with those obtained in the previous section when the losses in the network are sufficiently small.

Appendix

Proof that $[\delta]$ in equation (3.3.1) is the transpose of $[\beta]$ in equation (3.2.9) when the voltage relations are evaluated around the contours traversed by the loop currents. Consider a circuit element labelled m, with corresponding branch current I_m and voltage drop ϕ_m. Choose a set of loop currents, which for convenience we take always in a clockwise direction, and suppose the

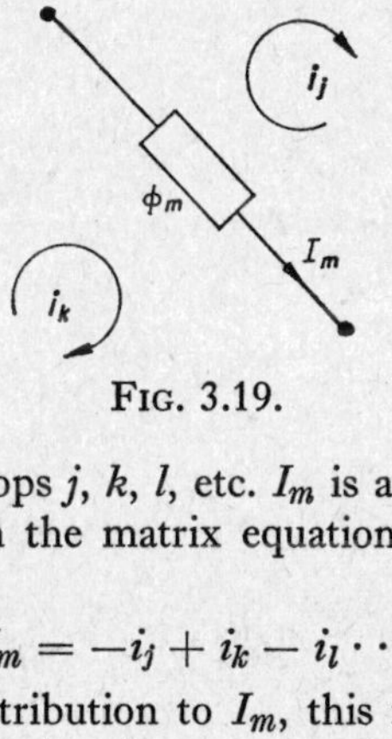

FIG. 3.19.

branch m is common to loops j, k, l, etc. I_m is a summation of the i_j, i_k, i_l, etc., and so one row from the matrix equation $[\delta].i] = I]$ would be, for example,

$$I_m = -i_j + i_k - i_l \cdots$$

If i_k makes a positive contribution to I_m, this means that I_m and i_k have the same direction through the element m and so ϕ_m will make a positive contribution to the voltage sum for the loop defined by i_k, for example,

$$\beta_{k1}\phi_1 + \beta_{k2}\phi_2 + \cdots (+1)\phi_m + \cdots = \mathscr{E}_k$$

Similarly if i_j makes a negative contribution to I_m, ϕ_m will make a negative contribution to the voltage sum equated to $\mathscr{E}_j$, and if there is no contribution to I_m there will be no contribution to the voltage sum.

The coefficients of ϕ_m, however, occur in the *columns* of the matrix $[\beta]$, whereas the coefficients of i_j, i_k, i_l . . . occurred in the *rows* of the matrix $[\delta]$. Hence the columns of $[\beta]$ are derived from the rows of $[\delta]$, and the mth row in $[\delta]$ gives the mth column in $[\beta]$. This proves that $[\delta]$ is the transpose of $[\beta]$, that is,

$$[\delta] = [\beta]^T$$

Bibliography

GUILLEMAN, E. A. *Introductory Circuit Theory*, John Wiley, 1953
GUILLEMAN, E. A. *Theory of Linear Physical Systems*, John Wiley, 1963
KUO, F. F. *Network Analysis and Synthesis*, John Wiley, 1962
PFEIFFER, P. E. *Linear Systems Analysis*, McGraw-Hill, 1961
SESHU, S. and BALABAMIAN, N. *Linear Network Analysis*, John Wiley, 1959

Problems

1. Calculate the branch current through R_A and the natural frequencies of the network E.1.

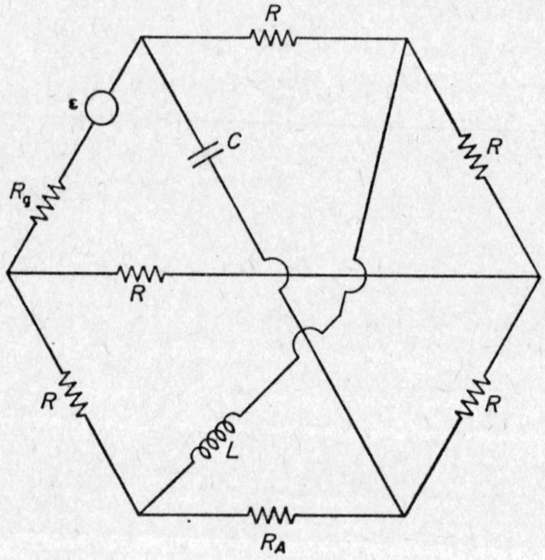

E.1

2. Calculate the driving point impedance functions and the resonance frequencies for the networks E.2 to E.5.

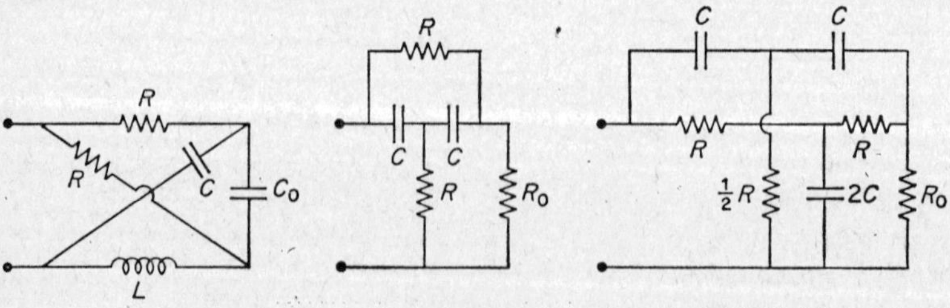

E.2 E.3 E.4

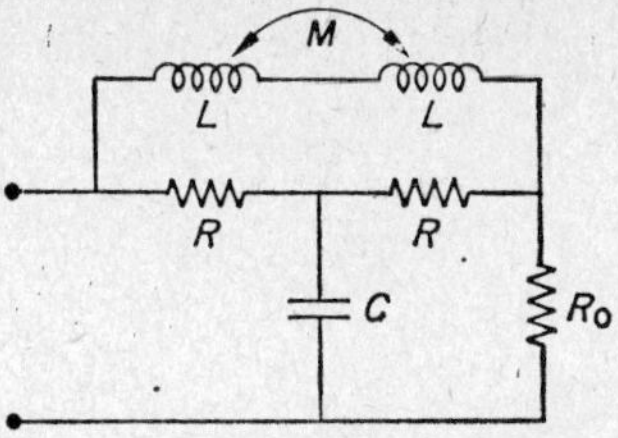

E.5

3. Calculate the output voltage $V(j\omega)$ for networks E.6, E.7.

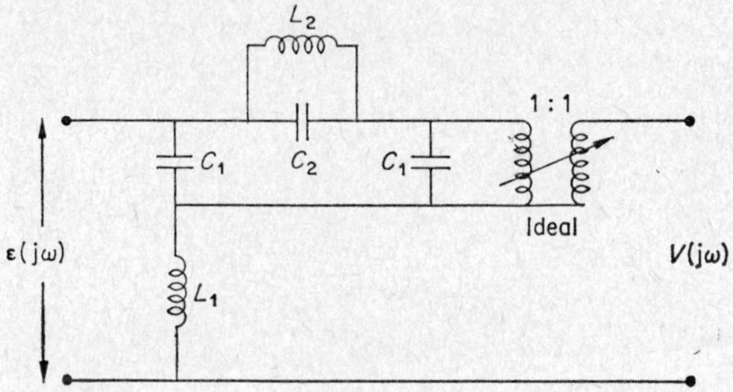

E.6

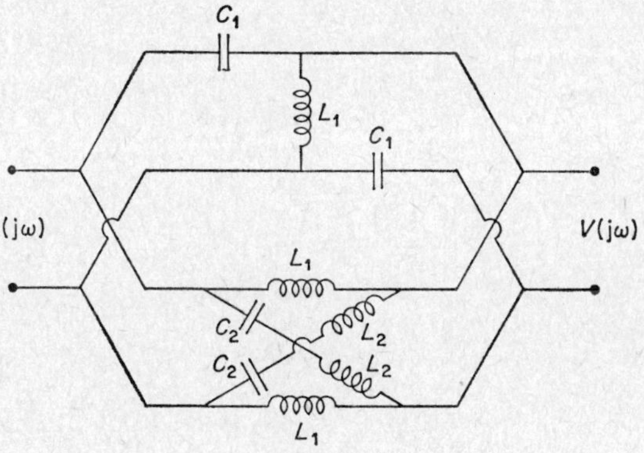

E.7

4. An ideal transformer is used to obtain maximum power transfer from a harmonic generator with internal impedance, R_g, to a load resistance, R_L. Calculate the turns ratio for an ideal transformer.

5. Calculate the total power dissipated in each of the networks E.2 to E.5, and the resonance frequencies as defined by the condition for unity power factor. Assume a harmonic generator, $\mathscr{E}(t) = \mathscr{E}_0 \, e^{j\omega t}$.

6. Calculate the driving point impedance for the circuit E.8.

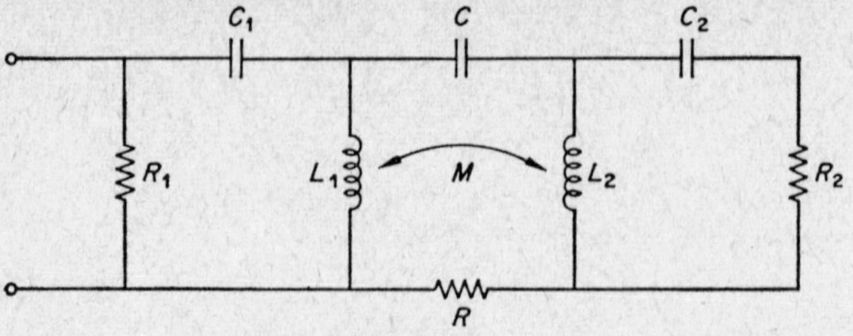

E.8

7. Show that a signal flow graph for the circuit E.9 is that given in E.10 and evaluate V_0 in terms of V_1 algebraically and graphically assuming the input impedance of the amplifier is infinite, and that it has a voltage gain $-K$.

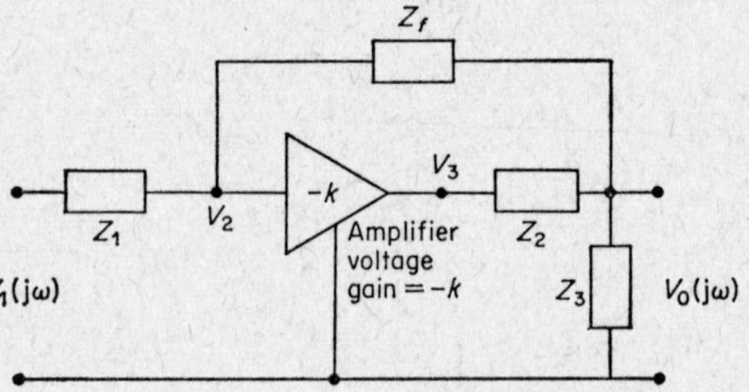

E.9

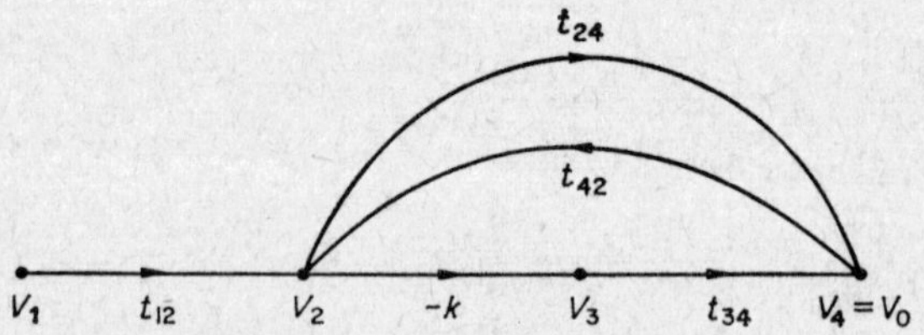

E.10

Chapter 4

Representation of Signals

It is usually necessary to write down an explicit mathematical representation for a signal when carrying through the analysis of an electrical circuit. The physical signal itself is normally visualized as a real function of the time variable which will have a more or less complicated form according to the type of signal involved. The mathematical representation, however, need not be so severely restricted to real time functions and may be constructed, for example, from complex functions which retain the time variable explicitly or, alternatively, from frequency distribution functions or correlation functions which do not in themselves contain the time variable at all.

The choice of a particular representation will be determined by the kind of signal involved and the particular circuit application under consideration. It is useful to distinguish the following classes of signal.

4.1 Classes of Signals

4.1.1 Periodic, Aperiodic Signals

A periodic signal repeats exactly after a definite time interval and may be described by an amplitude function

$$x(t) = x(t + t_0) \tag{4.1.1}$$

The smallest positive value of t_0 which satisfies this equation is called the *period*. If there is no value of t_0 satisfying equation (4.1.1) the signal is said to be *non-periodic* or *aperiodic*. Strictly speaking all signals in actual physical systems are *aperiodic* since they start and stop at finite times, but many signals are of a sufficiently long duration (that is, exist for a long time compared with t_0) for equation (4.1.1) to be an adequate approximation.

4.1.2 Random, Non-random Signals

A signal may also be classified as *random* or *non-random* (deterministic). The form of a random signal cannot be predicted at a time in the future whereas a non-random signal can be so predicted. A random signal will be characterized by probability functions or correlation functions. If these functions are independent of the particular time at which they are evaluated they describe a *stationary* or *time invariant* random process. We shall only consider such stationary random processes since they have a tractable mathematical formulation and correspond to the most common physical situations in which the process does not change appreciably during the time of observation. Thus, the thermal noise generated in a resistance of R ohms at a temperature of T K depends upon the values of R and T, but under normal conditions these are constant and the random process is stationary. The process will, however, be non-stationary if, for example, the temperature changes with time.

Random signals are random functions of the time variable. In order to describe the properties of such a signal we may either measure the signal at successive intervals of time (which in the limit becomes a continuous function of the time), or, alternatively, carry out a simultaneous set of measurements on a large number of equivalent signals. Thus we may either measure the noise voltage of a given resistor as a function of the time or measure the noise voltages generated by a set of identical resistors at a given instant in time. When the measurements are carried out in time sequence the resulting function is called a *sample function* of the random signal. The alternative set of equivalent signals constitutes an ensemble.

The time average of a particular property of a signal is not necessarily equal to the ensemble average. However, there is an important class of random signals for which each individual sample function eventually takes on all the values which occur for any other sample function. Such a random signal is said to be Ergodic and in these circumstances all types of ensemble average are equal to the corresponding time average.

4.1.3 Energy, Power Signals

Signals may be classified in terms of energy or power functions which are related to the square of the amplitude function $x(t)$. If

$$E \equiv \int_{-\infty}^{\infty} x^2(t)\, \mathrm{d}t \leqslant |M| \tag{4.1.2}$$

where M is a finite quantity, the signal may be regarded as including a finite amount of energy and is identified as an *Energy Signal.* This description should rigorously be applied to all actual physical signals since these are necessarily of finite duration. However, the mathematical representation chosen for a signal may not satisfy the inequality of equation (4.1.2). This situation arises in a number of important cases. For example, a simple illustration is provided by an infinite sinusoidal signal. In these circumstances, if the *mean* value of the energy function remains finite, we may identify the signal as a *power signal*, for which

$$\langle P \rangle = \operatorname*{Lim}_{T_0 \to \infty} \frac{1}{2T_0} \int_{-T_0}^{T_0} x^2(t)\, \mathrm{d}t \leqslant |M| \tag{4.1.3}$$

where again M is a finite quantity. Evidently $\langle P \rangle$ is zero for an energy signal.

4.1.4 Singularity Functions

There is an important class of idealized signal functions known as singularity functions. These functions are not strictly realizable in physical systems but they form significant limiting cases of actual signals which occur, for example, in pulse and switching circuits, and their mathematical properties enable circuit analysis to be carried through more easily than would be possible if an exact representation for the physical signal was attempted.

The *unit impulse* singularity function or *Dirac delta* function may be

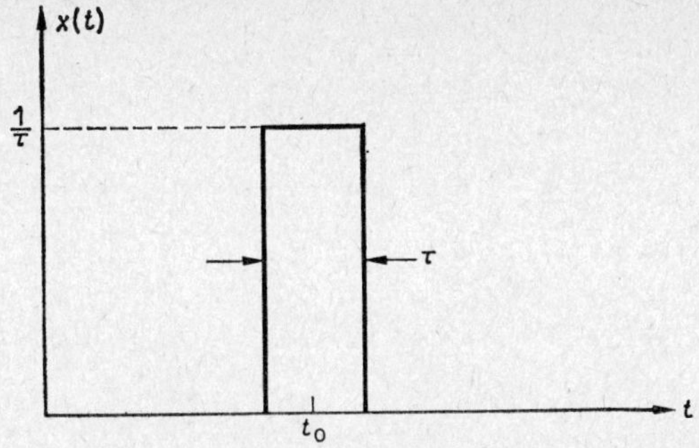

FIG. 4.1 A graphical representation for the unit impulse function.

visualized as the limiting case of a rectangular pulse having a time duration τ and height $1/\tau$, see Fig. 4.1. As τ tends to zero the pulse becomes taller and narrower but the area remains finite and equal to unity. The impulse function, denoted by $\delta(t - t_0)$ is therefore defined to have the following properties

$$\delta(t - t_0) = 0 \quad \text{for} \quad t \neq t_0 \tag{4.1.4}$$

$$\int_{-\infty}^{\infty} \delta(t - t_0)\, dt = 1 \tag{4.1.5}$$

Evidently

$$\int_{-\infty}^{\infty} f(t)\, \delta(t - t_0)\, dt = f(t_0) \tag{4.1.6}$$

Note that equation (4.1.5) implies that either $\delta(t)$ has the dimensions t^{-1} or that the right-hand side has the dimension t.

The impulse function is also required to be an even function of its argument, so that

$$\delta(t - t_0) = \delta(t_0 - t) \tag{4.1.7}$$

$$\int_{-\infty}^{t_0} \delta(t - t_0)\, dt = \int_{t_0}^{\infty} \delta(t - t_0)\, dt = \tfrac{1}{2} \tag{4.1.8}$$

$$\int_{-\infty}^{t} \delta(t - t_0)\, dt \equiv u(t - t_0) = \begin{cases} 0 & \text{if} \quad t < t_0 \\ \frac{1}{2} & \text{if} \quad t = t_0 \\ 1 & \text{if} \quad t > t_0 \end{cases} \tag{4.1.9}$$

Equation (4.1.9) defines another singularity function called the *unit step function*. This function is zero for $t < t_0$ and unity thereafter. Evidently the unit impulse may be regarded as the derivative of the unit step function. In addition the integral of the unit step function generates the *unit ramp function* $r(t)$.

$$r(t - t_0) \equiv \int_0^t u(t - t_0)\, dt = \begin{cases} t & \text{if} \quad t > t_0 \\ 0 & \text{if} \quad t < t_0 \end{cases} \tag{4.1.10}$$

Again, the unit step function may be regarded as the derivative of the unit ramp function.

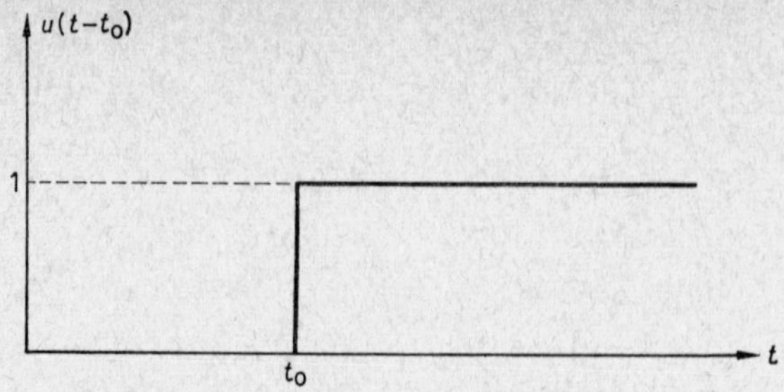

FIG. 4.2 A graphical representation for the unit step function.

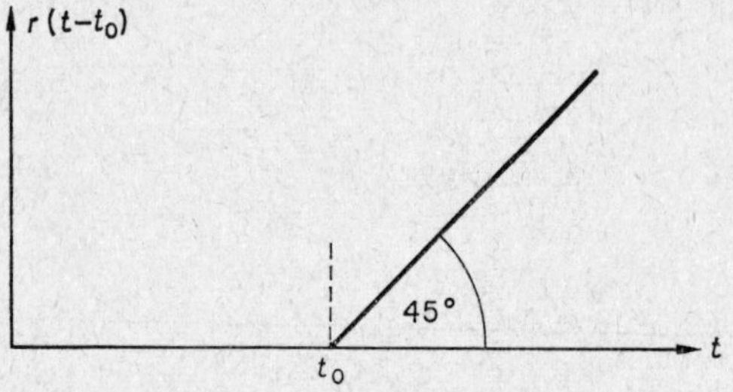

FIG. 4.3 A graphical representation for the unit ramp function.

A *unit doublet function* may also be defined by specifying a set of conditions similar to equations (4.1.4)–(4.1.6). It is denoted by $\delta'(t - t_0)$ and has the following properties:

$$\delta'(t - t_0) = 0 \quad \text{for} \quad t \neq t_0 \tag{4.1.11}$$

$$\int_{-\infty}^{\infty} \delta'(t - t_0)\, \mathrm{d}t = 0 \tag{4.1.12}$$

$$\int_{-\infty}^{\infty} f(t)\, \delta'(t - t_0)\, \mathrm{d}t = -f'(t_0) \tag{4.1.13}$$

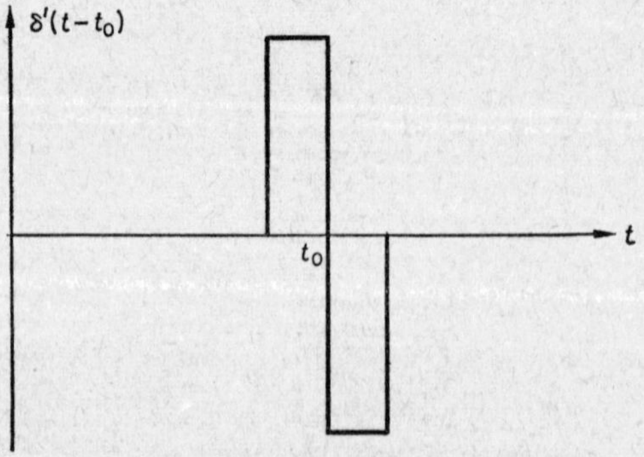

FIG. 4.4 A graphical representation for the unit doublet function.

The unit doublet function may be regarded as the derivative of the unit impulse function.

The singularity functions described above have been based on a visual model for the delta function as the limiting case of a rectangular pulse. Such a model function is essentially discontinuous and does not possess derivatives in the usual sense. It is, therefore, sometimes more convenient to use an alternative representation in terms of the appropriate limit of a continuous and differentiable function. We require a function $f(t - t_0)$ which can be made as sharply peaked as necessary in the vicinity of $t - t_0$, and which has the property that

$$\int_{-\infty}^{\infty} f(t - t_0)\,dt = 1 \tag{4.1.14}$$

Examples of suitable functions are

(a) The Gaussian pulse function, with $a > 0$

$$\delta(t - t_0) = \lim_{a\to\infty} \frac{a}{\sqrt{\pi}} \exp\{-a^2(t - t_0)^2\} \tag{4.1.15}$$

(b) The Sine function $\omega > 0$

$$\delta(t - t_0) = \lim_{\omega\to\infty} \frac{\sin \omega(t - t_0)}{\pi(t - t_0)} \tag{4.1.16}$$

(c) The Lorentz function

$$\delta(t - t_0) = \lim_{b\to\infty} \frac{(b/\pi)}{(t - t_0)^2 + b^2} \tag{4.1.17}$$

(d) The Fourier Transform representation is also of considerable importance,

$$\int_{-\infty}^{\infty} \exp\{j(\omega_0 - \omega)t\}\,dt = 2\pi\,\delta(\omega - \omega_0) \tag{4.1.18}$$

(c) Note also

$$\int \exp(jk_x x)\,dk_x = 2\pi\,\delta(x) \tag{4.1.19}$$

$$\int e^{j\mathbf{k}.\mathbf{r}}\,d^3\mathbf{k} = (2\pi)^3\,\delta(\mathbf{r}) \tag{4.1.20}$$

$$\nabla^2(1/r) = -4\pi\,\delta(r) \tag{4.1.21}$$

4.2 Representation by means of Fourier Series

It is often convenient to represent a complicated periodic signal $x(t)$ in terms of a linear combination of harmonic time functions. Such a representation enables steady state harmonic analysis techniques to be extended to periodic signals of a more complicated form since the response of a linear circuit to such a signal will be the sum of the responses appropriate to each individual sinusoidal component (see Chapter 5.) The choice of sinusoidal basis function for this type of representation is a convenience and not a fundamental necessity but it is the choice most frequently made in practice and gives rise to the Fourier Series representation. However, in order to emphasize the point that the harmonic representation is only one particular form, we shall first develop the theory in terms of a more general set of functions and then specialize to the Fourier Series at a later stage.

4.2.1 Orthogonal Function Representation

Let a set of basis functions be denoted by

$$\phi_{-N}(t), \ldots \phi_{-1}(t), \phi_0(t), \phi_1(t), \ldots \phi_N(t) \tag{4.2.1}$$

where N extends to infinity, and $\phi(t)$ may be a complex function of the time variable. We write

$$x(t) = \sum_{n=-\infty}^{\infty} a_n \phi_n(t) \tag{4.2.2}$$

and a_n may also be complex. The requirement that $x(t)$ be a real function of time is satisfied if

$$a_{-n} \phi_{-n}(t) = a_n^* \phi_n^*(t) \tag{4.2.3}$$

For all the functions we shall consider, we shall have

$$\phi_{-n}(t) = \phi_n^*(t) \tag{4.2.4}$$

and so we require that

$$a_{-n} = a_n^* \tag{4.2.5}$$

In practice it is desirable to be able to determine any coefficient a_n, in the expansion of equation (4.2.2), without reference to any other coefficient in the series. This feature is achieved if the basis functions are orthogonal over the fundamental time interval, t_0, characteristic of the representation. That is, we require

$$\int_{t_1}^{t_1+t_0} \phi_n(t) \phi_k^*(t)\, \mathrm{d}t = \begin{Bmatrix} 0 & \text{if} \quad n \neq k \\ \lambda_k & \text{if} \quad n = k \end{Bmatrix} = \lambda_k \delta_{kn} \tag{4.2.6}$$

where λ_k is a real constant and δ_{kn} is the Kronecker δ symbol which is equal to unity if $n = k$ and is zero otherwise. A particular coefficient a_j may now be obtained by multiplying $x(t)$ by $\phi_j^*(t)$ and integrating

$$\int_{t_1}^{t_1+t_0} x(t) \phi_j^*(t)\, \mathrm{d}t = \int_{t_1}^{t_1+t_0} \{\Sigma a_n \phi_n(t)\} \phi_j^*(t)\, \mathrm{d}t = a_j \lambda_j \tag{4.2.7}$$

Hence

$$a_j = \frac{1}{\lambda_j} \int_{t_1}^{t_1+t_0} x(t) \phi_j^*(t)\, \mathrm{d}t \tag{4.2.8}$$

or alternatively

$$a_j = \frac{\displaystyle\int_{t_1}^{t_1+t_0} x(t) \phi_j^*(t)\, \mathrm{d}t}{\displaystyle\int_{t_1}^{t_1+t_0} \phi_j(t) \phi^*(t)\, \mathrm{d}t} \tag{4.2.9}$$

4.2.2 Finite Sum Representation

If we can make use of an infinite number of terms in the summation of equation (4.2.2) we may expect to obtain an exact representation for $x(t)$. It is not usually possible to do this, however, and so the expression resulting from a finite (or partial) sum is an approximation to $x(t)$. Consider the partial sum

$$s_m(t) = \sum_{-m}^{m} a_n \phi_n(t) \tag{4.2.10}$$

We wish to choose a set of coefficients a_n to give the closest approximation of $s_m(t)$ to $x(t)$. A measure of the closeness of the approximation may be derived from the error function

$$\varepsilon(t) = x(t) - s_m(t) \tag{4.2.11}$$

The specific requirement is that the mean squared error averaged over the fundamental range should be a minimum. This is not the only possible choice for error determination but it is widely accepted as an adequate measure in circuit analysis.

We require a minimum value for the function

$$\langle \varepsilon^2(t) \rangle = \frac{1}{t_0} \int_{t_1}^{t_1+t_0} \{x(t) - s_m(t)\}^2 \, dt = \frac{1}{t_0} \int_{t_1}^{t_1+t_0} \{x(t) - \sum_{-m}^{m} a_n \phi_n(t)\}^2 \, dt \tag{4.2.12}$$

Expanding the bracket and interchanging the order of integration and summation gives

$$\langle \varepsilon^2(t) \rangle = \frac{1}{t_0} \int_{t_1}^{t_1+t_0} x^2(t) \, dt - \frac{2}{t_0} \sum_{-m}^{m} a_n \int_{t_1}^{t_1+t_0} \phi_n(t) \, x(t) \, dt$$

$$+ 1 \sum_{n=-m}^{m} \sum_{k=-m}^{m} a_n a_k \int_{t_1}^{t_1+t_0} \phi_n(t) \, \phi_k(t) \, dt \tag{4.2.13}$$

In this equation

$$\frac{1}{t_0} \int_{t_1}^{t_1+t_0} x^2(t) \, dt = \text{a constant} = C \tag{4.2.14}$$

and from equations (4.2.4) and (4.2.6)

$$\sum_{n=-m}^{m} \sum_{k=-m}^{m} a_n a_k \int_{t_1}^{t_1+t_0} \phi_n(t) \, \phi_k(t) \, dt = \sum_{-m}^{m} a_n a_{-n} \lambda_n = \sum_{-m}^{m} |a_n|^2 \lambda_n \tag{4.2.15}$$

We may write $x(t)$ in the exact representation using a set of coefficients labelled $a_n(\infty)$ which refer to the expansion of equation (4.2.2) where an infinite summation is used, and substitute this expression in the second term on the right-hand side of equation (4.2.13).

Hence in equation (4.2.13)

$$\sum_{-m}^{m} a_n \int_{t_1}^{t_1+t_0} \phi_n(t) \, x(t) \, dt = \sum_{-m}^{m} a_n a_{-n}(\infty) . \lambda_n$$

which is also
$$= \sum_{-m}^{m} a_n a^*(\infty) . \lambda_n \tag{4.2.16}$$

and also
$$= \sum_{-m}^{m} a^* a_n(\infty) . \lambda_n$$

The complete equation (4.2.13) may now be rewritten as

$$\langle \varepsilon^2(t) \rangle = C - \frac{2}{t_0} . \frac{1}{2} \sum_{-m}^{m} \{a_n a^*(\infty) + a^* a_n(\infty)\} \lambda_n + \frac{1}{t_0} \sum_{-m}^{m} |a_n|^2 \lambda_n \tag{4.2.17}$$

Add and subtract from the right-hand side of this equation a term

$$\frac{1}{t_0} \sum_{-m}^{m} |a_n(\infty)|^2 \lambda_n$$

so that,

$$\langle \varepsilon^2(t) \rangle = C - \frac{1}{t_0} \sum_{-m}^{m} |a_n(\infty)|^2 \lambda_n + \frac{1}{t_0} \sum_{-m}^{m} \{a_n - a_n(\infty)\}\{a_n^* - a_n^*(\infty)\} \lambda_n$$

$$= C - \frac{1}{t_0} \sum_{-m}^{m} |a_n(\infty)|^2 \lambda_n + \frac{1}{t_0} \sum_{-m}^{m} |(a_n - a_n(\infty))|^2 \lambda_n \tag{4.2.18}$$

The coefficients a_n referring to the partial sum occur only in the third term on the right-hand side of this equation and this particular term can never be negative. Hence $\langle \varepsilon^2(t) \rangle$ will have its minimum value when the a_n are chosen so that

$$a_n = a_n(\infty) \tag{4.2.19}$$

Hence, the closest approximation of $s_m(t)$ to $x(t)$, when measured by the mean squared error criterion, is achieved by using coefficients in the partial sum equal to those which are calculated as for the infinite sum. It is a great advantage to be able to use the same coefficients in both the partial sum and infinite sum since this means that terms can be added to the partial sum without modifying those which have been determined previously.

The closeness of the approximation which has been achieved using a given value for m may be judged by comparing the mean energy in the error signal with that in the actual signal.

$$\frac{\langle \text{Error signal energy} \rangle}{\langle \text{Signal energy} \rangle} = \frac{\langle \varepsilon^2(t) \rangle}{\dfrac{1}{t_0} \displaystyle\int_{t_1}^{t_1+t_0} x^2(t)\, dt} = \frac{C - \dfrac{1}{t_0} \sum_{-m}^{m} |a_n(\infty)|^2 \lambda_n}{C}$$

$$= 1 - \frac{\sum_{-m}^{m} |a_n(\infty)|^2 \lambda_n}{E} \tag{4.2.20}$$

where

$$E = C \times t_0 = \int_{t_1}^{t_1+t_0} x^2(t)\,\mathrm{d}t$$

is the total signal energy over the fundamental range.

4.2.3 Fourier Series

The particular choice of basis functions characteristic of the Fourier Series representation is a set of sinusoids whose arguments are integer multiples of a fundamental. The time function reproduced by such a series is periodic with a period equal to that of the fundamental term in the expansion. For the Fourier Series representation, equation (4.2.2) becomes

$$x(t) = \sum_{n=-\infty}^{\infty} a_n \exp(\mathrm{j}n\omega_0 t) \tag{4.2.21}$$

and

$$a_n = a^*_{-n} = \frac{\omega_0}{2\pi}\int_{t_1}^{t_1+2\pi/\omega_0} x(t)\exp(-\mathrm{j}n\omega_0 t)\,\mathrm{d}t \tag{4.2.22}$$

The validity of this series expansion for a given $x(t)$ is usually adequately covered for physical systems by the Dirichlet theorem.† This theorem states that if $x(t)$ is a bounded function with period $2\pi/\omega_0$, which in any one period has a finite number of maxima and minima and a finite number of discontinuities, then the Fourier Series converges to $x(t)$ at all points where $x(t)$ is continuous and to the average value (that is, $\frac{1}{2}\{x(t_+) + x(t_-)\}$) at the points where $x(t)$ is discontinuous. Thus even if, in a given physical situation, $x(t)$ consists of a set of discontinuous arcs, it can still be represented by a Fourier Series. This covers most physical cases of interest. The Fourier coefficients for the expansion of such a discontinuous function are obtained by breaking

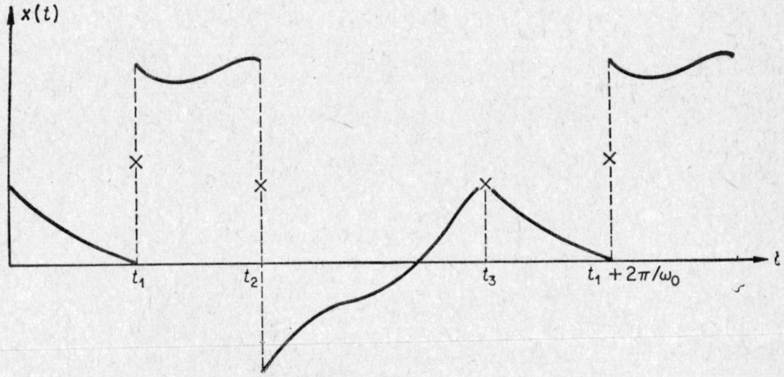

Fig. 4.5 A periodic function with three discontinuities within a period.

† See, for example, Sokolnikoff, I. S. and Redheffer, R. M. *Mathematics of Physics and Modern Engineering*, McGraw-Hill, 1958.

up the fundamental period into segments bounded by the discontinuities. For the curve shown in Fig. 4.5

$$a_n = \frac{\omega_0}{2\pi}\left\{\int_{t_1}^{t_2} \exp(-jn\omega_0 t)\, x(t)\, dt + \int_{t_2}^{t_3} \exp(-jn\omega_0 t)\, x(t)\, dt \right.$$
$$\left. + \int_{t_3}^{t_1+2\pi/\omega_0} \exp(-jn\omega_0 t)\, x(t)\, dt\right\} \tag{4.2.23}$$

The Fourier Series representation converges to the function $x(t)$ over the continuous curves shown in Fig. 4.5 and to the mean values, indicated by crosses, at the points, t_1, t_2, t_3, $t_1 + 2\pi/\omega_0$, etc.

4.2.4 Examples of Orthogonal Function Expansions

Consider the Fourier Series expansion of the saw-tooth waveform shown in Fig. 4.6

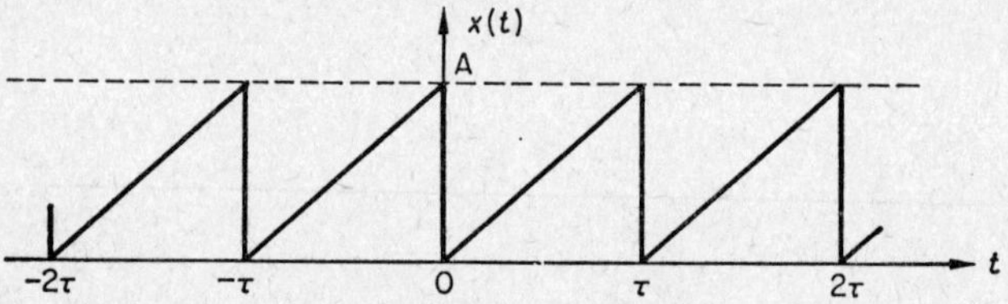

Fig. 4.6 A saw-tooth waveform.

For this wave form

$$x(t) = \frac{A}{\tau} t \quad \text{for} \quad 0 < t \leqslant \tau$$

$$\tau = \frac{2\pi}{\omega_0} \tag{4.2.24}$$

From Equation (4.2.22)

$$a_0 = \frac{\omega_0}{2\pi}\int_0^{2\pi/\omega_0} \frac{A}{\tau} t\, dt = \frac{A}{2} \tag{4.2.25}$$

and a general coefficient

$$a_n = \frac{\omega_0}{2\pi}\int_0^{2\pi/\omega_0} \frac{A}{\tau} t \exp(-jn\omega_0 t)\, dt \tag{4.2.26}$$

$$a_n = \left(\frac{\omega_0}{2\pi}\right)^2 \frac{A}{(-jn\omega_0)}\int_0^{2\pi/\omega_0} t d\{\exp(-jn\omega_0 t)\} = -\frac{A}{j2\pi n} \tag{4.2.27}$$

Taking pairs of terms a_n, a_{-n} together, we obtain

$$x(t) = \frac{A}{2} - \sum_{n=+1}^{\infty} \frac{A}{\pi n} \sin n\omega_0 t \tag{4.2.28}$$

Note that all the sine terms are zero when $t = 2\pi n/\omega_0$, where $x(t)$ has discontinuities. In agreement with Dirichlet's theorem, the Fourier expansion has the mean value $A/2$ at these points.

Moving the time origin of the representation simply modifies the phase angle of each coefficient a_n in the Fourier expansion by an amount proportional to n. Thus, from equation (4.2.21)

$$x(t') = \sum_{n=-\infty}^{\infty} a_n \exp(jn\omega_0 t') \equiv x(t + T_0)$$

$$= \sum_{n=-\infty}^{\infty} a_n \exp(jn\omega_0 T_0) \exp(jn\omega_0 t) \quad (4.2.29)$$

so that $x(t')$ is obtained immediately from $x(t)$ by identifying

$$a_n' = a_n \exp(jn\omega_0 T_0),$$

and $t' = t$. Consider a change of time origin for the saw-tooth waveform in Fig. 4.6, to give the waveform shown in Fig. 4.7.

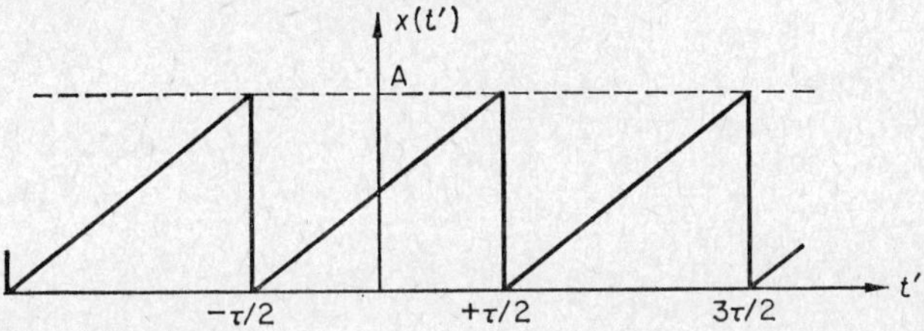

FIG. 4.7 The saw-tooth waveform from Fig. 4.6 with time origin shifted by $\tau/2$.

In this case $T_0 = \tau/2 = \pi/\omega_0$ and $a_n' = a_n\, e^{j\pi n} = (-1)^n a_n$. Hence

$$x(t') = \frac{A}{2} - \sum_{n=1}^{\infty} (-1)^n \frac{A}{\pi n} \sin n\omega_0 t' \quad (4.2.30)$$

As an alternative representation for the saw-tooth waveform shown in Fig. 4.6, consider the expansion in terms of a set of orthogonal rectangular basis functions. An orthogonal set of these functions may be constructed if any given function has one half the period of its immediate predecessor. From equation (4.2.6)

$$\lambda_k = \tau$$

and so,

$$a_0 = A/2\ a_n = \frac{1}{\tau}\int_0^{2\pi/\omega_0} \phi_n(t)\, x(t)\, dt \quad (4.2.31)$$

Hence, for the rectangular wave with period T, such that $nT = \tau$

$$a_n = \frac{1}{\tau}\cdot\frac{A}{\tau}\left[\left[\int_0^T (-1)t\, dt + \int_{T/2}^{T} t\, dt + \int_T^{3T/2} (-1)t\, dt \cdots \int_{(2n-1)T/2}^{nT} t\, dt\right]\right. \quad (4.2.32)$$

$$= \frac{A}{\tau^2}\left[\frac{1}{2}\cdot 2\left(\frac{T}{2}\right)^2 \{-1 + 2^2 - 3^2 + \cdots -(2n-1)^2 + \tfrac{1}{2}(2n)^2\}\right]$$

$$= \frac{A}{\tau^2}\left[\left(\frac{T}{2}\right)^2 \{-1 - 2 - 3 \cdots -(2n-1) + \tfrac{1}{2}(2n)^2\}\right] \quad (4.2.33)$$

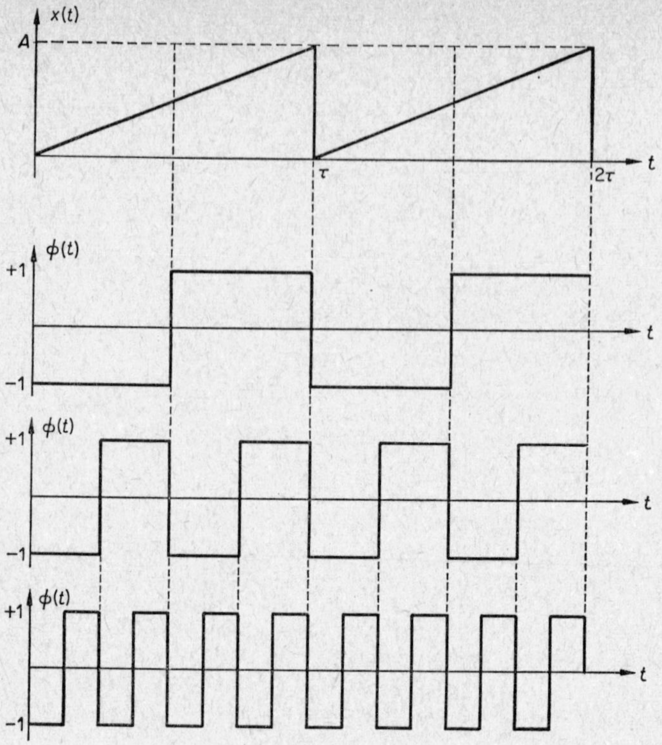

FIG. 4.8

that is

$$a_n = \frac{A}{4n} \tag{4.2.34}$$

$$x(t) = \frac{A}{2} + \sum_{n=1}^{\infty} \frac{A}{4n} \phi\left(\frac{\tau}{2^{n-1}}\right) \tag{4.2.35}$$

4.3 Fourier and Laplace Transforms

Periodic time functions which satisfy the Dirichlet conditions can, in principle, be represented by a Fourier Series. However, many of the signals which occur in electrical circuits are markedly transient and non-periodic. It is therefore necessary to seek a representation which can, if possible, include both periodic and non-periodic functions. One such representation can be developed by extending the previous discussion which used sinusoids as basis functions.

4.3.1 Fourier Transforms

Consider the Fourier Series representations for a periodic function. Rewriting equations (4.2.21) and (4.2.22) once more

$$x(t) = \sum_{n=-\infty}^{\infty} a_n \exp(jn\omega_0 t) \tag{4.3.1}$$

and

$$a_n = a_{-n}^* = \frac{\omega_0}{2\pi} \int_{t_1}^{t_1+2\pi/\omega_0} x(t) \exp(-jn\omega_0 t)\,dt \tag{4.3.2}$$

The fundamental period is $t_0 = 2\pi/\omega_0$ and so it is evident that a non-periodic function may formally be included in the representation if the range of t_0 can be extended to infinity. As $t_0 \to \infty$, then $\omega_0 \to 0$ and so the factor $n\omega_0$ becomes a continuous variable which can be written as $\omega(\equiv n\omega_0)$. Moreover, the spacing between the adjacent frequency components

$$\Delta\omega = \omega_{n+1} - \omega_n = \omega_0 \to d\omega$$

as $t_0 \to \infty$, and so equation (4.3.1) may be written

$$x(t) = \sum_{-\infty}^{\infty} \left\{ \frac{\Delta\omega}{2\pi} \int_{-\infty}^{\infty} x(t)\, e^{-j\omega t}\,dt \right\} \exp(jn\omega_0 t) \tag{4.3.3}$$

Proceeding to the limit of the summation

$$x(t) = \frac{1}{2\pi} \int_{-\infty}^{\infty} \left\{ \int_{-\infty}^{\infty} x(t)\, e^{-j\omega t}\,dt \right\} e^{j\omega t}\,d\omega \tag{4.3.4}$$

that is,

$$x(t) = \frac{1}{2\pi} \int_{-\infty}^{\infty} X(j\omega)\, e^{j\omega t}\,d\omega \tag{4.3.5}$$

where

$$X(j\omega) = \int_{-\infty}^{\infty} x(t)\, e^{-j\omega t}\,dt \tag{4.3.6}$$

Equations (4.3.5) and (4.3.6) constitute a *Fourier Transform Pair.* It should be noted that in these equations the factor $1/2\pi$ has arbitrarily been assigned to the integral representation for $x(t)$ given in equation (4.3.5). This is a widely accepted, but not a universal, convention and care must be exercised when using tables of Fourier Transforms.

The function $X(j\omega)$ may be regarded as a representation of the signal $x(t)$. Evidently $X(j\omega)$ does not contain the time variable explicitly and the complete representation is in terms of a continuum of sinusoids with infinitesimal amplitudes $(1/2\pi)X(j\omega)\,d\omega$ in the range ω to $\omega + d\omega$.† Since $d\omega$ was derived from $\Delta(n\omega_0)$ it is a uniform frequency interval. The finite quantity $X(j\omega)$ is proportional therefore to the complete amplitude function

† An exception occurs when there are discrete harmonic components for $x(t)$. See p. 157.

$(1/2\pi)X(j\omega)\,d\omega$ and itself gives the correct frequency distribution of the amplitude function.

The Fourier Transform of a function $x(t)$ is usually written in the abbreviated notation

$$F\{x(t)\} = X(j\omega) \tag{4.3.7}$$

This function is of considerable importance in circuit analysis. Its significance derives chiefly from the property that the Fourier Transform of the derivative of a time function is simply related to the Transform of the original time function. Thus

$$F\left\{\frac{d^n}{dt^n}x(t)\right\} = (j\omega)^n X(j\omega) \tag{4.3.8}$$

where $X(j\omega) = F\{x(t)\}$. This means that complicated differential equations can be reduced to an algebraic form for which there are usually adequate techniques for solution.

4.3.2 The Dirichlet Condition

Not all time functions can be represented by means of the Fourier Transform. When such a representation is possible, however, there is a unique correspondence between the time function and its Fourier Transform. There is only one Fourier Transform appropriate to a given time function and conversely only a single time function appropriate to a given Fourier Transform.

Since we require $X(j\omega)$ to be finite at all frequencies we may anticipate from the value for $X(j\omega)$ at $\omega = 0$ that $\int_{-\infty}^{\infty} x(t)\,dt$ is required to be finite. However, this is not of itself a sufficiently stringent condition to ensure that the complete Fourier Transform exists since for $\omega \neq 0$ the term $e^{-j\omega t}$ oscillates between ± 1. The Dirichlet condition states that the Fourier Transform exists if $x(t)$ has only a finite number of maxima and minima and only a finite number of discontinuities in any finite time interval and if

$$\int_{-\infty}^{\infty} |x(t)|\,dt \leqslant |M| \tag{4.3.9}$$

where M is a finite quantity.

These conditions are fulfilled by practically all energy signals but they exclude many important ideal signal types. Thus pure sinusoidal functions do not possess a straightforward Fourier Transform. This difficulty may be overcome, however, if the Transform function $X(j\omega)$ is allowed to include singularity functions. For example, if $x(t) = A$, where A is a constant, we may write by inspection

$$x(t) = A\int_{-\infty}^{\infty} \delta(\omega)\, e^{j\omega t}\, d\omega \tag{4.3.10}$$

and so we can identify the Fourier Transform of a constant term as

$$F\{x(t) = A\} = 2\pi A\,\delta\omega \tag{4.3.11}$$

If

$$x(t) = A \sin \omega_0 t = \frac{A}{2\mathrm{j}}\{\exp(\mathrm{j}\omega_0 t) - \exp(-\mathrm{j}\omega_0 t)\} \tag{4.3.12}$$

we may write by inspection

$$x(t) = \frac{A}{2j}\int_{-\infty}^{\infty} \{e^{j\omega t}\,\delta(\omega - \omega_0) - e^{j\omega t}\,\delta(\omega + \omega_0)\,d\omega\} \tag{4.3.13}$$

so that

$$F\{x(t) = A \sin \omega_0 t\} = \frac{\pi A}{\mathrm{j}}\{\delta(\omega - \omega_0) - \delta(\omega + \omega_0)\} \tag{4.3.14}$$

In particular, note that the Fourier Transform of the complex time function $x(t) = \exp(\mathrm{j}\omega_0 t)$ is

$$F\{x(t) = \exp(\mathrm{j}\omega_0 t)\} = 2\pi\,\delta(\omega - \omega_0) \tag{4.3.15}$$

Hence a *periodic* function, which may be also represented by means of a Fourier Series, can be described by the relation

$$F\left\{x(t) = \sum_{n=-\infty}^{\infty} a_n \exp(jn\omega_0 t)\right\} = \sum_{n=-\infty}^{\infty} a_n . 2\pi\,\delta(\omega - n\omega_0) \tag{4.3.16}$$

Since a_n can be written in terms of an integral over the fundamental period, this equation may be put in the form

$$F\left\{x(t) = \sum_{n=-\infty}^{\infty} a_n \exp(jn\omega_0 t)\right\}$$

$$= \omega_0 \sum_{n=-\infty}^{\infty} \left\{\int_{-\pi/\omega_0}^{\pi/\omega_0} x(t) \exp(-\mathrm{j}n\omega_0 t)\,\mathrm{d}t\right\} \delta(\omega - n\omega_0) \tag{4.3.17}$$

and the frequency distribution of the Fourier Transform is a sequence of pulses at harmonics of the fundamental with amplitudes proportional to the Fourier Series coefficients.

If $x(t)$ is itself a unit impulse train, then

$$\int_{-\pi/\omega_0}^{\pi/\omega_0} \delta(t) \exp(-\mathrm{j}n\omega_0 t)\,\mathrm{d}t = 1$$

and so

$$F\left\{x(t) = \sum_{n=-\infty}^{\infty} \delta(t - nt_0)\right\} = \omega_0 \sum_{n=-\infty}^{\infty} \delta(\omega - n\omega_0) \tag{4.3.18}$$

The unit impulse train in the time domain therefore has a Fourier Transform represented by a constant amplitude impulse train in the frequency domain.

4.3.3 Laplace Transforms

Although it is possible to extend the Fourier Transform method by making use of singularity functions this is not necessarily a convenient way of handling problems in circuit analysis where a representation which is continuous and differentiable may be preferable. The Laplace Transform which may be regarded as an extension of the Fourier Transform into the complex frequency domain provides a representation which avoids these difficulties and at the same time gives a powerful method for computing circuit response with the initial conditions included in the solution.

The Laplace Transform of the (real)time function $x(t)$ existing for positive times only is defined in terms of the complex frequency variable $s = q + j\omega$ to be

$$L\{x(t)\} = X(s) = \int_{0^-}^{\infty} x(t)\, e^{-st}\, dt \qquad (4.3.19)$$

and the inverse transform

$$x(t)\, u(t) = \frac{1}{2\pi j} \int_{q-j\infty}^{q+j\infty} X(s)\, e^{st}\, ds \qquad (4.3.20)$$

where $u(t)$ is the unit step function.

Note that the integration in equation (4.3.19) extends only from zero to plus infinity and the inverse transform of equation (4.3.20) gives rise to a positive-time function only. The lower limit in equation (4.3.19) is written 0^- so that discontinuities and impulses at $t = 0$ are included. This convention simplifies the inclusion of initial conditions.

The use of the Laplace Transform representation enables a linear differential equation with constant coefficients to be reduced to an algebraic form and is therefore of considerable importance for linear circuit analysis. The following relation is used in order to being about this reduction:

$$L\left\{\frac{d^n}{dt^n}x(t)\right\} = s^n L\{x(t)\} - s^{n-1}x(0) - s^{n-2}x'(0) \cdots x^{(n-1)}(0)$$

The resulting algebraic equation describing the linear circuit will now specifically include the initial conditions.

4.3.4 Existence of the Laplace Transform

The Laplace Transform is comparable to the Fourier Transform of a function $x(t)\, e^{-qt}$ and so the condition for the existence of the Laplace Transform is weaker than that set out in equation (4.3.9). We only require that,† for some real and positive value of q

$$\int_{0^-}^{\infty} |x(t)|\, e^{-qt}\, dt \leqslant |M| \qquad (4.3.21)$$

where M is a finite quantity.

4.3.5 Evaluation of Laplace Transforms

Equation (4.3.20) involves integration in the complex domain whereas the equations for the Fourier Transform pair involve only real integration. However, this is not a serious disadvantage. The usefulness of the Laplace Transform method, and indeed also of the Fourier Transform method in practice, is based on the availability of extensive tables of transforms. A short table of such transforms is given on pages 174–179. It is instructive, however, to

† Wylie, C. R. *Advanced Engineering Mathematics*, McGraw-Hill, 1960.

derive from first principles some of the simpler Laplace Transforms by means of equation (4.3.19) (which requires only real integration), and then to compare these with the corresponding Fourier Transforms

The Laplace Transforms of the singularity functions $\delta(t)$, $u(t)$, $r(t)$ are evidently

$$L\{\delta(t)\} = \int_{0^-}^{\infty} \delta(t)\, e^{-st}\, dt = 1 \tag{4.3.22}$$

$$L\{u(t)\} = \int_{0^-}^{\infty} u(t)\, e^{-st}\, dt = \frac{1}{s} \tag{4.3.23}$$

$$L\{r(t)\} = \int_{0^-}^{\infty} t\, u(t)\, e^{-st}\, dt = \frac{1}{s^2} \tag{4.3.24}$$

The corresponding Fourier Transforms are

$$F\{\delta(t)\} = 1;\ F\{u(t)\} = \pi\,\delta(\omega) + \frac{1}{j\omega};\ F\{r(t)\} = j\pi\,\delta'(\omega) - \frac{1}{\omega^2} \tag{4.3.25}$$

For a sinusoidal function switched on at $t = 0$

$$x(t) = A \sin \omega_0 t\,.\,u(t) \tag{4.3.26}$$

and

$$L\{x(t) = A \sin \omega_0 t\,.\,u(t)\} = \frac{A}{2j}\int_{0^-}^{\infty} u(t)\{\exp(j\omega_0 t) - \exp(-j\omega_0 t)\}\exp(-st)\, dt$$

$$= \frac{A\omega_0}{\omega_0^2 + s^2} \tag{4.3.27}$$

This function should be compared with equation (4.3.14).

Consider a general periodic function which is switched on at $t = 0$, repeating after a fundamental time period t_0. In the first period the function may be represented by $x_1(t)$ which is defined for $0 \leqslant t \leqslant t_0$ and is zero outside this interval. In the second period the function is described by $x_1(t - t_0)$, and the complete signal is

$$x(t) = x_1(t) + x_1(t - t_0) + x_1(t - 2t_0) + \cdots \tag{4.3.28}$$

Let

$$L\{x_1(t)\} = X_1(s)$$

Then

$$L\{x_1(t - t_0)\} = \int_{0^-}^{\infty} x_1(t - t_0)\, e^{-st}\, dt$$

$$= \exp(-st_0)\int_{0^-}^{\infty} x_1(t - t_0)\exp\{-s(t - t_0)\}\, d(t - t_0)$$

$$= \exp(-st_0)\int_{0^-}^{\infty} x(\xi)\, e^{-s\xi}\, d\xi = \exp(-st_0)X_1(s) \tag{4.3.29}$$

and so

$$L\{x(t)\} = X_1(s) + \exp(-st_0)\, X_1(s) + \exp(-2st_0)\, X_1(s) + \cdots \tag{4.3.30}$$

$$= \frac{X_1(s)}{1 - \exp(-st_0)} \equiv \frac{L\{x_1(t)\}}{1 - \exp(-st_0)} \tag{4.3.31}$$

This expression for the Laplace Transform of a general periodic function involves the Transform of the function in the first fundamental period only and should be compared with equation (4.3.17) for the corresponding Fourier Transform.

4.4 Representation in terms of Spectral Density and Correlation Functions

4.4.1 Periodic Signals

The periodic function $x(t)$ in equation (4.2.21) is determined when the coefficients a_n are known. These coefficients therefore provide a representation for $x(t)$.

The use of the mean square error criterion as a measure of the convergence of the Fourier Series representation for periodic signals implies that (compare equation (4.2.14)),

$$\int_{t}^{t_1+t_0} \{x(t)\}^2 \, dt \leqslant |M| \tag{4.4.1}$$

where t_0 is the fundamental period associated with the representation and M is a finite quantity. This equation is characteristic of a power signal since it requires that the energy in a period (that is, the power in the signal) be finite. When equation (4.4.1) is satisfied, the mean square error function $\langle \varepsilon^2(t) \rangle$ tends to zero as the number of terms in the Fourier expansion tends to infinity.

For the infinite series expansion, with $a_n = a_{-n}^*$ and $\phi_n = \phi_{-n}^*$

$$\int_{t_1}^{t_1+t_0} \{x(t)\}^2 \, dt = \int_{t_1}^{t_1+t_0} \left[\sum_{n=-\infty}^{\infty} a_n \phi_n \right]^2 dt = \sum_{n=-\infty}^{\infty} \lambda_n |a_n|^2 \tag{4.4.2}$$

The Fourier series has $\lambda_n = t_0$ so that

$$\frac{1}{t_0} \int_{t_1}^{t_1+t_0} \{x(t)\}^2 \, dt = \sum_{n=-\infty}^{\infty} |a_n|^2 \tag{4.4.3}$$

This result is Parseval's theorem. The average energy in the signal (that is, the signal power) is equal to the sum of the average energies for each individual harmonic component.

4.4.2 Power Spectral Density

Each term in the summation of equation (4.4.3) is associated with one frequency component. We may therefore define a power spectral density function

$$S_x(\omega) = 2\pi \sum_{n=-\infty}^{\infty} |a_n|^2 \, \delta(\omega - n\omega_0) \tag{4.4.4}$$

where the factor 2π is introduced because we are writing $S_x(\omega)$ in terms of the angular frequency $\omega = 2\pi\nu$ in radians per second rather than the frequency, ν in hertz (cycles per second). Evidently equation (4.4.3) for the signal power gives

$$\frac{1}{t_0}\int_{t_1}^{t_1+t_0} \{x(t)\}^2\,\mathrm{d}t = \sum_{n=-\infty}^{\infty} |a_n|^2 = \frac{1}{2\pi}\int_{-\infty}^{\infty} S_x(\omega)\,\mathrm{d}\omega \qquad (4.4.5)$$

The function $S_x(\omega)$ offers an alternative representation for the periodic signal $x(t)$. In this form, however, all information about the phases of the various frequency components has been lost. Two functions $x(t)$ and $x'(t)$ which have Fourier coefficients of the same magnitude but different phases will have the same power spectral density function $S_x(\omega)$, and the same representation. Note that, since $a_n = a^*_{-n}$, $|a_n|^2 = |a_{-n}|^2$, $S_x(\omega)$ is an even function of frequency and from equation (4.4.4) is always positive or zero (that is, $S_x(\omega)$ is a non-negative function). In addition $S_x(\omega)$ is always real.

4.4.3 Energy Spectral Density

A non-periodic signal in an actual physical system belongs to the class of energy signals (see page 144) for which

$$\int_{-\infty}^{\infty} \{x(t)\}^2\,\mathrm{d}t \leqslant |M| \qquad (4.4.6)$$

since it will be switched on and off at finite times.

If $X(\mathrm{j}\omega)$ is the Fourier Transform of $x(t)$, then,

$$\int_{-\infty}^{\infty} \{x(t)\}^2\,\mathrm{d}t = \int_{-\infty}^{\infty} x(t).x(t)\,\mathrm{d}t = \int_{-\infty}^{\infty} x(t)\left[\frac{1}{2\pi}\int_{-\infty}^{\infty} X(\mathrm{j}\omega)\,\mathrm{e}^{\mathrm{j}\omega t}\,\mathrm{d}\omega\right]\mathrm{d}t \qquad (4.4.7)$$

Interchanging the order of integration gives

$$\int_{-\infty}^{\infty} \{x(t)\}^2\,\mathrm{d}t = \frac{1}{2\pi}\int_{-\infty}^{\infty} X(\mathrm{j}\omega)\left[\int_{-\infty}^{\infty} x(t)\,\mathrm{e}^{\mathrm{j}\omega t}\,\mathrm{d}t\right]\mathrm{d}\omega \qquad (4.4.8)$$

However, $x(t)$ is a real function of the time variable, so that

$$\int_{-\infty}^{\infty} x(t)\,\mathrm{e}^{\mathrm{j}\omega t}\,\mathrm{d}t = X(-\mathrm{j}\omega) = X^*(\mathrm{j}\omega) \qquad (4.4.9)$$

and so

$$\int_{-\infty}^{\infty} \{x(t)\}^2\,\mathrm{d}t = \frac{1}{2\pi}\int_{-\infty}^{\infty} |X(\mathrm{j}\omega)|^2\,\mathrm{d}\omega \qquad (4.4.10)$$

This is the analogue of Parseval's theorem for Transform pairs. The total energy in the signal is equal to $1/2\pi$ times the area under the curve representing the square magnitude of the Fourier Transform of the signal.

In this case we may, by analogy with equations (4.4.4) and (4.4.5) define an energy spectral density function

$$S_E(\omega) = |X(\mathrm{j}\omega)|^2 \qquad (4.4.11)$$

4.4.4 Correlation Functions

The argument leading to equation (4.4.10) may be generalized to include the case of two different time functions $x(t), y(t)$. Following through the steps of equation (4.4.7)–(4.4.10), we have

$$\int_{-\infty}^{\infty} x(t)\, y(t)\, \mathrm{d}t = \frac{1}{2\pi}\int_{-\infty}^{\infty} X(\mathrm{j}\omega)\, Y(-\mathrm{j}\omega)\, \mathrm{d}\omega = \frac{1}{2\pi}\int_{-\infty}^{\infty} X(-\mathrm{j}\omega) Y(\mathrm{j}\omega)\, \mathrm{d}\omega \tag{4.4.12}$$

Integrals of the form $\int_{-\infty}^{\infty} \{x(t)\}^2\, \mathrm{d}t$ and $\int_{-\infty}^{\infty} x(t)\, y(t)\, \mathrm{d}t$ lead to the consideration of a class of integrals called correlation functions.

For energy signals, the correlation functions are defined by the following relations:

Auto-correlation Function for an Energy Signal

$$R_x^{(E)}(\tau) = \int_{-\infty}^{\infty} x(t+\tau)\, x(t)\, \mathrm{d}t \tag{4.4.13a}$$

Cross-correlation Function for Two Energy Signals

$$R_{xy}^{(E)}(\tau) = \int_{-\infty}^{\infty} x(t+\tau)\, y(t)\, \mathrm{d}t \tag{4.4.13b}$$

Evidently

$$R_x^{(E)}(0) = \int_{-\infty}^{\infty} \{x(t)\}^2 \mathrm{d}t = E \tag{4.4.13c}$$

$$= \frac{1}{2\pi}\int_{-\infty}^{\infty} |X(\mathrm{j}\omega)|^2\, \mathrm{d}\omega = \frac{1}{2\pi}\int_{-\infty}^{\infty} S_E(\omega)\, \mathrm{d}\omega \tag{4.4.13d}$$

where E is the total energy in the signal as discussed on page 144 and in equation (4.4.13d) it is assumed that the Fourier Transform $X(\mathrm{j}\omega)$ exists.

For power signals, time correlation functions are defined by the relations:

Time Auto-correlation Function for a Power Signal

$$R_x(\tau) = \lim_{T_0\to\infty} \frac{1}{2T_0}\int_{-T_0}^{T_0} x(t+\tau)\, x(t)\, \mathrm{d}t \tag{4.4.14a}$$

Time Cross-correlation Function for Two Power Signals

$$R_{xy}(\tau) = \lim_{T_0\to\infty} \frac{1}{2T_0}\int_{-T_0}^{T_0} x(t+\tau)\, y(t)\, \mathrm{d}t \tag{4.4.14b}$$

For stationary or time invariant processes (see page 143) these correlation functions depend only upon the time interval, τ, and not upon the specific times t and $(t+\tau)$ involved in these relationships. Evidently

$$R_x(0) = \lim_{T_0\to\infty} \frac{1}{2T_0}\int_{-T_0}^{T_0} \{x(t)\}^2\, \mathrm{d}t = \langle P\rangle \tag{4.4.14c}$$

where $\langle P\rangle$ is the power in the signal in accordance with the discussion of page 144.

Note that signals discussed above are not necessarily periodic or deterministic. The *correlation functions* may be formed for signals which represent random processes, as in the case of noise signals. In addition they need not necessarily possess Fourier Transforms.

In equations (4.4.13) and (4.4.14) we have implicitly assumed that $x(t)$, $y(t)$ are real functions such as arise in physical applications.† For such functions, substituting $t' = t + \tau$, yields

$$R_x^{(E)}(\tau) = R_x^{(E)}(-\tau) \quad R_{xy}^{(E)}(\tau) = R_{yx}^{(E)}(-\tau) \tag{4.4.15}$$

$$R_x(\tau) = R_x(-\tau) \qquad R_{xy}(\tau) = R_{yx}(-\tau) \tag{4.4.16}$$

4.4.5 Auto-correlation Function for a Periodic Signal

Consider a periodic signal such as we have considered previously on page 151, which can be represented by a Fourier Series. The auto-correlation function of

$$x(t) = \sum_{n=-\infty}^{\infty} a_n \exp(jn\omega_0 t)$$

is given by

$$R_x(\tau) = \operatorname*{Lim.}_{T_0\to\infty} \frac{1}{2T_0}\int_{-T_0}^{T_0}\left[\sum_{n=-\infty}^{\infty} a_n \exp\{jn\omega_0(t+\tau)\}\right] \times \left[\sum_{m=-\infty}^{\infty} a_m^* \exp(-jm\omega_0 t)\right] \mathrm{d}t \tag{4.4.17}$$

that is,

$$R_x(\tau) = \sum_{n=-\infty}^{\infty} |a_n|^2 \exp(jn\omega_0\tau) \tag{4.4.18}$$

since

$$\operatorname*{Lim.}_{T_0\to\infty} \frac{1}{2T_0}\int_{-T_0}^{T_0} \exp\{j(n-m)\omega_0 t\}\,\mathrm{d}t = \operatorname*{Lim.}_{T_0\to\infty} \frac{\sin(n-m)\omega_0 T_0}{(n-m)\omega_0 T_0} = \begin{cases} 1 \text{ if } n = m \\ 0 \text{ if } n \neq m \end{cases} \tag{4.4.19}$$

† The more general definitions for mathematical functions which may be complex are of the form

$$\int x(t+\tau)\, y^*(t)\,\mathrm{d}t.$$

Hence the auto-correlation function of a periodic signal is itself periodic and the harmonic frequencies coincide with those of the original signal. In this case the Fourier Transform of $R_x(\tau)$ is evidently

$$F\{R_x(\tau)\} = F\left\{ \sum_{n=-\infty}^{\infty} |a_n|^2 \exp\,(jn\omega_0\tau) \right\} \tag{4.4.20}$$

$$= 2\pi \sum_{n=-\infty}^{\infty} |a_n|^2\, \delta(\omega - n\omega_0) \tag{4.2.21}$$

from equation (4.3.16).

4.4.6 Spectral Density

Comparing equation (4.4.21) with equation (4.4.4), we see that, for a periodic function, the power spectral density is the Fourier Transform of the auto-correlation function for the signal

$$F\{R_x(\tau)\} = S_x(\omega) \tag{4.4.22}$$

We therefore extend this reasoning and define a more general power spectral density function, which will now refer also to random processes, by the Fourier Transform relations

$$S_x(\omega) = \int_{-\infty}^{\infty} R_x(\tau)\, e^{-j\omega\tau}\, d\tau \tag{4.4.23}$$

with the inverse transform relation

$$R_x(\tau) = \frac{1}{2\pi} \int_{-\infty}^{\infty} S_x(\omega)\, e^{j\omega\tau}\, d\omega \tag{4.4.24}$$

It follows immediately from equation (4.4.24) that the spectral density function itself integrates to the mean power in the signal

$$R_x(0) = \frac{1}{2\pi} \int_{-\infty}^{\infty} S_x(\omega)\, d\omega = \lim_{T_0 \to \infty} .\frac{1}{2T_0} \int_{-T_0}^{T_0} \{x(t)\}^2\, dt \tag{4.4.25}$$

When two functions $x(t)$, $y(t)$ represent stationary processes the Fourier Transform of $R_{xy}(\tau)$ defines a cross-spectral density $S_{xy}(\omega)$. By analogy with equation (4.4.23)

$$S_{xy}(\omega) = \int_{-\infty}^{\infty} R_{xy}(\tau)\, e^{-j\omega\tau}\, d\tau \tag{4.4.26}$$

and so

$$R_{xy}(\tau) = \frac{1}{2\pi} \int_{-\infty}^{\infty} S_{xy}(\omega)\, e^{-j\omega\tau}\, d\omega \tag{4.4.27}$$

Similar relations may also be defined to give energy spectral density functions. These are the Fourier Transforms of the correlation functions for energy signals which were introduced in equations (4.4.13).

4.4.7 Properties of Spectral Density Functions

The power spectral density function $S_x(\omega)$ is a real, positive and even function of ω for real values of ω.

If $S_x(\omega)$ is real, then $S_x(\omega) = \{S_x(\omega)\}^*$. To prove this property we consider the defining relation for $S_x(\omega)$, together with $R_x(\tau) = R_x(-\tau)$ from equation (4.4.16)

$$S_x(\omega) = \int_{-\infty}^{\infty} R_x(\tau)\, e^{-j\omega\tau}\, d\tau = \int_{-\infty}^{\infty} R_x(-\tau)\, e^{j\omega\tau}\, d\tau$$

therefore

$$S_x(\omega) = \int_{-\infty}^{\infty} R_x(\tau)\, e^{j\omega\tau}\, d\tau = \{S_x(\omega)\}^* \text{ for } \omega \text{ real.} \qquad (4.4.28)$$

In addition, these equations show that $S_x(\omega) = S_x(-\omega)$ and so $S_x(\omega)$ is an even function of ω for ω real.

To show that $S_x(\omega) \geqslant 0$ for real values of ω, consider the output function from an ideal narrow band filter. Let the filter have an amplitude transfer function $|H(j\omega)|$ over the frequency range $\Delta\omega$ centred upon ω and be zero elsewhere. We shall show in equation (5.4.6), that if the power spectral density for the input function is $S_x(\omega)$ and that for the output function is $S_y(\omega)$ then

$$S_y(\omega) = |H(j\omega)|^2\, S_x(\omega) \qquad (4.4.29)$$

Hence

$$\frac{1}{2\pi}\int_{-\infty}^{\infty} S_y(\omega)\, d\omega = \langle y^2 \rangle = \lim_{T_0 \to \infty} \frac{1}{2T_0}\int_{-T_0}^{T_0} \{y(t)\}^2\, dt$$

$$= \frac{\Delta\omega}{2\pi}\, |H(j\omega)|^2\, S_x(\omega) \qquad (4.4.30)$$

But $\langle y^2 \rangle$ and $\dfrac{\Delta\omega}{2\pi}\, |H(j\omega)|^2$ are each $\geqslant 0$

so that

$$S_x(\omega) \geqslant 0 \qquad (4.4.31)$$

Note that equations (4.4.30) imply that $S_y(\omega)$ (and in general $S_x(\omega)$ also) is integrable over the complete range of real ω from $-\infty$ to $+\infty$.

These properties of the power spectral density function give rise to restrictions on the location of the poles and zeros of $S_x(\omega)$. When $S_x(\omega)$ can be written in the form of a rational function, we have

$$S_x(\omega) = a^2 \frac{(\omega - \Omega_1)(\omega - \Omega_2) \ldots (\omega - \Omega_N)}{(\omega - p_1)(\omega - p_2) \ldots (\omega - p_M)} \qquad (4.4.32)$$

where Ω, p are the roots of the numerator and denominator polynomials and may be complex.

$S_x(\omega)$ is real for real values of ω, and $S_x(\omega) = \{S_x(\omega)\}^*$. Hence a^2 is real, and the complex roots, both Ω and p, must occur as conjugate pairs.

$S_x(\omega)$ is integrable for real ω over the range $-\infty$ to $+\infty$. Hence no root of the denominator can be real and the degree of the numerator function must be less than the denominator, that is $N < M$.

$S_x(\omega)$ is greater than or equal to zero for real ω. Hence any real root of the numerator must occur with even multiplicity otherwise the numerator will change sign.

Although the power spectral densities which occur in physical problems may usually be expressed as rational functions, important exceptions arise. Thus rational functions do not represent periodic or d.c. components for which δ-functions must be used. Consider $x(t) = \sum_{n=-\infty}^{\infty} a_n \exp(jn\omega_0 t)$ which represents a periodic component in the time domain. Since $S_x(\omega)$ is the Fourier Transform of $R_x(\tau)$ we have from equation (4.4.21)

$$S_x(\omega) = 2\pi \sum_{n=-\infty}^{\infty} |a_n|^2 \,\delta(\omega - n\omega_0) \tag{4.4.33}$$

and for the particular case of the d.c. component

$$S_x(\omega) = 2\pi a_0^2 \,\delta(\omega) \tag{4.4.34}$$

Cross-spectral density functions are not necessarily real, positive, or even functions of ω for real values of ω. They do, however, have the following properties which arise directly from the definition of $S_{xy}(\omega)$ as the Fourier Transform of $R_{xy}(\tau)$.

(a) $S_{xy}(\omega) = \{S_{yx}(\omega)\}^*$.
(b) For real functions $x(t)$, $y(t)$ and for real values of ω, Re $[S_{xy}(\omega)]$ is an even function of ω and Im $[S_{xy}(\omega)]$ is odd function of ω.
(c) If two stationary processes are uncorrelated their cross-spectral density is zero at all frequencies and so the power spectral density of their sum is the sum of their individual power spectral densities (see equation (4.4.43)).

4.4.8 Properties of Correlation Functions

The following properties of auto-correlation and cross-correlation functions should be noted.

(a) $R_x(\tau) \leqslant R_x(0)$. Thus the maximum value of $R_x(\tau)$ occurs at $\tau = 0$. There may be other values of τ, where $R_x(\tau) = R_x(0)$, (as, for example, may occur for a periodic signal) but $R_x(\tau)$ can never exceed $R_x(0)$. Since†

$$\left|\int_{-T_0}^{T_0} x(t+\tau)\,x(t)\,\mathrm{d}t\right|^2 \leqslant \int_{-T_0}^{T_0} |x(t+\tau)|^2\,\mathrm{d}t \int_{-T_0}^{T_0} |x(t)|^2\,\mathrm{d}t \tag{4.4.35}$$

$$R_x(\tau) = \lim_{T_0\to\infty} \cdot\frac{1}{2T_0}\int_{-T_0}^{T_0} x(t+\tau)\,x(t)\,\mathrm{d}t \tag{4.4.36}$$

$$\leqslant \lim_{T_0\to\infty} \cdot\frac{1}{2T_0}\left\{\int_{-T_0}^{T_0} |x(t+\tau)|^2\,\mathrm{d}t \int_{-T_0}^{T_0} |x(t)|^2\,\mathrm{d}t\right\}^{1/2}$$

therefore

$$R_x(\tau) \leqslant \lim_{T_0\to\infty} \cdot\frac{1}{2T_0}\int_{-T_0}^{T_0} |x(t)|^2\,\mathrm{d}t = R_x(0) \tag{4.4.37}$$

† This is the Schwarz inequality. See, for example, Margenau and Murphy referred to on p. 70.

(b) If $x(t)$ has a time independent component then $R_x(\tau)$ will have a constant component.

In particular if $x(t) = A$, a d.c. signal

$$R_x(\tau) = A^2 \tag{4.4.38}$$

and if $x(t)$ is a d.c. signal together with a component having zero mean value

$$x(t) = A + n(t) \tag{4.4.39}$$

then

$$R_x(\tau) = A^2 + R_n(\tau) \tag{4.4.40}$$

More generally if

$$x(t) = y(t) \pm n(t)$$

$$R_x(\tau) = R_y(\tau) + R_n(\tau) \pm R_{yn}(\tau) \pm R_{ny}(\tau) \tag{4.4.41}$$

If the processes giving rise to $y(t)$, $n(t)$ are statistically independent, and one of them has zero mean value, the cross-correlation terms vanish. The auto-correlation function of $x(t)$ is now just the sum of the auto-correlation functions for $y(t)$ and $n(t)$

$$R_x(\tau) = R_y(\tau) + R_n(\tau) \tag{4.4.42}$$

In this case, also, the power spectral density for the complete process is the sum of the power spectral densities for the individual processes

$$S_x(\omega) = S_y(\omega) + S_n(\omega) \tag{4.4.43}$$

(c) If $x(t)$ has neither a time independent component nor a periodic component

$$\lim_{\tau \to \infty} R_x(\tau) = 0 \tag{4.4.44}$$

This is because, for large values of τ, $x(t)$ and $x(t + \tau)$ tend to become statistically independent since the effect of past events diminishes as the time progresses.

(d) The cross-correlation function $R_{xy}(\tau)$ does not necessarily have its maximum value at $\tau = 0$, but

$$R_{xy}(\tau) \leqslant \{R_x(0)R_y(0)\}^{1/2} \tag{4.4.45}$$

(e) If either $x(t)$ or $y(t)$ is a time function having zero mean value and the processes are statistically independent, then

$$R_{xy}(\tau) = R_{yx}(\tau) = 0 \tag{4.4.46}$$

4.4.9 Thermal Noise

We illustrate this discussion of power spectral density functions and auto-correlation functions by considering thermal noise. Since the individual noise voltage pulses last for only a very short length of time the voltage frequency distribution is effectively flat (the Fourier Transform of an impulse function is a constant) and so the power spectral density is also a constant. Consider the case of a linear circuit with an ideal rectangular band pass centred at $\omega = 0$. This is an ideal system because it is not possible to realize a physical circuit with a rectangular band pass (see subsection 5.2.6 in Chapter 5). The thermal noise at the output will have a power spectral density which is constant over a finite band width and zero outside this range, as shown in Fig. 4.9.

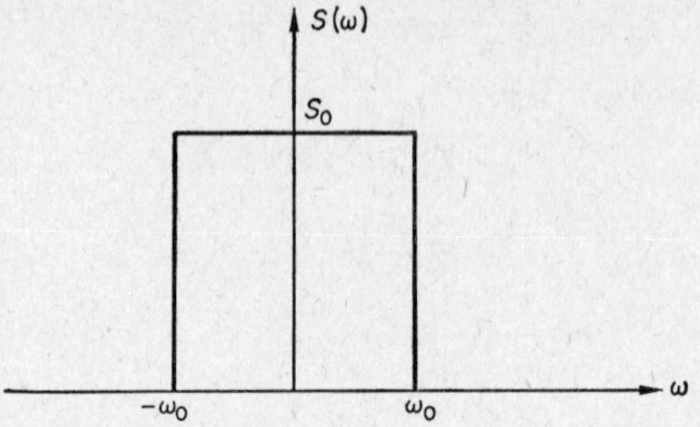

FIG. 4.9 Thermal noise power spectral density at the output of a linear circuit with a rectangular band passed centred on $\omega = 0$.

Let

$$\begin{aligned} S(\omega) &= S_0 \quad \text{for} \quad |\omega| \leqslant \omega_0 \\ S(\omega) &= 0 \quad \text{for} \quad |\omega| > \omega_0 \end{aligned} \tag{4.4.47}$$

Then

$$R_x(\tau) = \frac{1}{2\pi}\int_{-\infty}^{\infty} S_0 \, e^{j\omega\tau} \, d\omega = \frac{1}{2\pi}\int_{-\infty_0}^{\omega_0} S_0 \, e^{j\omega\tau} \, d\omega \tag{4.4.48}$$

$$= \frac{\omega_0}{2\pi} . S_0\left(\frac{2 \sin \omega_0\tau}{\omega_0\tau}\right) \tag{4.4.49}$$

This function is shown in Fig. 4.10.

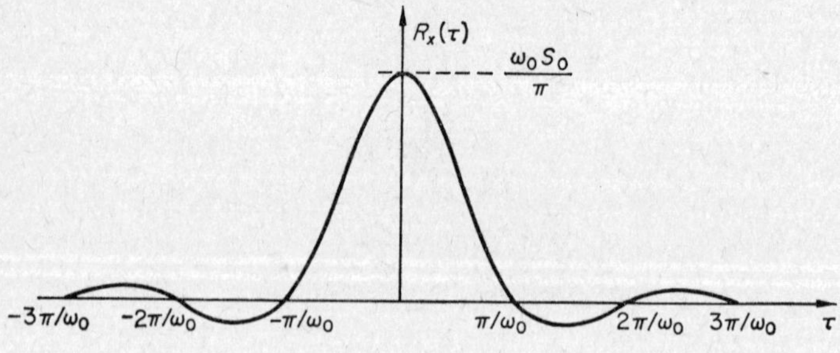

FIG. 4.10 $R_x(\tau)$ for band limited thermal noise.

As $\omega_0 \rightarrow \infty$ the band width increases without limit and equation (4.4.49) will describe the ideal case of 'white' thermal noise for which the spectral density is constant over the entire frequency range. This is again an ideal case—in this instance because the mean power given by $1/2\pi \int_{-\infty}^{\infty} S(\omega) \, d\omega$ does not converge and so $R_x(\tau)$ is infinite (compare equation (4.4.25). However, 'white noise' is a useful hypothetical concept for many circuit applications, and we obtain from equations (4.4.49) and (4.1.16) for 'white' noise

$$R_x(\tau) = S_0 \lim_{\omega_0 \to \infty} . \frac{\omega_0}{2\pi}\left(\frac{2 \sin \omega_0 \tau}{\omega_0 \tau}\right) \tag{4.4.50}$$

$$R_x(\tau) = S_0 \, \delta(\tau) \tag{4.4.51}$$

corresponding to

$$S(\omega) = S_0 \quad \text{for all } \omega \tag{4.4.52}$$

The singularity in $R_x(\tau)$ at $\tau = 0$, referred to above, is indicated by the δ-function in equation (4.4.51).

4.4.10 Periodic Signal in the Presence of Noise

Consider the extraction of a periodic signal from band limited noise. Let the output from an ideal band limited amplifier be

$$y(t) = s(t) + n(t) \tag{4.4.53}$$

where

$$s(t) = A_0 \cos(\omega t + \theta) \tag{4.4.54}$$

and

$$R_s(\tau) = \frac{A_0^2}{2} \cos \omega\tau \tag{4.4.55}$$

whilst from equation (4.4.49)

$$R_n(\tau) = \frac{\omega_0}{2\pi} S_0 \left[\frac{2 \sin \omega_0 \tau}{\omega_0 \tau}\right] \tag{4.4.56}$$

Evidently, since $n(t)$ is statistically independent of $y(t)$,

$$R_x(\tau) = \frac{A_0^2}{2} \cos \omega\tau + \frac{\omega_0}{2\pi} S_0 \left[\frac{2 \sin \omega_0 \tau}{\omega_0 \tau}\right] \tag{4.4.57}$$

and for large values of τ the auto-correlation function is derived almost entirely from the periodic signal component $s(t)$. Hence it should be possible to extract small periodic signals from a noisy background by measuring the auto-correlation function of the combined signal plus noise. A block diagram for a system capable of performing this operation is shown in Fig. 4.11(a).

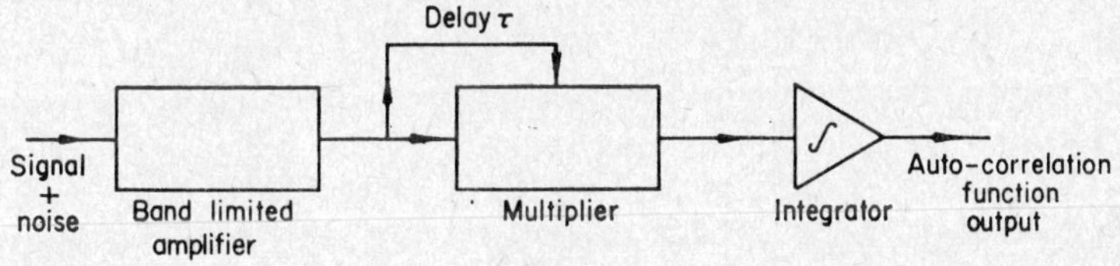

(a)

Fig. 4.11

(a) Block diagram of a system for detecting a signal by forming the auto-correlation function of the signal + noise.

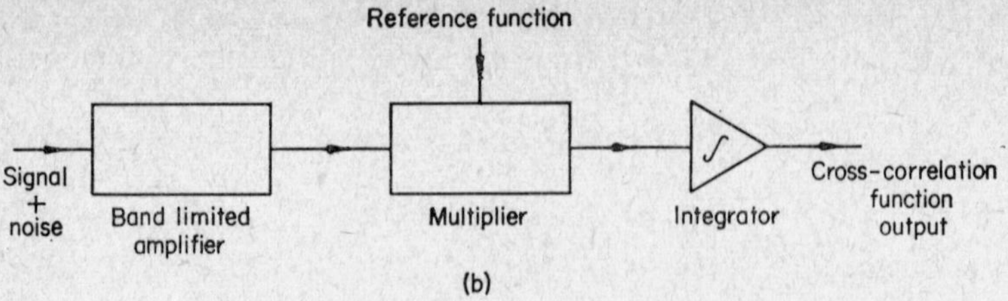

FIG. 4.11

(b) Block diagram of a system for detecting a signal by forming the cross-correlation function with an appropriate reference function.

This discussion can be extended immediately to demonstrate the principle underlying the correlation detector. Instead of forming an auto-correlation function, we now form a cross-correlation function. A system for performing this operation is illustrated in Fig. 4.11(b). Suppose the reference function is cosinusoidal having arbitrary but constant amplitude. Take the signal at the output of the amplifier to be

$$s(t) = A_0 \cos(\omega t + \theta) \tag{4.4.54b}$$

and the reference to be

$$f(t) = \alpha \cos(\omega t + \phi) \tag{4.4.54c}$$

then the cross-correlation function $R_{sf}(\tau)$ is given by

$$R_{sf}(\tau) = \frac{A_0 \alpha}{2} \cos(\omega\tau + \theta - \phi) \tag{4.4.57b}$$

and

$$R_{nf}(\tau) = 0 \tag{4.4.57c}$$

The output of the system is now a linear function of the amplitude of the signal which is to be extracted from the noise and, in practice, this amplitude carries the physical information. In a slight additional refinement of this procedure the reference function is taken as a rectangular waveform and the multiplier becomes a switch.

4.4.11 ERGODIC PROCESSES

A stationary random process is defined to be ergodic if all types of ensemble average are equal to the corresponding time average. Thus

$$\bar{x} = \int_{-\infty}^{\infty} x\, p(x)\, dx = \lim_{T_0 \to \infty} \frac{1}{2T_0} \int_{-T_0}^{T_0} x(t)\, dt = \langle x(t) \rangle \tag{4.4.58}$$

$$\overline{x^n} = \int_{-\infty}^{\infty} x^n\, p(x)\, dx = \lim_{T_0 \to \infty} \frac{1}{2T_0} \int_{-T_0}^{T_0} x^n(t)\, dt = \langle x^n(t) \rangle \tag{4.4.59}$$

$$\overline{xy} = \int_{-\infty}^{\infty} \int_{-\infty}^{\infty} xy\, p(x, y)\, dx\, dy = \lim_{T_0 \to \infty} \frac{1}{2T_0} \int_{-T_0}^{T_0} x(t)\, y(t)\, dt = \langle x(t)\, y(t) \rangle \tag{4.4.60}$$

In these expressions $p(x)\,\mathrm{d}x$ is the probability, derived from the ensemble distribution, of finding the signal value in the range x to $\mathrm{d}x$, whilst $p(x, y)\,\mathrm{d}x\,\mathrm{d}y$ is the joint probability of finding the values of two properties of the signal in the range x to $x + \mathrm{d}x$ and y to $y + \mathrm{d}y$.

We can see how these relations arise by considering the noise voltage $V(t)$ generated by a resistor. If we have a set of M identical resistors, the average value of $V(t)$ measured at time t for the ensemble is

$$\overline{V(t)} = \frac{1}{M}\{V_1(t) + V_2(t) + V_3(t) + \cdots V_M(t)\} \tag{4.4.61}$$

where M is a large number.

For a stationary random process this mean value is independent of the time at which the average is evaluated.

Hence, integrating both sides of equation (4.4.61) with respect to the time,

$$\begin{aligned}\overline{V(t)} &= \frac{1}{2T_0}\int_{-T_0}^{T_0} \overline{V(t)}\,\mathrm{d}t \\ &= \lim_{T_0\to\infty} .\frac{1}{2T_0}\int_{-T_0}^{T_0} \overline{V(t)}\,\mathrm{d}t \\ &= \frac{1}{M}.\lim_{T_0\to\infty} .\frac{1}{2T_0}\int_{-T_0}^{T_0} \{V_1(t) + V_2(t) + V_3(t) + \cdots V_M(t)\}\,\mathrm{d}t\end{aligned} \tag{4.4.62}$$

For an ergodic process each sample function eventually assumes all the values which occur for any other equivalent sample function and so

$$\begin{aligned}\lim_{T_0\to\infty} .\frac{1}{2T_0}\int_{-T_0}^{T_0} V_1(t)\,\mathrm{d}t &= \lim_{T_0\to\infty} .\frac{1}{2T_0}\int_{-T_0}^{T_0} V_2(t)\,\mathrm{d}t \\ = \ldots \qquad &= \lim_{T_0\to\infty} .\frac{1}{2T_0}\int_{-T_0}^{T_0} V_M(t)\,\mathrm{d}t \\ &\equiv \lim_{T_0\to\infty} .\frac{1}{2T_0}\int_{-T_0}^{T_0} V(t)\,\mathrm{d}t = \langle V(t)\rangle\end{aligned} \tag{4.4.63}$$

where $V(t)$ refers to any sample function.

Hence

$$\overline{V(t)} = \frac{1}{M}.M\langle V(t)\rangle = \langle V(t)\rangle \tag{4.4.64}$$

Similarly

$$\overline{xy} = \langle x(t)\,y(t)\rangle \tag{4.4.65}$$

Note that for ergodic processes this equation implies that the ensemble correlation functions $\overline{x(t + \tau)\,x(t)}$, $\overline{x(t + \tau)\,y(t)}$ are equal to the time correlation functions for the sample functions which were discussed in the previous section.

Some simple processes are not ergodic. Thus, if a set of voltages V_r are distributed randomly over the ensemble but individually are constant in time, the ensemble average V is not equal to the time average of each sample function (since $\langle V_r\rangle = V_r$ and $V_r \neq V_j$). Again, if

$$V_r(t) = V_r^0 \cos(\omega t + \phi_r) \tag{4.4.66}$$

where V_r^0 and ϕ_r are random variables with respect to the ensemble but constant in time, the two sets of averages are not necessarily equal. To examine this case, suppose ϕ is uniformly distributed over the ensemble with values between 0 and 2π and that V^0 and ϕ are statistically independent. Although

$$\begin{aligned}\overline{V} &= \int_{-\infty}^{\infty}\int_0^{2\pi} V^0 \cos(\omega t + \phi)\, \rho(V^0)\, \mathrm{d}V^0\, \rho(\phi)\, \mathrm{d}\phi \\ &= \int_{-\infty}^{\infty} V^0 \rho(V^0)\, \mathrm{d}V^0 \int_0^{2\pi} \cos(\omega t + \phi) . \frac{1}{2\pi}\, \mathrm{d}\phi \\ &= 0 \\ &= \langle V_r(t)\rangle \end{aligned} \tag{4.4.67}$$

The second moments are not necessarily equal

$$\begin{aligned}\overline{V^2} &= \int_{-\infty}^{\infty} (V^0)^2 \rho(V^0)\, \mathrm{d}V^0 \int_0^{2\pi} \cos^2(\omega t + \phi) . \frac{1}{2\pi}\, \mathrm{d}\phi \\ &= \frac{1}{2}\int_{-\infty}^{\infty} (V^0)^2 \rho(V^0)\, \mathrm{d}V^0 \end{aligned} \tag{4.4.68}$$

whilst for the rth individual sample function

$$\langle V_r^2(t)\rangle = \tfrac{1}{2}(V_r^0)^2, \text{ not necessarily} = \overline{V^2} \tag{4.4.69}$$

If, however, each individual sample function $V_r(t)$ is given by

$$V_r(t) = \sum_{n=0}^{N} V_{rn} \cos(\omega_n t + \phi_{rn}) \tag{4.4.70}$$

where $\omega_n = n\Delta\omega$ an almost continuous variable and V_{rn}, ϕ_{rn}, are independent random variables with zero mean, the process will be ergodic when the probability distribution over the ensemble and over the index n describes a random process. Consider

$$\overline{V_r^2(t)} = \left\{\sum_{n=0}^{N} V_{rn}\cos(\omega_n t + \phi_{rn})\right\}^2_{\text{average over ensemble}} \tag{4.4.71}$$

$$\begin{aligned} &= \left\{\sum_{n=0}^{N} V_{rn}\cos\phi_{rn}\right\}^2_{\text{average over ensemble}} \\ &= \Big\{V_{r1}^2\cos^2\phi_{r1} + V_{r2}^2\cos^2\phi_{r2}\cdots + \\ &\quad + \sum_i \sum_{\substack{j \\ i\neq j}} V_{ri}\cos\phi_{ri}\, V_{rj}\cos\phi_{rj}\Big\}_{\text{average over ensemble}} \\ &= \frac{N}{2}\overline{V_{r1}^2} = \frac{N}{2}\overline{V_{r2}^2}\cdots = \frac{N}{2}\overline{V_{rN}^2} \end{aligned} \tag{4.4.72}$$

The time average of the sample function $V_r(t)$ is given by

$$\langle V_r^2(t)\rangle = \operatorname*{Lim}_{T_0\to\infty} . \frac{1}{2T_0}\int_{-T_0}^{T_0}\left\{\sum_{n=0}^{N} V_{rn}\cos(\omega_n t + \phi_{rn})\right\}^2 \mathrm{d}t \tag{4.4.73}$$

and taking $T_0 = 2\pi/\Delta\omega$, the fundamental period, the cross-products are again zero.

$$\langle V_r^2(t)\rangle = \tfrac{1}{2}\sum_{n=0}^{N} V_{rn}^2 = \frac{N}{2}\{V_{rn}^2\} \text{ averaged over } n$$

Hence, since the distribution over the ensemble and over the index n is in each case random

$$\overline{V_r^2(t)} = \langle V_r^2(t)\rangle \tag{4.4.74}$$

The representation for the sample function given in equation (4.4.70) is of particular importance when discussing noise signals.

4.5 Summary

1. Fourier Series Representation

$$x(t) = \sum_{n=-\infty}^{\infty} a_n \exp(jn\omega_0 t)$$

$$a_n = a_{-n}^* = \frac{\omega_0}{2\pi}\int_{t_1}^{t_1+2\pi/\omega_0} x(t)\exp(-jn\omega_0 t)\,dt$$

2. Fourier Transform Representation

$$x(t) = \frac{1}{2\pi}\int_{-\infty}^{\infty} X(j\omega)\,e^{j\omega t}\,d\omega$$

$$F\{x(t)\} = X(j\omega) = \int_{-\infty}^{\infty} x(t)\,e^{-j\omega t}\,dt$$

3. Laplace Transform Representation

$$x(t)\,u(t) = \frac{1}{2\pi j}\int_{q-j\infty}^{q+j\infty} X(s)\,e^{st}\,ds$$

$$L\{x(t)\} = X(s) = \int_{0^-}^{\infty} x(t)\,e^{-st}\,dt$$

4. Power Spectral Density

$$S_x(\omega) = 2\pi\sum_{n=-\infty}^{\infty} |a_n|^2\,\delta(\omega - n\omega_0) \text{ for a periodic function}$$

5. Energy Spectral Density

$$S_E(\omega) = |X(j\omega)|^2$$

where

$$X(j\omega) = \int_{-\infty}^{\infty} x(t)\,e^{-j\omega t}\,dt = F\{x(t)\}.$$

6. Time Auto-correlation Function

$$R_x(\tau) = \operatorname*{Lim.}_{T_0\to\infty}\frac{1}{2T_0}\int_{-T_0}^{T_0} x(t+\tau)\,x(t)\,dt$$

7. Time Cross-correlation Function

$$R_{xy}(\tau) = \operatorname*{Lim}_{T_0 \to \infty} . \frac{1}{2T_0} \int_{-T_0}^{T_0} x(t+\tau)\, y(t)\, dt$$

8. Power Spectral Density

$$S_x(\omega) = \int_{-\infty}^{\infty} R_x(\tau)\, e^{-j\omega\tau}\, d\tau$$

$$R_x(\tau) = \frac{1}{2\pi} \int_{-\infty}^{\infty} S_x(\omega)\, e^{j\omega\tau}\, d\omega$$

$$R_x(0) = \frac{1}{2\pi} \int_{-\infty}^{\infty} S_x(\omega)\, d\omega = \operatorname*{Lim}_{T_0 \to \infty} . \frac{1}{2T_0} \int_{-T_0}^{T_0} \{x(t)\}^2\, dt$$

9. Cross-spectral Density

$$S_{xy}(\omega) = \int_{-\infty}^{\infty} R_{xy}(\tau)\, e^{-j\omega\tau}\, d\tau$$

$$R_{xy}(\tau) = \frac{1}{2\pi} \int_{-\infty}^{\infty} S_{xy}(\omega)\, e^{j\omega\tau}\, d\omega$$

We have now moved a considerable way from the initial attempt to represent a complicated signal $x(t)$ in terms of basis functions containing the time explicitly. It is evident from this discussion that a given signal may be as adequately described by a representation in the (complex) frequency domain as by functions in the more familiar time domain. There are advantages to be derived from a frequency domain representation—for example problems in linear circuit analysis are simplified. The use of correlation functions and power spectral density functions allows random signals to be included in the analysis and this is of particular importance for the design of optimum linear systems where both signal and noise have to be considered.

Bibliography

BROWN, W. M. *Analysis of Linear Time Invariant Systems*, McGraw-Hill, 1963

COOPER, G. R. and MCGILLEM, C. D. *Methods of Signal and System Analysis*, Holt, Rinehart and Winston Inc., 1967

DAVENPORT, W. B. and ROOT, W. L. *An Introduction to the Theory of Random Signals and Noise*, McGraw-Hill, 1958

HANCOCK, J. C. *The Principles of Communication Theory*, McGraw-Hill, 1961

PAPOULIS, A. *The Fourier Integral and its Applications*, McGraw-Hill, 1962

Tables of Transforms

TABLE 4.1

FOURIER TRANSFORMS

Time Domain	*Fourier Transform*
1. $f(t) = \frac{1}{2\pi} \int_{-\infty}^{\infty} F(j\omega)\, e^{j\omega t}\, d\omega$	$F(j\omega) = \int_{-\infty}^{\infty} f(t)\, e^{-j\omega t}\, dt$
2. $a_1 f(t) + a_2 f(t)$	$a_1 F(j\omega) + a_2 F_2(j\omega)$

	Time Domain	Fourier Transform
3.	$f(-t)$	$F(-j\omega) = F^*(j\omega)$
4.	$F(t)$	$2\pi f(-j\omega)$
5.	$f(at)$	$\frac{1}{a}F\left(\frac{j\omega}{a}\right)$
6.	$f(t-t_0)$	$e^{-j\omega t_0} F(j\omega)$
7.	$\frac{d^n}{dt^n} f(t)$	$(j\omega)^n F(j\omega)$
8.	$t^n f(t)$	$j^n \frac{d^n}{d\omega^n} F(j\omega)$
9.	$\int_0^t f_{even}(t)\,dt + \int_{-\infty} f_{odd}(t)\,dt$	$\frac{1}{j\omega} F(j\omega)$
10.	$\int_{-\infty}^t f(t)\,dt$	$j\omega F(j\omega) + \pi F(0)\,\delta(\omega)$
11.	$f_1(t) * f_2(t) \equiv \int_{-\infty}^{\infty} f_1(t-\tau) f_2(\tau)\,d\tau$	$F_1(j\omega)\,F_2(j\omega)$
12.	$f_1(t).f_2(t)$	$\frac{1}{2\pi}\int_{-\infty}^{\infty} F_1(j\omega - j\xi)\,F_2(j\xi)\,d\xi$
13.	$f(at)\,e^{jbt}$ $\quad a > 0$	$\frac{1}{a}F\left\{\frac{j(\omega-b)}{a}\right\}$
14.	$f(at)\cos bt$ $\quad a > 0$	$\frac{1}{2a}\left[F\left\{\frac{j(\omega-b)}{a}\right\} + F\left\{\frac{j(\omega+b)}{a}\right\}\right]$
15.	$f(at)\sin bt$ $\quad a > 0$	$\frac{1}{2aj}\left[F\left\{\frac{j(\omega-b)}{a}\right\} - F\left\{\frac{j(\omega+b)}{a}\right\}\right]$
16.	$e^{-\alpha t} u(t)$ $\quad \mathrm{Re}[\alpha] > 0$	$\frac{1}{j\omega + \alpha}$
17.	$\frac{1}{\beta - \alpha}(e^{-\alpha t} - e^{-\beta t})u(t)$ $\quad \mathrm{Re}[\alpha], \mathrm{Re}[\beta] > 0$	$\frac{1}{(j\omega+\alpha)(j\omega+\beta)}$
18.	$e^{-\alpha\|t\|}$ $\quad \mathrm{Re}[\alpha] > 0$	$\frac{2\alpha}{\alpha^2 + \omega^2}$
19.	$e^{-\alpha^2 t^2}$	$\frac{\sqrt{\pi}}{\alpha} \exp\left(-\frac{\omega^2}{4\alpha^2}\right)$
20.	$e^{-\alpha t}\sin\omega_0 t\, u(t)$	$\frac{\omega_0}{(\alpha + j\omega)^2 + \omega_0^2}$
21.	$e^{-\alpha t}\cos\omega_0 t\, u(t)$	$\frac{\alpha + j\omega}{(\alpha + j\omega)^2 + \omega_0^2}$

	Time Domain	Fourier Transform
22.	$\delta(t)$	1
23.	$u(t)$	$\pi\,\delta(\omega) + \dfrac{1}{j\omega}$
24.	A = constant	$2\pi A\,\delta(\omega)$
25.	$\exp(j\omega_0 t)$	$2\pi\,\delta(\omega - \omega_0)$
26.	$\sin\omega_0 t$	$-j\pi\{\delta(\omega - \omega_0) - \delta(\omega + \omega_0)\}$
27.	$\cos\omega_0 t$	$\pi\{\delta(\omega - \omega_0) + \delta(\omega + \omega_0)\}$

TABLE 4.2
LAPLACE TRANSFORMS

	Time Domain		Laplace Transforms
1.	$x(t)$		$X(s) = \int_{0^-}^{\infty} x(t)\,e^{-st}\,dt$
2.	$a_1\,x_1(t) + a_2\,x_2(t)$		$a_1\,X_1(s) + a_2\,X_2(s)$
3.	$\dfrac{d}{dt}\{x(t)\}$		$s\,X(s) - x(0^-)$
4.	$\dfrac{d^n}{dt^n}\{x(t)\}$		$s^n\,X(s) - \sum_{l=1}^{n} s^{n-l}\,x^{l-1}(0^-)$
5.	$\int_{0^-}^{t} x(\tau)\,d\tau$		$\dfrac{1}{s}X(s)$
6.	$(-t)^n\,x(t)$		$\dfrac{d^n}{ds^n}\{X(s)\}$
7.	$\dfrac{1}{t}x(t)$		$\int_{s}^{\infty} X(s)\,ds$
8.	$x(t - t_0)\,u(t - t_0)$		$\exp(-st_0)\,X(s)$
9.	$e^{\pm at}\,x(t)$	{a complex with non-negative real part}	$X(s \mp a)$
10.	$x(at)$	$a > 0$	$\dfrac{1}{a}X\left(\dfrac{s}{a}\right)$
11.	$\lim_{t\to\infty} x(t)$	{Poles of $X(s)$ in left half plane}	$\lim_{s\to 0} s\,X(s)$
12.	$\lim_{t\to 0} x(t)$		$\lim_{s\to\infty} s\,X(s)$
13.	$x_1(t) * x_2(t) = \int_{0^-}^{t} x_1(t - \tau, x_2(\tau)\,d\tau$		$X_1(s)\,X_2(s)$

	Time Domain	*Laplace Transforms*
14.	$\mathrm{Re}\,[x(t)]$	$\mathrm{Re}\,[X(s)]$
15.	$\mathrm{Im}\,[x(t)]$	$\mathrm{Im}\,[X(s)]$
16.	$\delta(t)$	1
17.	$\dfrac{\mathrm{d}^n}{\mathrm{d}t^n}\,\delta(t)$	s^n
18.	$u(t)$	$\dfrac{1}{s}$
19.	$t\,u(t)$	$\dfrac{1}{s^2}$
20.	$\mathrm{e}^{-at}\,u(t)$	$\dfrac{1}{s+a}$
21.	$t\,\mathrm{e}^{-at}\,u(t)$	$\dfrac{1}{(s+a)^2}$
22.	$\dfrac{1}{b-a}(\mathrm{e}^{-at}-\mathrm{e}^{-bt})\,u(t)$	$\dfrac{1}{(s+a)(s+b)}$
23.	$\sin \omega t \,.\, u(t)$	$\dfrac{\omega}{s^2+\omega^2}$
24.	$\cos \omega t \,.\, u(t)$	$\dfrac{s}{s^2+\omega^2}$
25.	$\mathrm{e}^{-at} \sin \omega t \,.\, u(t)$	$\dfrac{\omega}{(s+a)^2+\omega^2}$
26.	$\mathrm{e}^{-at} \cos \omega \,.\, u(t)$	$\dfrac{(s+a)}{(s+a)^2+\omega^2}$
27.	Staircase function	$\dfrac{1}{s \sinh as}$

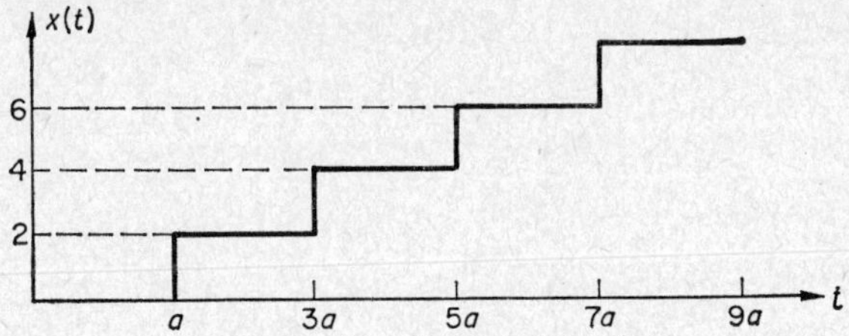

$$x(t) = 2\sum_{k=0}^{\infty} u\{t - 2(k+1)a\}$$

Time Domain	*Laplace Transforms*
28. Repeated pulse	
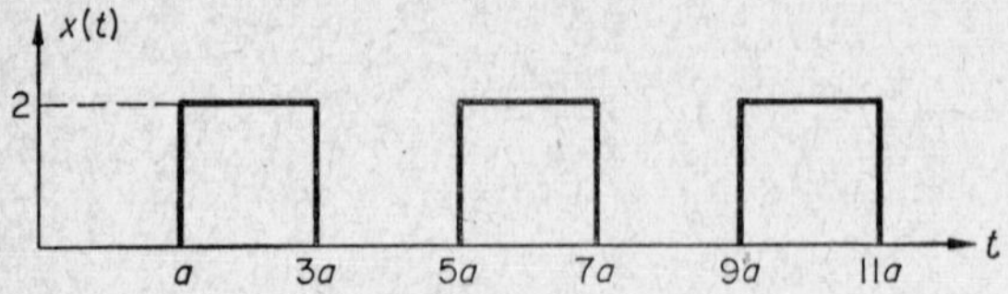	$\dfrac{1}{s \cosh as}$
$x(t) = 2\sum_{k=0}^{\infty}(-1)^k . u\{t-(2k+1)a\}$	
29. Triangular waveform	
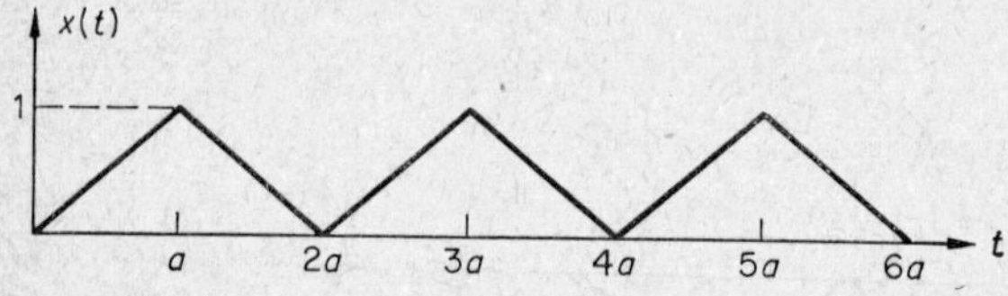	$\dfrac{1}{s^2}\tanh\left(\dfrac{as}{2}\right)$
$x(t) = \dfrac{1}{a}\left\{tu(t) + 2\sum_{k=1}^{\infty}(-1)^k(t-ka).u(t-ka)\right\}$	
30. Sawtooth waveform	
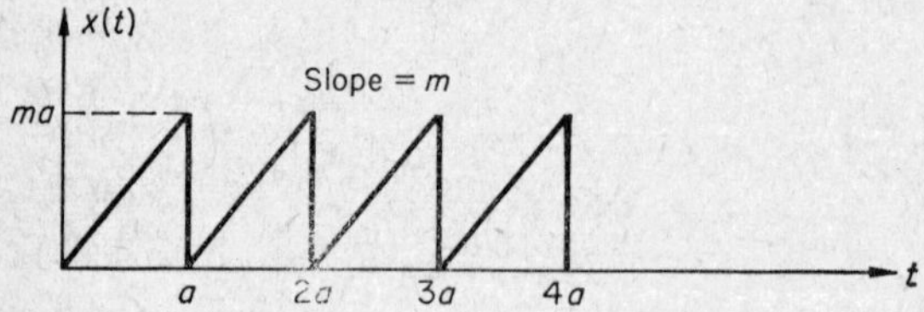	$\dfrac{m}{s^2} - \dfrac{ma}{2s}\left\{\coth\left(\dfrac{as}{2}\right) - 1\right\}$
$x(t) = mt\,u(t) - ma\sum_{k=1}^{\infty}u(t-ka)$	
31. Half-wave rectification of sine wave	
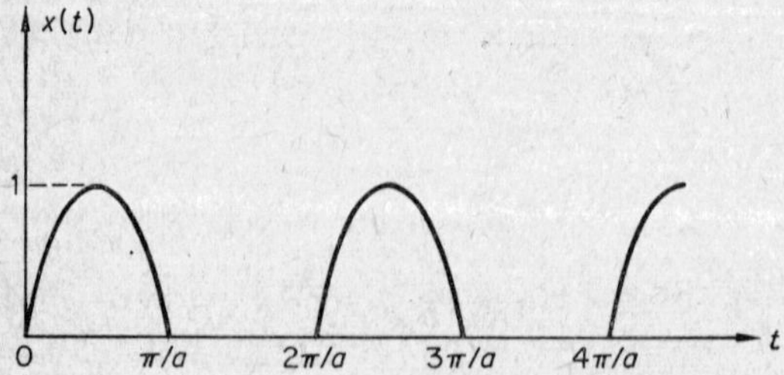	$\dfrac{a}{(s^2 + a^2)(1 - e^{-\pi s/a})}$
$x(t) = \sum_{k=0}^{\infty}\left\{\sin a\left(t - \dfrac{k\pi}{a}\right)\right\}u\left(t - \dfrac{k\pi}{a}\right)$	

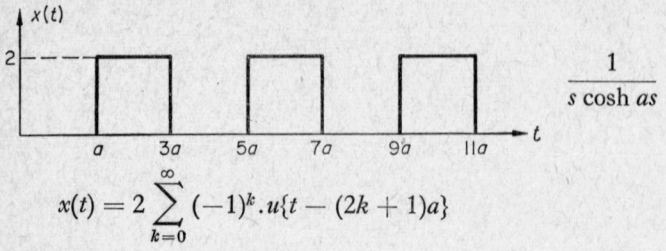

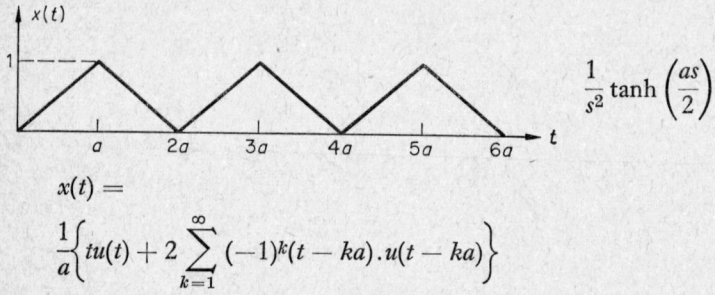

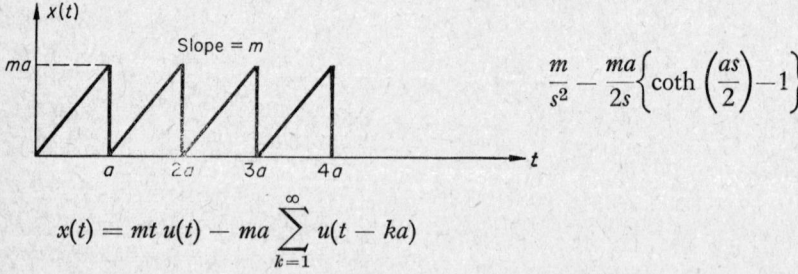

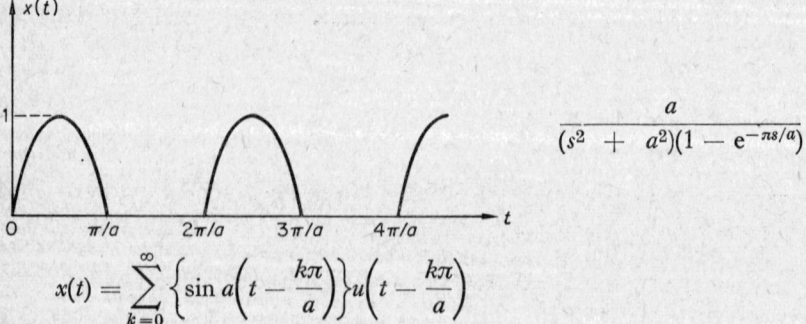

Time Domain	*Laplace Transforms*
32. Full wave rectification of a sine wave	
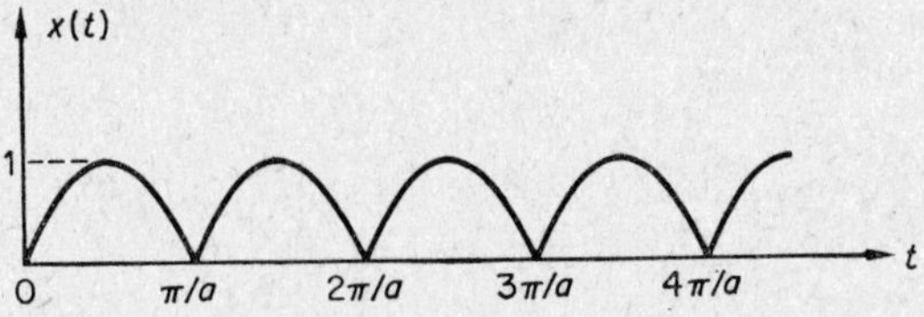	$\dfrac{a}{s^2 + a^2} \coth \dfrac{\pi s}{2a}$
$x(t) = \sin at . u(t) + 2 \sum_{k=1}^{\infty} \left\{ \sin a\left(t - \dfrac{k\pi}{a}\right) \right\} u\left(t - \dfrac{k\pi}{a}\right)$	
33. Single pulse	
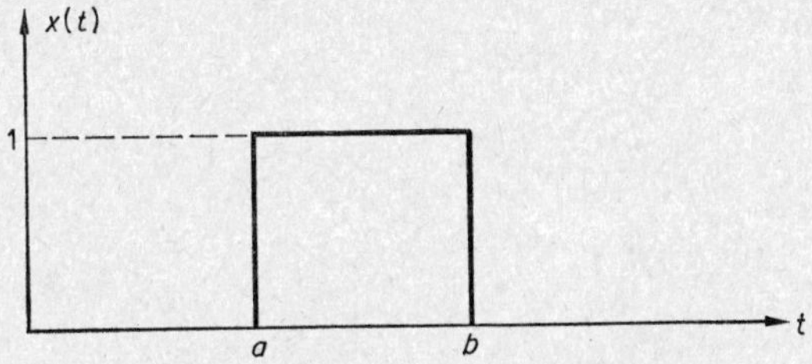	$\dfrac{1}{s}(e^{-as} - e^{-bs})$
$x(t) = u(t - a) - u(t - b)$	

Problems

1. Determine the Fourier Series representations for the waveforms shown below

(a) $x(t) = A\,|\sin \omega_0 t|$

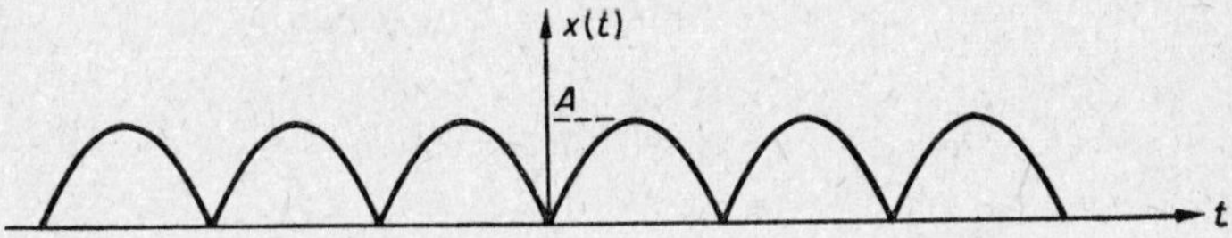

(b) $x(t)$ = square wave, amplitude A

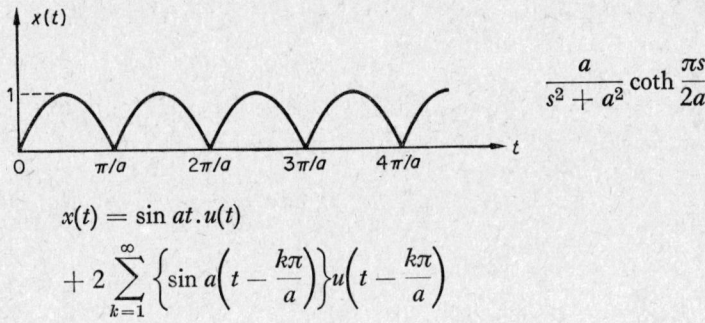

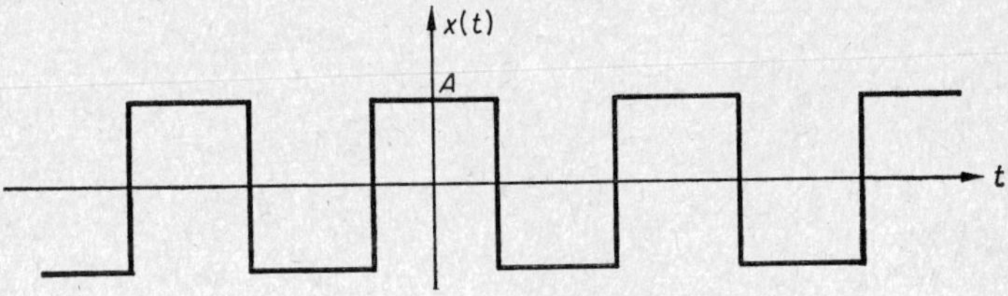

(c) $x(t)$ = spaced sawtooth, amplitude A

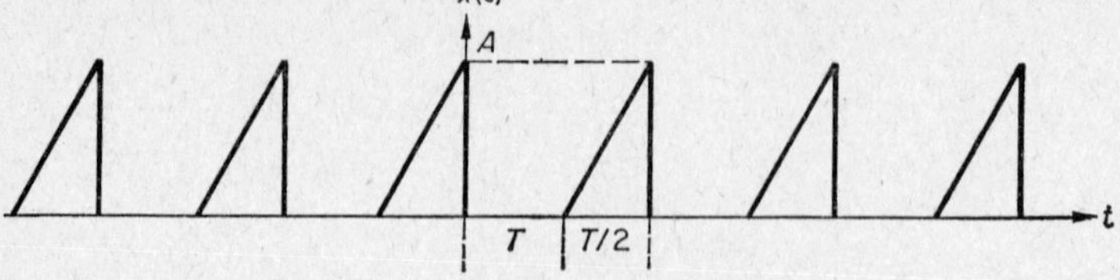

2. If a periodic signal $x(t)$ having no harmonics above $\omega = \omega_m$ is passed through an ideal low pass filter with cut-off ω_m, unit gain, and no time delay, the output is also $x(t)$. Show that a signal $y(t) = [1 + x(t)] \cos \omega_c t$, where $\omega_c \gg \omega_m$ will also pass through a similar ideal band pass filter, having a band pass $\omega_c - \omega_m$ to $\omega_c + \omega_m$ without significant distortion.

3. Evaluate the Fourier Transforms for the following time functions:

(a) $x(t) = A \sin \omega_0 t \, u(t)$

(b) $x(t) = [1 + m \cos \omega_1 t] \cos \omega_0 t \qquad -T \leqslant t \leqslant T$

(c) $x(t) = -1$ for $t < 0$
$= 0$ for $t = 0$
$= +1$ for $t > 0$

(d) $x(t) = 1 - \dfrac{2|t|}{T}$ for $|t| < T/2$
$= 0$ for all other values of t

(e) $x(t) = u(t + T/2) - u(t - T/2)$

(*f*) $x(t) = \cos \omega_0 t\{u(t + T/2) - u(t - T/2)\}$

4. A time function $f(t)$ has the ideal frequency spectrum

$$F(j\omega) = A \quad \text{for} \quad \omega \leqslant |\omega_0|, \quad F(j\omega) = 0 \quad \text{for} \quad \omega > |\omega_0|.$$

Evaluate $f(t)$.

5. Show that the Fourier Transform of a uniform sequence of doublets is a sequence of impulses, that is,

$$F\left\{\sum_{n=-\infty}^{\infty} \delta'(t - n\tau_0)\right\} = j\omega_0^2 \sum_{n=-\infty}^{\infty} n\,\delta(\omega - n\omega_0) \quad \text{where } \omega_0 = \frac{2\pi}{\tau_0}$$

and hence that

$$\sum_{n=-\infty}^{\infty} \delta'(t - n\tau_0) = -\frac{4\pi}{\tau_0^2} \sum_{n=1}^{\infty} n \sin\left(\frac{2\pi n t}{\tau_0}\right)$$

6. Prove the Sampling Theorem.

If the Fourier Transform $F(j\omega)$ of a time function $f(t)$ is zero for $|\omega| > |\omega_m|$ then $f(t)$ can be uniquely determined from sampled values $f_n \equiv f(n\pi/\omega_m)$ measured at a set of equal time intervals $\Delta t = \pi/\omega_m$ and

$$f(t) = \sum_{n=-\infty}^{\infty} f_n \frac{\sin(\omega_m t - n\pi)}{(\omega_m t - n\pi)}$$

7. Evaluate the Laplace Transforms for the following time functions

(a) $x(t) = 5u(t) - 5u(t-2)$
(b) $x(t) = t\,e^{-\gamma t}.u(t)$
(c) $x(t) = \cosh \gamma t.u(t)$
(d) $x(t) = e^{-\beta t} \cosh \gamma t.u(t)$
(e) $x(t) = e^{-\beta t^2}.u(t)$
(f) $x(t) = 3t^2.u(t-5)$

8. A sample function of an ergodic random process is passed through an ideal limiter for which the output is $+1$ whenever the input is positive and the output is -1 whenever the input is negative. Assuming that the number of transitions across zero at the input, and hence also the output is given by a Poisson distribution, for which the probability of n zeros occurring in T seconds is given by

$$P(n, T) = \frac{(\beta T)^n}{n!} e^{-\beta T}$$

where β is the average number of transitions occurring per second, show that the auto-correlation function at the output is

$$R_x(\tau) = e^{-2\beta|\tau|}$$

9. A random process has an auto-correlation function

$$\begin{aligned} R_x(\tau) &= 1 - |\tau| \quad \text{for } |\tau| \leqslant 1 \\ &= 0 \quad \text{for } |\tau| > 1 \end{aligned}$$

show that the spectral density is

$$S_x(\omega) = \left(\frac{\sin \omega/2}{\omega/2}\right)^2$$

10. Evaluate the auto-correlation functions and the spectral densities for the waveforms shown in Problem (1).

11. A wideband input includes a fixed frequency signal buried in noise. The complete input is passed through an ideal limiter for which the output is $+1$ whenever the input is positive and the output is -1 whenever the input is negative. Show that the auto-correlation function of the fixed frequency signal is given by

$$R_s(\tau) = \frac{A^2}{2} \sin\left\{\frac{\pi}{2} R_c(\tau)\right\}$$

where $R_c(\tau)$ is the auto-correlation function of the clipped signal, and discuss the use of this system for detecting signals in the presence of noise.

12. The spectral density for a stationary, band limited, white noise signal is given by

$$\begin{aligned} S_x(\omega) &= S_0 \quad \text{for} \quad |\omega \pm \omega_0| \leqslant \omega_m \\ &= 0 \quad \text{for} \quad |\omega \pm \omega_0| > \omega_m \end{aligned}$$

Show that the auto-correlation function is

$$R_x(\tau) = \frac{S_0}{\pi\tau}\{\sin(\omega_0 + \omega_m)\tau - \sin(\omega_0 - \omega_m)\tau\}$$

Chapter 5

Linear Systems

5.1 General Discussion of Linear Systems

In Chapter 3 we discussed the solution of network problems in terms of the detailed voltage and current pattern within a network. From this analysis a linear relation was derived between an input variable and an output variable (see for example equation (3.3.31)). If we consider the loop equations for a network which consists of a finite number of circuit elements, and write the time differential operator $d/dt = D$, we have the formal relation between an input generator, e.m.f. $= \mathscr{E}_r(t)$, and an output loop current $i_j(t)$

$$i_j(t) = \frac{\Delta_r(D)}{\Delta(D)}\mathscr{E}_r(t) \tag{5.1.1}$$

Here $\Delta(D)$ is a finite determinant composed of operator elements with real coefficients which are derived from the properties of the individual circuit elements set out in equations (3.1.1)–(3.1.4) and $\Delta_r(D)$ is the appropriate cofactor taken from $\Delta(D)$.

5.1.1 Linear Differential Equation

Equation (5.1.1) may be written quite generally as a relation between an excitation variable, $x(t)$, and a response variable $y(t)$ for the network. Since $\Delta(D)$ and $\Delta_r(D)$ are finite polynomials in the operator D, we may write

$$(b_m D^m + b_{m-1}D^{m-1} + \cdots b_0)y(t) = (a_n D^n + a_{n-1}D^{n-1} + \cdots a_0)x(t) \tag{5.1.2}$$

or

$$y(t) = \frac{a_n D^n + a_{n-1}D^{n-1} + \cdots a_0}{b_m D^m + b_{m-1}D^{m-1} + \cdots b_0}.x(t) \tag{5.1.3}$$

$$= H(D).x(t)$$

The function $H(D)$ is evidently a rational function of the operator D and depends only upon the particular configuration of circuit elements within the network; it is called the transfer function operator for the network.

Equation (5.1.2) may also be written as a set of coupled first-order differential equations. This representation has advantages when used with digital or analogue computers. Consider, for example, the nth order differential equation

$$y(t) = (a_n D^n + a_{n-1}D^{n-1} + \cdots a_0)x(t) \tag{5.1.4}$$

Writing

$$q_0 = x(t),\ q_1 = D\,x(t),\ q_2 = D^2\,x(t), \ldots \tag{5.1.5}$$

we have the following set of coupled first order differential equations

$$\begin{aligned} Dq_0 &= q_1 \\ Dq_1 &= q_2 \\ Dq_2 &= q_3 \\ &\vdots \\ Dq_{n-1} &= q_n \end{aligned} \tag{5.1.6}$$

and then,

$$y(t) = a_n q_n + a_{n-1} q_{n-1} + \cdots a_0 q_0$$

5.1.2 Discrete Parameter, Time Invariant, Linear Systems

Equations (5.1.1)–(5.1.6) describe a discrete parameter, time invariant, linear system. A discrete parameter system consists essentially of a finite number of circuit elements for which propagation effects can be neglected. This means that the differential equation relating the excitation and the response will be an ordinary differential equation as distinct from a partial differential equation. Examples of distributed parameter linear systems are provided by transmission lines and waveguides; these will not be described rigorously by equations (5.1.1)–(5.1.6) and the transfer function $H(D)$ need not be a rational function of the operator D.

A time invariant, or stationary, linear system is one such that if the response $y(t)$ arises from an excitation $x(t)$ then the response $y(t + \tau)$ will arise from the excitation $x(t + \tau)$. Thus the form of the response depends only upon the form of the excitation and does not depend upon the time of application.

A linear system is defined by the law of superposition. If $x_i(t)$ gives rise to the response $y_i(t)$ then the excitation

$$x(t) = \sum_i a_i \, x_i(t)$$

produces the response

$$y(t) = \sum_i a_i \, y_i(t)$$

The differential equation (5.1.2) describes a linear system because all derivatives of the excitation and response are raised to the first power only in $x(t)$ and $y(t)$ and there are no products of their derivatives.

Finally, the linear system which we have discussed is a fixed parameter system. For such a system the component elements do not change their characteristics with time nor with the magnitude of the excitation.

5.1.3 Linear Integral Equation

A linear system may also be described by an integral equation. We may remove the constraint that it must necessarily be a discrete parameter system but we now assume that it is initially inert. An initially inert, time invariant, linear system is described by the integral equation†

$$y(t) = \int_{-\infty}^{t} h(t - \tau)\, x(\tau)\, d\tau = \int_{0^-}^{\infty} h(\tau)\, x(t - \tau)\, d\tau \tag{5.1.7}$$

† This integral equation is an example of Volterra's equation. See for example Margenau, H. and Murphy, G. M. *The Mathematics of Physics and Chemistry*, Van Nostrand, 1944.

In this equation $y(t)$ is again the response to the excitation $x(t)$ and the function $h(\xi)$ with the variable $\xi = (t - \tau)$, τ, is called the *system weighting function.* Note that if $x(t)$ is a delta function excitation,

$$y(t) = \int_{-\infty}^{t} h(t-\tau)\,\delta(\tau)\,d\tau = \int_{0^-}^{\infty} h(\tau)\,\delta(t-\tau)\,d\tau = h(t) \tag{5.1.8}$$

so that $h(t)$, or correspondingly $h(\xi)$, is also referred to as the *impulse response of the system.*

The relations given in equation (5.1.8) arise as follows. The response of a linear circuit at a particular time, t, is the superposition of all responses present at time, t. These responses will be derived from all excitations which have occurred at times τ, previous to t, but will not be influenced by events which may occur at future times since the system is causal and non-anticipatory. This statement implies that $h(t - \tau)$ is zero for all $\tau > t$ in the first integral of equation (5.1.8) and sets the upper limit for that integral. It also sets the lower limit for the second integral, in which $h(\tau) = 0$ for $\tau < 0^-$.

Regard the continuous excitation as a sequence of impulses having amplitudes $x(t)$. If $x(\tau_i)$ is the amplitude of the instantaneous input at time τ_i the response remaining at the output of the system at time t will depend upon the time interval which has elapsed, that is, upon $(t - \tau_i)$. Consequently, if $y_i(t)$ is the response at time t arising from the impulse excitation of amplitude $x(\tau_i)$ at time τ_i, we write

$$y_i(t) = h(t-\tau_i)\,x(\tau_i)$$

where $h(t - \tau_i)$ is a weighting coefficient. The complete response for a time invariant linear system at time t is obtained by summing the responses from all the excitations

$$y(t) = \sum_i y_i(t) = \sum_{\tau_i=-\infty}^{t} h(t-\tau_i)\,x(\tau_i) \tag{5.1.9}$$

For continuous excitation, the summation becomes an integral

$$y(t) = \int_{-\infty}^{t} h(t-\tau)\,x(\tau)\,d\tau \tag{5.1.10}$$

In this equation the origin of the coordinate τ coincides with that for the time, t, as shown in Fig. 5.1. If, however, we define a new variable, λ, such that

$$\lambda = t - \tau \tag{5.1.11}$$

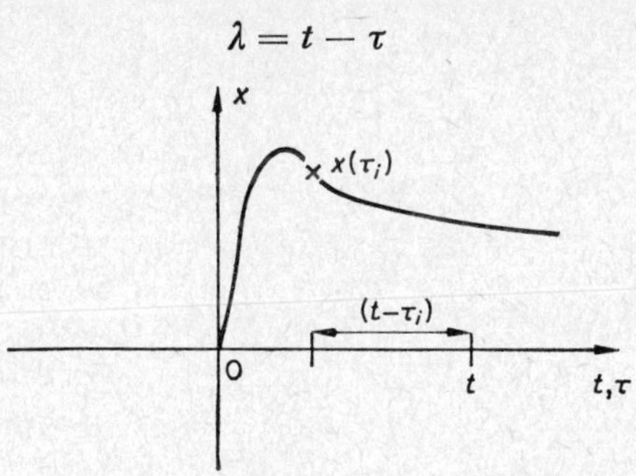

FIG. 5.1 The time axis t, τ, for equation (5.1.10).

the origin of λ is now at the point of observation in the time coordinate t, as shown in Fig. 5.2 and so,

$$y(t) = -\int_{\infty}^{0^-} h(\lambda)\, x(t-\lambda)\, d\lambda$$

$$= \int_{0^-}^{\infty} h(\lambda)\, x(t-\lambda)\, d\lambda \qquad (5.1.12)$$

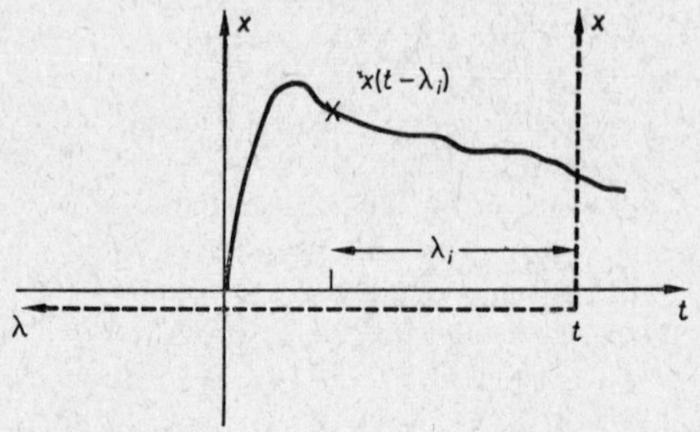

FIG. 5.2 The time axis t, λ, for equation (5.1.12).

Since equations (5.1.10) and (5.1.12) are definite integrals it is usual to write τ as the variable in each case. We then obtain the equations as written in equation (5.1.7).

In an actual physical system, the excitation is switched on at a particular time and it is convenient to take this point as the time origin. For this case, $x(t) = 0$ for $t < 0^-$ and so $x(\tau) = 0$ for $\tau < 0^-$. In addition, since the system is causal, $h(t-\tau) = 0$ for all $\tau > t$. The first integral of equation (5.1.7) can therefore be written with the same limits as the second integral and we have

$$y(t) = \int_{0^-}^{\infty} h(t-\tau)\, x(\tau)\, d\tau = \int_{0^-}^{\infty} h(\tau)\, x(t-\tau)\, d\tau \qquad (5.1.13)$$

for excitations of the form $x(t)\, u(t)$.

For an immediate illustration of the use of these equations consider the definitions of the fundamental circuit elements in terms of the system weighting function $h(t)$. A resistor is defined to be a two-terminal element for which the system weighting function relating a voltage response $\phi(t)$ to a current excitation $i(t)$ is given by,

$$h(t) = R\, \delta(t) \qquad (5.1.14a)$$

Hence, for a resistor,

$$\phi(t) = R \int_{0^-}^{\infty} \delta(\tau)\, i(t-\tau)\, u(t-\tau)\, d\tau = R\, i(t) \qquad (5.1.15a)$$

Similarly, for an inductance,

$$h(t) = L\, \delta'(t) \qquad (5.1.14b)$$

and

$$\phi(t) = L \int_{0}^{\infty} \delta'(\tau)\, i(t-\tau)\, u(t-\tau)\, d\tau = L \frac{di(t)}{dt} \qquad (5.1.15b)$$

whereas, for a capacitor,

$$h(t) = C^{-1}\,u(t) \tag{5.1.14c}$$

and

$$\begin{aligned}\phi(t) &= C^{-1}\int_{0^-}^{\infty} u(\tau)\,i(t-\tau)\,u(t-\tau)\,\mathrm{d}\tau \\ &= C^{-1}\int_{0^-}^{t} i(t-\tau)\,\mathrm{d}\tau = C^{-1}\int_{0^-}^{t} i(\xi)\,\mathrm{d}\xi\end{aligned}$$

$$\phi(t) = C^{-1}D^{-1}\,i(t) \tag{5.1.15c}$$

5.1.4 Laplace and Fourier Transform Solutions

Laplace Transform or Fourier Transform techniques may be used to convert either the differential equation (5.1.2) or the integral equation (5.1.13) into algebraic form. Consider the differential equation. If $L(s)$ is the Laplace Transform of $x(t)$, with the complex frequency variable, $s = q + j\omega$.

$$L\{D^n\,x(t)\} = s^n\,L(s) - s^{n-1}\,x(0^-) - s^{n-2}\,x'(0^-)\ldots x^{(n-1)}(0^-) \tag{5.1.16a}$$

Hence equation (5.1.2) reduces to

$$\begin{aligned}(b_m s^m + b_{m-1}s^{m-1} + \cdots b_0)\,Y(s) &- \sum_{j=0}^{m-1}\beta_j s^{m-1-j} y^j(0^-) \\ = (a_n s^n + a_{n-1}s^{n-1} + \cdots a_0)X(s) &- \sum_{i=0}^{n-1}\alpha_i s^{n-1-i}\,x^i(0^-)\end{aligned} \tag{5.1.16b}$$

where

$$L\{x(t)\} = X(s), \quad L\{y(t)\} = Y(s), \quad \{D^i\,x(t)\}_{t=0^-} = x^i(0^-)$$

and

$$\{D^j\,y(t)\}_{t=0^-} = y^j(0^-)$$

Equation (5.1.16b) may be written

$$Y(s) = \frac{P(s)}{Q(s)}.X(s) - \frac{\Gamma_0(s)}{Q(s)} = H(s)\,X(s) - \frac{\Gamma_0(s)}{Q(s)} \tag{5.1.16c}$$

where $P(s)$, $Q(s)$ are the polynomials

$$\sum_{l=0}^{n} a_l s^l \text{ and } \sum_{r=0}^{m} b_r s^r$$

respectively, and $\Gamma_0(s)$ is a polynomial in s with coefficients determined by the circuit parameters and the initial conditions for the system. The response of the system in the time domain, $y(t)$, therefore, may be obtained from the inverse Laplace Transform of the right-hand side of equation (5.1.16c) and will automatically include the initial conditions because of the term in $\Gamma_0(s)$. For an electrical network composed of circuit elements defined by the time domain relations set out in equations (3.1.1)–(3.1.4) the coefficients a_l, b_r in the polynomials $P(s)$, $Q(s)$ will be real numbers since R, L, C and M are themselves real quantities.

If Fourier Transforms had been used to solve equation (5.1.2) instead of Laplace Transforms, the initial conditions would not be included in the solution for $y(t)$. For the Fourier Transform, $F\{D^n x(t)\} = (j\omega)^n X(j\omega)$, where $X(j\omega)$ is the Fourier Transform of $x(t)$. Hence for this case

$$Y(j\omega) = \frac{P(j\omega)}{Q(j\omega)} . X(j\omega) \tag{5.1.17}$$

and the initial conditions are not specifically included. When the system is initially inert, however, both Fourier Transform and Laplace Transform techniques yield the same formal relations. From equation (5.1.16c), with $\Gamma_0(s)$ now zero for an initially inert system

$$Y(s) = \frac{P(s)}{Q(s)} . X(s) = H(s)\, X(s) \tag{5.1.18}$$

Hence, provided that the appropriate transforms exist, the choice between the Laplace or Fourier Transform can be made in this case on the grounds of mathematical convenience.

The Laplace Transform technique may also be applied to the integral equation for a linear system. Consider the integral from equation (5.1.13)

$$y(t) = \int_{0^-}^{\infty} h(t - \tau)\, x(\tau)\, d\tau \tag{5.1.19}$$

From the definition of the Laplace Transform given in Chapter 4 we have

$$L\{y(t)\} \equiv Y(s) = \int_{0^-}^{\infty} \left\{ \int_{0^-}^{\infty} h(t - \tau)\, x(\tau)\, d\tau \right\} e^{-st}\, dt \tag{5.1.20}$$

Changing the order of integration we find that

$$Y(s) = \int_{0^-}^{\infty} \int_{0^-}^{\infty} h(t - \tau)\, e^{-s(t-\tau)}\, dt . \{x(\tau)\, e^{-s\tau}\}\, d\tau$$

and putting $t - \tau = \xi$, with $h(\xi) = 0$ for $\xi < 0^-$ in the first integration for which τ is constant

$$Y(s) = \int_{0^-}^{\infty} h(\xi)\, e^{-s\xi}\, d\xi \int_{0^-}^{\infty} x(\tau)\, e^{-s\tau}\, d\tau$$

$$= L\{h(t)\} . L\{x(t)\} \tag{5.1.21}$$

where the variable can be taken as t, for each definite integral. Equation (5.1.21) may evidently be written as

$$Y(s) = H(s) . X(s) \tag{5.1.22}$$

This equation is of the same form as equation (5.1.18) but now $H(s)$ is not necessarily a rational function of s with real coefficients. When $H(s)$ refers to an initially inert, discrete parameter linear network, then those constraints will apply which led previously to the description in terms of an ordinary differential equation and $H(s)$ will be a rational function. On the other hand, the integral equation itself describes a more general class of linear systems and includes for example distributed parameter systems.

If the excitation function can be represented by a Fourier Transform (compare Chapter 4) a similar analysis can be carried through with the range of integration now extending from $-\infty$ to $+\infty$. For this case we obtain

$$Y(j\omega) = H(j\omega).X(j\omega) \tag{5.1.23}$$

which corresponds to equation (5.1.17).

5.1.5 Transfer Function

The function $H(s)$ is called the transfer function for the system. If the excitation is a delta function,

$$x(t) = \delta(t), \text{ then } X(s) = 1 \text{ and } y(t) = L^{-1}\{H(s)\} = h(t)$$

Hence the transfer function for a linear system is the Laplace Transform of the impulse response for the system.† The dimensions of $H(s)$ are not specified and will depend upon the choice of the input and response variables. Thus if $x(t)$ is a current and $y(t)$ a voltage, $H(s)$ is an impedance function. On the other hand if both $x(t), y(t)$ are voltages then $H(s)$ is dimensionless and is called the voltage ratio transfer function.

Since the transfer function is the transform of the impulse response of the system it is in principle sufficient to know the impulse response in the time domain to obtain the representation in the complex frequency domain. From this it follows that the constraints on $h(t)$ which arise in the description of physically realizable, causal and stable systems in the time domain will lead to corresponding constraints upon $H(s)$ and $H(j\omega)$ in the frequency domain.

5.1.6 Transform Network

In the previous two sections we have given a formal discussion of the properties of a linear system as derived from the general differential or integral relationships between the excitation and the response functions. When solving a practical problem in circuit analysis using transform techniques, it is often helpful to draw out the network in the transform representation initially so that the equations are given in an algebraic form at the outset. It is therefore necessary to find the representations for the different types of circuit element. We need to evaluate the voltage–current relationships in the complex frequency domain which correspond to the definitions of equations (3.1.1)–(3.1.4) in the time domain.

For a resistance,

$$\Phi(t) = R\,i(t) \tag{5.1.24}$$

and so on, taking the Laplace Transform, the definition of a resistance in the complex frequency domain is given by

$$\Phi(s) = RI(s) \tag{5.1.25}$$

For an inductance

$$\phi(t) = LD\,i(t) \tag{5.1.26}$$

and the transform relation is

$$\Phi(s) = sLI(s) - L\,i(0^-) \tag{5.1.27}$$

† Correspondingly in equation (5.1.23) $H(j\omega)$ is the Fourier Transform of $h(t)$.

or, alternatively,

$$I(s) = \frac{1}{sL}\Phi(s) + \frac{1}{s}i(0^-) \tag{5.1.28}$$

From equations (5.1.27) and (5.1.28) it can be seen that the transformed representation of an inductance is given by an inductive element together with an element representing the initial conditions. The description given by equation (5.1.27) requires an ideal generator having a transformed e.m.f. equal to $-Li(0^-)$ in series with the inductive element, whilst equation (5.1.28) requires a current source $i(0^-)/s$ in parallel with the element. These representations are shown in Fig. 5.3.

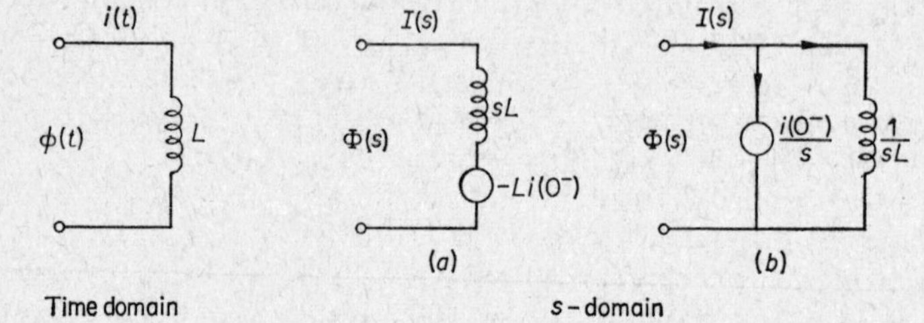

FIG. 5.3 Representation of an inductance.

For a capacitance

$$\phi(t) = C^{-1}D^{-1}\,i(t) \tag{5.1.29}$$

and the transform, relation is

$$\Phi(s) = \frac{1}{sC}I(s) + \frac{\phi(0^-)}{s} \tag{5.1.30}$$

or

$$I(s) = sC\,\Phi(s) - C\,\phi(0^-) \tag{5.1.31}$$

These representations are shown in Fig. 5.4.

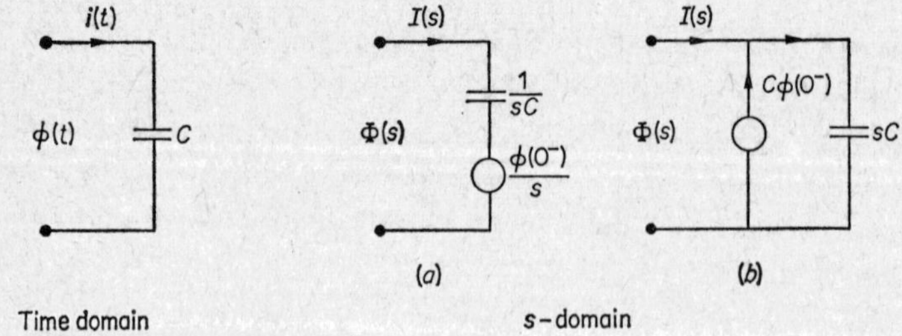

FIG. 5.4 Representation of a capacity.

For a mutual inductance

$$\phi(t) = MDI_1(t) \tag{5.1.32}$$

and so

$$\Phi(s) = sMI_1(s) - Mi_1(0^-) \tag{5.1.33}$$

Taking into account the self-inductances L_1, L_2, of the individual windings in a transformer circuit element with mutual inductance M, the output voltages will appear to be derived from generators with e.m.f.s $\Phi(s)$ in series with their appropriate internal impedances as shown in Fig. 5.5.

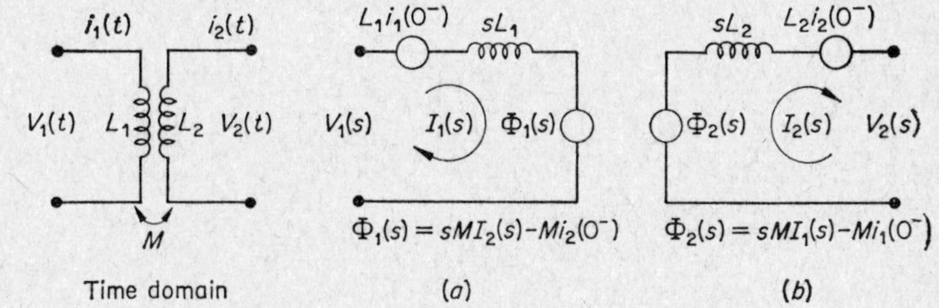

FIG. 5.5 Representation of a mutual inductance.

The transform representation of a network can now be drawn out as a set of impedances together with the appropriate generators to represent the initial conditions. The loop or node equations can be written down directly and the solution obtained either algebraically or graphically following the methods discussed in Chapter 3. These techniques will lead to an equation of the form of equation (5.1.16) with the terms in $H(s)$, $\Gamma_0(s)/Q(s)$, once more rational functions of the variable s. Note also that since the coefficients R, L, C are real numbers, the coefficients in these rational functions are also real numbers.

Example

As an example of this discussion, consider the circuit of Fig. 5.6.

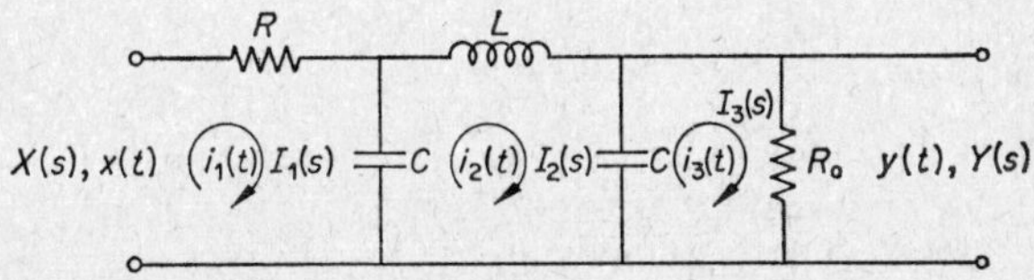

FIG. 5.6 Circuit for example discussed on page 191.

In order to simplify the algebra, assume that the system is initially inert. The loop equations are

$$
\begin{aligned}
RI_1(s) + \frac{1}{sC}I_1(s) - \frac{1}{sC}I_2(s) &= X(s) \\
-\frac{1}{sC}I_1(s) + \left(sL + \frac{2}{sC}\right)I_2(s) - \frac{1}{sC}I_3(s) &= 0 \\
-\frac{1}{sC}I_2(s) + \left(R_0 + \frac{1}{sC}\right) I_3(s) &= 0 \\
R_0 I_3(s) &= Y(s)
\end{aligned}
\tag{5.1.34}
$$

and so

$$Y(s) = \frac{R_0/s^2C^2}{\begin{vmatrix} \left(R + \dfrac{1}{sC}\right) & -\dfrac{1}{sC} & 0 \\ -\dfrac{1}{sC} & \left(sL + \dfrac{2}{sC}\right) & -\dfrac{1}{sC} \\ 0 & -\dfrac{1}{sC} & \left(R_0 + \dfrac{1}{sC}\right) \end{vmatrix}} . X(s) \tag{5.1.35}$$

and so $Y(s) = H(s)\, X(s)$ is given by

$$Y(s) = \frac{1/RLC^2}{s^3 + \dfrac{s^2}{C}\left(\dfrac{1}{R} + \dfrac{1}{R_0}\right) + s\left(\dfrac{2}{LC} + \dfrac{1}{RR_0C^2}\right) + \dfrac{1}{LC^2}\left(\dfrac{1}{R} + \dfrac{1}{R_0}\right)} . X(s) \tag{5.1.36}$$

This equation corresponds to the third-order differential equation

$$\frac{d^3y}{dt^3} + \frac{1}{C}\left(\frac{1}{R} + \frac{1}{R_0}\right)\frac{d^2y}{dt^2} + \left(\frac{2}{LC} + \frac{1}{RR_0C^2}\right)\frac{dy}{dt} + \frac{1}{LC^2}\left(\frac{1}{R} + \frac{1}{R_0}\right)y = \frac{1}{RLC^2}\, x(t) \tag{5.1.37}$$

which could have been obtained by the more tedious process of writing out and solving the set of simultaneous differential equations for the network.

The complete solution to the problem depends of course upon the prescribed form given for $x(t)$ and hence $X(s)$. When this is known, $Y(s)$ is a known rational function of s and $y(t)$ may then be obtained with the aid of a table of Laplace Transform functions.

5.2 Properties of Transfer Functions

The transfer function $H(s)$ for a discrete parameter, time invariant linear system (with discrete circuit elements as distinct from distributed elements) is a rational function of the complex frequency variable s. This is evident both from the discussion leading to equation (5.1.18) and from the analysis of a network in terms of the transformed representation discussed in the preceding section. Moreover, the coefficients in both the numerator and denominator polynomials are real. Hence, we may write

$$H(s) = \frac{a_n s^n + a_{n-1}s^{n-1} + \cdots a_0}{b_m s^m + b_{m-1}s^{m-1} + \cdots b_0} \equiv \frac{P(s)}{Q(s)} \tag{5.2.1}$$

where the a_i and b_j are real numbers. Evidently $H(s) = H(q + j\omega)$, becomes the real function $H(q)$ on the real axis of the s-domain. A function of a complex variable which is real on the real axis is called a *real function*, and so the transfer function $H(s)$ is a *real rational function.*

Consider the numerator polynomial $P(s)$, which can be written

$$P(s) \equiv P(q + j\omega) = U(q, \omega) + jV(q, \omega) \tag{5.2.2}$$

where $U(q, \omega)$ and $V(q, \omega)$ are both real functions. The function $U(q, \omega)$ is obtained from the even powers of $(j\omega)$, whilst $V(q, \omega)$ is obtained from the odd powers of $(j\omega)$ in $P(s)$, and so $U(q, \omega)$ is an even function of ω whilst $V(q, \omega)$ is an odd function of ω. Since j, ω, always occur as the product $(j\omega)$ in $P(s)$, changing the sign of j is equivalent to changing the sign of ω, hence,

$$P(s^*) = P(q - j\omega) = U(q, -\omega) + jV(q, -\omega)$$

or

$$P(s^*) = U(q, \omega) - jV(q, \omega) = P^*(s) \tag{5.2.3}$$

Similarly

$$Q(s^*) = Q^*(s)$$

and so

$$H(s^*) = \frac{P^*(s)}{Q^*(s)} = H^*(s) \tag{5.2.4}$$

The transfer function therefore assumes conjugate values at conjugate points in the complex frequency domain. This is the reflection property for $H(s)$, (and also for $P(s)$, $Q(s)$), which shows that a map of $H(s)$ in the U, V, plane is symmetrical about the real axis.

If $s_i = (q_i + j\omega_i)$ is a root of $P(s) = 0$, then s_i determines a zero of $H(s)$. For this value of s,

$$P(q_i + j\omega_i) = U(q_i, \omega_i) + jV(q_i, \omega_i) = 0 \tag{5.2.5}$$

and since $U(q_i, \omega_i)$, $V(q_i, \omega_i)$ are both real

$$U(q_i, \omega_i) = 0, \; V(q_i, \omega_i) = 0$$

Hence also

$$P(q_i - j\omega_i) = U(q_i, \omega_i) - jV(q_i, \omega_i) = 0 \tag{5.2.6}$$

and so, if $(q_i + j\omega_i)$ is a root of $P(s) = 0$, then $(q_i - j\omega_i)$ is also a root. The complex roots of $P(s)$ therefore occur in conjugate pairs, and $P(s)$ may be factored into the form

$$P(s) = P_0 \, \Pi(s - \alpha_i) \, \Pi(s - z_i)(s - z_i^*) \tag{5.2.7}$$

where P_0 is a real coefficient, and the product of linear factors gives real roots while the quadratic factors give conjugate complex roots. Some of the factors may of course be repeated.

The transfer function $H(s)$ may now be written

$$H(s) = \frac{P(s)}{Q(s)} = K \frac{\Pi(s - \alpha_i) \, \Pi(s - z_i)(s - z_i^*)}{\Pi(s - \beta_i) \, \Pi(s - w_i)(s - w_i^*)} \tag{5.2.8}$$

The roots α_i, z_i, z_i^* determine the zeros of $H(s)$ and the roots β_i, w_i, w_i^*, determine the poles of $H(s)$. The poles and zeros of $H(s)$ which are complex occur as conjugate pairs.

It can be seen from equation 5.2.8 that, apart from the scale factor, K, $H(s)$ is completely determined by the values of its poles and zeros. A pole-zero plot in the complex plane is therefore an important graphical method of displaying the transfer function. It is sufficient to know the scale factor, K, and the position of all the poles and zeros in order to be able to reconstruct $H(s)$. An example of such a plot for the function

$$H(s) = \frac{(s-1)(s^2+2s+2)}{(s+2)(s^2+2s+5)} \tag{5.2.9}$$

is given in Fig. 5.7.

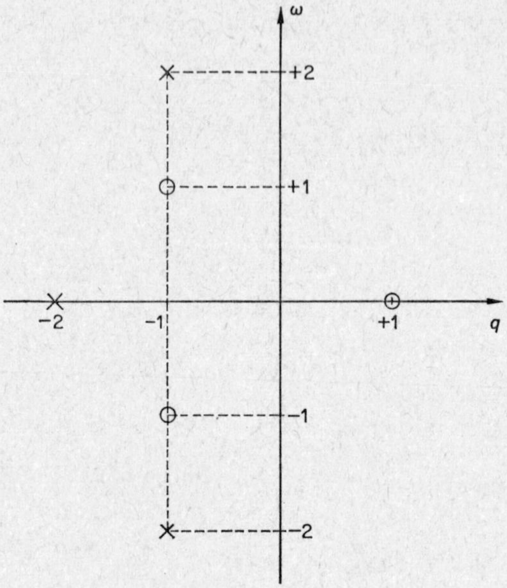

FIG. 5.7 Pole-zero plot for the function

$$\frac{(s-1)(s^2+2s+2)}{(s+2)(s^2+2s+5)} = \frac{(s-1)(s+1+j)(s+1-j)}{(s+2)(s+1+2j)(s+1-2j)}$$

A pole is denoted by ×; a zero by ○.

Note that since $H(s)$ is a rational function, the finite roots of $P(s)$, $Q(s)$ in equation (5.2.8) do not set out the complete set of poles and zeros for $H(s)$. It is necessary to examine the limit $\lim_{s\to\infty} H(s)$ in order to see whether there are also poles or zeros at infinity. From equation (5.2.1)

$$\lim_{s\to\infty} H(s) = Ks^{(n-m)} \tag{5.2.10}$$

If $n > m$ there is a pole of order $(n - m)$ at infinity whilst if $n < m$ there is a zero of order $(m - n)$ at infinity. When the poles and zeros at infinity are also counted, the number of poles of $H(s)$ is equal to the number of zeros.

5.2.1 Transfer Functions for Realizable, Stable Systems

In the previous section we have discussed properties of the transfer function $H(s)$ which arise from the general form of $H(s)$ as a rational function of the complex frequency variable. However, it does not follow that all rational functions of s are acceptable transfer functions. There are additional constraints upon $H(s)$ which are set by conditions for realizability and stability of the system.

A transfer function will be regarded as physically realizable in principle if it describes a system which is causal and non-anticipatory. A system which is initially inert may be described by the integral equation (5.1.12) with a system weighting function $h(\tau)$ which is zero for $\tau < 0^-$ and real for all $\tau \geqslant 0$. If a given rational function of s is a valid Laplace transform, then the system weighting function in the time domain is given by the inverse transformation

$$L^{-1}\{H(s)\} = h(\tau)\, u(\tau) \tag{5.2.11}$$

and realizability is automatically implied. This condition does not, however, necessarily require that the system is stable. On the other hand, a stable linear system is described by a transfer function which is a valid Laplace Transform and so the constraints imposed on the description of a stable system automatically guarantee realizability.

There are two definitions of system stability which are widely used. These are

(a) the response to an impulse excitation must be bounded excepting only impulse and doublet response terms,
(b) the response to any bounded excitation is also bounded.

Definition (a) is less restrictive than definition (b). Systems satisfying (a) are said to be conditionally stable, marginally stable, or simply stable, whereas systems satisfying (b) are said to be unconditionally stable, absolutely stable or strictly stable.

Consider the conditions imposed by definition (a). Since the Laplace Transform of the impulse function is unity

$$Y(s) = H(s)$$

and we require $y(t) = L^{-1}\{H(s)\}$ to be bounded except for impulse and doublet terms. The function $H(s)$ must therefore be a valid Laplace Transform and we may write it as

$$H(s) \equiv \frac{P(s)}{Q(s)} = \frac{P_1(s)}{Q(s)} + A + Bs \tag{5.2.12}$$

where the degree of the polynomial $P_1(s)$ is less than that of $Q(s)$ and the term A gives rise to an impulse term in $y(t)$ whilst Bs gives rise to a doublet.

Hence the degree of the complete numerator function $P(s)$, in the rational fraction expression for $H(s)$, cannot exceed the degree of the denominator function $Q(s)$ by more than unity. That is,

$$n - m \leqslant 1$$

or

$$n \leqslant m + 1 \tag{5.2.13}$$

The inverse Laplace Transform of $H(s)$ may be obtained from an expansion into partial fractions. The denominators of the individual terms in this expression will be obtained from the roots of $Q(s)$. From equations (5.2.8) and (5.2.12)

$$H(s) = \frac{P_1(s)}{Q_0 \Pi(s-\beta_i)\Pi(s-w_i)(s-w_i^*)} + A + Bs \tag{5.2.14}$$

$$= \sum_i \frac{\alpha_i}{(s-\beta_i)^p} + \sum_i \frac{\gamma_i s + \delta_i}{\{(s-q_i)^2+\omega_i^2\}^r} + A + Bs \tag{5.2.15}$$

In this equation the indices p, r, take care of the repeated roots of $Q(s)$. Thus, if there is a factor $(s-\beta_i)^p$ in $Q(s)$, the partial fraction expansion contains the terms

$$\frac{\alpha_p}{(s-\beta_j)^p} + \frac{\alpha_{p-1}}{(s-\beta_j)^{p-1}} + \cdots \frac{\alpha_1}{(s-\beta_j)} \tag{5.2.16}$$

whilst a repeated quadratic factor, $\{(s-q_i)^2+\omega_i^2\}^r$ gives rise to

$$\frac{\gamma_r s+\delta_r}{\{(s-q_i)^2+\omega_i^2\}^r} + \frac{\gamma_{r-1}s+\delta_{r-1}}{\{(s-q_i)^2+\omega_i^2\}^{r-1}} + \cdots \frac{\gamma_1 s+\delta_1}{\{(s-q_i)^2+\omega_i^2\}} \tag{5.2.17}$$

TABLE 5.1

	Form of term in $H(s)$	*Form of term in* $y(t)$
1.	Bs	Doublet $\frac{d}{dt}\{\delta(t)\}$
2.	A	Impulse $\delta(t)$
3.	$\frac{1}{s}$	$u(t)$
4.	$\frac{1}{s-\beta_i}$	$\exp(\beta_i t).u(t)$
5.	$\frac{\gamma_i s+\delta_i}{(s^2+\omega_i^2}$	$(\cos\omega_i t, \sin\omega_i t)\, u(t)$
6.	$\frac{\gamma_i s+\delta_i}{[(s-q_i)^2+\omega_i^2]}$	$\{\exp(q_i t).\cos\omega_i t, \exp(q_i t).\sin\omega_i t\}\, u(t)$
7.	$\frac{1}{s^n}, (n>1)$	$t^{n-1}\, u(t)$
8.	$\frac{1}{(s-\beta_i)^n}, (n>1)$	$t^{n-1}\exp(\beta_i t).u(t)$
9.	$\frac{\gamma_i s+\delta_i}{(s^2+\omega_i^2)}, (n>1)$	$(t^{n-1}\cos\omega_i t, t^{n-1}\, sin\, \omega_i t)\, u(t)$
10.	$\frac{\gamma_i s+\delta_i}{\{(s-q_i)^2+\omega_i^2\}^n}$	$\{t^{n-1}\exp(q_i t).\cos\omega_i t, t^{n-1}\exp(q_i t).\sin\omega_i t\}\, u(t)$ $n>1$

Equation (5.2.15) shows that the terms which can possibly occur in $y(t)$, without any further restrictions upon $H(s)$, are those set out in Table 5.1. It can be seen that terms of the form 1, 2, 3 and 5 are always allowed when stability is defined by definition (a). Terms 4, 6, 8 and 10, are only allowed if β_i is negative and q_i is negative. Terms 7, 9 are not allowed since these always become unbounded as $t \to \infty$. A transfer function, $H(s)$, which describes a conditionally stable linear system is therefore subject to the following constraints:

(a) The degree of the numerator polynomial cannot exceed the degree of the denominator polynomial by more than unity (from equation (5.2.13)).
(b) $H(s)$ cannot have poles in the right half complex plane (from β_i, q_i negative and terms 3, 5, of Table 5.1).
(c) $H(s)$ cannot have multiple poles on the $j\omega$ axis (from terms 7, 9 of Table 5.1).
(d) The poles and zeros of $H(s)$ which are complex occur as conjugate pairs (from the discussion leading to equation (5.2.8)).

The restrictions imposed on $H(s)$ by using definition (b) for stability may be derived by considering the integral form for the linear relation between excitation and response. For a bounded input $x(t)$, we have

$$|x(t)| \leqslant M_1 \quad \text{for} \quad 0 < t < \infty \tag{5.2.18}$$

and we require that

$$|y(t)| \leqslant M_2 \quad \text{for} \quad 0 < t < \infty \tag{5.2.19}$$

where M_1, M_2 are positive finite quantities. From the integral relation, equation (5.1.13).

$$\begin{aligned} |y(t)| &= \left|\int_{0^-}^{\infty} h(\tau)\, x(t-\tau)\, d\tau\right| \\ &\leqslant \int_{0^-}^{\infty} |h(\tau)| . |x(t-\tau)|\, d\tau \\ &\leqslant M_1 \int_{0^-}^{\infty} |h(\tau)|\, d\tau \end{aligned} \tag{5.2.20}$$

Hence, $y(t)$ is bounded for all bounded inputs, and so the system is unconditionally stable, if

$$\int_{0^-}^{\infty} |h(\tau)|\, d\tau \leqslant M \tag{5.2.21}$$

where M is a positive finite quantity.

Equation (5.2.21) is evidently sufficient to ensure that $y(t)$ is bounded when the excitation is bounded. In order to establish that this condition is also necessary (and therefore no more restrictive than is essential) we shall show that when equation (5.2.21) is not satisfied there exists a bounded excitation for which the response $y(t)$ is unbounded. Consider an excitation $x(t)$ which has the form $x(t-\tau) = +1$ whenever $h(\tau) > 0$ (that is, positive) and

$x(t-\tau) = -1$ whenever $h(\tau) < 0$ (that is, negative). The response $y(t)$ is given by

$$y(t) = \int_{0^-}^{\infty} h(\tau)\, x(t-\tau)\, \mathrm{d}\tau = \int_{0^-}^{\infty} |h(\tau)|\, \mathrm{d}\tau$$

and $y(t)$ will be unbounded if $\int_{0^-}^{\infty} |h(\tau)|\, \mathrm{d}\tau \rightarrow \infty$. Equation (5.2.21) is therefore both necessary and sufficient to ensure unconditional stability.

The system weighting function, $h(t)$, is also the impulse response function and so equation (5.2.21) imposes constraints upon $H(s)$ through the relation $h(t) = L^{-1}\{H(s)\}$ or $H(s) = L\{h(t)\}$. The transfer function for an unconditionally stable system will be more severely restricted than that for a conditionally stable system. Instead of requiring only that $h(t)$ be bounded except for impulse and doublet response terms, we now require that $h(t)$ be absolutely integrable. This requirement will still allow impulse terms to be present but excludes terms in the time domain of the form $\mathrm{d}/\mathrm{d}t\{\delta(t)\}$, $u(t)$, and $(\cos \omega_i t, \sin \omega_i t).u(t)$ which were acceptable for a conditionally stable system. It is immediately evident that the terms $u(t)$, $(\cos \omega_i t, \sin \omega_i t).u(t)$ are not absolutely integrable. The doublet term is excluded either by noting that the response to the bounded excitation $x(t) = u(t)$ is

$$\begin{aligned} y(t) &= \int_{0^-}^{\infty} \frac{\mathrm{d}}{\mathrm{d}t}\{\delta(\tau)\}\, u(t-\tau)\, \mathrm{d}\tau \\ &= u'(t) \\ &= \delta(t) \end{aligned} \tag{5.2.22}$$

and so is unbounded, or by observing that exclusion of terms

$$(\cos \omega_i t, \sin \omega_i t)\, u(t)$$

means that $H(s)$ cannot have any poles on the $\mathrm{j}\omega$ axis, including the points $s = \pm \mathrm{j}\infty$, so that $H(s)$ cannot contain a term of the form Bs which gives rise to the doublet response in the time domain. This condition also implies that the degree of the numerator function $P(s)$ in $H(s)$ cannot exceed the degree of the denominator function $Q(s)$.

From arguments similar to those used for the conditionally stable definition it is evident from equation (5.2.21) and Table 5.1, that terms of the form 1, 3, 5, 7, 9 are excluded from the transfer function for an unconditionally stable system and that terms 4, 6, 8, 10 are only allowed if β_i, q_i are each negative. An unconditionally stable system is therefore a more severely constrained type of stable system and the restrictions on the transfer function $H(s)$ may be summarized as follows:

(a) The degree of the numerator polynomial cannot exceed the degree of the denominator polynomial.
(b) All the poles of $H(s)$ lie in the left half plane and there are no poles on the $\mathrm{j}\omega$ axis.

(c) Conditions (a) and (b) arise directly from the constraint

$$\int_{0^-}^{\infty} |h(\tau)| \, d\tau \leqslant M \text{ and } H(s) = L^{-1}\{h(t)\}$$

(d) The poles and zeros of $H(s)$ which are complex occur as conjugate pairs.

5.2.2 Location of Poles for a Transfer Function

The conditions for stability discussed in the previous section impose restrictions on the location of the poles of the transfer function. These poles are determined by the roots of the denominator function $Q(s)$. Evidently $Q(s)$ must be a polynomial which has roots with real parts zero or negative and which is itself real when s is real. These conditions identify $Q(s)$ as a Hurwitz polynomial. A necessary condition for all the roots of the polynomial to have non-positive real parts is that all the coefficients in $Q(s)$ should have the same sign and that no coefficient should be zero. This, however, is not in itself a sufficient condition since some functions having coefficients all of the same sign and no coefficient zero may possess roots in the right half plane.† A more restrictive algebraic test is therefore usually applied which is both necessary and sufficient. This is the Routh–Hurwitz test which determines whether or not the roots have real parts which are strictly negative.

Consider a polynomial $Q(s)$ for which every coefficient is positive and no coefficient is zero.

$$Q(s) = b_m s^m + b_{m-1} s^{m-1} + \cdots b_0 \tag{5.2.23}$$

Construct the following determinants

$$\Delta_1 = |b_1| \quad \Delta_2 = \begin{vmatrix} b_1 & b_0 \\ b_3 & b_2 \end{vmatrix} \tag{5.2.24}$$

$$\Delta_3 = \begin{vmatrix} b_1 & b_0 & 0 \\ b_3 & b_2 & b_1 \\ b_5 & b_4 & b_3 \end{vmatrix} \quad \Delta_j = \begin{vmatrix} b_1 & b_0 & 0 & \ldots & 0 \\ b_3 & b_2 & b_1 & & \cdot \\ b_5 & b_4 & b_3 & & \cdot \\ \cdot & \cdot & \cdot & & \cdot \\ \cdot & \cdot & \cdot & & \cdot \\ \cdot & \cdot & \cdot & & \cdot \\ b_{2j-1} & b_{2j-2} & b_{2j-3} & \ldots & b_j \end{vmatrix}$$

When constructing the determinant of the general form Δ_j using the coefficients from the polynomial $Q(s)$, an element b_i is put equal to zero whenever the index i exceeds m. The Routh–Hurwitz criterion states that a necessary and sufficient condition for each root of $Q(s) = 0$ to have a strictly negative real part is that each Δ_j for $j = 1$ to $(m - 1)$ be positive.‡

An alternative graphical method for locating the poles of a transfer function can be developed from a basic mapping theorem for functions of a complex variable.‡ If $f(s)$ is analytic within and on a closed contour C except at a finite

† For example $s^4+s^3+s^2+11s+10 = 0$ has roots $s_i = -1, -2, (+1 \pm 2j)$.

‡ See, for example, Wylie, C. R., *Advanced Engineering Mathematics*, McGraw-Hill, 1960; and Pfeiffer, P. E., *Linear Systems Analysis*, McGraw-Hill, 1961.

number of poles, and if $f(s)$ has neither poles nor zeros on C, then

$$\frac{1}{2\pi}\{\text{variation of arg } f(s) \text{ around } C\} = N - P \tag{5.2.25}$$

where N is the number of zeros of $f(s)$ within C and P is the number of poles within C, multiple zeros and poles being counted the appropriate number of times.

When the contour is C_∞, chosen to be the $j\omega$ axis and the infinite semicircle enclosing the entire right half plane, this theorem provides the basis for the Nyquist criterion for stability. If the locus of $W = H(s)$ is mapped in the $W(= U, V)$-plane for values of s ranging round the contour C_∞ then the number of times this locus encircles the origin, counted with the appropriate signs, is equal to the difference in the number of poles and zeros in the right half plane. Moreover since $H(s) = 0$ requires $W = 0$ and so $U = 0$, $V = 0$ it is evident that any zero on C_∞ is demonstrated by the mapped curve passing through the origin in the W-plane.

This discussion only determines the difference between the number of poles and zeros of $H(s)$ which lie in the right half plane. It can, however, be extended to determine the number of poles separately by considering the denominator polynomial $Q(s)$. A polynomial equation has only a finite number of roots and so $Q(s)$, which has no poles,† has only a finite number of zeros (compare equation (5.2.8)). It is evident therefore that the number of zeros of $Q(s)$ which lie in the right half plane can be determined by considering again a contour consisting of the $j\omega$ axis and a semicircle of sufficiently large radius. Since $Q(s)$ is a polynomial, it is only necessary to retain the term in the highest power of s when considering points ranging round the semicircle. Thus, writing $s = R\,e^{j\theta}$ for this part of the contour, and taking R sufficiently large,

$$Q(s) \rightarrow b_m s^m = b_m R^m\,e^{jm\theta} \tag{5.2.26}$$

and as θ varies between $+\pi/2$ and $-\pi/2$, the argument of $Q(s)$ varies from $+m\pi/2$ to $-m\pi/2$ giving a total change of $m\pi$ in arg $Q(s)$. It is now only necessary to evaluate arg $Q(s)$ over the imaginary axis in order to complete the contour calculation. Hence if the net change in arg $Q(s)$ is determined for values of s ranging from $+j\infty$ to $-j\infty$ and this value is added (with its proper sign) to $m\pi$, the number of roots of $Q(s)$ lying in the right half plane is given by this total divided by 2π.

The labour of calculation can be further reduced by recalling that $Q(s^*) = Q^*(s)$ and so only positive values of $j\omega$ need be considered in order to derive the complete change in arg $Q(j\omega)$. When the degree m of $Q(s)$ is even, the number of encirclements of the origin for the map of $Q(j\omega)$ in the W-plane may be added to $m/2$ with appropriate sign in order to evaluate the total number of zeros in the right half plane. Note that if $Q(s) = 0$ for a value of s on the $j\omega$ axis, the mapped curve for $Q(j\omega)$ will pass through the origin of the W-plane.

† The poles at infinity are left outside the contour and so do not contribute.

Examples

As an example of this discussion consider the polynomial

$$Q(s) = s^5 + 7s^4 + 25s^3 + 47s^2 + 52s + 24 \tag{5.2.27}$$

All the coefficients have the same sign, and we apply the Routh–Hurwitz test by evaluating the determinants

$$\Delta_1 = +52 \quad \Delta_2 = \begin{vmatrix} 52 & 24 \\ 25 & 47 \end{vmatrix} = +1844$$

$$\Delta_3 = \begin{vmatrix} 52 & 24 & 0 \\ 25 & 47 & 52 \\ 1 & 7 & 25 \end{vmatrix} = +28420 \qquad \Delta_4 = \begin{vmatrix} 52 & 24 & 0 & 0 \\ 25 & 47 & 52 & 24 \\ 1 & 7 & 25 & 47 \\ 0 & 0 & 1 & 7 \end{vmatrix} = +257{,}712 \tag{5.2.28}$$

Since all the determinants are positive, the roots of $Q(s)$ all lie in the left half plane. The roots are, in fact $s = -1$, $s = -1 \pm j\sqrt{2}$, $s = -2 \pm 2j$.

We may consider the same function and use the graphical method.

$$Q(j\omega) = (7\omega^4 - 47\omega^2 + 24) + j(\omega^5 - 25\omega^3 + 52\omega) \tag{5.2.29}$$

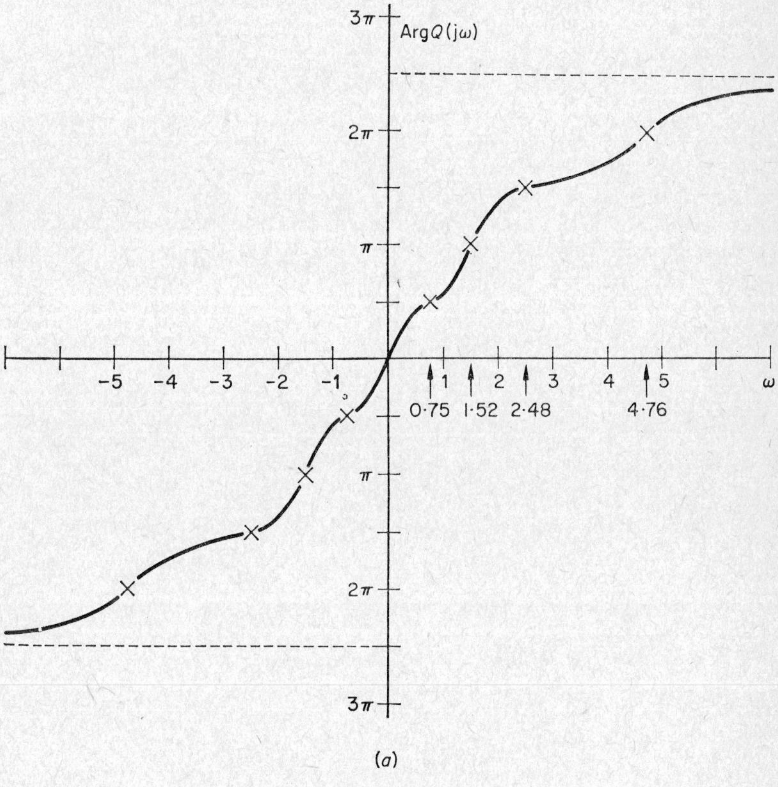

(a)

FIG. 5.8

(a) Sketch of arg $Q(j\omega)$ corresponding to equation (5.2.30b).

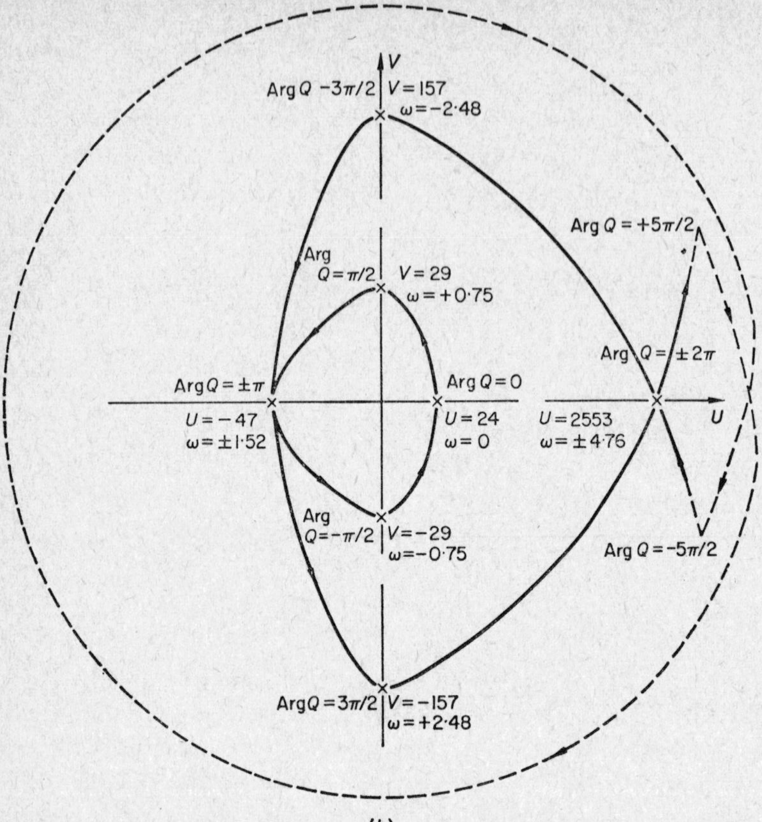

(b) Map of $Q(j\omega)$ in the W-plane, with rounded up values of the co-ordinates.

and so

$$\arg Q(j\omega) = \tan^{-1} \frac{\omega(\omega^4 - 25\omega^2 + 52)}{7\omega^4 - 47\omega^2 + 24} \tag{5.2.30a}$$

that is,

$$\arg Q(j\omega) = \tan^{-1} \frac{\omega(\omega - 1{\cdot}52)(\omega - 4{\cdot}76)(\omega + 1{\cdot}52)(\omega + 4{\cdot}76)}{7(\omega - 0{\cdot}75)(\omega - 2{\cdot}48)(\omega + 0{\cdot}75)(\omega + 2{\cdot}48)} \tag{5.2.30b}$$

This function is shown in Fig. 5.8(a).

It can be seen that, as ω ranges from $-\infty$ to $+\infty$, arg $Q(j\omega)$ varies from $-5\pi/2$ to $+5\pi/2$. On the semicircular curve, arg $Q(s)$ varies from $+5\pi/2$ to $-5\pi/2$ so that the net change in arg $Q(s)$ for the contour enclosing the right half plane is zero. Hence the roots of $Q(s)$ all lie in the left half plane.

The map of $Q(j\omega)$ in the W-plane is shown in Fig. 5.8(b). This curve does not pass through the origin in the W-plane and so $Q(s)$ has no zeros on the imaginary axis. From this curve it can again be seen that arg $Q(j\omega)$ changes from $-5\pi/2$ to $+5\pi/2$ as ω ranges from $-\infty$ to $+\infty$.

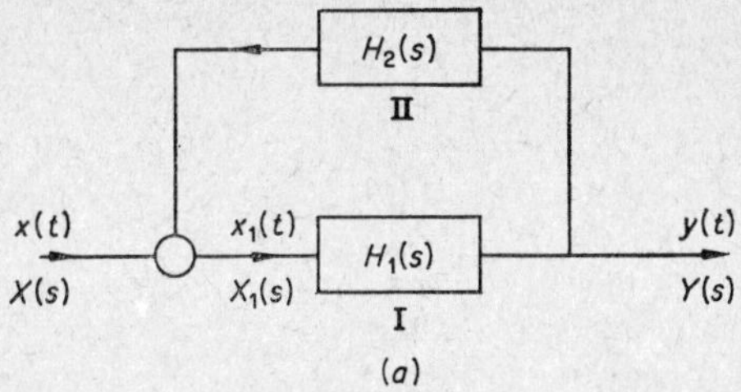

FIG. 5.9
(a) Representation of a feedback loop.

As an additional example consider the feedback loop shown in Fig. 5.9(a). This circuit consists of two linear systems having transfer functions $H_1(s)$, $H_2(s)$ together with a summing junction at the input to system 1. If $x_1(t)$ is the actual input to system 1,

$$Y(s) = H_1(s)\, X_1(s)$$

But

$$X_1(s) = X(s) + H_2(s)\, Y(s)$$

and so

$$Y(s) = \frac{H_1(s)}{1 - H_1(s)\, H_2(s)} . X(s)$$

Hence, the over-all transfer function for the system is

$$H(s) = \frac{H_1(s)}{1 - H_1(s)\, H_2(s)}$$

Assuming that system 1 is unconditionally stable when there is no feedback loop, $H_1(s)$ has no poles in the right half plane and so the stability of the complete system is determined by the zeros of the denominator function. The denominator will, however, generally be in the form of a rational function of s rather than a polynomial. We write

$$D(s) = 1 - H_1(s)\, H_2(s) \tag{5.2.31}$$

If the product $H_1(s)\, H_2(s)$ tends to a constant, A, or to zero as s tends to infinity, the map of $D(s)$ in the W-plane for values of s ranging round the infinite semicircle is a fixed point, either $(1 - A, 0)$ or $(1, 0)$, lying on the U-axis. Hence the product $H_1(s)\, H_2(s)$ does not change for these values of s and so there is no contribution to arg $D(s)$ from the semicircular part of the contour. The zeros of $D(s)$ are now determined completely from the variation of arg $D(j\omega)$, or correspondingly from any encirclements of the origin by the map of $D(j\omega)$ in the W-plane when $H_2(s)$ has no poles in the right half plane.

The function $D(j\omega) = 1 - H_1(j\omega)\, H_2(j\omega)$ is determined by the open loop transfer functions of the individual components of the feedback network, evaluated on the $j\omega$ axis. These transfer functions can in principle be derived from the steady state response of each component to a sinusoidal excitation and so the graphical procedure has a direct physical interpretation.

A simple operational amplifier with resistive feedback is shown in Fig.

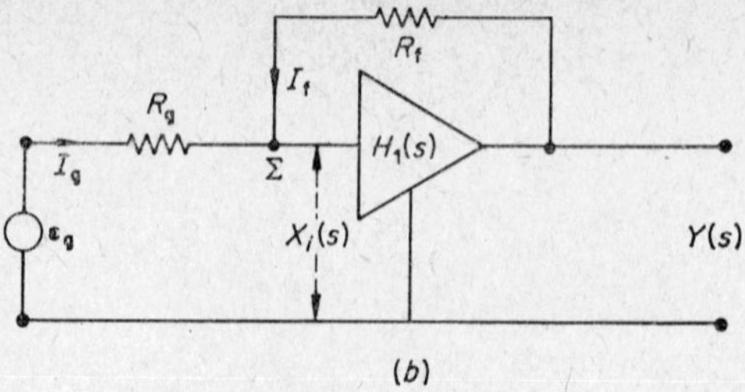

(b) A simple operational amplifier configuration with resistive feedback.

5.9(b). The circuit is excited by the voltage generator $\mathscr{E}_g$ and Σ is a current summing junction.

If $y(t)$ is the voltage output from the amplifier and $x_i(t)$ is the voltage at the amplifier input terminals,

$$Y(s) = H_1(s)\, X_i(s)$$

with

$$X_i(s) = \mathscr{E}_g - R_g\, I_g(s)$$

and

$$R_f I_f(s) = Y(s) - X_i(s) \simeq Y(s)$$

since we assume that the amplifier has sufficient gain for $X_i(s)$ to be negligible compared with $Y(s)$. Moreover, a good operational amplifier draws no significant input current, and so

$$I_g(s) + I_f(s) = 0$$

Consequently

$$Y(s) = H_1(s)\{\mathscr{E}_g - R_g I_g(s)\}$$

$$= H_1(s)\left\{\mathscr{E}_g + \frac{R_g}{R_f} Y(s)\right\}$$

$$Y(s) = \frac{H_1(s)}{1 - (R_g/R_f)\, H_1(s)} \cdot \mathscr{E}_g$$

Describe the operational amplifier by a transfer function† $H_1(s) = -|K(\omega)|\, e^{-s\tau_0}$, where we write $|K(\omega)|$ to identify a positive quantity, and put $R_g/R_f = \beta$. Then the overall transfer function is

$$H(s) = \frac{-|K(\omega)| \exp(-s\tau_0)}{1 + \beta |K(\omega)| \exp(-s\tau_0)}$$

The exponential factor $\exp(-s\tau_0)$ may be regarded as representing a polynomial in s and so our previous discussion of stability is valid. The function $H_1(s)$ has no poles in the right half plane and so the significant poles of $H(s)$ are determined by the zeros of the denominator function. That is,

$$1 + \beta\, |K(\omega)| \exp(-s\tau_0) = 0$$

† See equation (5.2.37).

Write $s = q + j\omega$ and equate the real and imaginary parts to zero.

$$U = 1 + \beta\,|K(\omega)|\,\exp(-q\tau_0)\cos\omega\tau_0 = 0$$
$$V = -\beta\,|K(\omega)|\,\exp(-q\tau_0)\sin\omega\tau_0 = 0$$

The term $\exp(-q\tau_0)$ is a positive quantity and we assume that $|K(\omega)|$ is well behaved having neither poles nor zeros over the range of interest. The first zero of the equation for $V = 0$, which has the appropriate sign, occurs when $\omega = \omega_i$ and

$$\omega_i\tau_0 = \pi$$

and so

$$\exp(-q_i\tau_0) = \frac{1}{\beta\,|K(\omega_i)|}$$
$$q_i = \frac{1}{\tau_0}\log_e \beta\,|K(\omega_i)|$$

which will lie in the right-half plane if $\beta\,|K(\omega_i)| > 1$. This amplifier configuration will therefore be unstable if $\beta\,|K(\omega_i)| > 1$ at that frequency ω_i for which the phase shift $\omega\tau_0$ is equal to π.

To derive this condition from a map in the U, V-plane, plot the function of ω,

$$U = 1 + \beta\,|K(\omega)|\cos\omega\tau_0$$
$$V = -\beta\,|K(\omega)|\sin\omega\tau_0$$

This function will encircle the origin if $\beta\,|K(\omega)| > 1$ when $\omega\tau_0 = \pi$ (or more generally when $\omega\tau_0 = (2n + 1)\pi$).

5.2.3 Particular Transfer Functions

The transfer functions which refer to the driving point terminals of a network and those which describe an ideal amplifier or delay line deserve some additional comments. We shall also introduce the minimum phase and the all pass transfer functions.

The driving point impedance function $H_z(s)$ is the reciprocal of the driving point admittance function $H_A(s)$ and so these transfer functions are subject to additional constraints. For example, both of the functions $H_z(s) = Z_{dr}(s)$ and

$$H_A(s) = \frac{1}{Z_{dr}(s)}\{= Y_{dr}(s)\}$$

must be valid transfer functions. Hence the driving point transfer function for an unconditionally stable system must have all its *poles* and *zeros* in the left half plane. This is more restrictive than the conditions which apply to a general transfer function since for this case the zeros may lie anywhere in the complex plane, subject only to the symmetry requirement that they occur as conjugate complex pairs.

The reciprocal relation between the driving point impedance and admittance functions also determines a relation between the degree of the polynomials $P(s)$ and $Q(s)$. All transfer functions when written in the form

$$H(s) = \frac{P(s)}{Q(s)}$$

require that (compare equation (5.2.13)),

$$n \leqslant m + 1 \tag{5.2.32}$$

if the system is to be stable. This is the case for all passive networks. However, for the driving point function,

$$\{H(s)\}^{-1} \equiv H_1(s) = \frac{Q(s)}{P(s)}$$

is also a valid transfer function and so we also require that

$$m \leqslant n + 1 \tag{5.2.33}$$

Hence from equations (5.2.32) and (5.2.33)

$$m - 1 \leqslant n \leqslant m + 1 \tag{5.2.34}$$

For a driving point transfer function, which refers to a stable network, the degree of the numerator function can differ from the degree of the denominator function by unity at the most. In addition $\mathrm{Re}\,[Z_{dr}(j\omega)] \geqslant 0$, $\mathrm{Re}\,[Y_{dr}(j\omega)] \geqslant 0$ for all ω, see page 138 Chapter 3, and so the driving point function is a positive real function.

An ideal amplifier or delay line produces at its output a response signal which is an exact replica of the input signal. The response may be time delayed with respect to the excitation and it may be scaled by a numerical factor K. If $|K|$ is greater than unity we may regard the system as a linear amplifier, whereas, if $|K|$ is less than or equal to unity we may regard the system as a delay line. Evidently, if the time delay is τ_0,

$$y(t) = Kx(t - \tau_0) \tag{5.2.35}$$

taking the Laplace Transform

$$Y(s) = K \exp(-s\tau_0) . X(s) \tag{5.2.36}$$

and so the transfer function for the ideal linear amplifier or the delay line is

$$H(s) = K \exp(-s\tau_0) \tag{5.2.37}$$

The corresponding impulse response of the system is

$$h(t) = K\, \delta(t - \tau_0) \tag{5.2.38}$$

Since $H(s)$ given by equation (5.2.37) is not a rational function of s, an ideal delay line cannot be constructed from discrete parameter R, L, C, networks. It may, however, in principle, be constructed from distributed elements such as arise, for example, in the ideal transmission line.

Note that the gain factor K is not necessarily a *dimensionless* constant in these equations. For example if $y(t)$ is the voltage response to an input current $x(t)$, then K will have the dimensions of impedance.

An arbitrary transfer function $H(s)$, for a stable system when given as a rational function of s, with all its poles in the left half plane can be written as the product of two functions $H_1(s)$, $H_2(s)$ which have the following properties:

$H_1(s)$ is a transfer function with all finite poles and zeros on the $j\omega$ axis or the left half plane.† This type of function is called a *minimum phase transfer function.*

† It may, however, have a pole or zero at infinity as discussed on page 194.

$H_2(s)$ is a transfer function with all of its poles in the left half plane and all of its zeros in the right half plane. The poles and zeros occur as mirror images in the jω axis. This type of function is called an *all pass function* since $|H_2(j\omega)| = 1$ for all ω.†

We can write an arbitrary transfer function, $H(s)$, as

$$H(s) = \frac{P_1(s)\, P_2(-s)}{Q(s)} \tag{5.2.39}$$

where $P_1(s)$ is a polynomial with all zeros on the jω axis or in the left half plane, $P_2(-s)$ is a polynomial with all zeros in the right half plane, and $Q(s)$ has zeros only in the left half plane from the conditions for stability.

Hence we may also write, if $P_2(s)$ is the image function of $P_2(-s)$

$$\begin{aligned} H(s) &= \frac{P_1(s)\, P_2(s)}{Q(s)} \cdot \frac{P_2(-s)}{P_2(s)} \\ &= H_1(s) . H_2(s) \end{aligned} \tag{5.2.40}$$

where $H_1(s)$, $H_2(s)$ are the functions described above.

Note that a driving point transfer function is a minimum phase transfer function.

5.2.4 Real and Imaginary Parts

Transfer functions for unconditionally stable linear systems have all their poles in the left half plane and we may apply the Kramers–Kronig relations derived in Section 1.3.6 of Chapter 1 to the real and imaginary parts of $H(s)$. There are no poles on the jω axis and so, if we write

$$H(s) = U(q, \omega) + jV(q, \omega)$$

we obtain from equations (1.3.68) and (1.3.69),

$$U(\Omega) = U(\infty) - \frac{2}{\pi} \mathrm{P} \int_0^\infty \frac{\omega V(\omega)}{\omega^2 - \Omega^2}\, d\omega \tag{5.2.41}$$

$$V(\Omega) = V(\infty) + \frac{2\Omega}{\pi} \mathrm{P} \int_0^\infty \frac{U(\omega)}{\omega^2 - \Omega^2}\, d\omega \tag{5.2.42}$$

A conditionally stable system may have simple poles on the jω axis and a pole at $s = \infty$. For this case the poles on the jω axis may be by-passed with small semicircular indentations in the contour integral as shown in Fig.

† We may write $H_2(s) = \dfrac{\Pi(s - q_r - j\omega_r)}{(\Pi s + q_r - j\omega_r)}$

and $H_2(j\omega) = \dfrac{\Pi\{-q_r + j(\omega - \omega_r)\}}{\Pi\{q_r + j(\omega - \omega_r)\}}$

that is $H_2(j\omega) = \dfrac{\Pi\{q_r^2 + (\omega - \omega_r)^2\}^{1/2} \exp(-j\phi_r)}{\Pi\{q_r^2 + (\omega - \omega_r)^2\}^{1/2} \exp(+j\phi_r)}$

with $\tan \phi_r = (\omega - \omega_r)/q_r$

Hence $|H_2(j\omega)| = 1$. Also $\arg H_2(j\omega) = -2\Sigma\, \phi_r$ and as ω ranges from 0 to ∞ $\arg H_2(j\omega)$ ranges from 0 to $-(r\pi)$.

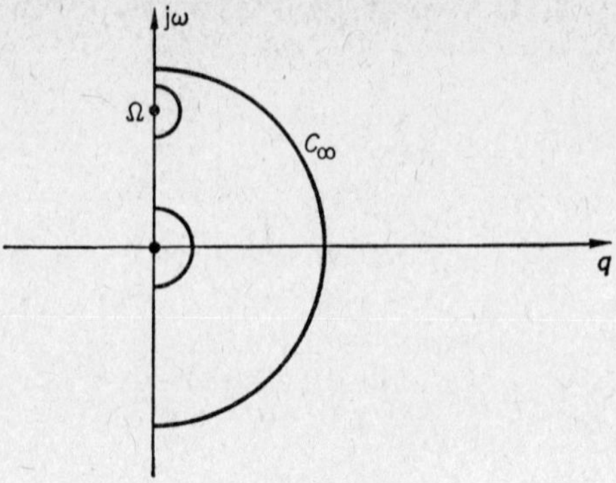

(a)

FIG. 5.10

(a) Curve for the contour integral leading to equations (5.2.43) and (5.2.44).

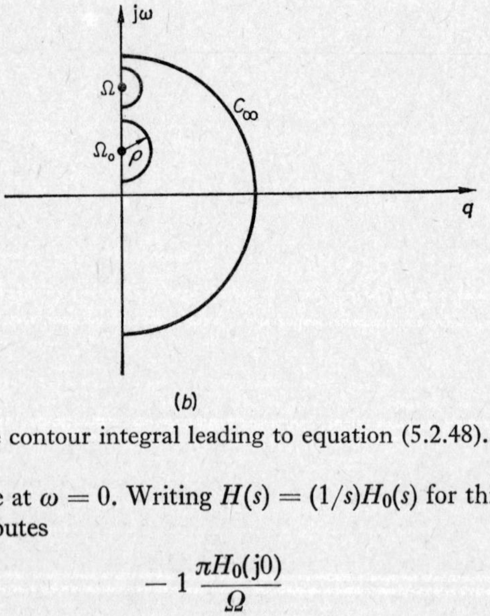

(b)

(b) Curve for the contour integral leading to equation (5.2.48).

5.10(a) for a pole at $\omega = 0$. Writing $H(s) = (1/s)H_0(s)$ for this case the pole at $\omega = 0$, contributes

$$-1\,\frac{\pi H_0(\mathrm{j}0)}{\Omega}$$

to the complete integral. Hence for a pole at $\omega = 0$, but no pole at $s = \infty$

$$U(\Omega) = U(\infty) + \frac{\mathrm{Im}\,[H_0(\mathrm{j}0)]}{\Omega} - \frac{1}{\pi}\mathrm{P}\int_{-\infty}^{\infty} \frac{V(\omega)}{\omega - \Omega}\,\mathrm{d}\omega \qquad (5.2.43\mathrm{a})$$

$$V(\Omega) = V(\infty) - \frac{\mathrm{Re}\,[H_0(\mathrm{j}0)]}{\Omega} + \frac{1}{\pi}\mathrm{P}\int_{-\infty}^{\infty} \frac{U(\omega)}{\omega - \Omega}\,\mathrm{d}\omega \qquad (5.2.44\mathrm{a})$$

But

$$H(\mathrm{j}\omega) = \frac{1}{\mathrm{j}\omega} H_0(\mathrm{j}\omega)$$

and so

$$H_0(j\omega) = j\omega\, U(\omega) - \omega V(\omega)$$

Consequently we may write equations (5.2.43a) and (5.2.44a) in the form

$$U(\Omega) = U(\infty) + \frac{\{\omega U(\omega)\}_{\omega=0}}{\Omega} - \frac{1}{\pi}\mathrm{P}\int_{-\infty}^{\infty} \frac{V(\omega)}{\omega - \Omega}\,d\omega \quad (5.2.43b)$$

$$V(\Omega) = V(\infty) + \frac{\{\omega V(\omega)\}_{\omega=0}}{\Omega} + \frac{1}{\pi}\mathrm{P}\int_{-\infty}^{\infty} \frac{U(\omega)}{\omega - \Omega}\,d\omega \quad (5.2.44b)$$

A pole at $\omega = 0$ may therefore be included in the Kramers–Kronig relations (and a similar technique may be used to include any pole on the $j\omega$ axis) but now the analysis is of more limited value because a knowledge of $H(j\omega)$ near $\omega = 0$ is required. This difficulty may be overcome by applying the Kramers–Kronig relations to the logarithm of the transfer function rather than to the transfer function itself, provided that it is a minimum phase function.

Consider a minimum phase transfer function having neither poles nor zeros in the right half plane. $\log_e H(s)$ is regular, therefore, in the right half plane and so the Kramers–Kronig relations may now be applied to the logarithm of the transfer function. This gives rise to amplitude-phase relationships. We write

$$H(j\omega) = |H(j\omega)|\, e^{j\beta(\omega)}$$

Excluding, for the moment, singular points on the $j\omega$ axis, we have

$$\log_e H(j\omega) = \log_e |H(j\omega)| + j\beta(\omega)$$

and, from equations (5.2.41), (5.2.42)

$$\log_e |H(j\Omega)| = \log_e |H(j\infty)| - \frac{1}{\pi}\mathrm{P}\int_{-\infty}^{\infty} \frac{\beta(\omega)}{\omega - \Omega}\,d\omega \qquad (5.2.45)$$

$$\beta(\omega) = \beta(\infty) + \frac{1}{\pi}\mathrm{P}\int_{-\infty}^{\infty} \frac{\log_e |H(j\omega)|}{\omega - \Omega}\,d\omega \qquad (5.2.46)$$

A singular point on the $j\omega$ axis may be by-passed with a small semicircular contour, as in Fig. 5.10(b). Let

$$H(s) = (s - j\Omega_0)^k H_0(s) \qquad (5.2.47)$$

where a pole is described by negative values of k and a zero by positive values of k.

We may write $(s - j\Omega_0) = |s - j\Omega_0|\, e^{j\theta} = \rho\, e^{j\theta}$, and $ds = \rho\, e^{j\theta}.jd\theta$ for points on the small semicircle by-passing $j\Omega_0$. The contribution from this element to the complete contour integral required for equations (5.2.45) and (5.2.46) is

$$\int_{\substack{\text{semicircle}\\ \text{Lim } \rho\to 0}} \log_e H(s)\, ds = \int_{\text{Lim } \rho\to 0} \log_e H_0(s)\, ds + \int_{\text{Lim } \rho\to 0} k \log_e (s - j\Omega_0)\, ds$$

$$= \int_{\text{Lim } \rho\to 0} \{\log_e H_0(j\Omega_0) + k \log_e \rho + k\, j\theta\}\rho\, e^{j\theta}\, j\, d\theta \qquad (5.2.48)$$

The term $\log_e H_0(j\Omega_0)$ is regular at $j\Omega_0$ and so the right-hand side of equation (5.2.48) tends to zero as ρ tends to zero. Hence a singular point on the $j\omega$ axis makes no contribution to the contour integral and so $H(s)$ can have a finite number of poles or zeros on the $j\omega$ axis without affecting the relations given in equations (5.2.45) and (5.2.46) for minimum phase transfer functions having $\log_e H(s)$ regular at infinity.

5.2.5 Amplitude Transfer Function and Linear Filters

For a large number of practical applications it is only necessary to require that the amplitude of the transfer function for a linear system should vary in a prescribed manner with frequency. In these circumstances $|H(j\omega)|$ is given but the phase of the transfer function, $\beta(\omega)$, is unspecified. The question arises as to whether the required amplitude function can be associated with an acceptable phase function so that the given characteristics can be reproduced using a realizable physical system. The phase function cannot be chosen arbitrarily since it will be related to the amplitude function through real and imaginary part relations such as have been derived in equations (5.2.41)–(5.2.46). For example, it can be seen from equation (5.2.46) that difficulties arise if $|H(j\omega)|$ is zero over a band of frequencies since $\beta(\omega)$ will become infinite. There will therefore be constraints imposed which will restrict the allowed form of $|H(j\omega)|$. These restrictions are of considerable importance in electrical applications since they will determine the types of filter circuit which can be used for the design of optimum linear systems. A filter is defined essentially as a linear network for which the amplitude transfer function varies in a specified way as a function of frequency. Usually $|H(j\omega)|$ is large over one or more bands of frequency and small over the rest of the $j\omega$ axis.

If $|H(j\omega)|$ is a valid amplitude transfer function, it must be derived from a complete transfer function $H(s)$ evaluated along the $j\omega$ axis. When $H(s)$ is itself a Laplace Transform, then $h(t) = L^{-1}\{H(s)\}$ is zero for $t < 0$ and so $|H(j\omega)|$ refers to a realizable causal system according to the discussion on page 195. There will be a number of transfer functions $H(s)$ which reduce to an acceptable form of $|H(j\omega)|$ but we need only consider those which fall within the class of minimum phase transfer functions. This is because an arbitrary function $H(s)$ can be written as the product of a minimum phase function $H_1(s)$ and an all pass function $H_2(s)$ in accordance with equation (5.2.40). Along the $j\omega$ axis

$$\begin{aligned}|H(s)|_{s=j\omega} = |H(j\omega)| &= |H_1(j\omega)|\,.\,|H_2(j\omega)| \\ &= |H_1(j\omega)| \end{aligned} \tag{5.2.49}$$

Hence the amplitude transfer function $|H(j\omega)|$ is sufficient to determine only the minimum phase part of a corresponding transfer function $H(s)$.

An acceptable amplitude transfer function $|H(j\omega)|$ must therefore satisfy conditions set by minimum phase functions evaluated along the $j\omega$ axis. A minimum phase function has all its finite poles and zeros in the left half plane and so $\log_e H(s)$ is regular throughout the right half plane except for the possibility of poles and zeros in $H(s)$ at infinity. Consider the integral of

$$\left\{\frac{(s+a)^{-1}.\log_e H(s)}{(s-a)}\right\}$$

over the contour consisting of the entire jω axis and the infinite semicircle in the right half plane. Let a be a point on the real axis, $a>0$, then the function $\{(s+a)^{-1}.\log_e H(s)\}$ is regular throughout the right half plane even when $H(s)$ has poles or zeros at infinity. This is because as $s \to \infty$, $H(s) \to As^{n-m}$ and

$$\lim_{s\to\infty}\left\{\frac{\log_e H(s)}{s+a}\right\} \to \lim_{s\to\infty}\left\{\frac{(n-m)\log_e s+\log_e A}{s}\right\} \to 0 \qquad (5.2.50)$$

Hence

$$\oint \frac{(s+a)^{-1}\log_e H(s)}{s-a}\,ds = 2\pi j(2a)^{-1}\log_e H(a) \qquad (5.2.51)$$

Treating the jω axis and the infinite semicircle separately, and using the limit in equation (5.2.50)

$$\int_{-\infty}^{\infty} \frac{(j\omega+a)^{-1}\log_e H(j\omega)}{(j\omega-a)}\,j\,d\omega = \int_{-\infty}^{\infty} \frac{\log_e H(j\omega)}{\omega^2+a^2}\,.\,jd\omega$$

$$= \frac{\pi j}{a}\log_e H(a) \qquad (5.2.52)$$

The function $\beta(\omega)$ in the expression

$$\log_e H(j\omega) = \log_e |H(j\omega)| + j\,\beta(\omega) \qquad (5.2.53)$$

is an odd function of ω, whilst $\log_e |H(j\omega)|$ is an even function of ω. Hence $\beta(\omega)$ does not contribute to the integral in equation (5.2.52), and so we may write

$$\int_{-\infty}^{\infty} \frac{\log_e |H(j\omega)|}{\omega^2+a^2}\,.a\,d\omega = \pi\log_e H(a) \qquad (5.2.54)$$

A minimum phase transfer function belongs to the class of positive real functions and so $H(a)$ is a real, finite number, greater than zero for all finite values of a along the real axis in the right half plane. Hence the integral in equation (5.2.54) is bounded and we may write

$$-C_1 \leqslant \int_{-\infty}^{\infty} \frac{\log_e |H(j\omega)|}{\omega^2+a^2}\,.a\,d\omega \leqslant C_2 \qquad (5.2.55)$$

where C_1, C_2 are positive numbers. Equation (5.2.55) is always satisfied by an amplitude function derived from a minimum phase function. However, an arbitrary function which satisfies equation (5.2.55) is not necessarily derived from a minimum phase function. For example, $\log_e |H(j\omega)|$ can take both positive and negative values arising from $|H(j\omega)| \lesseqgtr 1$ and so infinities may arise over a band of frequencies but cancel in the complete integral. To remove these types of singularities we may impose the stronger condition

$$\int_{-\infty}^{\infty} \frac{|\log_e|H(j\omega)||}{\omega^2+a^2}\,.a\,d\omega \leqslant |M| \qquad (5.2.56)$$

where M is a finite quantity. This will ensure that $|H(j\omega)|$ may be derived from a minimum phase transfer function. The additional constraints implied by equation (5.2.56) may be derived by writing

$$\log_e |H(j\omega)| = \log_e^+ |H(j\omega)| + \log_e^- |H(j\omega)| \tag{5.2.57}$$

where $\log_e^+ |H(j\omega)|$ represents all the positive parts for which $|H(j\omega)| \geqslant 1$ whilst $\log_e^- |H(j\omega)|$ represents all the negative parts for which

$$0 < |H(j\omega)| < 1.$$

Evidently, the positive function

$$|\log_e |H(j\omega)|| = \log_e^+ |H(j\omega)| - \log_e^- |H(j\omega)| \tag{5.2.58}$$

and so

$$\int_{-\infty}^{\infty} \frac{|\log_e |H(j\omega)||}{\omega^2 + a^2} . a \, d\omega = \int_{-\infty}^{\infty} \frac{2 \log_e^+ |H(j\omega)| - \log_e |H(j\omega)|}{\omega^2 + a^2} . a \, d\omega \tag{5.2.59}$$

and is always a positive quantity. Now

$$\int_{-\infty}^{\infty} \frac{2 \log_e^+ |H(j\omega)|}{\omega^2 + a^2} . d\omega < \int_{-\infty}^{\infty} 2 \log_e^+ |H(j\omega)| \, d\omega \leqslant \int_{-\infty}^{\infty} |H(j\omega)|^2 \, d\omega \tag{5.2.60}$$

Hence if

$$\int_{-\infty}^{\infty} |H(j\omega)|^2 \, d\omega < \infty \tag{5.2.61}$$

the inequality of equation (5.2.56) will also be satisfied because

$$\int_{-\infty}^{\infty} \frac{\log_e |H(j\omega)|}{\omega^2 + a^2} . a \, d\omega$$

is bounded according to equation (5.2.55).

Thus the additional constraint imposed by equation (5.2.56) is that

$$\int_{-\infty}^{\infty} |H(j\omega)|^2 \, d\omega = 2\pi \int_{-\infty}^{\infty} \{h(t)\}^2 \, dt < \infty \tag{5.2.62}$$

The impulse response of the circuit is therefore of integrable square and belongs to the class of energy response signals (see Chapter 4).

Equation (5.2.56) may be written in an alternative form by noting that the number, a, is simply a scale factor for ω in the definite integral. Upon writing, $a\omega = \omega$, we obtain the Paley–Wiener condition,† to be satisfied by an arbitrary amplitude transfer function which refers to a realizable, causal network.

$$\begin{aligned} &\int_{-\infty}^{\infty} \frac{|\log_e |H(j\omega)||}{\omega^2 + 1} \, d\omega < \infty \\ &\int_{-\infty}^{\infty} |H(j\omega)|^2 \, d\omega < \infty \end{aligned} \tag{5.2.63}$$

† *American Math. Soc. Colloquium Publ.* **19**, 16–20, 1934.

5.2.6 Particular Amplitude Transfer Functions for Linear Filters

The conditions set out in equations (5.2.63) for an acceptable amplitude transfer function exclude those functions for which $|H(j\omega)|$ is zero over a range of frequencies. This means that an ideal band pass filter with a characteristic such as is shown in Fig. 5.11(a) is not realizable. However, this limitation can be overcome for practical purposes by noting that $\log_e |H(j\omega)|$ will remain finite for $|H(j\omega)| = \varepsilon$, however small ε may be, provided that $\varepsilon > 0$. This is illustrated in Fig. 5.10(b). Equation (5.2.63) will be satisfied but the phase shifts associated with the filter will be large.

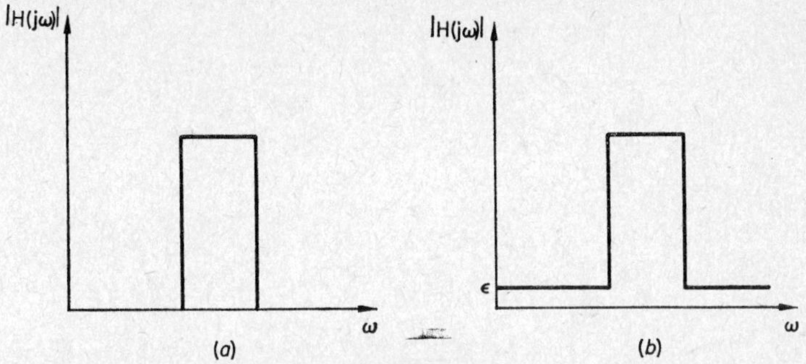

Fig. 5.11
(*a*) Band pass function which is not realizable.
(b) Realizable band pass function.

A more serious constraint is imposed on amplitude transfer functions which tend to zero as ω tends to infinity. Evidently equations (5.2.63) require that $\log_e |H(j\omega)|$ should not increase more rapidly than ω^2 and this means that $|H(j\omega)|$ cannot fall to zero more rapidly than $e^{-\omega^2}$. In particular, a band-pass filter with a Gaussian characteristic $|H(j\omega)| = \exp\{-\alpha^2(\omega - \omega_0)^2\}$ is not realizable since

$$\int_{-\infty}^{\infty} \frac{\alpha^2(\omega^2 - \omega_0^2)}{1 + \omega^2} \, d\omega$$

diverges.

The ideal band-pass filter is not realizable and so approximations have been devised which satisfy equation (5.2.63). We need only to discuss the particular case of a low-pass filter since frequency transformations can be applied to transform it into any other type.

Consider the amplitude transfer function

$$|H(j\omega)| = \frac{1}{\{1 + (\omega/\omega_0)^{2n}\}^{1/2}} \tag{5.2.64}$$

where n is a positive integer, and ω_0 is the half-power frequency. Evidently, for n sufficiently large, this function will approximate to an ideal low-pass filter with

$$|H(j\omega| = 1 \quad \text{for} \quad \omega < \omega_0$$
$$|H(j\omega)| = 0 \quad \text{for} \quad \omega > \omega_0$$

Equation (5.2.64) describes a realizable filter according to equation (5.2.63). It is called the Butterworth or 'maximally flat' approximation. This filter agrees most closely with the ideal case at $\omega = 0$. For small values of ω,

$$|H(j\omega)| = \{1 - \tfrac{1}{2}(\omega/\omega_0)^{2n} + \tfrac{3}{8}(\omega/\omega_0)^{4n} - \tfrac{5}{16}(\omega/\omega_0)^{6n} + \tfrac{35}{128}(\omega/\omega_0)^{8n} - \cdots\} \tag{5.2.65}$$

and so the first $(2n - 1)$ derivatives of $|H(j\omega)|$ are zero at $\omega = 0$. For $\omega/\omega_0 \gg 1$ the response falls off as ω^{-n}. A typical amplitude response curve for a Butterworth filter is shown in Fig. 5.12.

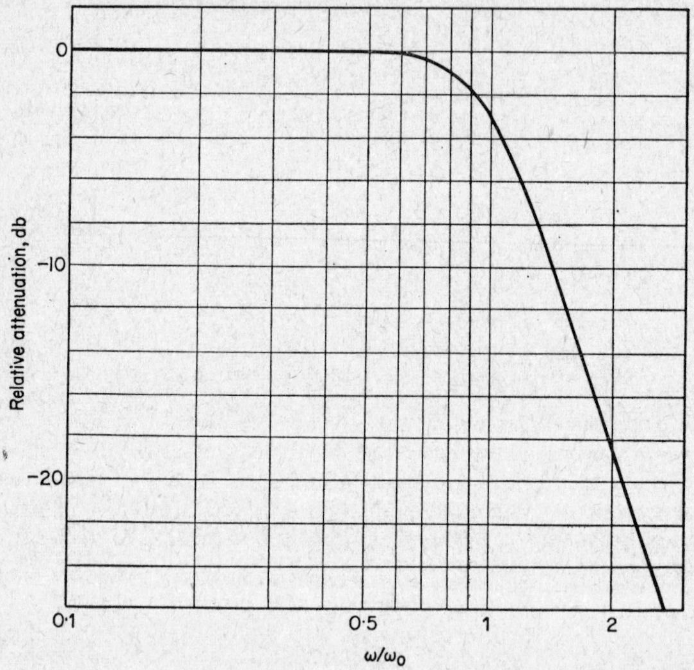

FIG. 5.12 A Butterworth response function for a low-pass later with $n = 3$.

Another possible amplitude transfer function which approximates to the ideal low-pass filter is given by

$$|H(j\omega)| = \frac{1}{[1 + \varepsilon^2\{C_n(\omega/\omega_0)\}^2]^{1/2}} \tag{5.2.66}$$

where $\varepsilon \leqslant 1$ and $C_n(\omega/\omega_0)$ is an nth degree polynomial. A particularly important polynomial used with this expression is the Chebyshev cosine polynomial which gives rise to the Chebyshev or equal-ripple filter. The Chebyshev polynomial is defined by

$$\begin{aligned} C_n(\omega/\omega_0) &= \cos\{n \cos^{-1}(\omega/\omega_0)\} \quad \text{for} \quad |\omega/\omega_0| \leqslant 1 \\ &= \cosh\{n \cosh^{-1}(\omega/\omega_0)\} \quad \text{for} \quad |\omega/\omega_0| \geqslant 1 \end{aligned} \tag{5.2.67}$$

When $\omega/\omega_0 \leqslant 1$, $C_n(\omega/\omega_0)$ reduces to the following polynomials

$$C_0(\omega/\omega_0) = 1$$
$$C_1(\omega/\omega_0) = \omega/\omega_0$$
$$C_2(\omega/\omega_0) = 2(\omega/\omega_0)^2 - 1$$
$$C_3(\omega/\omega_0) = 4(\omega/\omega_0)^3 - 3(\omega/\omega_0)$$
$$\cdot \quad \cdot \quad \cdot \quad \cdot \quad \cdot \quad \cdot \quad \cdot$$
$$C_{m+1}(\omega/\omega_0) = 2(\omega/\omega_0)C_m - C_{m-1} \tag{5.2.68}$$

The function $|H(j\omega)|$ oscillates between a maximum value of unity and a minimum value of $(1 + \varepsilon^2)^{-1/2}$ as ω ranges over the frequency interval $0 \leqslant |\omega/\omega_0| \leqslant 1$. Outside this range $|H(j\omega)|$ falls rapidly towards zero, and at high frequencies varies as $\{\varepsilon^2 2^{n-1}(\omega/\omega_0)^n\}^{-1/2}$. A typical amplitude response curve for a Chebyshev filter is shown in Fig. 5.13. The total number of ripple maxima and minima is equal to n.

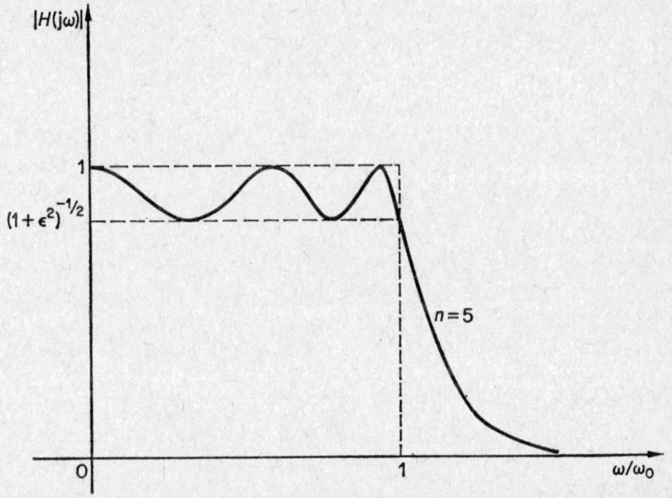

FIG. 5.13 A Chebyshev response function for a low-pass filter with $n = 5$.

The Butterworth filter is evidently a closer approximation to the ideal case than is the Chebyshev filter, when $|\omega/\omega_0| < 1$. However, near cut-off the Chebyshev response is the better and the initial roll-off is usually greater than that for a Butterworth filter. The attenuation characteristics for the two types of filter are given by:

Butterworth

$$20 \log_{10} |H(j\omega)| = -20n \log_{10} (\omega/\omega_0)$$

Chebyshev

$$20 \log_{10} |H(j\omega)|$$
$$= -\{20 \log_{10}\varepsilon + 20(n - 1) \log_{10} 2 + 20n \log_{10} (\omega/\omega_0)\} \tag{5.2.69}$$

when $|\omega/\omega_0| \gg 1$. Thus although the initial advantage usually lies with the Chebyshev filter,† both types eventually attenuate at the same rate of $20n$ decibels per decade.

A low-pass filter which transmits, without distortion, a signal having all its frequency components within the pass band has a transfer function along the $j\omega$ axis derived from equation (5.2.37).

$$\begin{aligned} H(j\omega) &= K \exp(-j\omega\tau_0) \quad \text{for} \quad |\omega| < \omega_0 \\ &= \varepsilon(\ll 1) \quad \text{for} \quad |\omega| > \omega_0 \end{aligned} \tag{5.2.70}$$

Evidently the phase of $H(j\omega)$ is

$$\beta(\omega) = -\omega\tau_0 \tag{5.2.71}$$

and so $\beta(\omega)$ is a linear function of the frequency. The transfer function for this filter, $H(s) = K \exp(-s\tau_0)$, is not a rational function of s and so it is not possible, in principle, to construct such a filter using a discrete parameter network. The Butterworth and Chebyshev filters may be regarded as polynomial approximations to the characteristics derived from equation (5.2.70). However, these filters are non-linear in their phase functions and so give rise to distortion of the signal. For example, the Butterworth filter with $n = 1$, gives rise to

$$H(j\omega).H(-j\omega) = \frac{1}{1 + (\omega/\omega_0)^2} = \frac{1}{\{1 + (j\omega/\omega_0)\}\{1 - (j\omega/\omega_0)\}} \tag{5.2.72}$$

and a minimum phase transfer function in the pass band,

$$H(s) = \frac{1}{1 + s/\omega_0}$$

for which

$$H(j\omega) = \frac{\omega_0}{(\omega_0^2 + \omega^2)^{1/2}} e^{-j\beta(\omega)} \tag{5.2.73}$$

with

$$\beta(\omega) = \tan^{-1}(\omega/\omega_0)$$

A better approximation to a linear phase filter is obtained using Bessel polynomials. Consider

$$\begin{aligned} H(s) &= K \exp(-s\tau_0) \\ &= \frac{K}{\sinh s\tau_0 + \cosh s\tau_0} \end{aligned} \tag{5.2.74}$$

Expanding the sinh and cosh functions

$$\cosh \phi = 1 + \frac{\phi^2}{2!} + \frac{\phi^4}{4!} + \frac{\phi^6}{6!} + \cdots$$

$$\sinh \phi = \phi + \frac{\phi^3}{3!} + \frac{\phi^5}{5!} + \frac{\phi^7}{7!} + \cdots$$

† Note that $\log_{10}\varepsilon$ is usually negative and so this term has to be compensated by choosing n sufficiently large.

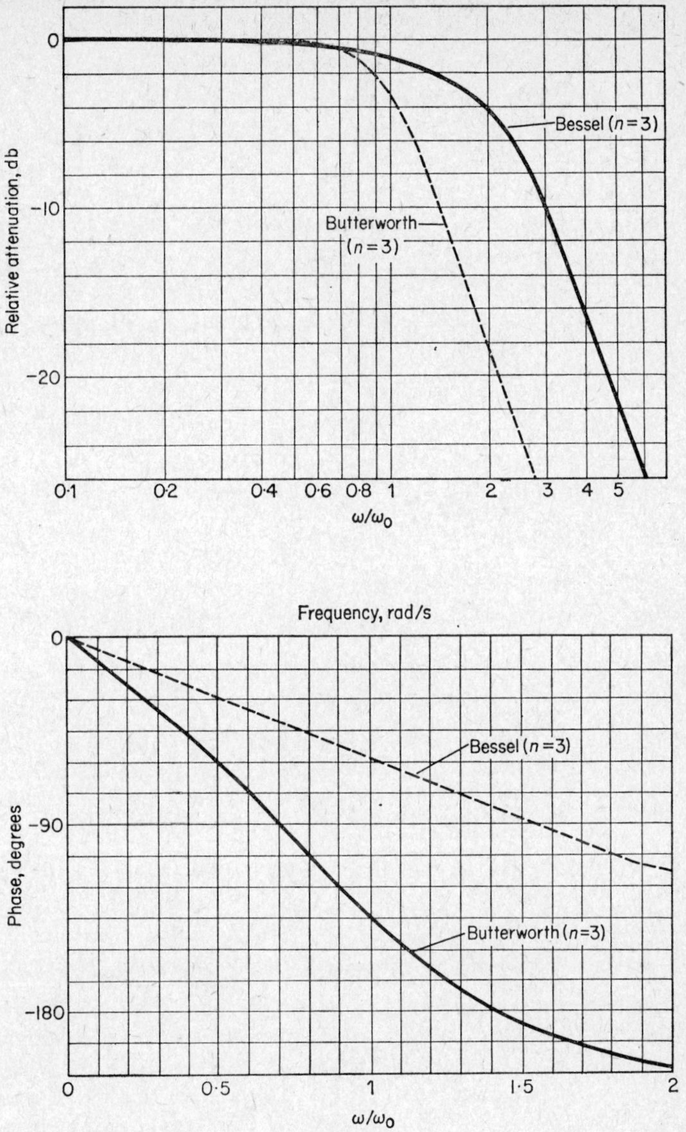

FIG. 5.14 Amplitude and phase response curves for $n = 3$ Butterworth and Bessel filters.

the denominator of equation (5.2.74) is approximated by the Bessel polynomials $B_n(\phi)$, for which

$$B_0(\phi) = 1$$

$$B_1(\phi) = \phi + 1$$

$$B_2(\phi) = \phi^2 + 3\phi + 3$$

$$B_3(\phi) = \phi^3 + 6\phi^2 + 15\phi + 15$$

$$\cdot \quad \cdot \quad \cdot \quad \cdot \quad \cdot \quad \cdot \quad \cdot \quad \cdot$$

$$B_n(\phi) = (2n - 1)B_{n-1}(\phi) + \phi^2 B_{n-2}(\phi) \tag{5.2.75}$$

The roots of the Bessel polynomials all lie in the left half plane and so these polynomials give rise to minimum phase transfer functions. The amplitude and phase functions for a Bessel filter are compared with those for a Butterworth filter in Fig. 5.14. It can be seen that although the Bessel filter has a more linear phase characteristic the cut-off function is much more gradual than that for the Butterworth filter.

5.2.7 Frequency Transformations

By means of suitable frequency transformations the transfer function for a low-pass filter can be used to generate transfer functions which describe high-pass, band-pass or band-reject filters. Particularly useful transformations are given by the relations

$$s = \frac{\omega_0^2}{S} \tag{5.2.76}$$

for the low-pass to high-pass transformation

$$s = \frac{\omega_0^2}{\Delta\omega_0}\left(\frac{\omega_0}{S} + \frac{S}{\omega_0}\right) \tag{5.2.77}$$

for the low-pass to band-pass transformation, and

$$s = \Delta\omega_0\left(\frac{\omega_0}{S} + \frac{S}{\omega_0}\right)^{-1} \tag{5.2.78}$$

for the low-pass to band-reject transformation.

In each of these equations, s, is the complex frequency variable in $H(s)$ describing the low-pass filter, whilst S is the transformed variable in $H(S)$ describing the derived filter. The terms ω_0, $\Delta\omega_0$, are constants for the transformation which are related to the cut-off frequency for the high-pass filter and to a 'mean' frequency and bandwidth for the band-pass/reject filters.

The use of the transformations given in equations (5.2.76)–(5.2.78) enables circuit elements in the original low-pass filter to be replaced by simple circuit elements in the derived filter. Thus for the high-pass transformation, an inductance L occurring in the low-pass filter, having an impedance sL, is converted into an element with impedance $\omega_0^2 L/S$, that is, into a capacitance element with capacity $C = 1/\omega_0^2 L$. Similarly a capacitance element in the low-pass filter is converted into an inductance $L = 1/\omega_0^2 C$. For the band-pass transformation, an inductance L is converted into a series combination of an inductance $\omega_0 L/\Delta\omega_0$ and a capacitance $\Delta\omega_0/\omega_0^3 L$, whilst a capacitance in the low-pass filter is converted into a parallel combination of an inductance $\Delta\omega_0/\omega_0^3 C$ and a capacitance $\omega_0 C/\Delta\omega_0$. Similar transformations can be carried through for the band-reject filter and the complete set is summarized in Fig. 5.15.

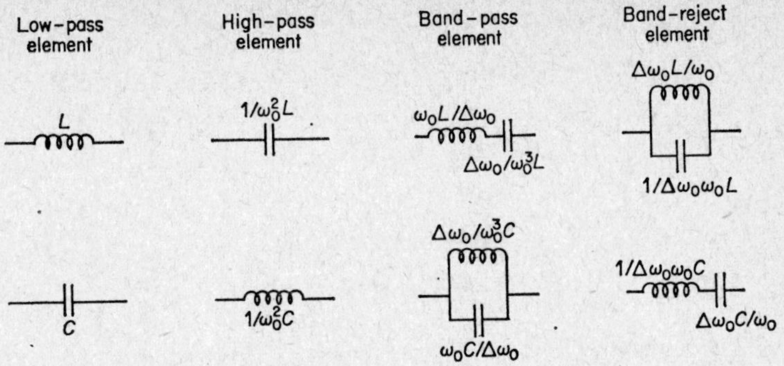

FIG. 5.15 Circuit element configurations corresponding to frequency transformations.

Example

As an example of these transformations, consider the low-pass filter formally described by

$$H(s) = H_P(s) \quad \text{for} \quad -\omega_c \leqslant \omega \leqslant \omega_c$$
$$H(s) = \varepsilon(\ll 1) \quad \text{for} \quad |\omega| > \omega_c \tag{5.2.79}$$

where ω_c is a positive quantity. The band-pass transformation will define frequency intervals where

$$H(S) = H_P(S)$$

and

$$H(S) = \varepsilon(\ll 1)$$

Evidently the pass bands are derived from the condition

$$-\omega_c \leqslant \frac{\omega_0}{\Delta\omega_0}\left(\frac{\Omega^2 - \omega_0^2}{\Omega}\right) \leqslant \omega_c \tag{5.2.80}$$

where we have used $s = q + j\omega$ and $S = Q + j\Omega$ to relate the frequency variables, and we have enquired how the $j\omega$ axis maps on to the $j\Omega$ axis.

From the lower bound

$$\frac{\omega_0}{\Delta\omega_0}\left(\frac{\Omega^2 - \omega_0^2}{\Omega}\right) + \omega_c \geqslant 0$$

or

$$\frac{\omega_0}{\Delta\omega_0}\left[\frac{\Omega^2 - \omega_0^2 + \Omega\{(\omega_c\Delta\omega_0)/\omega_0\}}{\Omega}\right] \geqslant 0 \tag{5.2.81}$$

Taking the transformation constants ω_0, $\Delta\omega_0$ to be positive quantities, this equation may be written

$$\frac{\omega_0}{\Delta\omega_0}\cdot\frac{(\Omega - \Omega_1)(\Omega + \Omega_2)}{\Omega} \geqslant 0 \tag{5.2.82}$$

with Ω_1, Ω_2 positive quantities and $\Omega_2 > \Omega_1$. Similarly, from the upper bound of equation (5.2.80)

$$\frac{\omega_0}{\Delta\omega_0}\cdot\frac{(\Omega + \Omega_1)(\Omega - \Omega_2)}{\Omega} \leqslant 0 \tag{5.2.83}$$

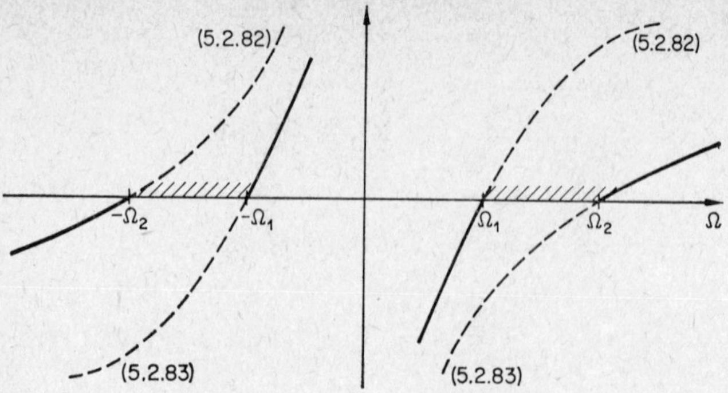

FIG. 5.16 Functions in equations (5.2.82) and (5.2.83)

The functions in equations (5.2.82) and (5.2.83) are shown in Fig. 5.16. From this figure it is evident that the inequalities in equations (5.2.82) and (5.2.83) are only satisfied simultaneously for Ω in the ranges

$$\Omega_1 \leqslant \Omega \leqslant \Omega_2$$
$$-\Omega_2 \leqslant \Omega \leqslant -\Omega_1 \tag{5.2.84}$$

These ranges define the band-pass range of the derived filter. From the relation between the functions in equations (5.2.81) and (5.2.82) we obtain

$$\sqrt{(\Omega_2\Omega_1)} = \omega_0$$
$$(\Omega_2 - \Omega_1) = \frac{\omega_c}{\omega_0}.\Delta\omega_0 \tag{5.2.85}$$

5.3 Operational Amplifier with Feedback

An ideal operational amplifier is a unidirectional circuit element having constant gain K over the entire frequency range $+\infty$ to $-\infty$ together with an infinite input impedance and zero output impedance. The output signal from the amplifier is a replica of the input and so, from equation (5.2.37), the transfer function is

$$H(s) = K \exp(-s\tau_0)$$

where the gain factor K may be either positive or negative and τ_0 is the associated time delay. In many applications τ_0 is taken to be zero.

The ideal operational amplifier is usually represented by a triangular element as shown in Fig. 5.17.

A practical operational amplifier will have a finite bandwidth, a finite input

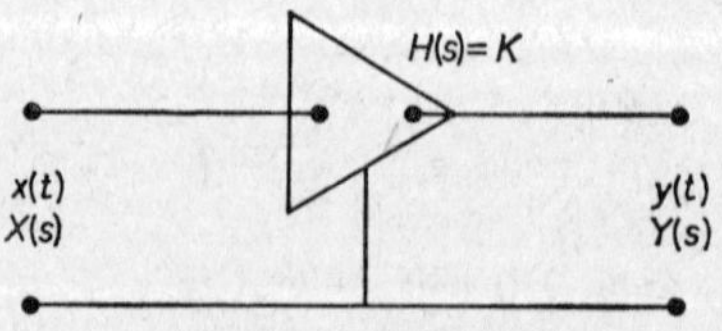

FIG. 5.17 Representation for an ideal amplifier.

impedance and non-zero output impedance. The overall transfer function will therefore only approximate to that for the ideal element. We shall assume that both the transfer function for the amplifier itself and these characteristic impedances are unaffected by any additional circuit connections (for example, a feedback loop). This assumption is referred to as the condition of 'no-loading' of the amplifier by external circuit elements.

A practical operational amplifier usually has two input terminals and one output terminal. One of the input terminals (labelled non-inverting, i/p^+) gives rise to a direct replica of the signal, the other input terminal (labelled inverting, i/p^-) gives rise to an inverted replica of the signal. Taken together these terminals form a voltage summing or subtracting junction. The practical operational amplifier is drawn as shown in Fig. 5.18.

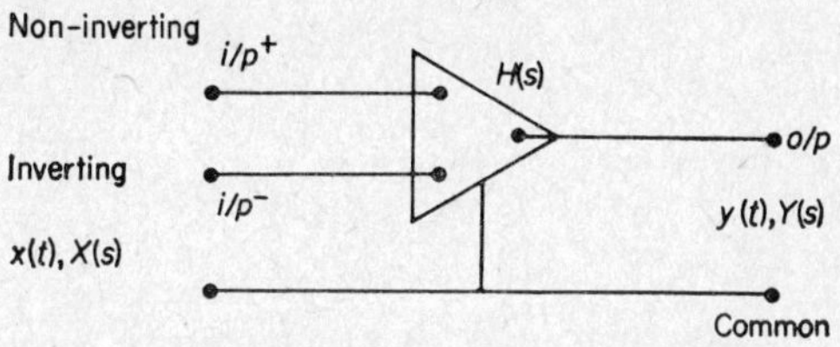

FIG. 5.18 Representation for a practical operation amplifier

For the purpose of circuit analysis an operational amplifier (whether ideal or practical) may be regarded as an element with a particular transfer function, just as is formally the case for a resistive, capacitive, mutual or self-inductive element. A feedback loop containing operational amplifiers is therefore included in the general discussion given previously on page 203, and the transfer functions will satisfy the conditions for stability discussed in Section 5.2, pages 195–199.

5.3.1 An Integrating Circuit

Consider first the ideal amplifier used to form an integrating circuit as shown in Fig. 5.19. The summing junction Σ at the input to the amplifier

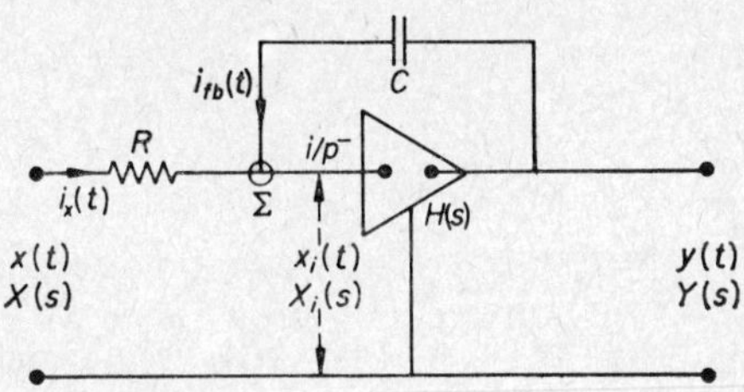

FIG. 5.19 An ideal amplifier forming an integrating circuit.

forms the summation of the currents $i_x(t)$ and $i_{fb}(t)$. However, the input impedance of the ideal amplifier is infinite, and so

$$i_x(t) = -i_{fb}(t) = i(t) \tag{5.3.1}$$

Hence,

$$R\,I(s) = X(s) - X_1(s)$$
$$-\frac{1}{sC}I(s) = Y(s) - X_1(s) \qquad (5.3.2)$$

with

$$Y(s) = H(s)\,X_1(s) \qquad (5.3.3)$$

so that

$$Y(s) = \frac{H(s)}{1 - sCR\{H(s) - 1\}}\,X(s) \qquad (5.3.4)$$

Writing $H(s) = K$, with $|K| \gg 1$,

$$Y(s) = \frac{K}{1 - sCRK}\,X(s) = H'(s)\;X(s) \qquad (5.3.5)$$

For stability, the poles of the overall transfer function $H'(s)$ must lie in the left half plane, and so K must be negative.† Hence, writing now $H(s) = -|K|$,

$$Y(s) = -\frac{1}{RC}.\frac{1}{s}.X(s) \qquad (5.3.6)$$

for $|K|.sRC \gg 1$.
That is,

$$y(t) = -\frac{1}{RC}\int_0^t x(t)\,\mathrm{d}t \qquad (5.3.7)$$

The circuit therefore acts as an integrating element.

We may compare the ideal amplifier circuit with the passive RC integrating circuit shown in Fig. 5.20.

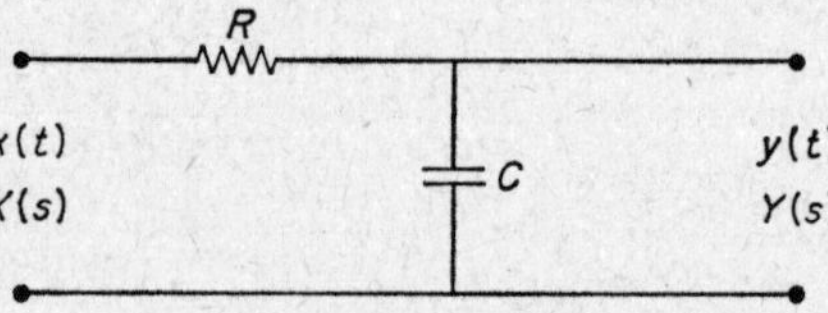

FIG. 5.20 An RC integrating circuit.

For this case

$$Y(s) = \frac{1}{1 + sRC}X(s) \qquad (5.3.8)$$

and for $sRC \gg 1$

$$y(t) = \frac{1}{RC}\int_0^t x(t)\,\mathrm{d}t \qquad (5.3.9)$$

Thus, in order to act as an integrator, the passive circuit requires $sRC \gg 1$ whilst the active circuit only imposes the weaker condition $|K|.sRC \gg 1$. The passive circuit therefore integrates over a time interval determined by RC, whilst the corresponding time interval for the active circuit is determined by $|K|.RC$.‡

† The excitation is therefore connected to i/p^- as shown in Fig. 5.19.

‡ Note that the complete response of the ideal amplifier circuit as given by equation (5.3.5) with K negative is exactly analogous to that for the RC circuit as given by equation (5.3.8).

5.3.2 Voltage Feedback with a Practical Operational Amplifier

Consider next the analysis of a feedback circuit using a practical operational amplifier which derives an input signal from a source with internal resistance R_s. For this analysis we assume that the input connections give rise to voltage subtraction. The feedback circuit is shown in Fig. 5.21.

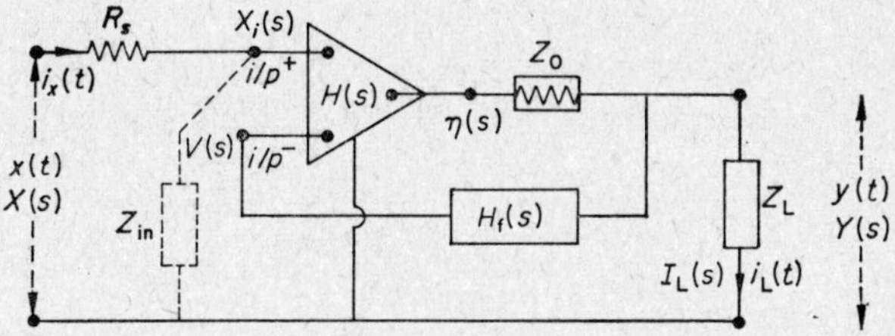

Fig. 5.21 A practical operational amplifier with voltage feedback.

The effective input impedance to the amplifier with feedback, Z_{in}, is indicated by broken lines in the diagram. This is the impedance seen by the signal source. It will reduce to Z_{in}^0, the input impedance without feedback (which is a constant for the amplifier) when $H_f(s) = 0$. Note also that Z_o is drawn inside the feedback loop and is the constant characteristic output impedance of the amplifier. The apparent overall output impedance when feedback is present will be calculated later.

Evidently

$$Y(s) = \frac{Z_L}{Z_L + Z_o}\,\eta(s)$$

$$\eta(s) = H(s)\{X_i(s) - V(s)\}$$

$$V(s) = H_f(s).Y(s)$$

$$X_i(s) = \frac{Z_{in}}{Z_{in} + R_s}.X(s) \tag{5.3.10}$$

The overall transfer function for the circuit is derived from

$$\frac{Z_L + Z_o}{Z_L}\,Y(s) = H(s)\left\{\frac{Z_{in}}{Z_{in} + R_s}\,X(s) - H_f(s)\,Y(s)\right\} \tag{5.3.11}$$

that is

$$Y(s) = \left[\frac{H(s)}{\{(Z_L + Z_o)/Z_L\} + H(s)\,H_f(s)}\cdot\frac{Z_{in}}{Z_{in} + R_s}\right].X(s) \tag{5.3.12}$$

For an idealized operational amplifier with an open loop gain factor $H(s) = K$, and a simple voltage ratio feedback loop for which $H_f(s) = \beta$

$$Y(s) = \left\{\frac{K}{1 + (Z_o/Z_L) + K\beta}\cdot\frac{Z_{in}}{Z_{in} + R_s}\right\}.X(s) \tag{5.3.13}$$

and for $K\beta \gg 1, (Z_o/Z_L)$; and $Z_{in} \gg R_s$

$$Y(s) = \frac{1}{\beta}\,X(s) \tag{5.3.14}$$

Equation (5.3.12) involves the effective input impedance of the amplifier with feedback.

$$Z_{in} = \frac{X_i(s)}{I_x(s)} = \frac{V(s) + \eta(s)/H(s)}{I_x(s)}$$

$$= \left\{\frac{Z_L}{Z_L + Z_o} H(s)\, H_f(s) + 1\right\}\left\{\frac{1}{H(s)}\cdot\frac{\eta(s)}{I_x(s)}\right\} \qquad (5.3.15)$$

But, $\eta(s)/H(s) = X_i^0(s)$ is the excitation at the input terminal of the amplifier which gives rise to $\eta(s)$ when there is no feedback (open loop conditions). Hence

$$\frac{\eta(s)}{H(s)}\cdot\frac{1}{I_x(s)} = \frac{X_i^0(s)}{I_x(s)} = Z_{in}^0 \qquad (5.3.16)$$

and so

$$Z_{in} = Z_{in}^0\left\{1 + \frac{Z_L}{Z_L + Z_o} H(s)\, H_f(s)\right\} \qquad (5.3.17)$$

For $H(s) = K$, $H_f(s) = \beta$, with $K\beta \gg 1$ and $Z_L \gg Z_o$

$$Z_{in} = K\beta Z_{in}^0 \qquad (5.3.18)$$

Hence voltage feedback increases the effective input impedance of the amplifier. The particular case of $\beta = 1$ gives rise to the voltage follower circuit for which $Y(s) = X(s)$ and $Z_{in} = KZ_{in}^0$. Note that

$$Y(s) = Z_L\, I_L(s) = \frac{Z_L}{Z_L + Z_o} H(s)\{X_i(s) - H_f(s)\, Z_L\, I_L(s)\}$$

so that

$$I_L(s) = \frac{H(s)}{\{1 + Z_L\,(Z_L + Z_o)^{-1} H(s)\, H_f(s)\}}\cdot\frac{X_i(s)}{Z_L + Z_o} \qquad (5.3.19)$$

but $X_i(s) = Z_{in}\, I_x(s)$, and hence

$$I_L(s) = \frac{Z_{in}^0}{Z_L + Z_o} H(s) . I_x(s) \qquad (5.3.20)$$

The current gain is therefore unaffected by voltage feedback.

The effective output impedance of the overall system, Z_{out}, is derived from

$$Y(s) = \frac{Z_L}{Z_L + Z_{out}}\; Y^{(\infty)}(s) \qquad (5.3.21)$$

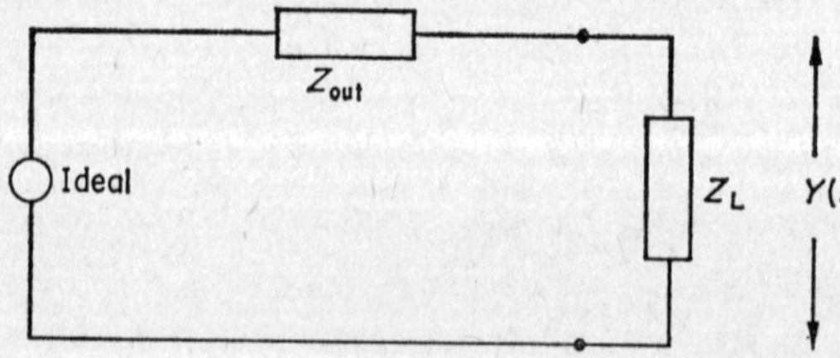

FIG. 5.22 An operational amplifier regarded as an ideal generator with impedance Z_{out}, working into a load Z_L.

where $Y^{(\infty)}(s)$ is the value of $Y(s)$ as $R_L \to \infty$. This is evident from Fig. 5.22.

Using equation (5.3.12)

$$1+\frac{Z_{out}}{Z_L}=\frac{Y^{(\infty)}(s)}{Y(s)}=\frac{1+(Z_o/Z_L)+H(s)\,H_f(s)}{1+H(s)\,H_f(s)} \tag{5.3.22}$$

Hence

$$Z_{out}=Z_o\{1+H(s)\,H_f(s)\}^{-1} \tag{5.3.23}$$

and now for $H(s)\,H_f(s)=K\beta \gg 1$

$$Z_{out}=\frac{1}{K\beta}Z_o \tag{5.3.24}$$

so that voltage feedback reduces the effective output impedance of the circuit.

5.3.3 Current Feedback with a Practical Operational Amplifier

We have discussed the case of the practical operational amplifier with voltage feedback. A similar analysis can be carried through when the input terminal acts as a current summing junction (see Problem 14). A typical circuit is shown in Fig. 5.23.

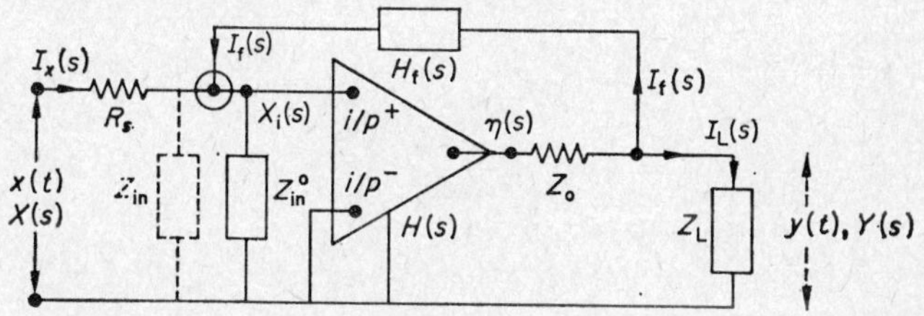

Fig. 5.23 A practical operational amplifier with current feedback. Note that the particular designation of the input terminals i/p^+, i/p^- is not significant.

For this case,

$$\begin{aligned} &Y(s)=\eta(s)-Z_o\{I_L(s)+I_f(s)\}=Z_L\,I_L(s)\\ &\eta(s)=H(s)\,X_i(s)=H(s)\,Z^0_{in}\{I_x(s)+I_f(s)\}\\ &I_f(s)=H_f(s)\,Y(s)\ \text{when}\ X_i(s)\ll Y(s)\\ &R_s\,I_x(s)=X(s)-Z^0_{in}\{I_x(s)+I_f(s)\} \end{aligned} \tag{5.3.25}$$

From these equations,

$$Y(s)\left\{1-\frac{R_s\,Z^0_{in}}{R_s+Z^0_{in}}H(s)\,H_f(s)+Z_o\left(\frac{1}{Z_L}+H_f(s)\right)\right\}=\frac{Z^0_{in}}{R_s+Z^0_{in}}H(s)\,X(s) \tag{5.3.26}$$

Assume that

$$I_f(s)=H_f(s)\,Y(s)\ll I_L(s) \tag{5.3.27}$$

Then

$$Z_o\,I_f(s)=Z_o\,H_f(s)\,Y(s)\ll Z_o\,I_L(s)=(Z_o/Z_L)Y(s)$$

Hence

$$Z_o H_f(s) \ll (Z_o/Z_L) \ll 1 \tag{5.3.28}$$

Consequently equation (5.3.26) gives,

$$Y(s)\left\{1 - \frac{R_s Z_{in}^0}{R_s + Z_{in}^0} H(s) H_f(s)\right\} = \frac{Z_{in}^0}{R_s + Z_{in}^0} H(s) X(s) \tag{5.3.29}$$

For a good operational amplifier $R_s \ll Z_{in}^0$. In this case

$$Y(s) = \frac{H(s)}{1 - R_s H(s) H_f(s)} X(s) \tag{5.3.30}$$

If $H(s) = -K$ and $|R_s H(s) H_f(s)| \gg 1$, then

$$Y(s) = -\frac{1}{R_s H_f(s)} X(s) \tag{5.3.31}$$

The overall current gain is

$$\frac{I_L(s)}{I_x(s)} = \frac{Y(s)/Z_L}{[\{X(s) - Z_{in}^0 I_f(s)\}/(R_s + Z_{in}^0)]} = \frac{R_s + Z_{in}^0}{Z_L} \cdot \frac{1}{\{X(s)/Y(s)\} - Z_{in}^0 H_f(s)}$$

Substitute for $\{X(s)/Y(s)\}$ from equation (5.3.30),

$$\frac{I_L(s)}{I_x(s)} = \frac{R_s + Z_{in}^0}{Z_L} \cdot \frac{H(s)}{1 - (R_s + Z_{in}^0) H(s) H_f(s)} \tag{5.3.32}$$

and when $R_s \ll Z_{in}^0$ the current gain is

$$\frac{I_L(s)}{I_x(s)} = \frac{(-1)}{Z_L H_f(s)} \tag{5.3.33}$$

The effective input impedance with feedback is

$$Z_{in} = \frac{X(s) - R_s I_x(s)}{I_x(s)} = \frac{X(s)(R_s + Z_{in}^0)}{X(s) - Z_{in}^0 H_f(s) Y(s)} - R_s$$

$$= \frac{Z_{in}^0[1 + R_s H_f(s) . \{Y(s)/X(s)\}}{1 - Z_{in}^0 H_f(s) \{Y(s)/X(s)\}} \tag{5.3.34}$$

upon using equation (5.3.30) we find that

$$Z_{in} = \frac{Z_{in}^0}{1 - (R_s + Z_{in}^0) H(s) H_f(s)} \tag{5.3.35}$$

and when $R_s \ll Z_{in}^0$; $1 \ll |Z_{in}^0 H(s) H_f(s)|$

$$Z_{in} \approx \frac{(-1)}{H(s) H_f(s)} \tag{5.3.36}$$

The effective output impedance with feedback is obtained from Fig. 5.21

$$\frac{Z_{out} + Z_L}{Z_L} = \frac{Y^\infty(s)}{Y(s)}$$

$$Z_{out} = \frac{Z_o}{1 - \{(R_s Z_{in}^0)/(R_s + Z_{in}^0)\} H(s) H_f(s) + Z_o H_f(s)} \tag{5.3.37a}$$

and is therefore dependent upon the source resistance.

For $R_s \ll Z^0_{\text{in}}$

$$Z_{\text{out}} = \frac{Z_o}{1 - R_s H(s) H_f(s) + Z_o H_f(s)} \tag{5.3.37b}$$

A simple configuration will have $H(s) = -K$, $H_f(s) = 1/R$ so that,

$$Z_{\text{out}} = \frac{Z_o}{1 + \{K(R_s/R)\} + Z_o/R} \tag{5.3.38a}$$

$$\approx \frac{Z_o}{K} \cdot \frac{R}{R_s} \tag{5.3.38b}$$

The examples of an operational amplifier with a feedback loop which have been discussed in this section have explicitly related the feedback signal to the output *voltage* through an appropriate transfer function. This is not a fundamental requirement. The feedback signal may, if so desired, be derived from the output *current*. In accordance with the general discussion on page 203 it is only necessary for the feedback signal to be the response to an appropriate excitation and this excitation can be either the output voltage or the output current. The analysis of an operational amplifier with feedback signal derived from the output current may therefore be carried through in the same manner as for the examples given above provided that the appropriate transfer functions are used.

5.4 Optimum Linear Systems

An important class of linear networks is used to extract wanted signals from an undesirable noise background. The design of a system to perform this function in an optimum manner requires the choice of an appropriate parameter to characterize the quality of the network. If it is sufficient to detect the presence of a signal, without regard to reproducing its original form, the criterion of maximum signal to noise ratio may be used. For this case, the impulse response of the optimum linear system is directly related to the original form of required signal, and so it is a good criterion to use when this signal is deterministic (non-random) and the form is known *a priori*. On the other hand, if it is necessary to reproduce the original signal as closely as possible, a more satisfactory criterion is obtained by reducing to a minimum the mean square error between the required signal excitation and response functions.

That part of the total excitation for the network, which arises from the noise background, will be in the form of a random function of the time. The analysis of linear networks in the complex frequency domain which has been discussed in the previous sections is therefore not directly applicable and it is necessary to consider how random excitations can be treated.

5.4.1 Excitation by a Random Signal

When the excitation $x(t)$ is a random function of the time special techniques are required for handling network problems because the Fourier and Laplace representations used in equations (5.1.17) and (5.1.18) no longer exist.

Relations between the excitation and response for a linear system for a stationary process may, however, be derived from the auto-correlation and spectral density functions. These representations have been introduced in Chapter 4. They may, of course, also be used, when necessary, with deterministic time functions.

For a causal system, the impulse response function $h(\tau)$ is zero for $\tau < 0^-$. and so equation (5.1.13) may be taken in the form

$$y(t) = \int_{-\infty}^{\infty} h(\tau)\, x(t-\tau)\, \mathrm{d}\tau \tag{5.4.1}$$

From equation (4.4.14a), the time auto-correlation function for $y(t)$, is

$$R_y(T) = \lim_{T_0\to\infty} \frac{1}{2T_0}\int_{-T_0}^{T_0} y(t)\, y(t+T)\, \mathrm{d}t \tag{5.4.2}$$

substituting from equation (5.4.1)

$$R_y(T) =$$

$$\lim_{T_0\to\infty} \frac{1}{2T_0}\int_{-T_0}^{T_0} \left\{\int_{-\infty}^{\infty} h(\tau_1)\, x(t-\tau_1)\, \mathrm{d}\tau_1 \int_{-\infty}^{\infty} h(\tau_2)\, x(t+T-\tau_2)\, \mathrm{d}\tau_2\right\} \mathrm{d}t$$

Interchange the order of integration,

$$R_y(T) = \int_{-\infty}^{\infty}\int_{-\infty}^{\infty} h(\tau_1)\, h(\tau_2)$$

$$\times \left\{\lim_{T_0\to\infty} \frac{1}{2T_0}\int_{T-0}^{T_0} x(t-\tau_1)\, x(t+T-\tau_2)\, \mathrm{d}t\right\} \mathrm{d}\tau_1\, \mathrm{d}\tau_2$$

or

$$R_y(T) = \int_{-\infty}^{\infty}\int_{-\infty}^{\infty} h(\tau_1)\, h(\tau_2)\, R_x(T-\tau_2+\tau_1)\, \mathrm{d}\tau_1\, \mathrm{d}\tau_2 \tag{5.4.3}$$

This is the integral equation which relates the excitation and response auto-correlation functions for a linear system. The derivation refers specifically to the time auto-correlation function but for an ergodic process the same relation will be true for ensemble averages.

The power spectral density representation for the random response function is obtained from equation (4.4.23)

$$S_y(\omega) = \int_{-\infty}^{\infty} R_y(T)\, \mathrm{e}^{-\mathrm{j}\omega T}\, \mathrm{d}T \tag{5.4.4}$$

That is

$$S_y(\omega) = \int_{-\infty}^{\infty}\int_{-\infty}^{\infty}\int_{-\infty}^{\infty} h(\tau_1)\, h(\tau_2)\, R_x(T-\tau_2+\tau_1)\, \mathrm{e}^{-\mathrm{j}\omega T}\, \mathrm{d}\tau_1\, \mathrm{d}\tau_2\, \mathrm{d}T$$

$$= \int_{-\infty}^{\infty} h(\tau_1)\, \mathrm{e}^{+j\omega\tau_1}\, \mathrm{d}\tau_1 \int_{-\infty}^{\infty} h(\tau_2)\, \mathrm{e}^{-j\omega\tau_2} \mathrm{d}\tau_2$$

$$\times \int_{-\infty}^{\infty} R_x(T-\tau_2+\tau_1)\, \exp\{-\mathrm{j}\omega(T-\tau_2+\tau_1)\}\, \mathrm{d}T \tag{5.4.5}$$

Recalling that $h(\tau_1)$, $h(\tau_2)$ are each zero for $\tau_1, \tau_2 < 0^-$, it is evident that the first two integrals in this equation are the transfer functions $H(-\mathrm{j}\omega)$,

$H(j\omega)$, for the network derived from $H(s)$ evaluated along the $(j\omega)$ axis. The third integral is the spectral density representation for the excitation. Hence we obtain the important relation,

$$S_y(\omega) = H(-j\omega)\, H(j\omega)\, S_x(\omega) = |H(j\omega)|^2\, S_x(\omega) \tag{5.4.6}$$

Example

As an example of this discussion, consider the output from the RC network shown in Fig. 5.24, when the excitation is white noise.

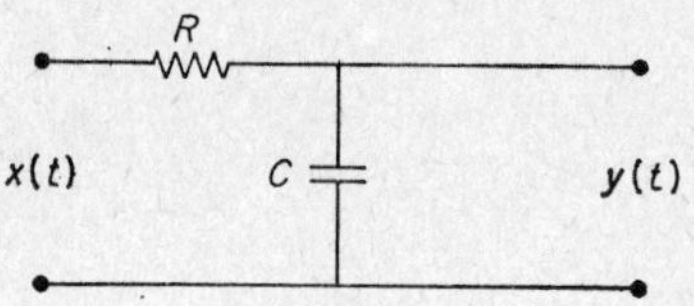

FIG. 5.24 A simple RC network: for the discussion on page 229, $x(t)$ is white noise.

The transform relation for this circuit is evidently,

$$Y(s) = \frac{(1/sC)}{R + (1/sC)} . X(s)$$

and so

$$H(s) = \frac{1}{RC} \cdot \frac{1}{\{s + (1/RC)\}} \tag{5.4.7}$$

and

$$h(t) = \frac{1}{RC}\, e^{-t/RC} . u(t) \tag{5.4.8}$$

The auto-correlation function for the white noise excitation is obtained from equation (4.4.51)

$$R_x(\tau) = S_0\, \delta(\tau) \tag{5.4.9}$$

so that

$$R_y(T) = \int_{-\infty}^{\infty} \int_{-\infty}^{\infty} h(\tau_1)\, h(\tau_2)\, S_0\, \delta(T - \tau_2 + \tau_1)\, d\tau_1\, d\tau_2$$

$$= S_0 \int_{-\infty}^{\infty} h(\tau_1)\, h(T + \tau_1)\, d\tau_1 \tag{5.4.10}$$

provided that $\tau_2 = (T + \tau_1) \geqslant 0^-$ from the lower limit on $h(t)$ set by $u(t)$ in equation (5.4.8). Moreover, $\tau_1 \geqslant 0^-$, again from the function $u(t)$ occurring in equation (5.4.8) and so we require $T \geqslant 0^-$, for which

$$R_y(T) = \frac{S_0}{R^2C^2} \int_{0^-}^{\infty} \exp\{-(T + 2\tau_1)/RC\}\, d\tau_1 \tag{5.4.11}$$

Hence

$$R_y(T) = \frac{S_0}{2RC}\, e^{-T/RC} \quad \text{for } T \geqslant 0^- \tag{5.4.12}$$

However, from equation (4.4.16), $R_y(T) = R_y(-T)$, so that

$$R_y(T) = \frac{S_0}{2RC}\,\mathrm{e}^{-|T|/RC} \quad \text{for } -\infty < T < \infty \tag{5.4.13}$$

The mean square output voltage is obtained using equation (4.4.14c)

$$\langle y^2 \rangle = R_y(0) = \frac{S_0}{2RC} \tag{5.4.14}$$

This relation can also be derived using the spectral density representation. From equations (5.4.6) and (5.4.7),

$$S_y(\omega) = \frac{1}{1 + \omega^2 R^2 C^2} S_x(\omega) \tag{5.4.15}$$

and now using equations (4.4.25) and (4.4.52)
we have,

$$\langle y^2 \rangle = \frac{1}{2\pi}\int_{-\infty}^{\infty} S_y(\omega)\,\mathrm{d}\omega = \frac{S_0}{2\pi}\int_{-\infty}^{\infty} \frac{1}{1 + \omega^2 R^2 C^2}\,\mathrm{d}\omega \tag{5.4.16}$$

and again obtain

$$\langle y^2 \rangle = \frac{S_0}{2RC} \tag{5.4.17}$$

Equations (5.4.14) and (5.4.17) draw attention to the interesting feature that the RC time constant for the circuit shown in Fig. 5.24 can in principle be determined by measuring the mean square output voltage when the circuit is excited by white noise.

5.4.2 Maximum Signal to Noise Ratio—the Matched Filter

The criterion of maximum signal to noise ratio is a particularly useful measure of the quality of a linear circuit when the principal objective is the detection of a signal of given deterministic form without any constraint relating this form to that of the output signal. Consider the linear system shown in Fig. 5.25 which is excited by a required signal function $s_i(t)$ together with a

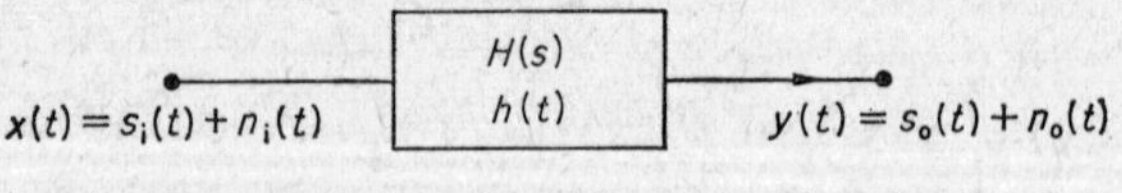

Fig. 5.25 A linear system excited by a signal and noise function.

noise background $n_i(t)$. The signal $s_i(t)$ is assumed to be a deterministic time function whilst the noise excitation $n_i(t)$ is a sample function of a stationary random process. We wish to find the transfer function (or impulse response function) for a linear system which maximizes the signal to noise ratio subject to the condition that the system be realizable.

The output signal to noise ratio at a specified time, t_0, is defined to be

$$\rho(t_0) = \frac{s_0^2(t_0)}{\langle n_0^2(t) \rangle} \tag{5.4.18}$$

where

$$\langle n_o^2(t)\rangle = \lim_{T_o\to\infty} \frac{1}{2T_o}\int_{-T_o}^{T_o} n_o^2(t)\,dt$$

is the mean output noise power. There is no difficulty in identifying the output signal and noise functions separately as $s_o(t)$ and $n_o(t)$, for a linear system which performs the general operation O, since upon taking,

$$x(t) = s_i(t) + n_i(t)$$

then

$$\begin{aligned} y(t) &= O\{x(t)\} = O\{s_i(t)\} + O\{n_i(t)\} \\ &= s_o(t) + n_o(t) \end{aligned}$$

and it is reasonable to define $s_o(t) = O\{s_i(t)\}$ and $n_o(t) = O\{n_i(t)\}$.

If $h(t)$ is the impulse response function for the desired linear system, the output signal at time t_o is obtained from equation (5.1.13)

$$s_o(t) = \int_{0^-}^{\infty} h(\tau_1)\, s_i(t_o - \tau_1)\, d\tau_1 \tag{5.4.19}$$

whilst $\langle n_o^2(t)\rangle$ is obtained from equations (4.4.14c) and (5.4.3), using the auto-correlation function representation,

$$\langle n_o^2(t)\rangle = R_{n_o}(0) = \int_{-\infty}^{\infty}\int_{-\infty}^{\infty} h(\tau_1)\, h(\tau_2)\, R_{ni}(\tau_1 - \tau_2)\, d\tau_1\, d\tau_2 \tag{5.4.20}$$

When the noise excitation arises from white noise, for which the spectral density function is equal to the constant S_0, equation (4.4.24) shows that

$$R_{ni}(\tau_1 - \tau_2) = S_0\, \delta(\tau_1 - \tau_2) \tag{5.4.21}$$

so that

$$\langle n_o^2(t)\rangle = S_0 \int_{-\infty}^{\infty} h^2(\tau_1)\, d\tau_1 = S_0 \int_{0^-}^{\infty} h^2(\tau_1)\, d\tau_1 \tag{5.4.22}$$

where the lower limit may be taken as 0^- because $h(\tau) = 0$ when $\tau < 0^-$ for a realizable system.

We wish to maximize

$$\rho(t_o) = \frac{s_o^2(t_o)}{\langle n_o^2(t)\rangle} = \frac{\left\{\int_{0^-}^{\infty} h(\tau_1)\, s_i(t_o - \tau_1)\, d\tau_1\right\}^2}{S_0 \int_{0^-}^{\infty} h^2(\tau_1)\, d\tau_1} \tag{5.4.23}$$

Using the Schwarz inequality,† since $h(t)$ and $s(t)$ are both real functions,

$$\left\{\int_{0^-}^{\infty} h(\tau_1)\, s_i(t_o - \tau_1)\, d\tau_1\right\}^2 \leqslant \int_{0^-}^{\infty} h^2(\tau_1)\, d\tau_1 \int_{0^-}^{\infty} s_i^2(t_o - \tau_1)\, d\tau_1 \tag{5.4.24}$$

so that

$$\rho(t_o) \leqslant \frac{1}{S_0}\int_{0^-}^{\infty} s_i^2(t_o - \tau_1)\, d\tau_1 \tag{5.4.25}$$

† See, for example, MARGENAU, H. and MURPHY, G. M., *The Mathematics of Physics and Chemistry*, Van Nostrand, 1944.

Hence the maximum value which $\rho(t_0)$ can achieve is

$$\rho_{\text{max}}(t_0) = \frac{1}{S_0} \int_{0-}^{\infty} s_i^2(t_0 - \tau_1)\, d\tau_1 \tag{5.4.26}$$

and this value will evidently be given by equation (5.4.23) when

$$h(\tau_i) = s_i(t_0 - \tau_i) \tag{5.4.27}$$

and, specifically for a realizable optimum linear network, the impulse response is given by

$$h(t) = s_i(t_0 - t)\, u(t) \tag{5.4.28}$$

Hence $h(t)$ has the form of the signal run backward starting from the particular time, t_0, at which $\rho_{\text{max}}(t_0)$ is evaluated. A linear system described by equation (5.4.28) is called a *matched filter* since the impulse response function is determined by the form of the required incoming signal to which it is 'matched'.

For a signal of finite duration, covering the range in time $0 \leqslant t \leqslant T$, $\rho_{\text{max}}(t_0)$ will reach its greatest value for $t_0 = T$ and thereafter remain constant. This condition arises because, from equation (5.4.26), we now have

$$\rho_{\text{max}}(t_0) = \frac{1}{S_0} \int_{0-}^{t_0} s_i^2(t_0 - \tau_1)\, d\tau_1 \tag{5.4.29}$$

but for $t_0 \geqslant T$, $s_i(t_0 - \tau_1)$ is only non-zero over the range $t_0 - T \leqslant \tau_1 \leqslant t_0$, so that

$$\begin{aligned} \rho_{\text{max}}(t_0) &= \frac{1}{S_0} \int_{t_0-T}^{t_0} s_i^2(t_0 - \tau_1)\, d\tau_1 \quad \text{for} \quad t_0 \geqslant T \\ &= \frac{1}{S_0} \int_{0}^{T} s_i^2(T - \xi)\, d\xi = \rho_{\text{max}}(T) \end{aligned} \tag{5.4.30}$$

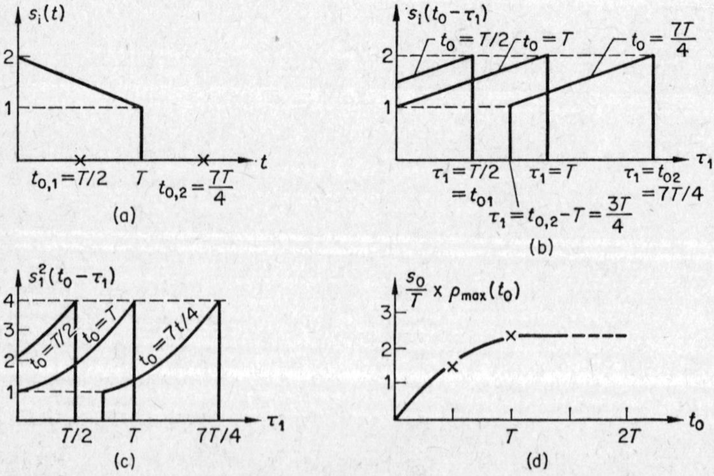

FIG. 5.26 Graphical illustrations of the functions involved in the derivation of $\rho_{\text{max}}(t_0)$ when the particular signal form is $s_i(t) = (2 - t/T)$ for $0 \leqslant t \leqslant T$ and $s_i(t)$ is zero elsewhere.

where $\xi = \tau_1 - (t_0 - T)$. This discussion is illustrated graphically for a particular signal form in Fig. 5.26. The transfer function evaluated along the $j\omega$ axis, corresponding to equation (5.4.28), with $t_0 \geqslant T$, is

$$\begin{aligned} H_T(j\omega) &= F\{s_i(t_0 - t)\, u(t)\} \\ &= F\{s_i(\eta)\, u(T - \eta)\} = e^{-j\omega T}\, S_i^*(j\omega) \end{aligned} \tag{5.4.31}$$

It can be seen from equation (5.4.30) that, for a signal of finite duration, the greatest value of the signal to noise ratio is

$$\rho_{max}(T) = \frac{E}{S_0} \tag{5.4.32}$$

where E is the total energy in the signal. Hence the only way to improve the detection of such a signal of known form in a background of white noise is to increase the total energy of the signal.

The complete signal $y(t)$ at the output of a matched filter is

$$\begin{aligned} y(t) &= \int_{0^-}^{\infty} h(\tau)\, x(t - \tau)\, d\tau \\ &= \int_{-\infty}^{\infty} s_i(t_0 - \tau)\, u(\tau)\, x(t - \tau)\, d\tau \end{aligned} \tag{5.4.33}$$

Changing the variable of integration to $\xi = t_0 - \tau$

$$y(t) = \int_{-\infty}^{\infty} s_i(\xi)\, x\{\xi + (t - t_0)\}\, u(t_0 - \xi)\, d\xi \tag{5.4.34}$$

Evidently the matched filter is forming a cross-correlation function of the desired signal with the total input excitation $x(t) = s_i(t) + n_i(t)$.

The useful output signal is

$$\begin{aligned} s_0(t) &= \int_{0^-}^{\infty} h(\tau)\, s_i(t - \tau)\, d\tau \\ &= \int_{-\infty}^{\infty} s_i(\xi)\, s_i\{\xi + (t - t_0)\}\, u(t_0 - \xi)\, d\xi \end{aligned} \tag{5.4.35}$$

When $t_0 > T$, the duration of the signal, the function $u(t_0 - \xi)$ need not be included specifically in the integral since $s_i(\xi) = 0$ for $\xi > T$. Equation (5.4.35) may now be written

$$s_0(t) = \int_{-\infty}^{\infty} s_i(\xi)\, s_i\{\xi + (t - t_0)\}\, d\xi = R_{s_1}^{(E)}(t - t_0) \tag{5.4.36}$$

which is an auto-correlation function having a shape which may bear little resemblance to that of the input signal. Hence this type of system has little value when it is necessary to reproduce the shape of the original signal as closely as possible.

5.4.3 Minimum Mean Square Error Filter

The mean square error criterion provides a measure of the degree to which the complete output function from the linear network reproduces the shape of the required input signal. Both the signal and noise excitations are assumed to be stationary (time invariant) processes, and we seek an impulse response function $h(t)$ for a realizable system which makes the mean square error a minimum. The error function is taken in the form

$$\epsilon(t) = y(t) - K\, s_1(t+\alpha) \tag{5.4.37}$$

which allows for an overall gain factor K and a delay $\alpha \geqslant 0$. Substituting

$$y(t) = \int_{-\infty}^{\infty} h(\tau)\{s_1(t-\tau) + n_1(t-\tau)\}\, d\tau = \int_{-\infty}^{\infty} h(\tau)\, x(t-\tau)\, d\tau \tag{5.4.38}$$

with $h(\tau) = 0$, when $\tau < 0^-$ for a realizable system, we shall require that the mean square error averaged over time shall be a minimum.
That is

$$\langle \epsilon^2(t) \rangle = \left\langle \left\{ \int_{-\infty}^{\infty} h(\tau)\, x(t-\tau)\, d\tau - K\, s_1(t+\alpha) \right\}^2 \right\rangle \tag{5.4.39}$$

be a minimum.

This equation may be written as

$$\begin{aligned} \langle \epsilon^2(t) \rangle &= \langle K^2\, s_1^2(t+\alpha) \rangle \\ &\quad + \left\langle \int_{-\infty}^{\infty} \int_{-\infty}^{\infty} h(\tau_1)\, h(\tau_2)\, x(t-\tau_1)\, x(t-\tau_2)\, d\tau_1\, d\tau_2 \right\rangle \\ &\quad - 2\left\langle \int_{-\infty}^{\infty} h(\tau)\, x(t-\tau)\, K s_1(t+\alpha)\, d\tau \right\rangle \end{aligned} \tag{5.4.40}$$

and since the average is taken over the time variable, t, for stationary processes,

$$\begin{aligned} \langle \epsilon^2(t) \rangle &= \langle K^2\, s_1^2(t+\alpha) \rangle \\ &\quad + \int_{-\infty}^{\infty} \int_{-\infty}^{\infty} h(\tau_1)\, h(\tau_2)\, \langle x(t-\tau_1)\, x(t-\tau_2) \rangle\, d\tau_1\, d\tau_2 \\ &\quad - 2\int_{-\infty}^{\infty} h(\tau)\, K \langle x(t-\tau)\, s_1(t+\alpha) \rangle\, d\tau \end{aligned} \tag{5.4.41}$$

$$\begin{aligned} &= K^2 R_{s_1}(0) + \int_{-\infty}^{\infty} \int_{-\infty}^{\infty} h(\tau_1)\, h(\tau_2)\, R_x(\tau_2 - \tau_1)\, d\tau_1\, d\tau_2 \\ &\quad - 2\int_{-\infty}^{\infty} h(\tau)\, K R_{x,s_1}(\tau + \alpha)\, d\tau \end{aligned} \tag{5.4.42}$$

In order to find the function $h(\tau)$ which makes $\langle \epsilon^2(t) \rangle$ a minimum, suppose that $g(t)$ is the impulse response function for any realizable linear network. Then $h(t) + \lambda\, g(t)$ is also the impulse response function of a realizable linear network, and if $h(t)$ corresponds to the function giving the minimum squared error in equation (5.4.42), the right-hand side of this equation must increase in value (or at least remain constant) when $h(\tau)$ is replaced by $\{h(\tau) + \lambda\, g(\tau)\}$. Making this substitution in equation (5.4.42) we obtain

$$\begin{aligned} \langle \epsilon^2(t) \rangle_g &= K^2 R_{s_1}(0) + \int_{-\infty}^{\infty} \int_{-\infty}^{\infty} h(\tau_1)\, h(\tau_2)\, R_x(\tau_2 - \tau_1)\, d\tau_1\, d\tau_2 \\ &\quad + \lambda^2 \int_{-\infty}^{\infty} \int_{-\infty}^{\infty} g(\tau_1)\, g(\tau_2)\, R_x(\tau_2 - \tau_1)\, d\tau_1\, d\tau_2 \end{aligned} \tag{5.4.43}$$

$$+2\lambda\int_{-\infty}^{\infty}\int_{-\infty}^{\infty} h(\tau_1)\, g(\tau_2)\, R_x(\tau_2 - \tau_1)\, d\tau_1\, d\tau_2$$

$$-\, 2K\int_{-\infty}^{\infty} h(\tau)\, R_{x,s_1}(\tau + \alpha)\, d\tau - 2\lambda K\int_{-\infty}^{\infty} g(\tau)\, R_{x,s_1}(\tau + \alpha)\, d\tau$$

and we require

$$\langle \epsilon^2(t)\rangle_g - \langle \epsilon^2(t)\rangle \geqslant 0 \tag{5.4.44}$$

for all values of the parameter λ.

That is, we require,

$$\lambda^2\int_{-\infty}^{\infty}\int_{-\infty}^{\infty} g(\tau_1)\, g(\tau_2)\, R_x(\tau_2 - \tau_1)\, d\tau_1\, d\tau_2$$
$$+\, 2\lambda\left\{\int_{-\infty}^{\infty}\int_{-\infty}^{\infty} h(\tau_1)\, g(\tau_2)\, R_x(\tau_2 - \tau_1)\, d\tau_1\, d\tau_2\right.$$
$$\left. -\, K\int_{-\infty}^{\infty} g(\tau)\, R_{x,s_1}(\tau + \alpha)\, d\tau\right\} \geqslant 0 \tag{5.4.45}$$

The first term on the left-hand side of this inequality can be seen to be $\left\langle\left\{\lambda\int_{-\infty}^{\infty} g(\tau)\, x(t - \tau)\, d\tau\right\}^2\right\rangle$ by comparison with equations (5.4.39) and (5.4.42). This term is therefore greater than or equal to zero. However, if the second term, enclosed in brackets {}, differs from zero a choice of λ, either positive or negative, can be made for which the complete expression on the left-hand side will be less than zero. We therefore require

$$\int_{-\infty}^{\infty}\int_{-\infty}^{\infty} h(\tau_1)\, g(\tau_2)\, R_x(\tau_2 - \tau_1)\, d\tau_1\, d\tau_2$$
$$-\, K\int_{-\infty}^{\infty} g(\tau)\, R_{x,s_1}(\tau + \alpha)\, d\tau = 0$$

that is,

$$\int_{-\infty}^{\infty}\left\{\int_{-\infty}^{\infty} h(\tau_1)\, R_x(\tau - \tau_1)\, d\tau_1 - K\, R_{x,s_1}(\tau + \alpha)\right\} g(\tau)\, d\tau = 0 \tag{5.4.46}$$

This equation will be satisfied for all possible $g(\tau)$ if

$$\int_{-\infty}^{\infty} h(\tau_1)\, R_x(\tau - \tau_1)\, d\tau_1 = K\, R_{x,s_1}(\tau + \alpha) \quad \text{for} \quad \tau \geqslant 0^- \tag{5.4.47}$$

where we have included the constraint that $h(t)$ and $g(t)$ are zero for $t < 0^-$. Equation (5.4.47) is the integral equation which must be satisfied by $h(t)$ in order that the corresponding realizable linear network should give a minimum mean square error between the complete output function and the desired input signal.

The technique for solving equation (5.4.47) involves factorizing the spectral density function for the entire input excitation into two parts, one of which is the Fourier Transform of a function $\xi_+(t)$ which vanishes for $t \leqslant 0$ and the other is the Fourier Transform of a function $\xi_-(t)$ which vanishes for $t \geqslant 0$.

This factorization is certainly possible when $S_x(\omega)$ is a rational function and this will be assumed to be the case. In these circumstances we can write

$$S_x(\omega) = a^2 \frac{(\omega - \Omega_1)(\omega - \Omega_2) \cdots (\omega - \Omega_N)}{(\omega - p_1)(\omega - p_2) \cdots (\omega - p_M)}. \tag{5.4.48}$$

where the roots Ω, p may be complex. $S_x(\omega)$ has particular properties which restrict the location of the poles and zeros. These have been discussed on pages 164–166 where it was shown that

(a) $S_x(\omega)$ is real for real values of ω, and so $S_x(\omega) = S_x^*(\omega)$. Hence a^2 is real and all the roots Ω, p with non-zero imaginary parts occur as conjugate pairs.
(b) $S_x(\omega)$ is integrable over the range $-\infty$ to $+\infty$ for ω real and so no root of the denominator can be real. This also implies $N < M$
(c) $S_x(\omega) \geqslant 0$ for real values of ω and hence any real root of the numerator must occur with even multiplicity.

The spectral density function may therefore be split into two factors, $S_x^{(+)}(\omega)$ containing all the poles and zeros with positive imaginary parts, $S_x^{(-)}(\omega)$ containing all the poles and zeros with negative imaginary parts. The even number of real root terms and the constant a^2 are shared equally into $S_x^{(+)}(\omega)$, $S_x^{(-)}(\omega)$ so that we may write

$$S_x(\omega) = S_x^{(+)}(\omega)\, S_x^{(-)}(\omega) \tag{5.4.49}$$

with

$$\{S_x^{(+)}(\omega)\}^* = S_x^{(-)}(\omega) \tag{5.4.50}$$

The Fourier Transforms of the functions are

$$\xi_+(t) = \frac{1}{2\pi}\int_{-\infty}^{\infty} e^{j\omega t}\, S_x^{(+)}(\omega)\, d\omega \quad \text{and} \quad \xi_-(t) = \frac{1}{2\pi}\int_{-\infty}^{\infty} e^{j\omega t}\, S_x^{(-)}(\omega)\, d\omega \tag{5.4.51}$$

where $\xi_+(t) = 0$ for all $t < 0$ and $\xi_-(t) = 0$ for all $t > 0$. These conditions arise since, for example $S_x^{(+)}(\omega)$ is a function of ω having all its poles in the upper half plane.† The auto-correlation function $R_x(\eta)$ is given by

$$R_x(\eta) = \frac{1}{2\pi}\int_{-\infty}^{\infty} e^{j\omega\eta}\, S_x(\omega)\, d\omega = \frac{1}{2\pi}\int_{-\infty}^{\infty} e^{j\omega\eta}\, S_x^{(+)}(\omega)\, S_x^{(-)}(\omega)\, d\omega$$

$$= \int_{-\infty}^{0} \xi_+(\eta - \lambda)\, \xi_-(\lambda)\, d\lambda \tag{5.4.52}$$

where the upper limit is taken to be zero because $\xi_-(\lambda) = 0$ for $\lambda > 0$. This equation provides an expression for $R_x(\tau - \tau_1)$ occurring in equation (5.4.47) in terms of the functions $\xi_+(t)$ and $\xi_-(t)$. It is now necessary to find a similar relation for $R_{xs_1}(\tau + \alpha)$.

† From Table 4.1. $F\{e^{-\alpha t}\, u(t)\} = \dfrac{1}{j\omega + \alpha}$ for Re $\alpha > 0$, and a typical partial fraction from $S_x^{(-)}(\omega)$ is

$$\frac{1}{\omega - (a + jb)} = \frac{j}{j(\omega - a) + b}$$

The cross-spectral density function $S_{x,s_1}(\omega)$ is related to $R_{x,s_1}(\tau + \alpha)$ by equation (4.4.27)

$$R_{x,s_1}(\eta) = \frac{1}{2\pi}\int_{-\infty}^{\infty} e^{j\omega\eta} S_{x,s_1}(\omega)\, d\omega \tag{5.4.53}$$

Introducing a new function $A(\omega)$ by the equation

$$S_{x,s_1}(\omega) \quad A(\omega)\, S_x^{(-)}(\omega) \tag{5.4.54}$$

we obtain

$$\begin{aligned} R_{x,s_1}(\eta) &= \frac{1}{2\pi}\int_{-\infty}^{\infty} e^{j\omega\eta}\, A(\omega)\, S_x^{(-)}(\omega)\, d\omega \\ &= \int_{-\infty}^{0} a(\eta - \lambda)\, \xi_-(\lambda)\, d\lambda \end{aligned} \tag{5.4.55}$$

where again the upper limit is taken to be zero because $\xi_-(\lambda) = 0$ for $\lambda > 0$, and the function $a(t)$, is defined by the Fourier Transform relation

$$a(t) = \frac{1}{2\pi}\int_{-\infty}^{\infty} e^{j\omega t}\, A(\omega)\, d\omega \tag{5.4.56}$$

Equation (5.4.47) may now be written

$$\begin{aligned} &\int_{0^-}^{\infty} h(\tau_1)\left\{\int_{-\infty}^{0} \xi_+(\tau - \tau_1 - \lambda)\, \xi_-(\lambda)\, d\lambda\right\} d\tau_1 \\ &\qquad = K\int_{-\infty}^{0} a(\tau + \alpha - \lambda).\xi_-(\lambda)\, d\lambda \quad \text{for } \tau \geqslant 0 \end{aligned} \tag{5.4.57}$$

that is

$$\begin{aligned} &\int_{-\infty}^{0} \xi_-(\lambda)\left\{K.a(\tau + \alpha - \lambda) - \int_{0^-}^{\infty} h(\tau_1).\xi_+(\tau - \tau_1 - \lambda)\, d\tau_1\right\} d\lambda \\ &\qquad = 0 \quad \text{for } \tau \geqslant 0 \end{aligned} \tag{5.4.58}$$

This equation is satisfied if the expression in brackets {} is zero for all $\lambda < 0$ and $\tau > 0$, which is the case if

$$K.a(\tau + \alpha) = \int_{0^-}^{\infty} h(\tau_1).\xi_+(\tau - \tau_1)\, d\tau_1 \quad \text{for } \tau \geqslant 0$$

This equation can be solved by taking Fourier Transforms because $\xi_+(\tau_1 - \tau)$ vanishes for $(\tau_1 - \tau) < 0$.

$$\begin{aligned} &\int_{0^-}^{\infty} e^{-j\omega\tau} K.a(\tau + \alpha)\, d\tau \\ &\quad = \int_{0^-}^{\infty} e^{-j\omega\tau_1} h(\tau_1)\, d\tau_1 \int_{0^-}^{\infty} e^{\{-j\omega(\tau-\tau_1)\}}\, \xi_+(\tau - \tau_1)\, d(\tau - \tau_1) \end{aligned} \tag{5.4.59}$$

so that the transfer function $H_0(j\omega)$ evaluated along the $j\omega$ axis for a linear network giving a minimum mean square error between the complete output function and the desired signal input is

$$H_0(j\omega) = \frac{K}{S_x^{(+)}(\omega)}\int_{0^-}^{\infty} e^{-j\omega\tau}\, a(\tau + \alpha)\, d\tau \tag{5.4.60}$$

Using equations (5.4.54) and 5.4.56) we obtain,

$$H_0(j\omega) = \frac{K}{S_x^{(+)}(\omega)} \int_{0^-}^{\infty} e^{-j\omega\tau} \left\{ \frac{1}{2\pi} \int_{-\infty}^{\infty} e^{j\omega'(\tau+\alpha)} \frac{S_{x,s_i}(\omega')}{S_x^{(-)}(\omega')} \, d\omega' \right\} d\tau \quad (5.4.61)$$

With $\alpha = 0$, the particular case for which there is no delay of the signal, equation (5.4.61) describes a *smoothing filter*. Recall that for a Fourier Transform (see, for example, Table 4.1)

$$F(j\omega) = \int_{-\infty}^{\infty} e^{-j\omega\tau} \left\{ \frac{1}{2\pi} \int_{-\infty}^{\infty} e^{j\omega'\tau} F(j\omega') \, d\omega' \right\} d\tau \quad (5.4.62)$$

and if the expression in brackets {} vanishes for negative values of τ, which is the case when $F(j\omega) \equiv F_+(j\omega)$ is a rational function with all its poles in the upper half of the complex ω-plane,

$$F_+(j\omega) = \int_{0^-}^{\infty} e^{-j\omega\tau} \left\{ \frac{1}{2\pi} \int_{-\infty}^{\infty} e^{j\omega'\tau} F_+(j\omega') \, d\omega' \right\} d\tau \quad (5.4.63)$$

Evidently, if $F(j\omega) \equiv F_-(j\omega)$ is a rational function with all its poles in the lower half of the complex ω-plane

$$\int_{0^-}^{\infty} e^{-j\omega\tau} \left\{ \frac{1}{2\pi} \int_{-\infty}^{\infty} e^{j\omega'\tau} F_-(j\omega') \, d\omega' \right\} d\tau = 0 \quad (5.4.64)$$

Consider the application of these relations to equation (5.4.61) with $\alpha = 0$, when the signal and noise input functions are uncorrelated. For this case

$$\begin{aligned} S_x(\omega) &= S_{si}(\omega) + S_{ni}(\omega) \\ S_{x,si}(\omega) &= S_{si}(\omega) \end{aligned} \quad (5.4.65)$$

and

$$H_0(j\omega) = \frac{K}{S_x^{(+)}(\omega)} \left\{ \frac{S_s(\omega)}{S_x^{(-)}(\omega)} \right\}_+ \quad (5.4.66)$$

and so $H_0(j\omega)$ can be obtained from a partial fraction expansion of the function $\{S_{si}(\omega)/S_x^{(-)}(\omega)\}$ without having to carry through the integrations in equation (5.4.61).

5.4.4 Examples of Optimum Linear Networks

Consider the design of the optimum linear network which maximizes the signal to noise ratio when the input signal is $s_i(t) = A \sin^2 \beta t . u(t)$ and the input noise is described by the power spectral density function

$$S_{ni}(\omega) = \frac{1}{a^2 + \omega^2}.$$

Since the input noise is not white, it is necessary to insert a shaping network to convert the noise to white noise before applying equation (5.4.28). This shaping network will of course, also change the form of the input signal. The complete network is shown in Fig. 5.27.

The transfer function for the shaping network is derived using equation (5.4.6). We require

$$|H_1(j\omega)|^2 S_{n_i}(\omega) = S_0 \quad (5.4.67)$$

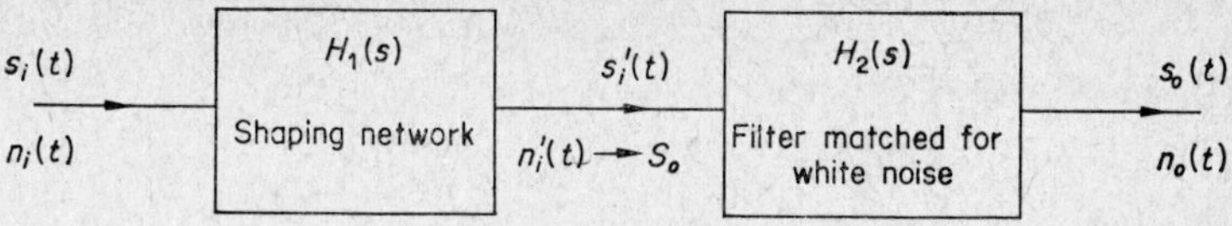

FIG. 5.27 Optimum linear circuit arrangement for maximum signal to noise ratio when input noise is not white.

so that

$$\begin{aligned} H_1(j\omega).H_1(-j\omega) &= \frac{S_0}{S_{n_i}(\omega)} \\ &= (a^2 + \omega^2).S_0 \\ &= (a + j\omega)(a - j\omega).S_0 \end{aligned} \qquad (5.4.68)$$

and so an acceptable transfer function is

$$H_1(s) = (s + a)\sqrt{S_0} \qquad (5.4.69)$$

The transfer function for the matched filter section is obtained from equation (5.4.28)

$$h_2(t) = s_i'(t_0 - t)\, u(t) \qquad (5.4.70)$$

with

$$\begin{aligned} L\{s_i'(t)\} &= H_1(s).L\{s_i(t)\} \\ &= (s + a)\sqrt{S_0}\cdot\frac{A}{2}\cdot\frac{4\beta^2}{s(s^2 + 4\beta^2)} \end{aligned} \qquad (5.4.71)$$

and hence

$$s_i'(t) = \frac{A}{2}(\sqrt{S_0})\{2\beta \sin 2\beta t + a(1 - \cos 2\beta t)\}\, u(t) \qquad (5.4.72)$$

which gives

$$h_2(t) = \frac{A}{2}(\sqrt{S_0})[2\beta \sin 2\beta(t_0 - t) + a\{1 - \cos 2\beta(t_0 - t)\}]\, u(t_0 - t).u(t) \qquad (5.4.73)$$

Assume that the signal to noise ratio is evaluated at $t_0 = 2\pi n/\beta$. This simplifies the calculation. Note also that

$$u(t_0 - t)\, u(t) = u(t) - u(t - t_0) \qquad (5.4.74)$$

For this case

$$\begin{aligned} h_2(t) &= \frac{A}{2}(\sqrt{S_0})\{-2\beta \sin 2\beta t + a(1 - \cos 2\beta t)\}\, u(t) \\ &\quad - \frac{A}{2}(\sqrt{S_0})[-2\beta \sin 2\beta(t - t_0) + a\{1 - \cos 2\beta(t - t_0)\}]\, u(t - t_0) \end{aligned} \qquad (5.4.75)$$

and

$$H_2(s) = \frac{A}{2}(\sqrt{S_0})\left[\frac{4\beta^2(a - s)}{s(s^2 + 4\beta^2)}\{1 - \exp(-st_0)\}\right], \quad t_0 = \frac{2\pi n}{\beta} \qquad (5.4.76)$$

The complete transfer function for the optimum linear network is therefore, from equations (5.4.69) and (5.4.76)

$$H(s) = H_1(s)\,H_2(s) = \frac{A}{2}S_0\left[\frac{4\beta^2(a^2 - s^2)}{s(s^2 + 4\beta^2)}\left\{1 - \exp\left(-s.\frac{2\pi n}{\beta}\right)\right\}\right] \tag{5.4.77}$$

Next consider the design of an optimum linear network which minimizes the mean square error between the complete output function and the input signal. Suppose that the power spectral density function for the input signal is $S_{s_1}(\omega) = 1/(1 + \omega^2)$ and the input noise is white noise for which $S_{n_1}(\omega) = S_0$. The power spectral density for the complete input function, assuming the signal and noise to be uncorrelated, is

$$\begin{aligned} S_x(\omega) = S_{s_1}(\omega) + S_{n_1}(\omega) &= \frac{(1 + S_0) + S_0\omega^2}{1 + \omega^2} \\ &= \frac{\{\omega\sqrt{S_0} + j\sqrt{(1 + S_0)}\}\{\omega\sqrt{S_0} - j\sqrt{(1 + S_0)}\}}{(\omega + j)(\omega - j)} \end{aligned} \tag{5.4.78}$$

From equation (5.4.66), for the filter with no delay, we require

$$\left\{\frac{S_{s_1}(\omega)}{S_x^{(-)}(\omega)}\right\}_+ = \left[\frac{1}{(\omega + j)(\omega - j)}.\frac{(\omega + j)}{\{\omega\sqrt{S_0} + j\sqrt{(1 + S_0)}\}}\right]_+ \tag{5.4.79}$$

that is

$$\begin{aligned} \left\{\frac{S_{s_1}(\omega)}{S_x^{(-)}(\omega)}\right\}_+ &= \frac{1}{j\{\sqrt{(1 + S_0)} + \sqrt{S_0}\}}\left[\frac{1}{(\omega - j)} - \frac{\sqrt{S_0}}{\{\omega\sqrt{S_0} + j\sqrt{(1 + S_0)}\}}\right]_+ \\ &= \frac{1}{j\{\sqrt{(1 + S_0)} + \sqrt{S_0}\}}.\frac{1}{(\omega - j)} \end{aligned} \tag{5.4.80}$$

and so

$$\begin{aligned} H_0(j\omega) &= \frac{K.(\omega - j)}{\{\omega\sqrt{S_0} - j\sqrt{(1 + S_0)}\}}.\frac{1}{j\{\sqrt{(1 + S_0)} + \sqrt{S_0}\}}.\frac{1}{(\omega - j)} \\ &= \frac{K}{\{\sqrt{(1 + S_0)} + \sqrt{S_0}\}}.\frac{1}{\{\sqrt{(1 + S_0)} + j\omega\sqrt{S_0}\}} \end{aligned} \tag{5.4.81}$$

Note the similarity between this transfer function and $H(j\omega)$ describing the low-pass *RC* filter shown in Fig. 5.24.

5.4.5 Thermal Noise

In this section we shall consider briefly the noise voltage which is developed between the terminals of a resistor because of the thermally excited random motion of the electrons. This noise voltage consists of individual pulses lasting for very short intervals of time and so the voltage spectral density can be assumed to be essentially constant

$$S_{V_n}(\omega) = S_0 \tag{5.4.82}$$

The resistor is therefore a white noise source. We shall not consider other sources of noise which may be of importance in practical applications for example shot noise (which is also white noise) and '1/f' noise in semiconductor devices, or atmospheric noise in communication systems.

It is useful to represent the noisy resistance as an ideal noise voltage generator in series with a noiseless resistance. This equivalent circuit is shown in Fig. 5.28(a). The value of S_0 in equation (5.4.82) may be determined by examining the thermal equilibrium behaviour of the circuit shown in Fig. 5.28(c) which is composed of a noisy resistance connected in parallel with an ideal lossless inductance. Assuming that the theorem of equipartition of energy applies to this system we assign $\frac{1}{2}kT$ to the mean energy stored

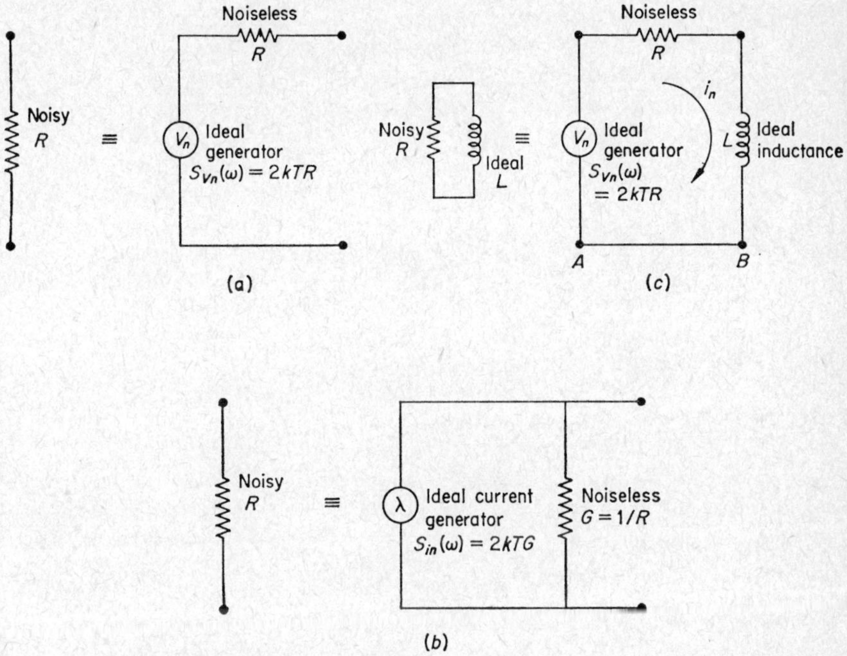

FIG. 5.28 Equivalent noise circuits for a resistance; in (c) the noisy resistance is connected in parallel with an ideal inductance.

in the inductance. The instantaneous energy in the inductance is $\frac{1}{2}L\,i_n^2$ and so

$$\tfrac{1}{2}L\,\langle i_n^2\rangle = \tfrac{1}{2}kT \tag{5.4.83}$$

The mean square value $\langle i_n^2\rangle$ is related to the current spectral density function

$$\langle i_n^2\rangle = \frac{1}{2\pi}\int_{-\infty}^{\infty} S_{i_n}(\omega)\,\mathrm{d}\omega \tag{5.4.84}$$

and so, if $H(\mathrm{j}\omega)$ is the transfer function relating i_n, V_n, from equation (5.4.6)

$$\begin{aligned} S_{i_n}(\omega) &= |H(\mathrm{j}\omega)|^2\,S_{V_n}(\omega) \\ &= \frac{1}{R^2+\omega^2L^2}\,S_0 \end{aligned}$$

Hence

$$\langle i_n^2 \rangle = \frac{S_0}{2\pi} \int_{-\infty}^{\infty} \frac{d\omega}{R^2 + \omega^2 L^2} = \frac{S_0}{2RL} \tag{5.4.85}$$

and using equation (5.4.83)

$$S_0 = 2kTR \tag{5.4.86}$$

This discussion has assigned $\frac{1}{2}kT$ to the mean energy stored in the inductance and has therefore assumed that the degree of freedom is fully excited. This assumption will not be valid at sufficiently low temperatures or high frequencies. In practice, this is not a significant limitation because the frequencies involved are of the order of 10^{14} c/s at room temperature, well above the range where the concept of a resistor as a discrete parameter (lumped-constant) circuit element is valid†. It does, however, imply that equation (5.4.86) can be used rigorously only over a finite frequency range when evaluating the mean square noise voltage $\langle V_n^2 \rangle$. Thus in a frequency range $\Delta\omega$,

$$\langle V_n^2 \rangle_{\Delta\omega} = \frac{1}{2\pi} \int_{-(\omega+\Delta\omega)}^{-\omega} S_0 \, d\omega + \frac{1}{2\pi} \int_{\omega}^{\omega+\Delta\omega} S_0 \, d\omega \tag{5.4.87}$$

where both positive and negative frequencies must be included, and so the r.m.s. noise voltage across a resistance is

$$\sqrt{(\langle V_n^2 \rangle_{\Delta\nu})} = 2\sqrt{(kTR\,\Delta\nu)} \tag{5.4.88}$$

where $\Delta\nu$ is the frequency bandwidth in cycles per second.

Note that the noisy resistance can also be represented by an ideal current generator as shown in Fig. 5.28(b). Evidently the r.m.s. short circuit current between the terminals of the equivalent circuit in Fig. 5.28(a) is

$$\sqrt{(\langle i_n^2 \rangle_{\Delta\nu})} = 2\sqrt{(kT\,\Delta\nu/R)} \tag{5.4.89}$$

and so the power spectral density for the equivalent noise current generator is

$$S_{i_n}(\omega) = \frac{S_0}{R^2} = 2kTG \tag{5.4.90}$$

A similar analysis can be carried through for an ideal capacitor in parallel with a noisy resistance. For this case the electric energy stored in the capacitor at any instant is $\frac{1}{2}C\,V_c^2$ and so we write

$$\tfrac{1}{2}C\langle V_c^2 \rangle = \tfrac{1}{2}kT \tag{5.4.91}$$

From Fig. 5.29(a) it is evident that

$$S_{V_c}(\omega) = \frac{1}{1 + \omega^2 R^2 C^2} S_0 \tag{5.4.92}$$

and so $S_0 = 2kTR$ again.

The analysis leading to equation (5.4.86) also implies that an actual lossy inductive element with total impedance $Z_L = R_L + j\omega L$ will act as a noise

† For a detailed discussion see VALLEY, G. E. and WALLMAN, H. 'Vacuum Tube Amplifiers' *M.I.T. Rad. Lab. Series* **18**, McGraw-Hill, 1948.

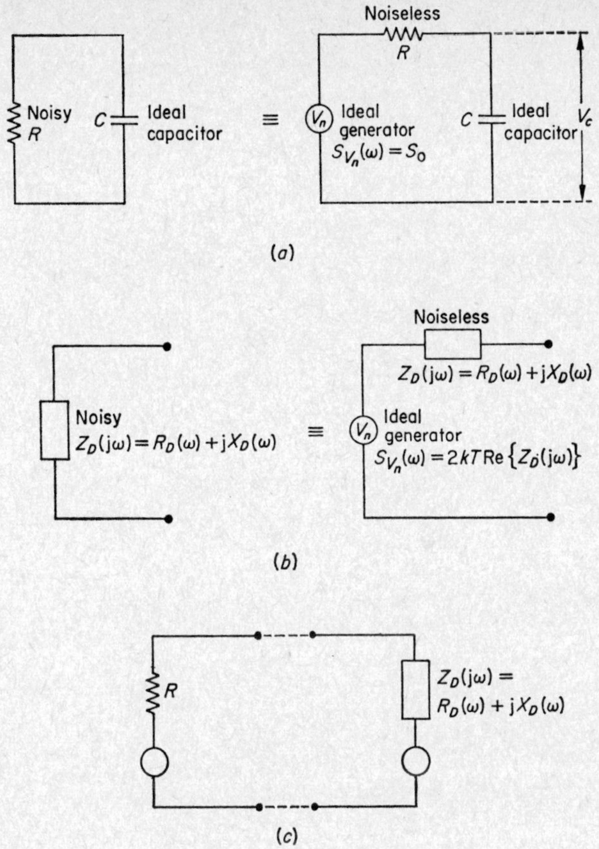

FIG. 5.29 Equivalent noise circuits.

generator having an internal impedance Z_L in series with an ideal voltage source for which the power spectral density is

$$S_{V_n}(\omega) = 2kTR_L = 2kT \text{ Re } [Z_L].$$

This is evident from Fig. 5.28(c) when the terminals A, B are open circuited. Similarly, for the case of a lossy capacitor, the noise voltage appearing at the terminals is V_c in Fig. 5.29(a) and so the element may be represented as an impedance

$$Z_c = \left(\frac{1}{R_c} + j\omega C\right)^{-1}$$

in series with an ideal voltage generator for which

$$S_{V_n}(\omega) = \left|\frac{1/j\omega C}{R_c + 1/j\omega C}\right|^2 .2kTR_c = 2kT \text{ Re } [Z_c]$$

In general, a passive two terminal network with driving point impedance $Z_D(j\omega)$ will act as a noise generator having internal impedance $Z_D(j\omega)$ in series with an ideal voltage source for which $S_{V_n}(\omega) = 2kT\,\text{Re}\,[Z_D(j\omega)]$. We may show this as follows. Suppose that $Z_D(j\omega)$ is in thermal equilibrium with a resistance R as in Fig. 5.29(c). Then the average thermal noise power absorbed by Z_D and arising from R in any frequency interval $\Delta\omega$ must be equal to the average noise power absorbed by R and arising from Z_D in the same frequency interval.

Thus

$$\tfrac{1}{2}\,\text{Re}\,[Z_D(j\omega)]\cdot\frac{2kTR\,\Delta\omega}{(R+R_D(\omega))^2+X_D^2(\omega)} = \tfrac{1}{2}R\cdot\frac{S_{V_n}(\omega)\,\Delta\omega}{(R+R_D(\omega))^2+X_D(\omega)^2} \tag{5.4.93}$$

$$S_{V_n}(\omega) = 2kT\,\text{Re}\,[Z_D(j\omega)] \tag{5.4.94}$$

which gives rise to the equivalent noise circuit shown in Fig. 5.29(b).

Bibliography

Cooper, G. R. and McGillem, C. D., *Methods of Signal and System Analysis*, Holt, Rinehart and Winston, 1967

Davenport, W. B., and Root, W. L., *An Introduction to the Theory of Random Signals and Noise*, McGraw-Hill, 1958

Hancock, J. C., *An Introduction to the Principles of Communication Theory*, McGraw-Hill, 1961

Kuo, F. F., *Network Analysis and Synthesis*, Wiley, 1962

Papoulis, A., *The Fourier Integral and its Applications*, McGraw-Hill, 1962

Pfeiffer, P. E., *Linear Systems Analysis*, McGraw-Hill, 1961

Ruston, H. and Bordogna, J., *Electric Networks, Functions and Filters Analysis*, McGraw-Hill, 1966

Weinberg, L., *Network Analysis and Synthesis*, McGraw-Hill, 1962

Operational Amplifier Circuit Configurations

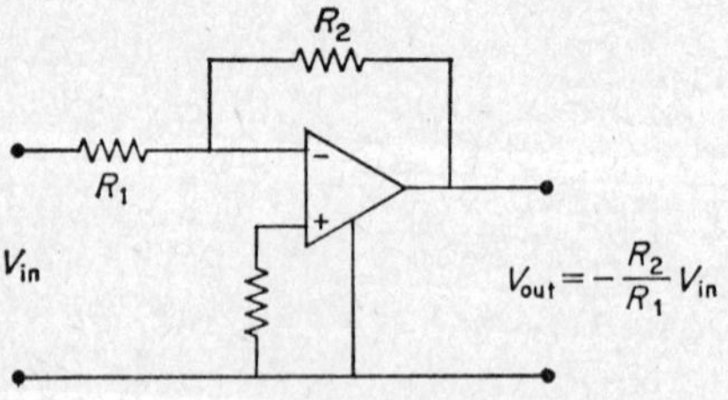

(*a*) Inverting Amplifier

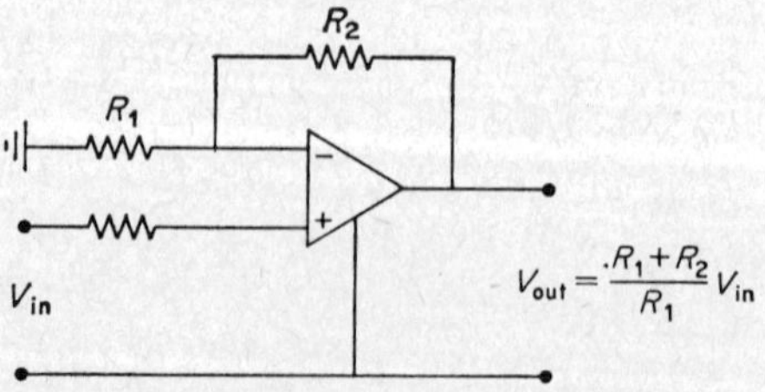

(*b*) Non-inverting Amplifier

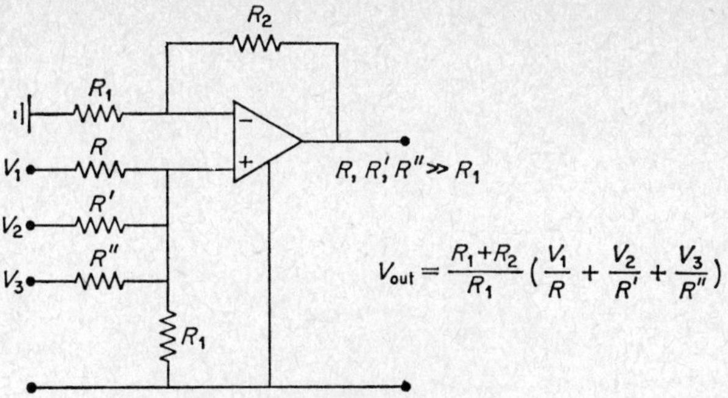

(*c*) Summing Amplifier

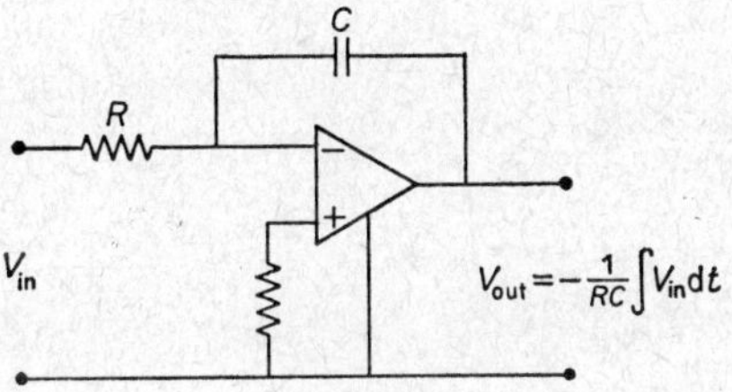

(*d*) Integrator

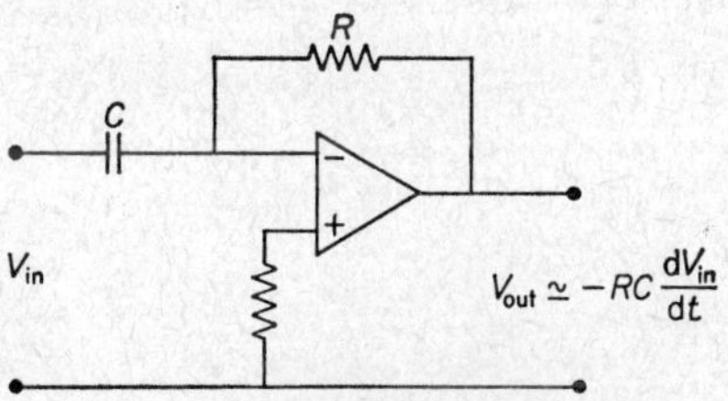

(*e*) Differentiator

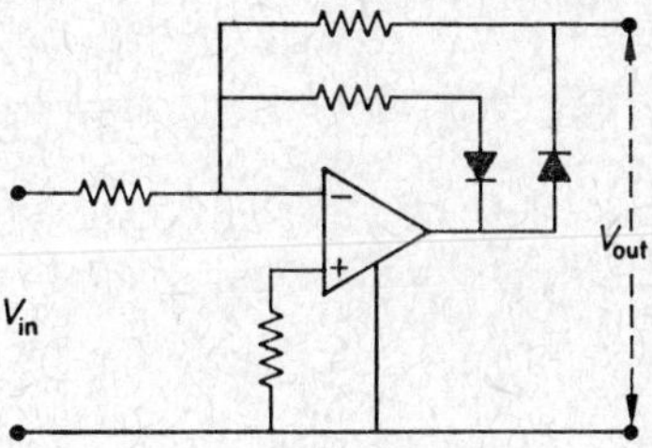

(*f*) Rectifier

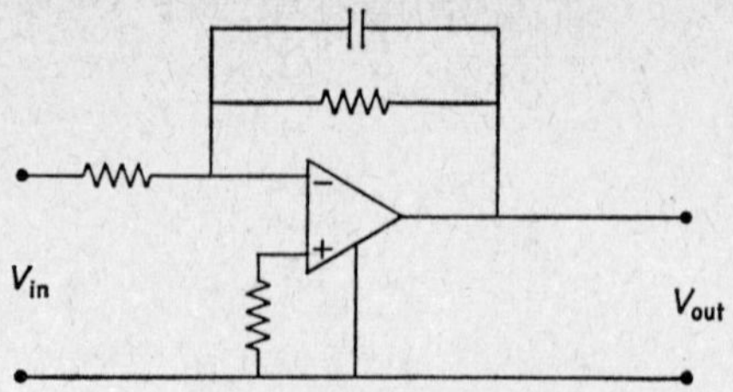

(*g*) Low pass Filter

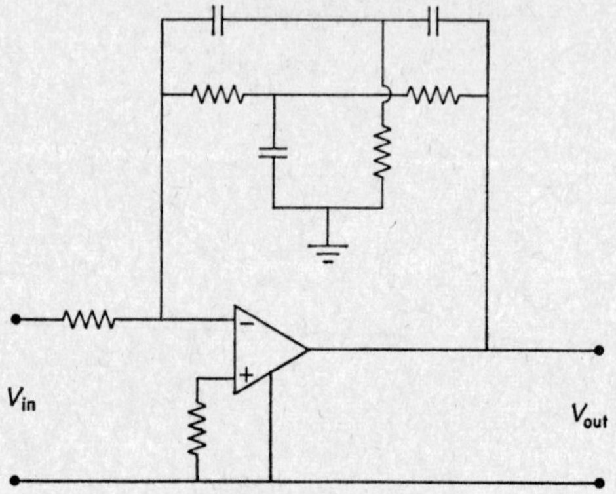

(*h*) Band pass Filter

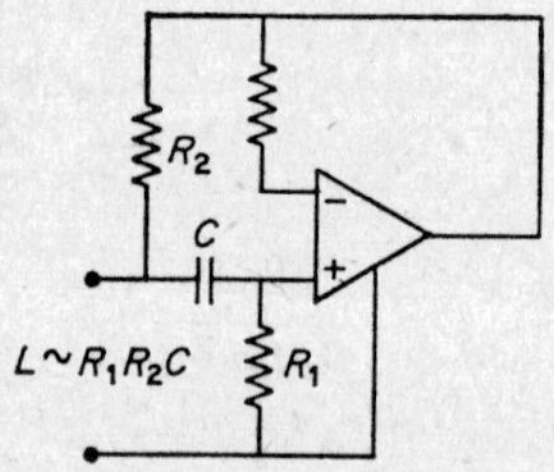

(*i*) Simulated Inductance

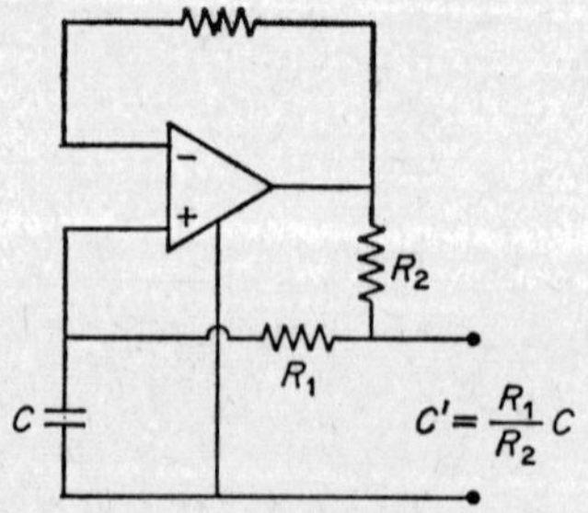

(*j*) Capacitance Multiplier

Problems

1. The switch in the circuit shown in Fig. 5.P.1 is moved from position A to position B at time $t = t_0$. Find (a) the voltage $V_0(t)$, (b) the total energy dissipated in the resistor R_2, when the input voltage is $V_i(t) = V_0\, u(t)$ and both condensers are uncharged at $t = 0^-$.

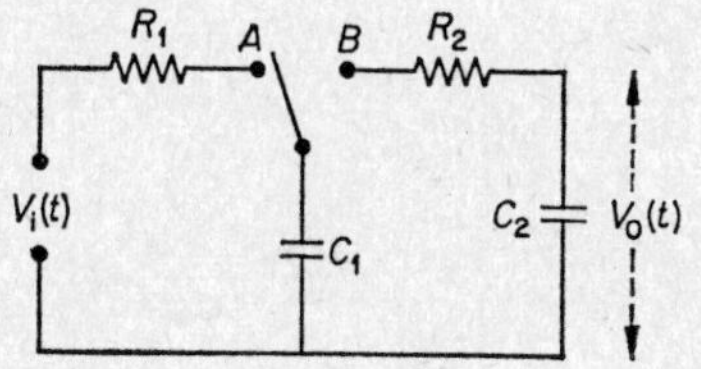

Fig. 5.P.1

2. Evaluate $V_2(t)$ for the network shown in Fig. 5.P.2, when

$$V_1(t) = \sum_{n=0}^{\infty} (-1)^n\, u(t - n\tau)$$

represents a rectangular input wave form, and the system is initially inert.

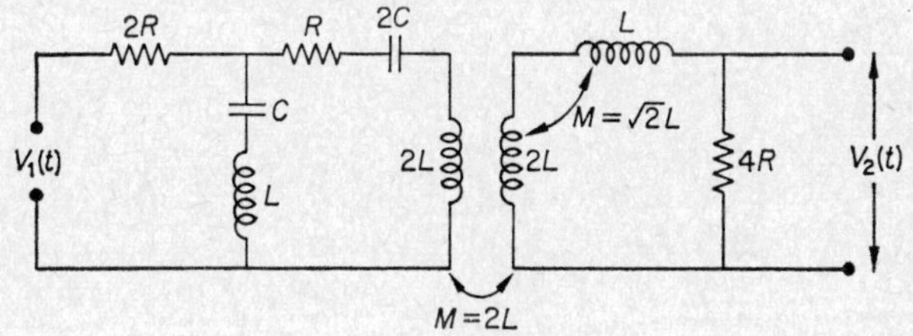

Fig. 5.P.2

3. Determine V(t) for $i(t) = \delta(t)$ assuming the circuit shown in Fig. 5.P.3 is initially inert.

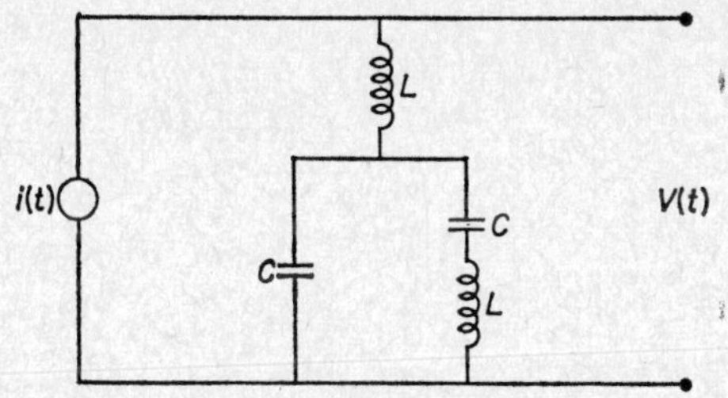

Fig. 5.P.3

4. Find expressions for the currents $i_1(t)$, $i_2(t)$, $i_3(t)$ given that $V(t) = t\, e^{-t}\, u(t)$ and that the circuit shown in Fig. 5.P.4 is initially inert.

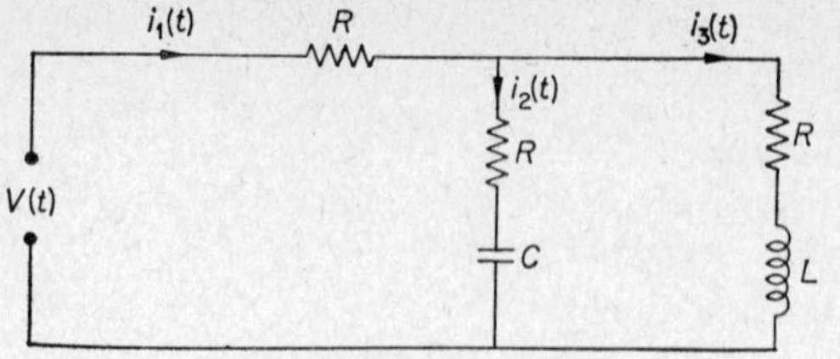

Fig. 5.P.4

5. Plot the poles and zeros for the following transfer functions, and comment on the stability of the networks described by these functions.

(a) $H(s) = \dfrac{s^2 + 4s + 3}{s^2 + 2s}$

(b) $H(s) = \dfrac{(s^2 + 3s)(s + 1)}{s(s^2 - 2s + 5)}$

(c) $H(s) = \dfrac{(s^2 + 4)(s - 2)}{(s^2 - 1)(s^2 + 2s + 2)}$

(d) $H(s) = \dfrac{s^2 + s - 2}{s^3 - s^2 + s + 1}$

(e) $H(s) = \dfrac{s^2 + 4}{s^3 + 2s^2 + 3s + 1}$

(f) $H(s) = \dfrac{2s^2 + 2s + 4}{(s + 1)^2(s^2 + 2)^2}$

(g) $H(s) = \dfrac{2s^2 + 2s + 4}{(s + 1)(s^2 + 2)}$

(h) $H(s) = \dfrac{(s + 1)(s + 2)}{s(s^2 + 3s + 2)}$

6. Determine whether the following polynomials have real parts which are strictly negative

(a) $s^3 + s^2 + 2s + 2$

(b) $s^4 + 2s^2 + 3s + 1$

(c) $7s^7 + 5s^5 + 3s^3 + 2s$

(d) $s^5 + 6s^4 + 3s^3 + s^2 + s + 1$

(e) $3s^3 + 2s^2 + s + 2$

(f) $4s^4 + 3s^3 + 2s^2 + s + 1$

(g) $s^6 + s^5 + s^4 + s^3 + s^2 + s + 1$

7. Construct a complete transfer function $H(s)$ corresponding to each of the following functions

(a) $\text{Re}\,[H(j\omega)] = \dfrac{1}{1 + \omega^6}$

(b) $\text{Re}\,[H(j\omega)] = \dfrac{1 + \omega^4}{(1 + \omega^2)^2}$

(c) $\text{Im}[H(j\omega)] = \dfrac{\omega^5 - 6\omega^3 + 5\omega}{1 + \omega^6}$

(d) $\text{Im}\,[H(j\omega)] = \dfrac{\omega(1 - \omega^2)}{(1 + \omega^2)^2}$

(e) $|H(j\omega)|^2 = \dfrac{1}{1 + \omega^6}$

(f) $|H(j\omega)|^2 = \dfrac{1 - \omega^2 + \omega^4}{1 + \omega^4}$

(g) $\beta(j\omega) = \tan^{-1} \dfrac{\omega}{\omega^2 - 1}$

(h) $\beta(j\omega) = \tan^{-1}(-\omega^3)$

8. Sketch the amplitude and phase response for the stable networks from Problem 5.

9. An ideal low-pass filter has an amplitude transfer function

$$|H(j\omega)| = K \quad \text{for } |\omega| < |\omega_0|$$
$$= 0 \quad \text{for } |\omega| > |\omega_0|$$

If the associated delay time is τ_0, discuss the response of this filter to a unit step function excitation $u(t)$.

10. The phase lag of minimum phase transfer function is given by

$$\beta(\omega) = 0 \quad \text{for } |\omega| < \omega_c$$
$$= \beta_0 \quad \text{for } \omega > \omega_c$$

Determine the amplitude transfer function $|H(j\omega)|$. Is this filter realizable?

11. Convert the low-pass filter shown in Fig. 5.P.5
(a) into a high-pass filter with cut-off ω_h;
(b) into a band-pass filter, with bandwidth B and characteristic frequency ω_c;
(c) into a band reject filter with bandwidth B and characteristic frequency ω_c.

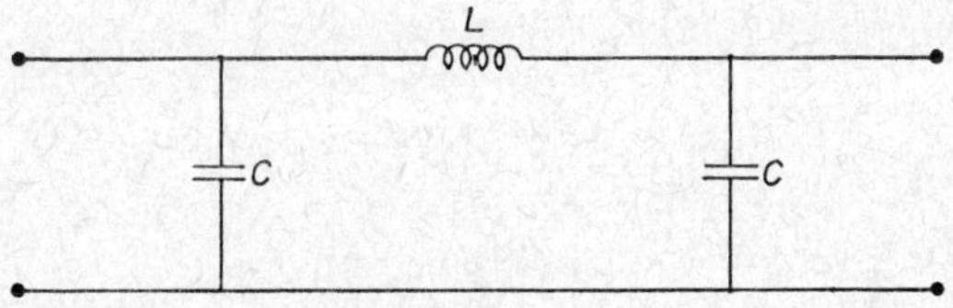

Fig. 5.P.5

12. Determine the impulse response $h(t)$ for each of the filters in Problem 11.

13. An operational amplifier with voltage feedback has an open loop transfer function

$$H(s) = \frac{A}{1 + s/\omega_0}$$

where ω_0 defines the bandwidth. If $H_f(s) = \beta$ and $Z_0 \ll Z_L$ show that the effective transfer function becomes

$$H_e(s) = \frac{A}{1 + A\beta}\left(\frac{1}{1 + s/\Omega}\right)$$

where $\Omega = \omega_0(1 + A\beta)$, and that the product of the d.c. gain $\times$ bandwidth is a constant.

14. Derive expressions for the overall current gain, overall voltage gain, effective input and effective output impedances for the circuit shown in Figure 5.P.6.

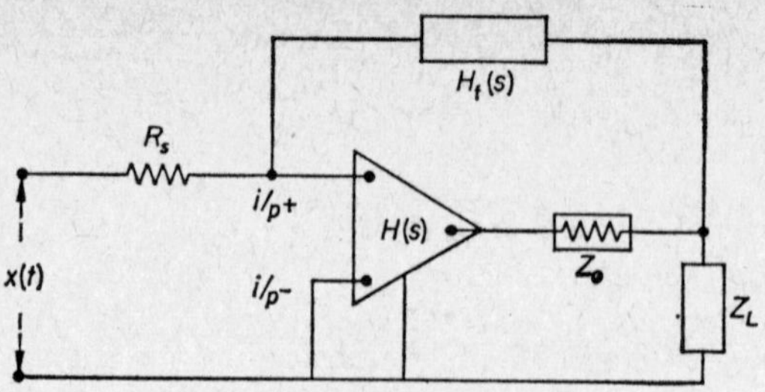

Fig. 5.P.6

15. The input to an RC low-pass filter consists of a signal

$$s_i(t) = V_0 \cos \beta t \, u(t)$$

and white noise with power spectral density S_0. Determine the output signal to noise ratio as a function of the time constant RC.

16. A signal has the form $s_i(t) = V_0 \cos \beta t \, u(t)$. Discuss the design of (a) the matched filter, (b) the minimum mean square error filter for this signal if the power spectral density of the associated noise is

$$(1)\ S_{ni}(\omega) = S_0, \quad (2)\ S_{ni}(\omega) = \frac{1}{a^2 + \omega^2}$$

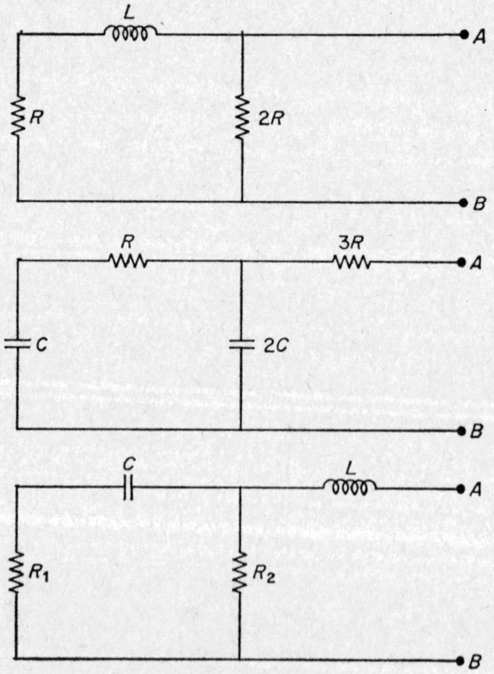

Fig. 5.P.7

17. A signal has power spectral density

$$S_{si}(\omega) = \frac{A}{a^2 + \omega^2}$$

Discuss the design of a minimum mean square error filter for this signal using an *RC* low-pass network when the input noise is white noise.

18. Determine the spectral density of the noise voltage appearing at the terminals *A*, *B* of the circuits shown in Fig. 5.P.7.

Chapter 6

Two-Port Networks

The transfer function for a linear system relates the response to the excitation when the complete circuit conditions are given. This method of description is entirely satisfactory for driving point functions which refer essentially to a network with one pair of accessible terminals. It is not, however, always the most convenient technique for describing circuits with two pairs of accessible terminals since the overall transfer function $H(s)$ will be dependent upon the terminating conditions. This is evident for example from the discussion on page 191 for the transformer mutual inductance element. There, the overall voltage–current relations were not simply the defining relation for mutual inductance, equation (5.1.33), but involved also the currents in the primary and secondary circuits through the self-inductances L_1, L_2 shown in Fig. 5.5.

An important class of networks can be regarded as linear systems with two pairs of accessible terminals. These will be called two-port systems (networks, circuit elements) and we will develop a technique for describing their behaviour which does not involve specific terminating conditions. This is desirable when it is required to construct a network for a particular application. It is then convenient to attempt to seek a solution using, as basic units, two-port elements which can be interconnected to produce the desired overall characteristics.

6.1 Coupling Matrices

Consider the two-port circuit element shown in Fig. 6.1. The terminals

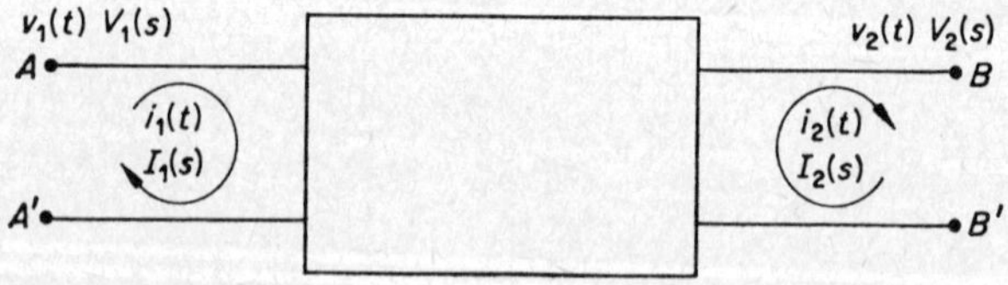

FIG. 6.1. The representation for a two-port circuit element.

A, A' form one pair (port), the terminals B, B' form the other pair (port). Two terminals may be identified as a pair (port) by the fact that the current flowing into one of the terminals must equal that flowing out at the associated terminal. There are four variables involved which may be represented as functions of time, t, or complex frequency variable s. These are $v_1(t)$, $i_1(t)$, $v_2(t)$, $i_2(t)$, or the corresponding transforms $V_1(s)$, $I_1(s)$, $V_2(s)$, $I_2(s)$. Only two from a set of four variables are independent variables. We shall see that the specification of any two determines the values of the remaining two. Thus if $v_1(s)$, $v_2(s)$ are specified then $I_1(s)$, $I_2(s)$ are determined by the properties of the circuit element.

It will be assumed that there are no internal independent voltage or current sources and so the two-port element is essentially a passive device. (This requirement does not however exclude operational amplifier circuit elements.) We may therefore regard the two accessible terminals as being connected by an array of linear circuit elements and write down a set of transform relations which will take the general form:

$$\begin{aligned} V_1(s) &= H_{11}(s)\,I_1(s) + H_{12}(s)\,I_2(s) + H_{13}(s)\,I_3(s) + \cdots H_{1n}(s)\,I_n(s) \\ -V_2(s) &= H_{21}(s)\,I_1(s) + H_{22}(s)\,I_2(s) + H_{23}(s)\,I_2(s) + \cdots H_{2n}(s)\,I_n(s) \\ 0 &= H_{31}(s)\,I_1(s) + H_{32}(s)\,I_2(s) + H_{33}(s)\,I_3(s) + \cdots H_{3n}(s)\,I_n(s) \\ &\cdot \quad \cdot \quad \cdot \quad \cdot \quad \cdot \quad \cdot \quad \cdot \quad \cdot \\ 0 &= H_{n1}(s)\,I_1(s) + H_{n2}(s)\,I_2(s) + H_{n3}(s)\,I_3(s) + \cdots H_{nn}(s)\,I_n(s) \end{aligned} \tag{6.1.1}$$

when there are n significant elements within the complete two-port system and the $H_{ij}(s)$ are appropriate transfer functions.

6.1.1 Admittance and Impedance Matrices

The set of simultaneous equations (6.1.1) may be solved to give $I_1(s)$ and $I_2(s)$.

$$\Delta . I_1(s) = \begin{vmatrix} V_1(s) & H_{12}(s) & H_{13}(s) \ldots H_{1n}(s) \\ -V_2(s) & H_{22}(s) & H_{23}(s) \ldots H_{2n}(s) \\ \vdots & \vdots & \vdots \quad\quad \vdots \\ 0 & H_{n2}(s) & H_{n3}(s) \ldots H_{nn}(s) \end{vmatrix}$$

$$\Delta . I_2(s) = \begin{vmatrix} H_{11}(s) & V_1(s) & H_{13}(s) \ldots H_{1n}(s) \\ H_{21}(s) & -V_2(s) & H_{23}(s) \ldots H_{2n}(s) \\ \vdots & \vdots & \vdots \quad\quad \vdots \\ H_{n1}(s) & 0 & H_{n3}(s) \ldots H_{nn}(s) \end{vmatrix} \tag{6.1.2}$$

$$\Delta = \begin{vmatrix} H_{11}(s) & H_{12}(s) & H_{13}(s) \ldots & H_{1n}(s) \\ H_{21}(s) & H_{22}(s) & H_{23}(s) \ldots & H_{2n}(s) \\ \vdots & \vdots & \vdots & \vdots \\ H_{n1}(s) & H_{n2}(s) & H_{n3}(s) \ldots & H_{nn}(s) \end{vmatrix}$$

Hence

$$\begin{aligned} I_1(s) &= \frac{\Delta_{11}}{\Delta} V_1(s) + \frac{\Delta_{21}}{\Delta} V_2(s) \\ I_2(s) &= -\left\{ \frac{\Delta_{12}}{\Delta} V_1(s) + \frac{\Delta_{22}}{\Delta} V_2(s) \right\} \end{aligned} \tag{6.1.3}$$

where Δ_{ij} is the cofactor of the term $H_{ij}(s)$ in the determinant Δ. Since Δ_{ij}/Δ is a transfer function giving a current response in terms of a voltage excitation, it will have the dimensions of admittance and so equations (6.1.3) may be written quite generally as

$$\begin{aligned} I_1(s) &= Y_{11}(s)\,V_1(s) + Y_{12}(s)\,V_2(s) \\ I_2(s) &= Y_{21}(s)\,V_1(s) + Y_{22}(s)\,V_2(s) \end{aligned} \tag{6.1.4}$$

or

$$\begin{bmatrix} I_1(s) \\ I_2(s) \end{bmatrix} = \begin{bmatrix} Y_{11}(s) & Y_{12}(s) \\ Y_{21}(s) & Y_{22}(s) \end{bmatrix} \begin{bmatrix} V_1(s) \\ V_2(s) \end{bmatrix} = [Y] \begin{bmatrix} V_1(s) \\ V_2(s) \end{bmatrix} \tag{6.1.5}$$

where $[Y]$ is called the admittance matrix for the two-port. If equations (6.1.4) are solved for $V_1(s)$, $V_2(s)$, the coefficients will have the dimensions of impedance. The equations now become

$$\begin{bmatrix} V_1(s) \\ V_2(s) \end{bmatrix} = \begin{bmatrix} Z_{11}(s) & Z_{12}(s) \\ Z_{21}(s) & Z_{22}(s) \end{bmatrix} \begin{bmatrix} I_1(s) \\ I_2(s) \end{bmatrix} = [Z] \begin{bmatrix} I_1(s) \\ I_2(s) \end{bmatrix} \tag{6.1.6}$$

where $[Z]$ is called the impedance matrix for the two-port. A relation between $[Y]$ and $[Z]$ is obtained by multiplying this equation from the left by $[Y]$ and using equation (6.1.5). We obtain

$$\begin{bmatrix} I_1(s) \\ I_2(s) \end{bmatrix} = [Y]\,[Z] \begin{bmatrix} I_1(s) \\ I_2(s) \end{bmatrix}$$

which is true for all $I_1(s)$, $I_2(s)$. Hence

$$[Y][Z] = [1] \tag{6.1.7}$$

and so the impedance matrix $[Z]$ for the two-port network is the reciprocal of the admittance matrix $[Y]$.

When the two-port network is constructed from passive circuit elements involving only resistance, capacitance and inductance, the relation $H_{ij}(s) = H_{ji}(s)$ holds in equation (6.1.1) (compare equation (3.3.7)). For this case there are no operational amplifiers or other strictly unidirectional elements (for example diodes) in the two-port. The determinant Δ_{21} is now equal to determinant Δ_{12} in equation (6.1.3) since the one is obtained from the other simply by interchanging all the rows and columns. Hence

$$Y_{12}(s) = -Y_{21}(s) \quad \text{and} \quad Z_{12}(s) = -Z_{21}(s)$$

so that the matrices $[Y]$ and $[Z]$ for a passive reciprocal two-port network are antisymmetric.

6.1.2 Transfer Matrices

There are other possible ways of relating pairs of terms from $V_1(s)$, $I_1(s)$, $V_2(s)$, $I_2(s)$, which will also provide useful techniques for describing two-port systems. Thus we may solve equations (6.1.3) for the input voltage and current in terms of the output voltage and current.† This will emphasize the

† For example an ideal transformer is described by $V_1(s) = (1/n)V_2(s)$, $I_1(s) = nI_2(s)$ and so

$$Z_{11}(s) = [V_1(s)/I_1(s)]_{I_2=0} = \infty,\ Z_{12}(s) = [V_1(s)/I_2(s)]_{I_1=0} = \infty, \text{ etc.}$$

Accordingly

$$[Z] = \begin{bmatrix} \infty & \infty \\ \infty & \infty \end{bmatrix}.$$

On the other hand

$$[A] = \begin{bmatrix} 1/n & 0 \\ 0 & n \end{bmatrix}.$$

transmission properties of the two-port. We obtain

$$\begin{bmatrix} V_1(s) \\ I_1(s) \end{bmatrix} = \begin{bmatrix} A_{11} & A_{12} \\ A_{21} & A_{22} \end{bmatrix} \begin{bmatrix} V_2(s) \\ I_2(s) \end{bmatrix} = [A] \begin{bmatrix} V_2(s) \\ I_2(s) \end{bmatrix} \tag{6.1.8}$$

The matrix $[A]$ is called the transfer matrix for the two-port and we note that the elements of $[A]$ do not all have the same dimensions. The elements A_{11}, A_{22}, are just numbers whilst A_{12} is an impedance and A_{21} is an admittance. The matrix $[A]$ does not possess the symmetry of the $[Y]$ and $[Z]$ matrices but instead has the important property that

$$\det [A] = -\frac{Z_{12}(s)}{Z_{21}(s)} \tag{6.1.9}$$

which can be derived by rearranging terms in equation (6.1.6). Hence

$$\det [A] = 1 \tag{6.1.10}$$

when $\Delta_{12} = \Delta_{21}$ which is the case for a simple passive system.

The relation complementary to equation (6.1.8) is

$$\begin{bmatrix} V_2(s) \\ I_2(s) \end{bmatrix} = \begin{bmatrix} B_{11} & B_{12} \\ B_{21} & B_{22} \end{bmatrix} \begin{bmatrix} V_1(s) \\ I_1(s) \end{bmatrix} = [B] \begin{bmatrix} V_1(s) \\ I_1(s) \end{bmatrix} \tag{6.1.11}$$

with

$$\det [B] = -\frac{Z_{21}(s)}{Z_{12}(s)} \tag{6.1.12}$$

and

$$\det [B] = 1 \tag{6.1.13}$$

when $\Delta_{12} = \Delta_{21}$.

The relation between the $[A]$ and $[B]$ matrices is

$$[A][B] = [1] \tag{6.1.14}$$

6.1.3 Hybrid Parameter Matrices

The remaining matrix relations between pairs of the variables $V_1(s)$, $I_1(s)$, $V_2(s)$, $I_2(s)$ are written in terms of 'hybrid' coefficients

$$\begin{bmatrix} V_1(s) \\ I_2(s) \end{bmatrix} = \begin{bmatrix} h_{11} & h_{12} \\ h_{21} & h_{22} \end{bmatrix} \begin{bmatrix} I_1(s) \\ V_2(s) \end{bmatrix} = \begin{bmatrix} h \end{bmatrix} \begin{bmatrix} I_1(s) \\ V_2(s) \end{bmatrix} \tag{6.1.15}$$

and

$$\begin{bmatrix} I_1(s) \\ V_2(s) \end{bmatrix} = \begin{bmatrix} g_{11} & g_{12} \\ g_{21} & g_{22} \end{bmatrix} \begin{bmatrix} V_1(s) \\ I_2(s) \end{bmatrix} = [g] \begin{bmatrix} V_1(s) \\ I_2(s) \end{bmatrix} \tag{6.1.16}$$

with

$$[h][g] = [1] \tag{6.1.17}$$

The matrices $[h]$ and $[g]$ reflect the symmetry of the $[Z]$ and $[Y]$ matrices. When $\Delta_{12} = \Delta_{21}$ in equations (6.1.3)

$$h_{12} = h_{21} \quad \text{and} \quad g_{12} = g_{21} \tag{6.1.18}$$

The description of a two-port system based on $[h]$ and $[g]$ matrices is used extensively as a representation for a transistor.

6.1.4 Relations between the Matrix Elements

The various matrix representations of a two-port system are of course all interrelated through equations (6.1.3). The choice of any particular representation is simply determined by its convenience for the network problem under investigation. A table of relations between the different matrix elements is given in Table 6.1.

Table 6.1

[Z]	$Z_{11}(s)$	$Z_{12}(s)$	$Z_{21}(s)$	$Z_{22}(s)$
[Y]	$Y_{11} = \dfrac{Z_{22}}{\Delta_Z}$	$Y_{12} = \dfrac{-Z_{12}}{\Delta_Z}$	$Y_{21} = \dfrac{-Z_{21}}{\Delta_Z}$	$Y_{22} = \dfrac{Z_{11}}{\Delta_Z}$
[A]	$A_{11} = \dfrac{Z_{11}}{Z_{21}}$	$A_{12} = \dfrac{-\Delta_Z}{Z_{21}}$	$A_{21} = \dfrac{1}{Z_{21}}$	$A_{22} = \dfrac{-Z_{22}}{Z_{21}}$
[B]	$B_{11} = \dfrac{Z_{22}}{Z_{12}}$	$B_{12} = \dfrac{-\Delta_Z}{Z_{12}}$	$B_{21} = \dfrac{+1}{Z_{12}}$	$B_{22} = \dfrac{-Z_{11}}{Z_{12}}$
[h]	$h_{11} = \dfrac{\Delta_Z}{Z_{22}}$	$h_{12} = \dfrac{Z_{12}}{Z_{22}}$	$h_{21} = \dfrac{-Z_{21}}{Z_{22}}$	$h_{22} = \dfrac{1}{Z_{22}}$
[g]	$g_{11} = \dfrac{1}{Z_{11}}$	$g_{12} = \dfrac{-Z_{12}}{Z_{11}}$	$g_{21} = \dfrac{Z_{21}}{Z_{11}}$	$g_{22} = \dfrac{\Delta_Z}{Z_{11}}$

where $\Delta_Z \equiv \det [Z]$

6.2 Interconnection of Two-Port Networks

Two-port networks may be combined together to form another overall two-port circuit. There are five basic ways for doing this: (1) series, (2) parallel, (3) series–parallel, (4) parallel–series, (5) cascade. For the first four methods of interconnection, when certain validity conditions are satisfied (known as Brune tests), one of the matrices describing the overall circuit is derived simply by summing the elements from appropriate matrices which refer to the constituent two-port networks before interconnection. The cascade arrangement does not require any specific validity condition and for this case the transfer matrix for the complete circuit is the product of the individual two-port transfer matrices.

6.2.1 Series Connection

Consider the series connection of two elements labelled λ, μ, as shown in Fig. 6.2. If the currents which enter and leave any given port of the constituent elements remain equal to one another after interconnection that port will continue to be a valid terminal pair. When all the original ports continue to be valid terminal pairs each of the constituent elements will have input and output voltages and currents related through the matrices for the isolated elements. This will be the case for the circuit shown in Fig. 6.2 when the loop current $I(s) = 0$

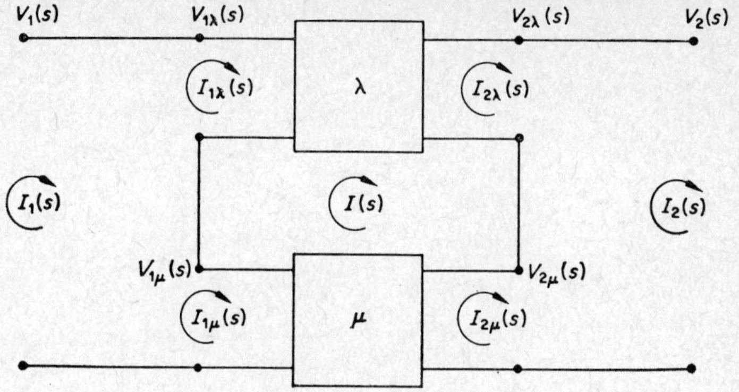

FIG. 6.2 Two elements connected in series.

For $I(s) = 0$

$$\begin{aligned} I_1(s) &= I_{1\lambda}(s) = I_{1\mu}(s) \\ I_2(s) &= I_{2\lambda}(s) = I_{2\mu}(s) \end{aligned} \tag{6.2.1}$$

Then

$$\begin{bmatrix} V_1(s) \\ V_2(s) \end{bmatrix} = \begin{bmatrix} V_{1\lambda}(s) \\ V_{2\lambda}(s) \end{bmatrix} + \begin{bmatrix} V_{1\mu}(s) \\ V_{2\mu}(s) \end{bmatrix} = \{[Z_\lambda] + [Z_\mu]\} \begin{bmatrix} I_1(s) \\ I_2(s) \end{bmatrix} \tag{6.2.2}$$

and so

$$[Z] = [Z_\lambda] + [Z_\mu] \tag{6.2.3}$$

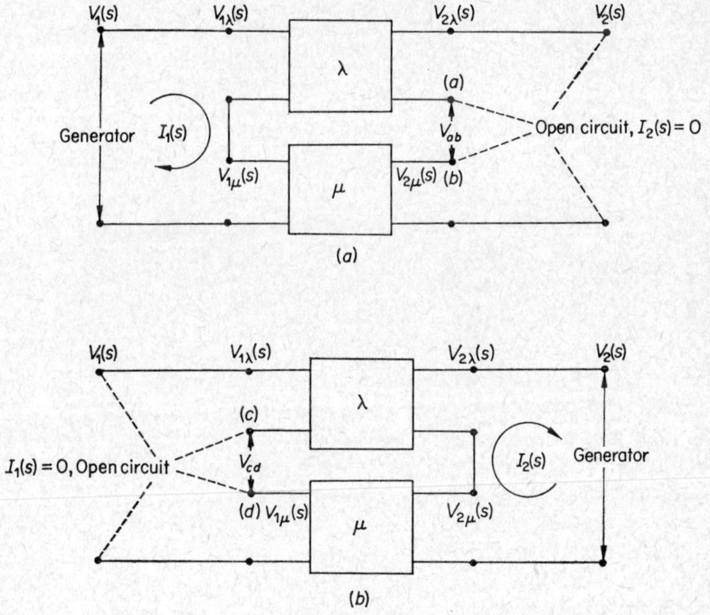

FIG. 6.3 Brune tests for the series connection of two-port networks.

This equation will be valid and the conditions of equation (6.2.1) satisfied if the open circuit voltages V_{ab}, V_{cd}, shown in Fig. 6.3 are zero. When the terminals are open circuited as shown in Fig. 6.3(a), $I(s) = 0$ and $I_2(s) = I_{2\lambda}(s) = I_{2\mu}(s) = 0$. But, from equation (6.1.6), and with $V_{ab} = 0$

$$Z_{11}(s) = \left\{\frac{V_1(s)}{I_1(s)}\right\}_{I_2=0} = \left\{\frac{V_{1\lambda}(s)}{I_{1\lambda}(s)}\right\}_{I_{2\lambda}=0} + \left\{\frac{V_{1\mu}(s)}{I_{1\mu}(s)}\right\}_{I_{2\mu}=0} = Z_{11,\lambda}(s) + Z_{11,\mu}(s) \tag{6.2.4}$$

$$Z_{21}(s) = \left\{\frac{V_2(s)}{I_1(s)}\right\}_{I_2=0} = \left\{\frac{V_{2\lambda}(s)}{I_{1\lambda}(s)}\right\}_{I_{2\lambda}=0} + \left\{\frac{V_{2\mu}(s)}{I_{1\mu}(s)}\right\}_{I_{2\mu}=0} = Z_{21,\lambda}(s) + Z_{21,\mu}(s)$$

Similarly for Fig. 6.3(b), when $V_{cd} = 0$

$$Z_{22}(s) = Z_{22,\lambda}(s) + Z_{22,\mu}(s)$$
$$Z_{12}(s) = Z_{12,\lambda}(s) + Z_{12,\mu}(s) \tag{6.2.5}$$

Hence if $V_{ab} = 0$ and $V_{cd} = 0$ for the open circuit conditions of Fig. 6.3(a) and (b) equation (6.2.3) is valid. However, in these circumstances, the pairs of terminals $(a)(b)$, and $(c)(d)$, may be connected together without affecting the circuit conditions, and equation (6.2.3) will be valid for the interconnected two-port system shown in Fig. 6.2.

The tests indicated by Fig. 6.3(a) and (b) are the Brune tests for the series connection of two-port networks.

6.2.2 Parallel Connection

The parallel connection of two elements labelled λ, μ, is shown in Fig. 6.4.

For this case the individual admittance matrices may be summed to give an overall admittance matrix for the system. The conditions for this to be valid may be derived by an argument similar to that used for series connection.

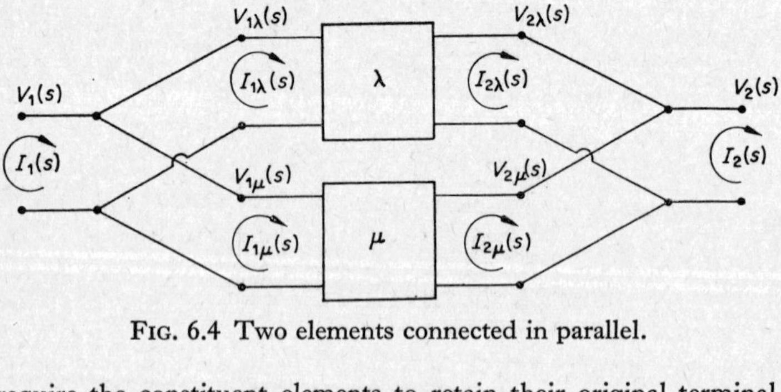

Fig. 6.4 Two elements connected in parallel.

We require the constituent elements to retain their original terminal pair configuration. The process of interconnection will not then give rise to perturbing currents in the additional loops which have been formed. When this is the case we have $I_1(s) = I_{1\lambda}(s) + I_{1\mu}(s)$ and $I_2(s) = I_{2\lambda}(s) + I_{2\mu}(s)$ together with

$$V_1(s) = V_{1\lambda}(s) = V_{1\mu}(s)$$
$$V_2(s) = V_{2\lambda}(s) = V_{2\mu}(s) \tag{6.2.6}$$

so that

$$\begin{bmatrix} I_1(s) \\ I_2(s) \end{bmatrix} = \begin{bmatrix} I_{1\lambda}(s) \\ I_{2\lambda}(s) \end{bmatrix} + \begin{bmatrix} I_{1\mu}(s) \\ I_{2\mu}(s) \end{bmatrix} = \{[Y_\lambda] + [Y_\mu]\} \begin{bmatrix} V_1(s) \\ V_2(s) \end{bmatrix} \quad (6.2.7)$$

and

$$[Y] = [Y_\lambda] + [Y_\mu] \quad (6.2.8)$$

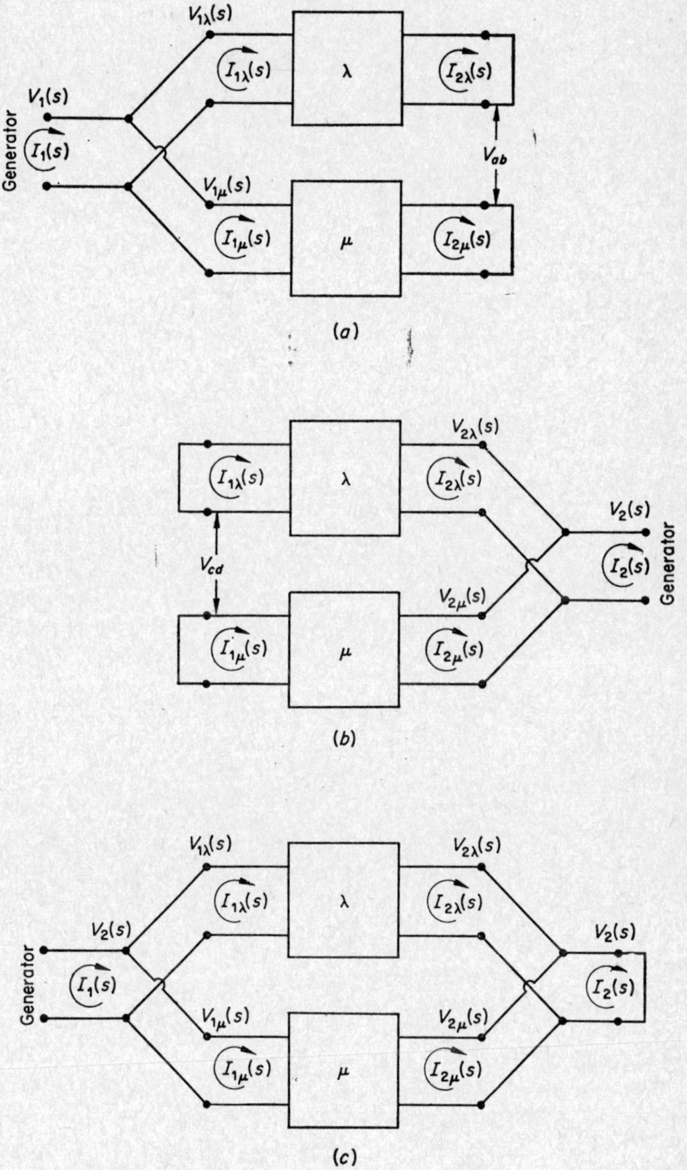

FIG. 6.5 Brune tests for the parallel connection of two-port networks.

This equation will be valid if the voltages V_{ab}, V_{cd}, shown in Figs. 6.5(a), (b) are zero when the various terminal pairs are short circuited. When $V_{ab} = 0$ in Fig. 6.5(a), the output terminals may be connected together as shown in Fig. 6.5(c) without affecting the circuit. But the two-port networks λ, μ in Fig. 6.5(a) have entirely separate current paths, the currents entering and leaving the various terminal pairs remain equal to one another after interconnection and so the original λ, μ, matrices will continue to provide a valid representation for the individual two-ports. From equation (6.1.4) and Fig. 6.5(c) the matrix elements Y_{11}, Y_{21}, for the overall admittance matrix are given by

$$Y_{11}(s) = \left\{\frac{I_1(s)}{V_1(s)}\right\}_{V_2=0} = \left\{\frac{I_{1\lambda}(s)}{V_{1\lambda}(s)}\right\}_{V_{2\lambda}=0} + \left\{\frac{I_{1\mu}(s)}{V_{1\mu}(s)}\right\}_{V_{2\mu}=0}$$
$$= Y_{11,\lambda}(s) + Y_{11,\mu}(s)$$
$$Y_{21}(s) = \left\{\frac{I_2(s)}{V_1(s)}\right\}_{V_2=0} = \left\{\frac{I_{2\lambda}(s)}{V_{1\lambda}(s)}\right\}_{V_{2\lambda}=0} + \left\{\frac{I_{2\mu}(s)}{V_{1\mu}(s)}\right\}_{V_{2\mu}=0} \quad (6.2.9)$$
$$= Y_{21,\lambda}(s) + Y_{21,\mu}(s)$$

Similarly, when $V_{cd} = 0$ for the conditions of Fig. 6.5(b)

$$Y_{22}(s) = Y_{22,\lambda}(s) + Y_{22,\mu}(s)$$
$$Y_{12}(s) = Y_{12,\lambda}(s) + Y_{12,\mu}(s) \quad (6.2.10)$$

Hence when $V_{ab} = 0$ and $V_{cd} = 0$ for the circuit configurations shown in Fig. 6.5(a) and (b) equation (6.2.8) will be valid for the overall two-port of Fig. 6.4. These circuit conditions are Brune's tests for the parallel connection of two-port networks.

6.2.3 Series–Parallel Connection

The series–parallel connection of two elements labelled λ, μ, is shown in Fig. 6.6. For this configuration the input ports are connected in series whilst the

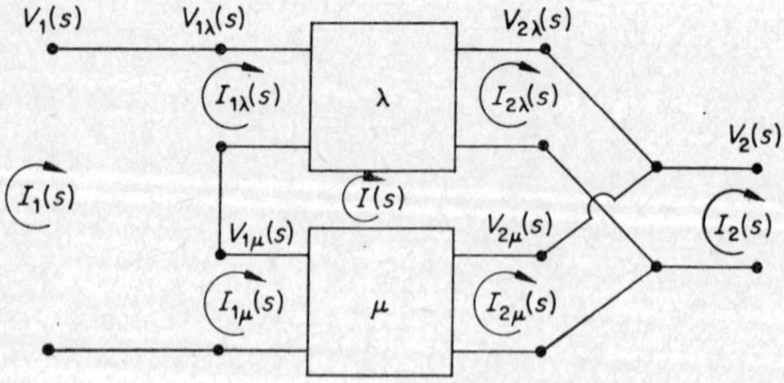

FIG. 6.6 Two elements connected in series parallel

output ports are connected in parallel. When the voltages V_{ab}, V_{cd} are zero for the circuit arrangements shown in Figs. 6.7(a) and (b) the following matrix equation is valid

$$[h] = [h_\lambda] + [h_\mu] \quad (6.2.11)$$

The validity conditions (Brune's tests for series–parallel connection) may be derived from the previous discussion of series connection and parallel connection.

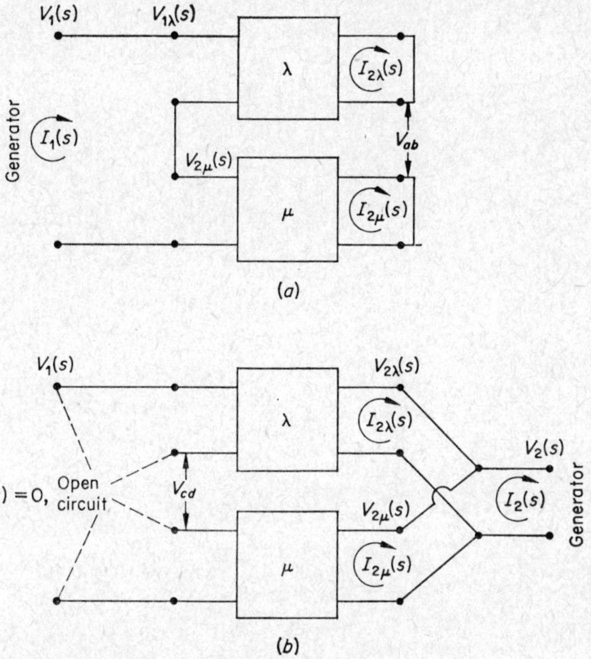

FIG. 6.7 Brune tests for the parallel connection of two-port elements.

6.2.4 Parallel–Series Connection

The parallel–series connection of two elements labelled γ, μ, is shown in Fig. 6.8(a). When the voltages V_{ab}, V_{cd} are zero for the circuit arrangements shown in Figs. 6.8(b), (c), the following matrix relation is valid

$$[g] = [g_\lambda] + [g_\mu] \tag{6.2.12}$$

The validity conditions (Brune's tests for parallel–series connection) may be derived from the previous discussion of series connection and parallel connection.

6.2.5 Examples of Brune's Tests

A set of circuit configurations illustrating the use of Brune's tests for the series connection of two networks is given in Fig. 6.9(a). A circuit satisfying Brune's test for parallel interconnection is shown in Fig. 6.9(b).

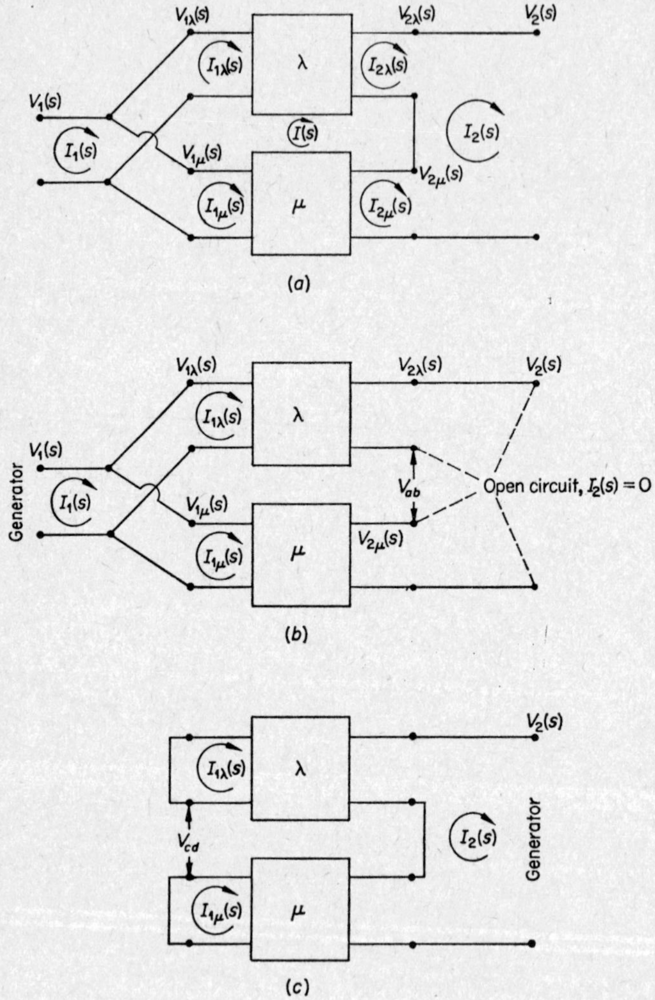

FIG. 6.8 Brune tests for the parallel-series connection of two-port elements.

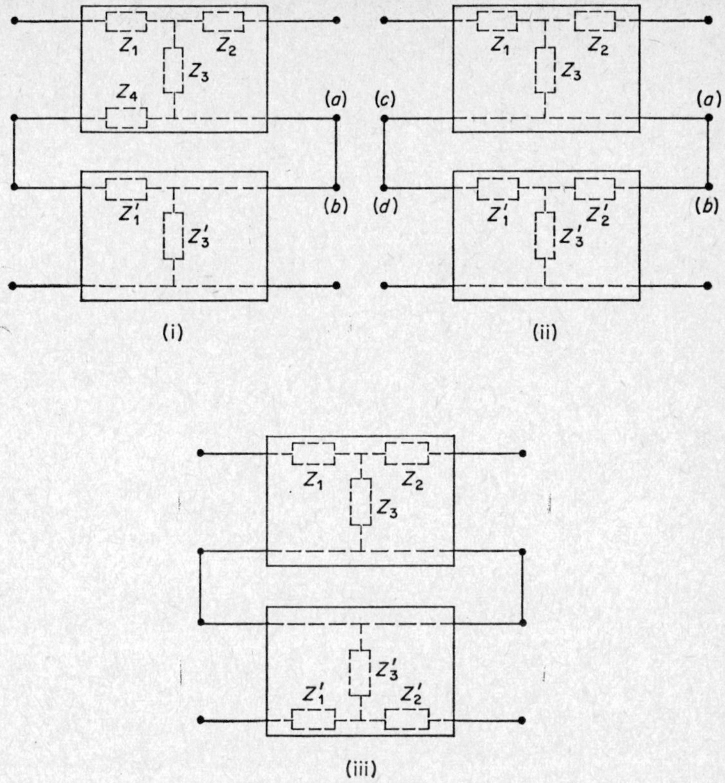

(a)

FIG. 6.9.

(a) Brune's tests for series connection:

(i) Brune's test is not satisfied for V_{ab},
(ii) Brune's test is not satisfied by either V_{ab} or V_{cd}
(iii) Brune's tests are satisfied.

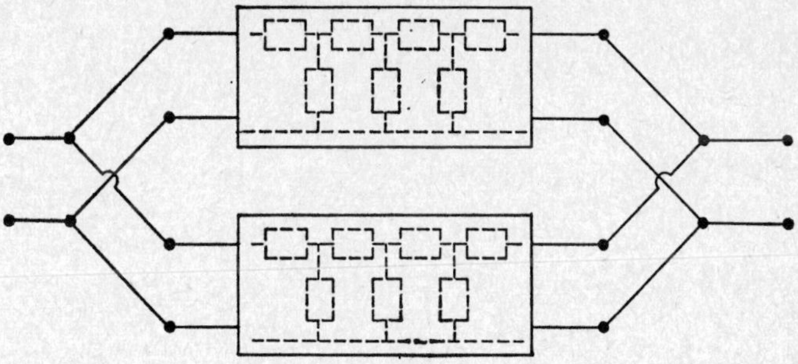

(b)

(b) Two elements satisfying Brune's tests for parallel connection.

6.2.6 Interconnection of Two-Port Networks using Transformers

When the validity conditions for the interconnection of two-port elements are not satisfied immediately, the circuits may be made to conform by adding transformer elements. This technique is illustrated in Fig. 6.10, where

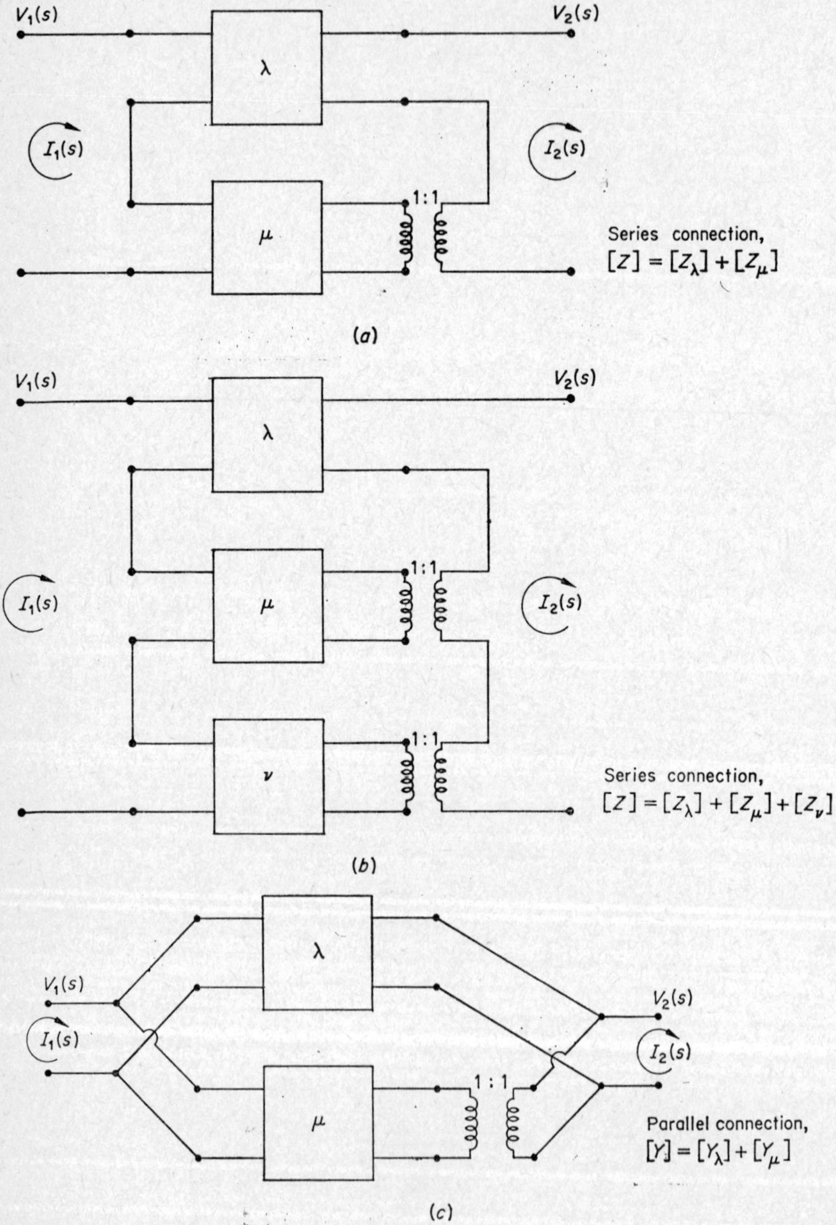

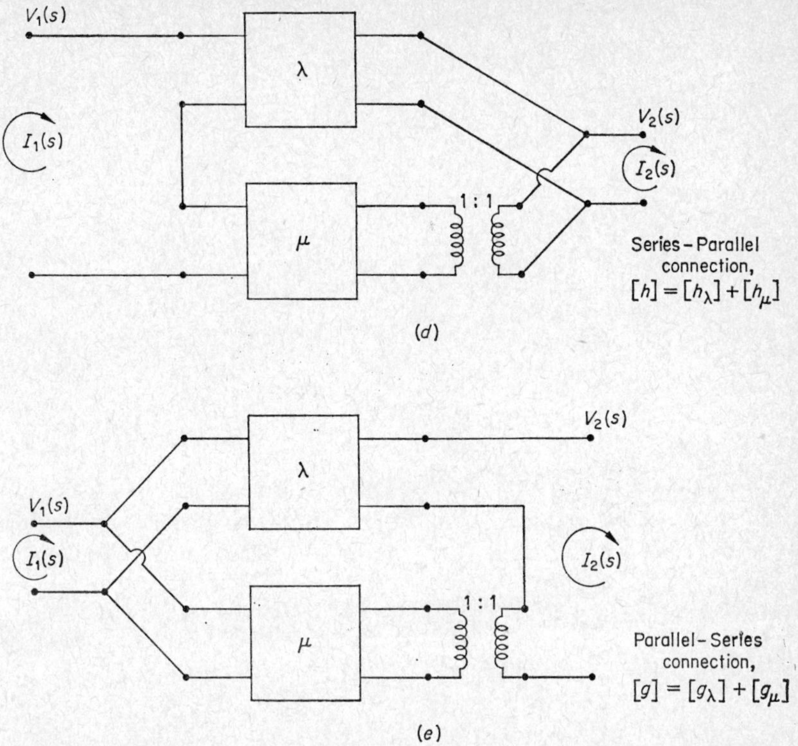

FIG. 6.10 The interconnection of two-port networks using transformers.

ideal 1 : 1 transformer elements are used. In general, the interconnection of a total of N two-port elements may require $(N - 1)$ transformers. Although the transformers are shown connected to the individual output terminal pairs in Fig. 6.10 they may of course be inserted at either input or output terminals in order to satisfy the Brune tests.

6.2.7 Cascade Connection

The fifth method of interconnection of two-port networks, cascade, is shown in Fig. 6.11. Evidently, the output voltage for the network labelled λ is the input voltage for the succeeding network labelled $(\lambda + 1)$, when this method of interconnection is used. Similarly the output current for the element λ is the input current for the element $(\lambda + 1)$.

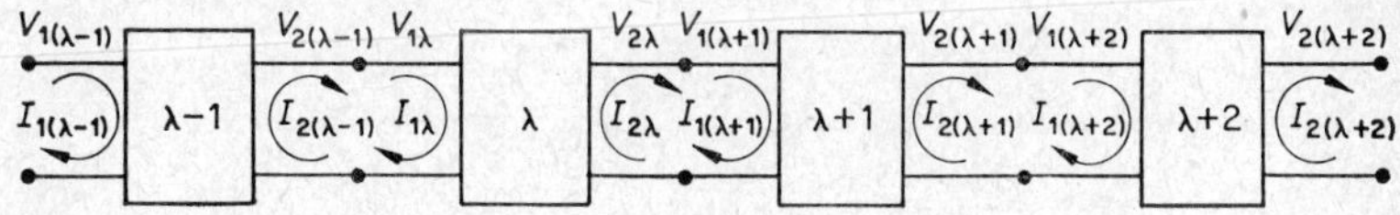

FIG. 6.11 Two-port elements connected in cascade

Hence, for a sequence of networks,

$$V_{2\lambda}(s) = V_{1(\lambda+1)}(s); \quad V_{2(\lambda+1)}(s) = V_{1(\lambda+2)}(s); \text{ etc}$$
$$I_{2\lambda}(s) = I_{1(\lambda+1)}(s); \quad I_{2(\lambda+1)}(s) = I_{1(\lambda+2)}(s); \text{ etc} \tag{6.2.13}$$

and

$$\begin{bmatrix} V_{1j}(s) \\ I_{1j}(s) \end{bmatrix} = [A_j]\begin{bmatrix} V_{1(j+1)}(s) \\ I_{1(j+1)}(s) \end{bmatrix} = [A_j][A_{j+1}]\begin{bmatrix} V_{1(j+2)}(s) \\ I_{1(j+2)}(s) \end{bmatrix}$$
$$= [A_j][A_{j+1}][A_{j+2}] \ldots [A_{j+r}]\begin{bmatrix} V_{2(j+r)}(s) \\ I_{2(j+r)}(s) \end{bmatrix} \tag{6.2.14}$$

or

$$\begin{bmatrix} V_{\text{in}}(s) \\ I_{\text{in}}(s) \end{bmatrix} = [A_1][A_2] \ldots [A_N]\begin{bmatrix} V_{\text{out}}(s) \\ I_{\text{out}}(s) \end{bmatrix} \tag{6.2.15}$$

so that the overall transfer matrix $[A]$ is equal to the product of the individual transfer matrices $[A_\lambda]$ without any validity condition being imposed.

$$[A] = [A_1][A_2] \ldots [A_N] \tag{6.1.16}$$

Multiplying equation (6.2.15) from the left by the complementary transfer matrices $[B]$ in the order $[B_1]$, $[B_2]$ . . . gives

$$\begin{bmatrix} V_{\text{out}}(s) \\ I_{\text{out}}(s) \end{bmatrix} = [B_N][B_{N-1}] \ldots [B_1]\begin{bmatrix} V_{\text{in}}(s) \\ I_{\text{in}}(s) \end{bmatrix} \tag{6.2.17}$$

and

$$[B] = [B_N][B_{N-1}] \ldots [B_1] \tag{6.2.18}$$

6.2.8 Two Examples

We shall consider two useful transistor circuit configurations in order to illustrate the general discussion set out in the previous sections of this chapter. The chosen circuits are (a) the small signal common emitter amplifier with an emitter resistance and (b) the small signal emitter coupled difference amplifier.

The circuit for the common emitter amplifier with an emitter resistance is shown in Fig. 6.12(a), it is redrawn in Fig. 6.12(b) as a pair of series connected two-port elements, which are combined together in Fig. 6.12(c).

The circuit configuration satisfies Brune's test for the series connection of two-port elements and so the impedance matrices may be added. For the combined elements

$$\begin{bmatrix} V_1 \\ V_2 \end{bmatrix} = \{[Z_\lambda] + [Z_\mu]\}\begin{bmatrix} I_1 \\ I_2 \end{bmatrix} \tag{6.2.19}$$

We derive $[Z_\lambda]$ from the $[h]$ matrix for the transistor,

$$\begin{bmatrix} V_{1\lambda} \\ I_{2\lambda} \end{bmatrix} = \begin{bmatrix} h_{11} & h_{12} \\ h_{21} & h_{22} \end{bmatrix}\begin{bmatrix} I_{1\lambda} \\ V_{2\lambda} \end{bmatrix} \tag{6.2.20}$$

From the equation for $I_{2\lambda}$, we may immediately write

$$V_{2\lambda} = -\frac{h_{21}}{h_{22}}I_{1\lambda} + \frac{1}{h_{22}}I_{2\lambda} \tag{6.2.21}$$

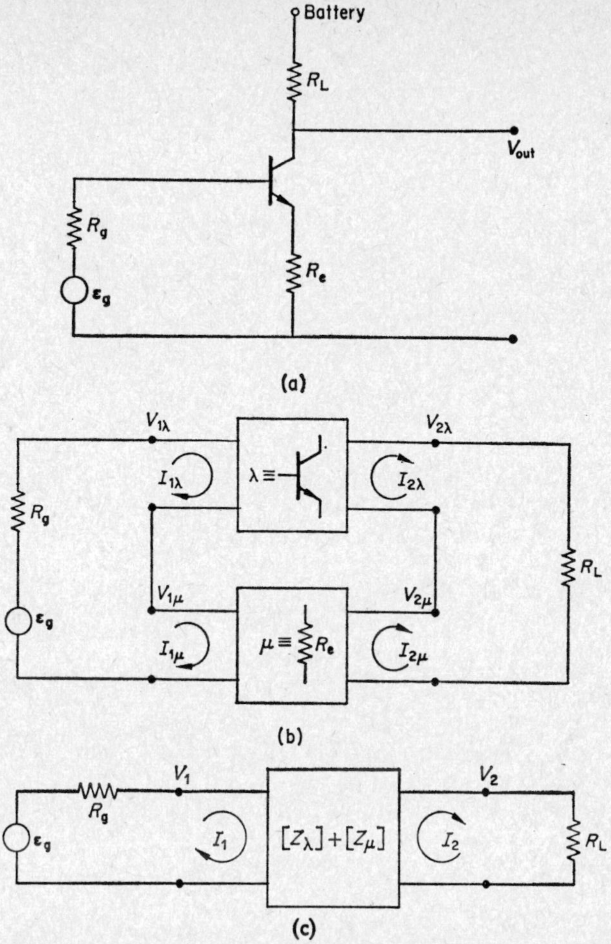

FIG. 6.12 The common emitter amplifier with an emitter resistance.

substituting this relation into the equation for $V_{1\lambda}$, gives

$$V_{1\lambda} = \left[h_{11} - \frac{h_{12}h_{21}}{h_{22}}\right]I_{1\lambda} + \frac{h_{12}}{h_{22}}I_{2\lambda} \tag{6.2.22}$$

Hence

$$\begin{bmatrix} V_{1\lambda} \\ V_{2\lambda} \end{bmatrix} = \begin{bmatrix} \dfrac{\Delta_h}{h_{22}} & \dfrac{h_{12}}{h_{22}} \\ -\dfrac{h_{21}}{h_{22}} & \dfrac{1}{h_{22}} \end{bmatrix} \begin{bmatrix} I_{1\lambda} \\ I_{2\lambda} \end{bmatrix} = [Z_\lambda] \begin{bmatrix} I_{1\lambda} \\ I_{2\lambda} \end{bmatrix} \tag{6.2.23}$$

with $\Delta_h = h_{11}h_{22} - h_{12}h_{21}$.

For the emitter resistance,

$$V_{1\mu} = V_{2\mu} = R_e(I_{1\mu} - I_{2\mu}) \tag{6.2.24}$$

so that

$$\begin{bmatrix} V_{1\mu} \\ V_{2\mu} \end{bmatrix} = \begin{bmatrix} R_e & -R_e \\ R_e & -R_e \end{bmatrix} \begin{bmatrix} I_{1\mu} \\ I_{2\mu} \end{bmatrix} = [Z_\mu] \begin{bmatrix} I_{1\mu} \\ I_{2\mu} \end{bmatrix} \tag{6.2.25}$$

We have $I_{1\lambda} = I_{1\mu} = I_1$ and $I_{2\lambda} = I_{2\mu} = I_2$. For the combined elements therefore,

$$\begin{bmatrix} V_1 \\ V_2 \end{bmatrix} = \begin{bmatrix} \dfrac{\Delta_h}{h_{22}} + R_e & \dfrac{h_{12}}{h_{22}} - R_e \\ -\dfrac{h_{21}}{h_{22}} + R_e & \dfrac{1}{h_{22}} - R_e \end{bmatrix} \begin{bmatrix} I_1 \\ I_2 \end{bmatrix} = [Z] \begin{bmatrix} I_1 \\ I_2 \end{bmatrix} \tag{6.2.26}$$

The current gain is obtained from

$$V_2 = R_L I_2 = \left(-\frac{h_{21}}{h_{22}} + R_e\right) I_1 + \left(\frac{1}{h_{22}} - R_e\right) I_2 \tag{6.2.27}$$

$$A_I = \frac{I_2}{I_1} = \frac{h_{21} - h_{22} R_e}{1 - h_{22}(R_L + R_e)} \tag{6.2.28a}$$

For many applications, A_I is given to sufficent accuracy as

$$A_I \approx h_{21} \tag{6.2.28b}$$

The voltage gain is

$$A_V = \frac{V_2}{V_1} = \frac{Z_{21} + Z_{22} A_I}{Z_{11} + Z_{12} A_I} \tag{6.2.29}$$

$$= \frac{A_I - h_{21} + h_{22} R_e(1 - A_I)}{\Delta_h + h_{12} A_I + h_{22} R_e(1 - A_I)} \tag{6.2.30}$$

but equation (6.2.28a) gives

$$A_I - h_{21} + h_{22} R_e(1 - A_I) = h_{22} R_L A_I \tag{6.2.31}$$

and so

$$A_V = \frac{h_{22} R_L A_I}{\Delta_h + h_{12} A_I + h_{22} R_e(1 - A_I)} \tag{6.2.32}$$

Substitute from equation (6.2.31) for $h_{12} A_I$ in the denominator of this expression, to obtain,

$$A_V = \frac{R_L A_I}{h_{11} + R_e(1 - h_{12}) + A_I\{h_{12}(R_L + R_e) - R_e\}} \tag{6.2.33}$$

Note that we could have written the voltage gain as

$$A_V = \frac{V_2}{V_1} = \frac{R_L I_2}{Z_{in} I_1} = \frac{R_L A_I}{Z_{in}} \tag{6.2.34}$$

where Z_{in} is the input impedance for the amplifier. Consequently equation (6.2.33) gives the input impedance directly,

$$Z_{in} = h_{11} + R_e(1 - h_{12}) + A_I\{h_{12}(R_L + R_e) - R_e\} \tag{6.2.35}$$

For many practical cases the term $A_I R_e$ dominates in this equation, so that

$$Z_{in} \approx A_I R_e \tag{6.2.36}$$

and correspondingly

$$A_V \approx -\frac{R_L}{R_e} \tag{6.2.37}$$

The input impedance is therefore considerably increased by the use of an emitter resistance. A typical n-p-n transistor† will have

$$A_I \approx h_{21} = -h_{fe} \sim -200$$

so that if $R_e = 1\ \text{k}\Omega$, $Z_{in} \sim 200\ \text{k}\Omega$. This value is to be compared with $h_{11} = h_{ie} \sim 5{\cdot}5\ \text{k}\Omega$ which is the order of magnitude for the input impedance without an emitter resistance. The introduction of the emitter resistance reduces the voltage gain, but in compensation the gain is determined by the elements R_L, R_e, which may lead to greater stability and bandwidth.

The output impedance may be derived by considering the overall two-port network to be a voltage generator with internal impedance Z_{out} operating into the load R_L. If the load R_L is an open circuit we measure $V_2 \equiv V_2^\infty$ whilst if the load is a short circuit ($R_L = 0$) we measure $I_2 \equiv I_2^s$. Then

$$V_2^\infty = Z_{out}\, I_2^s \tag{6.2.38}$$

but, from equation (6.2.26)

$$V_2^\infty = Z_{21}\, I_1^\infty = Z_{21} \frac{\mathscr{E}_g}{R_g + Z_{in}^\infty} \tag{6.2.39}$$

moreover,

$$I_2^s = A_I^s\, I_1^s = A_I^s \frac{\mathscr{E}_g}{R_g + Z_{in}^s} \tag{6.2.40}$$

Consequently,

$$Z_{out} = \frac{V_2^\infty}{I_2^s} = \frac{Z_{21}}{A_I^s}\left(\frac{R_g + Z_{in}^s}{R_g + Z_{in}^\infty}\right) \tag{6.2.41}$$

$$= \left(-\frac{h_{21}}{h_{22}} + R_e\right)\left(\frac{1-h_{22}R_e}{h_{21}-h_{22}R_e}\right)\frac{R_g+h_{11}+R_e(1-h_{12})+A_I^s R_e(h_{12}-1)}{R_g+Z_{in}^\infty}$$

$$= \left(-\frac{1}{h_{22}}\right)\frac{(1-h_{22}R_e)\{R_g+h_{11}+R_e(1-h_{12})\}+(h_{21}-h_{22}R_e)(h_{12}-1)R_e}{R_g+Z_{in}^\infty}$$

$$Z_{out} = \left(-\frac{1}{h_{22}}\right)\frac{(1-h_{22}R_e)(R_g+h_{11})+R_e(1-h_{12})(1-h_{21})}{\left(R_g+R_e+h_{11}-\dfrac{h_{12}h_{21}}{h_{22}}\right)} \tag{6.2.42}$$

This relation is usually dominated by the factor $h_{21} R_e$ in the numerator and by R_e in the denominator. In these circumstances

$$Z_{out} \approx \frac{h_{21}}{h_{22}} \tag{6.2.43}$$

The output impedance of the common emitter amplifier is therefore considerably increased by the introduction of the emitter resistance.

The circuit for an emitter coupled difference amplifier is shown in Fig. 6.13(a), it is redrawn, in Fig. 6.13(b), as a pair of two-port networks connected in parallel.

† For a typical n–p–n transistor, common emitter configuration,

$$\begin{bmatrix} h_{11} & h_{12} \\ h_{21} & h_{22} \end{bmatrix} = \begin{bmatrix} h_{ie} & h_{re} \\ -h_{fe} & -h_{oe} \end{bmatrix} \sim \begin{bmatrix} 5{\cdot}5\ \text{k}\Omega & 2{\cdot}10^{-4} \\ -250 & -25\ \mu\text{mho} \end{bmatrix}$$

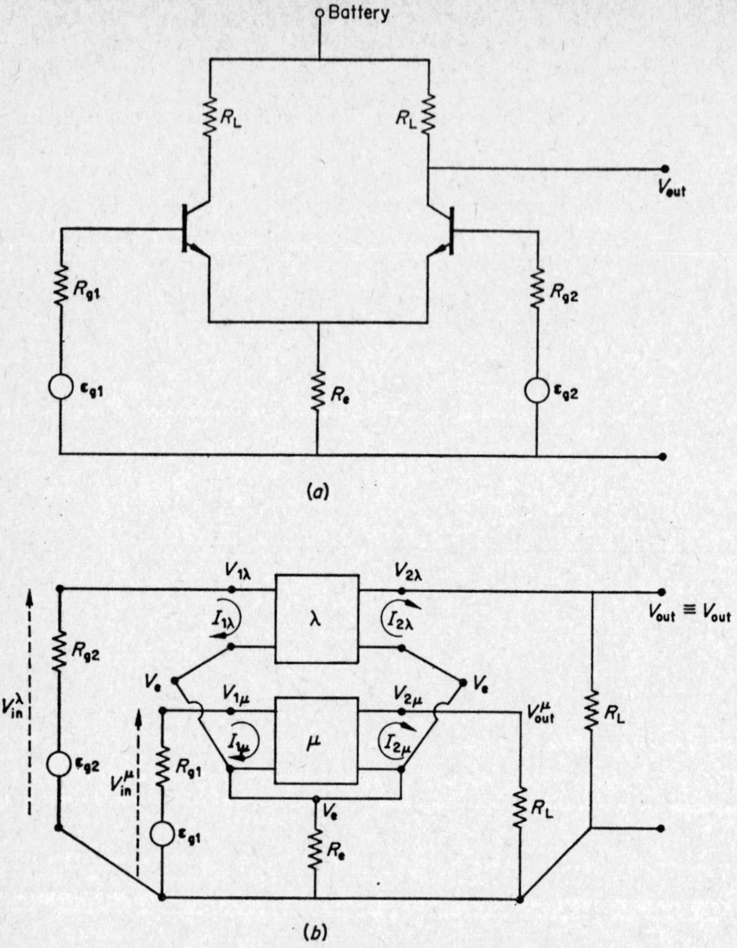

FIG. 6.13 An emitter coupled difference amplifier.

From Fig. 6.13(b) the output voltage V_{out} is taken from the collector of the transistor which is identified with the two-port network λ. The two transistors are assumed to have identical h parameters.

Evidently,

$$V_{\text{out}}^{\lambda} = V_{2\lambda} + V_e \tag{6.2.44}$$

But

$$\begin{bmatrix} V_{1\lambda} \\ I_{2\lambda} \end{bmatrix} = \begin{bmatrix} h_{11} & h_{12} \\ h_{21} & h_{22} \end{bmatrix} \begin{bmatrix} I_{1\lambda} \\ V_{2\lambda} \end{bmatrix} \tag{6.2.45}$$

from which

$$V_{2\lambda} = \frac{h_{11}}{\Delta_h} I_{2\lambda} - \frac{h_{21}}{\Delta_h} V_{1\lambda} \tag{6.2.46}$$

$$= \frac{h_{11}}{\Delta_h} \cdot \frac{V^{\lambda}_{\text{out}}}{R_{\text{L}}} - \frac{h_{21}}{\Delta_h} V_{1\lambda} \tag{6.2.47}$$

with $\Delta_h = h_{11} h_{22} - h_{21} h_{12}$.
Hence

$$V^{\lambda}_{\text{out}}\left(1 - \frac{h_{11}}{\Delta_h} \cdot \frac{1}{R_{\text{L}}}\right) = -\frac{h_{21}}{\Delta_h} V_{1\lambda} + V_{\text{e}} \tag{6.2.48}$$

In addition

$$V^{\lambda}_{\text{in}} = V_{1\lambda} + V_{\text{e}} \tag{6.2.49}$$

Consequently,

$$V^{\lambda}_{\text{out}}\left(1 - \frac{h_{11}}{\Delta_h} \cdot \frac{1}{R_{\text{L}}}\right) = -\frac{h_{21}}{\Delta_h} V^{\lambda}_{\text{in}} + V_{\text{e}}\left(1 + \frac{h_{21}}{\Delta_h}\right) \tag{6.2.50}$$

From the symmetry of the circuit we may immediately write down the corresponding relation for V^{μ}_{out}

$$V^{\mu}_{\text{out}}\left(1 - \frac{h_{11}}{\Delta_h} \cdot \frac{1}{R_{\text{L}}}\right) = -\frac{h_{21}}{\Delta_h} V^{\mu}_{\text{in}} + V_{\text{e}}\left(1 + \frac{h_{21}}{\Delta_h}\right) \tag{6.2.51}$$

We also have

$$V_{\text{e}} = R_{\text{e}}(I_{1\lambda} + I_{1\mu} - I_{2\lambda} - I_{2\mu}) \tag{6.5.52}$$

or

$$V_{\text{e}} = \left(\frac{1}{A_I} - 1\right) \frac{R_{\text{e}}}{R_{\text{L}}} (V^{\lambda}_{\text{out}} + V^{\mu}_{\text{out}}) \tag{6.2.53}$$

We wish to solve the simultaneous equations (6.2.50), (6.2.51) and (6.2.53) for $V_{\text{out}} \equiv V^{\lambda}_{\text{out}}$. Since we also wish to obtain the final expression as a function of the sum and difference voltages, $(V^{\lambda}_{\text{in}} + V^{\mu}_{\text{in}})$, $(V^{\lambda}_{\text{in}} - V^{\mu}_{\text{in}})$, it is convenient to proceed by adding and subtracting equations (6.2.50) and (6.2.51). Thus,

$$(V^{\lambda}_{\text{out}} + V^{\mu}_{\text{out}})\left(1 - \frac{h_{11}}{\Delta_h} \cdot \frac{1}{R_{\text{L}}}\right) = -\frac{h_{21}}{\Delta_h}(V^{\lambda}_{\text{in}} + V^{\mu}_{\text{in}}) + 2V_{\text{e}}\left(1 + \frac{h_{21}}{\Delta_h}\right) \tag{6.2.54}$$

$$(V^{\lambda}_{\text{out}} - V^{\mu}_{\text{out}})\left(1 - \frac{h_{11}}{\Delta_h} \cdot \frac{1}{R_{\text{L}}}\right) = -\frac{h_{21}}{\Delta_h}(V^{\lambda}_{\text{in}} - V^{\mu}_{\text{in}}) \tag{6.2.55}$$

Upon using equation (6.2.53), the sum voltage becomes

$$(V^{\lambda}_{\text{out}} + V^{\mu}_{\text{out}})\left\{1 - \frac{h_{11}}{\Delta_h} \cdot \frac{1}{R_{\text{L}}} - 2\left(\frac{1}{A_I} - 1\right) \frac{R_{\text{e}}}{R_{\text{L}}}\left(1 + \frac{h_{21}}{\Delta_h}\right)\right\}$$

$$= -\frac{h_{21}}{\Delta_h}(V^{\lambda}_{\text{in}} + V^{\mu}_{\text{in}}) \tag{6.2.56}$$

Add together equations (6.2.55) and (6.2.56), to obtain

$$2V_{\text{out}} \equiv 2V^{\lambda}_{\text{out}}$$

$$= \left(-\frac{h_{21}}{\Delta_h}\right)\frac{(V^{\lambda}_{\text{in}} - V^{\mu}_{\text{in}})}{\{1-(h_{11}/\Delta_h)R_{\text{L}}^{-1}\}}$$

$$+\left(-\frac{h_{21}}{\Delta_h}\right)\frac{(V^{\lambda}_{\text{in}} + V^{\mu}_{\text{in}})}{[1-(h_{11}/\Delta_h).R_{\text{L}}^{-1}-2(A_I^{-1}-1)(R_{\text{e}}/R_{\text{L}})\{1+(h_{21}/\Delta_h)\}]} \tag{6.2.57}$$

Thus we may write

$$V_{\text{out}} = A_{\text{d}}(V^{\lambda}_{\text{in}} - V^{\mu}_{\text{in}}) + A_{\text{c}}\cdot\tfrac{1}{2}(V^{\lambda}_{\text{in}} + V^{\mu}_{\text{in}}) \tag{6.2.58}$$

where A_{d} is the voltage gain for the difference signal and A_{c} is the voltage gain for the common mode signal.

$$A_{\text{d}} = -\frac{1}{2}\cdot\frac{h_{21}}{\Delta_h}\{1-(h_{11}/\Delta_h)R_{\text{L}}^{-1}\}^{-1} \tag{6.2.59}$$

$$A_{\text{c}} = -\frac{h_{21}}{\Delta_h}[1-(h_{11}/\Delta_h)R_{\text{L}}^{-1} - 2(A_I^{-1}-1)(R_{\text{e}}/R_{\text{L}})\{1+(h_{21}/\Delta_h\}]^{-1} \tag{6.2.60}$$

For many practical cases these equations reduce to

$$A_{\text{d}} \sim \frac{h_{21}\,R_{\text{L}}}{2h_{11}} \tag{6.2.61}$$

$$A_{\text{c}} \sim -\frac{R_{\text{L}}}{2R_{\text{e}}} \tag{6.2.62}$$

To this accuracy the common mode rejection ratio is

$$\rho = \left|\frac{A_{\text{d}}}{A_{\text{c}}}\right| \sim \left|\frac{h_{21}\,R_{\text{e}}}{h_{11}}\right| \tag{6.2.63}$$

Taking† $R_{\text{e}} \sim 50$ kΩ, $|h_{21}| = 250$, $h_{11} = 5{\cdot}5$ kΩ, we obtain $\rho \sim 2500$ and so the common mode rejection ratio exceeds 60 db.

The input impedance for the difference amplifier is derived from

$$Z^{\lambda}_{\text{in}} = \frac{V^{\lambda}_{\text{in}}}{I^{\lambda}_1} = \frac{V^{\lambda}_{\text{in}}\,A_I}{I_{2\lambda}} = \frac{V^{\lambda}_{\text{in}}}{V^{\lambda}_{\text{out}}}\,A_I\,R_{\text{L}} \tag{6.2.64}$$

Similarly

$$Z^{\mu}_{\text{in}} = \frac{V^{\mu}_{\text{in}}}{V^{\mu}_{\text{out}}}\,A_I\,R_{\text{L}} \tag{6.2.65}$$

But from the symmetry of the circuit $Z_{\text{in}} = Z^{\lambda}_{\text{in}} = Z^{\mu}_{\text{in}}$, so that

$$Z_{\text{in}}(V^{\lambda}_{\text{out}} - V^{\mu}_{\text{out}}) = A_I\,R_{\text{L}}(V^{\lambda}_{\text{in}} - V^{\mu}_{\text{in}}) \tag{6.2.66}$$

† With large values of emitter resistance it is customary to take the lower end of R_{e} to a negative tapping on the voltage supply.

and upon using equation (6.2.55)

$$Z_{\text{in}} = -\frac{h_{21}}{\Delta_h} \cdot \frac{A_I R_L}{\{1 - (h_{11}/\Delta_h)(1/R_L)\}} \tag{6.2.67}$$

$$\sim \frac{(h_{21} R_L)^2}{h_{11}} \tag{6.2.68}$$

Finally we remark that, when it is only required to calculate small signal voltage gain or current gain for a circuit using silicon n–p–n transistors, it is usually sufficiently accurate to take $h_{12} = h_{22} = 0$. In these circumstances the algebraic relations are simplified. The matrix equation

$$\begin{bmatrix} V_1 \\ I_2 \end{bmatrix} = \begin{bmatrix} h_{11} & h_{12} \\ h_{21} & h_{22} \end{bmatrix} \begin{bmatrix} I_1 \\ V_2 \end{bmatrix} \tag{6.2.69}$$

becomes

$$\begin{bmatrix} V_1 \\ I_2 \end{bmatrix} = \begin{bmatrix} h_{11} & 0 \\ h_{21} & 0 \end{bmatrix} \begin{bmatrix} I_1 \\ V_2 \end{bmatrix} \tag{6.2.70}$$

that is

$$V_1 = h_{11} I_1, \quad I_2 = h_{21} I_1 \tag{6.2.71}$$

The silicon n–p–n transistor may now be described by a *mutual conductance* parameter, g_m, according to the relation,

$$g_m V_1 = I_2 \tag{6.2.72}$$

with

$$g_m = \frac{h_{21}}{h_{11}} \tag{6.2.73}$$

Taking $h_{21} = -250$ and $h_{11} = 5{\cdot}5$ kΩ we obtain $|g_m| = 45$ mA/V.

For a field effect transistor, h_{11} is extremely large and I_1 correspondingly small. The characteristic equation for a (common source) field effect transistor is therefore obtained by solving equation (6.2.69) for I_2 in terms of V_1, V_2. Evidently,

$$I_2 = \frac{h_{21}}{h_{11}} V_1 + \left(h_{22} - \frac{h_{12} h_{21}}{h_{11}}\right) V_2 \tag{6.2.74}$$

and so,

$$I_2 = g_m V_1 + \frac{1}{r_d} V_2 \tag{6.2.75}$$

The parameter

$$\frac{1}{r_d} \equiv \left(h_{22} - \frac{h_{12} h_{21}}{h_{11}}\right)$$

is called the *drain* (or *output*) *resistance*. For a junction field effect transistor $g_m \sim 1$ mA/V and $r_d \sim 1$ MΩ.

6.3 Reversible Two-Port Networks

A two-port circuit element is said to be reversible if it does not matter which of the terminal pairs is used as entrance or exit port. However, a particular restriction is imposed on the procedure for reversing an element. It is assumed that the terminals retain their overall relation to one another. Thus reversing the element shown in Fig. 6.14 is equivalent to switching from (*a*) to (*a*′) and (*b*) to (*b*′), or rotating the element about the *z*-axis.

When the excitation is connected to the left-hand terminals of the two-port element in Fig. 6.14, the current $I_{in}(s)$ flows in the same direction as

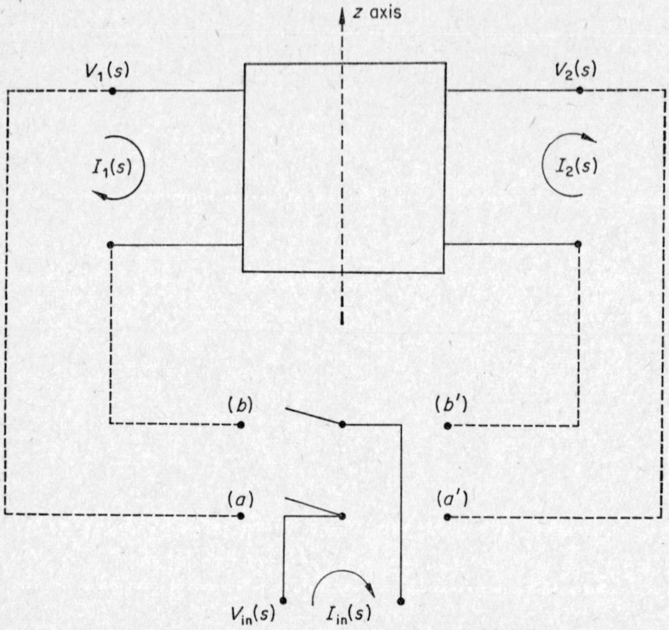

FIG. 6.14 Reversing a two-port element.

$I_1(s)$ and the transfer matrix is $[A]$. However, if the excitation is switched to the terminals (*a*′), (*b*′) for the reversed element, the current $I_{in}(s)$ is now in the opposite direction from $I_2(s)$. Thus the process of reversing a two-port element involves reversing the terminal currents also. The transfer matrix which represents the reversed element is not $[B]$ therefore but is a matrix which takes account of these reversed current relations.

With the excitation connected to the left-hand terminals in Fig. 6.14

$$\begin{bmatrix} V_{in}(s) \\ I_{in}(s) \end{bmatrix} = \begin{bmatrix} V_1(s) \\ I_1(s) \end{bmatrix} = [A]\begin{bmatrix} V_2(s) \\ I_2(s) \end{bmatrix} = [A]\begin{bmatrix} V_{out}(s) \\ I_{out}(s) \end{bmatrix} \tag{6.3.1}$$

For the reversed element, the excitation is connected to the right-hand terminal pair and the overall transfer matrix is given by

$$\begin{bmatrix} V'_{in}(s) \\ I'_{in}(s) \end{bmatrix} = [R]\begin{bmatrix} V'_{out}(s) \\ I'_{out}(s) \end{bmatrix} \tag{6.3.2}$$

But, in the notation of Fig. 6.14

$$\begin{bmatrix} V'_{\text{in}}(s) \\ I'_{\text{in}}(s) \end{bmatrix} = \begin{bmatrix} V_2(s) \\ -I_2(s) \end{bmatrix} = \begin{bmatrix} 1 & 0 \\ 0 & -1 \end{bmatrix} \begin{bmatrix} V_2(s) \\ I_2(s) \end{bmatrix} = \begin{bmatrix} 1 & 0 \\ 0 & -1 \end{bmatrix} [B] \begin{bmatrix} V_1(s) \\ I_1(s) \end{bmatrix}$$
$$= \begin{bmatrix} 1 & 0 \\ 0 & -1 \end{bmatrix} [B] \begin{bmatrix} 1 & 0 \\ 0 & -1 \end{bmatrix} \begin{bmatrix} V_1(s) \\ -I_1(s) \end{bmatrix}$$
$$= \begin{bmatrix} 1 & 0 \\ 0 & -1 \end{bmatrix} [B] \begin{bmatrix} 1 & 0 \\ 0 & -1 \end{bmatrix} \begin{bmatrix} V'_{\text{out}}(s) \\ I'_{\text{out}}(s) \end{bmatrix} \tag{6.3.3}$$

and so

$$[R] = \begin{bmatrix} 1 & 0 \\ 0 & -1 \end{bmatrix} [B] \begin{bmatrix} 1 & 0 \\ 0 & -1 \end{bmatrix} = \begin{bmatrix} B_{11} & -B_{12} \\ -B_{21} & B_{22} \end{bmatrix} \tag{6.3.4}$$

If the two-port element is reversible it does not matter which terminal pair is used as entrance or exit port and so we require

$$[R] = [A] \tag{6.3.5}$$

for a reversible two-port. Identifying the corresponding matrix elements, we have

$$B_{11} = A_{11}; \ -B_{12} = A_{12}; \ -B_{21} = A_{21}; \text{ and } B_{22} = A_{22} \tag{6.3.6}$$

Using the relations given in Table 6.1, we obtain

$$A_{11} = A_{22} \quad \text{and} \quad Z_{12}(s) = -Z_{21}(s) \tag{6.3.7}$$

The relation between the impedance matrix elements means that the co-factors Δ_{12}, Δ_{21} are equal, and so there must be no strictly unidirectional elements in the network. Note also that in the impedance matrix for a reversible two-port, equation (6.3.7) implies that

$$Z_{11}(s) = -Z_{22}(s) \tag{6.3.8}$$

Since $Z_{12}(s) = -Z_{21}(s)$ and det $[A] = 1$ for a reversible two-port network, we can write

$$A_{11} = A_{22} = \sqrt{1 + A_{12} A_{21}}$$
$$B_{11} = B_{22} = \sqrt{1 + B_{12} B_{21}} \tag{6.3.9}$$

6.4 Two-Port Network as a Matching Element

Consider a two-port network with a load. The load is described by a driving point impedance function $Z_L(s)$ and is connected across the output terminals as shown in Fig. 6.15.

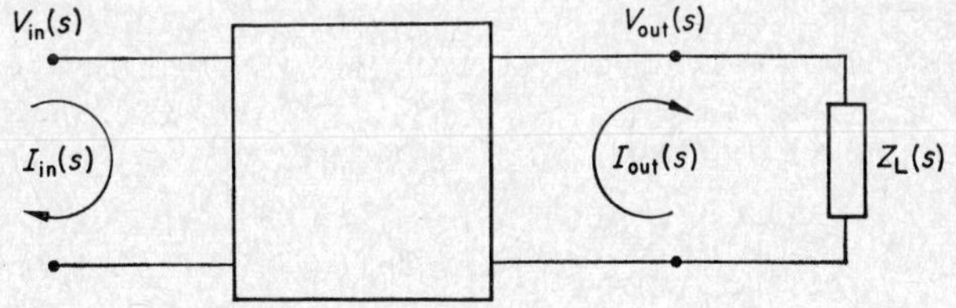

FIG. 6.15 A two-port element with a load.

The transfer matrix representation for the two-port gives

$$\frac{V_{\text{in}}(s)}{I_{\text{in}}(s)} = \frac{A_{11}\,V_{\text{out}}(s) + A_{12}\,I_{\text{out}}(s)}{A_{21}\,V_{\text{out}}(s) + A_{22}\,I_{\text{out}}(s)} \tag{6.4.1}$$

and so the driving point impedance function at the input terminals is given by

$$Z_{\text{in}}(s) = \frac{A_{11}\,Z_{\text{L}}(s) + A_{12}}{A_{21}\,Z_{\text{L}}(s) + A_{22}} \tag{6.4.2}$$

A two-port network may therefore be used to transform a driving point impedance function. In particular a two-port may be used to match a load to a generator.

For a reversible two-port network (see Section 6.3, above), $A_{11} = A_{22}$ and $\det [A] = 1$. A knowledge of two matrix elements is, therefore, sufficient to determine the complete transfer matrix. These may be determined experimentally by driving the network with a harmonic generator and measuring the input impedance (a) with the output terminals short circuited (b) with the output terminals open circuited. Let $Z_{\text{insc}}(j\omega)$ and $Z_{\text{inoc}}(j\omega)$ be the input impedances for conditions (a) and (b) respectively. Then, the complete transfer matrix is given by

$$\begin{aligned} A_{12}^2 &= \frac{Z_{\text{insc}}^2(j\omega)\,Z_{\text{inoc}}(j\omega)}{Z_{\text{inoc}}(j\omega) - Z_{\text{insc}}(j\omega)} \\ A_{21}^2 &= \frac{1}{Z_{\text{inoc}}(j\omega)[Z_{\text{inoc}}(j\omega) - Z_{\text{insc}}(j\omega)]} \\ A_{11}^2 &= 1 + A_{12}\,A_{21} \end{aligned} \tag{6.4.3}$$

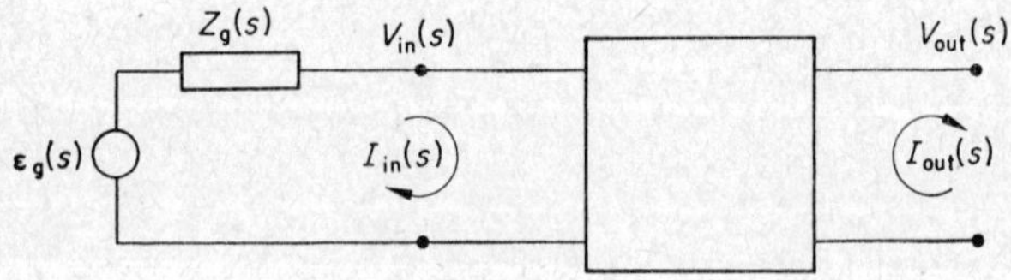

(a)

FIG. 6.16.

(a) A two-port element used to transform the characteristics of a generator.

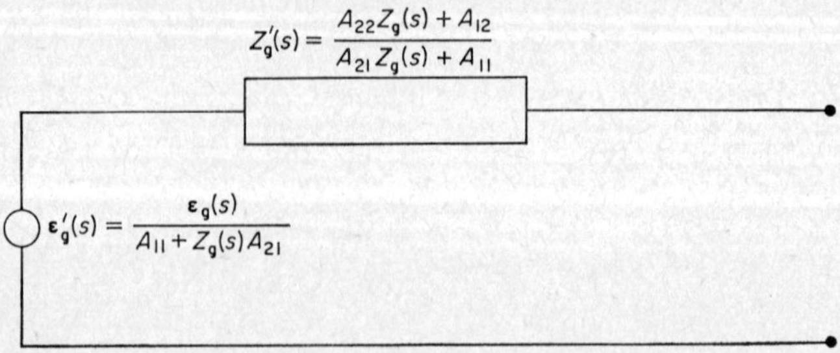

(b)

(b) The transformed generator.

A two-port network may also be used to transform the internal impedance of a generator. Consider the circuit shown in Fig. 6.16(a).
Evidently

$$V_{in}(s) = \mathscr{E}_g(s) - Z_g(s)\, I_{in}(s) = A_{11}\, V_{out}(s) + A_{12}\, I_{out}(s)$$
$$I_{in}(s) = A_{21}\, V_{out}(s) + A_{22}\, I_{out}(s) \tag{6.4.4}$$

When the output terminals are open circuit, $V_{out}(s)$ is the transformed e.m.f. of the generator, $\mathscr{E}'_g(s)$,

$$\mathscr{E}'_g(s) = \frac{\mathscr{E}_g(s)}{A_{11} + Z_g(s)\, A_{21}} \tag{6.4.5}$$

When the output terminals are short circuited, the short circuit current $I_{out}(s) \equiv I'(s)$ enables the transformed impedance of the generator $Z'_g(s)$ to be derived from the ratio $\mathscr{E}'_g(s)/I'(s)$

$$Z'_g(s) = \frac{A_{22}\, Z_g(s) + A_{12}}{A_{21}\, Z_g(s) + A_{11}} \tag{6.4.6}$$

The circuit of Fig. 6.16(a) may therefore be drawn in the equivalent form shown in Fig. 6.16(b)

6.4.1 Optimum Noise Figure

As an illustration of the need to match a generator and a load consider the noise factor of an amplifier (or indeed any linear system). This parameter is a measure of the quality of an amplifier with respect to its own internally generated noise. The integrated noise figure F_i is defined to be the ratio of the total available noise power at the output of the amplifier to the total available noise power at the output arising solely from the signal source itself.† The noise figure at a particular frequency (called the single frequency noise figure), denoted by $F(\omega)$ is defined to be the ratio of the available noise output power of the amplifier in an infinitesimal bandwidth, $\Delta\omega$ centred on ω, to the available noise output power of the amplifier in the same bandwidth arising solely from the signal generator

$$F(\omega) = \frac{\text{Entire noise output power of amplifier in } \Delta\omega}{\text{Noise output power of amplifier arising solely from signal generator in } \Delta\omega} \tag{6.4.7}$$

The noise figure of an amplifier is therefore a measure of the noisiness of the amplifier relative to the noisiness of the source.

A noisy amplifier may be represented as a noiseless two-port network with a particular transfer function $H(s)$ together with a noise current generator and a noise voltage generator connected to the input terminals. The circuit is shown in Fig. 6.17. The signal generator is assumed to have an internal impedance R_g which is purely resistive.

Suppose that the current spectral density for the amplifier noise generator is $S_{ia}(\omega)$ and the voltage spectral density is $S_{va}(\omega)$. The current source will

† Available power is defined to be the maximum power which can be delivered to a load. For a sinusoidal generator this requires that the load impedance is equal to the complex conjugate of the generator internal impedance.

give rise to an additional noise voltage in the signal generator impedance R_g, which will be described by a voltage spectral density function $S'_{va}(\omega)$. Using equation (5.4.6) we obtain

$$S'_{va}(\omega) = R_g\, S_{ia}(\omega) \tag{6.4.8}$$

The total output voltage spectral density from the amplifier, assuming the different noise sources to be uncorrelated, is therefore given by

$$S_{out}(\omega) = |H(j\omega)|^2[S_{vg}(\omega) + S_{va}(\omega) + R_g^2\, S_{ia}(\omega)] \tag{6.4.9}$$

In this equation $S_{vg}(\omega)$ is the voltage spectral density function describing

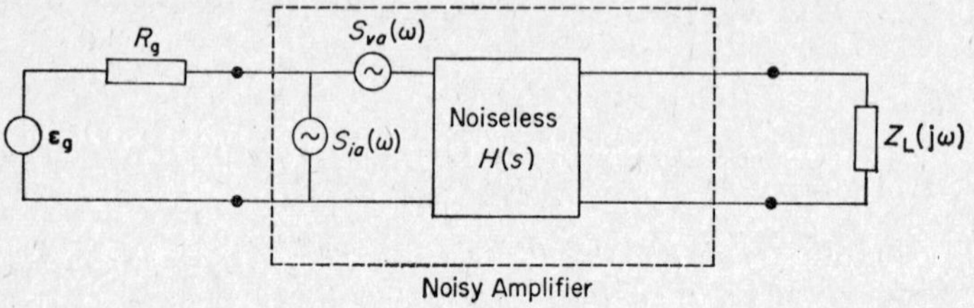

FIG. 6.17 The representation for a noisy amplifier.

the noise from the internal impedance of the signal generator. For the particular case we are considering, $S_{vg}(\omega)$ describes thermal noise in R_g, so that from Section 5.4, page 242

$$S_{vg}(\omega) = 2kTR_g \tag{6.4.10}$$

The noise power in an infinitesimal bandwidth $\Delta\omega$ centred on ω is proportional to $\langle V^2\rangle_{\Delta\omega} = 2 \times \left(\frac{1}{2\pi} S_V(\omega)\, \Delta\omega\right)$ from equation (5.4.87) and so the single frequency noise figure is given by

$$F(\omega) = \frac{|H(j\omega)|^2\{S_{Vg}(\omega) + S_{Va}(\omega) + R_g^2\, S_{ia}(\omega)\}(\Delta\omega/\pi)}{|H(j\omega)|^2\, S_{Vg}(\omega)\,(\Delta\omega/\pi)} \tag{6.4.11}$$

$$F(\omega) = 1 + \frac{S_{Va}(\omega) + R_g^2\, S_{ia}(\omega)}{2\,kT\,R_g} \tag{6.4.12}$$

The optimum value of the signal generator resistance R_g, which will make the noise figure a minimum, is obtained from $\partial F(\omega)/\partial R_g = 0$

$$R_g(\text{opt}) = \left\{\frac{S_{Va}(\omega)}{S_{ia}(\omega)}\right\}^{1/2} \tag{6.4.13}$$

When this equation is satisfied, the minimum noise figure is given by

$$F(\omega)_{min} = 1 + \frac{\sqrt{\{S_{Va}(\omega)\, S_{ia}(\omega)\}}}{kT} \tag{6.4.14}$$

In order to achieve this minimum noise figure a two-port network may be used to transform the source impedance so that equation (6.4.13) is satisfied.

Note that the integrated noise figure, F_i, involves a knowledge of the transfer characteristic of the amplifier. From equation (6.4.11) we see that

$$F_i = \frac{\int_{-\infty}^{\infty} |H(j\omega)|^2\{S_{Vg}(\omega) + S_{Va}(\omega) + R_g^2\, S_{ia}(\omega)\}\, d\omega}{\int_{-\infty}^{\infty} |H(j\omega)|^2\, S_{Vg}(\omega)\, d\omega} \tag{6.4.15}$$

$$= \frac{\int_{-\infty}^{\infty} |H(j\omega)|^2\{2kT\, R_g + S_{Va}(\omega) + R_g^2 S_{ia}(\omega)\}\, d\omega}{2kT\, R_g \int_{-\infty}^{\infty} |H(j\omega)|^2\, d\omega}$$

$$= \frac{\int_{-\infty}^{\infty} |H(j\omega|^2\, F(\omega)\, d\omega}{\int_{-\infty}^{\infty} |H(j\omega)|^2\, d\omega} \tag{6.4.16}$$

since both numerator and denominator are even functions of frequency.

$$F_i = \frac{\int_{0}^{\infty} |H(j\omega)|^2\, F(\omega)\, d\omega}{\int_{0}^{\infty} |H(j\omega)|^2\, d\omega} \tag{6.4.17}$$

6.5 Equivalent Circuits for Two-Port Networks

Two-port networks may be formally represented by T or π or lattice networks. The T or π representations have two terminals connected by a common line and this may have advantages for some circuit applications. However, it is not always possible to construct these particular equivalent circuits using physically realizable elements, and when these circumstances arise the equivalent circuit representation has value only for simplifying the mathematical analysis. The lattice network does not possess the advantage of a common line but it is of great importance in network synthesis because if a reversible two-port element is realizable at all it can certainly be realized by a lattice network.

6.5.1 The T Representation

The equivalent T representation for a two-port network is derived by considering the impedance matrix $[Z]$. When the network contains unidirectional elements, so that $Z_{12}(s) \neq -Z_{21}(s)$ the equivalent circuit is shown in Fig. 6.18(a). The non-reciprocal feature of the two-port is represented by a voltage generator. When the network contains no unidirectional elements so that $Z_{12}(s) = -Z_{21}(s)$, only three impedance parameters are required to represent the two-port and the equivalent circuit is shown in Fig. 6.18(b).

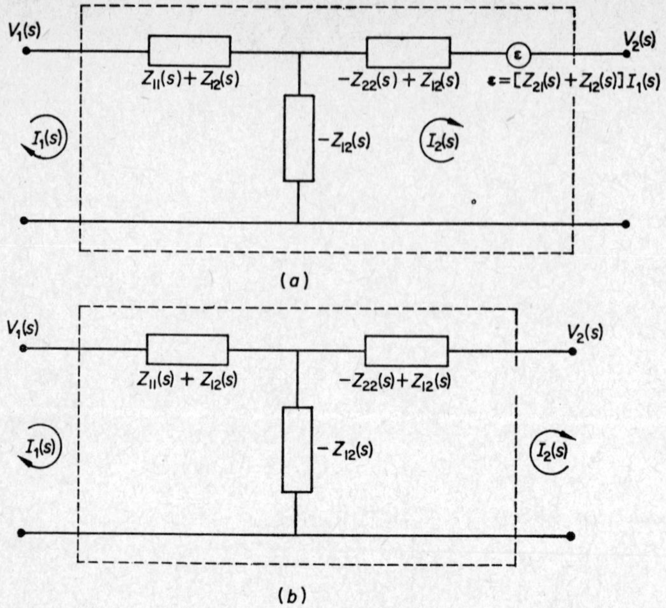

FIG. 6.18 T network equivalent circuit representations for a two-port element.

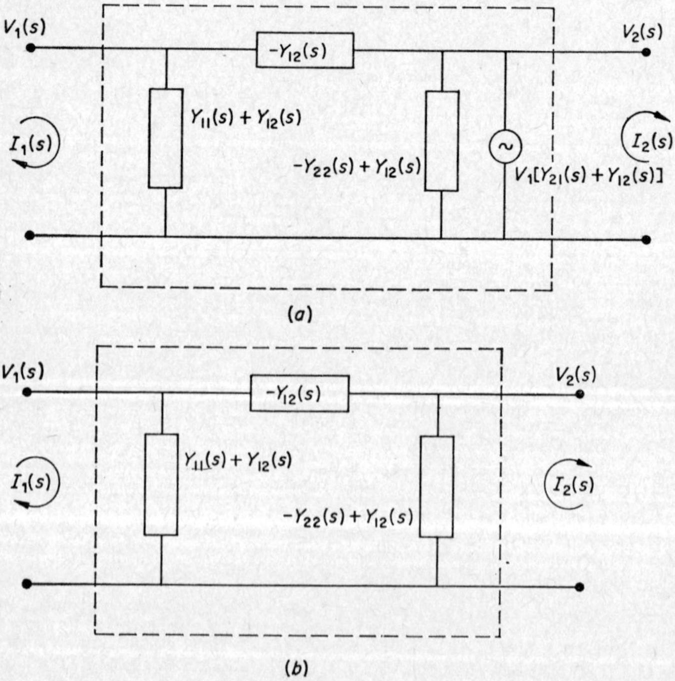

FIG. 6.19 π network equivalent circuit representations for a two-port element.

6.5.2 The π Representation

The equivalent π representation for a two-port network is derived by considering the admittance matrix $[Y]$. The general representation for a two-port which contains unidirectional elements is shown in Fig. 6.19(a) and that for a network composed entirely of reciprocal elements (so that $Y_{12} = -Y_{21}$) is shown in Fig. 6.19(b).

6.5.3 The Lattice Representation

A two-port circuit element drawn as a lattice network is shown in Fig. 6.20(a). This general form is called an *unsymmetrical lattice.*

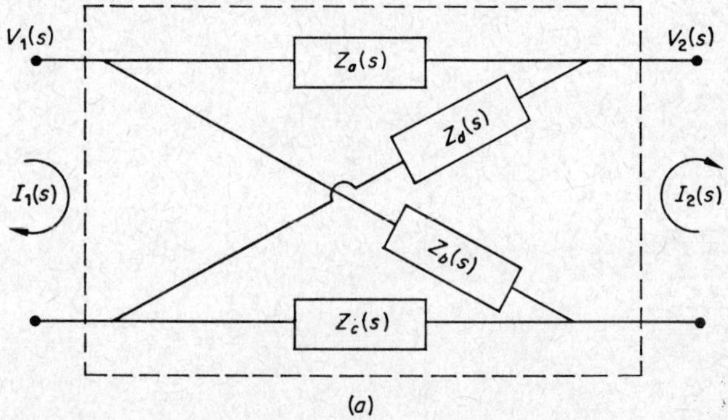

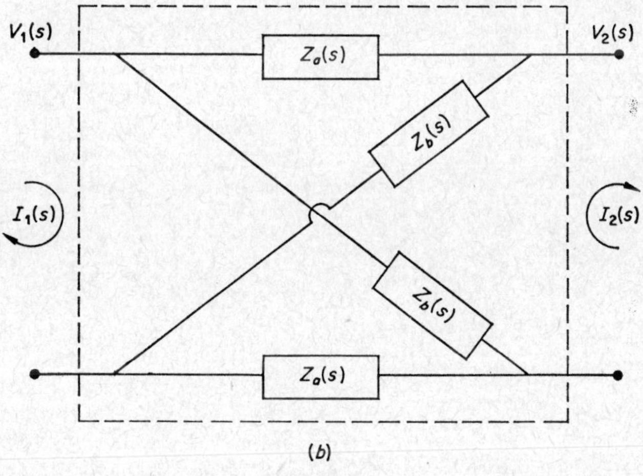

FIG. 6.20 Lattice network equivalent circuit representations for a two-port element:

(a) Unsymmetrical lattice,

(b) Symmetrical lattice,

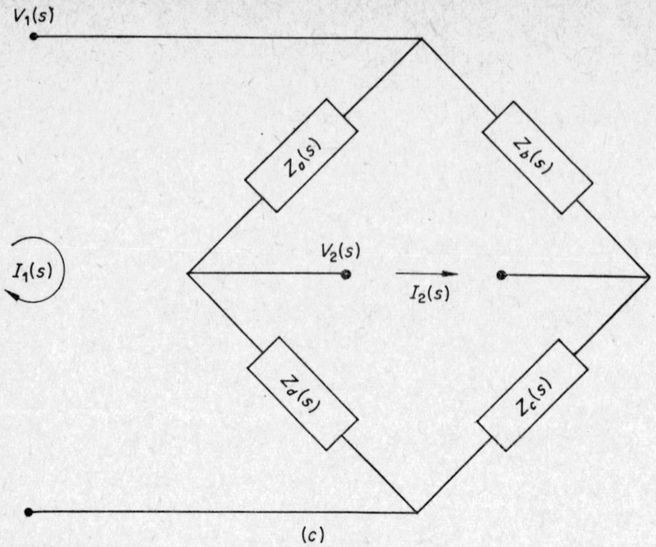

FIG. 6.20

(c) Bridge.

When $Z_a = Z_c$ and $Z_b = Z_d$ the resulting lattice, shown in Fig. 6.20(b) is called a *symmetrical lattice* or frequently just a *lattice*. Note that a lattice network may be redrawn as a bridge circuit as shown in Fig. 6.20(c).

The lattice network is of considerable importance in network synthesis. The coupling matrices may be derived by nodal analysis (see Section 3.4, Chapter 3) and for the symmetrical lattice those most frequently used are given by

$$[Z] = \tfrac{1}{2}\begin{bmatrix} Z_a(s) + Z_b(s) & Z_a(s) - Z_b(s) \\ -(Z_a(s) - Z_b(s)) & -(Z_a(s) + Z_b(s)) \end{bmatrix} \tag{6.5.1}$$

$$[Y] = \tfrac{1}{2}\begin{bmatrix} \dfrac{1}{Z_a} + \dfrac{1}{Z_b} & -\left(\dfrac{1}{Z_a} - \dfrac{1}{Z_b}\right) \\ \left(\dfrac{1}{Z_a} - \dfrac{1}{Z_b}\right) & -\left(\dfrac{1}{Z_a} + \dfrac{1}{Z_b}\right) \end{bmatrix} \tag{6.5.2}$$

$$[A] = \begin{bmatrix} -\left(\dfrac{Z_a + Z_b}{Z_a - Z_b}\right) & -\left(\dfrac{2Z_a\, Z_b}{Z_a - Z_b}\right) \\ -\dfrac{2}{(Z_a - Z_b)} & -\left(\dfrac{Z_a + Z_b}{Z_a - Z_b}\right) \end{bmatrix} \tag{6.5.3}$$

with†

$$\det[Z] = -(Z_a\, Z_b),\ \det[Y] = -(Z_a\, Z_b)^{-1} \quad \text{and} \quad \det[A] = 1. \tag{6.5.4}$$

† Note that $\det \alpha\,[Z] = \begin{vmatrix} \alpha Z_{11} & \alpha Z_{12} \\ \alpha Z_{21} & \alpha Z_{22} \end{vmatrix}$.

6.6 Matrix for a Realizable Two-Port

In this section we shall examine the conditions which a given 2×2 matrix must satisfy if it is to represent a realizable two-port network. Evidently the matrix must correspond to a set of circuit elements which are themselves realizable. We shall assume that the circuit elements are to be discrete parameter, reciprocal, and passive.

Consider the impedance matrix for a two-port element,

$$[Z] = \begin{bmatrix} Z_{11}(s) & Z_{12}(s) \\ -Z_{12}(s) & Z_{22}(s) \end{bmatrix} \tag{6.6.1}$$

The matrix elements

$$Z_{11}(s) = \left[\frac{V_1(s)}{I_1(s)}\right]_{I_2=0} \quad \text{and} \quad -Z_{22}(s) = \left[-\frac{V_2(s)}{I_2(s)}\right]_{I_1=0}$$

are driving point impedance functions and must satisfy the conditions set out in Chapter 5, Section 5.2. Driving point impedance functions have neither poles nor zeros in the right half complex frequency plane. For the special case of an ideal lossless network simple poles on the $j\omega$ axis are allowed, but for an actual network constructed from physical elements, which always involve some loss, the poles and zeros all lie in the left half complex frequency plane. Since the driving point impedance is a particular example of a general transfer function $H(s)$, it is also a real rational function of the complex frequency, s, it is real when s is real and both poles and zeros which are complex occur as conjugate pairs. Since the power dissipated in a driving point impedance is always greater than or equal to zero, we know from equation (3.8.14) that Re $[Z_{dr}(j\omega)] \geqslant 0$. However, Re $[Z_{dr}(s)]$ has neither poles nor zeros in the right half complex frequency plane, and so Re $[Z_{dr}(s)]$ cannot change sign for Re $(s) \equiv q \geqslant 0$. Hence not only is Re $[Z_{dr}(j\omega)] \geqslant 0$, but also

$$\text{Re}\,[Z_{dr}(s)] \equiv \text{Re}\,[Z_{11}(s)], \quad \text{Re}\,[-Z_{22}(s)] \geqslant 0 \quad \text{for} \quad q \geqslant 0 \tag{6.6.2}$$

These considerations impose no immediate constraints upon the matrix element $Z_{21}(s)$. However, the matrix element $Y_{11}(s)$ in the admittance matrix for a two-port must be a valid driving point admittance function and so $Y_{11}(s)$ has neither poles nor zeros in the right half complex frequency plane. From Table 6.1, $Y_{11}(s) = Z_{22}(s)/\Delta_Z$ and since $-Z_{22}(s)$ already satisfies the pole-zero condition, $-\Delta_Z$ must satisfy the condition also. Since

$$\Delta_Z = Z_{11}(s)\,Z_{22}(s) + Z_{12}^2(s)$$

it follows that $Z_{12}(s)$ can have no poles in the right half plane although it may have zeros and be either positive or negative in sign. Moreover, from the form of equations (6.1.6) and (6.1.4) it can be seen that the denominator function for $Z_{11}(s)$, $Z_{22}(s)$, $Z_{12}(s)$ is in each case Δ_Y ($\equiv$ det $[Y]$) and so the poles of $Z_{12}(s)$ will be included in the poles of $Z_{11}(s)$, $Z_{22}(s)$.

A further constraint is imposed upon the matrix elements for a passive two-port network because the power loss in the complete element must be greater than or equal to zero. The power loss in the element, $P(\omega)$ is the difference between the input power and the transmitted power. Consider a network driven by a harmonic generator. From equation (3.8.9)

$$P(\omega) = \tfrac{1}{2}\,\mathrm{Re}\,[I_1^*(j\omega)\,V_1(j\omega) - I_2^*(j\omega)\,V_2(j\omega)]$$

$$= \tfrac{1}{2}\,\mathrm{Re}\,[I_1^*(j\omega) \quad I_2^*(j\omega)]\begin{bmatrix} V_1(j\omega) \\ -V_2(j\omega) \end{bmatrix} \tag{6.6.3}$$

$$= \tfrac{1}{2}\,\mathrm{Re}\,[I_1^*(j\omega) \quad I_2^*(j\omega)]\begin{bmatrix} Z_{11}(j\omega) & Z_{12}(j\omega) \\ -Z_{21}(j\omega) & -Z_{22}(j\omega) \end{bmatrix}\begin{bmatrix} I_1(j\omega) \\ I_2(j\omega) \end{bmatrix} \tag{6.6.4}$$

But, $Z_{12}(j\omega) = -Z_{21}(j\omega)$ for a passive network. Consequently,

$$P(\omega) = \tfrac{1}{2}\,\mathrm{Re}\,[Z_{11}(j\omega)|I_1(j\omega)|^2 - Z_{22}(j\omega)|I_2(j\omega)|^2 + 2Z_{12}(j\omega)\,\mathrm{Re}\,\{I_1^*(j\omega)\,I_2(j\omega)\}] \tag{6.6.5}$$

Writing $R_{11}(\omega)$, $R_{22}(\omega)$, $R_{12}(\omega)$ for the real parts of the functions $Z_{11}(j\omega)$, $-Z_{22}(j\omega)$, $Z_{12}(j\omega)$, equation (6.6.5) is of the form

$$P(\omega) = \tfrac{1}{2}\,\{R_{11}(\omega)\,X_1^2 + R_{22}(\omega)\,X_2^2 + 2R_{12}(\omega)\,X_1\,X_2\cos\delta\} \tag{6.6.6}$$

where all the quantities $R_{ij}(\omega)$, X_i, δ, are real. We require

$$P(\omega) \geqslant 0 \tag{6.6.7}$$

for all values of X_1, X_2, δ.

Thus taking $X_2 = 0$ equation (6.5.7) gives $R_{11}(\omega) \geqslant 0$, whilst taking $X_1 = 0$ gives $R_{22}(\omega) \geqslant 0$. If $R_{11}(\omega) = R_{22}(\omega) = 0$ then $R_{12}(\omega) = 0$ since $\cos\delta$ may be either positive or negative.

Suppose $R_{11}(\omega) > 0$ (if there is a particular case for which $R_{11}(\omega) = 0$ then $R_{22}(\omega)$ must be used in the following argument), $P(\omega)$ may be written as

$$P(\omega) = \tfrac{1}{2}R_{11}(\omega)\left\{X_1 + \frac{R_{12}(\omega)\cos\delta}{R_{11}(\omega)}X_2\right\}^2 + \tfrac{1}{2}\left\{R_{22}(\omega) - \frac{R_{12}^2(\omega)\cos^2\delta}{R_{11}(\omega)}\right\}X_2^2 \tag{6.6.8}$$

which is of the form

$$P(\omega) = \tfrac{1}{2}R_{11}(\omega)\eta_1^2 + \tfrac{1}{2}\left\{R_{22}(\omega) - \frac{R_{12}^2(\omega)\cos^2\delta}{R_{11}(\omega)}\right\}\eta_2^2 \tag{6.6.9}$$

where

$$\eta_1 = X_1 + \frac{R_{12}(\omega)\cos\delta}{R_{11}(\omega)}X_2 \quad \eta_2 = X_2$$

Using the inequality of equation (6.6.7), it follows that

$$R_{11}(\omega)\,R_{22}(\omega) - R_{12}^2(\omega)\cos^2\delta \geqslant 0 \tag{6.6.10}$$

and since the maximum value for $\cos^2\delta$ is unity we require

$$R_{11}(\omega)\,R_{22}(\omega) - R_{12}^2(\omega) \geqslant 0 \tag{6.6.11a}$$

This equation is called the *real part condition* for the matrix elements of a 2×2 matrix which represents a realizable two-port.

For a reversible two-port $Z_{11}(s) = -Z_{22}(s)$ and equation (6.6.11a) becomes

$$R_{11}^2(\omega) - R_{12}^2(\omega) \geqslant 0 \tag{6.6.11b}$$

The conditions which we have derived may be summarized as

1. $Z_{11}(s)$, $-Z_{22}(s)$ must be valid driving point impedance functions.

2. $Y_{11}(s) = Z_{22}(s)/\Delta_Z$; $-\ Y_{22}(s) = -\ (Z_{11}(s)/\Delta_Z)$ must be valid driving point admittance functions
3. $R_{11}(\omega) \geqslant 0$; $R_{22}(\omega) \geqslant 0$; $\{R_{11}(\omega)\, R_{22}(\omega) - R_{12}^2(\omega)\} \geqslant 0$.

For a symmetrical lattice two-port network, equation (6.5.1) gives

$$Z_{11}(s) = -Z_{22}(s) = \tfrac{1}{2}\{Z_a(s) + Z_b(s)\} \qquad (6.6.12)$$
$$Z_{12}(s) = \tfrac{1}{2}\{Z_a(s) - Z_b(s)\}$$

If $Z_{11}(s)$ for a given 2×2 matrix satisfies condition (1) above then neither $Z_a(s)$ nor $Z_b(s)$ will have poles in the right half complex frequency plane. Moreover, for a lattice network $\Delta_Z = -Z_a(s)\, Z_b(s)$ and so, since $Z_{11}(s)$ is given as having neither poles nor zeros in the right half plane, it follows from condition (2) that Δ_Z and hence each of $Z_a(s)$, $Z_b(s)$ will have no zeros in the right half plane. (Note that there can be no cancellation of zeros because this would involve a pole of one function and a zero of the other but poles in the right half plane are excluded by condition (1).) Thus the individual functions $Z_a(s)$, $Z_b(s)$ have neither poles nor zeros in the right half plane. Condition (3) gives

$$R_a(\omega) + R_b(\omega) \geqslant 0 \qquad (6.6.13)$$
$$\tfrac{1}{4}\{R_a(\omega) + R_b(\omega)\}^2 - \tfrac{1}{4}\{R_a(\omega) - R_b(\omega)\}^2 = R_a(\omega)\, R_b(\omega) \geqslant 0$$

so that

$$R_a(\omega) \geqslant 0 \quad \text{and} \quad R_b(\omega) \geqslant 0 \qquad (6.6.14)$$

Since the individual functions $Z_a(s)$, $Z_b(s)$ have neither poles nor zeros in the right half plane they cannot change sign there and so equation (6.6.14) leads to

$$\text{Re}\,[Z_a(s)] \geqslant 0;\ \text{Re}\,[Z_b(s)] \geqslant 0; \quad \text{for} \quad \text{Re}\,(s) \equiv q \geqslant 0 \qquad (6.6.15)$$

Consequently $Z_a(s)$, $Z_b(s)$ are themselves valid driving point impedance functions describing realizable circuit elements. Thus if the elements of a given 2×2 matrix satisfy the three conditions set out above, this matrix can always be synthesized using a symmetrical lattice network. Note that an alternative discussion can be given starting from the admittance matrix $[Y]$ which will give rise to a set of conditions complementary to those given above, and again the matrix can be synthesized by a symmetrical lattice network.

6.6.1 Introduction to Synthesis Using Lattice Networks

As an introduction to network synthesis we consider the use of lattice networks to realize a given transfer function $H(s)$. The present discussion will be based on the impedance matrix representation which has been examined in some detail in the preceding paragraphs. A transfer function relates the current and voltage transforms at the ports of the network and for the impedance matrix representation the open circuit transfer functions are most directly relevant. (If short circuit transfer functions are prescribed it may be more convenient to base the analysis on the admittance matrix.) Of particular importance are the functions

$$H_1(s) = \left[\frac{V_1(s)}{I_1(s)}\right]_{I_2=0} = Z_{11}(s) \qquad H_2(s) = \left[\frac{V_1(s)}{I_2(s)}\right]_{I_1=0} = Z_{12}(s)$$
$$H_3(s) = \left[\frac{V_2(s)}{I_1(s)}\right]_{I_2=0} = Z_{21}(s) \qquad H_4(s) = \left[\frac{V_2(s)}{V_1(s)}\right]_{I_2=0} = \frac{Z_{21}(s)}{Z_{11}(s)} \qquad (6.6.16)$$

When the network synthesis is to be carried out using a symmetrical lattice structure, we know that $Z_{12}(s) = -Z_{21}(s)$; $Z_{11}(s) = -Z_{22}(s)$. If the matrix elements $Z_{11}(s)$ and $Z_{12}(s)$ can be derived from the given transfer function, the lattice network can be constructed with impedances $Z_a(s) = Z_{11}(s) + Z_{12}(s)$ and $Z_b(s) = Z_{11}(s) - Z_{12}(s)$. When forming the matrix elements $Z_{11}(s)$, $Z_{12}(s)$ it is necessary to remember that, although $Z_{12}(s)$ need not itself be a driving point impedance function, the functions

$$Z_{11}(s), \{Z_{11}(s) + Z_{12}(s)\} \quad \text{and} \quad \{Z_{11}(s) - Z_{12}(s)\}$$

must be realizable driving point impedance functions.

A useful technique for rewriting a given transfer function $H(s)$ in a form which is suitable for identifying $Z_{11}(s)$, $Z_{12}(s)$ makes use of the known properties of the driving point impedance functions for combinations of two elements, LC, RC, or RL. In the case of a general driving point impedance function the degree of the numerator polynomial, n, can differ from the degree of the denominator polynomial, m, by unity at the most (see equation (5.2.34) Section 5.2). That is

$$m - 1 \leqslant n \leqslant m + 1 \tag{6.6.17}$$

For an LC circuit, all the L's are open circuits and all the C's are short circuits as $\omega \longrightarrow \infty$. Hence the driving point impedance function $Z_{LC}(j\omega)$ for a network constructed only from L, C, elements, tends either to infinity or to zero as $\omega \longrightarrow \infty$. Thus the relation $n = m$ in equation (6.6.17) is not possible for an LC network and the degree of the numerator polynomial in $Z_{LC}(s)$ must differ by unity from the degree of the denominator polynomial. The function $Z_{LC}(s)$ is therefore the ratio of an even polynomial to an odd polynomial or vice versa. A similar discussion for the other two element network types shows that for $Z_{RC}(s)$, $n = m$ or $n = m - 1$, whilst for $Z_{RL}(s)$, $n = m$ or $n = m + 1$. When rewriting a given $H(s)$, it is therefore useful to form ratios of polynomials which fall into these two element categories so that relatively simple network constructions can be recognized.

Suppose an open circuit transfer function $H_4(s)$ is given to be

$$H_4(s) = \left[\frac{V_2(s)}{V_1(s)}\right]_{I_2=0} = \frac{s^2 + 1}{s^2 + 2s + 2} \tag{6.6.18}$$

Dividing numerator and denominator by s, leads to a possible identification. From equation (6.6.16) we may write

$$Z_{21}(s) = s + \frac{1}{s}$$

$$Z_{11}(s) = s + 2 + \frac{2}{s} \tag{6.6.19}$$

and

$$Z_a(s) = 2 + \frac{1}{s}$$

$$Z_b(s) = 2 + 2s + \frac{3}{s} \tag{6.6.20}$$

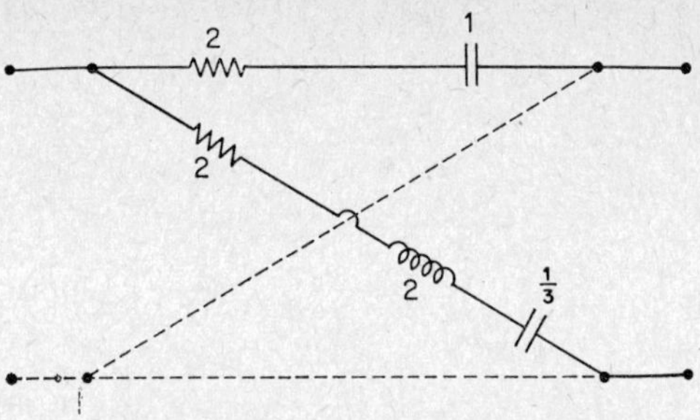

FIG. 6.21 A lattice network corresponding to $H_4(s) = \dfrac{s^2+1}{s^2+2s+2}$. The broken lines represent elements necessary to construct a symmetrical lattice. The units may be taken as ohms, henrys, farads.

The corresponding lattice network is shown in Fig. 6.21.

As a second example, suppose that it is required to synthesize a given $H_3(s)$ with an RC network. If we are given

$$H_3(s) = \left[\frac{V_2(s)}{I_1(s)}\right]_{I_2=0} = \frac{1+s^2}{s(1+s)} = Z_{21}(s) \tag{6.6.21}$$

$Z_{11}(s)$ is not specified and can be chosen arbitrarily to provide valid driving point impedance functions.† Equation (6.6.21) may be rewritten as

$$Z_{12}(s) = -\left[1 + \frac{1}{s} - \frac{2}{1+s}\right] \tag{6.6.22}$$

and if we take for $Z_{11}(s)$ the valid driving point function

$$Z_{11}(s) = 1 + \frac{1}{s} + \frac{2}{1+s} \tag{6.6.23}$$

then

$$Z_a(s) = \frac{4}{1+s}$$

$$Z_b(s) = 2 + \frac{2}{s} \tag{6.6.24}$$

and the corresponding lattice network is shown in Fig. 6.22.

As an example of a network terminated by a load, suppose the given transfer function refers to a terminating resistance R. If the circuit is that shown in Fig. 6.23, and the given transfer function is

$$H_3'(s) = \left[\frac{V_2(s)}{I_1(s)}\right]_{I_2=V_2/R} = \frac{s^2+1}{s^3+3s^2+4s+2} \tag{6.6.25}$$

† Note that in this example $-Z_{12}$ is not a valid driving point impedance function and the transfer function cannot be synthesized by the T network of Fig. 6.18(b). A T-network synthesis is possible for equation (6.6.18).

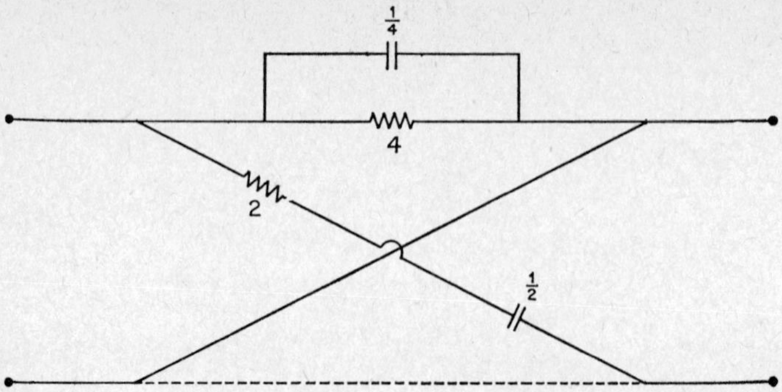

FIG. 6.22 A symmetrical lattice network corresponding to $H_3(s) = \frac{1+s^2}{s(1+s)}$. The units may be taken as ohms, farads.

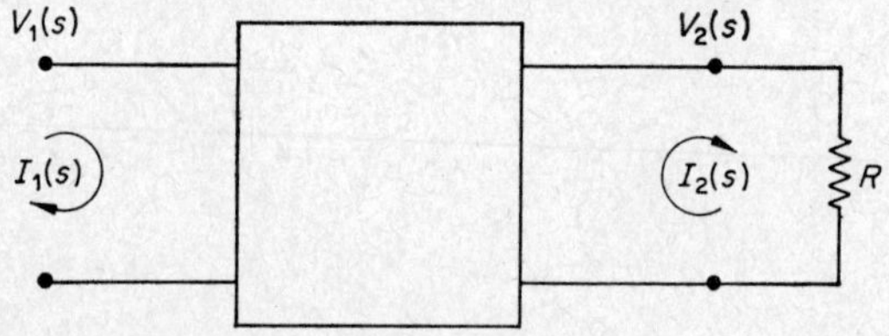

FIG. 6.23 A network terminated by a load resistance R.

we obtain from the matrix equation for the two-port

$$\left[\frac{V_2(s)}{I_1(s)}\right]_{I_2=V_2/R} = \frac{RI_2(s)}{I_1(s)} = \frac{RZ_{21}(s)}{R - Z_{22}(s)} = \frac{-RZ_{12}(s)}{R + Z_{11}(s)} \tag{6.6.26}$$

and so we may write

$$-\frac{Z_{12}(s)}{1 + \{Z_{11}(s)/R\}} = \frac{s^2+1}{s^3+3s^2+4s+2} \tag{6.6.27}$$

$$= \frac{(s^2+1)/(s^3+4s)}{1+(3s^2+2)/(s^3+4s)} \tag{6.6.28}$$

and identify

$$Z_{12}(s) = -\frac{(s^2+1)}{(s^3+4s)} = -\frac{\frac{1}{4}}{s} + \frac{\frac{3}{4}s}{s^2+4}$$

$$\frac{Z_{11}(s)}{R} = \frac{3s^2+2}{s^3+4s} = \frac{\frac{1}{2}}{s} + \frac{5s/2}{s^2+4} \tag{6.6.29}$$

so that

$$Z_a(s) = \frac{(R/2 - \frac{1}{4})}{s} + \frac{(5R/2 - \frac{3}{4})s}{s^2+4}$$

$$Z_b(s) = \frac{(R/2 + \frac{1}{4})}{s} + \frac{(5R/2 + \frac{3}{4})s}{s^2+4} \tag{6.6.30}$$

In particular, if $R = 1$ ohm, which is a value frequently specified,

$$Z_a(s) = \frac{\frac{1}{4}}{s} + \frac{\frac{7}{4}s}{s^2 + 4}$$
$$Z_b(s) = \frac{\frac{3}{4}}{s} + \frac{\frac{13}{4}s}{s^2 + 4} \qquad (6.6.31)$$

and the lattice network is as shown in Fig. 6.24.

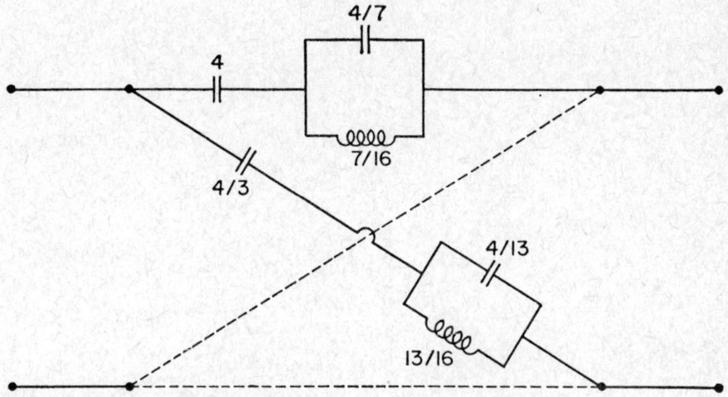

FIG. 6.24 A symmetrical lattice network corresponding to

$$H_3'(s) = \frac{s^2 + 1}{s^3 + 3s^2 + 4s + 2}$$

The units may be taken as henrys, farads.

Constant resistance networks are of particular importance in network synthesis. A constant resistance two-port network has the property that when one port is terminated by a resistance, R, the input impedance at the other port is also R. When a series of such networks are connected in cascade the overall voltage ratio transfer function is the product of the individual voltage ratio transfer functions. For the purpose of network synthesis therefore a prescribed transfer function may be broken down into a product of simpler functions which can be realized separately as constant resistance two-ports and subsequently connected together in cascade to yield the overall transfer function.

From equation (6.4.2) the input impedance for a two-port network terminated in a resistance R is given by

$$Z_{\text{in}}(s) = \frac{A_{11}R + A_{12}}{A_{21}R + A_{22}} \qquad (6.6.32)$$

For a reversible two-port $A_{11} = A_{22}$, so that if $Z_{\text{in}}(s) = R$

$$\frac{A_{12}}{A_{21}} = R^2 \qquad (6.6.33)$$

But from Table 6.1, and equation (6.5.4) we have, for a lattice

$$\frac{A_{12}}{A_{21}} = -\Delta_Z = Z_a(s)\, Z_b(s) \tag{6.6.34}$$

Thus the additional constraint on the network impedances when a lattice is to be realized in the form of a constant resistance network is

$$Z_a(s)\, Z_b(s) = R^2 \tag{6.6.35}$$

The voltage ratio transfer function for a two-port network terminated with a resistance R is given by

$$H_4'(s) = \left[\frac{V_2(s)}{V_1(s)}\right]_{I_2=V_2/R} = \frac{RI_2(s)}{[-\{Z_{11}(s)(Z_{22}(s) - R)/Z_{21}(s)\} + Z_{12}(s)]I_2(s)} \tag{6.6.36}$$

For a constant resistance lattice network, equations (6.6.34)–(6.6.36) give

$$\left[\frac{V_2(s)}{V_1(s)}\right]_{I_2=V_2/R} = \frac{Z_{21}(s)}{R + Z_{11}(s)} \tag{6.6.37}$$

Substituting for $Z_{21}(s)$, $Z_{11}(s)$ the impedance matrix elements from equation (6.5.1) we obtain

$$H_4'(s) = \left[\frac{V_2(s)}{V_1(s)}\right]_{I_2=V_2/R} = \frac{Z_b(s) - R}{Z_b(s) + R} \tag{6.6.38}$$

As an example of the use of this relation consider the synthesis of the voltage ratio transfer function $H_4'(s)$,

$$H_4'(s) = \left[\frac{V_2(s)}{V_1(s)}\right]_{I_2=V_2/R} = \frac{s^2 - 2s + 2}{s^2 + 2s + 2} \tag{6.6.39}$$

by a constant resistance lattice network with $R = 1$ ohm. Write equation (6.6.39) in the form

$$\left[\frac{V_2(s)}{V_1(s)}\right]_{I_2=V_2/R} = \frac{\{(s^2 + 2)/2s\} - 1}{\{(s^2 + 2)/2s\} + 1} \tag{6.6.40}$$

and identify

$$Z_b(s) = \frac{s^2 + 2}{2s} = \frac{s}{2} + \frac{1}{s} \tag{6.6.41}$$

then it follows that

$$Z_a(s) = \frac{2s}{s^2 + 2} \tag{6.6.42}$$

and the corresponding lattice network is shown in Fig. 6.25

6.7 Infinite Series of Identical Two-Ports in Cascade

An infinite series of identical two-port elements connected together in cascade has two important properties which are of immediate relevance for our present discussion:

(a) the input impedance $Z_{\text{in}}(s)$ at any pair of terminals is independent of the particular choice of terminals.

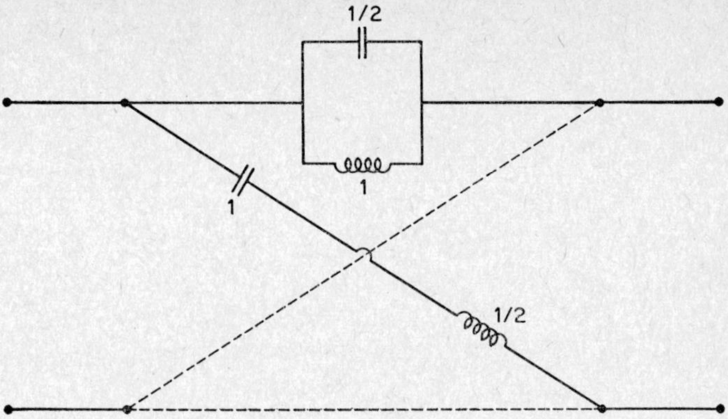

FIG. 6.25 A constant resistance symmetrical lattice network corresponding to

$$H_4'(s) = \frac{(s^2 - 2s + 2)}{(s^2 + 2s + 2)}.$$

The units may be taken as henrys, farads.

(b) the ratios of input to output voltage and of input to output current for any two-port in the series is independent of the particular choice of two-port.

Consider the three elements, from an infinite series, shown in Fig. 6.26.

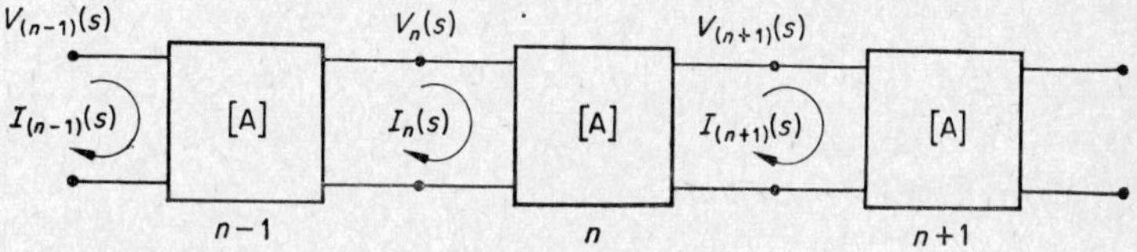

FIG. 6.26 Three elements from an infinite series of identical two-ports in cascade.

The elements are numbered from the left. The property (a) above, arises because the input impedance at any particular terminal pair always refers to an infinite sequence of two-port elements. If this input impedance is $Z_0(s)$, we have

$$\frac{V_{(n-1)}(s)}{I_{(n-1)}(s)} = \frac{V_n(s)}{I_n(s)} = \frac{V_{(n+1)}(s)}{I_{(n+1)}(s)} = Z_0(s) \tag{6.7.1}$$

and from equation (6.4.2)

$$A_{21}\, Z_0^2(s) + (A_{22} - A_{11})Z_0(s) - A_{12} = 0 \tag{6.7.2}$$

so that

$$Z_0(s) = \frac{-(A_{22} - A_{11}) \pm \sqrt{\{(A_{22} - A_{11})^2 + 4A_{21}\, A_{12}\}}}{2A_{21}} \tag{6.7.3}$$

If the two-port elements are reversible, $A_{11} = A_{22}$, and

$$Z_0(s) = \pm\sqrt{\frac{A_{12}}{A_{21}}} \tag{6.7.4}$$

The driving point impedance function $Z_0(s)$ is called the *characteristic impedance* for the sequence. A choice of sign occurs in equations (6.7.3) and (6.7.4) because we have not stated explicitly whether the amplitude of $V_n(s)$ is greater than or less than the amplitude of $V_{(n-1)}(s)$. If the excitation is at the left-hand side of the series, then, in a physical system which inevitably involves dissipative elements, the amplitude of the voltage must decrease along the series towards the right. The sign of $Z_0(s)$ must then be chosen to be consistent with

$$\left|\frac{V_{(n-1)}(s)}{V_n(s)}\right| > 1 \tag{6.7.5}$$

On the other hand, if the excitation is from the right-hand side, we require

$$\left|\frac{V_{(n-1)}(s)}{V_n(s)}\right| < 1 \tag{6.7.6}$$

and this will necessitate the alternative choice of sign in equations (6.7.3) and (6.7.4).

Note that when a finite series of two-port networks is terminated with an impedance equal to $Z_0(s)$ this finite set will have the transmission properties of the corresponding infinite sequence. However, this termination is not always realizable, even in principle, with a finite number of discrete parameter circuit elements because the square root factors in equations (6.7.3) and (6.7.4) may not allow $Z_0(s)$ to be a rational function of s.

The property (b) on page 291 states that

$$\frac{V_{(n-1)}(s)}{V_{(n-2)}(s)} = \frac{V_n(s)}{V_{(n-1)}(s)} = \frac{V_{(n+1)}(s)}{V_n(s)} = \cdots = \text{constant} \equiv \frac{1}{\xi} \tag{6.7.7}$$

where the constant has been written as $1/\xi$ in order to simplify the following mathematical analysis. Since

$$\frac{V_n(s)}{I_n(s)} = Z_0(s)$$

we also have

$$\frac{I_{(n-1)}(s)}{I_{(n-2)}(s)} = \frac{I_n(s)}{I_{(n-1)}(s)} = \frac{I_{(n+1)}(s)}{I_n(s)} = \cdots = \frac{1}{\xi} \tag{6.7.8}$$

and so

$$\begin{bmatrix} V_n(s) \\ I_n(s) \end{bmatrix} = \begin{bmatrix} \xi V_{(n+1)}(s) \\ \xi I_{(n+1)}(s) \end{bmatrix} \tag{6.7.9}$$

But

$$\begin{bmatrix} V_n(s) \\ I_n(s) \end{bmatrix} = [A] \begin{bmatrix} V_{(n+1)}(s) \\ I_{(n+1)}(s) \end{bmatrix} \tag{6.7.10}$$

Consequently, if there exists a set of $V(s)$, $I(s)$, which are not all identically zero, equations (6.7.9) and (6.7.10) lead to

$$\begin{vmatrix} A_{11} - \xi & A_{12} \\ A_{21} & A_{22} - \xi \end{vmatrix} = 0 \tag{6.7.11}$$

or, when det $[A] = 1$,

$$\xi^2 - (A_{11} + A_{22})\xi + 1 = 0 \tag{6.7.12}$$

and

$$\xi = \frac{(A_{11} + A_{22}) \pm \sqrt{\{(A_{11} + A_{22})^2 - 4\}}}{2} \tag{6.7.13}$$

Denoting the two roots of equation (6.7.12) by ξ_1, ξ_2 we note that

$$\xi_1 \, \xi_2 = 1 \tag{6.7.14}$$

and we may write

$$\begin{aligned} \xi_1 &= \mathrm{e}^{\Gamma} \equiv \mathrm{e}^{\alpha + \mathrm{j}\beta} \\ \xi_2 &= \mathrm{e}^{-\Gamma} \equiv \mathrm{e}^{-\alpha - \mathrm{j}\beta} \end{aligned} \tag{6.7.15}$$

with α, β, real quantities.

The discussion of the choice of sign for the characteristic impedance $Z_0(s)$ given above showed that $|V_{(n-1}(s)/V_n(s)| > 1$ for excitation from the left-hand side. For this case therefore we must choose $\xi = \xi_1 = \mathrm{e}^{\,\alpha + \mathrm{j}\beta}$ when α is positive. Hence ξ_1 represents transmission towards the right-hand side in our notation whilst ξ_2 represents transmission towards the left-hand side. The characteristic impedance may be written in terms of ξ. From equation 6.7.3, when det $[A] = 1$,

$$Z_0(s) = \frac{[(A_{11} + A_{22}) \pm \sqrt{\{(A_{11} + A_{22})^2 - 4\}}] - 2A_{22}}{2A_{21}} \tag{6.7.16}$$

Writing ξ_+, ξ_- for the two roots corresponding with the choice of sign in equation (6.7.13) we obtain

$$Z_0(s)^{(\pm)} = \frac{\xi_\pm - A_{22}}{A_{21}} \tag{6.7.17}$$

This equation again emphasizes that the choice of sign for $Z_0(s)$ is determined by the direction of transmission along the sequence of two-ports.

6.7.1 The Pass Band

When $\alpha = 0$ in equations (6.7.15) there is no attenuation along the sequence of two-ports. For this case, $\xi_1 = \mathrm{e}^{\mathrm{j}\beta}$, $\xi_2 = \mathrm{e}^{-\mathrm{j}\beta}$ and β is a real quantity. The amplitude ratio $|V_n(s)/V_{(n-1)}(s)|$ is therefore unity. Since β is real, $(\xi_1 + \xi_2) = 2\cos\beta$ must lie in the range

$$-2 \leqslant (\xi_1 + \xi_2) \leqslant +2 \tag{6.7.18}$$

or, from equation (6.7.12)

$$-2 \leqslant (A_{11} + A_{22}) \leqslant +2 \tag{6.7.19}$$

For an ideal two-port, containing no resistive elements, A_{11}, A_{22}, are real functions of the frequency ω and so it is possible, in principle, to find a frequency range where equation (6.7.19) is satisfied. This frequency range is the pass band for the sequence of two-ports. When resistive elements are present, however, A_{11} and A_{22} will in general be complex functions and a true pass band will not necessarily occur.

When there are no resistive elements in a two-port A_{21} is a pure imaginary function of the frequency ω. Within the pass band $(A_{11} + A_{22})^2 \leqslant 4$, and if the two-port is reversible $A_{11} = A_{22}$. For this case, it can be seen from equation (6.7.16) that $Z_0(j\omega)$ is real for frequencies within the pass band and pure imaginary for frequencies outside.

In the case of an ideal constant resistance lattice two-port, having no resistive elements, $Z_a(j\omega)$ and $Z_b(j\omega)$ are pure imaginary functions of ω. From equations (6.5.3) and (6.6.35).

$$A_{11} = \frac{R^2 + Z_a^2(j\omega)}{R^2 - Z_a^2(j\omega)} \tag{6.7.20}$$

and since $Z_a^2(j\omega)$ must be negative real,

$$-1 \leqslant A_{11} \leqslant 1 \tag{6.7.21}$$

at all frequencies. Hence this type of lattice two-port generates an all-pass system. The characteristic impedance for the system is real at all frequencies and given by $Z_0(s) = R$.

6.7.2 Attenuation Outside the Pass Band

For frequencies outside the pass band of the two-port system, we have $(A_{11} + A_{22}) > 2$ or $(A_{11} + A_{22}) < -2$. If the system is constructed from ideal two-port elements with no resistive components, A_{11} and A_{22} are real functions of ω even outside the pass band. Hence $(A_{11} + A_{22})$ is real and so $\cosh(\alpha + j\beta)$ is real and either positive or negative depending upon the sign of $(A_{11} + A_{22})$. However,

$$\cosh(\alpha + j\beta) = \cosh\alpha\cos\beta + j\sinh\alpha\sin\beta \tag{6.7.22}$$

and so if $\cosh(\alpha + j\beta)$ is to be real, we have

$$\beta = 0, \pm n\pi \tag{6.7.23}$$

The voltage change per element of the system is

$$\left|\frac{V_n(s)}{V_{(n-1)}(s)}\right| = e^{-\alpha} = \left|\frac{(A_{11} + A_{22}) \pm \sqrt{\{(A_{11} + A_{22})^2 - 4\}}}{2}\right| \tag{6.7.24}$$

and the sign must be chosen to give $\left|\frac{V_n(s)}{V_{(n-1)}(s)}\right| < 1$ when transmission is towards the right.

6.7.3 Characteristics of Simple Filters

A series of identical two-port networks connected together in cascade may be used to realize a linear filter. A simple filter system is drawn out formally in Fig. 6.27(a) and the corresponding T representation is shown in Fig. 6.27(b)

Assuming that the network is constructed from reciprocal passive components, the matrix elements for the impedance matrix can be derived from Fig. 6.18(b). Evidently

$$[Z] = \begin{bmatrix} \frac{1}{2}Z_1(s) + Z_2(s) & -Z_2(s) \\ Z_2(s) & -(\frac{1}{2}Z_1(s) + Z_2(s)) \end{bmatrix} \tag{6.7.25}$$

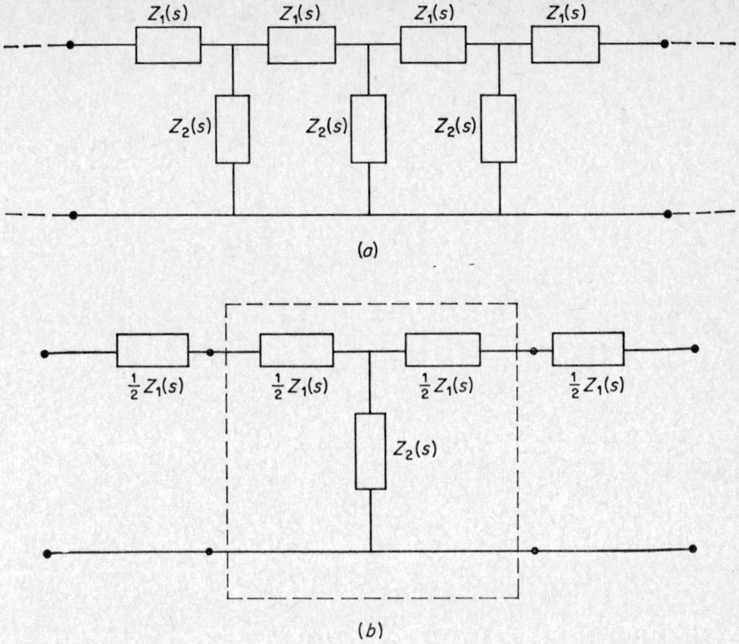

FIG. 6.27 A simple filter realized as a combination of T networks.

and the corresponding transfer matrix may be obtained from Table 6.1

$$[A] = \begin{bmatrix} \left(1 + \dfrac{Z_1(s)}{2Z_2(s)}\right) & Z_1(s)\left(1 + \dfrac{Z_1(s)}{4Z_2(s)}\right) \\ \dfrac{1}{Z_2(s)} & \left(1 + \dfrac{Z_1(s)}{2Z_2(s)}\right) \end{bmatrix} \tag{6.7.26}$$

The pass band of the filter is therefore determined by

$$-1 \leqslant 1 + \frac{Z_1(j\omega)}{2Z_2(j\omega)} \leqslant +1 \tag{6.7.27}$$

When the filter is realized in the form of π networks, as shown in Fig. 6.28(a) and 6.28(b), the transfer matrix is

$$[A] = \begin{bmatrix} \left(1 + \dfrac{Y_2(s)}{2Y_1(s)}\right) & \dfrac{1}{Y_1(s)} \\ Y_2(s)\left(1 + \dfrac{Y_2(s)}{4Y_1(s)}\right) & \left(1 + \dfrac{Y_2(s)}{2Y_1(s)}\right) \end{bmatrix} \tag{6.7.28}$$

and the pass band is determined by

$$-1 \leqslant 1 + \frac{Y_2(j\omega)}{2Y_1(j\omega)} \leqslant +1 \tag{6.7.29}$$

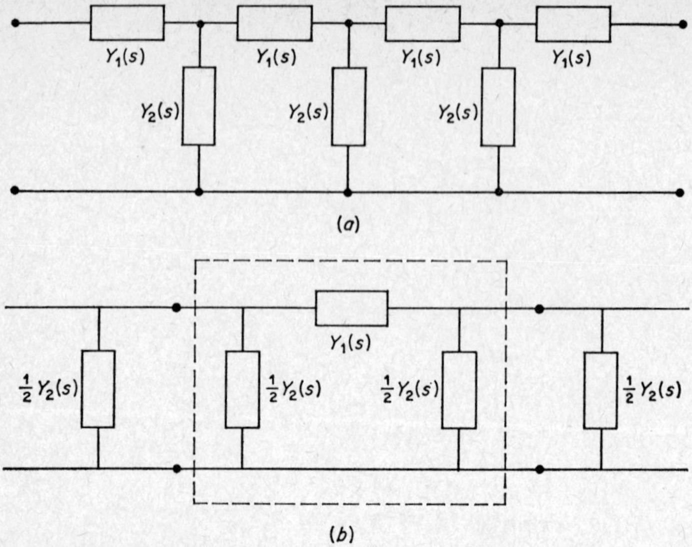

FIG. 6.28 A simple filter realized as a combination of π networks.

For the ideal low pass filter shown in Fig. 6.29(a), we may identify

$$Z_1(s) = sL \quad \text{and} \quad Z_2(s) = \frac{1}{sC}.$$

In this case the pass band is given by

$$-1 \leqslant \left(1 - \frac{\omega^2 LC}{2}\right) \leqslant 1 \tag{6.7.30}$$

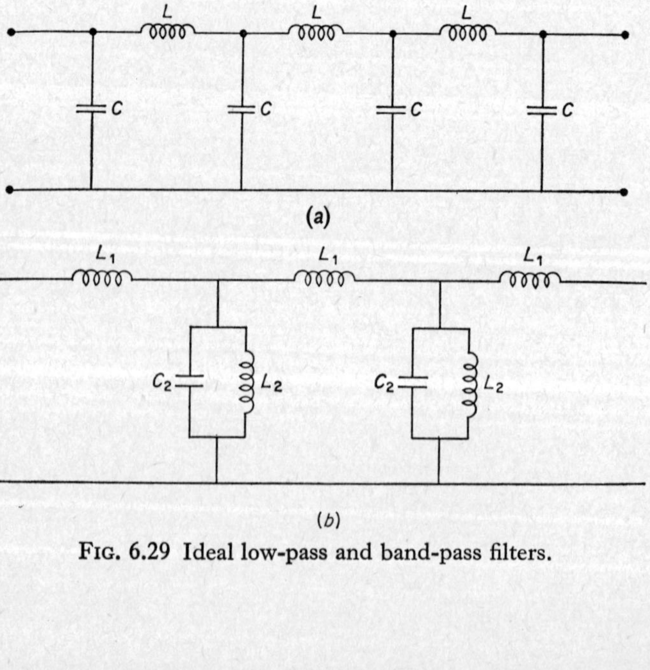

FIG. 6.29 Ideal low-pass and band-pass filters.

or

$$0 \leqslant \omega \leqslant \frac{2}{\sqrt{LC}} \tag{6.7.31}$$

For this representation, the characteristic impedance is

$$Z_0(s) = \sqrt{\left\{\frac{L}{C}\left(1 + \frac{s^2LC}{4}\right)\right\}} \tag{6.7.32}$$

In the case of the ideal band pass filter shown in Fig. 6.29(b) we may identify $Y_1(s) = \dfrac{1}{sL_1}$, $Y_2(s) = \dfrac{1}{sL_2} + sC_2$ and the pass band is given by

$$\frac{1}{\sqrt{(L_2C_2)}} \leqslant \omega \leqslant \sqrt{\left(\frac{4}{L_1C_2} + \frac{1}{L_2C_2}\right)} \tag{6.7.33}$$

The characteristic impedance for the π representation is

$$Z_0(s) = \frac{2s}{C_2[\{s^2 + (L_2C_2)^{-1}\}\{(4/L_1C_2) + (L_2C_2)^{-1} + s^2\}]^{\frac{1}{2}}} \tag{6.7.34}$$

6.7.4 Two-Ports in Cascade as a Transmission Line

An infinite series of identical and reversible two-ports connected together in cascade may be used to provide a representation for a transmission line. Consider the section of transmission line, length z, shown in Fig. 6.30 drawn as a sequence of two-ports.

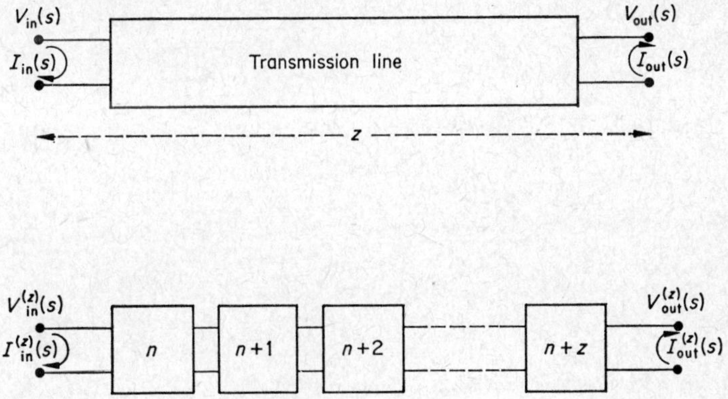

FIG. 6.30 A transmission line may be represented as a series of two-port elements in cascade.

Let each two-port network define a unit length along the sequence so that Γ in equation (6.7.15) also refers to unit length.

For each two-port element

$$\frac{V^{(n+r)}_{out}(s)}{V^{(n+r)}_{in}(s)} = \frac{I^{(n+r)}_{out}(s)}{I^{(n+r)}_{in}(s)} = \frac{1}{\xi} = e^{\pm\Gamma} \tag{6.7.35}$$

from equations (6.7.7), (6.7.8) and (6.7.15). For the sequence of length z.

$$\frac{V^{(z)}_{\text{out}}(s)}{V^{(z)}_{\text{in}}(s)} = \frac{I^{(z)}_{\text{out}}(s)}{I^{(z)}_{\text{in}}(s)} = \frac{1}{\xi_z} = \left(\frac{1}{\xi}\right)^z = e^{\pm\Gamma z} \tag{6.7.36}$$

and

$$\begin{bmatrix} V^{(z)}_{\text{in}}(s) \\ I^{(z)}_{\text{in}}(s) \end{bmatrix} = \begin{bmatrix} \xi_z V^{(z)}_{\text{out}}(s) \\ \xi_z I^{(z)}_{\text{out}}(s) \end{bmatrix} = [A^{(z)}] \begin{bmatrix} V^{(z)}_{\text{out}}(s) \\ I^{(z)}_{\text{out}}(s) \end{bmatrix} \tag{6.7.37}$$

which implies

$$\begin{vmatrix} A^{(z)}_{11} - \xi_z & A^{(z)}_{12} \\ A^{(z)}_{21} & A^{(z)}_{22} - \xi_z \end{vmatrix} = 0 \tag{6.7.38}$$

Since the complete sequence of two-ports is regarded as being constructed from reciprocal circuit elements and is reversible, $\det [A^{(z)}] = 1$ and $A^{(z)}_{11} = A^{(z)}_{22}$. Hence

$$\xi_z^2 - 2A^{(z)}_{11}\, \xi_z + 1 = 0 \tag{6.7.39}$$

If the roots of this equation are ξ_{1z}, ξ_{2z}, or equally well $e^{+\Gamma z}$, $e^{-\Gamma z}$, from equation (6.7.36), we have

$$\begin{aligned} \xi_{1z} + \xi_{2z} &= 2A^{(z)}_{11} = 2 \cosh \Gamma z \\ \xi_{1z}\, \xi_{2z} &= 1 = \{A^{(z)}_{11}\}^2 - A^{(z)}_{12}\, A^{(z)}_{21} \end{aligned} \tag{6.7.40}$$

Writing Z_0 for the characteristic impedance, we see that

$$Z_0 = \sqrt{\frac{A^{(z)}_{12}}{A^{(z)}_{21}}} = \sqrt{\frac{A_{12}}{A_{21}}} \tag{6.7.41}$$

which is characteristic of each two-port element and independent of the length of the sequence. The transfer matrix representation for a transmission line of length z, is therefore

$$\begin{bmatrix} V_{\text{in}}(s) \\ I_{\text{in}}(s) \end{bmatrix} = \begin{bmatrix} \cosh \Gamma z & Z_0 \sinh \Gamma z \\ \dfrac{1}{Z_0} \sinh \Gamma z & \cosh \Gamma z \end{bmatrix} \begin{bmatrix} V_{\text{out}}(s) \\ I_{\text{out}}(s) \end{bmatrix} \tag{6.7.42}$$

If the line is terminated with a load $Z_{\text{L}}(s)$, we obtain the standard equation for the transformation of an impedance by a transmission line

$$Z_{\text{in}}(s) = Z_0 \frac{Z_{\text{L}}(s) + Z_0 \tanh \Gamma z}{Z_0 + Z_{\text{L}}(s) \tanh \Gamma z} \tag{6.7.43}$$

It is frequently useful to describe a transmission line in terms of a series resistance R and inductance L per unit length together with a shunt conductance G and capacitance C per unit length. The transmission line may then be drawn as a sequence of π networks as shown in Fig. 6.31 with

$$Y_1(s) = \frac{1}{(R + sL)\,\delta z}$$

$$Y_2(s) = (G + sC)\,\delta z$$

From equation (6.7.28), the transfer matrix for a length δz of the transmission line is given, to first order in δz, by

$$[A_{(\delta_z)}] = \begin{bmatrix} 1 & (R + sL)\,\delta z \\ (G + sC)\delta z & 1 \end{bmatrix} \tag{6.7.44}$$

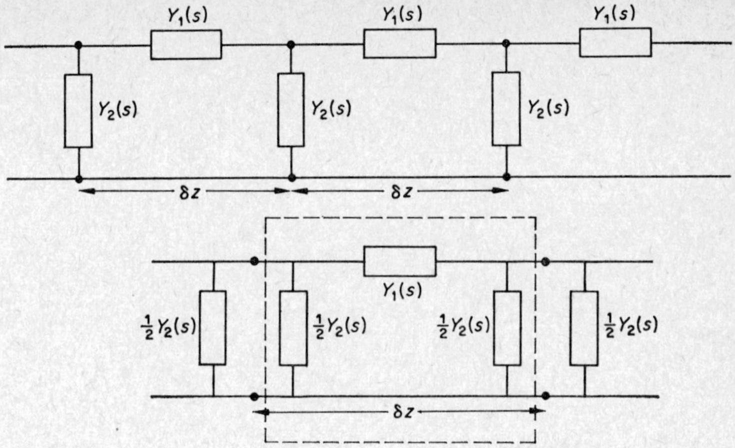

FIG. 6.31 A representation for a transmission line using π networks;

$$Y_1(s) = \frac{1}{(R + sL\ \delta_z}\quad Y_2(s) = (G + sC)\delta_z$$

with det $[A^{(\delta z)}] = 1$ also to first order in δz. The characteristic impedance is

$$Z_0(s) = \sqrt{\left(\frac{R + sL}{G + sC}\right)} \tag{6.7.45}$$

and the propagation constant Γ may be derived from equation (6.7.40), where δz now replaces z and $A_{11}^{(\delta z)}$ must be evaluated to second order in δz. We have

$$\begin{aligned} \cosh \Gamma\ \delta z &= 1 + \tfrac{1}{2}(\Gamma\ \delta z)^2 = A_{11}^{(\delta z)} \\ &= 1 + \tfrac{1}{2}(G + sC)(R + sL)(\delta z)^2 \end{aligned} \tag{6.7.46}$$

or

$$\Gamma = \pm\sqrt{\{(G + sC)(R + sL)} \tag{6.7.47}$$

Note that in the case of an ideal lossless transmission line, for which $G = R = 0$

$$\Gamma = \pm s\sqrt{(LC)} \tag{6.7.48}$$

From equation (6.7.36)

$$V_{\text{out}}^{(z)}(s) = \mathrm{e}^{-sz\sqrt{(LC)}} V_{\text{in}}^{(z)}(s) \tag{6.7.49}$$

and, taking Laplace Transforms,

$$V_{\text{out}}^{(z)}(t) = V_{\text{in}}^{(z)}(t - \tau) \tag{6.7.50}$$

with

$$\tau = z\sqrt{(LC)} \tag{6.7.51}$$

The lossless transmission line acts therefore as an ideal delay network (compare this discussion with that of Section 5.2.3.)

6.8 Summary

1. Two-port networks with reciprocal elements have

$$Z_{12}(s) = -Z_{21}(s) \qquad Y_{12}(s) = -Y_{21}(s)$$
$$h_{12} = h_{21} \qquad g_{12} = g_{21}$$
$$\det[A] = 1 \qquad \det[B] = 1$$

2. When Brune's tests are satisfied

(a) Series connection $[Z] = \sum_{\lambda} [Z_{\lambda}]$

(b) Parallel connection $[Y] = \sum_{\lambda} [Y_{\lambda}]$

(c) Series–parallel connection $[h] = \sum_{\lambda} [h_{\lambda}]$

(d) Parallel–series connection $[g] = \sum_{\lambda} [g_{\lambda}]$

3. Cascade connection

$$[A] = [A_1][A_2] \ldots [A_N]$$
$$[B] = [B_N][B_{N-1}] \ldots [B_1]$$

4. For a reversible two-port

$$A_{11} = A_{22};\ Z_{12}(s) = -Z_{21}(s),\ Z_{11}(s) = -Z_{22}(s).$$

5. Input impedance for a terminated two-port

$$Z_{\text{in}}(s) = \frac{A_{11}\, Z_{\text{L}}(s) + A_{12}}{A_{21}\, Z_{\text{L}}(s) + A_{22}}$$

6. For a symmetrical lattice

$$[Z] = \tfrac{1}{2}\begin{bmatrix} Z_a(s) + Z_b(s) & Z_a(s) - Z_b(s) \\ -(Z_a(s) - Z_b(s)) & -(Z_a(s) + Z_b(s)) \end{bmatrix}$$

7. For a constant resistance lattice

$$\det[Z] = -Z_a(s)\, Z_b(s) = -R^2$$

8. For an infinite series of identical two-ports in cascade, when $\det[A] = 1$

(a) $$Z_0(s) = \frac{(A_{11} - A_{22}) \pm \sqrt{\{(A_{11} + A_{22})^2 - 4\}}}{2A_{21}}$$

(b) $$\xi = \frac{(A_{11} + A_{22}) \pm \sqrt{\{(A_{11} + A_{22})^2 - 4\}}}{2}$$

(c) $-2 \leqslant A_{11} + A_{22} \leqslant +2$ defines the frequency pass band

9. Frequency pass band for a simple T representation

$$-1 \leqslant 1 + \frac{Z_1(j\omega)}{2Z_2(j\omega)} \leqslant 1$$

10. Frequency pass band for a simple π representation

$$-1 \leqslant 1 + \frac{Y_2(j\omega)}{2Y_1(j\omega)} \leqslant 1$$

11. Transfer matrix for a transmission line of length z

$$[A^{(z)}] = \begin{bmatrix} \cosh \Gamma z & Z_0 \sinh \Gamma z \\ \dfrac{1}{Z_0} \sinh \Gamma z & \cosh \Gamma z \end{bmatrix}$$

in terms of R, L, G, C, per unit length,

$$[A^{(\delta z)}] = \begin{bmatrix} 1 & (R + sL)\,\delta z \\ (G + sC)\,\delta z & 1 \end{bmatrix}$$

$$Z_0(s) = \sqrt{\left[\frac{R + sL}{G + sC}\right]}$$

$$\Gamma = \pm \sqrt{\{(R + sL)(G + sC)\}}$$

Bibliography

BRILLOUIN, L., *Wave Propagation in Periodic Structures*, McGraw-Hill, 1946.

DAVENPORT, W. B. and ROOT, W. L., *An Introduction to the Theory of Random Signals and Noise*, McGraw-Hill, 1958.

KUO, F. F., *Network Analysis and Synthesis*, Wiley, 1962.

RUSTON, H. and BORDOGNA, J., *Electric Networks, Functions, Filters, Analysis*, McGraw-Hill, 1966.

Problems

1. Determine the matrices $[Y]$, $[Z]$ and $[A]$ for the two-port networks shown in Fig. 6.P.1

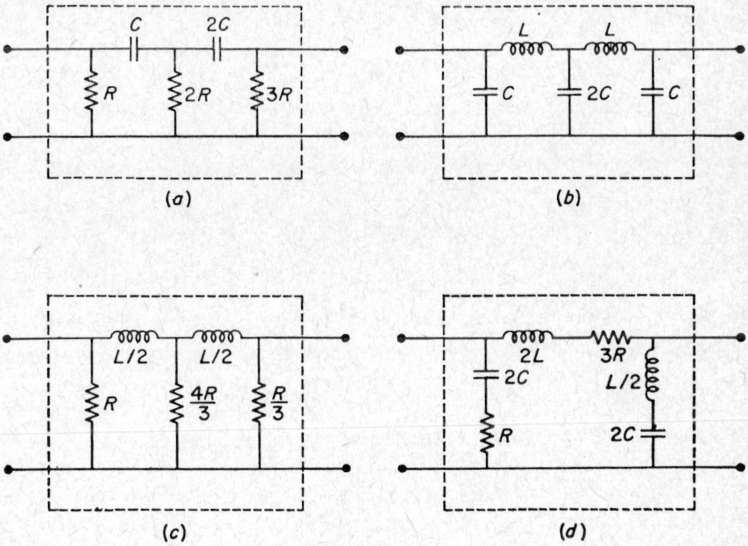

FIG. 6.P.1

2. Determine the matrices [Y], [Z] and [A] for the non-ideal transformer shown in Fig. 6.P.2. Draw out the equivalent T and π networks in terms of L_1, L_2 and M.

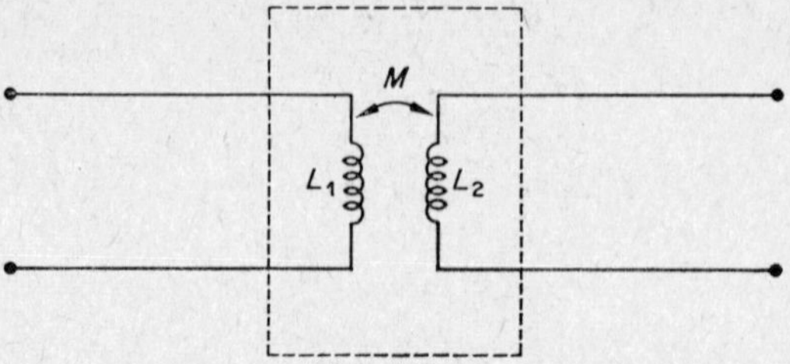

Fig. 6.P.2

3. Evaluate the functions $Z_{11}(s)$, $Z_{22}(s)$ and $Z_{in}(s)$ for the two-ports shown in Fig. 6.P.3.

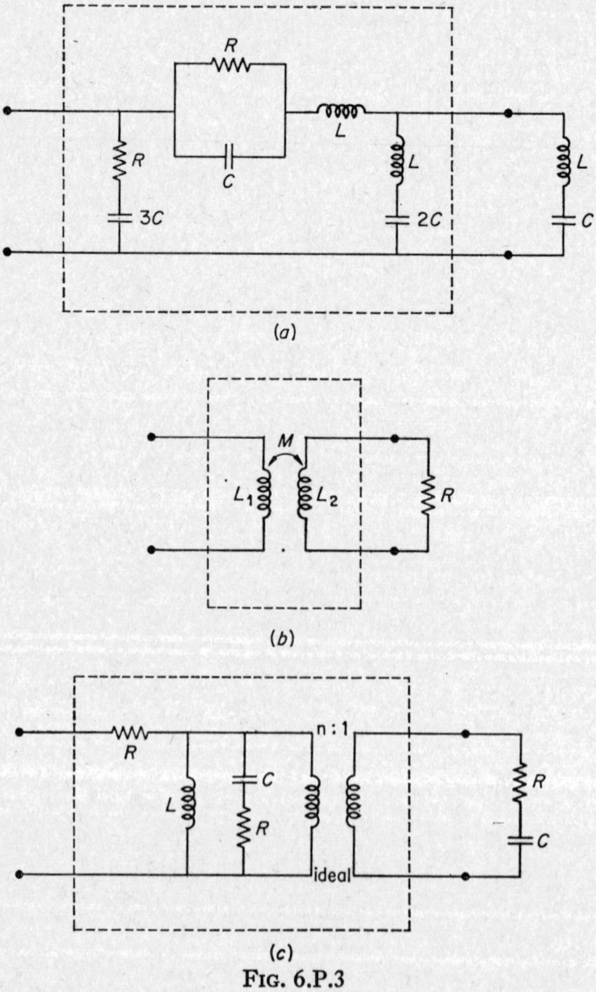

Fig. 6.P.3

4. Determine the $[Y]$, $[Z]$ and $[A]$ matrices for the frequency selective networks shown in Fig. 6.P.4. Find the frequency condition for $V_2(j\omega) = 0$ when the output termination is open circuit.

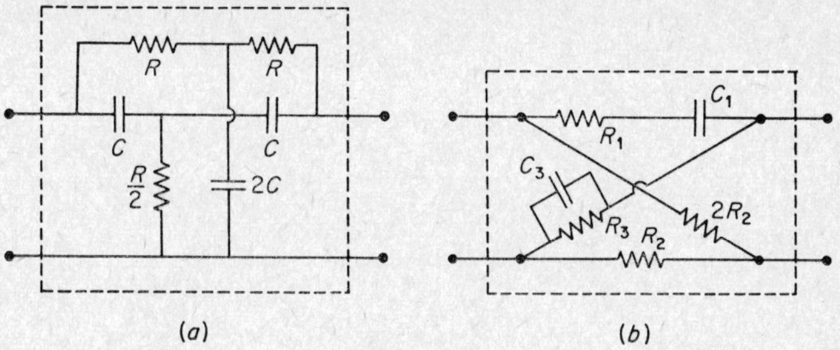

FIG. 6.P.4

5. Derive equations (6.2.61) and (6.2.62) for the small signal, emitter coupled, difference amplifier by using the simplified h matrix representation for a silicon n–p–n transistor given in equation (6.2.70).

6. Determine the matrix $[h]$ for the complete amplifier shown in Fig. 6.P.5 assuming that the two transistors are identical.

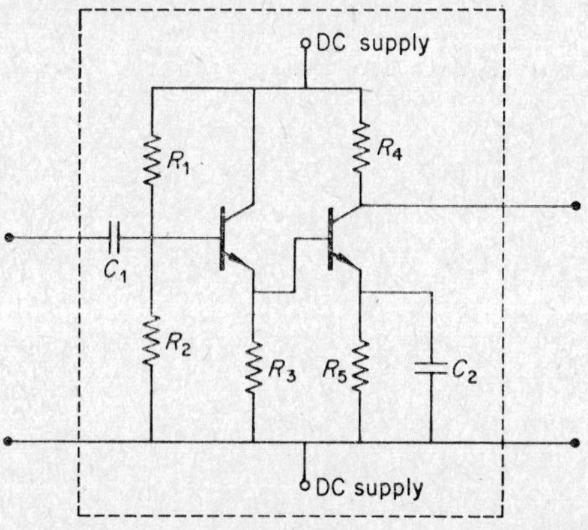

FIG. 6.P.5

7. If two networks are given in the forms shown in Fig. 6.P.6 determine the matrix $[Z]$ for series connection and the matrix $[Y]$ for parallel connection, arranging the networks so that Brune's tests are satisfied without using transformers.

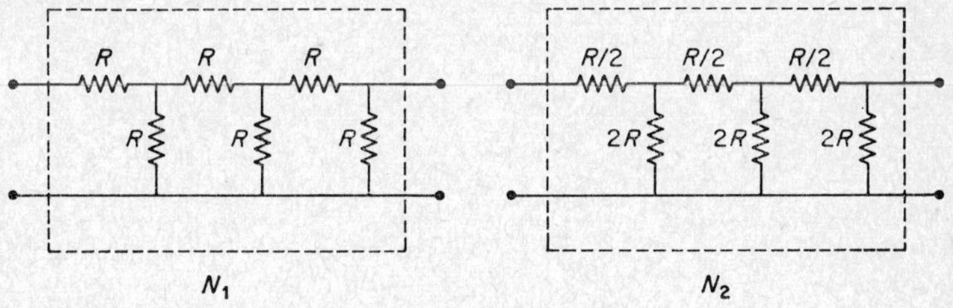

FIG. 6.P.6

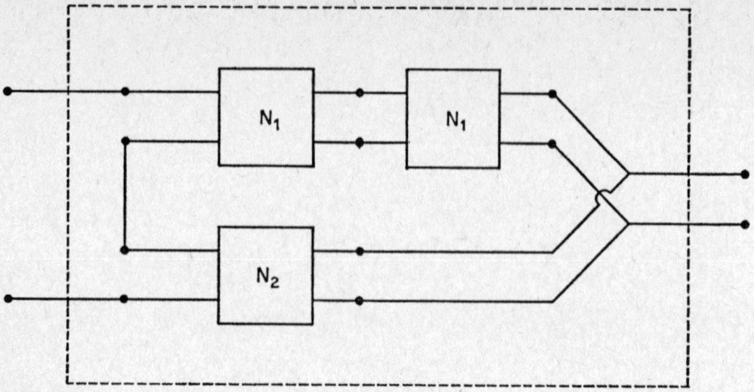

FIG. 6.P.7

8. Derive a matrix representation for the overall network configuration shown in Fig. 6.P.7. Assume that the individual networks are given by Fig. 6.P.6 and arrange the networks to satisfy Brune's tests without the use of transformers.

9. Determine the single frequency noise figure $F(\omega)$ and the integrated noise figure F_i for the two-port attenuator circuit shown in Fig. 6.P.8. What value

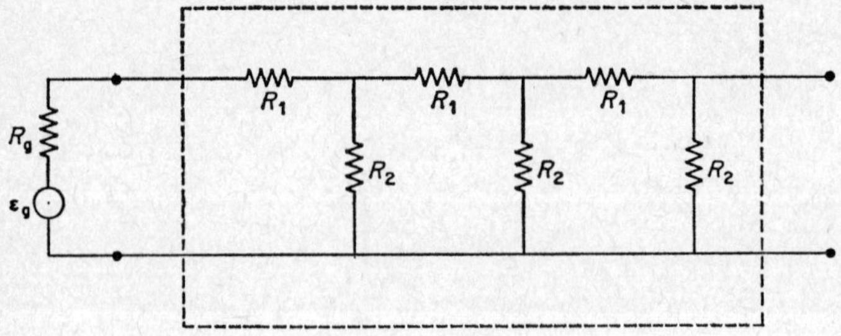

FIG. 6.P.8

of R_g will achieve a minimum noise figure? Derive the noise figures $F(\omega)$ and F_i for an attenuator constructed from a sequence of these two-ports connected together in cascade.

10. Synthesize networks for which

$$Z_{11}(s) = \frac{(s+2)(s+4)}{(s+1)(s+3)}$$

and $Z_{12}(s)$ is given by

(a) $Z_{12}(s) = \dfrac{Ks^2}{(s+1)(s+3)}$

(b) $Z_{12}(s) = \dfrac{Ks}{(s+1)(s+3)}$

(c) $Z_{12}(s) = \dfrac{K}{(s+1)(s+3)}$

Determine the value of K in each case.

11. Synthesize RC lattice networks for which

(a) $\left[\dfrac{V_2(s)}{V_1(s)}\right]_{I_2=0} = \dfrac{K}{(s+1)(s+3)}$

(b) $\left[\dfrac{V_2(s)}{V_1(s)}\right]_{I_2=0} = \dfrac{Ks}{(s+1)(s+3)}$

(c) $\left[\dfrac{V_2(s)}{V_1(s)}\right]_{I_2=V_2/R} = \dfrac{K(s^2+3)}{2s^2+2s+6}$

12. Synthesize a lattice network to give

$$\frac{V_2(s)}{\mathscr{E}_g(s)} = \frac{Ks^2}{15s^2+7s+2}$$

using the circuit arrangement shown in Fig. 6.P.9. Can this lattice be realized as a T network?

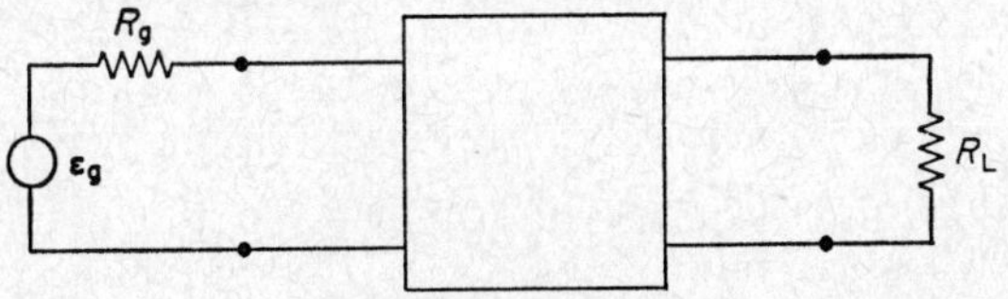

FIG. 6.P.9

13. Identify the following functions as LC, RC, or RL driving point functions

(a) $\dfrac{s^3+2s}{s^4+4s+3}$

(b) $\dfrac{s^2+6s+8}{s^2+4s+3}$

(c) $\dfrac{s^4+5s^2+6}{3s^3+3s}$

(d) $\dfrac{s^2+5s+6}{3s^2+3s}$

14. Synthesize the following functions using an LC network terminated in a resistance R

(a) $\left[\dfrac{V_2(s)}{I_1(s)}\right]_{I_2=V_2/R} = \dfrac{K}{s^3+3s^2+3s+2}$

(b) $\left[\dfrac{V_2(s)}{I_1(s)}\right]_{I_2=V_2/R} = \dfrac{Ks}{s^3+3s^2+3s+2}$

15. Synthesize the following functions using a constant-resistance lattice terminated with a resistance $R = 1$ ohm.

(a) $$\left[\frac{V_2(s)}{V_1(s)}\right]_{I_2=V_2/R} = K\,\frac{s^2 - s + 1}{s^2 + s + 1}$$

(b) $$\left[\frac{V_1(s)}{I_2(s)}\right]_{I_1=0} = \frac{K(s^2 + 3s + 2)}{s^2 - 3s + 2}$$

(c) $$\left[\frac{V_2(s)}{V_1(s)}\right]_{I_2=V_2/R} = \frac{K(s^3 - 20s^2 + 5s - 20)}{s^3 + 20s^2 + 5s + 20}$$

16. Discuss whether the transfer functions given in Problem (13) can be realized by equivalent T networks.

Chapter 7

Microscopic Theory of the Dielectric Constant

7.1 Introduction

The electromagnetic theory outlined in Chapters 1 and 2 provides a technique for describing the properties of electromagnetic waves and material media in terms of quantities averaged over a macroscopic volume. No attempt was made to account in detail for the *actual* values of the characteristic parameters such as ε, μ, σ, since this would have required a microscopic model of the underlying physical processes which gave rise to these quantities. We shall survey some aspects of a simplified microscopic theory for the dielectric constant in the following sections. A completely rigorous theory is not available but a useful approach is possible with a simple atomic model which regards material media as being composed of weakly interacting atoms or molecules distributed in free space. These fundamental units (the atoms and molecules) are capable of being polarized by an electromagnetic field and may also themselves possess permanent electric or magnetic moments. In the following discussion we shall for convenience refer simply to 'molecules' as the fundamental units, but this description is intended to include atoms also. We shall usually be considering interaction with a plane electromagnetic wave and will consider non-polar materials first of all.

7.2 Classical Theory of Non-polar Materials

Assume that each molecule is composed of charged particles (electrons and nuclei) which act as though they are held in their equilibrium positions by forces of an elastic nature. For this model, restoring forces, $\mathbf{F}^{(r)}$, which are proportional to the displacement from an equilibrium position will act on the charges. For a charge with mass m_i, we may write

$$\mathbf{F}_i^{(r)} = -m_i\omega_i^2\mathbf{r}_i \tag{7.2.1}$$

where the factor $m_i\omega_1^2$ is just the constant of proportionality. In addition dissipative damping forces $\mathbf{F}^{(d)}$ may be included. On classical theory these are assumed to be proportional to the particle velocity.

$$\mathbf{F}_i^{(d)} = -m_i\gamma_i\dot{\mathbf{r}}_i \tag{7.2.2}$$

The classical equation of motion for a charge, q_i, under the action of an electromagnetic field is therefore

$$m_i\ddot{\mathbf{r}}_i + m_i\gamma_i\dot{\mathbf{r}}_i + m_i\omega_i^2\mathbf{r}_i = q_i\left(\mathbf{E}' + \frac{\dot{\mathbf{r}}_i \wedge \mathbf{B}'}{c}\right) \tag{7.2.3}$$

where $\mathbf{E}'$, $\mathbf{B}'$ are the local fields acting on the charge q_i.

For a plane wave in a material for which ε, $\mu \sim 1$, $\sigma \sim 0$, the ratio of the electric to the magnetic forces can readily be estimated by making use of equation (2.3.51).

$$\frac{F_m}{F_e} \approx \frac{\dot{r}_i B'}{cE'} = \frac{\mu \dot{r}_i}{c} \frac{H'}{E'} = \dot{r}_i \frac{\sqrt{(\varepsilon\mu)}}{c} \approx \frac{\dot{r}_i}{c} \tag{7.2.4}$$

The velocity of the charged particles under the action of the electromagnetic field is assumed to be very much less than the wave velocity c. Hence the magnetic forces are small compared with the electric forces and may be neglected in the discussion of first order dielectric phenomena. With only $\mathbf{E}'$ significant in equation (7.2.3), the vector $\mathbf{r}_i$ is parallel to $\mathbf{E}'$ and the motion may be discussed in terms of the one-dimensional scalar equation.

$$m\ddot{\xi} + m\gamma\dot{\xi} + m\omega_0^2\xi = qE' \tag{7.2.5}$$

This is the basic equation underlying the derivation of first order dielectric phenomena on the classical microscopic theory.

7.2.1 Gases

The molecules in a gas are assumed to be sufficiently far apart so that interactions between the molecules can be neglected. The local field acting on the charges in a molecule is therefore assumed to be that of the incident electromagnetic wave. The wavelength of the radiation in which we are interested will always be long compared with the linear dimensions of a molecule and so the electric field will be approximately constant over this region. We may therefore assume the same value of $\mathbf{E}$, at a given time, for all the charges in a molecule, retardation effects are not significant and only the time variation of the field need be included in this part of the calculation. From equation (7.2.5) we write

$$m\ddot{\xi} + m\gamma\dot{\xi} + m\omega_0^2\xi = qE_0\, e^{j\omega t} \tag{7.2.6}$$

and obtain the steady state solution

$$\xi = \frac{(q/m)}{(\omega_0^2 - \omega^2) + j\gamma\omega} E_0\, e^{j\omega t} \tag{7.2.7}$$

Each electron in a molecule will contribute to the polarization a moment $\mathbf{p} = e\boldsymbol{\xi}$. There will be a contribution from the nuclei also but since the nuclear masses are heavy in comparison with the electron mass their contribution can be neglected in a first approximation. The oscillatory motion of a given electron is therefore equivalent to an induced electric dipole along the direction of the field with moment

$$\begin{aligned} \mathbf{p} &= \frac{(q^2/m)}{(\omega_0^2 - \omega^2) + j\gamma\omega} \mathbf{E}_0\, e^{j\omega t} \\ &= \frac{(q^2/m)}{(\omega_0^2 - \omega^2) + j\gamma\omega} \mathbf{E} \end{aligned} \tag{7.2.8}$$

The electric polarization $\mathbf{P}$ is the total moment per unit volume and this is obtained by summing $\mathbf{p}$ over all the electrons in unit volume. In general, the different electrons in a molecule will have different characteristic frequencies

ω_0 and also different damping factors γ. In the notation of equation (7.2.3), if N_i electrons per unit volume are characterized by the constants ω_i, γ_i, the polarization†

$$\mathbf{P} = \mathbf{E} \sum_i \frac{N_i(q^2/m)}{(\omega_i^2 - \omega^2) + \mathrm{j}\gamma_i\omega} \tag{7.2.9}$$

From equations (1.3.17) and (1.3.19)

$$\mathbf{P} = \frac{\varepsilon(\omega) - 1}{4\pi}\mathbf{E} \tag{7.2.10}$$

and so

$$\varepsilon(\omega) = 1 + \frac{4\pi q^2}{m} \sum_i \frac{N_i}{(\omega_i^2 - \omega^2) + \mathrm{j}\gamma_i\omega} \tag{7.2.11}$$

Evidently the static dielectric constant is

$$\varepsilon(0) = 1 + \frac{4\pi q^2}{m} \sum_i \frac{N_i}{\omega_i^2} \tag{7.2.12}$$

The static dielectric constant is directly related to the natural frequencies of the electrons and is a real quantity greater than unity.

The refractive index is given by equation (2.10.7),

$$\eta - \mathrm{j}\kappa = \sqrt{\varepsilon} \tag{7.2.13}$$

For gases the summation in equation (7.2.11) is usually small compared to unity and so we may expand the square root to obtain

$$\eta = 1 + \frac{2\pi q^2}{m} \sum_i \frac{N_i(\omega_i^2 - \omega^2)}{(\omega_i^2 - \omega^2)^2 + \gamma_i^2\omega^2} \tag{7.2.14}$$

$$\kappa = \frac{2\pi q^2}{m} \sum_i \frac{N_i\gamma_i\omega}{(\omega_i^2 - \omega^2)^2 + \gamma_i^2\omega^2} \tag{7.2.15}$$

The form of the frequency dependence of η, κ, predicted by equations (7.2.14) and (7.2.15) is shown in Fig. 7.1. The curves in this figure are called 'dispersion curves'. For the purpose of this illustration it has been assumed that only one set of constants ω_0, γ, are significant. In the general case the complete function will consist of a superposition of curves like those of Fig. 7.1, there will be rapid variations in the refractive index centred on each ω_i. Note that in the spectral region for which ω is less than the smallest of the ω_i equation (7.2.14) shows that the real part of the refractive index is always greater than unity. Since the characteristic frequencies for free atoms usually lie in the ultraviolet region of the spectrum, the real part of the refractive index of gases for visible light is usually greater than unity.

† Unit volume must be taken small enough for retardation effects to be negligible but yet large enough to contain sufficient molecules for a macroscopic dielectric constant to be defined.

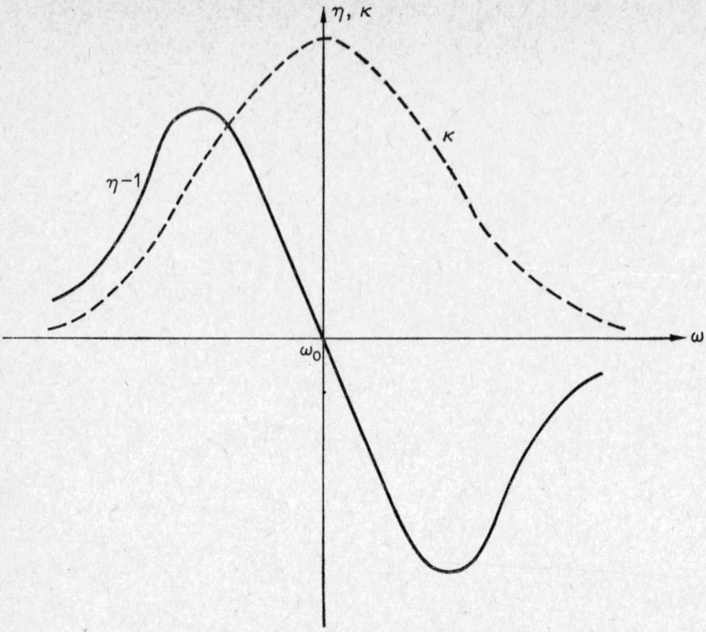

FIG. 7.1 Dispersion curves for a single pair of parameters ω_0, γ.

7.2.2 Insulating Liquids and Solids

In these materials the molecules are sufficiently close together for their interactions to be significant. The local field, $\mathbf{E}'$, $\mathbf{H}'$ which acts upon a given charge will now have a contribution arising from the polarization of the material. It is necessary therefore to distinguish $\mathbf{E}'$, $\mathbf{H}'$ from the mean field $\mathbf{E}$, $\mathbf{H}$ which refers to an average over a region containing a large number of molecules.

To estimate the local field $\mathbf{E}'$ consider a particular molecule and imagine that it is surrounded by a small sphere centred at the molecule. The radius of the sphere is taken to be small but yet sufficiently large to contain many other molecules. The effects produced by the material inside and outside this sphere are to be treated separately. Inside the sphere the discrete nature of the material must be taken into account whereas, if the dimensions are chosen correctly, the molecular structure may be disregarded for the region outside the sphere and the material may be considered to be continuous. We shall show that for a particular model the molecules within the sphere produce no resultant field at the central molecule. This molecule may therefore be regarded as being situated in a spherical hole within which is free space and outside which is a uniformly polarized continuous medium.

Consider a material having a simple cubic crystal structure with all the molecules identical. Assume that the molecules within the spherical volume can be represented by induced point dipoles oriented parallel to one another.

Taking the dipolar direction as the z-axis, the field at the central molecule will be along z and is given by (compare Problem 12, page 55).

$$\mathbf{E}^{(\mathrm{i})} = \sum_i \left\{ \frac{3(\mathbf{p}.\mathbf{r}_i)\mathbf{r}_i}{r_i^5} - \frac{\mathbf{p}}{r_i^3} \right\} \tag{7.2.16}$$

which for our case reduces to

$$\mathbf{E}^{(\mathrm{i})} = \hat{\mathbf{k}} \sum_i \frac{3pz_i^2 - pr_i^2}{r_i^5} \tag{7.2.17}$$

For a simple cubic lattice and a spherical cavity

$$\sum_i \frac{z_i^2}{r_i^5} = \sum_i \frac{x_i^2}{r_i^5} = \sum_i \frac{y_i^2}{r_i^5} = \frac{1}{3} \sum_i \frac{r_i^2}{r_i^5} \tag{7.2.18}$$

Hence

$$\mathbf{E}^{(\mathrm{i})} = 0 \tag{7.2.19}$$

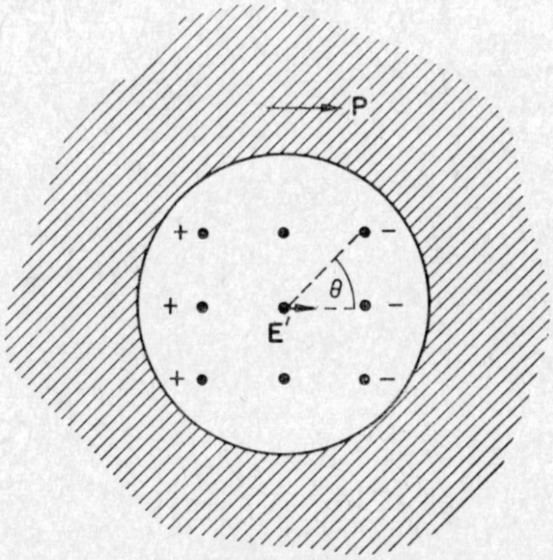

FIG. 7.2 The local field at the molecule at the centre of the spherical cavity in the dielectric.

This calculation involves a very particular model but the result that $\mathbf{E}^{(\mathrm{i})} = 0$ is also valid for simple body-centred cubic and face-centred cubic crystal lattices. It is also correct for a random distribution of molecules. It is therefore usually assumed for the purposes of calculation that $\mathbf{E}^{(\mathrm{i})}$ always vanishes. Such an assumption must, however, be treated with caution and more rigorously equation (7.2.16) should be used to evaluate $\mathbf{E}^{(\mathrm{i})}$ for a particular problem.

To calculate the contribution $\mathbf{E}^{(\mathrm{o})}$ arising from the material outside the spherical region note first that from the symmetry of the problem only field components parallel to $\mathbf{E}$ (and hence $\mathbf{P}$) need to be taken into account. If θ is the polar angle referred to $\mathbf{P}$ as axis, the surface density of charge on the surface of the sphere is $-P\cos\theta$. Hence, if a is the radius of the sphere and $\mathbf{n}$ is unit vector in the direction of $\mathbf{P}$, as in Fig. 7.2

$$\mathbf{E}^{(0)} = \mathbf{n}\int_0^{\pi} \cos\theta \frac{P\cos\theta\, 2\pi a^2 \sin\theta\, d\theta}{a^2} \tag{7.2.20}$$

$$= -2\pi\mathbf{P}\int_1^{-1} \cos^2\theta\, d(\cos\theta) \tag{7.2.21}$$

$$= \frac{4\pi}{3}\mathbf{P} \tag{7.2.22}$$

When equation (7.2.19) is satisfied, the local field $\mathbf{E}'$ at a molecule is therefore

$$\mathbf{E}' = \mathbf{E} + \frac{4\pi}{3}\mathbf{P} \tag{7.2.23}$$

Equation (7.2.9), which was derived for non-interacting molecules in gas, may now be rewritten, with $\mathbf{E}'$ replacing $\mathbf{E}$ and applied to liquids and solids. We obtain

$$\mathbf{P} = \left(\mathbf{E} + \frac{4\pi}{3}\mathbf{P}\right)\sum_i \frac{N_i(q^2/m)}{(\omega_i^2 - \omega^2) + j\gamma_i\omega} \tag{7.2.24}$$

using equation (7.2.10)

$$\frac{\mathbf{P}}{\mathbf{E} + (4\pi/3)\mathbf{P}} = \frac{3}{4\pi}\cdot\frac{\varepsilon(\omega) - 1}{\varepsilon(\omega) + 2} \tag{7.2.25}$$

so that

$$\frac{\varepsilon(\omega) - 1}{\varepsilon(\omega) + 2} = \frac{(\eta - j\kappa)^2 - 1}{(\eta - j\kappa)^2 + 2} = \frac{4\pi}{3}\sum_i \frac{N_i(q^2/m)}{(\omega_i^2 - \omega^2) + j\gamma_i\omega} \tag{7.2.26}$$

In this equation N_i is the number of charges of type i in unit volume. We may usefully rewrite the right-hand side of equation (7.2.26) as

$$\frac{4\pi}{3}\sum_i \frac{N_i(q^2/m)}{(\omega_i^2 - \omega^2) + j\gamma_i\omega} = \frac{4\pi}{3}N\langle\alpha\rangle \tag{7.2.27}$$

with now N the number of *molecules* per unit volume and $\langle\alpha\rangle$ an averaged quantity called the mean polarizability per molecule. Using this notation the dipole moment of each molecule is (compare equation (7.2.8))

$$\mathbf{p} = \langle\alpha\rangle\mathbf{E}' \tag{7.2.28}$$

Equation (7.2.26) now becomes

$$\frac{\varepsilon(\omega) - 1}{\varepsilon(\omega) + 2} = \frac{(\eta - j\kappa)^2 - 1}{(\eta - j\kappa)^2 + 2} = \frac{4\pi}{3}N\langle\alpha\rangle \tag{7.2.29}$$

In the case of a transparent material for which $\kappa = 0$

$$\langle\alpha\rangle = \frac{3}{4\pi N}\left(\frac{\varepsilon(\omega) - 1}{\varepsilon(\omega) + 2}\right) = \frac{3}{4\pi N}\left(\frac{\eta^2 - 1}{\eta^2 + 2}\right) \tag{7.2.30}$$

This relation is known as the Lorentz–Lorentz formula. The limit $\omega \to 0$ gives the Clausius–Mossotti formula

$$\frac{\varepsilon(0) - 1}{\varepsilon(0) + 2} = \frac{4\pi N}{3}\langle\alpha(0)\rangle \tag{7.2.31}$$

7.2.3 METALS

On classical theory a simple description of electrical conduction in metals can be given by assuming that the current is carried by a set of identical 'model' electrons each moving freely (or almost freely) through the lattice and each being subject to a dissipative damping force proportional to its velocity. A 'model' electron represents the average behaviour of the complete set of mobile electrons; it is no longer bound to an equilibrium position and so we may write its equation of motion in the form

$$m\ddot{\xi} + m\gamma\dot{\xi} = qE_0\,\mathrm{e}^{\mathrm{j}\omega t} \tag{7.2.32}$$

from which

$$\dot{\xi} = \frac{(q/m)E_0\,\mathrm{e}^{\mathrm{j}\omega t}}{\gamma + \mathrm{j}\omega} \tag{7.2.33}$$

The current density is (from equations (1.1.10), and (1.5.2)),

$$\mathbf{J} = \frac{1}{c}\rho\mathbf{v} = \frac{Nq}{c}\dot{\boldsymbol{\xi}} = \frac{\sigma}{c}\mathbf{E} \tag{7.2.34}$$

where N is the number of 'model' electrons per unit volume. Evidently

$$\sigma(\omega) = \frac{(Nq^2/\gamma m)}{1 + (\mathrm{j}\omega/\gamma)} \equiv \frac{\sigma_0}{1 + (\mathrm{j}\omega/\gamma)} \tag{7.2.35}$$

The constant σ_0 is evidently the d.c. conductivity.

From equation (2.1.7), the effective dielectric constant for a metal is

$$\varepsilon_c(\omega) = \varepsilon(\omega) - \frac{\mathrm{j}4\pi\sigma(\omega)}{\omega} \tag{7.2.36}$$

and so for a non-magnetic metal we may combine equations (7.2.11) and (7.2.35) and write

$$(\eta - \mathrm{j}\kappa)^2 = \varepsilon_c(\omega) = 1 + \frac{4\pi q^2}{m}\sum_i \frac{N_i}{(\omega_i^2 - \omega^2) + \mathrm{j}\gamma_i\omega} - \mathrm{j}\frac{4\pi}{\omega}\cdot\frac{(Nq^2/\gamma m)}{1 + \mathrm{j}\omega/\gamma} \tag{7.2.37}$$

In this equation the term involving $\sum_i$ represents the effect of the electrons which are bound to the atomic cores. Equation (7.2.11) has been used rather than equation (7.2.24) because of the screening effect of the conduction electrons.

The relationship for the refractive index given in equation (7.2.37) is evidently quite complicated but some limiting cases are of interest. First separate equation (7.2.37) into its real and imaginary parts,

$$\eta^2 - \kappa^2 = 1 + \frac{4\pi q^2}{m}\sum_i \frac{N_i(\omega_i^2 - \omega^2)}{(\omega_i^2 - \omega^2)^2 + \gamma_i^2\omega^2} + \frac{4\pi(Nq^2/\gamma^2 m)}{1 + (\omega/\gamma)^2} \tag{7.2.38}$$

$$2\eta\kappa = \frac{4\pi q^2}{m}\sum_i \frac{N_i\gamma_i\omega}{(\omega_i^2 - \omega^2)^2 + \gamma_i^2\omega^2} + \frac{(4\pi/\omega)(Nq^2/\gamma m)}{1 + (\omega/\gamma)^2} \tag{7.2.39}$$

In the low-frequency region where $4\pi\sigma/\omega \gg 1$, $\omega/\gamma \ll 1$, we have

$$\eta^2 - \kappa^2 \approx 1 + \frac{4\pi\sigma_0}{\gamma} \tag{7.2.40}$$

$$\eta\kappa \approx \frac{2\pi}{\omega}\left(\frac{Nq^2}{\gamma m}\right) = \frac{2\pi\sigma_0}{\omega} \gg 1 \tag{7.2.41}$$

Hence the product $\eta\kappa$ is very large whilst the difference $\eta^2 - \kappa^2$ is of order unity and small compared with $\eta\kappa$ and so

$$\eta \sim \kappa \sim \sqrt{\left(\frac{2\pi\sigma_0}{\omega}\right)} \tag{7.2.42}$$

This is the relation found previously (compare page 97) on the macroscopic theory.†

Equation (7.2.42) shows that at low frequencies the propagation effects are determined by the mobile electrons and the bound electrons make no significant contribution. At high frequencies, however, the bound electrons are important because of their resonant effects. Thus when $\omega \gg \gamma$ the metal begins to behave like an insulating dielectric. At sufficiently high frequencies (usually beyond the ultraviolet) when ω is greater than any resonant frequency ω_i,

$$\begin{aligned} \eta^2 - \kappa^2 &\rightarrow 1 \\ \eta\kappa &\rightarrow 0 \end{aligned} \tag{7.2.43}$$

so that

$$\eta \rightarrow 1 \quad \text{and} \quad \kappa \rightarrow 0$$

the metal becomes transparent.

These effects are exhibited by the alkali metals. For low frequencies they are opaque and highly reflecting whereas above some critical frequency they

TABLE 7.1

Critical wavelengths and frequencies for the alkali metals

Metal	*Wavelength* (Å)	*Frequency* (c/s)
Lithium	2050	$1{\cdot}46 . 10^{15}$
Sodium	2100	$1{\cdot}43 . 10^{15}$
Potassium	3150	$0{\cdot}95 . 10^{15}$
Rubidium	3600	$0{\cdot}83 . 10^{15}$
Caesium	4400	$0{\cdot}68 . 10^{15}$

At longer wavelengths (lower frequencies) the metals are highly reflecting and strongly absorbing whereas at shorter wavelengths (higher frequencies) the metals become transparent.

† We have taken the limit $\omega \rightarrow 0$ and so $\sigma(\omega) \rightarrow \sigma_0$ in equation (7.2.42).

become reasonably transparent and have comparatively low absorption. Table 7.1 shows the frequencies and wavelengths at which these transitions occur.

7.3 Classical Theory of Polar Molecules

Consider an assembly of a large number of identical molecules each having a permanent dipole moment $\mathbf{p}_0$. Assume that interactions between the molecules are so weak that they may be neglected in so far as they may correlate the orientations of the dipoles but are sufficient (for example in the form of collision processes in gases or lattice interactions in solids) to ensure thermal equilibrium. Without an external field the molecules are oriented randomly and the assembly has no resultant polarization. In an applied field $\mathbf{E}$ the molecules tend to orient along $\mathbf{E}$ but are also subject to the randomizing effects of thermal agitation processes.

Consider the thermal equilibrium state of an assembly of molecules at temperature T K in an applied electric field $\mathbf{E}$. The potential energy of a molecule is, from equation (1.2.46),

$$U = -\mathbf{p}_0 . \mathbf{E} = -p_0 E \cos\theta \tag{7.3.1}$$

where θ is the angle between the moment and the field direction. The polarization for the assembly is

$$\mathbf{P} = N p_0 \langle\cos\theta\rangle\, \mathbf{n} \tag{7.3.2}$$

Here N is the number of molecules per unit volume, $\mathbf{n}$ is unit vector along $\mathbf{E}$, and $\langle\cos\theta\rangle$ is the average of $\cos\theta$ over the distribution in thermal equilibrium. According to the Boltzmann distribution law the relative probability of a molecule being oriented in the range θ to $\theta + \mathrm{d}\theta$ is proportional to

$$\mathrm{e}^{-U/kT} . 2\pi \sin\theta\, \mathrm{d}\theta$$

Hence

$$\langle\cos\theta\rangle = \frac{\displaystyle\int_0^{\pi} [\exp\{(p_0 E \cos\theta)/kT\}] 2\pi \sin\theta \cos\theta\, \mathrm{d}\theta}{\displaystyle\int_0^{\pi} [\exp\{(p_0 E \cos\theta)/kT\}] 2\pi \sin\theta\, \mathrm{d}\theta} \tag{7.3.3}$$

To evaluate these integrals write

$$x = \cos\theta \qquad \frac{p_0 E}{kT} = a \tag{7.3.4}$$

$$\langle\cos\theta\rangle = \frac{\displaystyle\int_{-1}^{1} x\, \mathrm{e}^{ax}\, \mathrm{d}x}{\displaystyle\int_{-1}^{1} \mathrm{e}^{ax}\, \mathrm{d}x} \tag{7.3.5}$$

$$= \frac{\mathrm{e}^{a} + \mathrm{e}^{-a}}{\mathrm{e}^{a} - \mathrm{e}^{-a}} - \frac{1}{a} \tag{7.3.6}$$

$$= \coth a - \frac{1}{a} \equiv L(a) \tag{7.3.7}$$

The function $L(a)$ is called the Langevin function. It was originally derived by Langevin and used to account for the susceptibility of an assembly of magnetic dipoles which, on classical theory are described by the same formal equations used above for the electric case.†

Experimentally the most important applications of equation (7.3.7) occur when

$$a \ll 1 \qquad p_0 E \ll kT \tag{7.3.8}$$

This condition is usually easily satisfied since dipole moments are of order 10^{-18} e.s.u. so that for $E = 300$ V cm^{-1}, $p_0 E \sim 10^{-18}$ ergs, but $kT \sim 4.10^{-14}$ ergs at room temperature and so $a = p_0 E/kT \sim 0.25.10^{-4} \ll 1$. For $a \ll 1$

$$L(a) = \frac{a}{3} = \frac{p_0 E}{3kT} = \langle \cos \theta \rangle \tag{7.3.9}$$

and the polarization is obtained from equation (7.3.2)

$$\mathbf{P} = \frac{Np_0^2}{3kT}\mathbf{E} \tag{7.3.10}$$

7.3.1 Polar Gases

Equation (7.3.10) may be applied directly to polar gases. In a static field the total polarization will be the sum of the electronic contribution from equation (7.2.9) together with the dipolar contribution from equation (7.3.10)

$$\mathbf{P} = \left\{\sum_i \frac{N_i(q^2/m)}{\omega_i^2} + \frac{Np_0^2}{3kT}\right\}\mathbf{E}(0) \tag{7.3.11}$$

so that

$$\varepsilon(0) = 1 + \frac{4\pi q^2}{m}\sum_i \frac{N_i}{\omega_i^2} + \frac{4\pi N}{3kT}p_0^2 \tag{7.3.12}$$

The dipole moment p_0 may be determined by plotting $\varepsilon(0)$ as a function of $1/T$. Typical values for dipole moments are:‡

molecule	p_0 *e.s.u.*	*SI* (cm)
H_2O	$1{\cdot}82 \times 10^{-18}$	$6{\cdot}07 \times 10^{-30}$
HCl	$1{\cdot}07 \times 10^{-18}$	$3{\cdot}57 \times 10^{-30}$
NH_3	$1{\cdot}47 \times 10^{-18}$	$4{\cdot}90 \times 10^{-30}$
N_2O	$0{\cdot}18 \times 10^{-18}$	$0{\cdot}60 \times 10^{-30}$
SO_2	$1{\cdot}61 \times 10^{-18}$	$5{\cdot}37 \times 10^{-30}$
CO	$0{\cdot}13 \times 10^{-18}$	$0{\cdot}43 \times 10^{-30}$
CO_2	$0{\cdot}18 \times 10^{-18}$	$0{\cdot}60 \times 10^{-30}$
LiH	$5{\cdot}88 \times 10^{-18}$	$19{\cdot}61 \times 10^{-30}$

10^{-18} e.s.u. $= 1$ Debye unit $= 3{\cdot}335 \times 10^{-30}$ cm

† See Section 8.2.

‡ See, for example, McClellan, A. L., *Tables of Experimental Dipole Moments*, Freeman, 1963.

Evidently dipole moments are of order 10^{-18} e.s.u. which corresponds to the displacement of one electronic charge ($4{\cdot}8 \, . \, 10^{-18}$ e.s.u.) by about $0{\cdot}25 \, . \, 10^{-8}$ cm.

Since the molecules in a gas may be regarded as being essentially free, the dipolar contribution to the dielectric constant should exhibit no significant frequency dependence. The losses should be correspondingly small up to the infra red absorption frequencies of the molecules. In this respect polar gases differ considerably from polar liquids and solids.

7.3.2 Polar Liquids and Solids

The general discussion leading to equation (7.3.10) provides the basis for a microscopic model of polar liquids and solids. In these materials, however, it is necessary to evaluate the appropriate local field and also to take account of relaxation processes. We shall consider these two effects separately.

The local field is no longer adequately described by the Lorentz expression given in equation (7.2.23) for induced electric moments. In deriving that equation it was assumed that the dipole moments within the spherical cavity were all parallel. Although this is a good assumption for induced moments it is evident that it will not be satisfactory for the case of permanent dipole moments in the limit we are considering $\{(p_0E/kT) \ll 1\}$. For this case the dipoles are oriented more or less randomly. Onsager has suggested that the local field should be evaluated on the assumption that a particular dipole can be regarded as being situated within a real spherical cavity just large enough to contain one molecule. Outside the cavity the material is assumed to be continuous and at large distances from the cavity the field is assumed to be uniform and equal to $\mathbf{E}_0$.† Take as potential functions inside and outside the sphere of radius a,

$$\phi_i = Ar\cos\theta \qquad r \leqslant a \tag{7.3.13}$$

$$\phi_0 = \left(Br + \frac{b}{r^2}\right)\cos\theta \qquad r \geqslant a \tag{7.3.14}$$

where θ is the angle between $\mathbf{r}$ and $\mathbf{E}_0$. The boundary conditions at the surface of the sphere require that the normal component of $\mathbf{D}$ and the tangential component of $\mathbf{E}$ should be continuous (equations (2.5.18)). Assuming free space within the cavity we obtain, upon writing $\mathbf{E}_0 = \hat{\mathbf{k}}E_0$,

$$B = -E_0$$

$$A = -\frac{3\varepsilon}{2\varepsilon + 1}E_0$$

$$b = -a^3\frac{(\varepsilon - 1)}{2\varepsilon + 1}E_0 \tag{7.3.15}$$

Thus

$$\phi_i = -\frac{3\varepsilon}{2\varepsilon + 1}E_0 z \tag{7.3.16}$$

† In an actual experiment $\mathbf{E}_0$ will be related to the external field through depolarization factors dependent upon the sample shape.

where the z coordinate axis is taken along $\mathbf{E}_0$. The local field is therefore along $\mathbf{E}_0$ and given by

$$\mathbf{E}' = \frac{3\varepsilon}{2\varepsilon + 1}\mathbf{E}_0 \tag{7.3.17}$$

The polarization is now obtained from equation (7.3.10)

$$\mathbf{P} = \frac{Np_0^2}{3kT}\mathbf{E}' = \frac{Np_0^2}{3kT}\cdot\frac{3\varepsilon}{2\varepsilon + 1}\mathbf{E}_0 \tag{7.3.18}$$

But

$$\mathbf{P} = \frac{\varepsilon - 1}{4\pi}\mathbf{E}_0 \tag{7.3.19}$$

from which

$$(\varepsilon - 1) = \frac{3\varepsilon x}{2\varepsilon + 1}$$

with

$$x = \frac{4\pi N}{3kT}p_0^2$$

Hence

$$2\varepsilon^2 - \varepsilon(1 + 3x) - 1 = 0 \tag{7.3.20}$$

$$\varepsilon(0) = \tfrac{1}{4}\{1 + 3x + \sqrt{(9x^2 + 6x + 9)}\} \tag{7.3.21}$$

This calculation for ε involves an extreme assumption in the introduction of a real free space cavity at the site of a molecule. Although this is an over-simplified model it does have the advantage of removing in a straightforward manner the cooperative transition which arises if the Lorentz field is used. Thus if we had used the local field $\mathbf{E}'$ from equation (7.2.23), in the equation

$$\mathbf{P} = \frac{Np_0^2}{3kT}\mathbf{E}' \tag{7.3.22}$$

then

$$\frac{Np_0^2}{3kT} = \frac{\mathbf{P}}{\mathbf{E}'} = \frac{\mathbf{P}}{\mathbf{E} + (4\pi/3)\mathbf{P}} = \frac{3}{4\pi}\cdot\frac{\varepsilon - 1}{\varepsilon + 2} \tag{7.3.23}$$

and so,

$$\varepsilon = 1 + \frac{Np_0^2/3k}{T - T_c} \tag{7.3.24}$$

with $T_c = Np_0^2/9k$.

The dipole moment of water is $1{\cdot}87.10^{-18}$ e.s.u. and so $T_c \sim 1000$ K which is in contradiction with experiment.

On the other hand, if the dielectric constant is near to unity the Onsager and Lorentz fields become equivalent. Thus, from equation (7.3.17)

$$\mathbf{E}' = \frac{3\varepsilon}{2\varepsilon + 1}\mathbf{E}_0 \tag{7.3.25}$$

and writing $\varepsilon = 1 + \Delta$ where Δ is a small quantity

$$\mathbf{E}' = \left(1 + \frac{\Delta}{3}\right)\mathbf{E}_0 \qquad \text{Onsager} \qquad (7.3.26)$$

For the Lorentz field, from equation (7.2.23)

$$\mathbf{E}' = \left(1 + \frac{\varepsilon - 1}{3}\right)\mathbf{E}_0 \qquad (7.3.27)$$

$$\mathbf{E}' = \left(1 + \frac{\Delta}{3}\right)\mathbf{E}_0 \qquad \text{Lorentz} \qquad (7.3.28)$$

which is the same value as given above for the Onsager local field.

In polar liquids and solids there is usually a considerable frequency dependence of the dielectric constant. For liquids the real part of the dielectric constant decreases over a broad frequency range usually in the region of 10^{10} c/s whereas in solids the decrease may take place at much lower frequencies, say 10^6 c/s (in ice at $-20°$C the decrease is observed in the kilocycle range). Considerable dielectric losses occur in the frequency range where the real part of the dielectric constant is changing rapidly.

These effects are usually explained in terms of relaxation processes. They were first investigated by Debye and the theory is named after him. The dielectric is to be regarded as a dilute distribution of dipoles for which interactions between the dipoles can be neglected. In this case the local electric field acting on a molecule is equal to the macroscopic field **E** within the material. Consider a simple model in which the dipole moment has two allowed energy states. We suppose that the dipole moment may be oriented (a) parallel to **E** (b) antiparallel to **E**. If, at any instant of time there are $N_1(t)$, $N_2(t)$ molecules per unit volume in the two groups and the probability that a dipole in group (1) makes a transition to group (2) in a time δt is $a_{12}\,\delta t$, whilst the probability for the reverse process is $a_{21}\,\delta t$, the rate equations for $N_1(t)$ and $N_2(t)$ are,

$$\frac{\mathrm{d}N_1(t)}{\mathrm{d}t} = -N_1 a_{12} + N_2 a_{21}$$
$$\frac{\mathrm{d}N_2(t)}{\mathrm{d}t} = N_1 a_{12} - N_2 a_{21} \qquad (7.3.29)$$

For equilibrium,

$$\frac{\mathrm{d}N_1}{\mathrm{d}t} = \frac{\mathrm{d}N_2}{\mathrm{d}t} = 0 \qquad (7.3.30)$$

and so

$$\frac{N_1}{N_2} = \frac{a_{21}}{a_{12}} \qquad \text{(equilibrium)} \qquad (7.3.31)$$

But N_1, N_2 are also related by the Boltzmann distribution, for thermal equilibrium,

$$N_1 = A \exp(p_0 E/kT); \quad N_2 = A \exp(-p_0 E/kT) \qquad (7.3.32)$$

we may, therefore, write

$$a_{12} = \left(\frac{1}{2\beta}\right) \exp\,(-p_0E/kT); \quad a_{21} = \left(\frac{1}{2\beta}\right) \exp\,(p_0E/kT) \tag{7.3.33}$$

where β is a constant which will later be seen to play the role of a relaxation time. Taking $p_0E \ll kT$, equations (7.3.29) give

$$\begin{aligned} 2\beta\frac{\mathrm{d}N_1}{\mathrm{d}t} &= -(N_1 - N_2) + \frac{p_0E}{kT}(N_1 + N_2) \\ 2\beta\frac{\mathrm{d}N_2}{\mathrm{d}t} &= (N_1 - N_2) - \frac{p_0E}{kT}(N_1 + N_2) \end{aligned} \tag{7.3.34}$$

$$\beta\frac{\mathrm{d}}{\mathrm{d}t}(N_1 - N_2) + (N_1 - N_2) = \frac{N_0p_0}{kT}E \tag{7.3.35}$$

where $N_0 \equiv (N_1 + N_2)$ is the total number of dipoles per unit volume.

We may solve equation (7.3.35) using the Laplace Transform method (compare Chapters 4 and 5), the field $\mathbf{E}$ being switched on at time $t = 0$ and the initial conditions being $N_1(0^-) = N_2(0^-) = N_0/2$. Upon writing $n(t) = (N_1 - N_2)$, with $n(0^-) = 0$, and taking the Laplace Transform of equation (7.3.35) we obtain

$$(\beta s + 1)n(s) = \frac{N_0p_0}{kT}E(s)$$

$$n(s) = \frac{N_0p_0}{kT}\cdot\frac{E(s)}{1 + \beta s} \tag{7.3.36}$$

If $E(t)$ is a steady field, $E(t) = E_0\,u(t)$ and $E(s) = (1/s)E_0$. For this case

$$n(s) = \frac{N_0p_0}{kT}\,E_0\left\{\frac{1}{s} - \frac{1}{s + (1/\beta)}\right\}$$

$$n(t) = N_1 - N_2 = \frac{N_0p_0}{kT}E_0\left(1 - \mathrm{e}^{-t/\beta}\right)u(t) \tag{7.3.37a}$$

The polarization is proportional to $(N_1 - N_2)$ and so approaches its equilibrium value exponentially with a characteristic time constant β. When the field is switched off the polarization decays exponentially with the same characteristic time factor. If the field $E(t)$ is an impulse at $t = 0$, $E(t) = E_0\,\delta(t)$, and $E(s) = E_0$. For this case

$$n(s) = \frac{N_0p_0}{kT}\cdot\frac{E_0}{\beta}\cdot\frac{1}{s + (1/\beta)}$$

$$n(t) = (N_1 - N_2) = \frac{1}{\beta}\cdot\frac{N_0p_0}{kT}E_0\,\mathrm{e}^{-t/\beta}\,u(t) \tag{7.3.37b}$$

Again the polarization varies exponentially with the characteristic time constant β. This form of time dependence was used previously when discussing the Debye equations on page 27. If $E(t)$ has a harmonic time dependence $E(t) = E_0\, e^{j\omega t}\, u(t)$ and $E(s) = E_0/(s - j\omega)$

$$n(s) = \frac{N_0 p_0}{kT} E_0 \left(\frac{1/\beta}{s - j\omega} - \frac{1}{1 + \beta s} \right) \cdot \frac{1}{(1/\beta) + j\omega}$$

which leads to the steady state solution

$$n(t) = N_1 - N_2 = \frac{(N_0 p_0/kT)}{1 + j\omega\beta} E_0\, e^{j\omega t} \tag{7.3.37c}$$

upon using relation 20 of Table 4.2.

The electric susceptibility $\chi^e(\omega)$ is obtained directly from the polarization **P**,

$$\mathbf{P} = (N_1 - N_2) p_0 \mathbf{n}$$

where **n** is unit vector along **E**. From equation (7.3.37c), the complex susceptibility is,

$$\chi^{(e)}(\omega) = \frac{\mathbf{P}}{\mathbf{E}} = \frac{\varepsilon(\omega) - 1}{4\pi} = \frac{N_0 p_0^2/kT}{1 + j\omega\beta} \tag{7.3.38}$$

This is the Debye equation for the electric susceptibility given previously (equation (1.3.52)). From this equation we may write

$$\varepsilon(\omega) = 1 + \frac{\varepsilon(0) - 1}{1 + j\omega\beta} \tag{7.3.39}$$

but

$$\varepsilon(\omega) = \varepsilon_1(\omega) - j\varepsilon_2(\omega)$$

and consequently

$$\varepsilon_1(\omega) = 1 + \frac{\varepsilon(0) - 1}{1 + \omega^2\beta^2} \tag{7.3.40a}$$

$$\varepsilon_2(\omega) = \frac{\{\varepsilon(0) - 1\}\omega\beta}{1 + \omega^2\beta^2} \tag{7.3.40b}$$

These functions are sketched in Fig. 7.3.

It should be noted that $\varepsilon_1(\omega)$, $\varepsilon_2(\omega)$ depend on at least two parameters; the angular frequency ω and the temperature T. The frequency dependence is indicated explicitly in equations (7.3.40) but the temperature dependence only appears implicitly through $\varepsilon(0)$ and β. The function $\beta(T)$ is usually unknown and has to be determined experimentally. This is most directly carried out by evaluating the frequency at which $\varepsilon_2(\omega)$ is a maximum. If this frequency is $\omega_m(T)$ at temperature T, we obtain from equation (7.3.40b), with $\partial\varepsilon_2/\partial\omega = 0$,

$$\omega_m(T) = \frac{1}{\beta(T)} \tag{7.3.41}$$

As a further refinement of the theory we should observe that when effects due to the induced moments are included the appropriate high-frequency limit for equation (7.3.39) is not unity but ε_i. Equation (7.3.39) should now be rewritten using the relation

$$\varepsilon(\omega) = 1 + 4\pi\chi_i^{(e)} + 4\pi\chi_{dip}^{(e)} \tag{7.3.42}$$

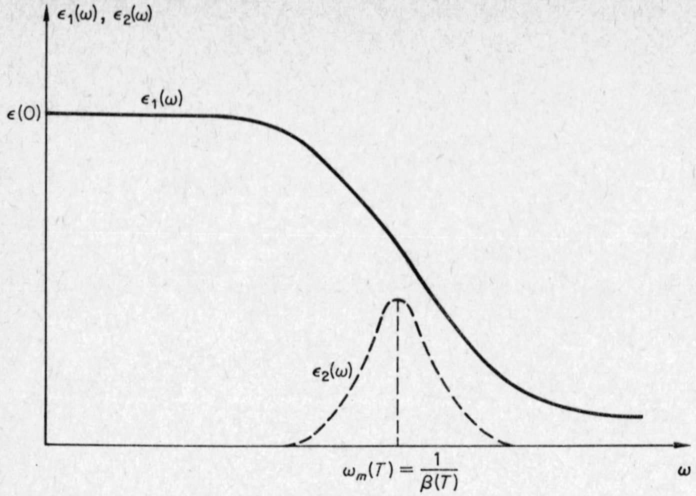

FIG. 7.3 Sketch of the Debye functions for $\varepsilon_1(\omega)$, $\varepsilon_2(\omega)$.

Here, $\chi_{\mathrm{i}}^{(\mathrm{e})}$, refers to the susceptibility at frequencies where the dipolar contribution is essentially zero (for example in the visible region) but below the range where absorption lines occur. We therefore write

$$\varepsilon(\omega) = \varepsilon_{\mathrm{i}} + 4\pi\chi_{\mathrm{dip}}^{(\mathrm{e})}(\omega) \tag{7.3.43a}$$

but

$$\varepsilon(0) = \varepsilon_{\mathrm{i}} + 4\pi\chi_{\mathrm{dip}}^{(\mathrm{e})}(0) \tag{7.3.43b}$$

and so equation (7.3.39) becomes

$$\varepsilon(\omega) = \varepsilon_{\mathrm{i}} + \frac{\varepsilon(0) - \varepsilon_{\mathrm{i}}}{1 + \mathrm{j}\omega\beta} \tag{7.3.44a}$$

Equations (7.3.40) may now be rewritten as

$$\omega^2\beta^2 = \frac{\varepsilon(0) - \varepsilon_{\mathrm{i}}}{\varepsilon_1(\omega) - \varepsilon_{\mathrm{i}}} - 1 \tag{7.3.44b}$$

$$\varepsilon_2(\omega) = \sqrt{[\{\varepsilon(0) - \varepsilon_1(\omega)\}\{\varepsilon_1(\omega) - \varepsilon_{\mathrm{i}}\}]} \tag{7.3.44c}$$

a representation which shows that when the Debye equations are valid β and $\varepsilon_2(\omega)$ can be obtained directly from measurements of real parts of the dielectric constant.

Finally note that this simplified discussion cannot be extended immediately to the situation where the dipoles interact with one another. When this is the case the transition probabilities a_{12}, a_{21} in equation (7.3.33) depend upon the local field $\mathbf{E}'$ which itself will depend upon $(N_1 - N_2)$. Equation (7.3.35) may no longer be a straightforward linear differential equation.

7.4 Quantum Theory of the Dielectric Constant

In this section we shall consider a microscopic theory of the dielectric constant from the standpoint of quantum theory. The discussion is based upon the simple atomic model for a material medium which was used previously in the classical theory but we now use quantum mechanics to describe the properties of the individual molecules. This is not therefore a comprehensive quantum mechanical treatment of the problem. The radiation field is still described by the classical electromagnetic potentials and the molecules are regarded as essentially isolated units whose translational motion can be ignored. In particular the contributions to the dielectric constant from the lattice waves in a solid are not treated in this discussion (see Section 7.6).

The technique for solving the Schrödinger equation for a molecule in an electromagnetic field assumes that in the absence of the field there exists a time independent equation which can be solved exactly. The presence of the electromagnetic field gives rise to a time dependent feature which may be regarded as a small disturbance of the initial state. The new equilibrium solution may now be sought by a perturbation method. The particular perturbation method used is sometimes called 'the method of variation of constants'.

7.4.1 Time Dependent Perturbation Theory

Consider the Schrödinger equation including the time†

$$\mathscr{H}(\mathbf{r}, \mathbf{p}, t)\Psi = \mathrm{j}\hbar \frac{\partial \Psi}{\partial t} \tag{7.4.1}$$

Perturbation theory requires that we are able to write the Hamiltonian operator $\mathscr{H}(\mathbf{r}, \mathbf{p}, t)$ as the sum of two terms

$$\mathscr{H}(\mathbf{r}, \mathbf{p}, t) = \mathscr{H}_0(\mathbf{r}, \mathbf{p}) + \mathscr{H}'(\mathbf{r}, \mathbf{p}, t) \tag{7.4.2}$$

Here $\mathscr{H}_0$ is the Hamiltonian for the unperturbed system whilst $\mathscr{H}'$ is small. The unperturbed eigenfunctions $\Psi^0(\mathbf{r}, t)$ satisfy

$$\mathscr{H}_0 \Psi^0 = \mathrm{j}\hbar \frac{\partial \Psi^0}{\partial t} \tag{7.4.3}$$

which, upon taking $\Psi^0(\mathbf{r}, t) = \psi^0(r) f(t)$, has solutions of the form

$$\Psi_n^0(\mathbf{r}, t) = \psi_n^0(\mathbf{r}) \exp\{-\mathrm{j}(E_n/\hbar)t\} \tag{7.4.4}$$

with E_n representing the nth energy eigenvalue of $\mathscr{H}_0$.

To find a solution for equation (7.4.1), expand Ψ in terms of Ψ_n^0 with expansion coefficients which depend upon the time,

$$\Psi(\mathbf{r}, t) = \sum_n a_n(t)\, \psi_n^0(\mathbf{r}) \exp\{-\mathrm{j}(E_n/\hbar)t\} \tag{7.4.5}$$

From equation (7.4.1),

$$\sum_n a_n(\mathscr{H}_0 + \mathscr{H}')\, \psi_n^0(\mathbf{r}) \exp\{-\mathrm{j}(E_n/\hbar)t\}$$

$$= \sum_n (\mathrm{j}\hbar \dot{a}_n(t) + a_n(t)E_n)\, \psi_n^0(\mathbf{r}) \exp\{-\mathrm{j}(E_n/\hbar)t\} \tag{7.4.6}$$

† Throughout the discussion we shall assume that the Ψ's and the ψ's are correctly normalized. For a general reference see, for example, Schiff, L. I., *Quantum Mechanics*, McGraw-Hill, 1949.

Since the eigenvalues E_n are obtained from

$$\mathscr{H}_0\, \psi_n^0(\mathbf{r}) = E_n\, \psi_n^0(\mathbf{r}) \tag{7.4.7}$$

equation (7.4.6) reduces to

$$\sum_n a_n(t)\, \mathscr{H}'\, \psi_n^0(r) \exp\{-\mathrm{j}(E_n/\hbar)t\} = \mathrm{j}\hbar \sum_n \dot{a}_n(t)\, \psi_n^0(\mathbf{r}) \exp\{-\mathrm{j}(E_n/\hbar)t\} \tag{7.4.8}$$

Multiply this equation throughout from the left by $(\Psi_m^0)^*$ and integrate over all coordinate space. Using the normalization and orthogonality of the $\psi_n^0(\mathbf{r})$ we obtain

$$\begin{aligned} \mathrm{j}\hbar\, \dot{a}_m(t) &= \sum_n a_n(t) \exp[\mathrm{j}\{(E_m - E_n)/\hbar\}t] \int \psi_m^{0*}\, \mathscr{H}'\, \psi_n^0\, \mathrm{d}^3\mathbf{r}, \\ &= \sum_n a_n(t) \exp[\mathrm{j}\{(E_m - E_n)/\hbar\}t] \langle m \mid \mathscr{H}' \mid n\rangle \end{aligned} \tag{7.4.9}$$

In a particular problem the perturbation approximation assumes that initially all the a_n are zero except for one, namely a_k, which we take to be unity. Because of the smallness of $\mathscr{H}'$, the $\dot{a}_m(t)$ and hence the $a_m(t)$ are also small whilst a_k will remain close to unity. To a first approximation therefore we may solve equation (7.4.9) for $a_m(t)$ by putting $a_k = 1$ and all other a_n zero. From equation (7.4.9)

$$\mathrm{j}\hbar\, \dot{a}_m(t) = \exp[\mathrm{j}\{(E_m - E_k)/\hbar\}t] \langle m |\mathscr{H}'| k\rangle \tag{7.4.10}$$

$$a_m(t) = \frac{1}{\mathrm{j}\hbar} \int \exp[\mathrm{j}\{(E_m - E_n)/\hbar\}t] \langle m|\mathscr{H}'|k\rangle dt \tag{7.4.11}$$

These equations will be satisfactory until the terms as calculated become large. They are entirely satisfactory for our purposes.

Our particular problem requires the evaluation of the electric dipole moment for a 'molecule'. On classical theory the electric dipole moment for a discrete charge distribution (compare equation (1.3.8) is given by

$$\mathbf{p}_e = \sum_i q_i \mathbf{r}_i \tag{7.4.12}$$

where $\mathbf{r}_i$ is the position vector for the ith charge. The position operator in quantum mechanics takes the same form as in the classical representation and so the electric dipole moment for a 'molecule' according to quantum mechanics is

$$\mathbf{p}_e = \int \Psi^* \Big(\sum_i q_i \mathbf{r}_i\Big) \Psi\, \mathrm{d}^3\mathbf{r} \tag{7.4.13}$$

Writing

$$\Psi = \Psi_k^0 + \sum_m a_m(t)\, \Psi_m^0 \tag{7.4.14}$$

we obtain

$$\mathbf{p}_e = \int \Psi_k^{0*}\Big(\sum_i q_i\mathbf{r}_i\Big)\,\Psi_k^0\,\mathrm{d}^3\mathbf{r}$$
$$+\int \Psi_k^{0*}\Big(\sum_i q_i\mathbf{r}_i\Big)\Big(\sum_m a_m(t)\,\Psi_m^0\Big)\,\mathrm{d}^3\mathbf{r}$$
$$+\int\Big(\sum_m a_m^*(t)\,\Psi_m^{0*}\Big)\Big(\sum_i q_i\mathbf{r}_i\Big)\Psi_k^0\,\mathrm{d}^3\mathbf{r} \qquad (7.4.15)$$

Here the small term involving products of the form $a_m\, a_l$ has been neglected because we are using a perturbation technique.

In order to use equation (7.4.15) it is necessary to determine the coefficients $a_m(t)$ from equation (7.4.11). To do this, however, we require the perturbation Hamiltonian $\mathscr{H}'$ which describes the effect of the electromagnetic field. This function will be discussed in the next section.

7.4.2 The Hamiltonian for a Particle in an Electromagnetic Field

The electromagnetic field may be described by the vector and scalar potential functions $\mathbf{A}(\mathbf{r}, t)$, $\phi(\mathbf{r}, t)$ which were introduced in Section 2.11, Chapter 2. Write

$$\mathbf{H} = \nabla \wedge \mathbf{A}(\mathbf{r}, t) \qquad (7.4.16)$$

$$\mathbf{E} = -\frac{1}{c}\frac{\partial}{\partial t}\mathbf{A}(\mathbf{r}, t) - \nabla\phi(\mathbf{r}, t) \qquad (7.4.17)$$

The classical Hamiltonian function for a particle with charge q is

$$\mathscr{H} = \frac{\{\mathbf{p} - (q\mathbf{A}/c)\}^2}{2m} + V(\mathbf{r}) + q\phi \qquad (7.4.18)$$

where $V(\mathbf{r})$ is that part of the potential energy which is non-electromagnetic in origin. The quantum-mechanical Hamiltonian which will operate on a wave function Ψ is obtained from equation (7.4.18). Recall that $\mathbf{p}$ is now a differential operator,

$$\mathscr{H}\Psi = \left(\frac{\mathbf{p}.\mathbf{p}}{2m} - \frac{q}{2mc}\mathbf{A}.\mathbf{p} - \frac{q}{2mc}\mathbf{p}.\mathbf{A} + \frac{q^2}{2mc^2}\mathbf{A}.\mathbf{A} + V(\mathbf{r}) + q\phi\right)\Psi \qquad (7.4.19)$$

Write $\mathbf{p} = -\mathrm{j}\hbar\nabla$

$$\mathscr{H}\Psi = \left\{-\frac{\hbar^2}{2m}\nabla^2 + V(\mathbf{r}) + q\phi + \frac{\mathrm{j}q\hbar}{mc}\mathbf{A}.\nabla + \frac{\mathrm{j}q\hbar}{2mc}(\nabla.\mathbf{A}) + \frac{q^2\mathbf{A}.\mathbf{A}}{2mc^2}\right\}\Psi \qquad (7.4.20)$$

For an electromagnetic wave we may take

$$\nabla.\mathbf{A} = 0 \quad \phi = 0 \qquad (7.4.21)$$

(compare the discussion on page 101) and also neglect the term in $\mathbf{A}.\mathbf{A}$ since we regard the field as a small perturbation. Equation (7.4.20) reduces to

$$\mathscr{H}\Psi = \left\{-\frac{\hbar^2}{2m}\nabla^2 + V(\mathbf{r}) + \frac{\mathrm{j}q\hbar}{mc}\mathbf{A}.\nabla\right\}\Psi \qquad (7.4.22)$$

This is the basic relation required for our discussion of dielectric phenomena. We are, of course, neglecting spin in this treatment. The effects of spin will be discussed later in connection with the Faraday effect. In the notation of equation (7.4.2) we shall evidently write the Hamiltonian operator as

$$\mathscr{H}_0(\mathbf{r}, \mathbf{p}) = -\frac{\hbar^2}{2m}\nabla^2 + V(\mathbf{r})$$
$$\mathscr{H}'(\mathbf{r}, \mathbf{p}, t) = \frac{\mathrm{j}q\hbar}{mc}\mathbf{A}.\nabla \tag{7.4.23}$$

7.4.3 Induced Moments

The electric susceptibility and the dielectric constant for a material medium may now be derived using the simple atomic model outlined previously in connection with the classical theory for these quantities. The moments induced in the fundamental molecular units by the electromagnetic field can be evaluated using the time dependent perturbation theory leading to equations (7.4.11) and (7.4.15). We shall be interested in integrals of the form

$$\langle m|\, \mathscr{H}'\, |n\rangle = \int \psi_m^{0}{}^* \,\mathscr{H}'\, \psi_n^0 \,\mathrm{d}^3\mathbf{r} \tag{7.4.24}$$

with

$$\mathscr{H}' = \frac{\mathrm{j}q\hbar}{mc}\mathbf{A}.\nabla$$

The wave functions $\psi_m^0(\mathbf{r})$, $\psi_n^0(\mathbf{r})$ are large only in a region of the order of the molecular dimensions whereas $\mathbf{A}$, which we shall take in the form $\mathbf{A}_0\, \mathrm{e}^{\mathrm{j}(\omega t - \mathbf{k}.\mathbf{r})}$, changes over a distance of the order of the wavelength. Thus the $\psi^0(\mathbf{r})$ are significant only over a distance of order 10^{-8} cm whilst, even in the optical range, $\mathbf{A}$ is changing much more slowly with a characteristic length of order 10^{-5} cm. The value of $\mathbf{A}$ at any point in a molecule may therefore be expressed by a Taylor expansion about an origin fixed in a molecule. Retaining only the first term in this expansion, we may write for the x component of $\mathbf{A}$ at the position x_i, y_i, z_i occupied by the ith electron in the molecule,

$$A_x^{(i)} = A_x^0 + x_i\left(\frac{\partial A_x}{\partial x}\right)_0 + y_i\left(\frac{\partial A_x}{\partial y}\right)_0 + z_i\left(\frac{\partial A_x}{\partial z}\right)_0 \tag{7.4.25}$$

The expansion for the vector $\mathbf{A}$ is

$$\mathbf{A}^{(i)} = \mathbf{A}^0 + \{(\mathbf{r}^{(i)}.\nabla)\mathbf{A}\}_0 \tag{7.4.26}$$

We are therefore interested in evaluating

$$\langle m|\, \mathscr{H}'\, |n\rangle = \frac{\mathrm{j}q_i\hbar}{m_i c}\int \psi_m^{0}{}^*[\mathbf{A}^0 + \{(\mathbf{r}^{(i)}.\nabla)\mathbf{A}\}_0].\nabla_i\, \psi_n^0\, \mathrm{d}^3\mathbf{r} \tag{7.4.27}$$

Consider first the integral involving $\mathbf{A}^0$

$$\frac{\mathrm{j}q_i\hbar}{m_i c}\langle m|\, A^0.\nabla_i\, |n\rangle = \frac{\mathrm{j}q_i\hbar}{m_i c}\int \psi_m^{0}{}^*\, \mathbf{A}^0.\nabla_i\, \psi_n^0\, \mathrm{d}^3\mathbf{r} \tag{7.4.28}$$

the contribution from A_x^0 is

$$\frac{\mathrm{j}q_i\hbar}{m_ic}\langle m|\, A_x^0\frac{\partial}{\partial x_i}\,|n\rangle = \frac{\mathrm{j}q_i\hbar}{m_ic}A_x^0\int \psi_m^{0*}\frac{\partial}{\partial x_i}\psi_n^0\,\mathrm{d}^3\mathbf{r} \qquad (7.4.29)$$

This integral may be written in an alternative form which is more significant for our purposes by making use of the momentum representation. Equation (7.4.29) can evidently be written as

$$\left(-\frac{q_i}{m_ic}\right)\langle m|\, A_x^0 p_{ix}\,|n\rangle = \left(-\frac{q_i}{m_ic}\right)A_x^0\int \psi_m^{0*}\,p_{ix}\,\psi_n^0\,\mathrm{d}^3\mathbf{r} \quad (7.4.30)$$

where p_{ix} is the x component of the momentum of the ith electron. But the Heisenberg equation of motion for a dynamical variable leads to,

$$\frac{1}{m_i}p_{ix} = \frac{\mathrm{d}}{\mathrm{d}t}x_i = \frac{1}{\mathrm{j}\hbar}(x_i\mathscr{H}_0 - \mathscr{H}_0x_i) \qquad (7.4.31)$$

and†

$$\begin{aligned}\langle m|\,\frac{\mathrm{d}}{\mathrm{d}t}x_i\,|n\rangle &= \langle m|\,\frac{1}{\mathrm{j}\hbar}(x_i\mathscr{H}_0 - \mathscr{H}_0x_i)\,|\,n\rangle \\ &= \frac{\mathrm{j}}{\hbar}(E_m - E_n)\,\langle m|\,x_i\,|n\rangle \qquad (7.4.32)\end{aligned}$$

since $|m\rangle$, $|n\rangle$, are eigenfunctions of $\mathscr{H}_0$.
Hence equation (7.4.29) becomes

$$\begin{aligned}\frac{\mathrm{j}q_i\hbar}{m_ic}\langle m|\,A_x^0\frac{\partial}{\partial x_i}\,|n\rangle &= \left(-\frac{q_i}{m_ic}\right)A_x^0\,\langle m|\,p_{ix}\,|n\rangle \\ &= -\frac{\mathrm{j}A_x^0(E_m - E_n)}{\hbar c}\int \psi_m^{0*}\,q_ix_i\,\psi_n^0\,\mathrm{d}^3\mathbf{r}\end{aligned} \qquad (7.4.33)$$

The integral on the right-hand side of this equation is the matrix element for the x component of the electric dipole moment arising from the ith electron. Taking account of the y and z components, equation (7.4.28) is therefore

$$\begin{aligned}\frac{\mathrm{j}q_i\hbar}{m_ic}\langle m|\,\mathbf{A}^0.\nabla_i\,|n\rangle &= \left(-\frac{q_i}{m_ic}\right)\langle m|\,\mathbf{A}^0.\mathbf{p}_i\,|n\rangle \\ &= -\frac{\mathrm{j}(E_m - E_n)}{\hbar c}\mathbf{A}^0.\langle m|\,q_i\mathbf{r}^{(i)}\,|n\rangle \qquad (7.4.34)\end{aligned}$$

Thus the zero order term $\mathbf{A}^0$, in the expansion of the vector potential $\mathbf{A}$ gives rise to electric dipole moments in the molecule.

The term $\{(\mathbf{r}^i.\nabla)\mathbf{A}\}_0$ in equation (7.4.26) gives rise to magnetic dipole and electric quadripole moments in the molecule. Electric quadripole moments take the form $\langle m|\,q_ix_iy_i\,|n\rangle$. Working back through the sequence of equations (7.4.32), (7.4.31), (7.4.30) and (7.4.29) we can see that matrix elements of the

† See, for example, Schiff, L. I., *Quantum Mechanics*, McGraw-Hill, 1949

electric quadripole moment will arise through relations of the form

$$\langle m|\, x_i y_i \,|n\rangle \rightarrow \langle m|\, \frac{\mathrm{d}}{\mathrm{d}t}(x_i y_i)\,|n\rangle \rightarrow \langle m|\, x_i \frac{\mathrm{d}y_i}{\mathrm{d}t} + y_i \frac{\mathrm{d}x_i}{\mathrm{d}t}\,|n\rangle$$

$$\rightarrow \langle m|\, x_i p_{iy} + y p_{ix} \,|n\rangle \rightarrow \langle m|\, x_i \frac{\partial}{\partial y_i} + y_i \frac{\partial}{\partial x_i}\,|n\rangle \quad (7.4.35)$$

We must therefore separate out terms of this form in the integral of equation (7.4.27). Consider

$$\{(\mathbf{r}^{(i)}\,.\,\nabla)\mathbf{A}\}_0\,.\,\nabla_i\psi$$

$$= \left\{x_i\left(\frac{\partial A_x}{\partial x}\right)_0 + y_i\left(\frac{\partial A_x}{\partial y}\right)_0 + z_i\left(\frac{\partial A_x}{\partial z}\right)_0\right\}\frac{\partial\psi}{\partial x_i}$$

$$+ \left\{x_i\left(\frac{\partial A_y}{\partial x}\right)_0 + y_i\left(\frac{\partial A_y}{\partial y}\right)_0 + z_i\left(\frac{\partial A_y}{\partial z}\right)_0\right\}\frac{\partial\psi}{\partial y_i}$$

$$+ \left\{x_i\left(\frac{\partial A_z}{\partial x}\right)_0 + y_i\left(\frac{\partial A_z}{\partial y}\right)_0 + z_i\left(\frac{\partial A_z}{\partial z}\right)_0\right\}\frac{\partial\psi}{\partial z_i} \quad (7.4.36)$$

This may be written in matrix form as

$$\{(\mathbf{r}^{(i)}\,.\,\nabla)\mathbf{A}\}\,.\,\nabla_i\psi$$

$$= [x_i, y_i, z_i]\begin{bmatrix} \left(\frac{\partial A_x}{\partial x}\right)_0 & \left(\frac{\partial A_y}{\partial x}\right)_0 & \left(\frac{\partial A_z}{\partial x}\right)_0 \\ \left(\frac{\partial A_x}{\partial y}\right)_0 & \left(\frac{\partial A_y}{\partial y}\right)_0 & \left(\frac{\partial A_z}{\partial y}\right)_0 \\ \left(\frac{\partial A_x}{\partial z}\right)_0 & \left(\frac{\partial A_y}{\partial z}\right)_0 & \left(\frac{\partial A_z}{\partial z}\right)_0 \end{bmatrix}\begin{bmatrix} \frac{\partial\psi}{\partial x_i} \\ \frac{\partial\psi}{\partial y_i} \\ \frac{\partial\psi}{\partial z_i} \end{bmatrix} \quad (7.4.37)$$

or

$$\{(\mathbf{r}^{(i)}\,.\,\nabla)\mathbf{A}\}\,.\,\nabla_i\psi$$

$$= [x_i, y_i, z_i][A]\begin{bmatrix} \frac{\partial\psi}{\partial x_i} \\ \frac{\partial\psi}{\partial y_i} \\ \frac{\partial\psi}{\partial z_i} \end{bmatrix} \quad (7.4.38)$$

The matrix $[A]$ can be written as the sum of two matrices one of which $[A_1]$ is symmetric the other $[A_2]$ is antisymmetric.

$$[A] =$$

$$\begin{bmatrix} \left(\frac{\partial A_x}{\partial x}\right)_0 & \frac{1}{2}\left\{\left(\frac{\partial A_y}{\partial x}\right)_0 + \left(\frac{\partial A_x}{\partial y}\right)_0\right\} & \frac{1}{2}\left\{\left(\frac{\partial A_z}{\partial x}\right)_0 + \left(\frac{\partial A_x}{\partial z}\right)_0\right\} \\ \frac{1}{2}\left\{\left(\frac{\partial A_y}{\partial x}\right)_0 + \left(\frac{\partial A_x}{\partial y}\right)_0\right\} & \left(\frac{\partial A_y}{\partial y}\right)_0 & \frac{1}{2}\left\{\left(\frac{\partial A_y}{\partial z}\right)_0 + \left(\frac{\partial A_z}{\partial y}\right)_0\right\} \\ \frac{1}{2}\left\{\left(\frac{\partial A_z}{\partial x}\right) + \left(\frac{\partial A_x}{\partial z}\right)_0\right\} & \frac{1}{2}\left\{\left(\frac{\partial A_y}{\partial z}\right)_0 + \left(\frac{\partial A_z}{\partial y}\right)_0\right\} & \left(\frac{\partial A_z}{\partial z}\right)_0 \end{bmatrix}$$

(7.4.39)

$$+\begin{bmatrix} 0 & \frac{1}{2}\left\{\left(\frac{\partial A_y}{\partial x}\right)_0-\left(\frac{\partial A_x}{\partial y}\right)_0\right\} & \frac{1}{2}\left\{\left(\frac{\partial A_z}{\partial x}\right)_0-\left(\frac{\partial A_x}{\partial z}\right)_0\right\} \\ -\frac{1}{2}\left\{\left(\frac{\partial A_y}{\partial x}\right)_0-\left(\frac{\partial A_x}{\partial y}\right)_0\right\} & 0 & \frac{1}{2}\left\{\left(\frac{\partial A_z}{\partial y}\right)_0-\left(\frac{\partial A_y}{\partial z}\right)_0\right\} \\ -\frac{1}{2}\left\{\left(\frac{\partial A_z}{\partial x}\right)_0-\left(\frac{\partial A_x}{\partial z}\right)_0\right\} & -\frac{1}{2}\left\{\left(\frac{\partial A_z}{\partial y}\right)_0-\left(\frac{\partial A_y}{\partial z}\right)_0\right\} & 0 \end{bmatrix}$$

$$=[A_1]+[A_2] \tag{7.4.40}$$

Equation (7.4.38) is now

$$\{(\mathbf{r}^{(i)}.\nabla)\mathbf{A}\}_0.\nabla_i\psi = [x_i, y_i, z_i]\{[A_1]+[A_2]\}\begin{bmatrix}\frac{\partial\psi}{\partial x_i}\\ \frac{\partial\psi}{\partial y_i}\\ \frac{\partial\psi}{\partial z_i}\end{bmatrix} \tag{7.4.41}$$

The symmetric matrix $[A_1]$ gives rise to terms of the type

$$\left(x_i\frac{\partial}{\partial y_i}+y_i\frac{\partial}{\partial x_i}\right)\psi.$$

These represent the electric quadripole moments. The antisymmetric matrix $[A_2]$, however, gives terms containing

$$\left(x_i\frac{\partial}{\partial y_i}-y_i\frac{\partial}{\partial x_i}\right)\psi.$$

These are angular momentum operators. Moreover the elements of $[A_2]$ are components of $\frac{1}{2}(\nabla\wedge\mathbf{A})_0$ which is a magnetic field. The matrix $[A_2]$ therefore is associated with the magnetic dipole terms. Writing out equation (7.4.41) in algebraic form so far as $[A_2]$ is concerned, we find that

$$[x_i, y_i, z_i][A_2]\begin{bmatrix}\frac{\partial\psi}{\partial x_i}\\ \frac{\partial\psi}{\partial y_i}\\ \frac{\partial\psi}{\partial z_i}\end{bmatrix} = \tfrac{1}{2}(\nabla\wedge\mathbf{A})_0.(\mathbf{r}^{(i)}\wedge\nabla_i\psi) \tag{7.4.42}$$

From equation (7.4.27) and the subsequent discussion the perturbation Hamiltonian for the ith electron now takes the form

$$\mathscr{H}' = \frac{jq_i\hbar}{m_ic}\left\{\mathbf{A}^0.\nabla_i+\tfrac{1}{2}(\nabla\wedge\mathbf{A})_0.(\mathbf{r}^i\wedge\nabla_i)+\begin{matrix}\text{electric quadripole}\\ \text{terms}\end{matrix}\right\} \tag{7.4.43}$$

$$=\begin{matrix}\text{electric}\\ \text{dipole}\\ \text{terms}\end{matrix}+\begin{matrix}\text{magnetic}\\ \text{dipole}\\ \text{terms}\end{matrix}+\begin{matrix}\text{electric}\\ \text{quadripole}\\ \text{terms}\end{matrix} \tag{7.4.44}$$

The electric quadripole terms are of no importance for our discussion and so we write

$$\langle m|\,\mathscr{H}'\,|n\rangle = -\left\{\frac{\mathrm{j}(E_m - E_n)}{\hbar c}\mathbf{A}^0.\langle m|\mathbf{R}|n\rangle + (\nabla \wedge \mathbf{A})_0.\langle m|\,\mathbf{M}\,|n\rangle\right\} \tag{7.4.45}$$

with

$$\langle m|\,\mathbf{R}\,|n\rangle = \int \psi_m^{0\,*}\left(\sum_i q_i \mathbf{r}^{(i)}\right)\psi_n^0\,\mathrm{d}^3\mathbf{r} \tag{7.4.46}$$

$$\langle m|\,\mathbf{M}\,|n\rangle = \int \psi_m^{0\,*}\left(\sum_i \frac{\mathrm{j}q_i\hbar}{2m_ic}(\mathbf{r}^{(i)} \wedge \nabla_i)\right)\psi_n^0\,\mathrm{d}^3\mathbf{r} \tag{7.4.47}$$

and we have now summed over all the electrons in a molecule to obtain the complete perturbation.

We may now evaluate the electric dipole moment for a molecule. The electric dipole moment for a molecule which is in a state described by an eigenfunction Ψ, is defined in equation (7.4.13), to be

$$\mathbf{p}_e = \int \Psi^*\left(\sum_i q\mathbf{r}_i\right)\Psi\,\mathrm{d}^3\mathbf{r} \tag{7.4.48}$$

$$= \int\left(\sum_i q\mathbf{r}_i\right)|\Psi|^2\,\mathrm{d}^3\mathbf{r} \tag{7.4.49}$$

here $\mathbf{r}_i$ is the position vector for a charge in the molecule (the charges are now all taken to be equal). Evidently $\mathbf{p}_e$ is a real quantity.

We do not know the exact eigenfunction for the molecule in the presence of the electromagnetic field and so we expand Ψ in terms of the unperturbed functions Ψ_l^0. Assume that in the absence of the electromagnetic field the molecule is in the state Ψ_k^0. Write

$$\Psi = \Psi_k^0 + \sum_l a_l(t)\,\Psi_l^0 \tag{7.4.50}$$

Substituting this expression into equation (7.4.48)

$$\begin{aligned}\mathbf{p}_e = &\int \Psi_k^{0*}\left\{\sum_i q\mathbf{r}^{(i)}\right\}\Psi_k^0\,\mathrm{d}^3\mathbf{r}\\ &+\int\left(\sum_i q\mathbf{r}^{(i)}\right)\left\{\Psi_k^{0*}\sum_l a_l(t)\,\Psi_l^0 + \Psi_k^0\sum_l a_l^*(t)\Psi_l^{0*}\right\}\mathrm{d}^3\mathbf{r}\\ &+ \text{a term in } a_m(t)\,a^*(t)\end{aligned} \tag{7.4.51}$$

The third term in equation (7.4.51) may be neglected because we are using perturbation theory and the $a_m(t)$, $a_l(t)$ are therefore small. The second integral in equation (7.4.51) contains the function

$$\left\{\Psi_k^{0*}\sum_l a_l(t)\,\Psi_l^0 + \Psi_k^0\sum_l a^*(t)\Psi_l^{0*}\right\} \tag{7.4.52}$$

which has the mathematical form

$$\{\Phi + \Phi^*\} \equiv 2\,\mathrm{Re}\,\Phi \tag{7.4.53}$$

Hence to first order we may write

$$\mathbf{p}_e = \int \Psi_k^{0*}\left(\sum_i q\mathbf{r}^{(i)}\right)\Psi_k^0\,\mathrm{d}^3\mathbf{r}$$
$$+ 2\,\mathrm{Re}\int \Psi_k^{0*}\left(\sum_i q\mathbf{r}^{(i)}\right)\left(\sum_l a_l(t)\Psi_l^0\right)\mathrm{d}^3\mathbf{r} \qquad (7.4.54)$$

In the notation of equation (7.4.46), this relation may be written as

$$\mathbf{p}_e = \langle k|\,\mathbf{R}\,|k\rangle + 2\,\mathrm{Re}\sum_{l\neq k} a_l(t)\exp\,[\{\mathrm{j}(E_k - E_l)/\hbar\}t]\,\langle k|\,\mathbf{R}\,|l\rangle \qquad (7.4.55)$$

where we have now indicated explicitly that the summation in the second term does not include the index k. The states $|k\rangle$, $|l\rangle$, are unperturbed states for the molecule. Hence the first term in equation (7.4.55) represents the permanent dipole moment $\mathbf{p}_0$ of the molecule whilst the second term, which involves the perturbing field through the coefficients $a_l(t)$, represents the induced dipole moment, $\mathbf{p}^{(\mathrm{i})}$. For many molecules $\mathbf{p}_0$ is zero.

We must now evaluate the coefficients $a_l(t)$ using equations (7.4.11) and (7.4.45). We have

$$a_l(t) = \frac{1}{\mathrm{j}\hbar}\int \exp\,[\{\mathrm{j}(E_l - E_k)/\hbar\}t](-1)\left\{\frac{\mathrm{j}(E_l - E_k)}{\hbar c}\mathbf{A}_0.\langle l|\,\mathbf{R}\,|k\rangle\right.$$
$$\left. + (\nabla\wedge\mathbf{A})_0.\langle l|\,\mathbf{M}\,|k\rangle\right\}\mathrm{d}t \qquad (7.4.56)$$

The vector potential $\mathbf{A}$ for a plane wave propagating along the z-direction with the electric field vector $\mathbf{E}$ along the x-axis may be written

$$\mathbf{A} = A_x\hat{\mathbf{i}} = \frac{A_{0x}}{2}\{\mathrm{e}^{\mathrm{j}(\omega t - kz)} + \mathrm{e}^{-\mathrm{j}(\omega t - kz)}\}\hat{\mathbf{i}} \qquad (7.4.57)$$

where we have written $\mathbf{A}$ explicitly in its real form to avoid confusion with the real part operations involved in equation (7.4.55). For this vector potential

$$\mathbf{E} = -\frac{1}{c}\frac{\partial\mathbf{A}}{\partial t} = -\frac{\mathrm{j}\omega}{c}\frac{A_{0x}}{2}\{\mathrm{e}^{\mathrm{j}(\omega t - kz)} - \mathrm{e}^{-\mathrm{j}(\omega t - kz)}\}\hat{\mathbf{i}} \qquad (7.4.58)$$
$$= \frac{\omega}{c}A_{0x}\sin(\omega t - kz)\hat{\mathbf{i}} = E_{0x}\sin(\omega t - kz)\hat{\mathbf{i}}$$

$$\mathbf{H} = \nabla\wedge\mathbf{A} = \left(\frac{\partial A_x}{\partial z}\right)\hat{\mathbf{j}} = (-\mathrm{j}k)\frac{A_{0x}}{2}\{\mathrm{e}^{\mathrm{j}(\omega t - kz)} - \mathrm{e}^{-\mathrm{j}(\omega t - kz)}\}\hat{\mathbf{j}} \qquad (7.4.59)$$
$$= kA_{0x}\sin(\omega t - kz)\hat{\mathbf{j}} = H_{0y}\sin(\omega t - kz)\hat{\mathbf{j}}$$

$$\nabla.\mathbf{A} = 0 \qquad (7.4.60)$$

$$\frac{\mathrm{d}\mathbf{E}}{\mathrm{d}t} = \frac{\omega^2}{c}A_{0x}\cos(\omega t - kz)\hat{\mathbf{i}} \qquad (7.4.61)$$

$$\frac{\mathrm{d}\mathbf{H}}{\mathrm{d}t} = \omega k\,A_{0x}\cos(\omega t - kz)\hat{\mathbf{j}} \qquad (7.4.62)$$

In equation (7.4.56) we shall need

$$\mathbf{A}_0 = \frac{A_{0x}}{2}(\mathrm{e}^{\mathrm{j}\omega t} + \mathrm{e}^{-\mathrm{j}\omega t})\hat{\mathbf{\imath}} \tag{7.4.63}$$

and

$$(\nabla \wedge \mathbf{A})_0 = (-\mathrm{j}k)\frac{A_{0x}}{2}(\mathrm{e}^{\mathrm{j}\omega t} - \mathrm{e}^{-\mathrm{j}\omega t})\hat{\mathbf{\jmath}} \tag{7.4.64}$$

Hence

$$\begin{aligned} a_l(t) = \frac{1}{\mathrm{j}\hbar}\int \exp\left[\left\{\frac{\mathrm{j}(E_l - E_k)}{\hbar}\right\}t\right] \\ \times (-1)\left\{\frac{\mathrm{j}(E_l - E_k)}{\hbar c}\frac{\mathrm{A}_{0x}}{2}(\mathrm{e}^{\mathrm{j}\omega t} + \mathrm{e}^{-\mathrm{j}\omega t})\hat{\mathbf{\imath}}.\langle l|\,\mathbf{R}\,|k\rangle \right. \\ \left. -\mathrm{j}k\frac{A_{0x}}{2}[\mathrm{e}^{\mathrm{j}\omega t} - \mathrm{e}^{-\mathrm{j}\omega t}]\hat{\mathbf{\jmath}}.\langle l|\,\mathbf{M}\,|k\rangle\right\}\mathrm{d}t \end{aligned} \tag{7.4.65}$$

which can be written more simply in the form

$$\begin{aligned} a_l(t) = \frac{(-1)}{\mathrm{j}\hbar}\int\left[\!\left[\alpha\left[\exp\left\{\frac{\mathrm{j}(E_{lk} + \hbar\omega)}{\hbar}t\right\} + \exp\left\{\frac{\mathrm{j}(E_{lk} - \hbar\omega)}{\hbar}t\right\}\right]\right.\right. \\ \left.\left. -\beta\left[\exp\left\{\frac{\mathrm{j}(E_{lk} + \hbar\omega)}{\hbar}t\right\} - \exp\left\{\frac{\mathrm{j}(E_{lk} - \hbar\omega)}{\hbar}t\right\}\right]\right]\!\right]\mathrm{d}t \end{aligned} \tag{7.4.66}$$

the function α is the first term in the integral of equation (7.4.65) and carries the electric field terms whilst the function β is the second term from the integral and carries the magnetic field terms.

Integrate equation (7.4.66).

$$a_l(t) = \left(\frac{-1}{\mathrm{j}\hbar}\right)\left\{\frac{\hbar}{\mathrm{j}}\frac{\alpha\exp\{\mathrm{j}(E_{lk}/\hbar)t\}}{E_{lk}^2 - \hbar^2\omega^2}(2E_{lk}\cos\omega t - 2\mathrm{j}\hbar\omega\sin\omega t)\right.$$
$$\left. - \left(\frac{\hbar}{j}\right)\frac{\beta\exp\{\mathrm{j}(E_{lk}/\hbar)t\}}{E_{lk}^2 - \hbar^2\omega^2}(2\mathrm{j}E_{lk}\sin\omega t + 2\hbar\omega\cos\omega t)\right\} \tag{7.4.67}$$

$$= \frac{2\exp\{\mathrm{j}(E_{lk}/\hbar)t\}}{E_{lk}^2 - \hbar^2\omega^2}\{\alpha(E_{lk}\cos\omega t - \mathrm{j}\hbar\omega\sin\omega t) - \beta(\mathrm{j}E_{lk}\sin\omega t + \hbar\omega\cos\omega t)\} \tag{7.4.68}$$

The constant of integration has been taken to be zero without loss of generality because it does not contain any functions of the perturbing fields.

The induced electric dipole moment $\mathbf{p}^{(\mathrm{i})}$ may now be obtained from equation (7.4.55)

$$\mathbf{p}^{(\mathrm{i})} = 2\,\mathrm{Re}\left[\sum_{l\neq k}\langle k|\,\mathbf{R}\,|l\rangle a_l(t)\exp\{\mathrm{j}(E_{kl}/\hbar)t\}\right] \tag{7.4.69}$$

$$\mathbf{p}^{(\mathrm{i})} = 2\,\mathrm{Re}\left[\sum_{l\neq k}\frac{2\langle k|\,\mathbf{R}\,|l\rangle}{E_{lk}^2 - \hbar^2\omega^2}\left\{\alpha(E_{lk}\cos\omega t - \mathrm{j}\hbar\omega\sin\omega t) - \beta(\mathrm{j}E_{lk}\sin\omega t + \hbar\omega\cos\omega t)\right\}\right] \tag{7.4.70}$$

Substituting the expressions for α and β, and making use of the relations in equations (7.4.58), (7.4.59), (7.4.61) and (7.4.62) we obtain

$$\mathbf{p}^{(i)} = 2\,\mathrm{Re}\left[\sum_{l\neq k}\frac{E_{lk}}{E_{lk}^2-\hbar^2\omega^2}\langle k|\,\mathbf{R}\,|l\rangle\{\langle l|\,\mathbf{R}\,|k\rangle.\mathbf{E}_{(z=0)}\}\right.$$
$$+\sum_{l\neq k}\frac{\mathrm{j}E_{lk}^2}{E_{lk}^2-\hbar^2\omega^2}\cdot\frac{1}{\hbar\omega^2}\langle k|\,\mathbf{R}\,|l\rangle\left\{\langle l\,|\mathbf{R}\,|k\rangle.\left(\frac{\mathrm{d}\mathbf{E}}{\mathrm{d}t}\right)_{z=0}\right\}$$
$$+\sum_{l\neq k}\frac{E_{lk}}{E_{lk}^2-\hbar^2\omega^2}\langle k|\,\mathbf{R}\,|l\rangle\{\langle l|\,\mathbf{M}\,|k\rangle.\mathbf{H}_{(z=0)}\}$$
$$\left.+\sum_{l\neq k}\frac{(-1)\mathrm{j}\hbar}{E_{lk}^2-\hbar^2\omega^2}\langle k|\,\mathbf{R}\,|l\rangle\left\{\langle l|\,\mathbf{M}\,|k\rangle.\left(\frac{\mathrm{d}\mathbf{H}}{\mathrm{d}t}\right)_{z=0}\right\}\right] \tag{7.4.71}$$

We may expect an electric dipole matrix element to be of the order of magnitude of the electronic charge multiplied by the Bohr radius whilst a magnetic dipole matrix element is probably of the order of the Bohr magneton. Hence we shall have

$$\langle l|\,\mathbf{R}\,|k\rangle \sim 2{\cdot}5\,.10^{-18}\text{ c.g.s. units}$$
$$\langle l|\,\mathbf{M}\,|k| \sim 9\,.10^{-21}\text{ c.g.s. units}$$

In first order therefore the magnetic field terms can be neglected in equation (7.4.71). This is also in agreement with the classical discussion on page 308.

For an isotropic material the first two terms in equation (7.4.71) involve matrix elements (see the discussion leading to equation (7.4.78))

$$\langle k|\,\mathbf{R}\,|l\rangle.\langle l|\,\mathbf{R}\,|k\rangle = |\langle k|\,\mathbf{R}\,|l\rangle|^2 \tag{7.4.72}$$

These are real functions. The term containing $\mathbf{E}_{(z=0)}$ is therefore purely real whilst that containing $(\mathrm{d}\mathbf{E}/\mathrm{d}t)_{z=0}$ is purely imaginary (recall that we used a real function representation for **E**). Hence, upon taking the real part of equation (7.4.71), the expression for the induced electric dipole moment to first order is

$$\mathbf{p}^{(i)} = 2\sum_{l\neq k}\frac{E_{lk}}{E_{lk}^2-\hbar^2\omega^2}\langle k|\,\mathbf{R}\,|l\rangle\{\langle l|\,\mathbf{R}\,|k\rangle.\mathbf{E}_{(z=0)}\} \tag{7.4.73}$$

In order to determine the polarization for an isotropic material we shall require the component of the induced moment in the direction of the electric field. This was the quantity we evaluated previously on a classical model leading to equation (7.2.9) where we were writing

$$\mathbf{P} = \sum_i N_i\mathbf{p}_i^{(i)} = \chi^{(e)}\mathbf{E} = \frac{\varepsilon(\omega)-1}{4\pi}\mathbf{E} \tag{7.4.74}$$

In our present discussion we have identified the x coordinate axis with the direction of **E** and so we require $p_{ix}^{(i)}$. Consider the function

$$\langle k|\,\mathbf{R}\,|l\rangle(\langle l|\,\mathbf{R}\,|k\rangle.\mathbf{E})$$

from equation (7.4.73). In our case we may write this quantity formally as

$$\mathbf{R}^*(\mathbf{R}.\mathbf{E}) = \mathbf{R}^*(R_xE_x) \tag{7.4.75}$$

The x component is

$$\{\mathbf{R}^*(\mathbf{R}.\mathbf{E})\}_x = \hat{\mathbf{\imath}}|R_x|^2E_x = |R_x|^2\mathbf{E} \tag{7.4.76}$$

But the identification of the x coordinate axis is arbitrary in an isotropic medium where also there are no cross terms. Thus

$$|R_x|^2 = |R_y|^2 = |R_z|^2 = \tfrac{1}{3}(\mathbf{R}^*.\mathbf{R}) \tag{7.4.77}$$

Hence we may write

$$\{\mathbf{R}^*(\mathbf{R}.\mathbf{E})\}_x = \tfrac{1}{3}(\mathbf{R}^*.\mathbf{R})\mathbf{E} \tag{7.4.78}$$

and

$$\{\langle k|\,\mathbf{R}\,|l\rangle[\langle l|\mathbf{R}\,|k\rangle.\mathbf{E}]\}_x = \tfrac{1}{3}|\,\langle k|\,\mathbf{R}\,|l\rangle\,|^2\mathbf{E} \tag{7.4.79}$$

The induced polarization is now obtained from equation (7.4.74). If there are N molecules per unit volume

$$\mathbf{P} = \frac{2N}{3}\left\{\sum_{l\neq k}\frac{E_{lk}|\,\langle k|\,\sum_i q\mathbf{r}^{(i)}|l\rangle\,|^2}{E_{lk}^2 - \hbar^2\omega^2}\right\}\mathbf{E} \tag{7.4.80}$$

$$= \frac{2Nq^2}{3}\left\{\sum_{l\neq k}\frac{(\omega_{lk}/\hbar)\,|\,\langle k|\,\sum_i \mathbf{r}^{(i)}\,|l\rangle\,|^2}{\omega_{lk}^2 - \omega^2}\right\}\mathbf{E} \tag{7.4.81}$$

$$= \frac{\varepsilon(\omega) - 1}{4\pi}\mathbf{E}$$

Note that the factor $(\omega_{lk}/\hbar)\,|\,\langle k|\,\sum_i \mathbf{r}^{(i)}\,|l\rangle\,|^2$ has the dimensions $(\text{mass})^{-1}$.

It is evident that the quantum mechanical expression for the induced polarization has the same form as that derived from classical theory in equation (7.2.9) when damping is neglected. The rest of the theory will therefore continue as in Sections 7.2 and 7.3, with $\mathbf{E}$ the local field at a molecule. Polar materials will also be described by the general classical treatment given previously. The permanent electric dipole moment $\mathbf{p}_0$ is now, however, given by

$$\mathbf{p}_0 = \langle k|\,\mathbf{R}\,|k\rangle \tag{7.4.82}$$

in accordance with the discussion following equation (7.4.55).

7.5 Optical Rotation Phenomena

Some materials which would otherwise be regarded as isotropic cause the plane of polarization of an electromagnetic wave to rotate as it propagates. These materials need not be in the form of single crystals and so the effect does not arise from a straightforward anisotropic dielectric constant. Thus sugar solutions exhibit this phenomenon in the absence of any perturbing factor (this is an example of natural optical activity) whilst materials which are paramagnetic or show a Zeeman splitting of absorption lines can cause the plane of polarization to rotate when a magnetic field is applied (this is the Faraday effect).

These materials are evidently exhibiting a particular form of anisotropy and the electric displacement vector $\mathbf{D}$ is no longer simply related to the field vector $\mathbf{E}$ by the relation $\mathbf{D} = \varepsilon(\omega)\mathbf{E}$. Instead we must write a tensor relation

$$\begin{bmatrix} D_x \\ D_y \\ D_z \end{bmatrix} = \begin{bmatrix} \varepsilon_{11} & \delta_{12} & \delta_{13} \\ \delta_{21} & \varepsilon_{22} & \delta_{23} \\ \delta_{31} & \delta_{32} & \varepsilon_{33} \end{bmatrix}\begin{bmatrix} E_x \\ E_y \\ E_z \end{bmatrix} \tag{7.5.1}$$

where the off diagonal elements in the dielectric constant matrix have been written δ_{ij} to emphasize that they differ from the usual components representing crystal anisotropy.

Consider first some general properties of any anisotropic material which can be derived from the fundamental electromagnetic field equations for an insulating medium. We have

$$\nabla \wedge \mathbf{E} + \frac{1}{c}\frac{\partial \mathbf{B}}{\partial t} = 0 \tag{7.5.2}$$

$$\nabla \wedge \mathbf{H} - \frac{1}{c}\frac{\partial \mathbf{D}}{\partial t} = 0 \tag{7.5.3}$$

$$\nabla . \mathbf{D} = 0 \qquad \nabla . \mathbf{B} = 0 \tag{7.5.4}$$

Take

$$\begin{aligned} \mathbf{D} &= \mathbf{D}_0\, e^{j(\omega t - \mathbf{k}.\mathbf{r})} \\ \mathbf{E} &= \mathbf{E}_0\, e^{j(\omega t - \mathbf{k}.\mathbf{r})} \\ \mathbf{H} &= \mathbf{H}_a + \mathbf{H}_0\, e^{j(\omega t - \mathbf{k}.\mathbf{r})} \end{aligned} \tag{7.5.5}$$

$$\mathbf{B} = \mathbf{H}_a + 4\pi \mathbf{M}_a + \mu(\omega)\mathbf{H}_0\, e^{j(\omega t - \mathbf{k}.\mathbf{r})} \tag{7.5.6}$$

Equations (7.5.5) and (7.5.6) describe the electromagnetic field of the wave together with a steady field $\mathbf{H}_a$ which gives rise to the magnetization $\mathbf{M}_a$. The wave field $\mathbf{H}_0\, e^{j(\omega t - \mathbf{k}.\mathbf{r})}$ is assumed to be related to $\mathbf{B}$ through the straightforward linear equation involving the isotropic permeability factor $\mu(\omega)$.

Substituting equations (7.5.5) and (7.5.6) into equations (7.5.2)–(7.5.4) we obtain,

$$\mathbf{k} \wedge \mathbf{E}_0 = \frac{\omega\mu}{c}\mathbf{H}_0 \tag{7.5.7}$$

$$-\mathbf{k} \wedge \mathbf{H}_0 = \frac{\omega}{c}\mathbf{D}_0 \tag{7.5.8}$$

$$\mathbf{k}.\mathbf{D}_0 = 0 \qquad \mathbf{k}.\mathbf{H}_0 = 0 \tag{7.5.9}$$

From equation (7.5.9) $\mathbf{k}$ is perpendicular to $\mathbf{D}_0$ and to $\mathbf{H}_0$. In addition, from equation (7.5.8), $\mathbf{D}_0$ is perpendicular to $\mathbf{k}$ and to $\mathbf{H}_0$. The vectors $\mathbf{D}_0$, $\mathbf{H}_0$, $\mathbf{k}$ form a right-handed set. The field vector $\mathbf{E}_0$ is no longer required to be parallel to $\mathbf{D}_0$, but from equation (7.5.7) $\mathbf{H}_0$ is perpendicular to both $\mathbf{k}$ and $\mathbf{E}_0$. The vector $\mathbf{E}_0$ therefore lies in the $\mathbf{D}_0$, $\mathbf{k}$ plane. These relations between the field vectors are illustrated in Fig. 7.4.

Evidently $\mathbf{D}_0$, $\mathbf{H}_0$ are transverse to the direction of propagation, whilst $\mathbf{E}_0$ can have a component along $\mathbf{k}$. The vectors $\mathbf{D}_0$ and $\mathbf{E}_0$ are formally related through equations (7.5.7) and (7.5.8)

$$\mathbf{D}_0 = -\frac{c}{\omega}(\mathbf{k} \wedge \mathbf{H}_0) = -\frac{c^2}{\omega^2\mu}\mathbf{k} \wedge (\mathbf{k} \wedge \mathbf{E}_0) \tag{7.5.10}$$

$$= \frac{c^2}{\omega^2\mu}\{(\mathbf{k}.\mathbf{k})\mathbf{E}_0 - (\mathbf{k}.\mathbf{E}_0)\mathbf{k}\} \tag{7.5.11}$$

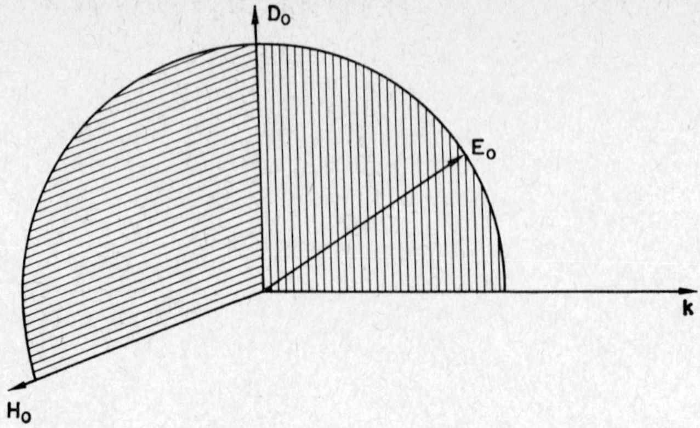

FIG. 7.4 The field vectors in an anisotropic material.

7.5.1 Natural Optical Activity

The specific anisotropic properties of a material giving rise to optical rotation effects depend upon the particular form of the off diagonal matrix elements in equation (7.5.1). Suppose that in equation (7.4.71) for the electric dipole moment the term involving $d\mathbf{H}/dt$ makes a small but significant contribution. We may write a simple formal relation for the electric displacement vector,

$$\mathbf{D} = \varepsilon(\omega)\mathbf{E} + \beta\frac{\partial \mathbf{H}}{\partial t} \tag{7.5.12}$$

Here we have taken $\varepsilon(\omega)$ corresponding to a straightforward real isotropic dielectric constant arising from permanent and induced dipole moments as described by the first terms of equations (7.4.55) and (7.4.71). The term in $\partial\mathbf{H}/\partial t$ arises from the fourth term in equation (7.4.71). Since we are writing down a formal relation in equation (7.5.12) we may use the complex representation for the field vectors. This will not lead to confusion in our discussion but care must be exercised if the actual matrix elements involved in $\varepsilon(\omega)$, β, need to be related to the measured values.

Assume that no external magnetic field is present, so that $\mathbf{H}_a = \mathbf{M}_a = 0$, and that the electromagnetic wave is propagating along the z-direction.† From equations (7.5.12), and (7.5.7)

$$\mathbf{D} = \varepsilon(\omega)\mathbf{E} + \frac{\mathrm{j}c\beta}{\mu}\mathbf{k}\wedge\mathbf{E} \tag{7.5.13}$$

From this equation $E_z = 0$, since $D_z = 0$ and $\varepsilon(\omega) \neq 0$, is an isotropic

† We shall eventually conclude that $\mathbf{E}$ has the form $E_0(z)\,e^{\mathrm{j}(\omega t - kz)}$. We assume that $E_0(z)$ is very slowly varying so that $\partial E_0/\partial z$ can be neglected. Similar remarks apply to the field vectors $\mathbf{D}$, $\mathbf{H}$. These assumptions are consistent with $\beta(\partial\mathbf{H}/\partial t)$ being a small term in equation (7.5.12).

dielectric constant. Hence

$$\begin{bmatrix} D_{0x} \\ D_{0y} \end{bmatrix} = \begin{bmatrix} \varepsilon(\omega) & -\dfrac{jc\beta}{\mu}k_z \\ \dfrac{jc\beta}{\mu}k_z & \varepsilon(\omega) \end{bmatrix} \begin{bmatrix} E_{0x} \\ E_{0y} \end{bmatrix} \tag{7.5.14}$$

The general field relation given by equation (7.5.11) must also be satisfied

$$\begin{bmatrix} D_{0x} \\ D_{0y} \end{bmatrix} = \begin{bmatrix} \dfrac{c^2k_z^2}{\omega^2\mu}E_{0x} \\ \dfrac{c^2k_z^2}{\omega^2\mu}E_{0y} \end{bmatrix} \tag{7.5.15}$$

Hence if our model is consistent with the field equations

$$\begin{bmatrix} \varepsilon(\omega) & -\dfrac{jc\beta}{\mu}k_z \\ \dfrac{jc\beta}{\mu}k_z & \varepsilon(\omega) \end{bmatrix} \begin{bmatrix} E_{0x} \\ E_{0y} \end{bmatrix} = \begin{bmatrix} \dfrac{c^2k_z^2}{\omega^2\mu}E_{0x} \\ \dfrac{c^2k_z^2}{\omega^2\mu}E_{0y} \end{bmatrix} \tag{7.5.16}$$

and

$$\begin{vmatrix} \varepsilon(\omega) - \dfrac{c^2k_z^2}{\omega^2\mu} & -\dfrac{jc\beta}{\mu}k_z \\ \dfrac{jc\beta}{\mu}k_z & \varepsilon(\omega) - \dfrac{c^2k_z^2}{\omega^2\mu} \end{vmatrix} = 0 \tag{7.5.17}$$

or

$$\frac{c^2k_z^2}{\omega^2\mu} = \varepsilon(\omega) \pm \frac{c\beta}{\mu}k_z \tag{7.5.18}$$

This is the (ω, k) dispersion relation for the material, it can be written as

$$\omega^2 = \frac{c^2}{\mu\varepsilon_\pm}k_z^2 \tag{7.5.19}$$

with an effective dielectric constant $\varepsilon_\pm$ given by

$$\varepsilon_\pm = \varepsilon(\omega) \pm \frac{c\beta}{\mu}k_z \tag{7.5.20}$$

In this equation k_z can be treated as a constant since the second term on the right-hand side is small.

The two values of the effective dielectric constant characterize the propagation of right-handed and left-handed circularly polarized waves. Write

$$E_{0y} = +jE_{0x} \tag{7.5.21}$$

for a right-handed circularly polarized wave.

From equation (7.5.14)

$$D_{0x} = \left\{\varepsilon(\omega) + \frac{c\beta}{\mu}k_z\right\}E_{0x} \tag{7.5.22}$$

$$D_{0y} = j\left\{\varepsilon(\omega) + \frac{c\beta}{\mu}k_z\right\}E_{0x}$$

so that

$$D_{0x} = \mathrm{j}D_{0y} \tag{7.5.23}$$

and we may write

$$\mathbf{D} = \varepsilon_+ \mathbf{E} \tag{7.5.24a}$$

with

$$\varepsilon_+ = \varepsilon(\omega) + \frac{c\beta}{\mu} k_z \tag{7.5.24b}$$

Similarly, for a left-handed circularly polarized wave with $E_{0y} = -\mathrm{j}E_{0x}$,

$$\mathbf{D} = \varepsilon_- \mathbf{E} \tag{7.5.24c}$$

and

$$\varepsilon_- = \varepsilon(\omega) - \frac{c\beta}{\mu} k_z \tag{7.5.24d}$$

From equation (7.5.19), the phase velocities for these circularly polarized waves are,

$$Vp_+ = \frac{c}{\sqrt{(\mu\varepsilon_+)}} \tag{7.5.25a}$$

$$Vp_- = \frac{c}{\sqrt{(\mu\varepsilon_-)}} \tag{7.5.25b}$$

The waves may be represented by functions having the form $\exp\{\mathrm{j}(\omega t - k_+ z)\}$, $\exp\{\mathrm{j}(\omega t - k_- z)\}$ with k_+ and k_- also derived from equation (7.5.19),

$$k_+ = \frac{\omega}{c}\sqrt{(\mu\varepsilon_+)} \quad k_- = \frac{\omega}{c}\sqrt{(\mu\varepsilon_-)} \tag{7.5.26}$$

A convenient representation makes use of a matrix notation for the field vectors. Thus we write

$$\begin{bmatrix} E_x \\ E_y \end{bmatrix} = \begin{bmatrix} 1 \\ \mathrm{j} \end{bmatrix} E_0 \exp\{\mathrm{j}(\omega t - k_+ z)\} \tag{7.5.27a}$$

for a right-handed circularly polarized wave, and

$$\begin{bmatrix} E_x \\ E_y \end{bmatrix} = \begin{bmatrix} 1 \\ -\mathrm{j} \end{bmatrix} E_0 \exp\{\mathrm{j}(\omega t - k_- z)\} \tag{7.5.27b}$$

for a left-handed circularly polarized wave. In this notation a linearly polarized wave at the plane $z = 0$ is written

$$\begin{bmatrix} E_x \\ E_y \end{bmatrix} = \begin{bmatrix} E_{0x} \\ E_{0y} \end{bmatrix} \mathrm{e}^{\mathrm{j}\omega t} \tag{7.5.27c}$$

Evidently, if the x coordinate axis is taken along the direction of the electric vector at the plane $z = 0$, the linearly polarized wave may be described directly, at this plane, as the sum of two circularly polarized waves by using the form,

$$\begin{aligned} \begin{bmatrix} E_x \\ E_y \end{bmatrix} &= \begin{bmatrix} 1 \\ 0 \end{bmatrix} E_0 \, \mathrm{e}^{\mathrm{j}\omega t} \\ &= \tfrac{1}{2}\begin{bmatrix} 1 \\ \mathrm{j} \end{bmatrix} E_0 \, \mathrm{e}^{\mathrm{j}\omega t} + \tfrac{1}{2}\begin{bmatrix} 1 \\ -\mathrm{j} \end{bmatrix} E_0 \, \mathrm{e}^{\mathrm{j}\omega t} \end{aligned} \tag{7.5.28}$$

As such a wave propagates away from the plane $z = 0$ in an optically active medium the plane of polarization will rotate because of the differing velocities of the component circular waves. The angular rotation of the plane of polarization may be evaluated quite easily by making use of the matrix notation. Consider a wave which has propagated a distance l beyond the plane $z = 0$. From equations (7.5.27) and (7.5.28), we may write

$$\begin{bmatrix} E_x(l) \\ E_y(l) \end{bmatrix} = \tfrac{1}{2}\begin{bmatrix} 1 \\ \mathrm{j} \end{bmatrix} E_0 \exp\{\mathrm{j}(\omega t - k_+ l)\} + \tfrac{1}{2}\begin{bmatrix} 1 \\ -\mathrm{j} \end{bmatrix} E_0 \exp\{\mathrm{j}(\omega t - k_- l)\} \tag{7.5.29}$$

Write

$$k_+ = \tfrac{1}{2}(k_+ + k_-) + \tfrac{1}{2}(k_+ - k_-)$$
$$k_- = \tfrac{1}{2}(k_+ + k_-) - \tfrac{1}{2}(k_+ - k_-)$$

to obtain

$$\begin{bmatrix} E_x(l) \\ E_y(l) \end{bmatrix} = \left\{ \tfrac{1}{2}\begin{bmatrix} 1 \\ \mathrm{j} \end{bmatrix} \exp\{-\tfrac{1}{2}\mathrm{j}(k_+ - k_-)l\} + \tfrac{1}{2}\begin{bmatrix} 1 \\ -\mathrm{j} \end{bmatrix} \exp\{+\tfrac{1}{2}\mathrm{j}(k_+ - k_-)l\} \right\}$$
$$\times\, E_0 \exp[\![\mathrm{j}[\omega t - \{(k_+ + k_-)/2\}l]]\!] \tag{7.5.30a}$$

and so

$$\begin{bmatrix} E_x(l) \\ E_y(l) \end{bmatrix} = \begin{bmatrix} \cos\theta \\ \sin\theta \end{bmatrix} E_0\, \mathrm{e}^{\mathrm{j}(\omega t - \psi)} \tag{7.5.30b}$$

with $\theta = \tfrac{1}{2}(k_+ - k_-)l$ and $\psi = \tfrac{1}{2}(k_+ + k_-)l$.

As the field components E_x, E_y oscillate with the same phase, equation (7.5.30b) represents a linearly polarized wave with the electric vector inclined now at an angle θ to the x-axis previously defined at the plane $z = 0$. Evidently the plane of polarization has been rotated through the angle θ given by

$$\theta = \tfrac{1}{2}(k_+ - k_-)l = \tfrac{1}{2}\omega l\left(\frac{1}{Vp_+} - \frac{1}{Vp_-}\right) \tag{7.5.31a}$$

$$= \frac{1}{2}\frac{\omega l}{c}\{\sqrt{(\mu\varepsilon_+)} - \sqrt{(\mu\varepsilon_-)}\} \tag{7.5.31b}$$

Upon using equations (7.5.24b) and (7.5.24d), we obtain to first order

$$\theta = \frac{1}{2}\frac{\omega l}{c}\sqrt{(\mu\varepsilon)}\left\{\left(1 + \frac{c\beta}{2\mu\varepsilon}k_z\right) - \left(1 - \frac{c\beta}{2\mu\varepsilon}k_z\right)\right\}$$
$$= \frac{1}{2}\frac{\omega^2\beta l}{c} \tag{7.5.31c}$$

The rotation per unit path length is therefore

$$\frac{\theta}{l} = \frac{1}{2}\frac{\omega^2\beta}{c} \quad \text{radians per centimetre} \tag{7.5.31d}$$

The rotation of the plane of polarization depends explicitly therefore upon the parameter β which characterizes the fourth term in equation (7.4.71). If β is non-zero then the function $\mathrm{Re}\,[-\mathrm{j}\hbar\,\langle k|\,\mathbf{R}\,|l\rangle\langle l|\,\mathbf{M}\,|k\rangle]$ must also be non-zero. This function is in fact zero for all molecules which possess a centre of symmetry. For such molecules the wave functions may be classified as odd or even according as to whether or not they change sign when subjected to inversion at the centre of symmetry (this is equivalent to replacing each coordinate by its negative). Since the operator **R** changes sign upon inversion $\langle k|\, \mathbf{R}\, |l\rangle$ is only non-zero between odd and even states. The operator **M**, however, does not change sign upon inversion and can only couple two odd or two even states. The product $\langle k|\, \mathbf{R}\, |l\rangle\langle l|\, \mathbf{M}\, |k\rangle$ is therefore zero for all states of a molecule with a centre of symmetry and such a molecule cannot give rise to natural optical activity. This is also the case for a molecule which possesses a plane of symmetry. Natural optical activity does occur quite widely in crystals. In these materials, however, a further restriction must be imposed when mirror planes exist. A molecular unit must not be identical with its mirror image for if this is the case the net optical rotation will be zero.

7.5.2 Faraday Rotation

Faraday discovered that rotation of the plane of polarization occurs when light propagates through a block of glass along the direction of an applied magnetic field $\mathbf{H}_a$. This effect occurs, in principle, in all isotropic materials. In first order the angular rotation is linearly proportional to the field $\mathbf{H}_a = \hat{\mathbf{k}}H_z$, and to the distance, z, which the light travels in the material. We may therefore write the angular rotation ϕ, as

$$\phi = Vz\,H_z \tag{7.5.32}$$

where V, the constant of proportionality, is called the Verdet constant.

The Faraday rotation is not the only magneto-optical effect of this kind. Thus if light which is initially plane polarized propagates through a material in a direction perpendicular to the applied magnetic field $\mathbf{H}_a$ the emergent light is in general elliptically polarized. This is the Cotton–Mouton effect which depends upon H_z^2. Evidently a material in the presence of a magnetic field becomes in some sense anisotropic with an axis of symmetry along the field direction. We shall examine the physical origins of this induced anisotropy by considering the Faraday effect.

Consider first a simple classical model based on equation (7.2.3). Neglect local field effects and assume that all the electrons are identical. In order to avoid the mathematical difficulties which arise when the dielectric constant $\varepsilon(\omega)$ is a complex quantity in the absence of the magnetic field assume that the material is transparent in the frequency range of interest and neglect the damping term in equation (7.2.3). The equation of motion for an electron is

$$m\ddot{\mathbf{r}} + m\omega_0^2\mathbf{r} = -|e|\left(\mathbf{E} + \frac{\dot{\mathbf{r}} \wedge \mathbf{B}_a}{c}\right) \tag{7.5.33}$$

Seek a steady state solution with **E** and **r** varying as $e^{j\omega t}$, we shall eventually take $\mathbf{B}_a = \hat{\mathbf{k}}B_z$.

$$m(\omega_0^2 - \omega^2)\mathbf{r} = -|e|\mathbf{E} - \frac{j\omega|e|}{c}\,\mathbf{r} \wedge \mathbf{B}_a$$

The electric polarization $\mathbf{P} = -N|e|\mathbf{r}$ when there are N electrons per unit volume,

$$(\omega_0^2 - \omega^2)\mathbf{P} = \frac{Ne^2}{m}\mathbf{E} - \frac{\mathrm{j}\omega|e|}{mc}\mathbf{B}_a \wedge \mathbf{P} \tag{7.5.34}$$

In zero magnetic field

$$\mathbf{P} = \frac{Ne^2/m}{\omega_0^2 - \omega^2}\mathbf{E} = \chi_0\mathbf{E} \tag{7.5.35}$$

and since we assume the magnetic field to give rise to a small perturbation we may insert this value for $\mathbf{P}$ in the second term on the right-hand side of equation (7.5.34). Consequently,

$$\mathbf{P} = \chi_0\mathbf{E} - \frac{\mathrm{j}\omega|e|}{mc}\cdot\frac{\chi_0}{(\omega_0^2 - \omega^2)}\mathbf{B}_a \wedge \mathbf{E} \tag{7.5.36a}$$

and

$$\mathbf{D} = \varepsilon(\omega)\mathbf{E} - \mathrm{j}\frac{4\pi\omega|e|}{mc}\cdot\frac{\chi_0}{(\omega_0^2 - \omega^2)}\mathbf{B}_a \wedge \mathbf{E} \tag{7.5.36b}$$

This equation has the same mathematical form as equation (7.5.13), and so for $\mathbf{B}_a = \hat{\mathbf{k}}B_z$ and propagation along the z-direction there will be an angular rotation of the plane of polarization for linear polarized light given by

$$\frac{\phi}{z} = \frac{2\pi\omega^2|e|}{mc^2}\cdot\frac{\chi_0}{(\omega_0^2 - \omega^2)}\left(\frac{\mu}{\varepsilon}\right)^{\frac{1}{2}} B_z \tag{7.5.37a}$$

$$= \frac{2\pi N|e|^3\omega^2}{m^2c^2}\cdot\frac{1}{(\omega_0^2 - \omega^2)^2}\left(\frac{\mu}{\varepsilon}\right)^{\frac{1}{2}} B_z \tag{7.5.37b}$$

The angular rotation of the plane of polarization is therefore determined strictly by B_z rather than H_z. However, the present discussion is limited to essentially diamagnetic materials since we have taken no account of electron spin. We may therefore, in practice, write $B_z = H_z$ and the high frequency permeability $\mu = 1$. In these circumstances we may write

$$\phi = Vz\,H_z$$

where V is a proportionality factor which is dependent upon frequency.

Note that equation (7.5.37b) may also be written in terms of the dispersion of the refractive index η in zero magnetic field. We have

$$\eta^2 = \varepsilon(\omega)$$

$$\frac{\phi}{z} = \frac{\omega\omega_L}{2c}\frac{\mathrm{d}\eta}{\mathrm{d}\omega} \tag{7.5.38}$$

with

$$\omega_L = \frac{|e|B_z}{mc} = \frac{|e|H_z}{mc}$$

In order to describe the Faraday effect on quantum theory we consider a simple model for a material in which the ground state for each molecular unit is split into two levels by the action of an applied magnetic field $\mathbf{H}_a$. These levels are each non-degenerate after the application of $\mathbf{H}_a$. The excited

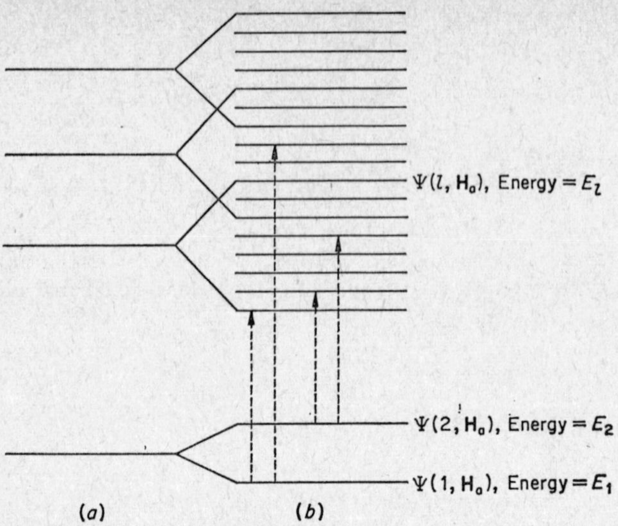

FIG. 7.5 Energy states of a molecule, (a) before application of the magnetic field $\mathbf{H}_a$, (b) after application of $\mathbf{H}_a$. The broken vertical lines indicate typical Zeeman transitions.

states are also split by the magnetic field and so the material will exhibit the usual Zeeman effects in its spectral lines as indicated in Fig. 7.5. Our discussion of this two level model will also include the case where the ground state is a singlet and non-degenerate for all values of $\mathbf{H}_a$.

Write the Hamiltonian operator for an electron in a molecule from equation (7.4.20) as

$$\mathscr{H} = \left\{-\frac{\hbar^2}{2m}\nabla^2 + V(\mathbf{r}) + \frac{jq\hbar}{mc}\mathbf{A}.\nabla + q^2\frac{\mathbf{A}.\mathbf{A}}{2mc^2}\right\} \tag{7.5.39a}$$

$$= \left\{-\frac{\hbar^2}{2m}\nabla^2 + V(\mathbf{r})\right\} + \left\{\frac{jq\hbar}{mc}\mathbf{A}_a.\nabla + q^2\frac{\mathbf{A}_a.\mathbf{A}_a}{2mc^2}\right\} + \frac{jq\hbar}{mc}\mathbf{A}_w.\nabla \tag{7.5.39b}$$

$$= \mathscr{H}_0 + \mathscr{H}_a + \mathscr{H}_w \tag{7.5.39c}$$

Here the vector potential $\mathbf{A}_a$ refers to the applied magnetic field $\mathbf{H}_a$. Thus if $\mathbf{H}_a$ is along the z-direction, $\mathbf{H}_a = \hat{\mathbf{k}}H_z$ and a typical vector potential for $\mathbf{A}_a$ will be

$$\mathbf{A}_a = (-\tfrac{1}{2}yH_z, \tfrac{1}{2}xH_z, 0) \tag{7.5.40a}$$

The vector potential $\mathbf{A}_w$ refers to the electromagnetic wave and we have neglected the terms in $(\mathbf{A}_w.\mathbf{A}_w)$, and $(\mathbf{A}_a.\mathbf{A}_w)$. We have also chosen $\mathbf{A}$ such that

$$\nabla.\mathbf{A}_a = 0, \quad \nabla.\mathbf{A}_w = 0, \quad \phi = 0 \tag{7.5.40b}$$

Assume that it is possible to first solve the Schrödinger equation for the Hamiltonian operator $(\mathscr{H}_0 + \mathscr{H}_a)$ and that the resulting eigenfunctions are

$\Psi(1, \mathbf{H}_a)$, $\Psi(2, \mathbf{H}_a)$, ... $\Psi(l, \mathbf{H}_a)$. We may now use the perturbation theory discussed previously in order to derive the electric polarization due to the electromagnetic wave. We neglect second order effects arising from the spatial variation of the electromagnetic wave over the molecular dimensions, and take

$$\mathbf{A}_W = \tfrac{1}{2}\mathbf{A}_0(e^{j\omega t} + e^{-j\omega t}) \tag{7.5.40c}$$

Consider the induced electric dipole moment arising from the state $\Psi(1, \mathbf{H}_a)$. From equation (7.4.55)

$$\mathbf{p}^{(i)}(1, \mathbf{H}_a) = 2\,\mathrm{Re}\left[\left[\sum_{l\neq 1} a(t, l, \mathbf{H}_a) \exp\left[\left\{\frac{j(E_1 - E_l)}{\hbar}\right\}t\right]\langle 1, \mathbf{H}_a|\,\mathbf{R}\,|l, \mathbf{H}_a\rangle\right]\right] \tag{7.5.41a}$$

with

$$a(t, l, \mathbf{H}_a)$$

$$= \frac{1}{j\hbar}\int \exp\left[\left\{\frac{j(E_l - E_1)}{\hbar}\right\}t\right] \frac{(-1)\,j(E_l - E_1)}{\hbar c}\,\mathbf{A}_W.\langle l, \mathbf{H}_a|\,\mathbf{R}\,|1, \mathbf{H}_a\rangle\, dt \tag{7.5.41b}$$

$$= \frac{(-1)}{\hbar^2 c}(E_l - E_1)\mathbf{A}_0.\langle l, \mathbf{H}_a|\mathbf{R}|1, \mathbf{H}_a\rangle$$

$$\times \int \exp\left[\left\{\frac{j(E_l - E_1)}{\hbar}\right\}t\right]\frac{(e^{j\omega t} + e^{-j\omega t})}{2}\, dt$$

$$= \frac{jE_{l1}}{\hbar c}\mathbf{A}_0.\langle l, \mathbf{H}_a|\,\mathbf{R}\,|1, \mathbf{H}_a\rangle \frac{(E_{l1}\cos\omega t - j\hbar\sin\omega t)}{E_{l1}^2 - \hbar^2\omega^2}\, e^{j(E_{l1}/\hbar)t} \tag{7.5.41c}$$

where $E_{l1} = (E_l - E_1)$.

Hence

$$\mathbf{p}^{(i)}(1, \mathbf{H}_a)$$

$$= 2\,\mathrm{Re}\left[\sum_{l\neq 1}\frac{E_{l1}}{E_{l1}^2 - \hbar^2\omega^2}\left(\frac{\omega}{c}\mathbf{A}_0.\langle l, \mathbf{H}_a|\,\mathbf{R}\,|1, \mathbf{H}_a\rangle\right)\langle 1, \mathbf{H}_a|\,\mathbf{R}\,|l, \mathbf{H}_a\rangle \sin\omega t\right]$$

$$+ 2\,\mathrm{Re}\left[\sum_{l\neq 1}\frac{jE_{l1}^2}{\hbar(E_{l1}^2 - \hbar^2\omega^2)}\left(\frac{1}{c}\mathbf{A}_0.\langle l, \mathbf{H}a|\,\mathbf{R}\,|1, \mathbf{H}a\rangle\right)\langle 1, \mathbf{H}a|\mathbf{R}\,|l, \mathbf{H}_a\rangle \cos\omega t\right] \tag{7.5.41d}$$

Using the relations,

$$\mathbf{E} = -\frac{1}{c}\frac{\partial \mathbf{A}}{\partial t} = \frac{\omega}{c}\mathbf{A}_0 \sin\omega t$$

$$\frac{\partial \mathbf{E}}{\partial t} = \frac{\omega^2}{c}\mathbf{A}_0\cos\omega t \tag{7.5.42}$$

We may write

$$\mathbf{p}^{(i)}(1, \mathbf{H}_a) = 2\,\mathrm{Re}\left[\sum_{l\neq 1}\frac{E_{l1}}{E_{l1}^2 - \hbar^2\omega^2}\langle 1, \mathbf{H}_a|\,\mathbf{R}\,|l, \mathbf{H}_a\rangle\Big(\langle l, \mathbf{H}_a|\,\mathbf{R}\,|1, \mathbf{H}_a\rangle.\mathbf{E}\Big)\right] + 2\,\mathrm{Re}\left[\sum_{l\neq 1}\frac{\mathrm{j}E_{l1}/\hbar\omega^2}{E_{l1}^2 - \hbar^2\omega^2}\langle 1, \mathbf{H}_a|\,\mathbf{R}\,|l, \mathbf{H}_a\rangle\Big(\langle l, \mathbf{H}_a|\,\mathbf{R}\,|1, \mathbf{H}_a\rangle.\frac{\partial\mathbf{E}}{\partial t}\Big)\right] \tag{7.5.43}$$

This relation corresponds with that derived previously in equation (7.4.71) if the second order terms in that equation are neglected. The discussion following equation (7.4.71), however, considered an isotropic material whereas we now wish to examine the implications of equation (7.5.43) without this restriction. Simplify the notation by writing

$$\mathbf{R}_l = \langle l, \mathbf{H}_a|\,\mathbf{R}\,|1, \mathbf{H}_a\rangle \tag{7.5.44}$$

and consider the x component of $\mathbf{p}^{(i)}(1, \mathbf{H}_a)$ as being a typical term in equation (7.5.43).

$$\mathbf{p}_x^{(i)}(1, \mathbf{H}_a) = 2\,\mathrm{Re}\left[\sum_{l\neq 1}\frac{E_{l1}}{E_{l1}^2 - \hbar^2\omega^2}\Big(|R_{lx}|^2\,E_x + R_{lx}^*\,R_{ly}\,E_y + R_{lx}^*R_{ez}E_z\Big)\right] + 2\,\mathrm{Re}\left[\sum_{l\neq 1}\frac{\mathrm{j}E_{l1}^2/\hbar\omega^2}{E_{l1}^2 - \hbar^2\omega^2}\Big(|R_{lx}|^2\,\frac{\partial E_x}{\partial t} + R_{lx}^*\,R_{ly}\,\frac{\partial E_y}{\partial t} + R_{lx}^*\,R_{lz}\,\frac{\partial E_z}{\partial t}\Big)\right] \tag{7.5.45}$$

In this equation $\mathbf{E}$ and $\partial\mathbf{E}/\partial t$ are real quantities and so we require the real parts of terms like $R_{lx}^*\,R_{ly}$. We may use the following notation to write equation (7.5.43) in terms of real quantities.

$$\mathrm{Re}\,[\{R_{lx}^*\,R_{ly}\}] = \frac{R_{lx}^*\,R_{ly} + R_{lx}\,R_{ly}^*}{2} \equiv \tfrac{1}{2}\{R_x\,R_y\}_l \tag{7.5.46}$$

$$\mathrm{Re}\,[\{\mathrm{j}R_{lx}^*\,R_{ly}\}] = \frac{\mathrm{j}R_{lx}^*\,R_{ly} - \mathrm{j}R_{lx}\,R_{ly}^*}{2} \equiv \frac{\mathrm{j}}{2}[R_x\,R_y]_l \tag{7.5.47}$$

Hence

$$p_x^{(i)}(1, \mathbf{H}_a) = 2\sum_{l\neq 1}\frac{E_{l1}}{E_{l1}^2 - \hbar^2\omega^2}[|R_{lx}|^2\,E_x + \tfrac{1}{2}\{R_x\,R_y\}_l\,E_y + \tfrac{1}{2}\{R_x\,R_z\}_l\,E_z] + 2\sum_{l\neq 1}\frac{E_{l1}^2/\hbar\omega^2}{E_{l1}^2 - \hbar^2\omega^2}\cdot\frac{\mathrm{j}}{2}\left[[R_x\,R_y]_l\,\frac{\partial E_y}{\partial t} + [R_x\,R_z]_l\,\frac{\partial E_z}{\partial t}\right] \tag{7.5.48}$$

Suppose that the electromagnetic wave propagates along the z-direction. From the transverse nature of $\mathbf{D}$, we have $D_z = 0$. We may also take $E_z = 0$ without losing any significant physical feature, so that

$$p_x^{(i)}(1, \mathbf{H}_a) = 2\sum_{l\neq 1}\frac{E_{l1}}{E_{l1}^2 - \hbar^2\omega^2}\left[R_{lx}|^2E_x + \tfrac{1}{2}\{R_x\,R_y\}_l\,E_y\right] + 2\sum_{l\neq 1}\frac{E_{l1}^2/\hbar\omega^2}{E_{l1}^2 - \hbar^2\omega^2}\cdot\frac{\mathrm{j}}{2}[R_x\,R_y]_l\,\frac{\partial E_y}{\partial t} \tag{7.5.49}$$

$$p_y^{(i)}(1, H_a) = 2\sum_{l\neq 1}\frac{E_{l1}}{E_{l1}^2 - \hbar^2\omega^2}\Big[|R_{ly}|^2\, E_y + \tfrac{1}{2}\{R_y\, R_x\}_l\, E_x\Big]$$
$$+ 2\sum_{l\neq 1}\frac{E_{l1}^2/\hbar\omega^2}{E_{l1}^2 - \hbar^2\omega^2}\cdot\frac{j}{2}[R_y\, R_x]_l\,\frac{\partial E_x}{\partial t} \qquad (7.5.50)$$

Since

$$\{R_x\, R_y\}_l = \{R_y\, R_x\}_l$$
$$[R_x\, R_y]_l = -[R_y\, R_x]_l \qquad (7.5.51)$$

we may evidently write

$$\begin{bmatrix} p_x^{(i)}(1, \mathbf{H}_a) \\ p_y^{(i)}(1, \mathbf{H}_a)\end{bmatrix} = \begin{bmatrix}\chi_{11} & \chi_{12} \\ \chi_{12} & \chi_{22}\end{bmatrix}\begin{bmatrix}E_x \\ E_y\end{bmatrix} + \begin{bmatrix}0 & \Delta_{12} \\ -\Delta_{12} & 0\end{bmatrix}\begin{bmatrix}\dfrac{\partial E_x}{\partial t} \\ \dfrac{\partial E_y}{\partial t}\end{bmatrix} \qquad (7.5.52)$$

In the matrix $[\chi]$, the elements χ_{12}, χ_{21} are equal. This matrix therefore describes the electric susceptibility of a straightforward anisotropic material. The matrix $[\Delta]$ gives rise to Faraday rotation of the plane of polarization. If the material is isotropic except for the Faraday rotation, we may put $\chi_{11} = \chi_{22} = \chi$ and $\chi_{12} = \chi_{21} = 0$, and now write equation (7.5.52) as

$$\mathbf{p}^{(i)}(1, \mathbf{H}_a) = \chi\mathbf{E} + \begin{bmatrix}0 & \Delta_{12} \\ -\Delta_{12} & 0\end{bmatrix}\begin{bmatrix}\hat{\mathbf{j}}\dfrac{\partial E_x}{\partial t} \\ \hat{\mathbf{i}}\dfrac{\partial E_y}{\partial t}\end{bmatrix} \qquad (7.5.53)$$

Recalling that χ and Δ_{12} are real because no dissipative damping terms have been included in our analysis, we may write equation (7.5.53) using the complex representation for $\mathbf{E}$.

$$\mathbf{p}^{(i)}(1, \mathbf{H}_a) = \chi(\omega)\mathbf{E} + j\omega\begin{bmatrix}0 & \Delta_{12} \\ -\Delta_{12} & 0\end{bmatrix}\begin{bmatrix}\hat{\mathbf{j}}E_x \\ \hat{\mathbf{i}}E_y\end{bmatrix} \qquad (7.5.54)$$

Take the applied magnetic field $\mathbf{H}_a$ to be directed along the z-axis also, $\mathbf{H}_a = \hat{\mathbf{k}}H_z$, and expand Δ_{12} in powers of H_z. Since the Faraday rotation goes to zero as H_z goes to zero and changes sign with H_z, the first term in this expansion can be written as $\alpha_1 H_z$. To first order therefore, for propagation along the magnetic field,

$$\mathbf{p}^{(i)}(1, \mathbf{H}_a) = \chi(\omega)\mathbf{E} + j\omega\begin{bmatrix}0 & \alpha_1 H_z \\ -\alpha_1 H_z & 0\end{bmatrix}\begin{bmatrix}\hat{\mathbf{j}}E_x \\ \hat{\mathbf{i}}E_y\end{bmatrix} \qquad (7.5.55)$$

$$= \chi(\omega)\mathbf{E} - j\omega\alpha_1\, \mathbf{H}_a \wedge \mathbf{E} \qquad (7.5.56)$$

If there are N_1 electrons per unit volume in the state $\Psi(1, \mathbf{H}_a)$, the corresponding electric displacement is

$$\mathbf{D} = \varepsilon(\omega)\mathbf{E} - j\omega 4\pi N_1\alpha_1\, \mathbf{H}_a \wedge \mathbf{E} \qquad (7.5.57)$$

This equation is of the same mathematical form as equation (7.5.13) if the factor $(jc\beta/\mu)\mathbf{k}$ is replaced by $-j\omega 4\pi N_1\alpha_1\mathbf{H}_a$. It will lead to a rotation of the plane of polarization just as was the case for natural optical activity. The (ω, k) dispersion relation corresponding to equation (7.5.18) is

$$\frac{c^2 k_z^2}{\omega^2 \mu} = \varepsilon(\omega) \pm 4\pi N_1 \alpha_1 \omega H_z \tag{7.5.58}$$

and the rotation of the plane of polarization arising from the ground state $\Psi(1, \mathbf{H}_a)$ is given by

$$\frac{\phi(1, H_z)}{zH_z} = \frac{2\pi N_1 \omega^2 \alpha_1}{c} \sqrt{\left(\frac{\mu}{\varepsilon}\right)} \tag{7.5.59}$$

The ratio (ϕ/zH_z) is Verdet's constant, typical values are given in Table 7.2. From equations (7.5.55), (7.5.52) and (7.5.49), we may identify α_1 through $\alpha_1 H_z = \Delta_{12}$ and derive

$$\frac{\phi(1, H_z)}{zH_z} = \frac{N_1}{c}\left(\frac{\mu}{\varepsilon}\right)^{\frac{1}{2}} \frac{2\pi\omega^2}{H_z} \sum_{l \neq 1} \frac{E_{l1}^2/\hbar\omega^2}{E_{l1}^2 - \hbar^2\omega^2} \mathrm{j}[R_x\, R_y]_l \tag{7.5.60}$$

Evidently this is also the rotation which will occur if the ground state is a singlet.

TABLE 7.2

Typical Values for the Verdet Constant for visible radiation (in angular minutes cm^{-1} $gauss^{-1}$)

Substance	*Verdet constant*
Glass (light flint)	0·032
Glass (dense flint)	0·118
Quartz	0·017
Sapphire	0·024
NaCl	0·041
H_2O	0·013
Acetone	0·011
Benzene	0·030

The induced electric dipole moment arising from the second state $\Psi(2, \mathbf{H}_a)$ will be of the same form as that from $\Psi(1, \mathbf{H}_a)$ and we shall evidently derive equations (7.5.56) and (7.5.57) but now with N_2 and α_2 referring to the state $\Psi(2, \mathbf{H}_a)$. The two states taken together will give rise to an electric displacement.

$$\mathbf{D} = \varepsilon(\omega)\mathbf{E} - \mathrm{j}\omega 4\pi(N_1\alpha_1 + N_2\alpha_2)\mathbf{H}_a \wedge \mathbf{E} \tag{7.5.61}$$

The rotation of the plane of polarization from the two ground states is therefore given by

$$\frac{\phi(1, 2, \mathbf{H}_a)}{zH_z} = \frac{2\pi\omega^2}{cH_z}\left(\frac{\mu}{\varepsilon}\right)^{\frac{1}{2}} \left\{ \sum_{l \neq 1} \frac{N_1 E_{l1}^2/\hbar\omega^2}{E_{l1}^2 - \hbar^2\omega^2} \mathrm{j}[R_x\, R_y]_l^{(1)} + \sum_{l \neq 2} \frac{N_2 E_{l2}^2/\hbar\omega^2}{E_{l2}^2 - \hbar^2\omega^2} \mathrm{j}[R_x\, R_y]_l^{(2)} \right\} \tag{7.5.62}$$

7.5.3 Examples on Faraday Rotation

To illustrate this general discussion of the quantum theory of the dielectric constant we consider two examples of Faraday rotation based on hydrogen-like wave functions. The first example ignores electron spin. It is therefore somewhat artificial but it provides a convenient illustration for a ground state singlet 'molecule' and corresponds closely to the explanation of Faraday rotation on the classical theory. The second example includes electron spin. This is more realistic but the mathematics is less straightforward.

Taking the applied magnetic field $\mathbf{H}_a = \hat{\mathbf{k}}H_z$, and the vector potential $\mathbf{A}_a$ from equation (7.5.40a), the Hamiltonian operator $\mathscr{H}_a$ in equation (7.5.39c) may be written as

$$\mathscr{H}_a = \frac{jq\hbar}{mc}\mathbf{A}_a . \nabla \tag{7.5.63}$$

$$= \frac{jq\hbar}{mc} \cdot \frac{H_z}{2}(-\hat{\mathbf{i}}y + \hat{\mathbf{j}}x) . \nabla$$

$$= \frac{jq\hbar}{2mc}H_z\left(x\frac{\partial}{dy} - y\frac{\partial}{dx}\right) = -\frac{qH_z}{2mc}L_z \tag{7.5.64}$$

Here the term involving $\mathbf{A}_a . \mathbf{A}_a$ in equations (7.5.39) has been neglected and the operator L_z represents the z component of angular momentum.

The unperturbed Hamiltonian $\mathscr{H}_0$ refers to the hydrogen atom without spin and we now seek eigenfunctions of $\mathscr{H}_0 + \mathscr{H}_a$

$$\mathscr{H}_0 + \mathscr{H}_a = \left[-\frac{\hbar^2}{2m}\nabla^2 + V(\mathbf{r})\right] - \frac{qH_z}{2mc}L_z \tag{7.5.65}$$

Base the calculation on the hydrogen-like function $\psi(1s)$ for the ground state and take the functions $\psi(2s)$, $\psi(2p)$ for the excited states as shown in Fig. 7.6. Assume that other possible excited states are sufficiently far away to give no significant contribution to the Faraday rotation. The eigenfunctions of equation (7.5.65) are†

$$\psi(1s) = R_{1s}(r)\, Y_{00} = R_{1s}(r)\frac{1}{(4\pi)^{\frac{1}{2}}} \tag{7.5.66}$$

$$\psi(2s) = R_{2s}(r)\, Y_{00} = R_{2s}(r)\frac{1}{(4\pi)^{\frac{1}{2}}} \tag{7.5.67}$$

$$\psi(2p_1) = R_{2p}(r)\, Y_{11} = R_{2p}(r)\left(\frac{3}{8\pi}\right)^{\frac{1}{2}} \sin\theta\, e^{j\phi} \tag{7.5.68}$$

$$\psi(2p_0) = R_{2p}(r)\, Y_{10} = R_{2p}(r)\left(\frac{3}{4\pi}\right)^{\frac{1}{2}} \cos\theta \tag{7.5.69}$$

$$\psi(2p_{-1}) = R_{2p}(r)\, Y_{1,-1} = R_{2p}(r)\left(\frac{3}{8\pi}\right)^{\frac{1}{2}} \sin\theta\, e^{-j\phi} \tag{7.5.70}$$

† See, for example, Schiff, L. I., *Quantum Mechanics*, pp. 70 and 73, McGraw-Hill, 1949.

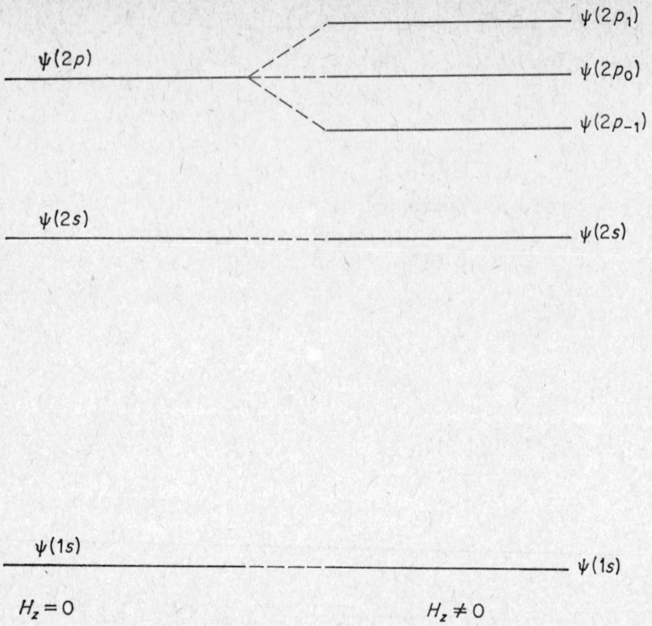

FIG. 7.6 Scheme of hydrogen-like energy states in a magnetic field H_z.

In these equations the functions $R(r)$ are solutions of the radial part of the Hamiltonian. We shall not need an explicit representation for these functions. The functions Y_{mn} are spherical harmonics which are written out explicitly.

It is evident from equation (7.5.60) that we shall need to evaluate integrals of the type $\int \psi^*(1s) x\, \psi(l)\, \mathrm{d}^3\mathbf{r}$ and $\int \psi^*(1s) y\, \psi(l)\, \mathrm{d}^3\mathbf{r}$. Using polar coordinates, take

$$x = r \sin\theta \cos\phi \qquad y = r \sin\theta \sin\phi \tag{7.5.71}$$

and note immediately that the integrals involving $\psi(2s)$ and $\psi(2p_0)$ are zero because the integration over ϕ vanishes. We therefore need to consider the function $[R_x\, R_y]_l$ for the two excited states $\psi(2p_1)$, $\psi(2p_{-1})$.

From equations (7.5.47), (7.5.44) and (7.4.46)

$$\begin{aligned}[R_x\, R_y]_l &= [R^*_{lx}\, R_{ly} - R_{lx}\, R^*_{ly}] \\ &= q^2[\langle 1s|\, x\, |l\rangle\langle l|y\, |1s\rangle - \langle 1s|\, y\, |l\rangle\langle l|\, x\, |1s\rangle]\end{aligned} \tag{7.5.72}$$

Consider the excited state $|l\rangle = \psi(2p_1)$

$$\langle 1s|\, x\, |l\rangle = \int R_{1s}(r) \frac{1}{(4\pi)^{\frac{1}{2}}} r \sin\theta \cos\phi\, R_{2p}(r) \left(\frac{3}{8\pi}\right)^{\frac{1}{2}} \sin\theta\, \mathrm{e}^{\mathrm{j}\phi}\, r^2 \sin\theta\, \mathrm{d}\theta\, \mathrm{d}\phi\, \mathrm{d}r \tag{7.5.73}$$

$$= C_{\mathrm{r}} \frac{1}{4\pi} \left(\frac{3}{2}\right)^{\frac{1}{2}} \int_0^{\pi} \sin^3\theta\, \mathrm{d}\theta \int_0^{2\pi} \frac{(\mathrm{e}^{\mathrm{j}\phi} + \mathrm{e}^{-\mathrm{j}\phi})}{2}\, \mathrm{e}^{\mathrm{j}\phi}\, \mathrm{d}\phi$$

$$\langle 1s|\, x\, |l\rangle = \frac{C_{\mathrm{r}}}{\sqrt{6}} \tag{7.5.74}$$

where C_r represents the integral over the radial part of the wave functions. Similarly

$$\langle l|\, y\, |1s\rangle = \int R_{2p}(r) \left(\frac{3}{8\pi}\right)^{\frac{1}{2}} \sin\theta\, e^{-j\phi}\, r\sin\theta \sin\phi \frac{R_{1s}(r)}{(4\pi)^{\frac{1}{2}}} r^2 \sin\theta\, d\theta\, d\phi\, dr$$

$$\langle l|\, y\, |1s\rangle = \frac{C_r}{j\sqrt{6}} \qquad (7.5.75)$$

$$\langle 1s|\, y\, |l\rangle = -\frac{C_r}{j\sqrt{6}} \qquad (7.5.76)$$

$$\langle l|\, x\, |1s\rangle = \frac{C_r}{\sqrt{6}} \qquad (7.5.77)$$

Hence, for the excited state $|l\rangle = \psi(2p_1)$

$$[R_x R_y]_l = q^2 \frac{C_r^2}{3j} \qquad (7.5.78)$$

Similarly, for the excited state $|l\rangle = \psi(2p_{-1})$

$$[R_x R_y]_l = -q^2 \frac{C_r^2}{3j} \qquad (7.5.79)$$

From equation (7.5.60), the Faraday rotation is given by

$$\frac{\phi}{z} = \frac{2\pi N\omega^2 q^2}{c}\left(\frac{\mu}{\varepsilon}\right)^{\frac{1}{2}} \frac{C_r^2}{3\hbar\omega^2}\left\{\frac{E^2(1)}{E^2(1)-\hbar^2\omega^2} - \frac{E^2(-1)}{E^2(-1)-\hbar^2\omega^2}\right\} \qquad (7.5.80)$$

$$= \frac{2\pi N\omega^2 q^2}{c}\left(\frac{\mu}{\varepsilon}\right)^{\frac{1}{2}} \frac{\hbar C_r^2}{3}\left\{\frac{-\{E^2(1)-E^2(-1)\}}{(E^2(1)-\hbar^2\omega^2)(E^2(-1)-\hbar^2\omega^2)}\right\} \qquad (7.5.81)$$

For a negatively charged electron, ignoring spin, we obtain from equations (7.5.65), (7.5.68) and (7.5.70)

$$E(1) = E_0 + \frac{|e|H_z}{2mc}\hbar = E_0 + \tfrac{1}{2}\hbar\omega_L$$

$$E(-1) = E_0 - \frac{|e|H_z}{2mc}\hbar = E_0 - \tfrac{1}{2}\hbar\omega_L \qquad (7.5.82)$$

with

$$\omega_L = \frac{|e|H_z}{mc}$$

Equation (7.5.81) therefore, gives, with $q = -|e|$,

$$\frac{\phi}{z} = -\frac{2\pi N\omega^2 e^2}{c}\left(\frac{\mu}{\varepsilon}\right)^{\frac{1}{2}} \times \frac{\hbar C_r^2}{3}\left[\frac{(E_0+\frac{1}{2}\hbar\omega_L)^2-(E_0-\frac{1}{2}\hbar\omega_L)^2}{\{(E_0+\frac{1}{2}\hbar\omega_L)^2-\hbar^2\omega^2\}\{(E_0-\frac{1}{2}\hbar\omega_L)^2-\hbar^2\omega^2\}}\right] \qquad (7.5.83a)$$

and the term linear in H_z gives

$$\frac{\phi}{z} = -\frac{2\pi N\omega^2 e^2}{c}\left(\frac{\mu}{\varepsilon}\right)^{\frac{1}{2}} \frac{\hbar C_r^2 E_0}{3} \cdot \frac{2\hbar\omega_L}{(E_0^2-\hbar^2\omega^2)^2} \qquad (7.5.83b)$$

We may evaluate the factor C^2E_0 for the states $\psi(1s)$, $\psi(2p)$ of this example.

$$C_r^2E_0 = \frac{\hbar^2}{m} \times \text{(a numerical constant)}$$

Hence, upon writing $\hbar\omega_0 = E_0$, $\omega_L = \frac{|e|H_z}{mc}$, the Faraday rotation becomes

$$\frac{\phi}{z} = -\frac{2\pi N\omega^2|e|^3}{m^2c^2}\cdot\frac{1}{(\omega_0^2-\omega^2)^2}\left(\frac{\mu}{\varepsilon}\right)^{\frac{1}{2}}H_z \times \text{(a numerical constant)} \quad (7.5.83c)$$

which has the same form as the classical relation given in equation (7.5.37b).

We now wish to include electron spin in the calculation of the Faraday rotation. The unperturbed Hamiltonian operator $\mathscr{H}_0$ in equation (7.5.65) will be modified by a term representing the spin-orbit interaction energy whilst $\mathscr{H}_a$ will include the spin energy in a magnetic field. The Hamiltonian operator $(\mathscr{H}_0 + \mathscr{H}_a)$ is therefore

$$\mathscr{H}_0 + \mathscr{H}_a = \left[-\frac{\hbar^2}{2m}\nabla^2 + V(\mathbf{r}) + \zeta\mathbf{L}.\mathbf{S}\right] + \frac{|e|H_z}{2mc}(L_z + 2S_z) \quad (7.5.84)$$

In this equation $\mathbf{L}$ is the orbital angular momentum operator and $\mathbf{S}$ is the spin angular momentum operator for the electron. Assuming that the magnetic field energy is small compared with the spin-orbit term (this corresponds to the weak field Zeeman approximation) the eigenfunctions of $(\mathscr{H}_0 + \mathscr{H}_a)$ are to a reasonable approximation also the eigenfunctions of $\mathbf{J}^2 = (\mathbf{L}+\mathbf{S})^2$ and of $J_z = (L_z + S_z)$. The relevant functions for our calculation based on hydrogen-like $\psi(1s)$ and $\psi(2p)$ functions are given by†

$$\begin{array}{lll} {}^2S_{1/2} & \left\{\begin{array}{l} J_z = \frac{1}{2} \\ J_z = -\frac{1}{2} \end{array}\right. & \begin{array}{l} \psi = (+)R_{1s}(r)\,Y_{0,0} \\ \psi = (-)R_{1s}(r)\,Y_{0,0} \end{array} \\ \text{Ground State} & & \end{array} \quad (7.5.85)$$

$$\begin{array}{ll} {}^2P_{1/2} \\ \text{Excited State} \end{array} \left\{\begin{array}{ll} J_z = \frac{1}{2} & \psi = \frac{1}{\sqrt{3}}R_{2p}(r)\{(+)\,Y_{1,0} - \sqrt{2}(-)\,Y_{1,1}\} \\ J_z = -\frac{1}{2} & \psi = \frac{1}{\sqrt{3}}R_{2p}(r)\{(-)\,Y_{1,0} - \sqrt{2}(+)Y_{1,-1}\} \end{array}\right. \quad (7.5.86)$$

$$\begin{array}{ll} {}^2P_{3/2} \\ \text{Excited State} \end{array} \left\{\begin{array}{ll} J_z = \frac{3}{2} & \psi = (+)R_{2p}(r)\,Y_{1,1} \\ J_z = \frac{1}{2} & \psi = \frac{1}{\sqrt{3}}R_{2p}(r)\{\sqrt{2}(+)Y_{1,0} + (-)Y_{1,1}\} \\ J_z = -\frac{1}{2} & \psi = \frac{1}{\sqrt{3}}R_{2p}(r)\{\sqrt{2}(-)Y_{1,0} + (+)Y_{1,-1}\} \\ J_z = -\frac{3}{2} & \psi = (-)R_{2p}(r)Y_{1,-1} \end{array}\right. \quad (7.5.87)$$

In these equations the functions $R(r)$, Y_{mn}, correspond to those given previously in equations (7.5.66)–(7.5.70), whilst the spin functions are represented by the symbols $(+)$ and $(-)$. From the orthogonality of the spin functions and the properties of the spherical harmonics discussed

† See, for example, SCHIFF, L. I., *Quantum Mechanics*, p. 281, McGraw-Hill, 1949.

following equation (7.5.71), it can be seen that the following functions will be coupled together by

$$x = r \sin \theta \cos \phi \qquad y = r \sin \theta \sin \phi$$

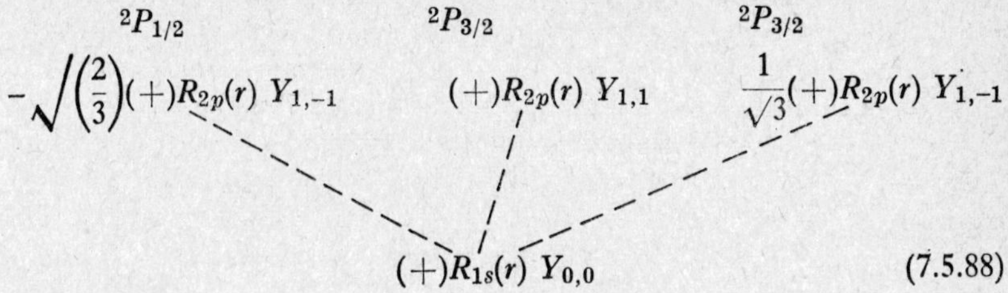

$$\begin{array}{ccc} {}^2P_{1/2} & {}^2P_{3/2} & {}^2P_{3/2} \\ -\sqrt{\left(\frac{2}{3}\right)}(+)R_{2p}(r)\,Y_{1,-1} & (+)R_{2p}(r)\,Y_{1,1} & \frac{1}{\sqrt{3}}(+)R_{2p}(r)\,Y_{1,-1} \end{array}$$

$$(+)R_{1s}(r)\,Y_{0,0} \qquad (7.5.88)$$

$$\begin{array}{ccc} {}^2P_{1/2} & {}^2P_{3/2} & {}^2P_{3/2} \\ -\sqrt{\left(\frac{2}{3}\right)}(-)R_{2p}(r)\,Y_{1,1} & \frac{1}{\sqrt{3}}(-)R_{2p}(r)\,Y_{1,1} & (-)R_{2p}(r)\,Y_{1,-1} \end{array}$$

$$(-)R_{1s}(r)\,Y_{0,0} \qquad (7.5.89)$$

We need to evaluate integrals of the form

$$\int R_{1s}(r)\,Y_{0,0}\binom{x}{y}R_{2p}(r)\,Y_{1,1}\,d^3\mathbf{r}$$

These integrals have already been calculated in equations (7.5.73)–(7.5.77). We obtain the following values for $[R_x R_y]_l$ by using equations (7.5.78) and (7.5.79).

TABLE 7.3

Ground state	Excited state	$[R_x R_y]_l$	Energy designation
$(+)R_{1s}(r)\,Y_{0,0}$	$-\sqrt{\left(\frac{2}{3}\right)}(+)R_{2p}(r)\,Y_{1,-1}$	$q^2\left(-\frac{C_r^2}{3j}\right)\frac{2}{3}$	E_{II}, E_a
$(+)R_{1s}(r)\,Y_{0,0}$	$(+)R_{2p}(r)\,Y_{1,1}$	$q^2\left(\frac{C_r^2}{3j}\right)$	E_{II}, E_f
$(+)R_{1s}(r)\,Y_{0,0}$	$\frac{1}{\sqrt{3}}(+)R_{2p}(r)\,Y_{1,-1}$	$q^2\left(-\frac{C_r^2}{3j}\right)\frac{1}{3}$	E_{II}, E_d
$(-)R_{1s}(r)\,Y_{0,0}$	$-\sqrt{\left(\frac{2}{3}\right)}(-)R_{2p}(r)\,Y_{1,1}$	$q^2\left(\frac{C_r^2}{3j}\right)\frac{2}{3}$	E_{I}, E_b
$(-)R_{1s}(r)\,Y_{0,0}$	$\frac{1}{\sqrt{3}}(-)R_{2p}(r)\,Y_{1,1}$	$q^2\left(\frac{C_r^2}{3j}\right)\frac{1}{3}$	E_{I}, E_e
$(-)R_{1s}(r)\,Y_{0,0}$	$(-)R_{2p}(r)\,Y_{1,-1}$	$q^2\left(-\frac{C_r^2}{3j}\right)$	E_{I}, E_c

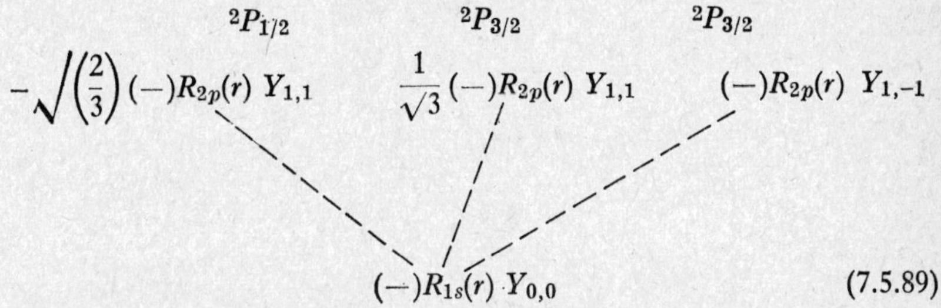

In Table 7.3 the energy designations correspond with the scheme shown in Fig. 7.7 and serve to identify the associated Zeeman Transitions.

The summations over the excited states required for equation (7.5.62) are evidently rather complicated. We may simplify the mathematics by making the approximation of neglecting the energy level splitting arising from the magnetic field in the terms E_{l1}, E_{l2}. This is consistent with our choice of eigenfunctions which refer to the weak field Zeeman region where the spin-orbit splitting is large compared with that due to the magnetic field. Collecting together the factors $[R_x R_y]_l$ from Table 7.3, we obtain from equation (7.5.62) with $q = -|e|$,

$$\frac{\phi(\mathrm{I}, \mathrm{II}, H_z)}{z} = \frac{2\pi\omega^2 e^2}{c}\left(\frac{\mu}{\varepsilon}\right)^{\frac{1}{2}} \frac{C_r^2}{3\hbar\omega^2} \times \left\{\left(-\frac{E_{\alpha 1}^2}{E_{\alpha 1}^2 - \hbar^2\omega^2} + \frac{E_{\beta 1}^2}{E_{\beta 1}^2 - \hbar^2\omega^2}\right)\frac{2}{3}N_{\mathrm{II}} + \left(\frac{E_{\alpha 1}^2}{E_{\alpha 1}^2 - \hbar^2\omega^2} - \frac{E_{\beta 1}^2}{E_{\beta 1}^2 - \hbar^2\omega^2}\right)\frac{2}{3}N_{\mathrm{I}}\right\} \tag{7.5.90}$$

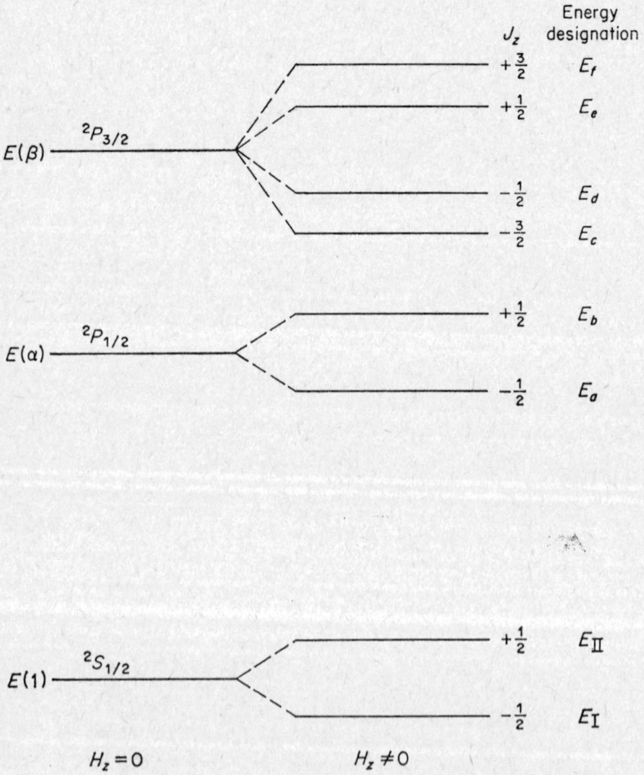

FIG. 7.7 Energy level scheme for $^2S_{\frac{1}{2}}$, $^2P_{\frac{1}{2}}$, $^2P_{\frac{3}{2}}$ states.

$$\frac{\phi(\mathrm{I}, \mathrm{II}, H_z)}{z} = \frac{2\pi\omega^2 e^2}{c}\left(\frac{\mu}{\varepsilon}\right)^{\frac{1}{2}} \frac{C_r^2}{3\hbar\omega^2} \times \frac{2}{3}\left(\frac{E_{\alpha 1}^2}{E_{\alpha 1}^2 - \hbar^2\omega^2} - \frac{E_{\beta 1}^2}{E_{\beta 1}^2 - \hbar^2\omega^2}\right)(N_{\mathrm{I}} - N_{\mathrm{II}}) \tag{7.5.91}$$

In this equation the magnetic field dependence is contained in the factor $(N_{\mathrm{I}} - N_{\mathrm{II}})$. For thermal equilibrium

$$N_{\mathrm{I}} - N_{\mathrm{II}} = N\left\{\frac{\exp(-E_{\mathrm{I}}/kT) - \exp(-E_{\mathrm{II}}/kT)}{\exp(-E_{\mathrm{I}}/kT) + \exp(-E_{\mathrm{II}}/kT)}\right\} = N\left\{\frac{\exp(+\mu_z H_z/kT) - \exp(-\mu_z H_z/kT)}{\exp(+\mu_z H_z/kT) + \exp(-\mu_z H_z/kT)}\right\} \tag{7.5.92}$$

$$N_{\mathrm{I}} - N_{\mathrm{II}} = N \tanh\left(\frac{\mu_z H_z}{kT}\right) \tag{7.5.93}$$

Here $N = (N_{\mathrm{I}} + N_{\mathrm{II}})$ is the total number of electrons per unit volume and $\mu_z = \beta$ is the magnetic moment of an electron along the field H_z. Equation (7.5.91) may therefore be written in the form

$$\frac{\phi}{z} = \rho \tanh\left(\frac{\beta H_z}{kT}\right) \tag{7.5.94}$$

where ρ is a constant factor. The term $\tanh(\beta_z H_z/kT)$ is proportional to the magnetization of the material which results from the splitting of the doubly degenerate ground state. For $\beta H_z \ll kT$ the function $\tanh(\beta_z H_z/kT)$ may be expanded, the first term being linear in H_z. In this approximation the Verdet constant is given by

$$\frac{\phi}{zH_z} = \rho\frac{\beta}{kT} \equiv \frac{C_0}{T} \tag{7.5.95}$$

and is proportional to the paramagnetic susceptibility of the material.

7.6 Lattice Waves and the Dielectric Constant

The discussion of the dielectric constant presented in the previous sections of this chapter has avoided any explicit representation of the motion of the fundamental molecular units in the material medium. These motions are particularly significant in the case of ionic crystals and we shall now consider the effects which arise from the lattice vibrations taking as a model a simple diatomic crystal which is optically isotropic. For example we may consider a crystal such as NaCl which has the simple cubic structure with Na^+ ions and Cl^- ions on alternate lattice sites.

7.6.1 Mechanical Vibrations

Consider first a very simple linear array formed from Na^+ ions each having charge $+q$, mass M_+, and Cl^- ions each having charge $-q$, and mass M_-. Let the Na^+ ions be situated near the even numbered lattice points with coordinates labelled $(2nd + x_{2n})$ and the Cl^- ions have coordinates labelled $[(2n + 1)d + x_{2n+1}]$ corresponding to the odd numbered lattice points. In this representation, for static equilibrium the adjacent Na^+, Cl^- ions are separated by the same distance, d, throughout the lattice, as shown in Fig. 7.8.

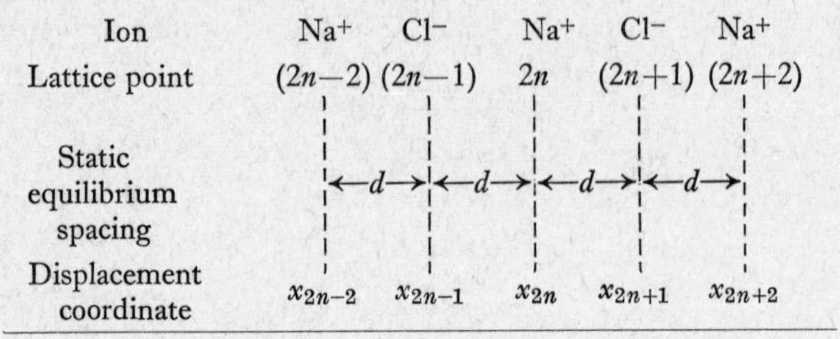

FIG. 7.8 A linear array of Na^+, Cl^- ions.

The mutual potential energy is assumed to depend only upon the distance between two ions. If nearest neighbour interactions alone are significant the potential energy of the Na^+ ion at lattice site $2n$ may be written

$$U_{2n} = U(|d + x_{2n} - x_{2n-1}|) + U(|d + x_{2n+1} - x_{2n}|) \tag{7.6.1}$$

Assuming that the displacements x_{2n}, x_{2n+1}, x_{2n-1} are small compared with d, we expand U_{2n} in a Taylor series to second order,

$$\begin{aligned} U_{2n} = U(2nd) &+ \{|x_{2n} - x_{2n-1}| + |x_{2n+1} - x_{2n}|\}U'(2nd) \\ &+ \tfrac{1}{2}\{(x_{2n} - x_{2n-1})^2 + (x_{2n+1} - x_{2n})^2\}U''(2nd) \end{aligned} \tag{7.6.2}$$

where $U(2nd)$ is the zero order term and $U'(2nd)$, $U''(2nd)$ are derivatives evaluated at the equilibrium coordinate 2nd. Evidently $U'(2nd)$ must be zero since it refers to the equilibrium position where no net force is acting on the ion. Hence the restoring force acting on the displaced Na^+ ion at the lattice point labelled $2n$, is

$$\begin{aligned} F = -\frac{\partial U_{2n}}{\partial x_{2n}} &= -2(2x_{2n} - x_{2n-1} - x_{2n+1})U''(2nd) \\ &= -\beta(2x_{2n} - x_{2n-1} - x_{2n+1}) \end{aligned} \tag{7.6.3}$$

with

$$\beta = 2U''(2nd)$$

The equation of motion for this Na^+ ion in the absence of applied electric fields is therefore

$$M_+\ddot{x}_{2n} = -\beta(2x_{2n} - x_{2n-1} - x_{2n+1}) \tag{7.6.4}$$

Similarly, for the Cl^- ion at the lattice point labelled $(2n + 1)$,

$$M_-\ddot{x}_{2n+1} = -\beta(2x_{2n+1} - x_{2n} - x_{2n+2}) \tag{7.6.5}$$

Equations (7.6.4) and (7.6.5) describe the usual acoustic waves. Solutions are obtained by writing

$$x_{2n} = \xi \, e^{j(\omega t - 2nkd)} \tag{7.6.6}$$

$$x_{2n+1} = \eta \, e^{j(\omega t - (2n+1)kd)} \tag{7.6.7}$$

so that

$$-\omega^2 M_+ \xi = \eta\beta(e^{jkd} + e^{-jkd}) - 2\beta\xi \tag{7.6.8}$$

$$-\omega^2 M_- \eta = \xi\beta(e^{jkd} + e^{-jkd}) - 2\beta\eta \tag{7.6.9}$$

These equations have a non-trivial solution if the determinant of the coefficients is zero,

$$\begin{vmatrix} (2\beta - M_+\omega^2) & -2\beta \cos kd \\ -2\beta \cos kd & (2\beta - M_-\omega^2) \end{vmatrix} = 0 \tag{7.6.10}$$

The (ω, k) dispersion relation is,

$$\omega^4 - 2\beta\omega^2\left(\frac{1}{M_+} + \frac{1}{M_-}\right) + \frac{4\beta^2}{M_+M_-}\sin^2 kd = 0 \tag{7.6.11}$$

giving two branches for the lattice vibrations as shown in Fig. 7.9.

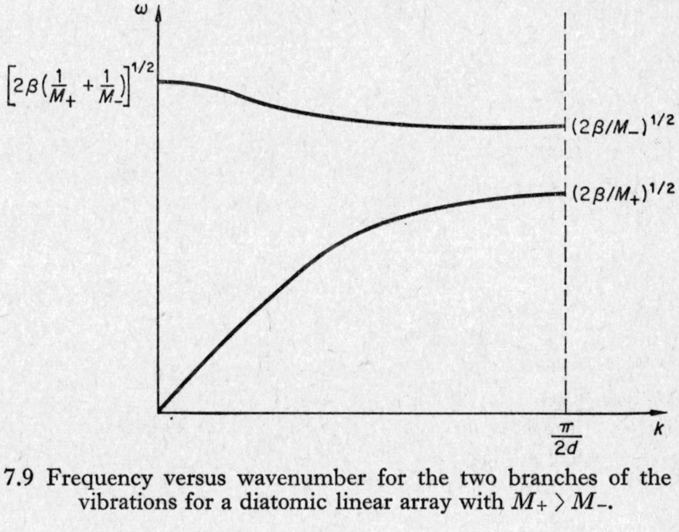

FIG. 7.9 Frequency versus wavenumber for the two branches of the lattice vibrations for a diatomic linear array with $M_+ > M_-$.

7.6.2 COUPLED WAVES

For our purposes, where we are considering the interaction of lattice vibrations with electromagnetic radiation, we may simplify the equations of motion for the ions given in equations (7.6.4) and (7.6.5). The electromagnetic radiation which is of interest usually has a wavelength of 10 micron or longer. Thus the wavenumber is $2\pi . 10^3$ cm^{-1} or smaller. On the other hand, the lattice vibrations have wavenumbers extending up to $\pi/2d \sim \pi . 10^8$ cm^{-1}. We shall therefore be seeking coupled solutions having finite frequencies ($\sim 10^{13}$ c/s) but with wavenumbers ($<10^3$ cm^{-1}) which are essentially zero on the lattice vibration scale. We may therefore take $k = 0$ in equations (7.6.6) and (7.6.7) and write $x_{2n} = x_{2n+2}$, $x_{2n-1} = x_{2n+1}$. Equations (7.6.4) and (7.6.5) now become

$$M_+\ddot{x}_{2n} = -2\beta(x_{2n} - x_{2n+1}) \tag{7.6.12}$$

$$M_-\ddot{x}_{2n+1} = -2\beta(x_{2n+1} - x_{2n}) \tag{7.6.13}$$

Multiply equation (7.6.12) by $M_-/(M_+ + M_-)$, equation (7.6.13) by $M_+/(M_+ + M_-)$ and subtract,

$$\frac{M_+M_-}{M_+ + M_-}(\ddot{x}_{2n} - \ddot{x}_{2n+1}) = -2\beta(x_{2n} - x_{2n+1}) \tag{7.6.14}$$

which may be written formally as

$$\ddot{x} = a_{11}x \tag{7.6.15}$$

with $x = (x_{2n} - x_{2n+1})$ and $a_{11} = -2\beta\left(\dfrac{1}{M_+} + \dfrac{1}{M_-}\right)$. Evidently the new coordinate x is the displacement of a positive ion relative to that of its neighbouring negative ion.

The presence of an electric field will add additional forces $\pm q\mathbf{E}'$ where $\mathbf{E}'$ is the local field at an ion. Writing these forces into equations (7.6.12) and 7.6.13) gives an equation of motion,

$$\ddot{\mathbf{x}} = a_{11}\mathbf{x} + q\left(\frac{1}{M_+} + \frac{1}{M_-}\right)\mathbf{E}' \tag{7.6.16}$$

The electric polarization of the material arising from the ionic displacement, together with the action of the electric field on the individual ions, is

$$\mathbf{P} = \tfrac{1}{2}Nq\mathbf{x} + N\langle\alpha\rangle\mathbf{E}' \tag{7.6.17}$$

where N is the number of ions per unit volume and $\langle\alpha\rangle$ is the mean polarizability per ion which was defined previously in equation (7.2.28). Equations (7.6.16) and (7.6.17) are the material equations which must be solved in conjunction with the electromagnetic field equations in order to describe the interaction of lattice vibrations and electromagnetic waves. We may write them in a more general and yet simplified notation by observing that the local field $\mathbf{E}'$ is linearly related to the macroscopic field $\mathbf{E}$ in the material either through the Onsager relation (equation (7.3.17)) or through the Lorentz polarization factor,

$$\frac{4\pi}{3}\mathbf{P} = \frac{4\pi}{3}\left\{\frac{\varepsilon(\omega) - 1}{4\pi}\right\}\mathbf{E}$$

We now write equations (7.6.16) and (7.6.17) as†

$$\ddot{\mathbf{x}} = a_{11}\mathbf{x} + a_{12}\mathbf{E} \tag{7.6.18}$$

$$\mathbf{P} = a_{21}\mathbf{x} + a_{22}\mathbf{E} \tag{7.6.19}$$

† The factor a_{11} in this equation is strictly not equal to a_{11} in equation (7.6.16) when we use $\mathbf{E}$ rather than $\mathbf{E}'$. This point is discussed in relation to equation (7.6.62) but can be ignored at present.

We use the equations in this form to describe a general diatomic isotropic crystal without restriction to the linear array which was the model system underlying equation (7.6.16). The solutions of interest must be consistent with the electromagnetic field equations where we shall take the field vectors to have a harmonic time dependence in the form $e^{j(\omega t - \mathbf{k}.\mathbf{r})}$. Equations (7.6.18) and (7.6.19) may now be written

$$-\omega^2\mathbf{x} = a_{11}\mathbf{x} + a_{12} \tag{7.6.20}$$

$$\mathbf{P} = \frac{\varepsilon(\omega) - 1}{4\pi}\mathbf{E} = a_{21}\mathbf{x} + a_{22}\mathbf{E} \tag{7.6.21}$$

From which

$$\varepsilon(\omega) = 1 + 4\pi\left(a_{22} + \frac{a_{12}a_{21}}{-a_{11} - \omega^2}\right) \tag{7.6.22}$$

and†

$$\varepsilon(0) = 1 + 4\pi\left(a_{22} + \frac{a_{12}a_{21}}{-a_{11}}\right) \tag{7.6.23}$$

$$\varepsilon(\infty) = 1 + 4\pi a_{22} \tag{7.6.24}$$

so that

$$\varepsilon(\omega) = \varepsilon(\infty) + \frac{\varepsilon(0) - \varepsilon(\infty)}{1 - (\omega^2/-a_{11})} \tag{7.6.25}$$

Consider solutions of the electromagnetic field equations for an insulating material,

$$\nabla \wedge \mathbf{E} + \frac{1}{c}\frac{\partial \mathbf{B}}{\partial t} = 0$$

$$\nabla \wedge \mathbf{H} - \frac{1}{c}\frac{\partial \mathbf{D}}{\partial t} = 0$$

$$\nabla.\mathbf{D} = 0 \quad \nabla.\mathbf{B} = 0$$

with $\mathbf{B} = \mu\mathbf{H}$, and the field vectors in the form $e^{j(\omega t - \mathbf{k}.\mathbf{r})}$, we derive

$$\mathbf{k} \wedge \mathbf{E} = \frac{\omega\mu}{c}\mathbf{H} \tag{7.6.27}$$

$$-\mathbf{k} \wedge \mathbf{H} = \frac{\omega}{c}\mathbf{D} \tag{7.6.28}$$

$$\mathbf{k}.\mathbf{D} = 0 \quad \mathbf{k}.\mathbf{H} = 0 \tag{7.6.29}$$

We assume that $\mu \neq 0$. In these circumstances the solution with $\mathbf{E} = 0$ is trivial for a wave process since it also implies $\mathbf{H} = 0$ from equation (7.6.27) and $\mathbf{x} = 0$ from equation (7.6.21). Distinguish two significant classes of solution (a) $\varepsilon(\omega) \neq 0$, (b) $\varepsilon(\omega) = 0$.

† $\varepsilon(\infty)$ refers to the dielectric constant for a freqency which is high compared with the characteristic lattice vibrations but low compared with the frequencies associated with the induced molecular polarization. Compare this discussion with page 322.

Case (*a*), $\varepsilon(\omega) \neq 0$ (Transverse Waves)

From equation (7.6.29), with $\mathbf{D} = \varepsilon(\omega)\mathbf{E}$ and $\mathbf{E}$ non-zero, it follows that $\mathbf{E}$ is perpendicular to $\mathbf{k}$. From equation (7.6.27), $\mathbf{H}$ is perpendicular to both $\mathbf{E}$ and $\mathbf{k}$. The vectors $\mathbf{E}$, $\mathbf{H}$, $\mathbf{k}$ form a right-handed set, the wave motion is transverse with $\mathbf{E}$, $\mathbf{H}$, perpendicular to the direction of propagation.

Take $\mathbf{E}$ along the x-direction which will also be the direction for $\mathbf{P}$ and $\mathbf{x}$. Then $\mathbf{H}$ is along the y-direction for propagation along z.

From equations (7.6.20), (7.6.21), (7.6.27) and (7.6.28)

$$\begin{aligned} (-a_{11} - \omega^2)x - a_{12}E_x &= 0 \\ a_{21}x + a_{22}E_x - P_x &= 0 \\ kE_x - \frac{\omega\mu}{c}H_y &= 0 \\ \frac{\omega}{c}E_x + \frac{4\pi\omega}{c}P_x - kH_y &= 0 \end{aligned} \tag{7.6.30}$$

For these equations to have a non-trivial solution

$$\begin{vmatrix} (-a_{11} - \omega^2) & -a_{12} & 0 & 0 \\ a_{21} & a_{22} & -1 & 0 \\ 0 & k & 0 & -\dfrac{\omega\mu}{c} \\ 0 & \dfrac{\omega}{c} & \dfrac{4\pi\omega}{c} & -k \end{vmatrix} = 0 \tag{7.6.31}$$

which is the (ω, k) dispersion relation for these waves. Using the relations in equations (7.6.23) and (7.6.24), we obtain

$$\omega^4 \frac{\mu\varepsilon(\infty)}{c^2} - \omega^2\left\{-a_{11}\frac{\mu\varepsilon(\omega)}{c^2} + k^2\right\} - a_{11}k^2 = 0 \tag{7.6.32}$$

The limiting solutions for $k \to 0$, are

$$\omega^2 = -a_{11}\frac{\varepsilon(0)}{\varepsilon(\infty)} \equiv \omega_l^2 \qquad \text{if } \omega_l^2 \gg 0 \tag{7.6.33}$$

and

$$-\omega^2\left(-a_{11}\frac{\mu\varepsilon(0)}{c^2}\right) - a_{11}k^2 = 0$$

$$\omega^2 = \frac{c^2}{\mu\varepsilon(0)}k^2 \qquad \text{if } \omega \to 0 \tag{7.6.34}$$

For large values of k the limiting solutions are

$$\omega^2 = \frac{c^2}{\mu\varepsilon(\infty)}k^2 \quad \text{and} \quad \omega^2 = -a_{11} \equiv \omega_t^2 \tag{7.6.35}$$

The form of the dispersion curve is sketched in Fig. 7.10. It can be seen that there are no solutions for transverse waves with frequencies between ω_t and

ω_l. This gives rise to a forbidden optical band and appears experimentally as a region of high reflectivity.

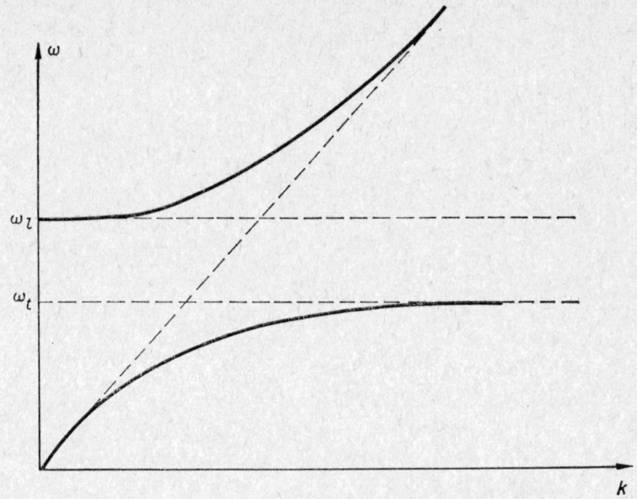

FIG. 7.10 The dispersion relation for coupled lattice vibrations and transverse electromagnetic waves.

Case (*b*), $\varepsilon(\omega) = 0$ (Longitudinal Waves)

There may be a frequency ω_l at which the dielectric constant, as given by equation (7.6.22), is zero. We assume μ and $\mathbf{E}$ to be non-zero, but evidently $\mathbf{D} = 0$. From equation (7.6.28) and (7.6.29)

$$\mathbf{k} \wedge \mathbf{H} = 0 \quad \mathbf{k} . \mathbf{H} = 0 \tag{7.6.36}$$

Since $\mathbf{k} \neq 0$, these equations require

$$\mathbf{H} = 0 \tag{7.6.37}$$

It now follows from equation (7.6.27) that

$$\mathbf{k} \wedge \mathbf{E} = 0 \tag{7.6.38}$$

but $\mathbf{k}.\mathbf{E} \neq 0$ and so $\mathbf{E}$ is parallel to the direction of propagation for these waves. Moreover, since $\mathbf{D} = 0$,

$$\mathbf{E} = -4\pi \mathbf{P} \tag{7.6.39}$$

From equation (7.6.25) the frequency of the wave with longitudinal $\mathbf{E}$, is given by

$$\omega_l^2 = (-a_{11}) \frac{\varepsilon(0)}{\varepsilon(\infty)} = \omega_t^2 \frac{\varepsilon(0)}{\varepsilon(\infty)} \tag{7.6.40}$$

where $\omega_t^2 = -a_{11}$ in accordance with the notation of equation (7.6.35). This solution is satisfied for all wave vectors $\mathbf{k}$, a continuous set of longitudinal waves is therefore available at a particular frequency ω_l. Equation (7.6.40) is usually known as the Lyddane–Sachs–Teller relation.

According to equation (7.6.40) the dielectric constant becomes zero at a frequency ω_l which is directly related to the frequency ω_t at which the dielectric constant becomes infinite in the absence of damping. This relation is of considerable generality and is not restricted to the particular model system which we have used. For example, in a physical system we expect to observe an absorption line at ω_t because many transverse waves can be excited at this frequency (as can be seen from Fig. 7.10) and the loss terms in the actual lattice vibrations will dissipate energy. Idealize this situation by an infinitely narrow absorption line at ω_t and write quite generally

$$\begin{aligned}\varepsilon(\omega) &= \varepsilon_1(\omega) - \mathrm{j}\varepsilon_2(\omega)\\ &= \varepsilon_1(\omega) - \mathrm{j}\alpha\{\delta(\omega-\omega_t) - \delta(\omega+\omega_t)\}\end{aligned} \tag{7.6.41}$$

where $\delta(\omega - \omega_t)$ is a delta function centred on ω_t and we require the two components because $\varepsilon_2(\omega)$ is an odd function of ω. From the Kramers–Kronig relations given in equations (1.3.76) and (1.3.77) we obtain

$$\begin{aligned}\varepsilon_1(\Omega) &= \varepsilon(\infty) + \frac{2}{\pi}\mathrm{P}\int_0^\infty \frac{\omega\varepsilon_2(\omega)}{\omega^2-\Omega^2}\,\mathrm{d}\omega\\ &= \varepsilon(\infty) + \frac{2\alpha}{\pi}\mathrm{P}\int_0^\infty \frac{\omega\delta(\omega-\omega_t)}{\omega^2-\Omega^2}\,\mathrm{d}\omega\end{aligned}$$

$$\varepsilon_1(\Omega) = \varepsilon(\infty) + \frac{2\alpha}{\pi}\cdot\frac{\omega_t}{\omega_t^2-\Omega^2} \tag{7.6.42}$$

This equation is evidently equivalent to equation (7.6.25) since

$$\varepsilon(0) - \varepsilon(\infty) = \frac{2\alpha}{\pi}\cdot\frac{1}{\omega_t} \tag{7.6.43}$$

The frequency, ω_l, at which $\varepsilon(\Omega)$ is zero is therefore given by

$$\varepsilon(\infty) + \frac{\varepsilon(0)-\varepsilon(\infty)}{\omega_t^2-\omega_l^2} = 0 \tag{7.6.44}$$

$$\omega_l^2 = \omega_t^2\,\frac{\varepsilon(0)}{\varepsilon(\infty)} \tag{7.6.45}$$

A Generalization

In an actual crystal there are usually a number of significant force constants a_{11} each giving rise to a different critical frequency ω_t. Equation (7.6.25) may be generalized to sum over these different frequencies and it is usual to write

$$\varepsilon(\mathrm{j}\omega) = \varepsilon(\infty) + \sum_i \frac{4\pi\rho_i\omega_{ti}^2}{\omega_{ti}^2-\omega^2+2\mathrm{j}\gamma_i\omega} \tag{7.6.46}$$

where ω_{ti} is the characteristic frequency associated with the transverse mode labelled i, γ_i is a formal damping factor and $4\pi\rho_i$ is a dimensionless factor called the strength of the mode. In this equation we have written the dielectric constant as $\varepsilon(\mathrm{j}\omega)$ because, when damping is included, $\varepsilon(\omega)$ is no longer real.† Note that the lattice contribution to the dielectric constant now has the

† $\varepsilon(\infty)$ is, however, real since the materials are usually transparent and non-absorbing in the visible range.

same mathematical form as the molecular contribution which was discussed in the previous sections (see for example equation (7.2.11)).

In the notation of equation (7.6.46) the static dielectric constant is

$$\varepsilon(0) = \varepsilon(\infty) + \sum_i 4\pi\rho_i \qquad (7.6.47)$$

This relation can be used to check the assignment of the mode strengths.

When damping is present there are no longer rigorous infinities and zeros for $\varepsilon(j\omega)$.† Instead there are maxima in the imaginary parts of $\varepsilon(j\omega)$ and $\varepsilon(j\omega)^{-1}$. A pole-zero representation can, however, usefully be introduced by rewriting equation (7.6.46) as

$$\varepsilon(j\omega) = \varepsilon(\infty) + \sum_i \frac{4\pi\rho_i\omega_{ti}^2}{(j\omega - w_i)(j\omega - w_i^*)} \qquad (7.6.48)$$

Here $w_i = -\gamma_i + j\omega_i$ and $\omega_{ti}^2 = \omega_i^2 + \gamma_i^2$. Bringing all the terms on the right-hand side over a common denominator and taking $\varepsilon(\infty)$ as a common multiplying factor, we may write

$$\varepsilon(j\omega) = \varepsilon(\infty)\frac{P(j\omega)}{\prod_i(j\omega - w_i)(j\omega - w_i^*)} \qquad (7.6.49)$$

where $P(j\omega)$ is a polynomial in $j\omega$ with real coefficients. Consider the polynomial $P(s)$ with $s \equiv (q + j\omega)$ replacing $j\omega$ in $P(j\omega)$. The function $P(s)$ is a 'real' function of the complex variable s and consequently the roots of $P(s) = 0$ occur in conjugate pairs.‡ Hence we may write

$$P(s) = \prod_n (s - z_n)(s - z_n^*) \quad \text{and} \quad P(j\omega) = \prod_n (j\omega - z_n)(j\omega - z_n^*).$$

Equation (7.6.49) may therefore be written

$$\varepsilon(j\omega) = \varepsilon(\infty)\frac{\prod_n (j\omega - z_n)(j\omega - z_n^*)}{\prod_i (j\omega - w_i)(j\omega - w_i^*)} \qquad (7.6.50)$$

Since we are assuming that the damping is small the 'poles' of $(j\omega)$ will occur close to $\omega = \mathrm{Im}\,[w_i]$ and the 'zeros' close to $\omega = \mathrm{Im}\,[z_n]$. These frequencies correspond to the transverse and longitudinal frequencies discussed previously. Equation (7.6.50) leads to a generalization of the Lyddane–Sachs–Teller relation, we now have instead of equation (7.6.40),

$$\varepsilon(0) = \varepsilon(\infty)\frac{\prod |z_n|^2}{\prod |w_i|^2} \qquad (7.6.51)$$

In order to determine the 'poles' and 'zeros' of $\varepsilon(j\omega)$ experimentally the lattice vibrations must be excited by the electromagnetic field. The transverse modes can be excited readily by an electromagnetic wave (which is itself usually transverse) and $\varepsilon(j\omega)$ can be derived from an analysis of the frequency variation of the absorption and reflection coefficients. Direct observation of

† ω is a real excitation frequency. The infinities and zeros occur for $\varepsilon(s)$ with $s = q + j\omega$, a complex variable.

‡ See Section 5.2.

the longitudinal modes at ω_l requires some care in the experimental arrangement. Plane polarized light incident on the crystal surface at an angle away from the normal can be polarized in a direction to provide a component of **E** perpendicular to the surface. This vector will drive the longitudinal modes through the continuity of the normal component of **D**.† Such a direct observation is not usually necessary however since complete information about the poles and zeros of $\varepsilon(j\omega)$ may in principle be derived from an analysis of the reflection coefficient for the material.

7.6.3 Reflection Coefficient

The dielectric constant $\varepsilon(j\omega)$ can be derived from a measurement of the reflection coefficient at normal incidence over a wide frequency range. The Fresnel equations, equations (2.10.6) and (2.10.8), are used to interpret the measurements. Taking the dielectric constant of air to be unity, we have

$$R = \left|\frac{1-\sqrt{\varepsilon}}{1+\sqrt{\varepsilon}}\right|^2 = \left|\frac{\eta - j\kappa - 1}{\eta - j\kappa + 1}\right|^2 \tag{7.6.52}$$

Write

$$r = \rho\, e^{-j\theta} = \frac{\eta - j\kappa - 1}{\eta - j\kappa + 1} \tag{7.6.53}$$

so that

$$R = rr^* = \rho^2 \tag{7.6.54}$$

and

$$\log_e r = \log_e \rho - j\theta \tag{7.6.55}$$

The Kramers–Kronig relations (see page 30) give a relation between θ and $\log_e \rho$,

$$\theta(\Omega) = -\frac{2\Omega}{\pi} P \int_0^\infty \frac{\log_e \rho}{\omega^2 - \Omega^2}\, d\omega \tag{7.6.56}$$

and so, from equation (7.6.54)

$$\theta(\Omega) = -\frac{\Omega}{\pi} P \int_0^\infty \frac{\log_e R}{\omega^2 - \Omega^2}\, d\omega \tag{7.6.57}$$

Hence measurements of R over a wide frequency range determine ρ and θ as functions of frequency.

The dielectric constant may now be derived using equation (7.6.53),

$$\frac{(\eta - j\kappa) - 1}{(\eta - j\kappa) + 1} = \rho\, e^{-j\theta}$$

$$(\eta - j\kappa) = \frac{1 + \rho\, e^{-j\theta}}{1 - \rho\, e^{-j\theta}} = \frac{(1 + \rho\, e^{-j\theta})(1 - \rho\, e^{-j\theta})^*}{(1 - \rho\, e^{-j\theta})(1 - \rho\, e^{-j\theta})^*}$$

$$(\eta - j\kappa) = \frac{1 - \rho^2 - 2j\rho \sin\theta}{1 + \rho^2 - 2\rho\cos\theta} \tag{7.6.58}$$

† See, for example, Berreman, D. W., *Phys. Rev.* **130**, 2193, 1963.

$$\eta = \frac{1 - \rho^2}{1 + \rho^2 - 2\rho \cos \theta} \qquad \kappa = \frac{2\rho \sin \theta}{1 + \rho^2 - 2\rho \cos \theta} \tag{7.6.59}$$

$$\begin{aligned} \varepsilon_1(\omega) &= \eta(\omega)^2 - \kappa(\omega)^2 \\ \varepsilon_2(\omega) &= 2\eta(\omega)\kappa(\omega) \\ \varepsilon(j\omega) &= \varepsilon_1(\omega) - j\varepsilon_2(\omega) \end{aligned} \tag{7.6.60}$$

Transverse and longitudinal mode frequencies derived for LaF_3 from an analysis of this type are given in the Table 7.4. The values refer to a temperature 78 K.

TABLE 7.4

$\omega_{tt}(\text{cm}^{-1})$†	$\omega_{ln}(\text{cm}^{-1})$†	$4\pi\rho_i$‡
100	110	2·36
127	131	0·20
142	143	
168	178	3·94
193	195	0·51
208	229	1·76
248	268	1·02
274	297	0·51
354	364	0·4
367	462	0·07

7.6.4 Ferroelectric Transition

Finally we observe that in certain circumstances the frequency of a transverse mode may fall to zero (indicating that this mode becomes unstable). For this to occur, the effective mechanical force constant must become zero under the action of the local electric field $\mathbf{E}'$, which depends upon the polarization of the material. Consider equations (7.6.16) and (7.6.17) with the local field taken as $\mathbf{E}' = \mathbf{E} + (4\pi/3)\mathbf{P}$. From equation (7.6.17)

$$\mathbf{P}\left(1 - \frac{4\pi N\langle\alpha\rangle}{3}\right) = \tfrac{1}{2}Nq\mathbf{x} + N\langle\alpha\rangle\mathbf{E} \tag{7.6.61}$$

Using this value for $\mathbf{P}$ in the local field of equation (7.6.16), we obtain

$$\ddot{\mathbf{x}} = \left\{a_{11} + \frac{(2\pi/3)Nq^2(M_+^{-1} + M_-^{-1})}{1 - (4\pi N\langle\alpha\rangle/3)}\right\}\mathbf{x} + \left\{\frac{q(M_+^{-1} + M_-^{-1})}{1 - (4\pi N\langle\alpha\rangle/3)}\right\}\mathbf{E} \tag{7.6.62}$$

The characteristic mode frequency is therefore given by

$$\omega^2 = -\left\{a_{11} + \frac{(2\pi/3)Nq^2(M_+^{-1} + M_-^{-1})}{1 - (4\pi N\langle\alpha\rangle/3)}\right\} \tag{7.6.63}$$

† Lowndes, R. P., Parrish, J. F., and Perry, C. H., *Phys. Rev.* **182**, 913, 1969.

‡ Rast, H. E., Caspers, H. H., Miller, S.A., and Buchanan, R. A., *Phys. Rev.* **171**, 1051, 1968.

and becomes zero if $\langle\alpha\rangle$ is sufficiently large, so that

$$\left\{\frac{(2\pi/3)Nq^2(M_+^{-1}+M_-^{-1})}{1-(4\pi N\langle\alpha\rangle/3)}\right\}=-a_{11} \tag{7.6.64}$$

In NaCl the left-hand side of equation (7.6.64) is about one half of $-a_{11}$ and so the transverse mode does not become unstable. In $SrTiO_3$, however, and similar materials with the perovskite crystal structure, this equation can be satisfied and the materials exhibit a ferroelectric transition. The explicit temperature dependence of the ferroelectric transition is not derived from the present discussion and probably arises from anharmonic components in the lattice vibrations.

Bibliography

BARKER, A. S., 'Infrared Dielectric Behaviour of Ferroelectric Crystals', in *Ferroelectricity*, Ed., E. F. Weller, Elsevier, 1967.

BORN, M. and HUANG, K., *Dynamical Theory of Crystal Lattices*, O.U.P., 1956.

BORN, M. and WOLF, E., *Principles of Optics*, Pergamon, 1969.

BROWN, W. F., JR., '*Dielectrics*' in *Handbuch der Physik* Volume 17, Springer-Verlag, 1956.

EYRING, H., WALTER, J., and KIMBALL, G. E., *Quantum Chemistry*, Wiley, 1949.

LANDAU, L. D., and LIFSHITZ, E. M., *Electrodynamics of Continuous Media*, Pergamon, 1963.

MARTIN, D. H., 'The Study of the Vibrations of Crystal Lattices by Far Infra Red Spectroscopy', in *Advances in Physics*, Ed. B. R. Coles, Volume 14, Taylor & Francis, 1965.

SMITH, J. W., *Electric Dipole Moments*, Butterworths, 1955.

SMYTH, C. P., *Dielectric Behaviour and Structure*, McGraw-Hill, 1955.

STERN, F., 'Elementary Theory of the Optical Properties of Solids', in *Solid State Physics*, Eds. F. Seitz and D. Turnbull, Volume 15, Academic Press, 1963.

FARADAY ROTATION

KRAMERS, H. A., *Collected Scientific Papers*, p. 522, North-Holland, 1956.

SERBER, R., *Phys. Rev.*, **41**, 489, 1932.

VAN VLECK, J. H., and HEBB, M. H., *Phys. Rev.*, **46**, 17, 1934.

EXPERIMENTAL METHODS

COLE, R. H., and GROSS, P. M., *Rev. Sci. Instrum.* **20**, 252, 1949.

HAGUE, B., *A.C. Bridge Measurements*, Pitman, 1946.

HARTSHORN, L., *Radiofrequency Measurements*, Wiley, 1941.

MAYBURG, S., *Phys. Rev.* **79**, 375, 1950.

Technique of Microwave Measurements, Ed. C. G. Montgomery, Chapters 8, 9 and 10, Radiation Laboratory Series, McGraw-Hill, 1947.

WILHELM, H. T., *Bell System Technical J.*, **31**, 999, 1952.

Also SMYTH C. P., and SMITH, J. W., referred to above.

MORT, J., LÜTY, F., and BROWN, F. C., *Phys. Rev.* **137**, A566, 1965 (Faraday Rotation).

BERREMAN, D. W., and UNTERWALD, F. C., *Phys. Rev.* **174**, 791, 1968 (Lattice Vibrations).

Chapter 8

The Microscopic Theory of Magnetic Materials

8.1 Introduction

When attempting to give a microscopic theory for the magnetic properties of materials it is necessary to distinguish at the outset two different model systems depending on whether the material is an insulator or a metallic conductor. Insulating materials may be described on the basis of the atomic model used previously in Chapter 7 to describe the properties of dielectrics. In these materials the magnetic properties arise from electrons in well-localized shells centred on particular lattice sites in solids and on particular nuclei in gases or liquids. For metals, on the other hand, the magnetic properties arise not only from the localized, 'core', electrons but also from mobile, 'itinerant', electrons which are taking part to a greater or lesser degree in the processes of electrical conduction. The model system for a metal must therefore take direct account of the band structure of the material. This is a formidable problem and so for the most part we shall restrict our discussion to insulating magnetic materials.

8.2 Diamagnetism and Paramagnetism in Insulators

We shall base this treatment on the simple atomic model which was used in Chapter 7 to describe the properties of insulating dielectric materials. We assume, in the first instance, that there are no significant interactions between the fundamental molecular units except for those processes which maintain thermal equilibrium within the system. These thermal interactions are sufficient to give rise to a definite temperature within the system but they are regarded as being so weak that they need not be included explicitly in the Hamiltonian operator which describes the interaction with an electromagnetic field.

From equation (7.4.20), the Hamiltonian operator for a system of charged particles in an electromagnetic field (neglecting spin at this stage) may be written,

$$\mathscr{H} = \mathscr{H}_0 + \mathscr{H}' \tag{8.2.1}$$

Here $\mathscr{H}_0$ is the Hamiltonian operator for the system in the absence of the electromagnetic field, and

$$\mathscr{H}' = \sum_i \left[\frac{jq\hbar}{m_i c}\mathbf{A}_i . \nabla_i + \frac{jq\hbar}{2m_i c}(\nabla_i . \mathbf{A}_i) + \frac{q^2}{2m_i c^2}(\mathbf{A}_i . \mathbf{A}_i) + q\phi_i\right] \tag{8.2.2}$$

represents the perturbation arising from the electromagnetic field. For the present discussion of magnetic properties we take the macroscopic electromagnetic field within the material to be a uniform steady magnetic field **H**, with a magnitude H_0. Choose the direction of the z coordinate axis for all the charges along the direction of **H** so that

$$\mathbf{H} = \hat{\mathbf{k}} H_0 \tag{8.2.3}$$

The vector and scalar potential functions for the charges may now be taken in the form

$$\mathbf{A}_i = \hat{\mathbf{i}}(-\tfrac{1}{2} y_i H_0) + \hat{\mathbf{j}}(\tfrac{1}{2} x_i H_0) \qquad A_{iz} = 0 \quad \phi_i = 0 \tag{8.2.4}$$

from which

$$\nabla_i . \mathbf{A}_i = 0 \quad \mathbf{A}_i . \mathbf{A}_i = \tfrac{1}{4} H_0 (x_i^2 + y_i^2)$$

$$\sum_i \mathbf{A}_i . \nabla_i = \tfrac{1}{2} H_0 \sum_i \left(x_i \frac{\partial}{\partial y_i} - y_i \frac{\partial}{\partial x_i} \right) \equiv \tfrac{1}{2} \frac{\mathrm{j} H_0 L_z}{\hbar} \tag{8.2.5}$$

with

$$L_z = \frac{\hbar}{\mathrm{j}} \sum_i \left(x_i \frac{\partial}{\partial y_i} - y_i \frac{\partial}{\partial x_i} \right)$$

Evidently L_z is the operator for the z component of the total orbital angular momentum.

The perturbation Hamiltonian $\mathscr{H}'$ for a system of electrons, for which $q = -|e|$, may now be written as

$$\mathscr{H}' = \frac{|e|}{2mc} H_0 L_z + \frac{|e|^2 H_0^2}{8mc^2} \sum_i (x_i^2 + y_i^2) \tag{8.2.6}$$

Each electron has an intrinsic magnetic moment along the direction of its spin axis. An additional term must therefore be written into equation (8.2.6) to take account of this contribution to the energy in a magnetic field. The electronic moment is

$$\mathbf{m} = -\frac{|e|}{mc} \mathbf{s}$$

and the extra energy is

$$-\mathbf{m} . \mathbf{H} = \frac{|e|}{mc} \mathbf{H} . \mathbf{s}$$

In equation (8.2.6), therefore, we shall require an additional term

$$\frac{|e|}{mc} \sum_i \mathbf{H} . \mathbf{s}_i = \frac{|e|}{mc} H_0 S_z$$

where S_z is the component of the total electronic spin in the direction of the magnetic field. The perturbation Hamiltonian for a system of electrons in a steady magnetic field, including the spin energy, is therefore given by

$$\mathscr{H}' = \frac{|e|}{2mc} H_0 (L_z + 2S_z) + \frac{|e|^2 H_0^2}{8mc^2} \sum_i (x_i^2 + y_i^2) = \mathscr{H}'_{\mathrm{p}} + \mathscr{H}'_{\mathrm{d}} \tag{8.2.7}$$

The first term in this Hamiltonian gives rise to paramagnetism in the system of localized electrons which we are considering at present. The second term gives rise to diamagnetism. All atoms and molecules exhibit diamagnetism regardless of whether or not they are also paramagnetic. Only those molecules for which $(\mathbf{L} + 2\mathbf{S})$ is non-zero will exhibit paramagnetism.

8.2.1 Diamagnetic Susceptibility

Consider the perturbation arising from the second term in equation (8.2.7). This term describes the diamagnetic response. We have

$$\mathscr{H}'_{\mathrm{d}} = \frac{|e|^2 H_0^2}{8mc^2} \sum_i (x_i^2 + y_i^2) \tag{8.2.8}$$

The first order energy change arising from this term is

$$\begin{aligned} E_{\mathrm{d}} &= \langle\psi|\, \mathscr{H}'_{\mathrm{d}}\, |\psi\rangle \\ &= \frac{|e|^2 H_0^2}{8mc^2} \sum_i \langle x_i^2 + y_i^2\rangle \end{aligned} \tag{8.2.9}$$

Here

$$\sum_i \langle x_i^2 + y_i^2\rangle = \sum_i \int \psi^*(x_i^2 + y_i^2)\psi \, \mathrm{d}^3\mathbf{r}$$

and is the expectation value for $\sum_i (x_i^2 + y_i^2)$ derived from the unperturbed wave function. Since all directions in space are regarded as equivalent before the magnetic field is applied, we shall have $\langle x_i^2\rangle = \langle y_i^2\rangle = \langle z_i^2\rangle = \frac{1}{3}\langle r_i^2\rangle$, where the function $\langle r_i^2\rangle$ is a mean square distance for the electron labelled i measured from the centre of gravity of the molecule. We may now write equation (8.2.9) as

$$E_{\mathrm{d}} = \frac{|e|^2 H_0^2}{8mc^2}\cdot\frac{2}{3} \sum_i \langle r_i^2\rangle \tag{8.2.10}$$

The induced magnetic moment per molecule arising from the perturbation is derived from the relation

$$\langle \mu_z^{\mathrm{d}}\rangle = -\frac{\partial E_{\mathrm{d}}}{\partial H_0} \tag{8.2.11a}$$

which gives

$$\langle \mu_z^{\mathrm{d}}\rangle = -\frac{|e|^2 H_0}{6mc^2} \sum_i \langle r_i^2\rangle \tag{8.2.11b}$$

Here $\langle \mu_z^{\mathrm{d}}\rangle$ is the magnetic moment in the direction of the magnetizing field.

The magnetic moment per unit volume $\mathscr{M}_z^{\mathrm{d}}$ is given by

$$\mathscr{M}^{\mathrm{d}} = N\langle \mu_z^{\mathrm{d}}\rangle = -\frac{N\,|e|^2 H_0}{6mc^2} \sum_i \langle r_i^2\rangle \tag{8.2.12a}$$

where N is the number of molecules per unit volume. The diamagnetic susceptibility per unit volume is

$$\chi^{\mathrm{d}} = \frac{\mathscr{M}_z^{\mathrm{d}}}{H_0} = -\frac{N\,|e|^2}{6mc^2} \sum_i \langle r_i^2\rangle \tag{8.2.12b}$$

Note that the diamagnetic response described by equations (8.2.11) and (8.2.12) gives rise to a magnetization which is oppositely directed to the applied field **H**.

8.2.2 Paramagnetic Susceptibility

The perturbation arising from the first term in equation (8.2.7) leads to paramagnetism. The first order perturbation energy is

$$E_{\mathrm{p}} = \langle\psi| \mathscr{H}'_{\mathrm{p}} |\psi\rangle \tag{8.2.13}$$

$$= \frac{|e|}{2mc} H_0 \langle L_z + 2S_z \rangle \tag{8.2.14}$$

and the magnetic moment in the direction of the field, $\hat{\mathbf{k}}H_0$, is

$$\langle \mu_z^{\mathrm{p}} \rangle = -\frac{\partial E_{\mathrm{p}}}{\partial H_0} = -\frac{|e|}{2mc}\langle L_z + 2S_z \rangle \tag{8.2.15}$$

In this first order approximation the moment is independent of the magnetizing field and corresponds to the z component of a permanent dipole, $\langle \boldsymbol{\mu}^{\mathrm{p}} \rangle$,

$$\langle \boldsymbol{\mu}^{\mathrm{p}} \rangle = -\frac{|e|}{2mc}\langle \mathbf{L} + 2\mathbf{S} \rangle \tag{8.2.16}$$

The evaluation of the expectation value $\langle L_z + 2S_z \rangle$ in equation (8.2.15) requires some care. It can be conveniently carried through in some special cases. Consider, for example, an atomic system which is characterized by a set of quantum numbers $|\mathbf{L}|^2$, $|\mathbf{S}|^2$, $|\mathbf{J}|^2$, J_z, associated with orbital, spin, and total angular momentum respectively. We call this the $|\mathbf{L}, \mathbf{S}, \mathbf{J}, J_z\rangle$ representation. In this representation the operators $\mathbf{L}^2$, $\mathbf{S}^2$, $\mathbf{J}^2$ and J_z are diagonal having eigenvalues $L(L+1)\hbar^2$, $S(S+1)\hbar^2$, $J(J+1)\hbar^2$ and $\langle J_z \rangle = M\hbar$, with M ranging over $J, (J-1) \ldots -J$. In order to evaluate $\langle L_z + 2S_z \rangle$ (which is not explicitly diagonal) we need to write this function in terms of the operators $\mathbf{L}^2$, $\mathbf{S}^2$, $\mathbf{J}^2$ and J_z. To carry this through we note that $\langle L_z + 2S_z \rangle$ is also the z component of the expectation value of $(\mathbf{L} + 2\mathbf{S})$ projected on to the total angular momentum $\mathbf{J}$. Thus we may first project $(\mathbf{L} + 2\mathbf{S})$ along $\mathbf{J}$ and then project this vector quantity along the z axis as shown in Fig. 8.1. By this technique we can obtain the diagonal representation for $\langle L_z + 2S_z \rangle$.†

Consider the expectation value of the projection of $(\mathbf{L} + 2\mathbf{S})$ along $\mathbf{J}$. We shall show that this vector is related to the total angular momentum operator $\mathbf{J}$ through a pure number denoted by g which is called the Landé g factor. The magnitude of the expectation value of the projection of $(\mathbf{L} + 2\mathbf{S})$ along $\mathbf{J}$ is evidently

$$\left\langle (\mathbf{L} + 2\mathbf{S}) \cdot \frac{\mathbf{J}}{|\mathbf{J}|} \right\rangle$$

† This is essentially first-order perturbation theory using the vector model for an atom.

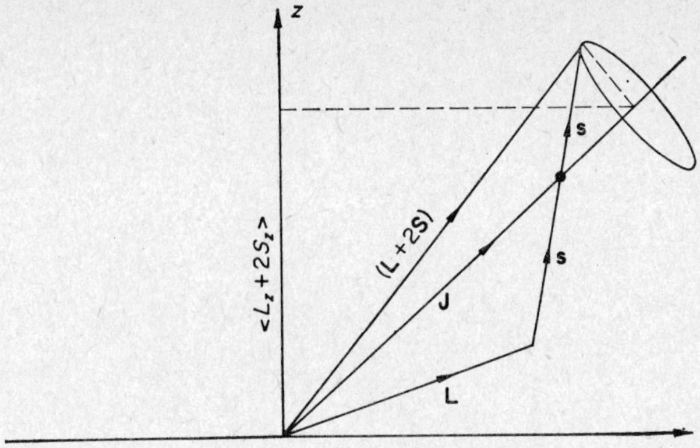

FIG. 8.1 Diagram illustrating the technique for deriving the quantity $\langle L_z + 2S_z \rangle$ when using the $|LSJJ_z\rangle$ representation.

where $\mathbf{J}/|\mathbf{J}|$ is unit vector in the direction of **J**. The explicit vector description for this projection along **J**, is

$$\langle \mathbf{L} + 2\mathbf{S} \rangle_{\mathbf{J}} \equiv \left\langle (\mathbf{L} + 2\mathbf{S}) . \frac{\mathbf{J}}{|\mathbf{J}|} \right\rangle \frac{\langle \mathbf{J} \rangle}{|\mathbf{J}|} \tag{8.2.17}$$

$$= \frac{1}{|\mathbf{J}|^2} \langle \mathbf{J}.\mathbf{J} + \mathbf{S}.\mathbf{J} \rangle \langle \mathbf{J} \rangle \tag{8.2.18}$$

$$= \frac{1}{|\mathbf{J}|^2} \langle \mathbf{J}^2 + \mathbf{S}^2 + \mathbf{S}.\mathbf{L} \rangle \langle \mathbf{J} \rangle \tag{8.2.19}$$

But

$$\langle \mathbf{J}^2 \rangle = \langle \mathbf{L}^2 + \mathbf{S}^2 + 2\mathbf{S}.\mathbf{L} \rangle \tag{8.2.20}$$

so that

$$\langle \mathbf{L} + 2\mathbf{S} \rangle_{\mathbf{J}} = \frac{1}{|\mathbf{J}|^2} \langle \tfrac{3}{2}\mathbf{J}^2 - \tfrac{1}{2}\mathbf{L}^2 + \tfrac{1}{2}\mathbf{S}^2 \rangle \langle \mathbf{J} \rangle \tag{8.2.21}$$

$$= \frac{3J(J+1) - L(L+1) + S(S+1)}{2J(J+1)} \langle \mathbf{J} \rangle \tag{8.2.22}$$

We may therefore write

$$\langle \mathbf{L} + 2\mathbf{S} \rangle_{\mathbf{J}} = g_J \langle \mathbf{J} \rangle \tag{8.2.23}$$

with

$$g_J = 1 + \frac{J(J+1) - L(L+1) + S(S+1)}{2J(J+1)} \tag{8.2.24}$$

and g_J is a pure number.

The magnitude of the expectation value of the z component of $\langle \mathbf{L} + 2\mathbf{S} \rangle_{\mathbf{J}}$, which we require for equation (8.2.15), is evidently given by $\langle \{ \langle \mathbf{L} + 2\mathbf{S} \rangle_{\mathbf{J}} . \hat{\mathbf{k}} \} \rangle$ where $\hat{\mathbf{k}}$ is the unit vector along the z-axis. From equation (8.2.23) we obtain

$$\langle L_z + 2S_z \rangle = \langle \{ \langle \mathbf{L} + 2\mathbf{S} \rangle_{\mathbf{J}} . \hat{\mathbf{k}} \} \rangle \tag{8.2.25}$$

$$= g_J \langle \mathbf{J} . \hat{\mathbf{k}} \rangle \tag{8.2.26}$$

so that

$$\langle L_z + 2S_z\rangle = g_J\langle J_z\rangle \tag{8.2.27}$$

with g_J given by equation (8.2.24). Hence, for an atomic system which can be described adequately by the $|\mathbf{L}, \mathbf{S}, \mathbf{J}, J_z\rangle$ representation, the dipole moment $\langle\mu_z^{\mathrm{p}}\rangle$ in the direction of the magnetic field is given by

$$\langle\mu_z^{\mathrm{p}}\rangle = -\frac{|e|}{2mc}\, g_J\, \langle J_z\rangle \tag{8.2.28}$$

and $\langle J_z\rangle$ will take the values $M\hbar$ with M ranging over $J, (J-1)\ldots -J$.

If there are N atoms per unit volume contributing to the magnetization of material, then in thermal equilibrium the number of these atoms with energy E_n is given by the Boltzmann relation

$$N(E_n) = N\frac{\exp(-E_n/kT)}{\sum\limits_n \exp(-E_n/kT)} \tag{8.2.29}$$

The magnetic moment per unit volume is therefore

$$\mathscr{M}_z^{\mathrm{p}} = \sum_n N(E_n)\,\langle\mu_z^{\mathrm{p}}\rangle_n \tag{8.2.30}$$

$$= -\frac{N|e|}{2mc}\,\frac{\sum\limits_n g_n\langle J_z\rangle_n \exp(-E_n/kT)}{\sum\limits_n \exp(-E_n/kT)} \tag{8.2.31}$$

This relation simplifies considerably if only one state with total angular momentum J is occupied. This is a realistic situation which arises when the excited states are separated from the ground state by an energy interval large compared with kT. In these circumstances the significant energy states for an atom in the presence of the magnetic field are obtained from equations (8.2.14) and (8.2.27)

$$E = E_{0J} + \frac{|e|}{2mc}H_0\langle L_z + 2S_z\rangle \tag{8.2.32}$$

$$= E_{0J} + \frac{|e|}{2mc}H_0 g_J\langle J_z\rangle \tag{8.2.33}$$

where E_{0J} is the unperturbed energy of the ground state. Using this relation for the energy terms in the exponential factors of equation (8.2.31), and cancelling through the common factor $\exp(-E_{0J}/kT)$, the magnetization of the material is

$$\mathscr{M}_z^{\mathrm{p}} = -\frac{N|e|}{2mc}g_J\,\frac{\sum\langle J_z\rangle\exp\{-(|e|/2mc).(H_0 g_J\langle J_z\rangle/kT)\}}{\sum\exp\{-(|e|/2mc).(H_0 g_J\langle J_z\rangle/kT)\}} \tag{8.2.34}$$

Write in the explicit eigenvalues for J_z, in the form $M\hbar$ with M ranging over

$J, (J-1)\ldots -J$, together with $\beta = (|e|\hbar)/2mc$ (the Bohr magneton), and obtain†

$$\mathscr{M}_z^{\mathrm{p}} = -Ng_J\beta \frac{\sum_{-J}^{J} M \exp\{-(g_J\beta MH_0)/kT\}}{\sum_{-J}^{+J} \exp\{-(g_J\beta MH_0)/kT\}} \tag{8.2.35}$$

For the condition $g_J\beta MH_0 \ll kT$, we may expand the exponential functions and retain only the first two terms.

$$\mathscr{M}_z^{\mathrm{p}} = -Ng_J\beta \frac{\sum_{-J}^{J} M - (g_J\beta H_0/kT)\sum_{-J}^{J} M^2}{\sum_{-J}^{J}\{1-(g_J\beta H_0/kT)M\}} \tag{8.2.36}$$

Since

$$\sum_{-J}^{J} M = 0, \qquad \sum_{-J}^{J} 1 = 2J+1 \tag{8.2.37}$$

and

$$\sum_{-J}^{J} M^2 = \frac{J(J+1)(2J+1)}{3}$$

we obtain

$$\mathscr{M}_z^{\mathrm{p}} = +\frac{Ng_J^2\beta^2 J(J+1)}{3kT} H_0 \tag{8.2.38}$$

The paramagnetic susceptibility is therefore, in this approximation,

$$\chi^{\mathrm{p}} = \frac{Ng_J^2\beta^2 J(J+1)}{3kT} \tag{8.2.39}$$

or

$$\chi^{\mathrm{p}} = \frac{C}{T} \tag{8.2.40}$$

where C is a constant independent of H. The relation given in equation (8.2.40) is known as Curie's law. It is obeyed qualitatively in this form by many materials even when the specific model used to derive equation (8.2.39) is not rigorously applicable. Note that the paramagnetic response described by equations (8.2.38)–(8.240) leads to a magnetization which is in the same direction as the applied field H_0.

† Note that for large values of **J** equation (8.2.35) gives the classical relation

$$\mathscr{M}_z^{\mathrm{p}} = -Ng_J\beta J \frac{\int \cos\theta \exp(-a\cos\theta)\, 2\pi \sin\theta \, d\theta}{\int \exp(-a\cos\theta)\, 2\pi\sin\theta\, d\theta}$$

with $a = (g_J\beta J/kT)H_0$. Hence $\mathscr{M}_z^{\mathrm{p}} = Ng_J\beta JL(a)$ where $L(a) \equiv \coth a - (1/a)$ is the Langevin function derived previously in connection with the dielectric constant (see Section 7.3).

8.2.3 Total Susceptibility

The complete magnetization of the material, derived from the perturbation Hamiltonian equation (8.2.7), will be given by the sum of the diamagnetic and paramagnetic response terms. In the previous sections we have evaluated the magnetization in the direction of the magnetic field $\hat{\mathbf{k}}H_0$ for each contribution separately. We may therefore write the total magnetization $(\mathscr{M}_z)_{\text{total}}$, using equations (8.2.12) and (8.2.38), as

$$(\mathscr{M}_z)_{\text{total}} = \mathscr{M}_z^{\text{d}} + \mathscr{M}_z^{\text{p}} \tag{8.2.41}$$

$$(\mathscr{M}_z)_{\text{total}} = \left\{\frac{N_{\text{d}}\,|e|^2}{6mc^2}\sum_i \langle r_i^2\rangle + \frac{N_{\text{p}}g_J^2\beta^2 J(J+1)}{3kT}\right\}H_0 \tag{8.2.42}$$

The total susceptibility is, therefore

$$\chi_{\text{total}} = \frac{(\mathscr{M}_z)_{\text{total}}}{H_0} \tag{8.2.43}$$

$$= \left\{-\frac{N_{\text{d}}\,|e|^2}{6mc^2}\sum_i \langle r_i^2\rangle + \frac{N_{\text{p}}g_J^2\beta^2 J(J+1)}{3kT}\right\} \tag{8.2.44}$$

8.2.4 Second Order Paramagnetic Terms

The diamagnetic contribution to the total magnetization is usually much smaller than the paramagnetic contribution. It is possible therefore that second order terms in the paramagnetic response may be comparable with the first order diamagnetic terms. The second order perturbation energy $E^{(2)}$ for a level labelled $|0_i\rangle$ in the ground-state multiplet is of the form†

$$E^{(2)} = -\sum_{n\neq 0}\frac{|\langle n|\,\mathscr{H}'_{\text{p}}\,|0_i\rangle|^2}{E_n - E_{0i}} \tag{8.2.45}$$

The perturbation $\mathscr{H}'_{\text{p}}$ is given by equation (8.2.7), so that

$$E^{(2)} = -\frac{|e|}{2mc}H_0^2\sum_{n\neq 0}\frac{|\langle n|L_z + 2S_z\,|0_i\rangle|^2}{E_n - E_{0i}} \tag{8.2.46}$$

The additional contribution to the magnetic moment of the molecule is

$$\langle\mu_z^{(2)}\rangle = -\frac{\partial E^{(2)}}{\partial H_0}$$

$$= +\frac{|e|}{2mc}2H_0\sum_{n\neq 0}\frac{|\langle n|\,L_z + 2S_z\,|0_i\rangle|^2}{E_n - E_{0i}} \tag{8.2.47}$$

We again assume that the excited states are separated from the ground state by an energy large compared with kT. We therefore have

$$E_n - E_{0i} \gg kT \gg g_J\beta H_0,$$

and so all the molecules are in the ground-state multiplet with effectively the same energy E_0 for this second order perturbation. Moreover, when summing the contributions from the different states of the ground-state multiplet in correspondence with equation (8.2.34) observe that the additional

† See, for example, Schiff, L. I., *Quantum Mechanics*, McGraw-Hill, 1949.

moment from equation (8.2.47) is derived from the quadratic function $|\langle n| L_z + 2S_z |0_i\rangle|^2$. This term does not change sign with $\langle J_z\rangle$ and so the second order effect simply reflects the total population in the ground-state multiplet rather than the differences in the populations over the various $\langle J_z\rangle$ states which arose in the case of equation (8.2.34). This means that there are no longer significant terms in $1/kT$ (these were derived previously from the differences in the Boltzmann factors) and this second order perturbation gives a temperature independent paramagnetic contribution to the magnetization. The population of each of the groundstate components will be closely $N/(2J+1)$, and so the second order contribution to the susceptibility takes the form

$$\chi^{(2)\mathrm{p}} = \frac{|e|}{2mc} \cdot \frac{2N}{2J+1} \sum_i \sum_{n \neq 0} \frac{|\langle n| L_z + 2S_z |0_i\rangle|^2}{E_n - E_0} \tag{8.2.48}$$

8.2.5 The Brillouin Function

At low temperatures the approxmation $g_J\beta M H_0 \ll kT$ may no longer be satisfactory and it is necessary to derive the complete relation for the first order paramagnetic susceptibility using equation (8.2.35). We wish to evaluate

$$\mathscr{M}_z^{\mathrm{p}} = -Ng_J\beta \frac{\sum_{-J}^{J} M \exp\{-(g_J\beta M H_0)/kT\}}{\sum_{-J}^{J} \exp\{-(g_J\beta M H_0)/kT\}} \tag{8.2.49}$$

Note that if we make use of the thermodynamic partition function

$$Z = \sum_{-J}^{J} \exp\{-(g_J\beta M H_0)/kT\} \tag{8.2.50}$$

we may write

$$\mathscr{M}_z^{\mathrm{p}} = NkT \frac{\partial}{\partial H_0}(\log_e Z) \tag{8.2.51}$$

But

$$Z = \exp\left(-\frac{g_J\beta J H_0}{kT}\right)\left\{1 + \exp\left(\frac{g_J\beta H_0}{kT}\right) + \exp\left(\frac{2g_J\beta H_0}{kT}\right) + \cdots \exp\left(\frac{2Jg_J\beta H_0}{kT}\right)\right\} \tag{8.2.52}$$

Using the summation relation for a Geometrical Progression,

$$\sum(\mathrm{GP}) = a\frac{(1-r^n)}{1-r}.$$

we have

$$Z = \exp\left(-\frac{g_J\beta J H_0}{kT}\right)\left[\frac{1 - \exp\{+(2J+1)\,(g_J\beta H_0/kT)\}}{1 - \exp(+\,g_J\beta H_0/kT)}\right] \tag{8.2.53}$$

Evidently,

$$Z = \frac{\sinh[\{(g_J\beta H_0)/2kT\}(2J+1)]}{\sinh\{(g_J\beta H_0)/2kT\}} \tag{8.2.54}$$

Carrying through the differentiation in equation (8.2.51), we obtain

$$\mathscr{M}_z^p = Ng_J\beta J\, B_J(y) \tag{8.2.55}$$

where $B_J(y)$ is the Brillouin function with $y = (g_J\beta JH_0)/kT$, and

$$B_J(y) = \frac{1}{2J}\left[(2J+1)\coth\left\{\left(\frac{2J+1}{2J}\right)y\right\} - \coth\left(\frac{y}{2J}\right)\right] \tag{8.2.56}$$

Note that $B_J(\infty) = 1$ and so in very high fields the magnetization approaches the value

$$\mathscr{M}_z^p(H_0 \to \infty) = Ng_J\beta J \tag{8.2.57}$$

This is just the maximum permissible value which can be achieved from the component of angular momentum along the magnetic field **H**.

Allowing J to go to infinity whilst retaining y finite in equation (8.2.56)

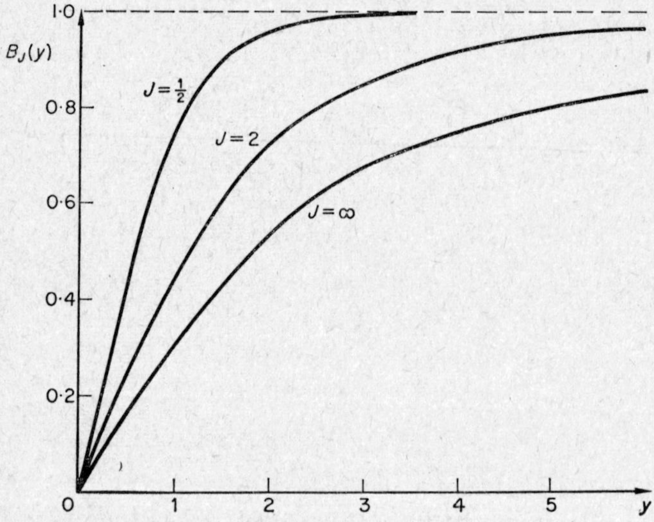

FIG. 8.2 The Brillouin function $B_J(y)$ for various values of J. Note that for $J = \infty$, $B_\infty(y)$ is the Langevin function $L(y)$

corresponds to taking the classical limit in which a continuous range of orientations is allowed for the total angular momentum vector. For this limit,

$$B_\infty(y) = L(y) = \coth y - \frac{1}{y} \tag{8.2.58}$$

This is the Langevin function which was discussed previously in connection with the classical theory of the dielectric constant (see Section 7.3). The function $B_J(y)$ is sketched for various values of J in Fig. 8.2.

8.2.6 SPECIAL CASE $\langle L_z\rangle = 0$

It frequently occurs that, in solids, the ground state configuration for a paramagnetic ion of the 3d transition group is described quite closely by an eigenfunction with $\langle L_z \rangle = 0$. This circumstance is of considerable practical importance, it simplifies the experimental determination of the total spin (and correspondingly the valency) of a paramagnetic ion. The condition $\langle L_z \rangle = 0$ arises because the distribution of charges on the lattice sites in a crystal gives rise to an electrostatic field which is of lower symmetry than that which refers to a free atom in a gas. The electrostatic field acting on an electron in a free atom is, in first order, the Coulomb field arising from the effective nuclear charge. The energy of interaction is $Z'e^2/r$ to this approximation and has spherical symmetry. In a crystal the additional contributions to the electrostatic field from charges on neighbouring lattice sites will lower the symmetry. The additional interaction energy may, for example, have cubic symmetry and may be large compared with the spin-orbit term $\lambda\mathbf{L}.\mathbf{S}$ which is the interaction giving rise to the $|\mathbf{L}, \mathbf{S}, \mathbf{J}, J_z\rangle$ representation for the free atom. In these circumstances a representation for the ground-state wave function in terms of a linear combination of free atom functions $|\mathbf{L}, L_z, \mathbf{S}, S_z\rangle$ is often more appropriate, and in many cases the particular linear combination required has $\langle L_z \rangle = 0$. When this situation occurs the ground-state orbital momentum is said to be quenched.

It is evident from equations (8.2.15) and (8.2.28) that when $\langle L_z \rangle = 0$ the paramagnetic response arises from the spin angular momentum alone and the factor g_J becomes $g_s = 2{\cdot}0$. The formula for the high temperature limit of the paramagnetic susceptibility, corresponding to equation (8.2.39), is now

$$\chi^{\mathrm{p}} = \frac{Ng^2\beta^2S(S+1)}{3kT} \qquad \text{spin only} \qquad (8.2.59)$$

In this equation we have written g rather than the explicit numerical value $2{\cdot}0$ because the spin-orbit coupling term $\lambda\mathbf{L}.\mathbf{S}$ may lead to a wave function for which $\langle L_z \rangle$ is not exactly zero. This perturbation can be conveniently included in equation (8.2.59) by allowing g to have values differing slightly from the spin-only value of $2{\cdot}0$. Typical values for g for a number of ions of the $3d$ transition group are given in Table 8.1.

In this table χ refers to one gram-mol of the material. From equation (8.2.59), we write

$$\chi(\mathrm{mol}) = \frac{N_A g^2\beta^2S(S+1)}{3kT} = \frac{C}{T} \qquad (8.2.60)$$

where N_A is Avogadro's number.

The number of effective Bohr magnetons per molecule, n_{eff}, is obtained by rewriting equation (8.2.60), as

$$\chi(\mathrm{mol}) = \frac{N_A(n_{\mathrm{eff}}\beta)^2}{3kT} \qquad (8.2.61)$$

from which

$$n_{\mathrm{eff}} = \left(\frac{3kC}{N_0\beta^2}\right)^{1/2} = 2{\cdot}83\sqrt{C} \qquad (8.2.62)$$

Salt	$\chi(293\ K) \times 10^6$ *c.g.s. units per mole*	*C per mole*	n_{eff} *per formula unit*	$\sqrt{\{4S(S+1)\}}$	*S*	*g*	θ K
(1) TiF_3	1,300	0·38	1·75	1·73	$\frac{1}{2}$	2·02	—
(2) $CrCl_3$	6,860	1·82	3·82	3·87	$\frac{3}{2}$	1·97	24
(3) $CrK(SO_4)_2.12H_2O$	6,350	1·86	3·86	3·87	$\frac{3}{2}$	2·0	0·004
(4) $MnCl_2$	14,500	4·17	5·78	5·92	$\frac{5}{2}$	1·95	3
(5) $Mn(NH_4)_2(SO_4)_2.6H_2O$	14,680	4·3	5·87	5·92	$\frac{5}{2}$	1·98	0·14
(6) $Fe(NH_4)(SO_4)_2.12H_2O$	14,640	4·3	5·86	5·92	$\frac{5}{2}$	1·98	0·043
(7) $Fe(SO_4).7H_2O$	11,930	3·52	5·3	4·91	2	2·16	—3
(8) $FeCl_2.4H_2O$	12,060	3·37	5·18	4·91	2	2·11	12
(9) $Co(NH_4)_2(SO_4)_2.6H_2O$	10,740	3·15	5·02	3·87	$\frac{3}{2}$	2·60	0·08
(10) $Co(SO_4).7H_2O$	9,780	2·94	4·85	3·87	$\frac{3}{2}$	2·51	—9
(11) $Ni(NH_4)_2(SO_4)_2.6H_2O$	4,240	1·24	3·15	2·83	1	2·23	—
(12) $Ni(NO_3)_2.6H_2O$	4,400	1·29	3·21	2·83	1	2·27	~—4
(13) $CuK_2(SO_4)_2.6H_2O$	1,520	0·45	1·89	1·73	$\frac{1}{2}$	2·18	—0·05
(14) $Cu(SO_4).5H_2O$	1,570	0·46	1·92	1·73	$\frac{1}{2}$	2·22	—0·7

using the values

$$k = 1{\cdot}380\,.\,10^{-16} \text{ erg deg}^{-1}$$
$$N_A = 6{\cdot}025\,.\,10^{23} \text{ g-mol}^{-1}$$
$$\beta = 0{\cdot}9273\,.\,10^{-20} \text{ erg gauss}^{-1}$$

8.2.7 A Generalization

The discussion given in the previous sections has drawn attention to particular cases in which relatively simple expressions can be derived for the magnetization and the susceptibility. These examples illustrate both the usefulness and the limitations of magnetic susceptibility measurements for identifying the ground state configuration of a molecule. We emphasize this aspect by a generalization of the previous discussion. The magnetic moment of a molecule which is in the energy state E_n is given by

$$\langle \mu_z \rangle_n = -\frac{\partial E_n}{\partial H_0} \tag{8.2.63}$$

Here E_n is the total energy of the state derived from complete Hamiltonian

$$\mathscr{H} = \mathscr{H}_0 + \mathscr{H}' \tag{8.2.64}$$

in the notation of equation (8.2.1).† The magnetization of the material is the sum of $\langle \mu_z \rangle_n$ over the thermodynamic distribution,

$$\mathscr{M}_z = \sum_n N(E_n) \langle \mu_z \rangle_n \tag{8.2.65}$$

$$= -N \frac{\sum (\partial E_n / \partial H_0) \exp(-E_n/kT)}{\sum \exp(-E_n/kT)} \tag{8.2.66}$$

$$= NkT \frac{\partial}{\partial H_0} (\log_e Z) \tag{8.2.67}$$

with

$$Z = \sum_n \exp(-E_n/kT) \tag{8.2.68}$$

For those states with $E_n \gg kT$ the occupation number $N(E_n)$ is negligible and their contribution to the magnetization can be neglected. If the remaining states all have $E_n \ll kT$ the exponential factor in equation (8.2.68) may be expanded to give

$$Z = \sum_n \left(1 - \frac{E_n}{kT} + \frac{E_n^2}{2k^2T^2} \cdots \right) \tag{8.2.69}$$

$$= A_0 + \frac{A_1}{kT} + \frac{A_2}{2k^2T^2} \tag{8.2.70}$$

with

$$A_0 = \sum_n 1, \ A_1 = -\sum_n E_n, \ A_2 = +\sum_n E_n^2, \text{ etc.} \tag{8.2.71}$$

† $\mathscr{H}_0$ and $\mathscr{H}'$ in equation (8.2.64), however, include the spin.

Hence

$$\mathscr{M}_z = NkT\frac{\partial}{\partial H_0}\log_e\left(A_0 + \frac{A_1}{kT} + \frac{A_2}{2k^2T^2}\cdots\right) \tag{8.2.72}$$

$$\simeq \frac{NkT\{(1/kT)(\partial A_1/\partial H_0) + (1/2k^2T^2)(\partial A_2/\partial H_0)\}}{A_0}$$

$$\mathscr{M}_z \simeq \frac{N}{A_0}\cdot\frac{\partial A_1}{\partial H_0} + \frac{N}{2kT}\cdot\frac{1}{A_0}\frac{\partial A_2}{\partial H_0} \tag{8.2.73}$$

The magnetic susceptibility is

$$\chi \simeq \frac{N}{A_0}\frac{\partial^2 A_1}{\partial H_0^2} + \frac{N}{2kT}\cdot\frac{1}{A_0}\frac{\partial^2 A_2}{\partial H_0^2} \tag{8.2.74}$$

which may be written

$$\chi = \alpha + \frac{\beta}{T}$$

Evidently there is, quite generally, a term in the magnetic susceptibility at high temperatures, of the form C/T which is also the form of the Curie law given in equation (8.2.40). The constant C is, however, no longer simply related to an angular momentum factor for the molecular unit but is now derived from the summation $\sum_n E_n^2$ over the occupied energy states. The fact that the susceptibility as measured experimentally varies with temperature according to the Curie law does not necessarily imply therefore that the ground state is adequately described by one of the simple models discussed in the previous sections.

The previous models are, however, included in the present discussion. Thus if we take the Hamiltonian to be

$$\mathscr{H} = \mathscr{H}_0 + g_J\beta J_z H_0 \tag{8.2.75}$$

and only one state of total angular momentum $\mathbf{J}$ is occupied

$$E_n = E_0 + g_J\beta\langle J_z\rangle H_0 \tag{8.2.76}$$

$$\sum E_n^2 = \sum_{-J}^{J}(E_0 + g_J\beta\langle J_z\rangle H_0)^2 \tag{8.2.77}$$

$$\frac{\partial A_2}{\partial H_0^2} = \frac{\partial^2}{\partial H_0^2}\left(\sum E_n^2\right) = 2g_J^2\beta^2\sum_{-J}^{J}\langle J_z\rangle^2 \tag{8.2.78}$$

From equations (8.2.37)

$$\frac{\partial^2 A_2}{\partial H_0^2} = 2g_J^2\beta^2\cdot\tfrac{1}{3}J(J+1)(2J+1) \tag{8.2.79}$$

and

$$A_0 = \sum_{-J}^{J} 1 = (2J+1) \tag{8.2.80}$$

Moreover

$$\frac{\partial^2 A_1}{\partial H_0^2} = \frac{\partial^2}{\partial H_0^2}\{\Sigma(E_0 + g_J\beta\langle J_z\rangle H_0)\} = 0 \tag{8.2.81}$$

The susceptibility obtained from equation (8.2.74) is

$$\begin{aligned}\chi &= \frac{N}{2kT}\cdot\frac{1}{(2J+1)}\cdot 2g_J^2\beta^2 . \tfrac{1}{3}J(J+1)(2J+1)\\ &= \frac{Ng_J^2\beta^2 J(J+1)}{3kT}\end{aligned} \tag{8.2.82}$$

in agreement with equation (8.2.39).

Equation (8.2.74) is of considerable interest since the susceptibility is derived from summations of energy terms over the occupied states. These summations are concerned with the diagonal elements of the energy matrix and can often be evaluated using the properties of the Spur (Trace) of the matrix. In order to carry the calculation through, however, it is necessary to know the form of the Hamiltonian operator for the molecular unit, and this usually has to be obtained from measurements (for example spectroscopic measurements) other than the susceptibility.

Example

As an example of the previous discussion we evaluate the paramagnetic susceptibility arising from Ni^{++} ions in a simple inorganic compound such as $Ni(NH_4)_2(SO_4)_2.6H_2O$ or $Ni(NO_3)_2.6H_2O$.

The free Ni^{++} ion has a lowest state electronic configuration designated $^3F_4$ so that the eigenfunction corresponds to three units of orbital angular momentum and one unit of spin angular momentum. The first order effect of the crystalline electric field is to split and combine the eigenstates labelled $|L = 3, L_z\rangle$ in such a way as to give a non-degenerate ground state orbital wave function

$$\psi(\text{orb}) = \frac{1}{\sqrt{2}}(|3, L_z = 2\rangle - |3, L_z = -2\rangle)$$

which is well separated from any of the excited states. The eigenfunctions, including the spin, which describe the ground state in this approximation, are therefore

$$\begin{aligned}\psi_{+1} &= \psi(\text{orb})\,|S_z = 1\rangle\\ \psi_0 &= \psi(\text{orb})\,|S_z = 0\rangle\\ \psi_{-1} &= \psi(\text{orb})\,|S_z = -1\rangle\end{aligned} \tag{8.2.83}$$

Note that $\langle\psi(\text{orb})|\, L_z\, |\psi(\text{orb})\rangle = 0$ so that we have an example of the special case $\langle L_z\rangle = 0$.

Taking into account the spin-orbit coupling, and second order effects of the crystalline electric field the perturbing Hamiltonian for the wave functions set out in equation (8.2.83) may be written in the form

$$\mathscr{H}' = D\{S_z^2 - \tfrac{1}{3}S(S+1)\} + E(S_x^2 - S_y^2) + g\beta\mathbf{S}.\mathbf{H} \tag{8.2.84}$$

Here the terms in D and E describe how the states are split in the absence of a magnetic field and g will differ slightly from the value 2·0 because of the spin-orbit coupling as described previously. The validity of equation (8.2.84) is most directly established by microwave resonance experiments.

In order to evaluate the susceptibility we first determine the energy states using the functions in equation (8.2.83) as basis functions. Take $\mathbf{H}$ along the z-axis, $\mathbf{H} = \hat{\mathbf{k}}H_0$, and rewrite equation (8.2.84) as

$$\mathscr{H}' = D(S_z^2 - \tfrac{2}{3}) + \tfrac{1}{2}E(S_+^2 + S_-^2) + g\beta S_z H_0 \tag{8.2.85}$$

Here we have used the fact that $S = 1$, and written

$$\begin{aligned} S_x + \mathrm{j}S_y &= S_+ \\ S_x - \mathrm{j}S_y &= S_- \end{aligned} \tag{8.2.86}$$

because there are useful operator relations, namely,

$$\begin{aligned} S_+ |S_z\rangle &= \{S(S+1) - S_z(S_z+1)\}^{1/2} |S_z + 1\rangle \\ S_- |S_z\rangle &= \{S(S+1) - S_z(S_z-1)\}^{1/2} |S_z - 1\rangle \end{aligned} \tag{8.2.87}$$

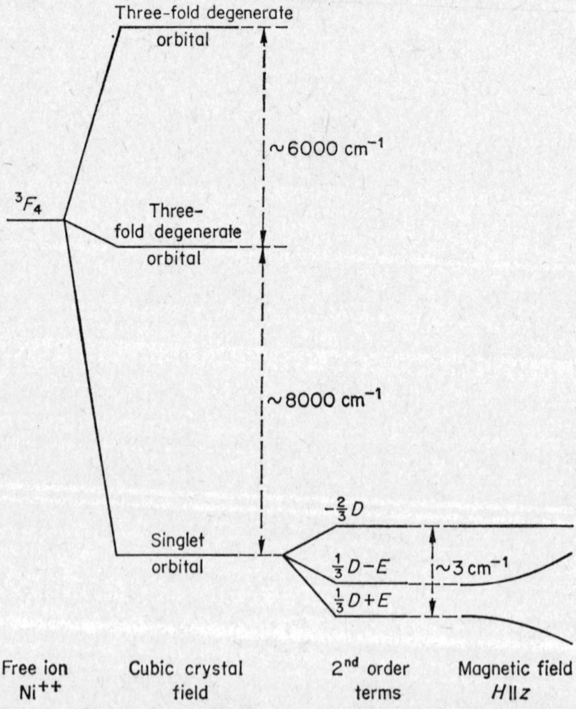

FIG. 8.3 Energy levels for an Ni^{++} ion in a simple inorganic crystal. For this diagram D and E are taken to be negative which is most commonly true for Ni^{++}. Note that transitions within the lowest triplet lie in the microwave frequency range and give rise to paramagnetic resonance.

The energy eigenvalues E_n are now given by the roots of the secular determinant

$$\begin{vmatrix} \frac{1}{3}D + g\beta H_0 - E_n & 0 & E \\ 0 & -\frac{2}{3}D - E_n & 0 \\ E & 0 & \frac{1}{3}D - g\beta H_0 - E_n \end{vmatrix} = 0 \qquad (8.2.88)$$

The roots of this equation are

$$\begin{aligned} E_1 &= \tfrac{1}{3}D + \sqrt{(E^2 + g^2\beta^2H_0^2)} \\ E_2 &= \tfrac{1}{3}D - \sqrt{(E^2 + g^2\beta^2H_0^2)} \\ E_3 &= -\tfrac{2}{3}D \end{aligned} \qquad (8.2.89)$$

The energy levels for the Ni^{++} ion are sketched in Fig. 8.3.

To first order, the magnetic moments are

$$\begin{aligned} \langle\mu_z\rangle_1 &= -\frac{\partial E_1}{\partial H_0} = -\frac{g^2\beta^2H_0}{E} \\ \langle\mu_z\rangle_2 &= -\frac{\partial E_2}{\partial H_0} = +\frac{g^2\beta^2H_0}{E} \\ \langle\mu_z\rangle_3 &= 0 \end{aligned} \qquad (8.2.90)$$

Note that for this particular example the magnetic moment of each level is zero in zero magnetic field. This arises because the wave functions appropriate to equation (8.2.85) with zero magnetic field take the form

$$\psi(\text{orb})\,\{|S_z = 1\rangle \pm |S_z = -1\rangle\}, \qquad \psi(\text{orb})\,|0\rangle.$$

The magnetization $\mathcal{M}_z$ is obtained from equation (8.2.66)

$$\begin{aligned} \mathcal{M}_z &= -\frac{Ng^2\beta^2H_0}{E}\cdot\frac{\{\exp(-E_1/kT) - \exp(-E_2/kT)\}}{\{\exp(-E_1/kT) + \exp(-E_2/kT) + \exp(-E_3/kT)\}} \\ &= \frac{2Ng^2\beta^2H_0}{E}\cdot\frac{\sinh(E/kT)}{\exp(D/kT) + 2\cosh(E/kT)} \end{aligned} \qquad (8.2.91)$$

In equation (8.2.91) we have assumed that terms involving H_0 occurring in the exponential factors can be neglected. Equation (8.2.91) leads therefore to the low field susceptibility

$$\chi^{\mathrm{p}} = \frac{2Ng^2\beta^2}{E}\cdot\frac{\sinh(E/kT)}{\exp(D/kT) + 2\cosh(E/kT)} \qquad (8.2.92)$$

At high temperatures $\sinh(E/kT) \to E/kT$, $\cosh(E/kT) \to 1$ and $\exp(D/kT) \to 1$. In this limit the susceptibility is

$$\chi^{\mathrm{p}} = \frac{2Ng^2\beta^2}{3kT} \qquad \text{(high } T\text{)} \qquad (8.2.93)$$

which is just Curie's law for molecules with spin $S = 1$,

$$\chi^{\mathrm{p}} = \frac{Ng^2\beta^2S(S+1)}{3kT} \qquad (8.2.94)$$

Note that at low temperatures the susceptibility will be anisotropic. We have oriented the magnetic field along the z-axis of coordinates when deriving equation (8.2.92); a different result will be obtained if **H** is directed along an x- or y-axis. The high temperature susceptibility given by equation (8.2.93) is, however, isotropic provided that g is isotropic (which has been assumed to be the case in the present calculation).

Consider next the derivation of the high temperature susceptibility using the Spur relation given in equation (8.2.74). The term A_1 is given by the sum of the diagonal elements in the determinant of equation (8.2.88), but omitting the factors E_n. Evidently

$$A_1 = 0 \tag{8.2.95}$$

The term A_2 is Spur $(\mathscr{H}')^2$,

$$\begin{aligned}\text{Spur}\,(\mathscr{H}')^2 = \text{Spur}\,\{&D^2(S_z^2 - \tfrac{2}{3})^2 + \tfrac{1}{4}E^2(S_+^2 + S_-^2)^2 + g^2\beta^2 S_z^2 H_0^2 \\ &+ DE(S_z^2 - \tfrac{2}{3})(S_+^2 + S_-^2) \\ &+ D(S_z^2 - \tfrac{2}{3})g\beta S_z H_0 \\ &+ E(S_+^2 + S_-^2)g\beta S_z H_0\}\end{aligned} \tag{8.2.96}$$

We shall require

$$\frac{\partial^2 A_2}{\partial H_0^2} = \frac{\partial^2}{\partial H_0^2}\{\text{Spur}\,(\mathscr{H}')^2\} \tag{8.2.97}$$

Evidently the only term in equation (8.2.96) which will contribute to $\partial^2 A_2/\partial H_0^2$ is that involving H_0^2. We therefore have

$$\frac{\partial^2 A_2}{\partial H_0^2} = \frac{\partial^2}{\partial H_0^2}\{\text{Spur}\,(g^2\beta^2 S_z^2 H_0^2)\} \tag{8.2.98}$$

$$= 2g^2\beta^2\,\text{Spur}\,(S_z^2) \tag{8.2.99}$$

$$= 2g^2\beta^2 . \tfrac{1}{3}S(S+1)(2S+1) \tag{8.2.100}$$

In addition

$$A_0 = \sum_{-S_z}^{S_z} 1 = (2S+1) \tag{8.2.101}$$

Hence equation (8.2.74) leads to

$$\begin{aligned}\chi^{\mathrm{p}} &= \frac{N}{2kT}\cdot\frac{1}{(2S+1)}\cdot\frac{2g^2\beta^2 S(S+1)(2S+1)}{3} \\ &= \frac{Ng^2\beta^2 S(S+1)}{3kT}\end{aligned} \tag{8.2.102}$$

in agreement with equation (8.2.94).

8.3 The Molecular Field

In solids and liquids there may be significant interactions which correlate the spin orientations throughout the medium. The spins may be coupled together more or less directly through the magnetic dipole fields of the molecular units or through exchange effects. They may, however, also be coupled together through more sophisticated and less direct mechanisms such as arise, for example, from interactions through the lattice vibrations or through mobile conduction electrons. To take account of these interactions in first order without requiring a microscopic model to describe the mechanism of the interaction, we use the 'molecular field' representation. We assume that the effect of all the interactions, whatever their origin, can be described by a single 'molecular' magnetic field $\mathbf{H}_m$. The total magnetic field, $\mathbf{H}$, experienced by a molecule is taken to be the vector sum of the field $\mathbf{H}_m$ and any macroscopic field $\mathbf{H}_a$ which is acting in the material. We write, therefore,

$$\mathbf{H} = \mathbf{H}_a + \mathbf{H}_m \tag{8.3.1}$$

The relation determining $\mathbf{H}_m$ is fundamental to the molecular field theory. The simplest assumption is that $\mathbf{H}_m$ is linearly proportional to the magnetization,

$$\mathbf{H}_m = \lambda\mathcal{M}$$

$$\mathbf{H} = \mathbf{H}_a + \lambda\mathcal{M} \tag{8.3.2}$$

Assume that in the high temperature regime the spin system obeys a Curie law referred to the total field of equation (8.3.1). In these circumstances

$$\mathcal{M} = \chi(\mathbf{H}_a + \mathbf{H}_m)$$

$$= \frac{C}{T}(\mathbf{H}_a + \lambda\mathcal{M}) \tag{8.3.3}$$

Hence

$$\mathcal{M} = \frac{C}{T - \lambda C}\mathbf{H}_a \tag{8.3.4}$$

and the susceptibility may be written as

$$\chi = \frac{C}{T - \theta} \tag{8.3.5}$$

with $\theta = \lambda C$, having the dimensions of temperature. Equation (8.3.5) is known as the Curie–Weiss Law, and the constant θ is called the Weiss constant. The constant θ can be determined experimentally by plotting the reciprocal of the high temperature susceptibility as a function of T. This should give a straight line graph which, when extrapolated to the point $1/\chi = 0$, determines the constant θ directly.

8.3.1 Spontaneous Magnetization

Athough equation (8.3.5) is satisfactory at high temperatures, it is evident that an anomaly will arise in the vicinity of $T = \theta$, if θ is positive. For this case the material will exhibit a spontaneous magnetization at temperatures less than θ K.† To see how this property arises it is necessary to write down a more general expression than that which is given by the simple Curie law for the magnetization as a function of temperature. We make use of the less restricted description which involves the Brillouin function introduced previously on page 373.

† Positive values of θ are usually called Curie temperatures and imply the existence of a uniform spontaneous magnetization throughout the material for $T < \theta$. They are sometimes written explicitly as θ_c.

Take $\mathbf{H}_a = \hat{\mathbf{k}} H_0$ and use equation (8.2.51),

$$\mathscr{M}_z = NkT \frac{\partial}{\partial H_0} (\log_e Z)$$

to obtain

$$\mathscr{M}_z = NkT \frac{\partial}{\partial H_0} \left[\!\left[\log_e \left[\exp\left\{ \frac{-g_J \beta J (H_0 + \lambda \mathscr{M}_z)}{kT} \right\} + \exp\left\{ \frac{-g_J \beta (J-1)(H_0 + \lambda \mathscr{M}_z)}{kT} \right\} + \cdots \exp\left\{ \frac{g_J \beta J (H_0 + \lambda \mathscr{M}_z)}{kT} \right\} \right] \right]\!\right] \quad (8.3.6)$$

$$= NkT \frac{\partial}{\partial H_0} \left\{ \log_e \frac{\sinh [\{g_J \beta (H_0 + \lambda \mathscr{M}_z)/2kT\}(2J+1)]}{\sinh [g_J \beta (H_0 + \lambda \mathscr{M}_z)/2kT]} \right\} \quad (8.3.7)$$

giving

$$\mathscr{M}_z = N g_J \beta J \, B_J(y) \equiv \mathscr{M}_0 \, B_J(y) \quad (8.3.8)$$

where $B_J(y)$ is the Brillouin function with $y = \{g_J \beta J (H_0 + \lambda \mathscr{M}_z)/kT\}$. As in equation (8.2.56),

$$B_J(y) = \frac{1}{2J} \left[(2J+1) \coth \left\{ \left(\frac{2J+1}{2J} \right) y \right\} - \coth \left(\frac{y}{2J} \right) \right] \quad (8.3.9)$$

The molecular field $\lambda \mathscr{M}_z$ can be two or three orders of magnitude larger than the usual laboratory fields. It is convenient, therefore, to discuss the solutions of equation (8.3.8) in the limit $H_0 = 0$. For this case we have

$$\mathscr{M}_z(H = 0) = N g_J \beta J \, B_J \left(\frac{g_J \beta J \lambda \mathscr{M}_z}{kT} \right) \quad (8.3.10)$$

This equation always has one solution with $\mathscr{M}_z = 0$. However, in the temperature range

$$0 \leqslant kT < \lambda \frac{N g_J^2 \beta^2 J(J+1)}{3} \quad (8.3.11)$$

there is an additional solution with $\mathscr{M}_z$ non-zero, which is the stable configuration for the magnetization. Consider equation (8.3.10) written as a pair of simultaneous equations

$$f_1 = \mathscr{M}_z \quad (8.3.12)$$

$$f_2 = N g_J \beta J \, B_J \left(\frac{g_J \beta J \lambda \mathscr{M}_z}{kT} \right) \quad (8.3.13)$$

Equation (8.3.12), when plotted graphically as a function of $\mathscr{M}_z$, gives a straight line through the origin with a slope

$$\frac{\partial f_1}{\partial \mathscr{M}_z} = 1 \quad (8.3.14)$$

Equation (8.3.13) describes a smooth curve which also passes through the origin. This curve tends to the limiting value $f_2 = N g_J \beta J$ as $\mathscr{M}_z$ tends to infinity. To demonstrate the possibility of two solutions for the pair of simultaneous equations we examine the slope $(\partial f_2/\partial \mathcal{M}_z)$ near $\mathcal{M}_z = 0$. We seek first a simplified expression for $B_J(y)$ near $y = 0$. Using an obvious formal notation, write equation (8.3.9), as

$$B_J(y) = \frac{1}{2J}\left\{(2J+1)\frac{e^{\alpha} + e^{-\alpha}}{e^{\alpha} - e^{-\alpha}} - \frac{e^{\beta} + e^{-\beta}}{e^{\beta} - e^{-\beta}}\right\} \tag{8.3.15}$$

For small values of the variable y,

$$B_J(y) = \frac{1}{2J}\left\{(2J+1)\frac{1 + (\alpha^2/2)}{\alpha + (\alpha^3/6)} - \frac{1 + (\beta^2/2)}{\beta + (\beta^3/6)}\right\} \tag{8.3.16}$$

$$= \frac{1}{2J}\left\{\frac{(2J+1)}{\alpha}\left(1 + \frac{\alpha^2}{3}\right) - \frac{1}{\beta}\left(1 + \frac{\beta^2}{3}\right)\right\} \tag{8.3.17}$$

$$= \frac{1}{2J}\left\{\frac{2J}{y}\left(1 + \frac{\alpha^2}{3}\right) - \frac{2J}{y}\left(1 + \frac{\beta^2}{3}\right)\right\} \tag{8.3.18}$$

$$= \frac{1}{3y}\left\{\left(\frac{y}{2J}\right)^2(2J+1)^2 - \left(\frac{y}{2J}\right)^2\right\} \tag{8.3.19}$$

so that,

$$B_J(y) = \frac{J+1}{3J}y \qquad (y \to 0) \tag{8.3.20}$$

Hence, near $\mathcal{M}_z = 0$

$$f_2 = \frac{Ng_J^2\beta^2 J(J+1)}{3kT}\lambda\mathcal{M}_z \tag{8.3.21}$$

and

$$\frac{\partial f_2}{\partial \mathcal{M}_z} = \lambda\frac{Ng_J^2\beta^2 J(J+1)}{3kT} \tag{8.3.22}$$

$$= \frac{\lambda C}{T} = \frac{\theta}{T} \tag{8.3.23}$$

where C is again the Curie constant, referring to the high temperature susceptibility, which was introduced in equation (8.2.40), and θ is the (positive) Weiss constant of equation (8.3.5). When

$$kT < \lambda\frac{Ng_J^2\beta^2 J(J+1)}{3} \tag{8.3.24}$$

and correspondingly

$$T < \lambda C \quad \text{or} \quad T < \theta \tag{8.3.25}$$

then,

$$\frac{\partial f_2}{\partial \mathcal{M}_z} > 1, \qquad \frac{\partial f_2}{\partial \mathcal{M}_z} > \frac{\partial f_1}{\partial \mathcal{M}_z} \qquad (\mathcal{M}_z \to 0) \tag{8.3.26}$$

and the simultaneous equations (8.3.12) and (8.3.13) have two solutions. On the other hand, for $T > \lambda C$, or $T > \theta$, there is only one solution, namely

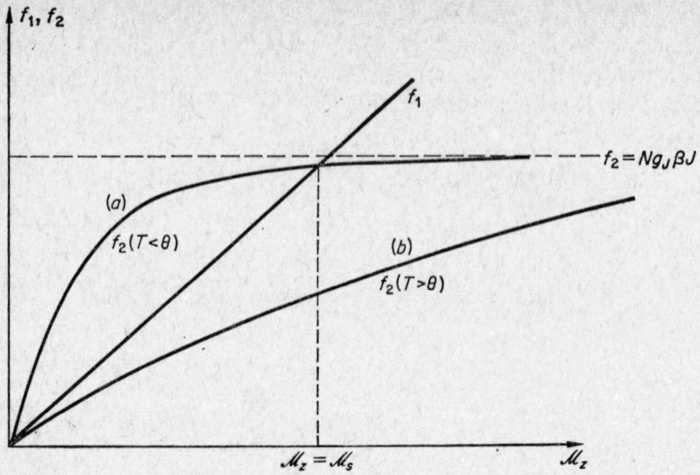

FIG. 8.4 Graphical solution for the molecular field equations (8.3.10), (8.3.12) and (8.3.13). Curve (*a*) intersects f_1 at the origin and at $\mathscr{M}_z = \mathscr{M}_s$; curve (*b*) intersects f_1 only at the origin.

$\mathscr{M}_z = 0$. The temperature $T_c = \lambda C = \theta$ is a magnetic 'critical' temperature for the material. Fig. 8.4 illustrates the technique we have used to establish the solutions for the magnetization.

To determine the stable configuration when $T < \theta$, consider the thermodynamic Free Energy function

$$F = -NkT \log_e Z \tag{8.3.27}$$

At a given temperature that configuration with the larger value of Z will have the lower energy value and will be the stable state. From equation (8.3.7), with $H_0 = 0$, we see that

$$Z = \frac{\sinh\{(g_J\beta\lambda\mathscr{M}_z/2kT)(2J+1)\}}{\sinh\{(g_J\beta\lambda\mathscr{M}_z/2kT)\}} \tag{8.3.28}$$

which we write as

$$Z = \frac{\sinh(2J+1)u}{\sinh u} \tag{8.3.29}$$

The solution $\mathscr{M}_z = 0$ for $H_0 = 0$, leads to

$$Z(\mathscr{M}_z = 0) = (2J+1) \tag{8.3.30}$$

On the other hand, the solution with $\mathscr{M}_z$ non-zero leads to

$$Z(\mathscr{M}_z = \mathscr{M}_s) = \frac{(2J+1)u + (3!)^{-1}(2J+1)^3u^3 + (5!)^{-1}(2J+1)^5u^5 + \cdots}{u + (3!)^{-1}u^3 + (5!)^{-1}u^5 + \cdots}$$
$$> \frac{(2J+1)[u + (3!)^{-1}u^3 + (5!)^{-1}u^5 + \cdots]}{u + (3!)^{-1}u^3 + (5!)^{-1}u^5 + \cdots} \tag{8.3.31}$$

and so

$$Z(\mathscr{M}_z = \mathscr{M}_s) > (2J+1)$$
$$Z(\mathscr{M}_z = \mathscr{M}_s) > Z(\mathscr{M}_z = 0) \tag{8.3.32}$$

Hence the configuration with $\mathcal{M}_z = \mathcal{M}_s$ has a lower free energy than that with $\mathcal{M}_z = 0$ and so it will be the stable state. Since the material has a magnetization $\mathcal{M}_s$ when there is no applied field H_0, the material is said to exhibit 'spontaneous magnetization'.

The present discussion of molecular field theory has resulted in a spontaneous magnetization which is uniform throughout the material. The molecular moments are all essentially parallel to one another for $T < \theta$. This arrangement is called a 'ferromagnetic' configuration as distinct from an 'antiferromagnetic' configuration in which spontaneous magnetization of different sublattices may occur but be so oriented as to give no overall magnetization. The critical temperature T_c for the onset of the ferromagnetic spontaneous magnetization

$$T_c = \lambda \frac{N g_J^2 \beta^2 J(J+1)}{3k} \tag{8.3.33}$$

is called the Curie temperature. In the case of an isotropic ferromagnetic material $T_c = \theta = \lambda C$.

8.3.2 Spontaneous Magnetization for $T < \theta$

It is evident from Fig. 8.4 that, for temperatures less than θ K, the spontaneous magnetization $\mathcal{M}_s$ is given by the point of intersection of the straight line representing f_1 with the curve representing f_2. There is unfortunately no simple analytical expression for $\mathcal{M}_s$ in this low temperature regime but a graph can be drawn without much difficulty from the numerical values.

The formal equation for the magnetization is given by equation (8.3.8),

$$\mathcal{M}_s = N g_J \beta J\, B_J(y) \equiv \mathcal{M}_0\, B_J(y) \tag{8.3.34a}$$

with

$$y = \frac{g_J \beta J \lambda \mathcal{M}_s}{kT} \tag{8.3.34b}$$

Here we have again taken $H_0 = 0$, to give the spontaneous magnetization. Write equations (8.3.34a) and (8.3.34b) as the pair of simultaneous equations.

$$\frac{\mathcal{M}_s}{\mathcal{M}_0} = B_J(y) \tag{8.3.35a}$$

$$\frac{\mathcal{M}_s}{\mathcal{M}_0} = \frac{J+1}{3J}\left(\frac{T}{\theta}\right) y \tag{8.3.35b}$$

Choose a particular value for the variable y and evaluate $\mathcal{M}_s/\mathcal{M}_0$ using equation (8.3.35a). Insert these values of y, $\mathcal{M}_s/\mathcal{M}_0$, in equation (8.3.35b) and determine (T/θ). Repetition of this process will enable the complete graph for $\mathcal{M}_s/\mathcal{M}_0$ versus (T/θ) to be constructed. Typical curves for different values of J are given in Fig. 8.5.

Note that the formal elimination of the variable y between the simultaneous equations (8.3.35) will lead to a relation

$$\frac{\mathcal{M}_s}{\mathcal{M}_0} = F_J(T/\theta) \tag{8.3.36}$$

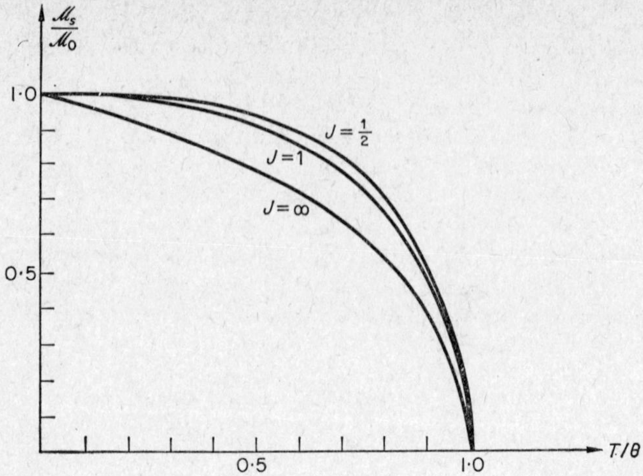

FIG. 8.5 The spontaneous magnetization plotted for (*a*) $J = \frac{1}{2}$, (*b*) $J = 1$, (*c*) $J = \infty$.

Although we do not have an analytical expression for the function F_J, equation (8.3.36) shows that $\mathcal{M}_s/\mathcal{M}_0$ should be a universal function of (T/θ), dependent upon the particular value of J but independent of other properties of the material once J is specified. Equation (8.3.36) is called the Law of Corresponding States.

Near $(T/\theta) = 1$, $\mathcal{M}_s \rightarrow 0$, and an approximate analytical relation can be derived for the ratio $\mathcal{M}_s/\mathcal{M}_0$. Expand $B_J(y)$ to terms in y^3, making use of equation (8.3.20) to give the linear term and noting that there will be no term in y^2. Equation (8.3.35a) may be written as

$$\frac{\mathcal{M}_s}{\mathcal{M}_0} = \frac{J+1}{3J}y - c(J)y^3 \qquad (y \rightarrow 0) \qquad (8.3.37)$$

where $c(J)$ is a rational function in J. Using equation (8.3.35b), we may now write

$$\frac{\mathcal{M}_s}{\mathcal{M}_0} = \left(\frac{\theta}{T}\right)\frac{\mathcal{M}_s}{\mathcal{M}_0} - c_1(J)\left(\frac{\theta}{T}\right)^3\left(\frac{\mathcal{M}_s}{\mathcal{M}_0}\right)^3 \qquad (\mathcal{M}_s \rightarrow 0) \qquad (8.3.38)$$

where $c_1(J) = \{3J/(J+1)\}^3\, c(J)$.

To this approximation, therefore,

$$\left(\frac{\mathcal{M}_s}{\mathcal{M}_0}\right)^2 = \frac{1}{c_1(J)}\left(\frac{T}{\theta}\right)^2\left(1 - \frac{T}{\theta}\right) \qquad (8.3.39)$$

so that, for temperatures just below θ K

$$\left(\frac{\mathcal{M}_s}{\mathcal{M}_0}\right) \propto \left(1 - \frac{T}{\theta}\right)^{1/2} \qquad (8.3.39a)$$

This relation is obeyed by some, but by no means all, materials.

8.3.3 Antiferromagnetism on the Molecular Field Model

When the high temperature susceptibility of a material follows a Curie–Weiss Law (compare equation (8.3.5)), in the form

$$\chi = \frac{C}{T+\theta} \tag{8.3.40}$$

the material will be antiferromagnetic at low temperatures. In equation (8.3.40) the constant θ is to be regarded as a positive quantity.

Molecular field theory can give a general description of antiferromagnetism. Consider a model for a crystalline material in which a system of identical molecular units, each characterized by an angular momentum $\mathbf{J}$, are distributed over lattice sites which may be labelled either type A or type B. One example of such a structure is the body-centred cubic crystal lattice where the sites A may be taken at the corners of the cubic unit cell and the sites B at the centre points. Another example is a simple layer structure with alternate crystal planes labelled A and B. Describe the total magnetization of the material in terms of two components, one component $\mathscr{M}_A$ arising from the molecules on the lattice sites labelled A (sublattice A) and the other component $\mathscr{M}_B$ arising from the sublattice B. If the applied magnetic field is $\mathbf{H}_a = \hat{\mathbf{k}}H_0$, the total field acting on a molecule in sublattice A is given by

$$\mathbf{H}_A = \hat{\mathbf{k}}H_0 - \lambda\mathscr{M}_A - \lambda'\mathscr{M}_B \tag{8.3.41}$$

and the total field acting on a molecule in sublattice B is,

$$\mathbf{H}_B = \hat{\mathbf{k}}H_0 - \lambda'\mathscr{M}_A - \lambda\mathscr{M}_B \tag{8.3.42}$$

The molecular field constants λ, λ' appear symmetrically in these equations because the sublattices are identical.

At high temperatures the magnetization of each sublattice will be in the direction of the applied field $\hat{\mathbf{k}}H_0$, and will be derived from a Curie Law referred to the *total* field.

$$\mathscr{M}_A = \frac{C'}{T}(H_0 - \lambda\mathscr{M}_A - \lambda'\mathscr{M}_B) \tag{8.3.43}$$

$$\mathscr{M}_B = \frac{C'}{T}(H_0 - \lambda'\mathscr{M}_A - \lambda\mathscr{M}_B) \tag{8.3.44}$$

The constant

$$C' = \tfrac{1}{2}\frac{Ng^2\beta^2 J(J+1)}{3k} = \tfrac{1}{2}C$$

because half of the molecules are on each type of lattice site. The total magnetization in the direction of the applied field is

$$\mathscr{M}_A + \mathscr{M}_B = \frac{C'}{T}\{2H_0 - (\lambda+\lambda')(\mathscr{M}_A + \mathscr{M}_B)\} \tag{8.3.45}$$

and the measured susceptibility for the material is

$$\chi = \frac{\mathscr{M}_A + \mathscr{M}_B}{H_0} = \frac{C}{T + \frac{1}{2}(\lambda+\lambda')C} \tag{8.3.46}$$

or

$$\chi = \frac{C}{T+\theta} \tag{8.3.47}$$

with

$$\theta = \tfrac{1}{2}(\lambda + \lambda')C \tag{8.3.48}$$

8.3.4 NÉEL TEMPERATURE (AXIAL MAGNETIZATION)

Below a particular temperature T_N, called the Néel temperature, an antiferromagnetic material exhibits spontaneous magnetization in each of the sublattices A and B but no net magnetic moment. A simple model system arises when the crystalline anisotropy field is uniaxial and determines a common *easy* direction along which the magnetizations $\mathscr{M}_A$ and $\mathscr{M}_B$ are aligned antiparallel. This anisotropy field need not be included explicitly in our theoretical account of antiferromagnetism, we only require that it should define a unique axis for the magnetization when no external field is present. The sign of the interaction constant λ' determines the ground state configuration to be that with $\mathscr{M}_A$ and $\mathscr{M}_B$ oppositely aligned.

Below the temperature T_N, the magnetization of each sublattice is a solution of the molecular field equation, the general form being given by equation (8.3.8). For the sublattice A, we write

$$\mathscr{M}_A = \mathscr{M}_A^0 B_J(y_A) \tag{8.3.49}$$

with

$$\mathscr{M}_A^0 = \tfrac{1}{2}Ng_J\beta J$$

$$y_A = \frac{g_J\beta J}{kT}(\mathrm{H}_0 - \lambda\mathscr{M}_A + \lambda'\mathscr{M}_B) \tag{8.3.50}$$

For sublattice B,

$$\mathscr{M}_B = \mathscr{M}_B^0 B_J(y_B) \tag{8.3.51}$$

with

$$\mathscr{M}_B^0 = \tfrac{1}{2}Ng_J\beta J = \mathscr{M}_A^0$$

$$y_B = \frac{g_J\beta J}{kT}(-H_0 + \lambda'\mathscr{M}_A - \lambda\mathscr{M}_B) \tag{8.3.52}$$

Note that in these equations the terms H_0, $\mathscr{M}_A$, $\mathscr{M}_B$, are the magnitudes of the corresponding vector quantities. We have taken $\hat{\mathbf{k}}H_0$ in the direction of $\mathscr{M}_A$ and included the signs appropriate to $\mathscr{M}_A$ and $\mathscr{M}_B$ for temperatures below T_N.

The spontaneous magnetization of each sublattice, $\mathscr{M}_A^s$, $\mathscr{M}_B^s$ refers to $H_0 = 0$. Thus

$$\mathscr{M}_A^s = \mathscr{M}_A^0 B_J(y_{A0}) \tag{8.3.53}$$

$$y_{A0} = \frac{g_J\beta J}{kT}(-\lambda\mathscr{M}_A^s + \lambda'\mathscr{M}_B^s) \tag{8.3.54}$$

In addition we seek solutions with the magnitude $\mathscr{M}_A^s = \mathscr{M}_B^s$ below T_N, and so we may write

$$y_{A0} = \frac{g_J\beta J}{kT}(\lambda' - \lambda)\mathscr{M}_A^s \tag{8.3.55}$$

Following the discussion given previously, in Section 8.3.1, there will be a solution to equation (8.3.53) with $\mathscr{M}^s_A$ non-zero at all temperatures below the particular temperature T_{N}, for which

$$\frac{\partial f_1}{\partial \mathscr{M}^s_A} = \frac{\partial f_2}{\partial \mathscr{M}^s_A} = 1 \tag{8.3.56}$$

with

$$f_1 = \mathscr{M}^s_A, \qquad f_2 = \mathscr{M}^0_A\, B_J(y_{A0})$$

Near T_{N} where $y_A \to 0$, use the expansion for $B_J(y)$ given in equation (8.3.20) to write

$$f_2 = \mathscr{M}^0_A\left(\frac{J+1}{3J}\right)y_{A0} \tag{8.3.57}$$

$$\frac{\partial f_2}{\partial \mathscr{M}^s_A} = \mathscr{M}^0_A\left(\frac{J+1}{3J}\right)\frac{g_J\beta J}{kT}(\lambda' - \lambda) \tag{8.3.58}$$

Hence the Néel temperature is

$$T_{\mathrm{N}} = \frac{\frac{1}{2}Ng_J^2\beta^2 J(J+1)}{3k}(\lambda' - \lambda) \tag{8.3.59}$$

$$T_{\mathrm{N}} = \tfrac{1}{2}C(\lambda' - \lambda) \tag{8.3.60}$$

Comparison with equation (8.3.48) gives

$$\frac{\theta}{T_{\mathrm{N}}} = \frac{\lambda' + \lambda}{\lambda' - \lambda} \tag{8.3.61}$$

Typical values of θ and T_{N} are given in Table 8.2.

TABLE 8.2

Magnetic parameters for simple antiferromagnetic materials

Substance	T_{N} (deg K)	θ (deg K)
CoF_2	37	48
FeF_2	80	120
MnF_2	67	117
NiF_2	73	100

8.3.5 SUSCEPTIBILITY BELOW T_{N} (AXIAL MAGNETIZATION)

At temperatures below T_{N} an applied magnetic field may induce a small magnetization in an antiferromagnetic material. Thus if $\hat{\mathbf{k}}H_0$ is directed along the common axis the sublattice magnetizations may no longer be equal in magnitude, whilst if $\hat{\mathbf{k}}H_0$ is perpendicular to the common axis the sublattice magnetizations may incline towards the direction of the applied field. These two configurations are illustrated in Fig. 8.6.

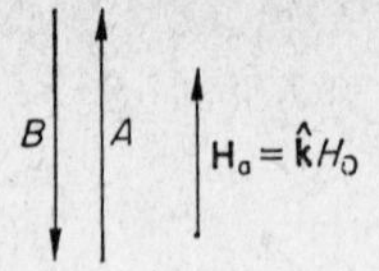

(a)

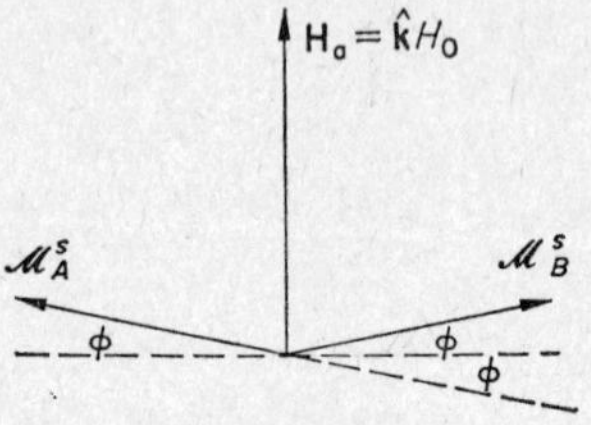

(b)

FIG. 8.6 The magnetization vectors for the sublattices A, B of an axial antiferromagnetic material in an applied magnetic field $\mathbf{H}_a$; along the easy axis of magnetization.

When the applied magnetic field $\hat{\mathbf{k}}H_0$ is along the common axis in the direction of $\mathscr{M}_A$, use equation (8.3.49) to write

$$\begin{aligned}\mathscr{M}_A &= \mathscr{M}_A^0\, B_J(y_A)\\ &= \mathscr{M}_A^0\{B_J(y_{A0}) + (y_A - y_{A0})\, B_J'(y_{A0})\}\end{aligned} \tag{8.3.62}$$

Here we have expanded $B_J(y_A)$, in a Taylor series about y_{A0}, assuming that H_0 is small.† From equations (8.3.50) and (8.3.54)

$$(y_A - y_{A0}) = \frac{g_J\beta J}{kT}\{H_0 - \lambda(\mathscr{M}_A - \mathscr{M}_A^s) + \lambda'(\mathscr{M}_B - \mathscr{M}_B^s)\} \tag{8.3.63}$$

Similarly, from equations (8.3.51), (8.3.52),

$$\mathscr{M}_B = \mathscr{M}_B^0\{B_J(y_{B0}) + (y_B - y_{B0})\, B_J'(y_{B0})\} \tag{8.3.64}$$

$$(y_B - y_{B0}) = \frac{g_J\beta J}{kT}\{-H_0 + \lambda'(\mathscr{M}_A - \mathscr{M}_A^s) - \lambda(\mathscr{M}_B - \mathscr{M}_B^s)\} \tag{8.3.65}$$

Recall that $\mathscr{M}_A^s = \mathscr{M}_B^s$, so that

$$y_{A0} = y_{B0} = (\lambda' - \lambda)\mathscr{M}_A^s\,\frac{g_J\beta J}{kT} \tag{8.3.66}$$

The net magnetization in the direction $\hat{\mathbf{k}}H_0$ is therefore

$$\mathscr{M}_A - \mathscr{M}_B = \mathscr{M}_A^0\, B_J'(y_{A0})\{(y_A - y_{A0}) - (y_B - y_{B0})\} \tag{8.3.67}$$

$$= \mathscr{M}_A^0\, B_J'(y_{A0})\,\frac{g_J\beta J}{kT}\{2H_0 - (\lambda + \lambda')(\mathscr{M}_A - \mathscr{M}_B)\} \tag{8.3.68}$$

† $B_J'(y_{A0}) = \left\{\dfrac{\partial}{\partial y_A} B_J(y_A)\right\}_{y_{A0}}$

Consequently,

$$\mathscr{M}_A - \mathscr{M}_B = H_0 \frac{Ng_J^2\beta^2J^2\, B_J'(y_{A0})}{kT + \frac{1}{2}Ng_J^2\beta^2J^2(\lambda + \lambda')\, B_J'(y_{B0})} \tag{8.3.69}$$

and the susceptibility is

$$\chi_{11} = \frac{Ng_J^2\beta^2J^2\, B_J'(y_{A0})}{kT + \frac{1}{2}Ng_J^2\beta^2J^2(\lambda + \lambda')\, B_J'(y_{A0})} \tag{8.3.70}$$

As $T \longrightarrow 0$, $B_J(y_{A0}) \longrightarrow 0$ more rapidly than T and so $\chi_{11}(T = 0)$ is zero.

When the applied magnetic field $\hat{\mathbf{k}}H_0$ is perpendicular to the common axis each sublattice magnetization experiences a torque which causes the magnetization to rotate to a new equilibrium position where the torque is zero. For sublattice A the equilibrium condition is,

$$\mathscr{M}_A^s \wedge (\hat{\mathbf{k}}H_0 - \lambda\mathscr{M}_A^s - \lambda'\mathscr{M}_B^s) = 0 \tag{8.3.71}$$

From Fig. 8.6,

$$\mathscr{M}_A^s H_0 \cos\phi - \lambda'\mathscr{M}_A^s\,\mathscr{M}_B^s \sin 2\phi = 0 \tag{8.3.72}$$

$$\sin\phi = \frac{H_0}{2\lambda'\mathscr{M}_B^s} \tag{8.3.73}$$

The resultant magnetization along $\hat{\mathbf{k}}H_0$ is

$$(\mathscr{M}_A^s + \mathscr{M}_B^s)\sin\phi = 2\mathscr{M}_B^s \sin\phi = \frac{H_0}{\lambda'} \tag{8.3.74}$$

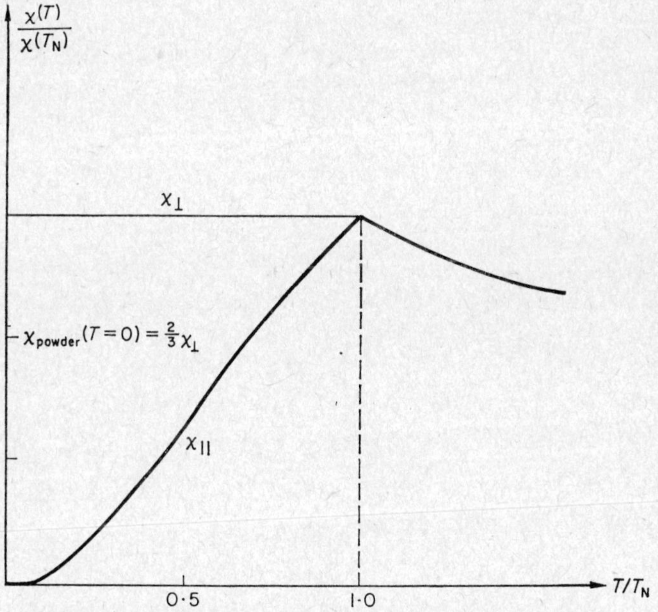

FIG. 8.7 The susceptibility of an axial antiferromagnetic crystal as a function of temperature according to the molecular field model.

The perpendicular susceptibility is

$$\chi_{\perp} = \frac{1}{\lambda'} \tag{8.3.75}$$

and so $\chi_{\perp}$ is independent of temperature for $T \leqslant T_{\mathrm{N}}$.

Note that for $T = T_{\mathrm{N}}$, we may write

$$B_J(y_{A0}) = \left(\frac{J+1}{3J}\right) y_{A0}$$

and so,

$$\chi_{\parallel}(T = T_{\mathrm{N}}) = \chi_{\perp}(T = T_{\mathrm{N}}) = \frac{1}{\lambda'} \tag{8.3.76}$$

This is also the value of χ derived from the high temperature susceptibility (equation (8.3.47)) at $T = T_{\mathrm{N}}$. The temperature dependence of the susceptibility is sketched in Fig. 8.7. For a powdered sample at $T = 0$,

$$\underset{\text{powder}}{\chi}(T = 0) = \tfrac{1}{3}(\chi_{\parallel} + 2\chi_{\perp})_{T=0} = \frac{2}{3\lambda'} \tag{8.3.77}$$

8.3.6 Antiferromagnetic Spiral

If the crystalline anisotropy is uniaxial and determines the *hard* axis of magnetization then, at temperatures below T_{N}, the sublattice magnetizations will lie in planes perpendicular to this axis. In such crystals a spiral configuration of the sublattice magnetization may occur. If the direction of the magnetization in a plane labelled A is taken as the reference, upon proceeding from plane to plane along the axis of anisotropy we may expect the magnetization vector to rotate by equal increments in angle. A simple antiparallel antiferromagnetic structure of the type $ABABA$ . . . corresponds to a turn angle π, whilst a ferromagnetic structure corresponds to a turn angle zero.

The spiral structure with an arbitrary turn angle ϕ is illustrated in Fig. 8.8. The molecular field acting on the spontaneous magnetization in a given plane may be written as

$$\mathbf{H}_{\mathrm{m}} = \lambda_0 \mathcal{M}_p^s + \lambda_1(\mathcal{M}_{p+1}^s + \mathcal{M}_{p-1}^s) + \lambda_2(\mathcal{M}_{p+2}^s + \mathcal{M}_{p-2}^s) + \cdots \tag{8.3.78}$$

Here the reference plane is labelled with the integer, p, and the subscripts indicate that the vector magnetizations, $\mathcal{M}^s$, refer to the nearest neighbour planes, the next nearest neighbour planes and so on. The energy per unit area of a plane arising from this interaction is

$$E = -\mathcal{M}_p^s.\mathbf{H}_{\mathrm{m}} = -(\mathcal{M}_p^s)^2(\lambda_0 + 2\lambda_1 \cos\phi + 2\lambda_2 \cos 2\phi + \cdots) \tag{8.3.79}$$

The equilibrium angle ϕ corresponds to the minimum value of E. Terminating the series at the term in 2ϕ (that is next nearest plane interactions) we require†

$$\frac{\partial E}{\partial \phi} = 0 \tag{8.3.80}$$

† For a discussion using Fourier Transform techniques see Nagamuja, T., 'Helical spin ordering' in *Solid State Physics*, Eds F. Seitz, D. Turnbull and H. Ehrenreich, Vol. 20, Academic Press, New York, 1967.

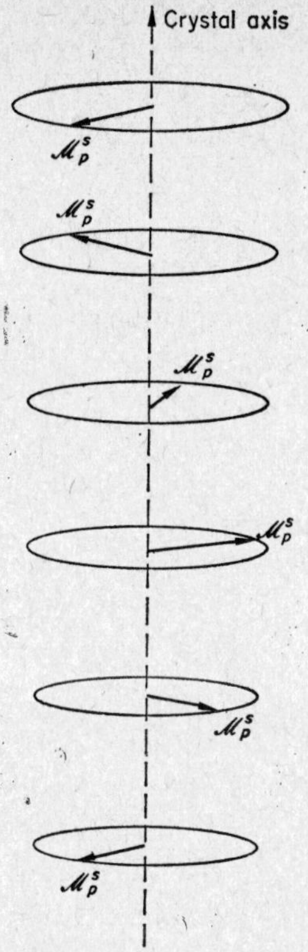

FIG. 8.8 The spontaneous magnetization for a spiral structure at temperatures below T_N,

so that either

$$\sin \phi = 0 \tag{8.3.81}$$

or

$$\cos \phi = -\frac{\lambda_1}{4\lambda_2} \tag{8.3.82}$$

A stable spiral configuration may occur with ϕ given by equation (8.3.82) provided that $4|\lambda_2| > |\lambda_1|$. Outside this range of values the solution is $\phi = 0$ or π, satisfying (8.3.81) and describing a simple ferromagnetic or antiparallel antiferromagnetic structure.

In order to identify the range of values for λ_1 and λ_2 over which the different configurations are stable we evaluate the minimum energy using equation (8.3.79). For the simple ferromagnetic structure $\phi = 0$, and so, writing $E_0 = -(\mathscr{M}_p^s)^2\lambda_0$, we find that

$$(E - E_0)_f = -2(\mathscr{M}_p^s)^2(\lambda_1 + \lambda_2) = -2(\mathscr{M}_p^s)^2\lambda_2\left(\frac{\lambda_1}{\lambda_2} + 1\right) \tag{8.3.83}$$

For the simple antiferromagnetic structure $\phi = \pi$,

$$(E - E_0)_a = -2(\mathscr{M}_p^s)^2(-\lambda_1 + \lambda_2) = -2(\mathscr{M}_p^s)^2\lambda_2\left(-\frac{\lambda_1}{\lambda_2} + 1\right) \tag{8.3.84}$$

For the spiral antiferromagnetic structure, $\cos \phi = -\lambda_1/4\lambda_2$,

$$(E - E_0)_{\text{sp}} = -2(\mathscr{M}_p^s)^2\left\{-\frac{\lambda_1^2}{4\lambda_2} + \lambda_2\left(2\frac{\lambda_1^2}{16\lambda_2^2} - 1\right)\right\}$$

$$= +2(\mathscr{M}_p^s)^2\lambda_2\left(\frac{\lambda_1^2}{8\lambda_2^2} + 1\right) \tag{8.3.85}$$

These functions are plotted in terms of the variable (λ_1/λ_2) in Figs. 8.9(a)

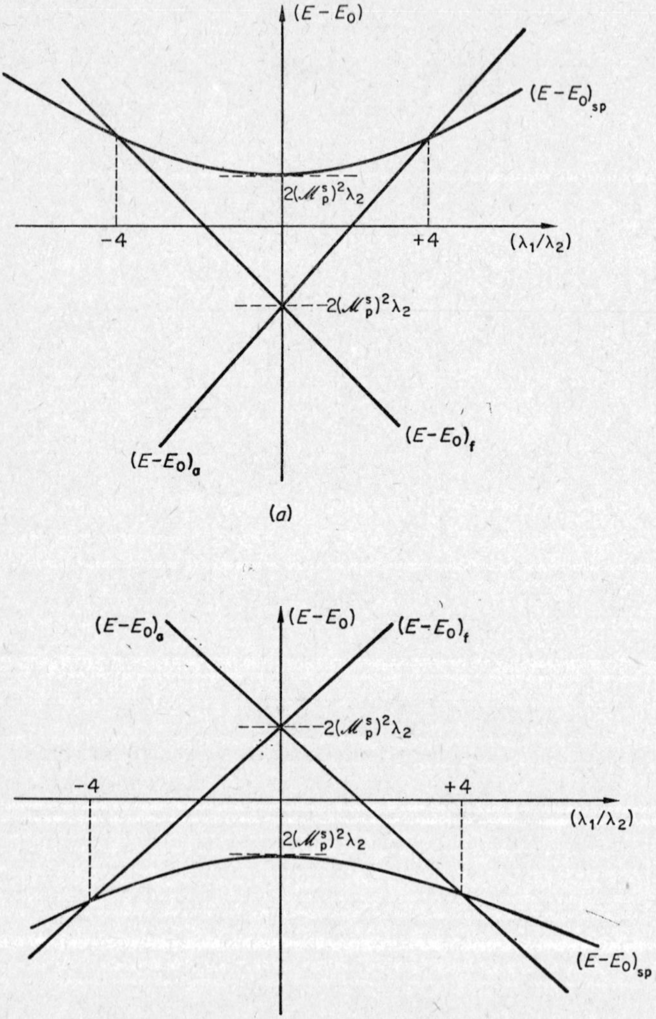

FIG. 8.9 Sketches of equations (8.3.83)–(8.3.85) for fixed values of λ_2.
(a) $\lambda_2 > 0$, a positive quantity
(b) $\lambda_2 < 0$, a negative quantity

and (b). These diagrams refer to a fixed value of λ_2. In Fig. 8.9(a) the value of λ_2 is positive whilst in Fig. 8.9(b) the value of λ_2 is negative. It is evident from these diagrams that the spiral configuration will only be stable if λ_2 is negative and satisfies the condition $4|\lambda_2| > |\lambda_1|$. If λ_2 is negative and $\lambda_1/\lambda_2 < -4$ the stable configuration has $\phi = 0$, whilst if λ_2 is negative and $\lambda_1/\lambda_2 > 4$ the stable configuration has $\phi = \pi$.

For the case $\lambda_1 > 0$, $\lambda_2 < 0$, an external magnetic field, applied perpendicular to the axis of the spiral at temperatures below T_N, can cause a transition to the ferromagnetic state, $\phi = 0$. For this transition to occur the external field must lower the energy of the ferromagnetic state below that of the spiral configuration. From equation (8.3.83), the energy of the ferromagnetic state, per unit volume, in a magnetic field is given by

$$E_f(H) = n(E - E_0)_f - n\mathcal{M}_p^s H_0 \tag{8.3.86}$$

Here n is the number of planes in the unit volume and H_0 is the magnetic field within the material. The energy of the spiral configuration is unchanged when a field is applied because there is no net moment in the direction of the field. This statement implies that the turn angle ϕ is not modified by the external field. It is not strictly correct but the induced magnetization is small and the corresponding energy term can be neglected.

The critical field at which the transition from spiral to ferromagnetic configuration takes place is obtained by equating $E_f(H)$ and $n(E - E_0)_{sp}$.

$$\mathcal{M}_p^s H_{0c} = (E - E_0)_f - (E - E_0)_{sp} \tag{8.3.87}$$

since $\lambda_1 > 0$, $\lambda_2 < 0$,

$$\mathcal{M}_p^s H_{0c} = 2(\mathcal{M}_p^s)^2|\lambda_2|\left\{\left(-\frac{\lambda_1}{|\lambda_2|} + 1\right) + \left(\frac{\lambda_1^2}{8\lambda_2^2} + 1\right)\right\} \tag{8.3.88}$$

and

$$H_{0c} = 4\mathcal{M}_p^s|\lambda_2|\left(1 - \frac{\lambda_1}{4|\lambda_2|}\right)^2 \tag{8.3.89}$$

or

$$H_{0c} = 4\mathcal{M}_p^s|\lambda_2|(1 - \cos\phi)^2 \tag{8.3.90}$$

8.3.7 Néel Temperature (Antiferromagnetic Spiral)

The temperature T_N, below which the spiral configuration is expected to be stable in zero magnetic field, can be estimated by identifying an effective molecular field term which can be written into the Brillouin function equation. On the simple molecular field model described above, the total field acting on a molecule is written

$$\mathbf{H} = \mathbf{H}_a + \lambda\mathcal{M} \tag{8.3.91}$$

and the energy of a molecule has the form

$$E(\mathbf{H}) = -\boldsymbol{\mu}.\mathbf{H}_a - \lambda\boldsymbol{\mu}.\mathcal{M} \tag{8.3.92}$$

so that below the transition temperature

$$E(\mathbf{H}) \simeq -\boldsymbol{\mu}.\mathbf{H}_a - \frac{\lambda}{N}(\mathcal{M}^s)^2 \tag{8.3.93}$$

From equation (8.3.79) the corresponding energy for a molecule in the spiral configuration is

$$E_{sp}(\mathbf{H}) = \boldsymbol{\mu}.\mathbf{H}_a - \frac{(\mathscr{M}_p^s)^2}{N'}(\lambda_0 + 2\lambda_1 \cos\phi + 2\lambda_2 \cos 2\phi + \cdots) \quad (8.3.94)$$

$$\equiv \boldsymbol{\mu}.\mathbf{H}_a - \frac{\lambda(\phi)}{N'}(\mathscr{M}_p^s)^2 \quad (8.3.95)$$

$$\simeq -\boldsymbol{\mu}.\mathbf{H}_a - \lambda(\phi)\boldsymbol{\mu}.\mathscr{M} \quad (8.3.96)$$

with

$$\lambda(\phi) = \lambda_0 + 2\lambda_1 \cos\phi + 2\lambda_2 \cos 2\phi \quad (8.3.97)$$

Comparing equation (8.3.96) with equation (8.3.92), we may make the formal substitution of $\lambda(\phi)$ for λ in the Brillouin function and obtain the critical temperature from equation (8.3.33)

$$T_N = \frac{N' g_J^2 \beta^2 J(J+1)}{3k}\lambda(\phi) \quad (8.3.98a)$$

Note that $\lambda(\phi)$ is a Fourier Transform.

$$\lambda(\phi) = \sum_n \lambda_n \, e^{jn\phi} = \sum_n \lambda_n \, e^{jnaQ} = \sum_n \lambda_n \, e^{j\mathbf{Q}\cdot\mathbf{r}_n} \equiv \lambda(\mathbf{Q})$$

where a is the interplanar spacing along the axis and $(2\pi/Q)$ is the pitch of the spiral. By comparison with equation (8.4.7) we see that, for a simple spin system,

$$T_N = \frac{S(S+1)}{3k} J(\mathbf{Q}) \quad (8.3.98b)$$

and $J(\mathbf{Q})$ is the Fourier Transform of the exchange interaction,

$$J(\mathbf{Q}) = \sum_j J_{ij} \exp(j\mathbf{Q}.\mathbf{r}_j)$$

8.3.8 Susceptibility below T_N (Antiferromagnetic Spiral)

When a small magnetic field, $\mathbf{H}_a$, is applied to an antiferromagnetic spiral a small net magnetization is induced along the direction of the field. We consider two particular orientations of the magnetic field,

(a) $\mathbf{H}_a$ parallel to the axis of the spiral,
(b) $\mathbf{H}_a$ perpendicular to the axis of the spiral.

For these cases the susceptibility can be calculated reasonably simply from the equilibrium configuration for the system in a magnetic field.

A field $\mathbf{H}_a$ parallel to the axis of the spiral will cause the magnetization vectors to tilt out from the basal planes towards the direction of the axis. Suppose the magnetization vectors are inclined at an angle θ to the field $\mathbf{H}_a = \hat{\mathbf{k}}H_0$ as shown in Fig. 8.10, and that the orientation in the basal plane is denoted by ϕ_p. Restrict the range of the interaction to next nearest planes

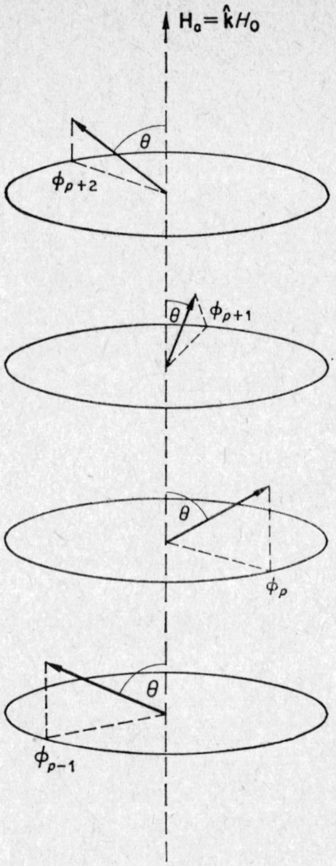

FIG. 8.10 The orientations of the magnetization vectors when the applied field $\mathbf{H}_a$ is along the axis of the spiral.

along the axis of the spiral and write the energy per unit area of the plane labelled p, from equations (8.3.78) and (8.3.79), as

$$E_p(\theta) = -\mathscr{M}_p^s.(\mathbf{H}_a + \mathbf{H}_m) - W_A \tag{8.3.99}$$

$$\begin{aligned} &= -\mathscr{M}_p^s H_0 \cos\theta \\ &\quad - (\mathscr{M}_p^s)^2[\lambda_0 + \lambda_1\{\cos(p, p+1) + \cos(p, p-1)\} \\ &\qquad + \lambda_2\{\cos(p, p+2) + \cos(p, p-2)\}] \\ &\quad - K_2 \sin^2\theta \end{aligned} \tag{8.3.100}$$

with

$$\begin{aligned} \cos(p, p+1) &= \sin\theta_p \sin\theta_{p+1} \cos\phi_p \cos\phi_{p+1} \\ &\quad + \sin\theta_p \sin\theta_{p+1} \sin\phi_p \sin\phi_{p+1} \\ &\quad + \cos\theta_p \cos\theta_{p+1} \qquad (8.3.101) \\ &= \sin^2\theta \cos(\phi_{p+1} - \phi_p) + \cos^2\theta \qquad (8.3.102) \end{aligned}$$

and the term $W_A = K_2 \sin^2\theta$, which describes the energy in the axial anisotropy field, has to be written explicitly into the present calculation because the magnetization is tilted out of the equilibrium plane. In equation (8.3.102) we have also written the angle θ without a subscript because we expect the tilt angle to be the same for all planes. We shall not expect the turn angle of the spiral to be significantly affected by a small axial magnetic field and so, for this case

$$\phi_{p+1} - \phi_p = \phi \tag{8.3.103}$$

with

$$\cos\phi = -\frac{\lambda_1}{4\lambda_2}$$

The equilibrium angle θ is determined by

$$\frac{\partial E_p(\theta)}{\partial\theta} = 0 \tag{8.3.104}$$

From equations (8.3.99) and (8.3.102)

$$\mathscr{M}_p^s H_0 \sin\theta - (\mathscr{M}_p^s)^2\{4\lambda_1 \sin\theta\cos\theta\,(\cos\phi - 1) + 4\lambda_2 \sin\theta\cos\theta\,(\cos 2\phi - 1)\} - 2K_2 \sin\theta\cos\theta = 0$$

$$H_0 - 4\mathscr{M}_p^s \cos\theta \left\{\lambda_1\left(-\frac{\lambda_1}{4\lambda_2} - 1\right) + \lambda_2\left(\frac{2\lambda_1^2}{16\lambda_2^2} - 2\right)\right\} - \frac{2K_2}{\mathscr{M}_p^s}\cos\theta = 0 \tag{8.3.105}$$

$$H_0 + \frac{\mathscr{M}_p^s}{2\lambda_2}\cos\theta\,(\lambda_1 + 4\lambda_2)^2 - \frac{2K_2}{\mathscr{M}_p^s}\cos\theta = 0 \tag{8.3.106}$$

The magnetization in the direction of the applied field is $\mathscr{M}_p^s \cos\theta$, where n_c is the number of planes in unit length along the c-axis. The susceptibility is therefore,

$$\chi_{\text{axis}} = \frac{n_c \mathscr{M}_p^s}{H_0} = \frac{n_c \mathscr{M}_p^s}{(2K_2/\mathscr{M}_p^s) - (\mathscr{M}_p^s/2\lambda_2)\,(\lambda_1 + 4\lambda_2)^2} \tag{8.3.107}$$

If the effective anisotropy field, $2K_2/\mathscr{M}_p^s$, is small,

$$\chi_{\text{axis}} = -\frac{2n_c\lambda_2}{(\lambda_1 + 4\lambda_2)^2}$$

or

$$\chi_{\text{axis}} = -\frac{n_c}{8\lambda_2(1 - \cos\phi)^2} \tag{8.3.108}$$

The calculation of the susceptibility is somewhat more difficult when the field is applied perpendicular to the axis of the spiral. In this case the magnetization vectors remain in the basal planes ($\theta = \pi/2$ in equations (8.3.101) and (8.3.102)), but the spiral is distorted and the turn angle ϕ, is modified. Write the turn angle ϕ_p at the plane labelled p as

$$\phi_p = \alpha_0 + p\phi + \varepsilon_p \tag{8.3.109}$$

where α_0 is simply a constant factor which is determined by the origin from which ϕ_p is measured and ε_p is the small angular displacement induced by the applied magnetic field $\mathbf{H}_a$. Evidently, in the notation of equation (8.3.100),

$$\begin{aligned}\cos(p, p+1) &= \cos(\phi_{p+1} - \phi_p) = \cos(\phi + \varepsilon_{p+1} - \varepsilon_p)\\ \cos(p, p-1) &= \cos(\phi_p - \phi_{p-1}) = \cos(\phi + \varepsilon_p - \varepsilon_{p-1})\end{aligned} \qquad (8.3.110)$$

and the energy per unit area of the plane labelled p, is

$$\begin{aligned}E_p(\varepsilon) = &- \mathscr{M}_p^s H_0 \cos(\alpha_0 + p\phi + \varepsilon_p)\\ &- (\mathscr{M}_p^s)^2[\lambda_0 + \lambda_1\{\cos(\phi + \varepsilon_{p+1} - \varepsilon_p) + \cos(\phi + \varepsilon_p - \varepsilon_{p-1})\}\\ &+ \lambda_2\{\cos(2\phi + \varepsilon_{p+2} - \varepsilon_p) + \cos(2\phi + \varepsilon_p - \varepsilon_{p-2})\}]\end{aligned} \qquad (8.3.111)$$

The total energy, W, is the summation of E_p. We shall only be interested in $\partial W/\partial \varepsilon_p$ and so we select from the complete summation those terms containing ε_p. Denote this partial sum by $W(\varepsilon_p)$. Then,

$$\begin{aligned}W(\varepsilon_p) = &-\mathscr{M}_p^s H_0 \cos(\alpha_0 + p\phi + \varepsilon_p)\\ &- 2(\mathscr{M}_p^s)^2[\lambda_1\{\cos(\phi + \varepsilon_{p+1} - \varepsilon_p) + \cos(\phi + \varepsilon_p - \varepsilon_{p-1})\}\\ &+ \lambda_2\{\cos(2\phi + \varepsilon_{p+2} - \varepsilon_p) + \cos(2\phi + \varepsilon_p - \varepsilon_{p-2})\}]\end{aligned}$$

Since the ε_p are small quantities, this equation may be written to sufficient accuracy as

$$\begin{aligned}W(\varepsilon_p) = &-\mathscr{M}_p^s H_0 \cos(\alpha_0 + p\phi + \varepsilon_p)\\ &- 2(\mathscr{M}_p^s)^2[2\lambda_1 \cos\phi \cos\tfrac{1}{2}(2\varepsilon_p - \varepsilon_{p+1} - \varepsilon_{p-1})\\ &+ 2\lambda_2 \cos 2\phi \cos\tfrac{1}{2}(2\varepsilon_p - \varepsilon_{p+2} - \varepsilon_{p-2})]\end{aligned} \qquad (8.3.112)$$

The equilibrium angle ε_p is determined by

$$\frac{\partial W}{\partial \varepsilon_p} = \frac{\partial W(\varepsilon_p)}{\partial \varepsilon_p} = 0$$

$$\begin{aligned}0 = &\mathscr{M}_p^s H_0 \sin(\alpha_0 + p\phi + \varepsilon_p)\\ &+ 4(\mathscr{M}_p^s)^2[\lambda_1 \cos\phi \sin\tfrac{1}{2}(2\varepsilon_p - \varepsilon_{p+1} - \varepsilon_{p-1})\\ &+ \lambda_2 \cos 2\phi \sin\tfrac{1}{2}(2\varepsilon_p - \varepsilon_{p+2} - \varepsilon_{p-2})]\end{aligned} \qquad (8.3.113)$$

To solve this difference equation approximate the sine terms

$$\begin{aligned}\sin\tfrac{1}{2}(2\varepsilon_p - \varepsilon_{p+1} - \varepsilon_{p-1}) &\approx \tfrac{1}{2}(2\varepsilon_p - \varepsilon_{p+1} - \varepsilon_{p-1})\\ \sin\tfrac{1}{2}(2\varepsilon_p - \varepsilon_{p+2} - \varepsilon_{p-2}) &\approx \tfrac{1}{2}(2\varepsilon_p - \varepsilon_{p+2} - \varepsilon_{p-2})\end{aligned} \qquad (8.3.114)$$

and expand the angle $\varepsilon_{p\pm r}$ about the value ε_p, with which we associate the coordinate z.

Thus

$$\begin{aligned}\varepsilon_{p+1} &= \varepsilon_p + a\frac{\partial \varepsilon}{\partial z} + \frac{a^2}{2}\frac{\partial^2 \varepsilon}{\partial z^2} + \frac{a^3}{3!}\frac{\partial^3 \varepsilon}{\partial z^3} + \frac{a^4}{4!}\frac{\partial^4 \varepsilon}{\partial z^4}\\ \varepsilon_{p-1} &= \varepsilon_p - a\frac{\partial \varepsilon}{\partial z} + \frac{a^2}{2}\frac{\partial^2 \varepsilon}{\partial z^2} - \frac{a^3}{3!}\frac{\partial^3 \varepsilon}{\partial z^3} + \frac{a^4}{4!}\frac{\partial^4 \varepsilon}{\partial z^4}\end{aligned} \qquad (8.3.115)$$

where a is the interplanar separation measured along the axis of the spiral. Hence

$$2\varepsilon_p - \varepsilon_{p+1} - \varepsilon_{p-1} = -2\left(\frac{a^2}{2}\frac{\partial^2\varepsilon}{\partial z^2} + \frac{a^4}{4!}\frac{\partial^4\varepsilon}{\partial z^4}\right)$$

$$2\varepsilon_p - \varepsilon_{p+2} - \varepsilon_{p-2} = -2\left(\frac{4a^2}{2}\frac{\partial^2\varepsilon}{\partial z^2} + \frac{16a^4}{4!}\frac{\partial^4\varepsilon}{\partial z^4}\right) \qquad (8.3.116)$$

From equation (8.3.113)

$$0 = \mathscr{M}_p^s H_0 \sin(\alpha_0 + p\phi) - 2(\mathscr{M}_p^s)^2\left\{(\lambda_1 \cos\phi + 4\lambda_2 \cos 2\phi)a^2 \frac{\partial^2\varepsilon}{\partial z^2} + (\lambda_1 \cos\phi + 16\lambda_2 \cos 2\phi)\frac{2a^4}{4!}\frac{\partial^4\varepsilon}{\partial z^4}\right\} \qquad (8.3.117)$$

$$(\lambda_1 \cos\phi + 4\lambda_2 \cos 2\phi)a^2 \frac{\partial^2\varepsilon}{\partial z^2} + (\lambda_1 \cos\phi + 16\lambda_2 \cos 2\phi)\frac{2a^4\partial^4\varepsilon}{4!\,\partial z^4} = \frac{H_0}{2\mathscr{M}_p^s}\sin\left(\alpha_0 + \frac{z}{a}\phi\right) \qquad (8.3.118)$$

On the right-hand side of this equation the integer p has been replaced by the z coordinate for the plane, $z = pa$. The solution for this differential equation is

$$\varepsilon = \frac{H_0}{2\mathscr{M}_p^s}\cdot\frac{\sin\{\alpha_0 + (z/a)\phi\}}{(\lambda_1 \cos\phi + 4\lambda_2 \cos 2\phi)(-\phi^2) + (\lambda_1 \cos\phi + 16\lambda_2 \cos 2\phi)(\phi^4/12)} \qquad (8.3.119)$$

Substitute $-(\lambda_1/4\lambda_2) = \cos\phi$ from equation (8.3.82)

$$\varepsilon = \frac{H_0}{2\mathscr{M}_p^s}\cdot\frac{1}{4\lambda_2}\cdot\frac{\sin\{\alpha_0 + (z/a)\phi\}}{(-\cos^2\phi + \cos 2\phi)(-\phi^2) + (-\cos^2\phi + 4\cos 2\phi)(\phi^4/12)} \qquad (8.3.120)$$

and, when ϕ is small, obtain

$$\varepsilon = \frac{H_0}{\mathscr{M}_p^s}\cdot\frac{1}{10\lambda_2\phi^4}\sin\left(\alpha_0 + \frac{z}{a}\phi\right) \qquad (8.3.121)$$

$$\equiv h \sin\left(\alpha_0 + \frac{z}{a}\phi\right) \qquad (8.3.122)$$

where

$$h = \frac{H_0}{\mathscr{M}_p^s}\cdot\frac{1}{10\lambda_2\phi^4}$$

is a small quantity since we are assuming H_0 to be small. The component of the magnetization, in the direction of the applied field, for the plane labelled p, with coordinate $z = pa$, is

$$\mathscr{M}_{H_0} = \mathscr{M}_p^s \cos\left(\alpha_0 + \frac{z}{a}\phi + \varepsilon\right) \qquad (8.3.123)$$

$$= \mathcal{M}_p^s \cos\left\{\alpha_0 + \frac{z}{a}\phi + h \sin\left(\alpha_0 + \frac{z}{a}\phi\right)\right\}$$

$$\approx \mathcal{M}_p^s\left\{\cos\left(\alpha_0 + \frac{z}{a}\phi\right) - h \sin^2\left(\alpha_0 + \frac{z}{a}\phi\right)\right\} \quad (8.3.124)$$

The net magnetization in the direction of the applied field is equal to the average value of $\mathcal{M}_{H_0}$ over one cycle of the spiral multiplied by the number of planes in unit length along the c-axis, n_c.

$$\{\mathcal{M}_{H_0}\}_{\text{net}} = \langle \mathcal{M}_{H_0} \rangle n_c \quad (8.3.125)$$

From equation (8.3.124)

$$\langle \mathcal{M}_{H_0} \rangle = \frac{\phi}{2\pi a} \int_0^{(2\pi/\phi)a} \mathcal{M}_{H_0}\, dz$$

$$= -\tfrac{1}{2} h \mathcal{M}_p^s \quad (8.3.126)$$

Hence the susceptibility for a small field H_0 perpendicular to the axis of the spiral is

$$\chi_{\text{perpendicular}} = \frac{\{\mathcal{M}_{H_0}\}_{\text{net}}}{H_0} = -\frac{n_c}{20\lambda_2\phi^4} \quad (8.3.127)$$

The most important examples of materials which exhibit a spiral antiferromagnetic structure are to be found among the rare earth metals which follow gadolinium in the Periodic Table. The magnetic properties of these metals are derived from the $4f$ electrons which are well localized within the atomic cores. The atomic model for a material medium, ignoring the band structure, is therefore adequate for describing the magnetism of the rare earth metals in a first order approximation. The elements terbium, dysprosium and holmium, and alloys of gadolinium with yttrium, most nearly conform with the simplified theory of the present section but even in these cases further transitions occur at low temperatures (thus terbium and

TABLE 8.3

Substance	*Critical Temp.* T_N K	*Order* $T_N > T > T^1$	*Turn Angle* $T_N > T > T^1$	*Critical Temp.* T^1 K	*Order* $T^1 > T$
Gd	—	—	—	293	Ferro
Tb	228	Spiral	17°–20°	219	Ferro
Dy	179	Spiral	26°–43°	87	Ferro
Ho	132	Spiral	36°–50°	20	Cone
Er	85, 53	Axial wave	—	20	Cone
Tm	56	Axial wave	—	40	Antiphase domains
Y/Gd (80 at %)	—	—	—	254	Ferro
Y/Gd (60 at %)	196	Spiral	14°–25°	84	Ferro
Y/Gd (40 at %)	145	Spiral	33°–40°	—	—
Y/Gd (10 at %)	59	Spiral	50°	—	—

dysprosium become ferromagnetic). These additional magnetic transformations may be interpreted on the present model if the theory is refined by including a more detailed description of the crystalline anisotropy fields and the magneto-elastic forces. A summary of the magnetic configurations for the heavy rare earth metals is given in Table 8.3.

8.4 Exchange Interaction

In the previous section we have represented the effect of all the magnetic interactions within the system by means of a molecular field $\mathbf{H}_m$. We shall now consider in more detail one of the most important contributing interactions which is called the *exchange interaction*. In conventional ferromagnetic materials the molecular field is believed to be derived almost entirely from these exchange terms and so other contributions such as arise for example from magnetic dipole interactions may be neglected in a first approximation.

The exchange interaction is usually written as a Hamiltonian operator, $\mathscr{H}_{ex}$,†

$$\mathscr{H}_{ex} = -\sum_{\substack{i,j \\ i>j}} J_{ij}\, \mathbf{S}_i \cdot \mathbf{S}_j \tag{8.4.1}$$

Here $\mathbf{S}_i$ is the total spin operator for the atom (molecular unit) at the position labelled i, $\mathbf{S}_j$ is the corresponding operator for the atom (molecular unit) at the position labelled j, and the coefficient, J_{ij}, is a function of the relative positions so that $J_{ij} = J(\mathbf{R}_{ij})$ where $\mathbf{R}_{ij}$ is the vector connecting the points i and j. Equation (8.4.1) is not to be regarded as a complete description of the exchange interaction. There may also be terms which are explicitly anisotropic and terms which involve higher powers of the operators $\mathbf{S}_i$, $\mathbf{S}_j$. It is, however, reasonable to make use of equation (8.4.1) if we understand it as representing the first (and possibly the dominant) term in a more general expansion for the complete representation involving product functions of the individual spin operators.

The Hamiltonian operator which leads to the magnetic energy of a spin system in an applied magnetic field $\mathbf{H}_a$, now takes the form

$$\mathscr{H}_{mag} = g\beta \mathbf{H}_a \cdot \sum_i \mathbf{S}_i - \sum_{\substack{i,j \\ i>j}} J_{ij}\, \mathbf{S}_i \cdot \mathbf{S}_j \tag{8.4.2}$$

This equation leads to an expression for the molecular field constant, λ, which was introduced in equation (8.3.2). Rewrite equation (8.4.2) in the molecular field approximation to give the interaction energy for one spin as,

$$\mathscr{H}^1_{mag} = \left(g\beta \mathbf{H}_a - \sum_j J_{ij}\, \mathbf{S}_j\right) \cdot \mathbf{S}_i \tag{8.4.3}$$

† The summation notation implies a summation over *distinct* pairs i, j.

Consider the situation at low temperatures where the magnetization is large and along $\mathbf{H}_a$ (= $\hat{\mathbf{k}}H_0$), and the transverse components may be neglected.

$$\mathscr{H}^1{}_{\text{mag}} \approx \left\{g\beta H_0 - S_j^z \sum_j J_{ij}\right\} S_i^z \tag{8.4.4}$$

The corresponding energy is

$$E^1 = \left\{g\beta H_0 - \langle S^z\rangle \sum_j J_{ij}\right\} \langle S_i^z\rangle \tag{8.4.5}$$

$$= g\beta\{H_0 + \lambda\mathscr{M}_z\}\ \langle S_i^z\rangle \tag{8.4.6}$$

with $\mathscr{M}_z = -Ng\beta\langle S^z\rangle$
and

$$\lambda = \frac{1}{Ng^2\beta^2}\sum_j J_{ij} \tag{8.4.7}$$

From equation (8.3.33), the Curie temperature is given by

$$T_c = \frac{1}{3}\frac{S(S+1)}{k}\sum_j J_{ij} \tag{8.4.8}$$

Equation (8.4.1) appears to indicate that the exchange energy arises from a direct interaction between pairs of spins. Such an interpretation is, however, too narrow and leads to considerable difficulties. Thus, whereas, within a single atom, calculations of J_{ij} based on the direct electrostatic interactions between electrons lead to the 'ferromagnetic' correlations described by Hund's rules, this is no longer the case when the electrons are associated with different atoms. The calculations for spins located on different atoms usually lead to negative values of J_{ij}, corresponding to an 'antiferromagnetic' ground state—which is indeed a characteristic feature of most molecules. A ferromagnetic ground state may, however, occur if equation (8.4.1) describes indirect as well as direct interactions between the spins. In the case of a ferromagnetic metal such indirect processes can readily arise via the conduction electrons. We shall now examine how this type of interaction can lead to ferromagnetism.

Consider a model for a ferromagnetic metal in which the essential magnetic properties arise from a set of spins $\mathbf{S}_i$ which are localized at particular lattice sites whilst electrical conduction takes place via a separate set of mobile electrons having associated spins $\mathbf{s}_k$. Extend the formalism of equation (8.4.1) to write down an interaction Hamiltonian for the whole system of electrons as

$$\mathscr{H}_{\text{int}} = -\sum_{i,j} J_{ij}^{(0)}\,\mathbf{S}_i.\mathbf{S}_j - \sum_{i,k} J_{ik}^{(1)}\,\mathbf{S}_i.\mathbf{s}_k - \sum_{k,l} J_{kl}^{(2)}\,\mathbf{s}_k.\mathbf{s}_l \tag{8.4.9}$$

The corresponding energy is

$$E_{\text{int}} = -\sum_{i,j} J_{ij}^{(0)}\,\langle\mathbf{S}_i.\mathbf{S}_j\rangle - \sum_{i,k} J_{ik}^{(1)}\,\langle\mathbf{S}_i.\mathbf{s}_k\rangle - \sum_{k,l} J_{kl}^{(2)}\,\langle\mathbf{s}_k.\mathbf{s}_l\rangle \tag{8.4.10}$$

For a ferromagnetic ordered state of the system at low temperatures where the transverse components can be neglected in accordance with the molecular field approximation discussed previously, the energy is

$$E_{\text{int}} \approx -\bar{J}^{(0)}\alpha N^2V^2\langle S^z\rangle^2 - \bar{J}^{(1)}nNV^2\langle S^z\rangle\langle s^z\rangle - \bar{J}^{(2)}n^2V^2\langle s^z\rangle^2 \quad (8.4.11)$$

Here, N is the number of localized spins per unit volume, n is the number of conduction electrons per unit volume, V is the volume of the sample and α is determined by the range of the interaction $J_{ij}^{(0)}$. The exchange parameters J have been replaced by mean values in equation (8.4.11). The equilibrium value of $\langle s^z\rangle$ for the mobile electrons is obtained by seeking a minimum energy value.

$$\frac{\partial E_{\text{int}}}{\partial\langle s^z\rangle} = -\bar{J}^{(1)}nNV^2\langle S^z\rangle - 2\bar{J}^{(2)}n^2V^2\langle s^z\rangle = 0 \quad (8.4.12)$$

$$n\langle s^z\rangle = -\frac{\bar{J}^{(1)}}{2\bar{J}^{(2)}}N\langle S^z\rangle \quad (8.4.13)$$

Substituting this value for $\langle s^z\rangle$ in equation (8.4.11)

$$E_{\text{int}} \approx -\bar{J}^{(0)}\alpha N^2V^2\langle S^z\rangle^2 + \frac{(\bar{J}^{(1)})^2}{4\bar{J}^{(2)}}N^2V^2\langle S^z\rangle^2 \quad (8.4.14)$$

$$= \left\{-\alpha\bar{J}^{(0)} + \frac{(\bar{J}^{(1)})^2}{4\bar{J}^{(2)}}\right\}N^2V^2\langle S^z\rangle^2 \quad (8.4.15)$$

which might be interpreted as being derived from an exchange Hamiltonian

$$\mathscr{H}_{\text{ex}} = -\sum_{\substack{i,j\\ i>j}}\left\{\alpha\bar{J}^{(0)} - \frac{(\bar{J}^{(1)})^2}{4\bar{J}^{(2)}}\right\}\mathbf{S}_i\cdot\mathbf{S}_j \quad (8.4.16)$$

having the form of equation (8.4.1), with

$$\bar{J}_{ij} = \alpha\bar{J}^{(0)} - \frac{(\bar{J}^{(1)})^2}{4\bar{J}^{(2)}} \quad (8.4.17)$$

The overall sign of the exchange interaction parameter may now be positive because of the indirect exchange term $(\bar{J}^{(1)})^2/4J^{(2)}$ and so favour ferromagnetism even though the direct exchange described by $\bar{J}^{(0)}$ would by itself lead to an antiferromagnetic ground state.

The model described above was first proposed by Zener (1951)† to account for ferromagnetism in metals. It contains the essential feature of an indirect interaction via the mobile conduction electrons. It can be seen from equation (8.4.13) that the localized spins $\mathbf{S}_i$ give rise to an internal field which magnetizes (polarizes) the conduction electrons and so generates the interaction.‡ Unfortunately even if doubts about the localized nature of the spins $\mathbf{S}_i$ can be overcome (as may for example be possible in the rare earth metals), this simple model does not appear to be entirely satisfactory. Thus the indirect interaction between two spins $\mathbf{S}_i$, $\mathbf{S}_j$, which is determined by the constants

† Zener, C., *Phys. Rev.* **81**, 440, 1951.

‡ This effect is similar in origin to the Knight shift produced by an external magnetic field.

$J^{(1)}, J^{(2)}$, through the coupling of $\mathbf{S}_i$, $\mathbf{S}_j$ separately to the conduction electrons, does not seem to depend upon the distance separating $\mathbf{S}_i$ from $\mathbf{S}_j$. Moreover, a simple estimate of $\bar{J}^{(2)}$ shows that this constant is dependent upon the *total* volume of the material. To estimate $\bar{J}^{(2)}$ we identify the energy term in equation (8.4.11) corresponding to the conduction electrons with that energy which arises in the internal field. Thus we may write

$$E_c = \tfrac{1}{2}\chi_c H_{int}^2 V \tag{8.4.18}$$

$$= \frac{1}{2}\frac{M_c^2}{\chi_c}V = \frac{1}{2}\frac{g^2\beta^2 n^2\langle s^z\rangle^2}{\chi_c}V \tag{8.4.19}$$

$$= -\bar{J}^{(2)}\, n^2 V^2 \langle s^z\rangle^2 \tag{8.4.20}$$

so that

$$\bar{J}^{(2)} = -\frac{g^2\beta^2}{\chi_c V} \tag{8.4.21}$$

Here χ_c is the susceptibility of the conduction electrons. Evidently we have obtained the anomalous result that the interaction constant, $\bar{J}^{(2)}$, depends upon the total volume of the material.

In order to resolve the difficulties which arise from this first order calculation it is necessary to consider the role of second order processes in the interactions between the localized spins and the conduction electrons. We shall examine a simplified model to show how a more realistic form of the indirect exchange interaction can be derived.

8.4.1 Ruderman–Kittel–Kasuya–Yosida Interaction

The role of the second order terms in the interaction via the conduction electrons was first investigated by Ruderman and Kittel (1954). These authors were primarily concerned with the indirect coupling between nuclear spins but their discussion is directly applicable to localized electron spins. The indirect interaction including second order effects is usually referred to as the Ruderman–Kittel–Kasuya–Yosida interaction when the coupled spins are electronic in origin.† We shall examine a simplified form of the mathematical arguments in order to bring out the essential physical features.

Consider a model of two localized electronic spins $\mathbf{S}_1$, and $\mathbf{S}_2$, situated at lattice points $\mathbf{R}_1$ and $\mathbf{R}_2$, respectively. The second term on the right-hand side of equation (8.4.9) becomes

$$\mathscr{H}^{(1)} = -\left\{\sum_k J_{1k}^{(1)}\,\mathbf{S}_1\cdot\mathbf{s}_k + \sum_k J_{2k}^{(1)}\,\mathbf{S}_2\cdot\mathbf{s}_k\right\} \tag{8.4.22}$$

We arbitrarily limit the range of interaction between the localized spin $\mathbf{S}$ and a conduction electron to the ideal case of a δ-function in the space coordinates.

† Ruderman M. A. and Kittel, C., *Phys. Rev.* **96**, 99, 1954; Kasuya, T., *Prog. theor. Phys.* (Kyoto), **16**, 45, 1956; Yosida, K., *Phys. Rev.*, **106**, 893, 1957.

We also assume that in these circumstances $J_{1k} = J_{2k} = J$. Equation (8.4.22) may therefore be written as

$$\mathscr{H}^{(1)} = -J\left\{\mathbf{S}_1 . \sum_k \mathbf{s}_k \, \delta(\mathbf{r}_k - \mathbf{R}_1) + \mathbf{S}_2 . \sum \mathbf{s}_k \, \delta(\mathbf{r}_k - \mathbf{R}_2)\right\} \tag{8.4.23}$$

$$= \mathscr{H}_1 + \mathscr{H}_2 \tag{8.4.24}$$

In order to determine the form of the Hamiltonian operator coupling $\mathbf{S}_1$, $\mathbf{S}_2$ in second order we examine the second order perturbation energy arising from $\mathscr{H}^{(1)}$. Take the unperturbed wave functions to be products of Bloch wave functions for the conduction electrons denoted by $\phi(\mathbf{k}, \mathbf{s})$ and appropriate wave functions for the localized spins denoted by $\psi(\mathbf{R})$. The second order perturbation energy is

$$E_2 = \sum_{\phi'\psi'} \frac{\langle\phi\psi| \mathscr{H}_1 + \mathscr{H}_2 |\phi'\psi'\rangle\langle\phi'\psi'| \mathscr{H}_1 + \mathscr{H}_2 |\phi\psi\rangle}{E(\phi, \psi) - E(\phi', \psi')} \tag{8.4.25}$$

$$= \sum_{\phi'\psi'} \frac{1}{E(\phi, \psi) - E(\phi', \psi')} \left\{\begin{matrix} \langle\phi\psi| \mathscr{H}_1 |\phi'\psi'\rangle\langle\phi'\psi'| \mathscr{H}_1 |\phi\psi\rangle \\ +\langle\phi\psi| \mathscr{H}_2 |\phi'\psi'\rangle\langle\phi'\psi'| \mathscr{H}_2 |\phi\psi\rangle \\ +\langle\phi\psi| \mathscr{H}_1 |\phi'\psi'\rangle\langle\phi'\psi'| \mathscr{H}_2 |\phi\psi\rangle \\ +\langle\phi\psi| \mathscr{H}_2 |\phi'\psi'\rangle\langle\phi'\psi'| \mathscr{H}_1 |\phi\psi\rangle \end{matrix}\right\} \tag{8.4.26}$$

In these equations $|\phi\psi\rangle$ describes the ground state wave function and $|\phi'\psi'\rangle$ denotes an excited state of the system. The first two terms within the brackets of equation (8.4.26) describe changes which occur for isolated spins $\mathbf{S}$, only the third and fourth terms refer to the energy change for the coupled pair. We are interested in the coupled pair and need consider only these latter terms. Thus we have

$$E_2^1 = \sum_{\phi'\psi'} \frac{\langle\phi\psi| \mathscr{H}_1 |\phi'\psi'\rangle\langle\phi'\psi'| \mathscr{H}_2 |\phi\psi\rangle}{E(\phi, \psi) - E(\phi', \psi')} + \begin{matrix}\text{complex} \\ \text{conjugate}\end{matrix} \tag{8.4.27}$$

The interaction Hamiltonian $\mathscr{H}^{(1)}$ of equation (8.4.23) contains the scalar products of the vector spin operators. Consider the products arising from the x components of the spins. Taking the wave functions to be properly normalized and noting that S_x operates only on ψ and s_x operates only on the spin part of $\phi \equiv \phi_s$, we may write

$$E_{2x}^1 = J^2 \sum_{\phi'} \frac{\sum_k \langle\phi_s|s_{kx}|\phi_s'\rangle\langle\phi_s'|s_{kx}|\phi_s\rangle\langle\phi(\mathbf{r})|\,\delta(\mathbf{r}_k - \mathbf{R}_1)\,|\phi'(\mathbf{r})\rangle\langle\phi'(\mathbf{r})|\,\delta(\mathbf{r}_k - \mathbf{R}_2)\,|\phi(\mathbf{r})\rangle}{E(\phi) - E(\phi')}$$

$$\times \sum_{\psi'} \langle\psi| \, S_{1x} \, |\psi'\rangle\langle\psi'| \, S_{2x} \, |\psi\rangle + \text{complex conjugate} \tag{8.4.28}$$

where we also assume that the energy $E(\phi, \psi)$ is not very sensitive to ψ. Adding together the contributions from the x, y, z components we see that the second order energy may be regarded as being derived from a Hamiltonian operator of the form

$$\mathscr{H}_{\text{int}} = A_{12} \, \mathbf{S}_1 . \mathbf{S}_2 \tag{8.4.29}$$

which describes a familiar exchange interaction (compare equation (8.4.1) with the coefficient A_{12} being composed of terms like

$$A_{12x} = J^2 \sum_{\phi'} \frac{\sum_k \langle \phi_s | s_{kx} | \phi_s' \rangle \langle \phi_s' | s_{kx} | \phi_s \rangle \langle \phi(\mathbf{r}) | \delta(\mathbf{r}_k - \mathbf{R}_1) | \phi'(\mathbf{r}) \rangle \langle \phi'(\mathbf{r}) | \delta(\mathbf{r}_k - \mathbf{R}_2) | \phi(\mathbf{r}) \rangle}{E(\phi) - E(\phi')} + \text{complex conjugate} \qquad (8.4.30)$$

The sum over k in this equation corresponds to a sum over all the occupied states of the Fermi distribution for the conduction electrons. The states ϕ' will correspond to unoccupied states at the temperature under consideration. We may therefore write

$$A_{12x} = J^2 \sum_{\substack{\phi \text{ occupied} \\ \phi' \text{ unoccupied}}} \frac{\langle \phi_s | s_x | \phi_s' \rangle \langle \phi_s' | s_x | \phi_s \rangle \langle \phi(\mathbf{r}) | \delta(\mathbf{r} - \mathbf{R}_1) | \phi'(\mathbf{r}) \rangle \langle \phi'(\mathbf{r}) | \delta(\mathbf{r} - \mathbf{R}_2) | \phi(\mathbf{r}) \rangle}{E(\phi) - E(\phi')} + \text{complex conjugate} \qquad (8.4.31)$$

The integrals involving the spin coordinates will give rise to a coefficient which we denote by C_s. The value of this coefficient is not of immediate interest since it does not depend upon the spatial coordinates. The integrals over the space coordinates are of considerable significance since these show that the interaction is dependent upon the distance between the localized spins.

Take the Bloch function $\phi(\mathbf{r})$ in the form†

$$\phi(\mathbf{r}) = u_{\mathbf{k}}(\mathbf{r}) \, e^{j\mathbf{k}.\mathbf{r}} \qquad (8.4.32)$$

We require

$$\langle \phi(\mathbf{r}) | \delta(\mathbf{r} - \mathbf{R}_1) | \phi'(\mathbf{r}) \rangle = \int u_{\mathbf{k}}^*(\mathbf{r}) \, e^{-j\mathbf{k}.\mathbf{r}} \, \delta(\mathbf{r} - \mathbf{R}_1) \, u_{\mathbf{k}'}(\mathbf{r}) \, e^{j\mathbf{k}'.\mathbf{r}} \, d^3\mathbf{r}$$
$$= u_k^*(\mathbf{R}_1) \, u_{\mathbf{k}'}(\mathbf{R}_1) \, e^{j(\mathbf{k}'-\mathbf{k}).\mathbf{R}_1} \qquad (8.4.33)$$

Hence

$$\langle \phi(\mathbf{r}) | \delta(\mathbf{r} - \mathbf{R}_1) | \phi'(\mathbf{r}) \rangle \langle \phi'(\mathbf{r}) | \delta(\mathbf{r} - \mathbf{R}_2) | \phi(\mathbf{r}) \rangle$$
$$= u_{\mathbf{k}}^*(\mathbf{R}_1) \, u_{\mathbf{k}}(\mathbf{R}_2) \, u_{\mathbf{k}'}^*(\mathbf{R}_2) \, u_{\mathbf{k}'}(\mathbf{R}_1) \, e^{j(\mathbf{k}-\mathbf{k}').(\mathbf{R}_2-\mathbf{R}_1)} \qquad (8.4.34)$$

The functions $u_{\mathbf{k}}(\mathbf{r})$ have the periodicity of the lattice and so

$$u_{\mathbf{k}}(\mathbf{R}_1) = u_{\mathbf{k}}(\mathbf{R}_2) = u_{\mathbf{k}}(0) \qquad (8.4.35)$$

$$u_{\mathbf{k}}^*(\mathbf{R}_1) \, u_{\mathbf{k}}(\mathbf{R}_2) = |u_{\mathbf{k}}(0)|^2 \qquad (8.4.36)$$

In addition we may expect $E(\phi)$ to be determined chiefly by the wave vector $\mathbf{k}$, so that, collecting together all the terms of the type set out in equation (8.4.31), we obtain

$$A_{12} = J^2 C_s \sum_{\substack{\mathbf{k} \text{ occupied} \\ \mathbf{k}' \text{ unoccupied}}} \frac{|u_{\mathbf{k}}(0)|^2 \, |u_{\mathbf{k}}(0)|^2 \cos (\mathbf{k} - \mathbf{k}').(\mathbf{R}_2 - \mathbf{R}_1)}{E(\mathbf{k}) - E(\mathbf{k}')} \qquad (8.4.37)$$

† See Chapter 9.

The evaluation of this sum is by no means straightforward. To obtain an indication of the form of the result make the following approximations,

$$|u_{\mathbf{k}}(0)|^2 = |u_{\mathbf{k}'}(0)|^2 = |u_{kF}(0)|^2$$

$$E(\mathbf{k}) = \frac{\hbar^2}{2m^*}k^2 \tag{8.4.38}$$

Here $|u_{kF}(0)|^2$ represents a function evaluated at the Fermi surface. Taking polar coordinates for $\mathbf{k}$ and $\mathbf{k}'$ with respect to the vector $\mathbf{R}_2 - \mathbf{R}_1 = \mathbf{R}_{12}$, and writing the sum as an integral, we obtain

$$A_{12} = J^2C_s'\,|u_{kF}(0)|^4 \iint\limits_{\substack{k \text{ occupied} \\ k' \text{ unoccupied}}} \frac{\begin{Bmatrix}\cos(kR_{12}\cos\theta)\cos(k'R_{12}\cos\theta') \\ +\sin(kR_{12}\cos\theta)\sin(k'R_{12}\cos\theta')\end{Bmatrix}}{k^2 - k'^2} \times (2\pi)^2 k^2k'^2 \sin\theta \, d\theta \sin\theta' \, d\theta' \, dk \, dk' \tag{8.4.39}$$

where C_s' is a constant related to C_s through a factor $\hbar^2/2m^*$. Integrating over θ, θ' gives

$$A_{12} = J^2C_s'\,|u_{kF}(0)|^4 \left(\frac{2\pi}{R_{12}}\right)^2 \iint\limits_{\substack{k \text{ occupied} \\ k' \text{ unoccupied}}} \frac{\sin(kR_{12})\sin(k'R_{12})\,kk'\,dk\,dk'}{k^2 - k'^2} \tag{8.4.40}$$

The integral over k' can be evaluated by a contour integral (compare Chapter 1, page 29) with k' extending over the range $-\infty$ to $+\infty$. The singularities are avoided by forming semicircles at the points $k' = \pm k$. Ignoring the numerical constants

$$A_{12} \sim |u_{kF}(0)|^4 \frac{1}{R_{12}^2} \int_{-k_F}^{k_F} k \sin(2kR_{12})\,dk \tag{8.4.41}$$

Integrate, by parts, to obtain

$$A_{12} \sim |u_{kF}(0)|^4 \frac{\{\sin(2k_FR_{12}) - 2k_FR_{12}\cos(2k_FR_{12})\}}{R_{12}^4} \tag{8.4.42}$$

This is the Ruderman–Kittel result.

The interaction constant is a decreasing oscillatory function of the separation between the localized spins, and can have the correct sign to give rise to ferromagnetism.

8.4.2 An Introduction to the Wave-Vector Dependence of the Susceptibility

We may now extend the discussion of the magnetic susceptibility, given in Chapter 1, to take account of a spatial variation in the magnetization. We shall use the molecular field model described by equation (8.3.2) and include the specific form of the exchange interaction given in equation (8.4.1).

The principles underlying the discussion in Chapter 5 suggest that, in the linear approximation, the magnetization of a material may be treated as the response arising from impressed magnetic fields according to the relation,

$$\mathscr{M}(\mathbf{r}, t) = \iint\limits_{\tau\ \boldsymbol{\rho}} h(\mathbf{r} - \boldsymbol{\rho}, t - \tau)\,\mathbf{H}(\boldsymbol{\rho}, \tau)\,d^3\boldsymbol{\rho}\,d\tau \tag{8.4.43}$$

This equation is an extension of equation (5.1.10) with the inclusion of both space and time variables. The function $h(\mathbf{r} - \boldsymbol{\rho}, t - \tau)$ is again the system weighting function, a function which now characterizes the magnetic behaviour of the material. The time dependent part of $h(\boldsymbol{\xi}, t)$ must conform to the conditions imposed by a causal system. These conditions have already been discussed in Chapter 5. Since they are not directly relevant for the present discussion we shall simply assume that they are satisfied.

Carry through a Fourier transformation in the time variable† to obtain from equation (8.4.43)

$$\mathcal{M}(\mathbf{r}, \omega) = \int_{-\infty}^{\infty} \mathcal{M}(\mathbf{r}, \mathbf{t})\, \mathrm{e}^{-\mathrm{j}\omega t}\, \mathrm{d}t$$

$$= \int_t\int_\tau\int_\rho h(\mathbf{r} - \boldsymbol{\rho}, t - \tau)\, \mathrm{e}^{-\mathrm{j}\omega(t-\tau)}\, \mathbf{H}(\boldsymbol{\rho}, \tau)\, \mathrm{e}^{-\mathrm{j}\omega\tau}\, \mathrm{d}^3\boldsymbol{\rho}\, \mathrm{d}\tau\, \mathrm{d}t \quad (8.4.44)$$

$$\mathcal{M}(\mathbf{r}, \omega) = \int h(\mathbf{r} - \boldsymbol{\rho}, \omega)\, \mathbf{H}(\boldsymbol{\rho}, \omega)\, \mathrm{d}^3\boldsymbol{\rho} \quad (8.4.45)$$

Carry through a corresponding Fourier transformation in the space variable

$$\mathcal{M}(\mathbf{q}, \omega) = \int \mathcal{M}(\mathbf{r}, \omega)\, \mathrm{e}^{-\mathrm{j}\mathbf{q}\cdot\mathbf{r}}\, \mathrm{d}^3\mathbf{r} \quad (8.4.46)$$

$$= \int_{\mathbf{r}}\int_\rho h(\mathbf{r} - \boldsymbol{\rho}, \omega)\, \mathrm{e}^{-\mathrm{j}\mathbf{q}\cdot(\mathbf{r}-\boldsymbol{\rho})}\, \mathbf{H}(\boldsymbol{\rho}, \omega)\, \mathrm{e}^{-\mathrm{j}\mathbf{q}\cdot\boldsymbol{\rho}}\, \mathrm{d}^3\boldsymbol{\rho}\, \mathrm{d}^3\mathbf{r} \quad (8.4.46)$$

Consequently, we may write

$$\mathcal{M}(\mathbf{q}, \omega) = \chi(\mathbf{q}, \omega)\, \mathbf{H}(\mathbf{q}, \omega) \quad (8.4.47)$$

with

$$\chi(\mathbf{q}, \omega) = \int h(\boldsymbol{\xi}, \omega)\, \mathrm{e}^{-\mathrm{j}\mathbf{q}\cdot\boldsymbol{\xi}}\, \mathrm{d}^3\boldsymbol{\xi} \quad (8.4.48)$$

Equation (8.4.47) introduces a susceptibility function $\chi(\mathbf{q}, \omega)$ which has a wave-vector and frequency dependence. There may indeed be a dispersion relation connecting the frequency and the wave vector, but this does not need to be stated explicitly. We shall illustrate the use of this generalized susceptibility function by examples which involve the space dependence of equation (8.4.43) directly and carry the time dependence in the calculations through the variable ω.

Consider the use of equation (8.4.47) for the description of a simple material which is uniformly magnetized by a uniform magnetic field. For this case, $\mathcal{M}(\mathbf{r}, \omega) = \mathcal{M}_0(\omega)$, $\mathbf{H}(\mathbf{r}, \omega) = \mathbf{H}_0(\omega)$ and $\mathcal{M}_0(\omega) = \chi_0(\omega)\, \mathbf{H}_0(\omega)$. Following the discussion of equation (8.4.46), we obtain,

$$\mathcal{M}(\mathbf{q}, \omega) = \int \mathcal{M}_0(\omega)\, \mathrm{e}^{-\mathrm{j}\mathbf{q}\cdot\mathbf{r}}\, \mathrm{d}^3\mathbf{r} = (2\pi)^3\, \mathcal{M}_0(\omega)\, \delta(\mathbf{q}) \quad (8.4.49a)$$

$$\mathbf{H}(\mathbf{q}, \omega) = \int \mathbf{H}_0(\omega)\, \mathrm{e}^{-\mathrm{j}\mathbf{q}\cdot\boldsymbol{\rho}}\, \mathrm{d}^3\boldsymbol{\rho} = (2\pi)^3\, \mathbf{H}_0(\omega)\, \delta(\mathbf{q}) \quad (8.4.49b)$$

† It would be preferable to use a Laplace Transform for the time variable to give $\mathcal{M}(\mathbf{r}, s)$ but we follow present convention in using a Fourier transformation.

so that,

$$\mathscr{M}(\mathbf{q}, \omega) = (2\pi)^3 \chi_0(\omega) \mathbf{H}_0(\omega) \delta(\mathbf{q}) = \chi_0(\omega) \mathbf{H}(\mathbf{q}, \omega) \qquad (8.4.50)$$

Hence,

$$\chi(\mathbf{q}, \omega) = \chi_0(\omega) = \int h(\boldsymbol{\xi}, \omega) \, e^{-j\mathbf{q}\cdot\boldsymbol{\xi}} \, d^3\boldsymbol{\xi} \qquad (8.4.51)$$

and

$$h(\boldsymbol{\xi}) = \chi_0(\omega) \, \delta(\boldsymbol{\xi}) \qquad (8.4.52)$$

Interactions within the spin system may be introduced by using the molecular field representation. Equation (8.3.2) gives,

$$\mathbf{H}(\mathbf{r}, \omega) = \mathbf{H}_a(\mathbf{r}, \omega) + \lambda \mathscr{M}(\mathbf{r}, \omega) \qquad (8.4.53)$$

with λ a constant in the present example. Equation (8.4.45) leads to

$$\mathscr{M}(\mathbf{r}, \omega) = \int_{\boldsymbol{\rho}} h(\mathbf{r} - \boldsymbol{\rho}, \omega)\{\mathbf{H}_a(\boldsymbol{\rho}, \omega) + \lambda \mathscr{M}(\boldsymbol{\rho}, \omega)\} \, d^3\boldsymbol{\rho} \qquad (8.4.54)$$

and

$$\mathscr{M}(\mathbf{q}, \omega) = \chi(\mathbf{q}, \omega)\{\mathbf{H}_a(\mathbf{q}, \omega) + \lambda \mathscr{M}(\mathbf{q}, \omega)\} \qquad (8.4.55)$$

where $\chi(\mathbf{q}, \omega)$ describes the non-interacting spin system. The Curie–Weiss law, which was previously derived from equation (8.3.3), may therefore be written as

$$\frac{\mathscr{M}(\mathbf{q}, \omega)}{\mathbf{H}_a(\mathbf{q}, \omega)} \equiv \tilde{\chi}(\mathbf{q}, \omega) = \frac{\chi(\mathbf{q}, \omega)}{1 - \lambda \chi(\mathbf{q}, \omega)} \qquad (8.4.56)$$

The exchange interaction may be introduced from equation (8.4.3) by writing the molecular field $\mathbf{H}_m(\mathbf{r}, \omega)$ as,

$$\mathbf{H}_m(\mathbf{r}, \omega) = -\frac{1}{g\beta} \sum_j J_{ij} \mathbf{S}_j \qquad (8.4.57)$$

Regarding the spins $\mathbf{S}_j$ as a continuous distribution, this equation may be written as

$$\mathbf{H}_m(\mathbf{r}, \omega) = \frac{1}{Ng^2\beta^2} \int_{\mathbf{r}'} J(\mathbf{r} - \mathbf{r}', \omega) \, \mathscr{M}(\mathbf{r}', \omega) \, d^3\mathbf{r}' \qquad (8.4.58)$$

Hence

$$\mathscr{M}(\mathbf{r}, \omega) = \int h(\mathbf{r} - \boldsymbol{\rho}, \omega) \Big\{ \mathbf{H}_a(\boldsymbol{\rho}, \omega) + \frac{1}{Ng^2\beta^2} \int J(\boldsymbol{\rho} - \mathbf{r}', \omega) \, \mathscr{M}(\mathbf{r}', \omega) \, d^3\mathbf{r}' \Big\} \, d^3\boldsymbol{\rho} \qquad (8.4.59)$$

and

$$\mathscr{M}(\mathbf{q}, \omega) = \chi(\mathbf{q}, \omega) \, \mathbf{H}_a(\mathbf{q}, \omega) + \frac{1}{Ng^2\beta^2} \int h(\mathbf{r} - \boldsymbol{\rho}, \omega) \, e^{-j\mathbf{q}\cdot(\mathbf{r} - \boldsymbol{\rho})}$$

$$\times J(\boldsymbol{\rho} - \mathbf{r}', \omega) \, e^{-j\mathbf{q}\cdot(\boldsymbol{\rho} - \mathbf{r}')} \, \mathscr{M}(\mathbf{r}', \omega) \, e^{-j\mathbf{q}\cdot\mathbf{r}'} \, d^3\mathbf{r} \, d^3\boldsymbol{\rho} \, d^3\mathbf{r}'$$

$$\mathscr{M}(\mathbf{q}, \omega) = \chi(\mathbf{q}, \omega) \Big\{ \mathbf{H}_a(\mathbf{q}, \omega) + \frac{1}{Ng^2\beta^2} J(\mathbf{q}, \omega) \, \mathscr{M}(\mathbf{q}, \omega) \Big\} \qquad (8.4.60)$$

so that

$$\frac{\mathscr{M}(\mathbf{q}, \omega)}{\mathbf{H}_a(\mathbf{q}, \omega)} \equiv \tilde{\chi}(\mathbf{q}, \omega) = \frac{\chi(\mathbf{q}, \omega)}{1-(Ng^2\beta^2)^{-1} J(\mathbf{q}, \omega)\, \chi(\mathbf{q}, \omega)} \tag{8.4.61}$$

Finally, note that the present discussion leads to a straightforward physical description of the RKKY indirect exchange interaction. Assume that a spin at the lattice site $\mathbf{R}_i$ gives rise to a polarizing field acting on the conduction electrons. Describe this localized spin by a density relation.

$$\mathbf{S}_i = \int \mathbf{S}_i\, \delta(\mathbf{r}_i - \mathbf{R}_i)\, \mathrm{d}^3\mathbf{r}_i = -\frac{1}{g\beta} \int \mathscr{M}(\mathbf{r}_i)\, \mathrm{d}^3\mathbf{r}_i \tag{8.4.62}$$

so that we may take

$$\mathscr{M}(\mathbf{r}_i) = -g\beta \mathbf{S}_i\, \delta(\mathbf{r}_i - \mathbf{R}_i) \tag{8.4.63}$$

and the molecular field at position $\mathbf{r}$, as

$$\mathbf{H}_m(\mathbf{r}, \omega) = -\frac{1}{g\beta} \int J(\mathbf{r} - \mathbf{r}_i)\, \mathbf{S}_i\, \delta(\mathbf{r}_i - \mathbf{R}_i)\, \mathrm{d}^3\mathbf{r}_i \tag{8.4.64}$$

$$\mathbf{H}_m(\mathbf{r}, \omega) = -\frac{1}{g\beta} J(\mathbf{r} - \mathbf{R}_i)\, \mathbf{S}_i \tag{8.4.65}$$

The conduction electron magnetization which results from this field is,

$$\boldsymbol{m}_c(\mathbf{r}, \omega) = \int h(\mathbf{r} - \boldsymbol{\rho}, \omega)\, \mathbf{H}_m(\boldsymbol{\rho}, \omega)\, \mathrm{d}^3\boldsymbol{\rho} \tag{8.4.66}$$

Take the conduction electron susceptibility to be a constant $\chi_c(\omega)$ so that $h(\boldsymbol{\xi}, \omega) = \chi_c(\omega)\, \delta(\boldsymbol{\xi})$, and derive the molecular field arising from the conduction electron magnetization,

$$\begin{aligned} \mathbf{H}_{mc}(\mathbf{r}, \omega) &= \int \lambda_c(\mathbf{r} - \mathbf{r}')\, \boldsymbol{m}_c(\mathbf{r}')\, \mathrm{d}^3\mathbf{r}' \\ &= \chi_c(\omega) \int \lambda_c(\mathbf{r} - \mathbf{r}')\, \mathbf{H}_m(\mathbf{r}', \omega)\, \mathrm{d}^3\mathbf{r}' \end{aligned} \tag{8.4.67}$$

The interaction energy for a second localized spin, situated at the lattice site $\mathbf{R}_j$ is

$$W_{ij} = g\beta \mathbf{S}_j . \mathbf{H}_{mc}(\mathbf{R}_j, \omega) \tag{8.4.68}$$

$$= -\mathbf{S}_i . \mathbf{S}_j\, \chi_c(\omega) \int \lambda_c(\mathbf{R}_j - \mathbf{r}')\, J(\mathbf{r}' - \mathbf{R}_i)\, \mathrm{d}^3\mathbf{r}' \tag{8.4.69}$$

This equation can be further simplified by writing

$$J(\mathbf{r}) = \frac{1}{(2\pi)^3} \int J(\mathbf{q}) \mathrm{e}^{\,\mathrm{j}\mathbf{q}.\mathbf{r}}\, \mathrm{d}^3\mathbf{q} \tag{8.4.70a}$$

$$\lambda_c(\mathbf{r}) = \frac{1}{(2\pi)^3} \int \lambda_c(\mathbf{q}')\, \mathrm{e}^{\mathrm{j}\mathbf{q}'.\mathbf{r}}\, \mathrm{d}^3\mathbf{q}' \tag{8.4.70b}$$

Then

$$W_{ij} = -\mathbf{S}_i . \mathbf{S}_j \frac{\chi_c(\omega)}{(2\pi)^6} \int_{\mathbf{q}} \int_{\mathbf{q}'} \int_{\mathbf{r}'} \lambda_c(\mathbf{q}')\, J(\mathbf{q})\, \mathrm{e}^{\mathrm{j}\mathbf{q}'.(\mathbf{R}_j - \mathbf{r}')} \times \mathrm{e}^{\mathrm{j}\mathbf{q}.(\mathbf{r}' - \mathbf{R}_i)}\, \mathrm{d}^3\mathbf{q}'\, \mathrm{d}^3\mathbf{q}\, \mathrm{d}^3\mathbf{r}'$$

$$= -\mathbf{S}_i . \mathbf{S}_j \frac{\chi_c(\omega)}{(2\pi)^6} \iiint \lambda_c(\mathbf{q}') J(\mathbf{q}) e^{j\mathbf{q}.(\mathbf{R}_j - \mathbf{R}_i)} e^{j(\mathbf{q}'-\mathbf{q}).\mathbf{R}_j} \times e^{j(\mathbf{q}-\mathbf{q}').\mathbf{r}'} d^3\mathbf{q}' \, d^3\mathbf{q} \, d^3\mathbf{r}' \quad (8.4.71)$$

Integration over $\mathbf{r}'$ yields a factor $(2\pi)^3 \delta(\mathbf{q} - \mathbf{q}')$, and subsequent integration over $\mathbf{q}'$ gives

$$W_{ij} = -\mathbf{S}_i . \mathbf{S}_j \frac{\chi_c(\omega)}{(2\pi)^3} \int \lambda_c(\mathbf{q}) J(\mathbf{q}) e^{j\mathbf{q}.(\mathbf{R}_j - \mathbf{R}_i)} d^3\mathbf{q} \quad (8.4.72)$$

$$= A(\mathbf{R}_j - \mathbf{R}_i) \, \mathbf{S}_i . \mathbf{S}_j \quad (8.4.73)$$

The interaction via the conduction electron magnetization has therefore the form of an exchange interaction and the factor $A(\mathbf{R}_j - \mathbf{R}_i)$ corresponds to the term A_{12} in the theory of Ruderman–Kittel–Kasuya–Yosida.

8.5 Spin Waves

The elementary modes of excitation for an interacting spin system may be represented as spin waves. We shall examine this form of representation using for a model system a set of identical localized spins which are situated at equivalent lattice points in a crystal and which are coupled together through an interaction described by equation (8.4.1). We shall assume that the interaction leads to ferromagnetism ($J(\mathbf{R}_{ij}) > 0$) so that we may make use of the known ground-state configuration for this case.

The ground state for an ideal ferromagnetic spin system has all the spin vectors parallel to one another. If an applied magnetic field, $\mathbf{H}_a = \hat{\mathbf{k}} H_0$, is present this simple spin system has its axis of magnetization along the direction of the applied field.† In order to be explicit, we shall assume that in the ground state all the spin vectors lie antiparallel to the vector $\hat{\mathbf{k}}$—a state for which all the spins are pointing 'down'. The eigenfunction for the ground state, $|0\rangle$, is represented so far as the spin variables are concerned by a product of the individual spin functions referred to each lattice site. Labelling the lattice sites by the subscripts $1, 2, \ldots i \ldots N$, and the individual spin functions by the z components of angular momentum (for the ground state each spin function will have its largest possible negative value $S_i^z = -S_i$) we may write for the ground state eigenfunction,

$$|0\rangle = |-S\rangle_1 |-S\rangle_2 |-S\rangle_3 \ldots |-S\rangle_i \ldots |-S\rangle_N \quad (8.5.1)$$

Evidently

$$S_i^z |0\rangle = -S_i |0\rangle = -S |0\rangle \quad (8.5.2)$$

and since S_i^z has its extreme negative value

$$S_i^- |0\rangle \equiv (S_i^x - jS_i^y)|0\rangle = 0 \quad (8.5.3)$$

The Hamiltonian operator describing the coupled system may be written in terms of the operators $S_i^+ = (S_i^x + jS_i^y)$ and $S_i^- = (S_i^x - jS_i^y)$. From equation (8.4.2)‡

$$\mathscr{H}_{mag} = g\beta H_0 \sum_i S_i^z - \sum_{\substack{i,j \\ i>j}} J_{ij} \{ S_i^z S_j^z + \tfrac{1}{2}(S_i^+ S_j^- + S_i^- S_j^+) \} \quad (8.5.4)$$

† Crystal anisotropy and demagnetizing effects are to be neglected.

‡ It is important to note that the summation is over *distinct* pairs of spins labelled i, j.

Since operators with different subscripts, i, j, will commute, the ground-state energy is obtained from

$$\mathscr{H}_{\text{mag}}\,|0\rangle = \Big(-Ng\beta H_0 S - S^2 \sum_{\substack{i,j\\ i>j}} J_{ij}\Big)\,|0\rangle = E_g\,|0\rangle \qquad (8.5.5)$$

$$E_g = -Ng\beta H_0 S - \tfrac{1}{2}NS^2 \sum_j J_{ij} \qquad (8.5.6)$$

Equation (8.5.5) shows that the state $|0\rangle$ is indeed an eigenfunction of $\mathscr{H}_{\text{mag}}$.

In order to derive a representation for the elementary excited states of the ferromagnetic spin system, we shall make use of the following commutation relations for the spin angular momentum operators,

$$\begin{aligned}
[S_i^+, S_i^-] &= S_i^+ S_i^- - S_i^- S_i^+ = 2S_i^z \\
[S_i^z, S_i^+] &= S_i^z\; S_i^+ - S_i^+\; S_i^z = S_i^+ \\
[S_i^z, S_i^-] &= S_i^z\; S_i^- - S_i^-\; S_i^z = -S_i^-
\end{aligned} \qquad (8.5.7)$$

whilst operators with different subscripts i, j, will commute. As an example of the use of these relations consider the z component of angular momentum at lattice site, λ, for a function derived by operating with S_λ^+ on the ground-state function $|0\rangle$.

$$S_\lambda^z\{S_\lambda^+\,|0\rangle\} = [S_\lambda^+ + S_\lambda^+\; S_\lambda^z]\,|0\rangle = (1 - S)\{S_\lambda^+\,|0\rangle\} \qquad (8.5.8)$$

Comparing equation (8.5.8) with equation (8.5.2) it is evident that the function $\{S_\lambda^+\,|0\rangle\}$ represents a state of the system with a single unit of spin deviation at the lattice site labelled λ. For this state the total z component of angular momentum will also be changed by one unit from $-NS$ to $(1 - NS)$.

$$\begin{aligned}
\Big(\sum_i S_i^z\Big)\{S_\lambda^+\,|0\rangle\} &= S_\lambda^+[(N-1)(-S)]\,|0\rangle + S_\lambda^z\, S_\lambda^+\,|0\rangle \\
&= (1 - NS)\{S_\lambda^+\,|0\rangle\}
\end{aligned} \qquad (8.5.9)$$

Unfortunately the function $\{S_\lambda^+\,|0\rangle$ is not an eigenfunction for the operator $\mathscr{H}_{\text{mag}}$ of equation (8.5.4). The function $\{S_\lambda^+\,|0\rangle\}$ is not therefore a valid representation for an elementary excitation of the coupled spin system. We can see how this arises by considering the operation $\mathscr{H}_{\text{mag}}\{S_\lambda^+\,|0\rangle\}$. Note that

$$\begin{aligned}
[\mathscr{H}_{\text{mag}}, S_\lambda^+]\,|0\rangle &= [\mathscr{H}_{\text{mag}}\, S_\lambda^+ - S_\lambda^+\, \mathscr{H}_{\text{mag}}]\,|0\rangle \\
&= \mathscr{H}_{\text{mag}}\{S_\lambda^+\,|0\rangle\} - E_g\{S_\lambda^+\,|0\rangle\}
\end{aligned} \qquad (8.5.10)$$

with E_g given by equation (8.5.6).

Using the commutation relations of equations (8.5.7) together with equation (8.5.3), we obtain

$$\begin{aligned}
[\mathscr{H}_{\text{mag}}, S_\lambda^+]\,|0\rangle = \Big\{ g\beta H_0[S_\lambda^z, S_\lambda^+] - \sum_j J_{\lambda j}\Big(S_j^z[S_\lambda^z, S_\lambda^+] \\
+ \tfrac{1}{2}S_j^+[S_\lambda^-, S_\lambda^+]\Big)\Big\}\,|0\rangle
\end{aligned} \qquad (8.5.11)$$

$$= \left\{g\beta H_0 + S\left(\sum_j J_{\lambda j}\right)\right\}\{S_\lambda^+ |0\rangle\} - S\sum_j J_{\lambda j} S_\lambda^+ |0\rangle \tag{8.5.12}$$

Hence, equation (8.5.10) leads to

$$\mathscr{H}_{\text{mag}}\{S_\lambda^+ |0\rangle\} = [g\beta H_0 + S\left(\sum_j J_{\lambda j}\right) + E_g]\{S_\lambda^+ |0\rangle\} - S\sum_j J_{\lambda j} S_j^+ |0\rangle \tag{8.5.13}$$

This equation does not represent an eigenfunction relation for $\{S_\lambda^+ |0\rangle\}$ because of the last term on the right-hand side, which describes a linear superposition of unit spin deviation states at the different lattice sites. The underlying physical mechanism for this effect is contained in the interaction term of the form $S_i^- S_j^+$ in $\mathscr{H}_{\text{mag}}$. This part of the interaction operating on a unit spin deviation function at the lattice site labelled i, $\{S_i^+ |0\rangle\}$, will turn it into a similar spin deviation function at the lattice site j. Thus

$$S_i^- S_j^+\{S_i^+ |0\rangle\} \sim S_j^+ |0\rangle,$$

and so the exchange interaction causes the spin deviations to move about within the coupled system leading to a representation which involves a linear superposition of single deviation functions. A particular eigenfunction representing a special elementary excitation of the spin system may in fact be written as

$$\psi_1^{(0)} = \sum_\lambda S_\lambda^+ |0\rangle \tag{8.5.14}$$

Add together all the equations of the type set out in equation (8.5.13) to obtain

$$\mathscr{H}_{\text{mag}}\, \psi_1^{(0)} = [g\beta H_0 + E_g]\psi_1^{(0)} \tag{8.5.15}$$

on making the reasonable assumption of translational invariance for the exchange parameter J_{ij}, that is,

$$\sum_j J_{\lambda j} = \sum_j J_{\mu j} \tag{8.5.16}$$

A general representation for an elementary excitation of the coupled spin system is obtained by introducing a suitable set of coefficients into the expansion of equation (8.5.14). Write

$$\psi_1 \equiv |\mathbf{k}\rangle = \frac{1}{\sqrt{(2SN)}}\sum_\lambda \{\exp(\mathrm{j}\mathbf{k}.\mathbf{r}_\lambda)\}\, S_\lambda^+ |0\rangle \tag{8.5.17}$$

In this equation $1/\sqrt{(2SN)}$ is a normalizing factor, $\mathbf{r}_\lambda$ is the lattice coordinate of the site labelled λ, and evidently $\psi_1(\mathbf{k} = 0) = |\mathbf{k} = 0\rangle = \psi_1^{(0)}$. Multiply each equation of the type given in equation (8.5.13) by $\exp(\mathrm{j}\mathbf{k}.\mathbf{r}_\lambda)$ and add the complete set,

$$\mathscr{H}_{\text{mag}}\left\{\sum_\lambda \{\exp(j\mathbf{k}.\mathbf{r}_\lambda)\}\, S_\lambda^+\,|0\rangle\right\}$$

$$= \left\{g\beta H_0 + S\left(\sum_j J_{\lambda j}\right) + E_g\right\}\left\{\sum_\lambda \{\exp(j\mathbf{k}.\mathbf{r}_\lambda)\}\, S_\lambda^+\,|0\rangle\right\}$$

$$-S\sum_\lambda\sum_j J_{\lambda j}\left[\exp\{j\mathbf{k}.(\mathbf{r}_\lambda - \mathbf{r}_j)\}\exp(j\mathbf{k}.\mathbf{r}_j)\right] S_j^+\,|0\rangle \quad (8.5.18)$$

Because of the translational invariance of the factor, $J_{\lambda j}\exp\{j\mathbf{k}.(\mathbf{r}_\lambda - \mathbf{r}_j)\}$, the summation over λ in the second line of equation (8.5.18) is independent of the index j. Hence

$$-S\sum_\lambda\sum_j J_{\lambda j}\left[\exp\{j\mathbf{k}.(\mathbf{r}_\lambda - \mathbf{r}_j)\}\exp(j\mathbf{k}.\mathbf{r}_j)\right] S_j^+\,|0\rangle$$

$$= -S\left(\sum_\lambda J_{\lambda j}\exp\{j\mathbf{k}.(\mathbf{r}_\lambda - \mathbf{r}_j)\}\right)\sum_j\{\exp(j\mathbf{k}.\mathbf{r}_j)\}S_j^+\,|0\rangle \quad (8.5.19)$$

The subscripts in the summations over all values of λ, j, may now be relabelled to conform with equation (8.5.18). Hence,

$$\mathscr{H}_{\text{mag}}\left\{\sum_\lambda \{\exp(j\mathbf{k}.\mathbf{r}_\lambda)\}S_\lambda^+\,|0\rangle\right\}$$

$$= \left[g\beta H_0 + E_g + S\sum_j J_{\lambda j}[1 - \exp\{j\mathbf{k}.(\mathbf{r}_\lambda - \mathbf{r}_j)\}]\right]$$

$$\times\left\{\sum_\lambda\{\exp(j\mathbf{k}.\mathbf{r}_\lambda)\}S_\lambda^+\,|0\rangle\right\} \quad (8.5.20)$$

or

$$\mathscr{H}_{\text{mag}}\,|\mathbf{k}\rangle = E\,|\mathbf{k}\rangle$$

with

$$E = E_g + g\beta H_0 + S\sum_j J_{\lambda j}[1 - \exp\{j\mathbf{k}.(\mathbf{r}_\lambda - \mathbf{r}_j)\}] \quad (8.5.21)$$

and the excitation energy is

$$\hbar\omega_{\mathbf{k}} = E - E_g = g\beta H_0 + S\sum_j J_{\lambda j}[1 - \exp\{j\mathbf{k}.(\mathbf{r}_\lambda - \mathbf{r}_j)\}] \quad (8.5.22)$$

This equation for the excitation energy is frequently written in the form

$$\hbar\omega_{\mathbf{k}} = g\beta H_0 + S\{J(0) - J(\mathbf{k})\} \quad (8.5.23)$$

$$J(\mathbf{k}) = \sum_j J_{\lambda j}\exp\{j\mathbf{k}.(\mathbf{r}_\lambda - \mathbf{r}_j)\} \quad (8.5.24)$$

The functions $\psi_1 \equiv |\mathbf{k}\rangle$ are the eigenfunctions for the elementary excitations. They are called Bloch spin waves.† Each eigenfunction represents a

† Spin waves were first suggested by F. Bloch in 1930. (*Z. Physik* **61,** 206, 1930.)

state with a characteristic wave vector **k** and one unit of total spin deviation. Thus, by comparison with equation (8.5.9),

$$\left(\sum_i S_i^z\right)|\mathbf{k}\rangle = \frac{1}{\sqrt{(2SN)}}\left(\sum_i S_i^z\right)\sum_\lambda \{\exp(\mathrm{j}\mathbf{k}.\mathbf{r}_\lambda)\}\, S_\lambda^+\,|0\rangle = (1 - NS)|\mathbf{k}\rangle \tag{8.5.25}$$

A particular spin wave may be visualized as a set of spin vectors, situated at the crystal lattice points, which are precessing about the direction of the magnetic field $\mathbf{H}_a = \hat{\mathbf{k}}H_0$. The phase of this precessional motion in the spin system varies in a periodic manner through the crystal and so gives rise to a wave.

8.5.1 Annihilation and Creation Operator Representation

An alternative description of the spin waves may be formulated in terms of operators $a_\mathbf{k}^+$, $a_\mathbf{k}^-$, which, when acting upon a state represented by $(a_\mathbf{k}^+)^n\,|0\rangle$, create and annihilate waves. Consider operators defined by the following relations.†

$$a_\mathbf{k}^+ = \frac{1}{\sqrt{(2SN)}}\sum_\lambda \{\exp(\mathrm{j}\mathbf{k}.\mathbf{r}_\lambda)\}\, S_\lambda^+;\quad S_\lambda^+ = \sqrt{\left(\frac{2S}{N}\right)}\sum_\mathbf{k} \{\exp(-\mathrm{j}\mathbf{k}.\mathbf{r}_\lambda)\}\, a_\mathbf{k}^+ \tag{8.5.26}$$

$$a_\mathbf{k}^- = \frac{1}{\sqrt{(2SN)}}\sum_\lambda \{\exp(-\mathrm{j}\mathbf{k}.\mathbf{r}_\lambda)\}\, S_\lambda^-;\quad S_\lambda^- = \sqrt{\left(\frac{2S}{N}\right)}\sum_\mathbf{k} \{\exp(\mathrm{j}\mathbf{k}.\mathbf{r}_\lambda)\}\, a_\mathbf{k}^- \tag{8.5.27}$$

Evidently the function $a_\mathbf{k}^+\,|0\rangle \equiv |\mathbf{k}\rangle$ and represents a spin wave with a characteristic wave vector **k** and one unit of total spin deviation.

The commutation rules for $a_\mathbf{k}^+$, $a_\mathbf{k}^-$ are obtained from equation (8.5.7),

$$\begin{aligned}[a_\mathbf{k}^+, a_\mathbf{k}^-] &= a_\mathbf{k}^+ a_\mathbf{k}^- - a_\mathbf{k}^- a_\mathbf{k}^+ \\ &= \frac{1}{2SN}\left[\left\{\sum_\lambda \{\exp(\mathrm{j}\mathbf{k}.\mathbf{r}_\lambda)\}\, S_\lambda^+\right\}\left\{\sum_\mu \{\exp(-\mathrm{j}\mathbf{k}.\mathbf{r}_\mu)\}\, S_\mu^-\right\}\right. \\ &\qquad \left. - \left\{\sum_\mu \{\exp(-\mathrm{j}\mathbf{k}.\mathbf{r}_\mu)\}\, S_\mu^-\right\}\left\{\sum_\lambda \{\exp(\mathrm{j}\mathbf{k}.\mathbf{r}_\lambda)\}\, S_\lambda^+\right\}\right] \\ &= \frac{1}{2SN}\sum_\lambda (S_\lambda^+ S_\lambda^- - S_\lambda^- S_\lambda^+)\end{aligned} \tag{8.5.28}$$

$$[a_\mathbf{k}^+, a_\mathbf{k}^-] = \frac{1}{SN}\sum_\lambda S_\lambda^z \tag{8.5.29}$$

† These relations are in fact approximations to the rigorous expressions used in the Holstein–Primakoff method of describing spin waves. Equations (8.5.26) and (8.5.27) do, however, provide a reasonable description of the elementary excitations and avoid some of the more difficult mathematical aspects of the theory.

We limit our discussion to the elementary excitations of the spin system for which $S^z|\mathbf{k}\rangle$ is very closely $(-S)|\mathbf{k}\rangle$ and replace the operator S^z by the eigenvalue $(-S)$ in equation (8.5.29). The commutation relation is now

$$[a_{\mathbf{k}}^-, a_{\mathbf{k}}^+] = 1 \tag{8.5.30}$$

In addition

$$[a_{\mathbf{k}}^-, a_{\mathbf{k}'}^+] \sim \frac{(-S)}{SN}\sum_{\lambda} \exp\{\mathrm{j}(\mathbf{k}-\mathbf{k}').\mathbf{r}_\lambda\} = 0 \tag{8.5.31}$$

since the lattice sum $\sum_{\lambda} \exp\{\mathrm{j}(\mathbf{k}-\mathbf{k}').\mathbf{r}_\lambda\}$ is zero for $\mathbf{k} \neq \mathbf{k}'$.

We also require a relation for the operator S^z which is more accurate than the eigenvalue $(-S)$ used in deriving equation (8.5.30). Use the operator relation

$$S_\lambda^2 = (S_\lambda^x)^2 + (S_\lambda^y)^2 + (S_\lambda^z)^2 \tag{8.5.32}$$
$$= \tfrac{1}{2}(S_\lambda^+ S_\lambda^- + S_\lambda^- S_\lambda^+) + (S_\lambda^z)^2$$

or

$$S_\lambda^2 = S_\lambda^+ S_\lambda^- + S_\lambda^z(S_\lambda^z - 1) \tag{8.5.33}$$

Approximate this equation first by substituting the eigenvalue $S(S+1)$ for S_λ^2 and then by writing $S(S+1) = S^2$ and $S_\lambda^z(S^z - 1) = (S_\lambda^z)^2$.

$$(S_\lambda^z)^2 \sim S^2 - S_\lambda^+ S_\lambda^- \tag{8.5.34}$$

To this approximation we can write

$$S_\lambda^z = -S + \frac{1}{2}\frac{S_\lambda^+ S_\lambda^-}{S} \tag{8.5.35}$$

where the sign of the square root has been chosen to give $S_\lambda^z|0\rangle = (-S)|0\rangle$. Substitute from equations (8.5.26) and (8.5.27) to obtain

$$S_\lambda^z = -S + \frac{1}{N}\left\{\sum_{\mathbf{k}} \{\exp(-\mathrm{j}\mathbf{k}.\mathbf{r}_\lambda)\}\, a_{\mathbf{k}}^+\right\}\left\{\sum_{\mathbf{k}'} \{\exp(\mathrm{j}\mathbf{k}'.\mathbf{r}_\lambda)\}\, a_{\mathbf{k}'}^-\right\} \tag{8.5.36}$$

$$\sum_{\lambda} S_\lambda^z = -NS + \sum_{\mathbf{k}} a_{\mathbf{k}}^+ a_{\mathbf{k}}^- \tag{8.5.37}$$

Use equations (8.5.26), (8.5.27) and (8.5.37) to write the exchange Hamiltonian operator in terms of the operators $a_{\mathbf{k}}^+$, $a_{\mathbf{k}}^-$. From equation (8.5.4)

$$\begin{aligned}\mathscr{H}_{\text{mag}} &= g\beta H_0\left(-NS + \sum_{\mathbf{k}} a_{\mathbf{k}}^+ a_{\mathbf{k}}^-\right)\\
&\quad - \sum_{\substack{i,j\\ i>}} J_{ij}\left[-S + \frac{1}{N}\sum_{\mathbf{kk}'} a_{\mathbf{k}}^+ a_{\mathbf{k}'}^- \exp\{\mathrm{j}(\mathbf{k}'-\mathbf{k}).\mathbf{r}_i\}\right]\\
&\qquad\qquad \times \left[-S + \frac{1}{N}\sum_{\mathbf{kk}'} a_{\mathbf{k}}^+ a_{\mathbf{k}'}^- \exp\{\mathrm{j}(\mathbf{k}'-\mathbf{k}).\mathbf{r}_j\right]\\
&\quad - \frac{1}{2}\sum_{\substack{i,j\\ i>}} J_{ij}\frac{2S}{N}\left[\left\{\sum_{\mathbf{k}} \{\exp(-\mathrm{j}\mathbf{k}.\mathbf{r}_i)\}\, a_{\mathbf{k}}^+\right\}\left\{\sum_{\mathbf{k}'} \{\exp(\mathrm{j}\mathbf{k}'.\mathbf{r}_j)\}\, a_{\mathbf{k}'}^-\right\}\right.\\
&\qquad\qquad \left. + \left\{\sum_{\mathbf{k}} \{\exp(\mathrm{j}\mathbf{k}.\mathbf{r}_i)\}\, a_{\mathbf{k}}^-\right\}\left\{\sum_{\mathbf{k}'} \{\exp(-\mathrm{j}\mathbf{k}'.\mathbf{r}_j)\}\, a_{\mathbf{k}'}^+\right\}\right]\end{aligned} \tag{8.5.38}$$

Neglecting the terms in $a_{\mathbf{k}}^{+}\,a_{\mathbf{k}}^{-}\,a_{\mathbf{k}'}^{+}\,a_{\mathbf{k}'}^{-}$,

$$\begin{aligned}\mathscr{H}_{\text{mag}} = {} & \Big(-Ng\beta H_0 S - \tfrac{1}{2}NS^2\sum_j J_{ij}\Big) \\ & + g\beta H_0 \sum_{\mathbf{k}} a_{\mathbf{k}}^{+}\,a_{\mathbf{k}}^{-} + S\sum_j J_{ij}\Big(\sum_{\mathbf{k}} a_{\mathbf{k}}^{+}\,a_{\mathbf{k}}^{-}\Big) \\ & - \frac{S}{N}\sum_{\substack{i,j \\ i>j}}\Big[\sum_{\mathbf{k}}\sum_{\mathbf{k}'} \exp\{\mathrm{j}(\mathbf{k}'-\mathbf{k}).\mathbf{r}_i\}\exp\{\mathrm{j}\mathbf{k}'.(\mathbf{r}_j-\mathbf{r}_i)\}\,a_{\mathbf{k}}^{+}\,a_{\mathbf{k}}^{-} \\ & + \sum_{\mathbf{k}}\sum_{\mathbf{k}'} \exp\{\mathrm{j}(\mathbf{k}-\mathbf{k}').\mathbf{r}_i\}\exp\{\mathrm{j}\mathbf{k}'.(\mathbf{r}_i-\mathbf{r}_j)\}\,a_{\mathbf{k}}^{-}\,a_{\mathbf{k}}^{+}\Big]\end{aligned} \tag{8.5.39}$$

The last term in this equation may be simplified by summing over the index i, whilst keeping $\mathbf{R}_{ij} = (\mathbf{r}_i - \mathbf{r}_j)$, $\mathbf{k}$, and $\mathbf{k}'$, constant, and using the relation $\sum_i \exp\{\mathrm{j}(\mathbf{k}'-\mathbf{k}).\mathbf{r}_i\} = 0$ unless $\mathbf{k} = \mathbf{k}'$. In addition, make use of equation (8.5.30) and the orthogonality relation for lattice sums, namely

$$\sum_{\mathbf{k}} \exp\{\mathrm{j}\mathbf{k}.(\mathbf{r}_i - \mathbf{r}_j)\} = 0 \quad \text{for} \quad \mathbf{r}_i \neq \mathbf{r}_j,$$

and obtain

$$\mathscr{H}_{\text{mag}} = E_g + \sum_{\mathbf{k}}\Big(g\beta H_0 + S\sum J_{ij}[1 - \exp\{\mathrm{j}\mathbf{k}.(\mathbf{r}_i - \mathbf{r}_j)\}]\Big)a_{\mathbf{k}}^{+}a_{\mathbf{k}}^{-} \tag{8.5.40}$$

In order to determine the eigenvalues of this equation we need to examine the properties of the operator $a_{\mathbf{k}}^{+}\,a_{\mathbf{k}}^{-}$. We know that

$$[a_{\mathbf{k}}^{-}, a_{\mathbf{k}}^{+}] = 1$$

and (8.5.41)

$$a_{\mathbf{k}}^{-}\,|0\rangle = 0$$

Consider the operator $a_{\mathbf{k}}^{+}\,a_{\mathbf{k}}^{-}$ acting on the function $a_{\mathbf{k}}^{+}\,|0\rangle$. Evidently

$$a_{\mathbf{k}}^{+}\,a_{\mathbf{k}}^{-}\{a_{\mathbf{k}}^{+}\,|0\rangle\} = a_{\mathbf{k}}^{+}(a_{\mathbf{k}}^{-}\,a_{\mathbf{k}}^{+})\,|0\rangle = a_{\mathbf{k}}^{+}\,|0\rangle \tag{8.5.42}$$

Hence the function $a_{\mathbf{k}}^{+}\,|0\rangle$ is an eigenfunction (unnormalized) of the operator $a_{\mathbf{k}}^{+}\,a_{\mathbf{k}}^{-}$ and the eigenvalue is unity. Moreover

$$a_{\mathbf{k}}^{+}\,a_{\mathbf{k}}^{-}\{(a_{\mathbf{k}}^{+})^{n_{\mathbf{k}}}\,|0\rangle\} = n_{\mathbf{k}}\{(a_{\mathbf{k}}^{+})^{n_{\mathbf{k}}}\,|0\rangle\} \tag{8.5.43}$$

and so the operator $a_{\mathbf{k}}^{+}a_{\mathbf{k}}^{-}$ may be regarded as a number operator with eigenvalues 0, 1, 2 . . . and eigenfunctions $|0\rangle$, $a_{\mathbf{k}}^{+}\,|0\rangle$, $(a_{\mathbf{k}}^{+})^2\,|0\rangle$ . . . ..

The normalized eigenfunctions may be denoted by

$$|n_{\mathbf{k}}\rangle = \alpha(n_{\mathbf{k}})(a_{\mathbf{k}}^{+})^{n_{\mathbf{k}}}\,|0\rangle.$$

We require

$$\langle n_{\mathbf{k}}|n_{\mathbf{k}}\rangle = |\alpha(n_{\mathbf{k}})|^2 \langle 0|\,(a_{\mathbf{k}}^-)^{n_{\mathbf{k}}}(a_{\mathbf{k}}^+)^{n_{\mathbf{k}}}|0\rangle = 1 \tag{8.5.44}$$

that is,

$$|\alpha(n_{\mathbf{k}})|^2 n_{\mathbf{k}}! = 1$$

$$\alpha(n_{\mathbf{k}}) = \frac{1}{\sqrt{(n_{\mathbf{k}}!)}} \tag{8.5.45}$$

and so the normalized eigenfunctions are

$$|n_{\mathbf{k}}\rangle = \frac{1}{\sqrt{(n_{\mathbf{k}}!)}}(a_{\mathbf{k}}^+)^{n_{\mathbf{k}}}\,|0\rangle \tag{8.5.46}$$

In addition, since $|n_{\mathbf{k}}+1\rangle = \dfrac{1}{\sqrt{\{(n_{\mathbf{k}}+1)!\}}}\,(a_{\mathbf{k}}^+)^{(n_{\mathbf{k}}+1)}\,|0\rangle$,

$$a_{\mathbf{k}}^+\,|n_{\mathbf{k}}\rangle = \sqrt{(n_{\mathbf{k}}+1)}\,|n_{\mathbf{k}}+1\rangle \tag{8.5.47}$$

and, since

$$a_{\mathbf{k}}^-\,|n_{\mathbf{k}}\rangle = a_{\mathbf{k}}^-\,\frac{a_{\mathbf{k}}^+}{\sqrt{(n_{\mathbf{k}})}}\,|n_{\mathbf{k}}-1\rangle$$

$$a_{\mathbf{k}}^-\,|n_{\mathbf{k}}\rangle = \sqrt{(n_{\mathbf{k}})}\,|n_{\mathbf{k}}-1\rangle \tag{8.5.48}$$

The eigenfunctions of the Hamiltonian operator in equation (8.5.40), which describe the *elementary* excitations, are therefore given by equation (8.5.46) as $|1_{\mathbf{k}}\rangle = a_{\mathbf{k}}^+|0\rangle$. These functions represent spin waves with characteristic wave vectors **k** and one unit of total spin deviation. The energy of an elementary spin wave having a wave vector **k**, is given by

$$\mathscr{H}_{\text{mag}}\{a_{\mathbf{k}}^+\,|0\rangle\} = \left(E_g + g\beta H_0 + S\sum J_{ij}[1 - \exp\{\mathrm{j}\mathbf{k}.(\mathbf{r}_i - \mathbf{r}_j\}]\right)\{a_{\mathbf{k}}^+\,|0\rangle\}$$

$$= E\{a_{\mathbf{k}}^+\,|0\rangle\} \tag{8.5.49}$$

The excitation energy is

$$\hbar\omega_{\mathbf{k}} = E - E_g = g\beta H_0 + S\sum J_{ij}[1 - \exp\{\mathrm{j}\mathbf{k}.(\mathbf{r}_i - \mathbf{r}_j)\}] \tag{8.5.50}$$

in agreement with equation (8.5.22).

Note that the Hamiltonian operator, given in equation (8.5.40), may be written in the form

$$\mathscr{H}_{\text{mag}} = E_g + \sum_{\mathbf{k}} \hbar\omega_{\mathbf{k}}\, a_{\mathbf{k}}^+\, a_{\mathbf{k}}^- \tag{8.5.51}$$

The energy of the state $|n_{\mathbf{k}}\rangle$ is

$$E(n_{\mathbf{k}}) = E_g + n_{\mathbf{k}}\,\hbar\omega_{\mathbf{k}} \tag{8.5.52}$$

and this state corresponds to a total spin deviation of $n_{\mathbf{k}}$ units since, from equation (8.5.37)

$$\left(\sum_{\lambda} S_{\lambda}^z\right)|n_{\mathbf{k}}\rangle = \left(-NS + \sum_{\mathbf{k}} a_{\mathbf{k}}^+\, a_{\mathbf{k}}^-\right)|n_{\mathbf{k}}\rangle \tag{8.5.53}$$

8.5.2 Low Temperature Magnetization

The description of spin waves in terms of the operators $a_{\mathbf{k}}^{+}$, $a_{\mathbf{k}}^{-}$ is particularly useful for deriving the low temperature variation of the magnetization. Each spin wave state will contribute to the total magnetization in accordance with equation (8.5.53) and so at a temperature T the complete magnetization, $\mathscr{M}(T)$, will be obtained by summing the eigenvalues of this equation over the Boltzmann distribution.

$$\mathscr{M}(T) = \mathscr{M}(0) - \left\langle \sum_{\mathbf{k}} n_{\mathbf{k}} \right\rangle_T g\beta \tag{8.5.54}$$

Here the term $\langle \sum_{\mathbf{k}} n_{\mathbf{k}} \rangle_T = \sum_{\mathbf{k}} \langle n_{\mathbf{k}} \rangle_T$ is to be evaluated from the Boltzmann relation,

$$\langle n_{\mathbf{k}} \rangle_T = N \frac{\sum n_{\mathbf{k}} \exp [-\{E(n_{\mathbf{k}})/kT\}]}{\sum \exp [-\{E(n_{\mathbf{k}})/kT\}]} \tag{8.5.55}$$

$$= N \frac{\sum n_{\mathbf{k}} \exp \{-(n_{\mathbf{k}}\hbar\omega_{\mathbf{k}} + E_g)/kT\}}{\sum \exp \{-(n_{\mathbf{k}}\hbar\omega_{\mathbf{k}} + E_g)/kT\}} = N \frac{\sum n_{\mathbf{k}} \exp (-n_{\mathbf{k}}\hbar\omega_{\mathbf{k}}/kT)}{\sum \exp (-n_{\mathbf{k}}\hbar\omega_{\mathbf{k}}/kT)}$$

$$= -\frac{NkT}{\hbar} \frac{\partial}{\partial \omega_{\mathbf{k}}} \log_e Z \tag{8.5.56}$$

with

$$Z = 1 + \exp (-\hbar\omega_{\mathbf{k}}/kT) + \exp (-2\hbar\omega_{\mathbf{k}}/kT) + \cdots \tag{8.5.57}$$

Hence

$$\langle n_{\mathbf{k}} \rangle_T = \frac{N}{\exp (\hbar\omega_{\mathbf{k}}/kT) - 1} \tag{8.5.58}$$

and

$$\sum_{\mathbf{k}} \langle n_{\mathbf{k}} \rangle_T \sim \int \frac{d^3\mathbf{k}}{\exp (\hbar\omega_{\mathbf{k}}/kT) - 1} \tag{8.5.59}$$

To estimate a value for this integral take H_0 to be zero in equation (8.5.50) and expand the exponential term with nearest neighbour only interactions noting also that the simple crystal lattice which we have considered has inversion symmetry. To first order we may write

$$\hbar\omega_{\mathbf{k}} = Ak^2 \tag{8.5.60}$$

so that $d^3\mathbf{k} = 4\pi k^2\, dk$, and

$$\sum_{\mathbf{k}} \langle n_{\mathbf{k}} \rangle_T \sim \int_0^\infty \frac{4\pi k^2\, dk}{\exp (Ak^2/k_B T) - 1} \sim \left(\frac{k_B T}{A}\right)^{3/2} \int_0^\infty \frac{x^{1/2}\, dx}{e^x - 1} \sim T^{3/2} \tag{8.5.61}$$

On introducing this result into equation (8.5.54) it can be seen that this simplified spin wave model of a ferromagnetic system suggests that the low temperature magnetization should vary according to a law of the form

$$\mathscr{M}(T) = \mathscr{M}(0)(1 - \alpha T^{3/2}) \tag{8.5.62}$$

The spin wave model which we have described in this section can only be applied to ferromagnetic systems at low temperatures. The Bloch spin waves are representations for the elementary excitations of the magnetic system whilst the operator relations set out in equations (8.5.26) and (8.5.27) are only acceptable when the spin deviations are small. The theory which we have developed from equations (8.5.26) and (8.5.27) places no upper bound on the spin deviation at a particular lattice site and can in principle generate spin waves corresponding to individual spin deviations greater than $2S$. This is clearly unrealistic. The rigorous transformations used by Holstein and Primakoff† avoid this difficulty by introducing an upper limit through a factor

$$\left(1 - \frac{a_\lambda^+ a_\lambda^-}{2S}\right)^{1/2}$$

which goes to zero when the eigenvalue of a $a_\lambda^+ a_\lambda^-$ is equal to $2S$. The Holstein–Primakoff transformations are

$$S_\lambda^+ = (2S)^{1/2}\left(1 - \frac{a_\lambda^+ a_\lambda^-}{2S}\right)^{1/2} a_\lambda^+ \quad \text{with} \quad a_\lambda^+ = \frac{1}{\sqrt{N}} \sum_{\mathbf{k}} \exp(-j\mathbf{k}.\mathbf{r}_\lambda)\, a_\mathbf{k}^+$$

$$S_\lambda^- = (2S)^{1/2} a_\lambda^- \left(1 - \frac{a_\lambda^+ a_\lambda^-}{2S}\right)^{1/2} \quad \text{with} \quad a_\lambda^- = \frac{1}{\sqrt{N}} \sum_{\mathbf{k}} \exp(j\mathbf{k}.\mathbf{r}_\lambda)\, a_\mathbf{k}^-$$

$$S_\lambda^z = -S + a_\lambda^+ a_\lambda^- \tag{8.5.63}$$

These transformations reduce to the more simple relations which we have used when the term $\{(a_\lambda^+ a_\lambda^-)/2S\}$ can be neglected and the problem of the upper bound can be ignored. This corresponds to the low temperature regime with small spin deviations. If one attempts to extend the temperature range of the spin wave model by using the rigorous transformations of equation (8.5.63) the mathematical complexities rapidly increase and the simplicity of equation (8.5.51) is lost because it is no longer allowable to linearize the Hamiltonian by neglecting the higher order terms involving $a_\mathbf{k}^+ a_\mathbf{k}^- a_{\mathbf{k}'}^+ a_{\mathbf{k}'}^-$.

8.6 Demagnetizing Fields

The magnetization of a ferromagnetic material gives rise to important demagnetizing fields. These fields are present both for the spin wave excitations and for the ground state configuration. In both cases they can in principle be calculated using the classical electromagnetic field equations which were discussed in some detail in Chapters 1 and 2.

8.6.1 Uniform Magnetization

In the case of the ground state configuration we assume that the material is uniformly magnetized so that $\nabla.\mathbf{M}$ is zero everywhere within the material excepting the actual boundary layer at the surface. From equations (1.1.2) and

† Holstein, T., and Primakoff, H., *Phys. Rev.* **58**, 1098 (1940).

(1.1.21), the magnetostatic equations for the field $\mathbf{H}(x, y, z)$ at the point (x, y, z) inside such a uniformly magnetized material, are

$$\left.\begin{aligned}\nabla \wedge \mathbf{H}(x, y, z) &= 0\\ \nabla . \mathbf{H}(x, y, z) &= 0\end{aligned}\right\} \quad \text{(except the boundary layer)} \tag{8.6.1}$$

The field $\mathbf{H}(x, y, z)$ may therefore be derived from a potential function $\phi^*(x, y, z)$

$$\mathbf{H}(x, y, z) = -\nabla\phi^*(x, y, z) \tag{8.6.2}$$

with $\phi^*(x, y, z)$ a solution of Laplace's equation

$$\nabla^2\phi^*(x, y, z) = 0 \quad \text{(except the boundary layer)} \tag{8.6.3}$$

These equations are mathematically equivalent to those used previously in Chapter 1 for describing the electrostatic properties of a dielectric material. The magnetization of the material may therefore be regarded as giving rise to a surface 'magnetic charge' distribution $\sigma(\xi, \eta, \zeta)$. From equation (1.3.7)

$$\sigma(\xi, \eta, \zeta) = -\mathbf{M}.\mathbf{n} \tag{8.6.4}$$

where $\mathbf{n}$ is a unit vector along the outward normal to the surface.

The potential function $\phi^*(x, y, z)$ at a point (x, y, z) due to the charge distribution $\sigma(\xi, \eta, \zeta)$ with coordinates (ξ, η, ζ) is given by (compare example (2), Chapter 1)

$$\phi^*(x, y, z) = \int_{\text{surface}} \frac{\sigma(\xi, \eta, \zeta)}{\{(x-\xi)^2 + (y-\eta)^2 + (z-\zeta)^2\}^{1/2}}\, \mathrm{d}S_{\xi,\eta,\zeta} \tag{8.6.5}$$

We seek now to formulate a solution for $\mathbf{H}(x, y, z)$ in terms of demagnetizing coefficients N_x, N_y, N_z, such that

$$\mathbf{H}(x, y, z) = -(\hat{\mathbf{i}}N_xM_x + \hat{\mathbf{j}}N_yM_y + \hat{\mathbf{k}}N_zM_z) \tag{8.6.6}$$

This is possible for a material which is ellipsoidal in shape and for which the (x, y, z), (ξ, η, ζ) coordinate axes coincide with the principal axes of the ellipsoid. We shall not attempt the difficult mathematical problem involved in solving equation (8.6.5) but shall instead derive a useful relation for the sum of the demagnetizing coefficients $(N_x + N_y + N_z)$. Consider an ellipsoidal specimen uniformly magnetized parallel to a principal axis of the ellipsoid which we identify as the x-axis of coordinates. From equations (8.6.2) and (8.6.5),

$$\begin{aligned}H_x(x, y, z) &= -\frac{\partial}{\partial x}\phi^*(x, y, z)\\ &= -\int_{\text{surface}} \frac{\partial}{\partial x}\cdot\frac{\sigma(\xi, \eta, \zeta)}{\{(x-\xi)^2 + (y-\eta)^2 + (z-\zeta)^2\}^{1/2}}\, \mathrm{d}S_{\xi,\eta,\zeta}\end{aligned} \tag{8.6.7}$$

Using equation (8.6.4) and writing $\mathbf{M} = \hat{\mathbf{i}}M_x$,

$$H_x(x, y, z) = +M_x \int_{\text{surface}} \frac{(x-\xi)(\hat{\mathbf{i}}.\mathbf{n})}{\{(x-\xi)^2 + (y-\eta)^2 + (z-\zeta)^2\}^{3/2}}\, \mathrm{d}S_{\xi,\eta,\zeta} \tag{8.6.8}$$

We may therefore identify the demagnetizing coefficient N_x as

$$N_x(x, y, z) = \int\limits_{\text{surface}} \frac{\{\hat{\mathbf{i}}(x-\xi).\mathbf{n}\}}{\{(x-\xi)^2+(y-\eta)^2+(z-\zeta)^2\}^{3/2}}\,\mathrm{d}S_{\xi,\eta,\zeta} \qquad (8.6.9)$$

Similarly

$$N_y(x, y, z) = \int\limits_{\text{surface}} \frac{\{\hat{\mathbf{j}}(y-\eta).\mathbf{n}\}}{\{(x-\xi)^2+(y-\eta)^2+(z-\zeta)^2\}^{3/2}}\,\mathrm{d}S_{\xi,\eta,\zeta} \qquad (8.6.10)$$

and

$$N_z(x, y, z) = \int\limits_{\text{surface}} \frac{\{\hat{\mathbf{k}}(z-\zeta).\mathbf{n}\}}{\{(x-\xi)^2+(y-\eta)^2+(z-\zeta)^2\}^{3/2}}\,\mathrm{d}S_{\xi,\eta,\zeta} \qquad (8.6.11)$$

Adding together equations (8.6.9), (8.6.10) and (8.6.11)

$$(N_x+N_y+N_z)_{x,y,z} = \int\limits_{\text{surface}} \frac{\{\hat{\mathbf{i}}(x-\xi)+\hat{\mathbf{j}}(y-\eta)+\hat{\mathbf{k}}(z-\zeta)\}.\mathbf{n}}{\{(x-\xi)^2+(y-\eta)^2+(z-\quad)^2\}^{3/2}}\,\mathrm{d}S_{\xi,\eta,\zeta} \qquad (8.6.12)$$

$$= -\int\limits_{\text{surface}} \{\nabla_{\xi,\eta,\zeta}[(x-\xi)^2+(y-\eta)^2+(z-\zeta)^2]^{-1/2}\} \,.\, \mathbf{n}\,\mathrm{d}S_{\xi,\eta,\zeta} \qquad (8.6.13)$$

This integral can be written as a volume integral taken throughout the material by using Green's transformation. We obtain

$$(N_x+N_y+N_z)_{x,y,z} =$$

$$-\int\limits_{\substack{\text{volume of}\\ \text{material}}} \nabla^2_{\xi,\eta,\zeta}\{(x-\xi)^2+(y-\eta)^2+(z-\zeta)^2\}^{-1/2}\,\mathrm{d}\xi\,\mathrm{d}\eta\,\mathrm{d}\zeta \qquad (8.6.14)$$

which may be written more neatly using the vector notation $\mathbf{r} \equiv (x, y, z)$, $\mathbf{r}' \equiv (\xi, \eta, \zeta)$

$$(N_x+N_y+N_z)_{\mathbf{r}} = -\int\limits_{\substack{\text{volume of}\\ \text{material}}} \nabla^2_{\mathbf{r}'}\left(\frac{1}{|\mathbf{r}-\mathbf{r}'|}\right)\mathrm{d}^3\mathbf{r}'$$

We now make use of the representation for the δ-function† (8.6.15)

$$\nabla^2\left(\frac{1}{|\mathbf{r}-\mathbf{r}'|}\right) = -4\pi\,\delta(\mathbf{r}-\mathbf{r}') \qquad (8.6.16)$$

† This relation may be derived by first considering the three-dimensional Fourier Transform of the screened Coulomb potential

$$\frac{1}{|\mathbf{r}|}\mathrm{e}^{-\alpha|\mathbf{r}|}.$$

Footnote continued on page 426.

where the differentiation may be taken with respect to either of the coordinates $\mathbf{r}$ or $\mathbf{r}'$. For our case differentiate with respect to $\mathbf{r}'$, and obtain immediately from equation (8.6.15) the important result that

$$N_x + N_y + N_z = 4\pi \tag{8.6.17}$$

for all points within the material.

For a spherical sample it is evident from symmetry that

$$N_x = N_y = N_z = \frac{4\pi}{3} \qquad \text{(sphere)} \tag{8.6.18}$$

whilst for an ellipsoid of revolution,

$$N_x = N_y = \frac{4\pi - N_z}{2} \qquad \text{(ellipsoid of revolution)} \tag{8.6.19}$$

leading to

$$N_x = N_y = 2\pi \qquad N_z = 0 \qquad \text{(cylinder)} \tag{8.6.20}$$

for a long cylinder, axis along z,
and

$$N_x = N_y = 0 \quad N_z = 4\pi \qquad \text{(disk)} \tag{8.6.21}$$

for a large diameter disk, normal to z.

8.6.2 The Spin Wave Field

Consider next the internal field which accompanies a spin wave. We assume that the time variations take place sufficiently slowly to allow propagation effects to be neglected and that the wavelengths of the spin waves are sufficiently short to allow boundary effects to be neglected. The magnetization is no longer uniform within the material, and so instead of equations

---

Write

$$\frac{1}{|\mathbf{r}|}e^{-\alpha|\mathbf{r}|} = \frac{1}{(2\pi)^3}\int_{-\infty}^{\infty} F(k_1\, k_2\, k_3) \exp\{j(k_1x + k_2y + k_3z)\}dk_1\, dk_2\, dk_3$$

$$= \frac{1}{(2\pi)^3}\int_{-\infty}^{\infty} F(\mathbf{k})\, e^{j\mathbf{k}.\mathbf{r}}\, d^3\mathbf{k},$$

with $F(\mathbf{k}) = 2\int_0^{\infty} \frac{1}{|\mathbf{r}|} e^{-\alpha|\mathbf{r}|}\, e^{-j\mathbf{k}.\mathbf{r}}\, d^3\mathbf{r} = \frac{4\pi}{|\mathbf{k}|^2 + \alpha^2}$.

Proceeding to the limit $\alpha = 0$,

$$\frac{1}{|\mathbf{r}|} = \frac{1}{(2\pi)^3}\int_{-\infty}^{\infty} \frac{4\pi}{|\mathbf{k}|^2}\, e^{j\mathbf{k}.\mathbf{r}}\, d^3\mathbf{k}$$

and

$$\nabla^2\frac{1}{|\mathbf{r}|} = -\frac{4\pi}{(2\pi)^3}\int_{-\infty}^{\infty} e^{j\mathbf{k}.\mathbf{r}}\, d^3\mathbf{k} = -4\pi\,\delta(\mathbf{r})$$

For this δ-function representation compare page 147 and entry number 25 in Table 4.1 For the evaluation of $F(\mathbf{k})$ compare page 486, Chapter 9.

(8.6.1) we have

$$\nabla \wedge \mathbf{H}(x, y, z) = 0$$
$$\nabla . \mathbf{H}(x, y, z) = -4\pi\nabla . \mathbf{M}(x, y, z) \qquad (8.6.22)$$

Seek plane wave solutions having the general form of a spin wave. With **r** representing the coordinate (x, y, z), we write

$$\mathbf{M}(\mathbf{r}) = \sum_{\mathbf{k}>\mathbf{k}_{\min}} \mathbf{M}(\mathbf{k})\, e^{j\mathbf{k}.\mathbf{r}}$$
$$\mathbf{H}(\mathbf{r}) = \sum_{\mathbf{k}'>\mathbf{k}_{\min}} \mathbf{H}(\mathbf{k}')\, e^{j\mathbf{k}'.\mathbf{r}} \qquad (8.6.23)$$

and the limiting value $\mathbf{k}_{\min}$ is necessary to allow the boundary effects to be neglected. Evidently

$$\nabla \wedge \mathbf{H}(\mathbf{r}) = j \sum_{\mathbf{k}'>\mathbf{k}_{\min}} \mathbf{k}' \wedge \mathbf{H}(\mathbf{k}')\, e^{j\mathbf{k}'.\mathbf{r}} = 0 \qquad (8.6.24)$$

and

$$\sum_{\mathbf{k}'>\mathbf{k}_{\min}} \mathbf{k}'.\mathbf{H}(\mathbf{k}')\, e^{j\mathbf{k}'.\mathbf{r}} = -4\pi \sum_{\mathbf{k}>\mathbf{k}_{\min}} \mathbf{k}.\mathbf{M}(\mathbf{k})\, e^{j\mathbf{k}.\mathbf{r}} \qquad (8.6.25)$$

Identifying coefficients of $e^{j\mathbf{k}.\mathbf{r}}$ on each side of equations (8.6.24) and (8.6.25),

$$\mathbf{k} \wedge \mathbf{H}(\mathbf{k}) = 0 \qquad (8.6.26)$$
$$\mathbf{k}.\mathbf{H}(\mathbf{k}) = -4\pi\mathbf{k}.\mathbf{M}(\mathbf{k}) \qquad (8.6.27)$$

and use the vector relation

$$\mathbf{a} \wedge (\mathbf{b} \wedge \mathbf{c}) = (\mathbf{a}.\mathbf{c})\mathbf{b} - (\mathbf{a}.\mathbf{b})\mathbf{c}$$

to obtain

$$\{\mathbf{k}.\mathbf{H}(\mathbf{k})\}\mathbf{k} = (\mathbf{k}.\mathbf{k})\mathbf{H}(\mathbf{k}) \qquad (8.6.28)$$
$$\mathbf{H}(\mathbf{k}) = -\frac{4\pi\{\mathbf{k}.\mathbf{M}(\mathbf{k})\}}{|\mathbf{k}|^2}\mathbf{k} \qquad (8.6.29)$$

The internal field accompanying the spin wave $\mathbf{M}(\mathbf{r})$ therefore takes the form

$$\mathbf{H}(\mathbf{r}) = -4\pi \sum_{\mathbf{k}>\mathbf{k}_{\min}} \frac{\{\mathbf{k}.\mathbf{M}(\mathbf{k})\}\mathbf{k}}{|\mathbf{k}|^2} e^{j\mathbf{k}.\mathbf{r}} \qquad (8.6.30)$$

8.6.3 Walker Modes

To neglect the boundary effects when deriving equation (8.6.30) is probably reasonable provided that the wavelength of the spin waves is small compared with the linear dimensions of the material. In these circumstances the surface distribution of 'magnetic charge' arising from $\mathbf{M}(\mathbf{k})\, e^{j\mathbf{k}.\mathbf{r}}$ will alternate rapidly in sign and the corresponding contribution to the internal field may be expected to be insignificant. The surface effects are, however, important for the long wavelength spin waves. The internal field is generally a complicated function of $\mathbf{M}(\mathbf{k})$ for these modes. We shall only examine how these modes can be treated in principle. The equations which we wish to solve inside the material are,

$$\nabla \wedge \mathbf{H} = 0 \quad \text{from which} \quad \mathbf{H} = -\nabla\phi^* \qquad (8.6.31)$$
$$\nabla . \mathbf{H} = -4\pi\nabla . \mathbf{M} \quad \text{from which} \quad \nabla^2\phi^* = 4\pi\nabla . \mathbf{M} \qquad (8.6.32)$$

together with the torque equation for the magnetization (see Section 8.7)

$$-\frac{d\mathbf{M}}{dt} = \gamma \mathbf{M} \wedge \mathbf{H} \tag{8.6.33}$$

Assume that the sample is in the form of an ellipsoid, magnetized to saturation along a principal axis which is identified as the z-axis, and that there may also be present an additional applied field $\hat{\mathbf{k}}H_0$. Take $\mathbf{M}(t)$ in the harmonic form $\mathbf{M}(t) = \mathbf{M}(\mathbf{r})\, e^{j\omega t}$, and write equation (8.6.33) in first order as

$$-\frac{j\omega}{\gamma} M_x = M_y(H_0 - N_z M_z^s) - M_z H_y \tag{8.6.34}$$

$$\frac{j\omega}{\gamma} M_y = M_x(H_0 - N_z M_z^s) - M_z H_x \tag{8.6.35}$$

$$M_z^s = \text{constant} \tag{8.6.36}$$

Equations (8.6.34), (8.6.35) may be solved for M_x, M_y, to give equations of the form

$$M_x = \alpha H_x + \beta H_y \tag{8.6.37}$$

$$M_y = -\beta H_x + \alpha H_y \tag{8.6.38}$$

from which

$$\nabla . \mathbf{M} = -\alpha\left(\frac{\partial^2 \phi^*}{\partial x^2} + \frac{\partial^2 \phi^*}{\partial y^2}\right) \tag{8.6.39}$$

Equation (8.6.32) may now be written

$$(1 + 4\pi\alpha)\left(\frac{\partial^2 \phi^*}{\partial x^2} + \frac{\partial^2 \phi^*}{\partial y^2}\right) + \frac{\partial^2 \phi^*}{\partial z^2} = 0 \tag{8.6.40}$$

This equation applies to the region inside the material. Outside the material

$$\nabla^2 \phi_0^* = 0 \tag{8.6.41}$$

The boundary conditions at the surface require that ϕ^* and

$$\mathbf{B}.\mathbf{n} \equiv (\nabla \phi^* + 4\pi \mathbf{M}).\mathbf{n}$$

should be continuous.

Solutions for ϕ^* have been obtained for samples which have the shape of an ellipsoid of revolution with the axis of symmetry along the direction of magnetization M_z^s. Even in these cases the solutions are complicated mathematical functions. The magnetic distributions described by these functions are usually referred to as Walker modes.†

8.7 Electromagnetic Waves in a Magnetized Material

The propagation of an electromagnetic wave in a magnetized material will be described by a simultaneous solution for the electromagnetic field equations,

$$\begin{aligned} \nabla \wedge \mathbf{E} + \frac{1}{c}\frac{\partial \mathbf{B}}{\partial t} = 0 \quad \nabla . \mathbf{D} = 0 \\ \nabla \wedge \mathbf{H} - \frac{1}{c}\frac{\partial \mathbf{D}}{\partial t} - 4\pi \mathbf{J} = 0 \quad \nabla . \mathbf{B} = 0 \end{aligned} \tag{8.7.1}$$

† Walker, L. R., *Phys. Rev.* **105**, 390, 1957.

taken together with the 'torque' equation which relates the magnetization **M** to the field vector **H**,

$$-\frac{1}{\gamma}\frac{\mathrm{d}\mathbf{M}}{\mathrm{d}t} = \mathbf{M} \wedge \mathbf{H} \tag{8.7.2}$$

When seeking a solution for these equations it is no longer possible to characterize the material by a simple isotropic permeability μ such as was used in the discussion of electromagnetic waves in Chapter 2. The magnetized material is essentially anisotropic because the direction of magnetization provides a unique axis within the material. In these circumstances it is usually most convenient to retain **B** as $(\mathbf{H} + 4\pi\mathbf{M})$ and to determine the relation between **H** and **M** from equation (8.7.2).

8.7.1 Torque Equation

Equation (8.7.2) may be regarded as a classical equation of motion. The left-hand side of this equation represents the rate of change of angular momentum for the spin system, derived from

$$\hbar\langle\mathbf{S}\rangle = -\frac{\mathbf{M}}{(g\beta/\hbar)}$$

whilst the right-hand side is the torque acting upon the magnetization. The most simple mode described by equation (8.7.2) may be represented by a vector **M** of constant magnitude which precesses about the direction of a uniform field $\mathbf{H} = \hat{\mathbf{k}}H_0$ with an angular frequency $\omega_0 = \gamma H_0$.

Equation (8.7.2) can, however, also be based upon the Hamiltonian operator for the spin system which was given in equation (8.4.2). Write equation (8.4.2) in the form

$$\mathscr{H}_{\text{mag}} = g\beta H_0 \sum_i S_i^z - \sum_{\substack{ij \\ i>j}} J_{ij}[S_i^x S_j^x + S_i^y S_j^y + S_i^z S_j^z] \tag{8.7.3}$$

The Heisenberg equation of motion for a dynamical variable gives†

$$\frac{\mathrm{d}}{\mathrm{d}t}\langle\mathbf{S}_i\rangle = -\frac{1}{\mathrm{j}\hbar}\langle[\mathscr{H}_{\text{mag}}\,\mathbf{S}_i - \mathbf{S}_i\,\mathscr{H}_{\text{mag}}]\rangle \tag{8.7.4}$$

Use the commutation relations

$$\begin{aligned}
[S_i^x, S_i^y] &= S_i^x S_i^y - S_i^y S_i^x = \mathrm{j}S_i^z \\
[S_i^x, S_i^z] &= S_i^x S_i^z - S_i^z S_i^x = -\mathrm{j}S_i^y \\
[S_i^y, S_i^z] &= S_i^y S_i^z - S_i^z S_i^y = \mathrm{j}S_i^x \\
[S_i^x, S_j^y] &= [S_i^x, S_j^z] = [S_i^y, S_j^z] = 0
\end{aligned} \tag{8.7.5}$$

to obtain

$$\frac{\mathrm{d}}{\mathrm{d}t}\langle S_i^x\rangle = -\frac{1}{\hbar}\left\langle\left[g\beta H_0\, S_i^y + \sum_j J_{ij}[S_j^y\, S_i^z - S_j^z\, S_i^y]\right]\right\rangle \tag{8.7.6}$$

† See, for example, Schiff, L. I., *Quantum Mechanics*, McGraw-Hill, 1949.

Add together all equations of this type,

$$\frac{\mathrm{d}}{\mathrm{d}t}\left\langle\sum_i S_i^x\right\rangle = -\frac{1}{\hbar}g\beta H_0\left\langle\sum_i S_i^y\right\rangle \tag{8.7.7}$$

or, writing

$$S^x = \sum_i S_i^x, \quad S^y = \sum_i S_i^y$$

we find that

$$\hbar\frac{\mathrm{d}}{\mathrm{d}t}\langle S^x\rangle = -g\beta H_0\langle S^y\rangle \tag{8.7.8}$$

Similarly

$$\hbar\frac{\mathrm{d}}{\mathrm{d}t}\langle S^y\rangle = g\beta H_0\langle S^x\rangle, \quad \frac{\mathrm{d}}{\mathrm{d}t}\langle S^z\rangle = 0 \tag{8.7.9}$$

Equations (8.7.8) and (8.7.9) may be written in vector notation with $\mathbf{H} = \hat{\mathbf{k}}H_0$, as

$$\hbar\frac{\mathrm{d}}{\mathrm{d}t}\langle\mathbf{S}\rangle = -g\beta\langle\mathbf{S}\rangle \wedge \mathbf{H} \tag{8.7.10}$$

Write $\mathbf{M} = -g\beta\langle\mathbf{S}\rangle$, $\gamma = g\beta/\hbar$, to obtain

$$-\frac{1}{\gamma}\frac{\mathrm{d}\mathbf{M}}{\mathrm{d}t} = \mathbf{M} \wedge \mathbf{H} \tag{8.7.11}$$

This equation appears to be equivalent to equation (8.7.2). However, it should be noted that we have included no phase factors as coefficients in the summation leading to equation (8.7.7). Equation (8.7.7) and equation (8.7.11) therefore refer only to the uniform precessional mode, or the $\mathbf{k} = 0$ spin wave mode of Section 8.5. The extension of equation (8.7.11) to more general modes of excitation presents a difficult mathematical problem. We may, however, make a reasonably convincing guess at the result by modifying the molecular field discussion leading from equation (8.4.3) to equation (8.4.6). The discussion of these equations in Section 8.4 was based on the ground state configuration with all the $\langle S_j^z\rangle = -S$ and the $\langle S_j^x\rangle = \langle S_j^y\rangle = 0$. To take account of a spin wave excitation with $\mathbf{k} \neq 0$ allow the transverse components $\langle S_j^x\rangle$, $\langle S_j^y\rangle$ to be non-zero so that, at any instant in time, the resultant vector $\langle\mathbf{S}_j\rangle$ will vary in direction at the different lattice sites. Assume that these variations are small so that $\langle\mathbf{S}_{j+m}\rangle$ may be expanded by Taylor's theorem about the value $\langle\mathbf{S}_j\rangle$. In addition, restrict the range of the interaction J_{ij} to nearest neighbours and consider a simple crystal lattice having inversion symmetry. The molecular field $\mathbf{H}_\mathrm{m}$ derived from equation (8.4.3) involves the summation $J_{nn}\sum\mathbf{S}_j$ taken over nearest neighbours and will have the form

$$\mathbf{H}_\mathrm{m} \sim J_{nn}[\langle\mathbf{S}_j\rangle + \alpha\nabla^2\mathbf{S}_j] \tag{8.7.12}$$

The first order terms in $\nabla S_j^{x,y,z}$ do not occur in this relation because the crystal has inversion symmetry. Equation (8.7.12) may be written in terms of the magnetization

$$\mathbf{H}_\mathrm{m} = \lambda(\mathbf{M} + \alpha'\nabla^2\mathbf{M}) \tag{8.7.13}$$

We may therefore expect the torque acting on the magnetization in a field $\mathbf{H}_a$ to take the form

$$\mathbf{M} \wedge (\mathbf{H}_a + \mathbf{H}_m) = \mathbf{M} \wedge (\mathbf{H}_a + \alpha'' \nabla^2 \mathbf{M}) \qquad (8.7.14)$$

Note that the zero order exchange term $\mathbf{M}$ does not contribute to the torque in accordance with our previous discussion leading to equation (8.7.11). The physical explanation is that the zero order molecular field component $\mathbf{M}$ has the same direction as the magnetization $\mathbf{M}$ and so does not give rise to any torque.

We may therefore expect to write equation (8.7.2) in the explicit form

$$-\frac{1}{\gamma}\frac{\mathrm{d}}{\mathrm{d}t}\mathbf{M}(\mathbf{r}, t) = \mathbf{M}(\mathbf{r}, t) \wedge \mathbf{H}'(\mathbf{r}, t) \qquad (8.7.15)$$

as quite a general classical equation of motion for the magnetization. In this equation $\mathbf{M}(\mathbf{r}, t)$ can vary both in space and time through the sample, and $\mathbf{H}'(\mathbf{r}, t)$ is the actual field experienced by the spin system including the term $\alpha'' \nabla^2 \mathbf{M}$.

8.7.2 Propagation through a Magnetized Disk

As an illustration of the technique used for obtaining a particular solution of equations (8.7.1) and (8.7.15), consider the propagation of a plane harmonic wave through a disk which is magnetized in the plane of the disk. Take the z-axis of coordinates along the direction of the magnetization and the direction of propagation to be y-axis, pointing into the disk, as shown in Fig. 8.11(a). In order to retain a tractable mathematical problem without losing any of the essential physical features make the following assumptions,

(a) that the disk may be regarded as an ellipsoid with good demagnetizing factors and that it is magnetized to saturation along a principal axis (the z-axis),

(b) that the linear dimensions of the disk are sufficiently large for all dimensional effects to be neglected so far as the electromagnetic wave is concerned, so that the propagation characteristics may be derived by treating the disk as a semi-infinite slab bounded by the plane $y = 0$.

In equations (8.7.1) take

$$\mathbf{J} = \frac{\sigma}{c}\mathbf{E}, \quad \mathbf{D} = \varepsilon\mathbf{E}, \quad \Lambda = \frac{\mathrm{j}\omega\varepsilon + 4\pi\sigma}{c} \qquad (8.7.16)$$

and assume a plane harmonic wave propagating along the y-axis,

$$\mathbf{H} = \mathbf{h}(\mathbf{r}, t) + \hat{\mathbf{k}}(H_0 - N_z M_z); \quad \mathbf{h}(\mathbf{r}, t) = \mathbf{h}^0 \exp\{\mathrm{j}(\omega t - k_y y)\} \qquad (8.7.17)$$

$$\mathbf{M} = \mathbf{m}(\mathbf{r}, t) + \hat{\mathbf{k}} M_z; \qquad \mathbf{m}(\mathbf{r}, t) = \mathbf{m}^0 \exp\{\mathrm{j}(\omega t - k_y y)\} \qquad (8.7.18)$$

$$\mathbf{E} = \mathbf{e}(\mathbf{r}, t) = \mathbf{e}^0 \exp\{\mathrm{j}(\omega t - k_y y)\} \qquad (8.7.19)$$

From

$$\nabla \wedge \mathbf{E} + \frac{1}{c}\frac{\partial \mathbf{B}}{\partial t} = 0$$

$$(\mathrm{j}k_y)[\hat{\mathbf{j}} \wedge \mathbf{e}^0] = \frac{\mathrm{j}\omega}{c}[\mathbf{h}^0 + 4\pi\mathbf{m}^0] \qquad (8.7.20)$$

and from $\nabla \wedge \mathbf{H} - \Lambda\mathbf{E} = 0$,

$$(-jk_y)(\hat{\mathbf{j}} \wedge \mathbf{h}^0) = \Lambda\mathbf{e}^0 \tag{8.7.21}$$

Substitute for $\mathbf{e}^0$ in equation (8.7.20),

$$\mathbf{h}^0\left(1 + \frac{ck_y^2}{j\omega\Lambda}\right) - \hat{\mathbf{j}}\left(\frac{ck_y^2}{j\omega\Lambda}\right)h_y^0 + 4\pi\mathbf{m}^0 = 0 \tag{8.7.22}$$

which, written out in component form, gives

$$h_x^0\left(1 + \frac{ck_y^2}{j\omega\Lambda}\right) + 4\pi m_x^0 = 0 \tag{8.7.23a}$$

$$h_y^0 + 4\pi m_y^0 = 0 \tag{8.7.23b}$$

$$h_z^0\left(1 + \frac{ck_y^2}{j\omega\Lambda}\right) + 4\pi m_z^0 = 0 \tag{8.7.23c}$$

In equation (8.7.15), write the effective field $\mathbf{H}'(\mathbf{r}, t)$ as

$$\mathbf{H}'(\mathbf{r}, t) = \mathbf{H} + \alpha''\nabla^2\mathbf{M} = \mathbf{h}(\mathbf{r}, t) + \hat{\mathbf{k}}(H_0 - N_zM_z) + \alpha''\nabla^2\mathbf{M} \tag{8.7.24}$$

so that

$$-\frac{j\omega}{\gamma}\mathbf{m}(\mathbf{r}, t) = \{\mathbf{m}(\mathbf{r}, t) + \hat{\mathbf{k}}M_z\} \wedge \{\mathbf{h}(\mathbf{r}, t) + \hat{\mathbf{k}}(H_0 - N_zM_z) + \alpha''\nabla^2\mathbf{M}\} \tag{8.7.25}$$

To first order, this equation may be written in component form as,

$$-\frac{j\omega}{\gamma}m_x^0 = m_y^0(H_0 - N_zM_z + \alpha''k_y^2M_z) - M_zh_y^0 \tag{8.7.26a}$$

$$-\frac{j\omega}{\gamma}m_y^0 = m_x^0(H_0 - N_zM_z + \alpha''k_y^2M_z) - M_zh_x^0 \tag{8.7.26b}$$

$$m_z^0 = 0 \tag{8.7.26c}$$

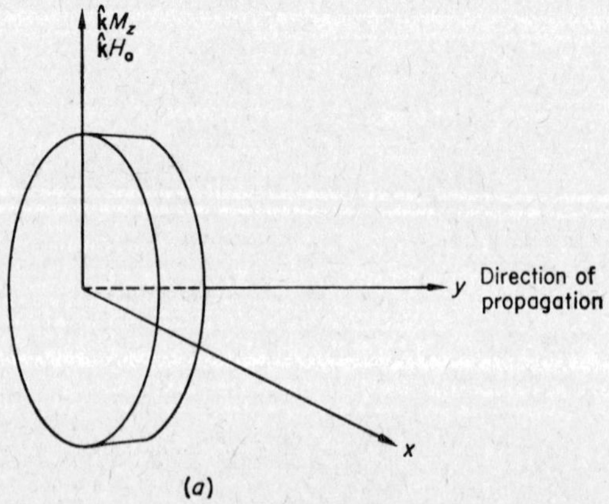

FIG. 8.11

(a) The coordinate axes for an electromagnetic wave propagating through a magnetized disk.

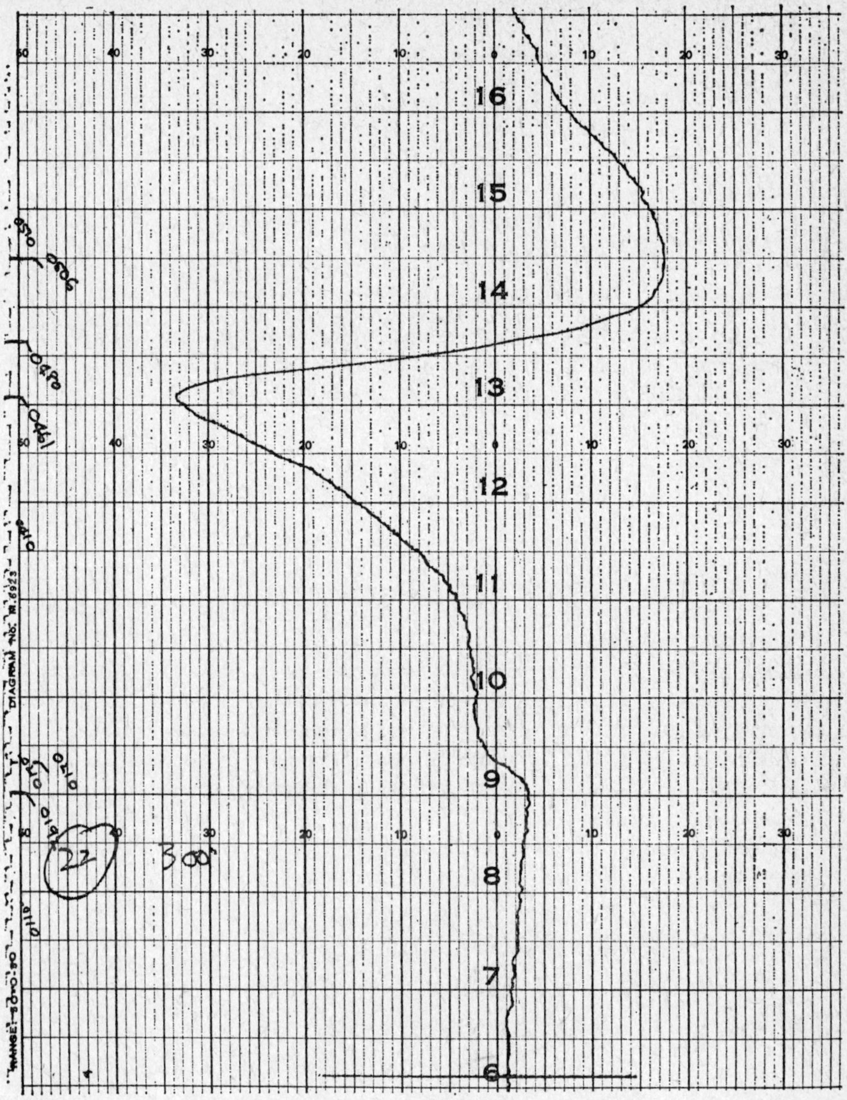

(b) An actual pen-recorder trace of ferromagnetic resonance in a single crystal of a gadolinium-yttrium alloy containing 64 at% Gd. The record is the differential of the power absorption in the sample and was obtained at a temperature of 181 K using radiation of 8·56 mm wavelength. Note that in addition to the principal resonance absorption corresponding to equation (8.7.37) there is a subsidiary absorption in low magnetic fields. This low field effect is called the 'anti-resonance' and arises from the solution of equation (8.7.34) for which $ky = 0$ and $\dfrac{\omega}{\gamma} = H_0 = N_z M_z + 4\pi M_z$.

These equations may be solved for m_x^0, m_y^0,

$$m_x^0 = \chi_1 h_x^0 + j\chi_2 h_y^0 \tag{8.7.27a}$$

$$m_y^0 = -j\chi_2 h_x^0 + \chi_1 h_y^0 \tag{8.7.27b}$$

$$m_z^0 = 0 \tag{8.7.27c}$$

with

$$\chi_1 = \frac{H_1 M_z}{\{(-\omega^2/\gamma^2) + H_1^2\}}; \quad \chi_2 = \frac{(\omega/\gamma)M_z}{\{(-\omega^2/\gamma^2) + H_1^2\}} \tag{8.7.28}$$
$$H_1 = (H_0 - N_z M_z + \alpha'' k_y^2 M_z)$$

The properties of the electromagnetic field may now be obtained by solving the set of algebraic equations given in equations (8.7.23) and (8.7.27). Consider a wave polarized such that $h_x^0 = h_y^0 = 0$, $h_z^0 \neq 0$. From equations (8.7.23c) and (8.7.27c).

$$1 + \frac{ck_y^2}{j\omega\Lambda} = 0 \tag{8.7.29}$$

or

$$k_y^2 = -\frac{j\omega\Lambda}{c}, \quad k_y \approx (1 - j)\delta^{-1} \tag{8.7.30}$$

where $\delta = \sqrt{(c^2/2\pi\sigma\omega)}$ is the usual skin depth parameter for a non-magnetic conducting material. This particular wave therefore will be attenuated as it propagates through the disk in the same manner as a wave in a simple conducting material and will exhibit no response to the magnetization. This is the first order description of the wave. In practice some response to the magnetization will arise because m_z^0 is not zero when higher order terms are included in equations (8.7.27).

The solutions for equations (8.7.23) and (8.7.27) having h_x^0, $h_y^0 \neq 0$, $h_z^0 = 0$, show a first order response to the magnetization. We shall have,

$$\begin{gathered}\left(1 + 4\pi\chi_1 + \frac{ck_y^2}{j\omega\Lambda}\right)h_x^0 + 4\pi j\chi_2 h_y^0 = 0 \\ -4\pi j\chi_2 h_x^0 + (1 + 4\pi\chi_1)h_y^0 = 0\end{gathered} \tag{8.7.31}$$

For these equations to have a non-trivial solution,

$$\left(1 + 4\pi\chi_1 + \frac{ck_y^2}{j\omega\Lambda}\right)(1 + 4\pi\chi_1) = 16\pi^2\chi_2^2 \tag{8.7.32}$$

or, from equations (8.7.28),

$$\frac{ck_y^2}{j\omega\Lambda} = -\left\{1 + \frac{4\pi M_z(H_1 + 4\pi M_z)}{H_1(H_1 + 4\pi M_z) - (\omega^2/\gamma^2)}\right\} \tag{8.7.33}$$

The effective field H_1 includes the exchange contribution $\alpha'' k_y^2 M_z$ and so this equation is cubic in k_y^2. A particular frequency, ω, will be associated with three distinct waves corresponding to the roots of this equation. The boundary conditions, at the surface $y = 0$, lead to complicated mathematical relations. We shall therefore simplify the analysis by assuming that the term

$\alpha'' k_y^2 M_z$ can be neglected. This will be a reasonable approximation for long wavelength excitations since, for these modes, $k \approx 0$.

Neglecting the exchange term in H_1, equation (8.7.33) becomes

$$\frac{ck_y^2}{j\omega\Lambda} = -\left\{1 + \frac{4\pi M_z(H_0 - N_z M_z + 4\pi M_z)}{(H_0 - N_z M_z)(H_0 - N_z M_z + 4\pi M_z) - (\omega^2/\gamma^2)}\right\} \tag{8.7.34}$$

This is quadratic in k_y, the positive root representing a wave propagating into the material.

The time average of the power flow into the material is obtained from the complex Poynting vector $\langle \mathbf{S} \rangle$ evaluated at the plane $y = 0$. From equation (2.7.14), Chapter 2,

$$\langle \mathbf{S} \rangle_y = \frac{c}{4\pi}\frac{1}{2}\,\mathrm{Re}\,[e_z h_x^*]_{y=0} \tag{8.7.35}$$

$$= \frac{c}{8\pi}\,|h_x^0|^2\,\mathrm{Re}\left[\frac{jk_y}{\Lambda}\right] \tag{8.7.36}$$

The power flow into the sample will have a singularity at the pole of k_y, which, according to equation (8.7.34), is given by

$$\frac{\omega^2}{\gamma^2} = (H_0 - N_z M_z)(H_0 - N_z M_z + 4\pi M_z) \tag{8.7.3}$$

This is the frequency relation for ferromagnetic resonance in the particular field configuration which we have chosen. A record of a ferromagnetic resonance line is shown in Fig. 8.11(b).

8.7.3 Resonance Frequency from the Torque Equation

Although we expect equation (8.7.37) to be reasonably correct for the long wavelength excitations this equation strictly refers to the uniform mode of precession since we have neglected the exchange field. The normal mode frequencies for uniform precession may, however, be derived from equation (8.7.2) alone and so they can be evaluated without undertaking a complete solution for the electromagnetic field. It is possible therefore to derive the resonance frequencies for the long wavelength excitations from the relatively simple vector equation, equation (8.7.2).

Consider an ellipsoidal sample, magnetized to saturation along a principal axis, which we take to be the z-axis of coordinates. Write the field vector $\mathbf{H}$ in terms of the applied field $\mathbf{H}_a = \hat{\mathbf{k}} H_0$ and the demagnetizing coefficients N_x, N_y, N_z,

$$\mathbf{H} = -\hat{\mathbf{i}} N_x M_x - \hat{\mathbf{j}} N_y M_y + \hat{\mathbf{k}}(H_0 - N_z M_z) \tag{8.7.38}$$

Equation (8.7.2) gives in first order

$$\begin{aligned} -\frac{j\omega}{\gamma} M_x &= M_y\{H_0 + (N_y - N_z)M_z\} \\ \frac{j\omega}{\gamma} M_y &= M_x\{H_0 + (N_x - N_z)M_z\} \\ \frac{dM_z}{dt} &= 0 \end{aligned} \tag{8.7.39}$$

from which

$$\frac{\omega^2}{\gamma^2} = \{H_0 + (N_x - N_z)M_z\}\{H_0 + (N_y - N_z)M_z\} \qquad (8.7.40)$$

This equation is usually referred to as the Kittel resonance condition.†

For a sample in the shape of a flat disk, magnetized in the plane of the disk, the field components of equation (8.7.38) are $(0, -4\pi M_y, H_0 - N_z M_z)$. This choice of field components corresponds with our previous discussion where dimensional effects in the x-direction were neglected. The demagnetizing coefficients in equation (8.7.40) are, therefore, $[0, 4\pi, N_z]$ and

$$\frac{\omega^2}{\gamma^2} = (H_0 - N_z M_z)(H_0 - N_z M_z + 4\pi M_z) \qquad (8.7.41)$$

in agreement with equation (8.7.37).

If the sample is a flat disk, magnetized perpendicular to the plane of the disk, the field components of equation (8.7.38) are $(0, 0, H_0 - 4\pi M_z)$. The normal mode frequency for this configuration is

$$\frac{\omega^2}{\gamma^2} = (H_0 - 4\pi M_z)^2 \qquad (8.7.42)$$

This equation may be verified by solving the coupled electromagnetic field equations for a wave propagating into the disk along the direction of magnetization.

Finally, we note that equation (8.7.15) is not an entirely satisfactory equation of motion for the magnetization because no damping terms have been included. Such terms are necessary for the spin system to become magnetized in the first instance and for the electromagnetic wave to continue to give up energy through the interaction with the magnetization. Damping may be incorporated into equation (8.7.15) by adding suitable terms to the component equations. We may write, for example,

$$\frac{\mathrm{d}M_x}{\mathrm{d}t} = -\gamma(\mathbf{M} \wedge \mathbf{H}')_x - \frac{M_x}{T_2}, \quad \frac{\mathrm{d}M_y}{\mathrm{d}t} = -\gamma(\mathbf{M} \wedge \mathbf{H}')_y - \frac{M_y}{T_2}$$

$$\frac{\mathrm{d}M_x}{\mathrm{d}t} = -\gamma(\mathbf{M} \wedge \mathbf{H}')_z - \frac{M_z - M^s}{T_1} \qquad (8.7.43)$$

Provided that these damping terms are small their effect can be taken into account by substituting $\{\omega - (\mathrm{j}/T_2)\}$ for ω in the solutions for the electromagnetic field. They do not modify significantly the discussion outlined in this section.

8.8 Frequency of Uniform Precession from Lagrange's Equations

A very useful technique for deriving the frequency of uniform precession for the magnetization is based on Lagrange's formulation of the equations of motion in classical dynamics. This method of calculation achieves an important advantage over equations (8.7.2) and (8.7.15) in replacing the effective field torque representation by an equation of motion which makes explicit

† Kittel, C., *Phys. Rev.* **73**, 155, 1948.

use of an energy function. The Lagrangian function is defined in terms of the kinetic energy T and the potential energy V of the system, to be,

$$L = T - V \tag{8.8.1}$$

The motion of a simple dynamical system is described by Lagrange's equations of motion†

$$\frac{\mathrm{d}}{\mathrm{d}t}\cdot\frac{\partial L}{\partial \dot{q}_p} - \frac{\partial L}{\partial q_p} = 0 \tag{8.8.2}$$

where q_p is a generalized coordinate for the system.

The magnetized system is represented by an ideal angular momentum which may be visualized as being derived from a symmetric top spinning about its axis. The top has a moment of inertia, C, about the axis of spin but no transverse components. A convenient set of coordinates for describing the

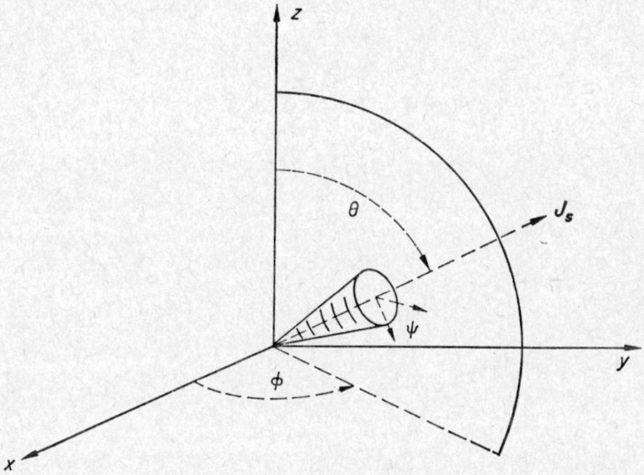

FIG. 8.12 The coordinate system for the spinning top.

motion of the top is known as the set of Eulerian angles. These are three in number, denoted by θ, ϕ, ψ. The angles θ, ϕ are the usual polar angles for the axis of the top with respect to a fixed set of rectangular axes, x, y, z, as shown in Fig. 8.12. The angle ψ refers to rotation of the top about its own axis and is chosen such that the angular momentum along this axis is

$$J_s = C(\dot{\psi} + \cos\theta\dot{\phi}) \tag{8.8.3}$$

The potential energy of the spin system is a function of (θ, ϕ) but not of ψ. Hence

$$\frac{\partial V}{\partial \psi} = \frac{\partial V}{\partial \dot{\psi}} = 0 \tag{8.8.4}$$

† See, for example, TER HAAR, D., *Elements of Hamiltonian Mechanics*, North-Holland, 1961, or SYNGE, J. L. and GRIFFITH, B. A., *Principles of Mechanics*, McGraw-Hill, 1959.

The kinetic energy function is

$$T = \tfrac{1}{2}C(\dot{\psi} + \cos\theta\dot{\phi})^2 \tag{8.8.5}$$

We may use equations (8.8.2) and (8.8.4) to derive a relation which will eliminate ψ, $\dot{\psi}$ from the equations of motion for the system. We have

$$\frac{\mathrm{d}}{\mathrm{d}t}\left(\frac{\partial T}{\partial \dot{\psi}}\right) - \frac{\partial T}{\partial \psi} = 0 \tag{8.8.6}$$

that is

$$\frac{\mathrm{d}}{\mathrm{d}t}\{C(\dot{\psi} + \cos\theta\dot{\phi})\} = 0 \tag{8.8.7}$$

or

$$C(\dot{\psi} + \cos\theta\dot{\phi}) = J_s = \text{constant} \tag{8.8.8}$$

Making use of this relation, equation (8.8.2), when evaluated with respect to the coordinates θ and ϕ, leads to the equations of motion

$$J_s \sin\theta\,\dot{\phi} + \frac{\partial V}{\partial \theta} = 0 \tag{8.8.9}$$

$$-J_s \sin\theta\,\dot{\theta} + \frac{\partial V}{\partial \phi} = 0 \tag{8.8.10}$$

Identify the magnetization per unit volume of material through the relation $M_s = -\gamma J_s$, to obtain the equations for uniform precession

$$-\frac{M_s}{\gamma}\sin\theta\,\dot{\phi} + \frac{\partial V}{\partial \theta} = 0 \tag{8.8.11}$$

$$\frac{M_s}{\gamma}\sin\theta\,\dot{\theta} + \frac{\partial V}{\partial \phi} = 0 \tag{8.8.12}$$

As an example of the use of these equations, consider the motion of the magnetization after being displaced by a small amount from a position of equilibrium. Expand $V(\theta, \phi)$ about its value at the equilibrium position θ_0, ϕ_0,

$$\begin{aligned} V(\theta, \phi) = V(\theta_0, \phi_0) &+ (\theta - \theta_0)\left(\frac{\partial V}{\partial \theta}\right)_{\theta_0\phi_0} + (\phi - \phi_0)\left(\frac{\partial V}{\partial \phi}\right)_{\theta_0,\phi_0} \\ &+ \tfrac{1}{2}(\theta - \theta_0)^2\left(\frac{\partial^2 V}{\partial \theta^2}\right)_{\theta_0,\phi_0} + \tfrac{1}{2}(\phi - \phi_0)^2\left(\frac{\partial^2 V}{\partial \phi^2}\right)_{\theta_0,\phi_0} \\ &+ (\theta - \theta_0)(\phi - \phi_0)\left(\frac{\partial^2 V}{\partial \theta\,\partial \phi}\right)_{\theta_0,\phi_0} + \cdots \end{aligned} \tag{8.8.13}$$

But

$$\left(\frac{\partial V}{\partial \theta}\right)_{\theta_0,\phi_0} = \left(\frac{\partial V}{\partial \phi}\right)_{\theta_0,\phi_0} = 0 \tag{8.8.14}$$

because the equilibrium position has coordinates θ_0, ϕ_0. Hence equations (8.8.11) and (8.8.12) give

$$-\frac{M_s}{\gamma}\sin\theta\,\dot{\phi} + (\theta - \theta_0)\left(\frac{\partial^2 V}{\partial \theta^2}\right)_{\theta_0,\phi_0} + (\phi - \phi_0)\left(\frac{\partial^2 V}{\partial \theta\,\partial \phi}\right)_{\theta_0,\phi_0} = 0 \tag{8.8.15}$$

$$\frac{M_s}{\gamma}\sin\theta\,\dot{\theta}+(\theta-\theta_0)\left(\frac{\partial^2 V}{\partial\theta\,\partial\phi}\right)_{\theta_0,\phi_0}+(\phi-\phi_0)\left(\frac{\partial^2 V}{\partial\phi^2}\right)_{\theta_0,\phi_0}=0 \quad (8.8.16)$$

Seek solutions of the form $(\theta-\theta_0)\sim e^{j\omega t}$, and $(\phi-\phi_0)\sim e^{j\omega t}$, with $\sin\theta=\sin\theta_0$ since the oscillations are assumed to be small in amplitude. The simultaneous equations, equations (8.8.15) and (8.8.16), will have a solution of this form if $\sin\theta_0$ is non-zero, and,

$$\left(\frac{\omega}{\gamma}\right)^2=\frac{1}{M_s\sin^2\theta_0}\left\{\left(\frac{\partial^2 V}{\partial\theta^2}\right)\left(\frac{\partial^2 V}{\partial\phi^2}\right)-\left(\frac{\partial^2 V}{\partial\theta\,\partial\phi}\right)\right\}_{\theta_0,\phi_0} \quad (8.8.17)$$

which determines the frequency of uniform precession.

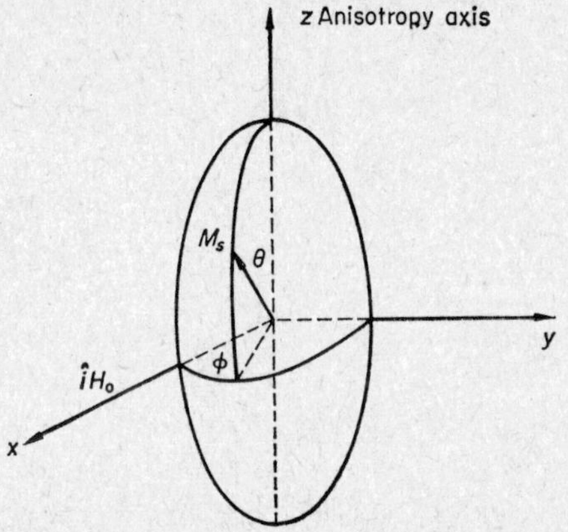

FIG. 8.13 The coordinate system for uniform precession in an ellipsoidal specimen with uniaxial anisotropy.

To illustrate the usefulness of this relation we shall derive the uniform precession frequency for an ellipsoidal specimen which has uniaxial anisotropy. Take the z-axis to be the axis of anisotropy and apply an external field along the x-direction as shown in Fig. 8.13. The external field is to be large enough to magnetize the sample to saturation in this direction, so that $\theta_0=\pi/2$, $\phi_0=0$.

The anisotropy energy V_A may be written, with K_2 a positive quantity,

$$V_A=K_2\sin^2\theta \quad (8.8.18)$$

which indicates that the easy direction of magnetization is along the z-axis. The total magnetic energy is

$$V=-\mathbf{M}_s.\mathbf{H}+\tfrac{1}{2}[N_xM_x^2+N_yM_y^2+N_zM_z^2]+V_A \quad (8.8.19)$$

$$=-M_sH_0\sin\theta\cos\phi+\frac{M_s^2}{2}(N_x\sin^2\theta\cos^2\phi+N_y\sin^2\theta\sin^2\phi+N_z\cos^2\theta)+K_2\sin^2\theta \quad (8.8.20)$$

and

$$\left(\frac{\omega}{\gamma}\right)^2=\left[H_0+(N_z-N_x)M_s-\frac{2K_2}{M_s}\right][H_0+(N_y-N_x)M_s] \quad (8.8.21)$$

This relation contains the interesting result that the uniform precession frequency will fall to zero at a value of the external field H_0 given by

$$H_0 + (N_z - N_x)M_s = \frac{2K_2}{M_s} \tag{8.8.22}$$

If the external field is directed along the z-axis in Fig. 8.13 the equilibrium condition is $\theta_0 = 0$ and equation (8.8.17) cannot be directly applied. For this special case make use of the underlying equations of motion given in equations (8.8.11) and (8.8.12). We have $\mathbf{H} = \hat{\mathbf{k}}H_0$, and

$$V = -M_sH_0 \cos\theta + \frac{M_s^2}{2}(N_x \sin^2\theta \cos^2\phi + N_y \sin^2\theta \sin^2\phi + N_z \cos^2\theta) + K_2 \sin^2\theta \tag{8.8.23}$$

$$\frac{\partial V}{\partial \theta} = M_sH_0 \sin\theta + \frac{M_s^2}{2}(N_x \cos^2\phi + N_y \sin^2\phi - N_z) \sin 2\theta + K_2 \sin 2\theta \tag{8.8.24}$$

Since θ is small, equation (8.8.11) gives

$$-\frac{1}{\gamma}\dot{\phi} + H_0 + M_s\{N_x \cos^2\phi + N_y \sin^2\phi - N_z\} + \frac{2K_2}{M_s} = 0 \tag{8.8.25}$$

or

$$\frac{1}{\gamma}\dot{\phi} = \left\{H_0 + (N_x - N_z)M_s + \frac{2K_2}{M_s}\right\}\cos^2\phi + \left\{H_0 + (N_y - N_z)M_s + \frac{2K_2}{M_s}\right\}\sin^2\phi \tag{8.8.26}$$

which we write in the form

$$\frac{1}{\gamma}\dot{\phi} = H_1 \cos^2\phi + H_2 \sin^2\phi$$

Divide through by $\cos^2\phi$,

$$\frac{1}{\gamma \cos^2\phi}\dot{\phi} = H_1\left(1 + \frac{H_2}{H_1}\tan^2\phi\right) \tag{8.8.27}$$

write

$$\tan\alpha = \sqrt{\left(\frac{H_2}{H_1}\right)}\tan\phi \tag{8.8.28}$$

to obtain

$$\dot{\alpha} = \gamma\sqrt{(H_1 H_2)} \tag{8.8.29}$$

and

$$\tan\phi = \sqrt{\left(\frac{H_1}{H_2}\right)}\tan(\omega t + \delta) \tag{8.8.30}$$

with the frequency of uniform precession given by

$$\left(\frac{\omega}{\gamma}\right)^2 = \left\{H_0 + (N_x - N_z)M_s + \frac{2K_2}{M_s}\right\}\left\{H_0 + (N_y - N_z)M_s + \frac{2K_2}{M_s}\right\} \tag{8.8.31}$$

which agrees with equation (8.7.40) when anisotropy is neglected.

Note that for this configuration, in which the field **H** is applied along the easy axis of magnetization, the frequency does not fall to zero. Consider the case of a sphere for which $N_x = N_y = N_z$. The equilibrium direction is still given by $\theta = 0$ even when H_0 is zero. The frequency of this uniform precession mode in zero field is therefore

$$\omega(H = 0) = \gamma \frac{2K_2}{M_s} \tag{8.8.32}$$

and indicates the existence of a spin wave energy gap.

8.9 Itinerant Electrons

The magnetic electrons in a metal are unlikely to be completely localized. In general we shall expect them to be taking part in the motion of charges through the crystal lattice which gives rise to electrical conduction. A rigorous account of magnetism in metals should, therefore, be based on their detailed band structures. We shall not attempt so complete a discussion but will consider a limiting form of this itinerant electron model which regards the electrons as being completely free and forming an ideal electron gas obeying Fermi–Dirac statistics.

For the ideal electron gas, the number of states in the range of kinetic energy between E_k and $E_k + dE_k$ is given by

$$g(E_k)\, dE_k = \frac{V}{2\pi^2(\hbar)^3} 2^{1/2} m^{3/2} E_k^{1/2}\, dE_k \tag{8.9.1}$$

Here V is the total volume of the material. The number of electrons in the range of kinetic energy between E_k and $E_k + dE_k$ is given by

$$N(E_k)\, dE_k = g(E_k) f(E)\, dE_k \tag{8.9.2}$$

where $f(E)$ is the Fermi–Dirac function,

$$f(E) = \frac{1}{\exp\{(E - \eta)/kT\} + 1} \tag{8.9.3}$$

and E is the total energy—kinetic and potential—for an electron. The parameter η is called the Fermi potential. It is determined from the total number of electrons,

$$N_{\text{total}} = 2\int_0^\infty g(E_k) f(E)\, dE_k \tag{8.9.4}$$

In this equation the factor of two occurs because the distribution function in equation (8.9.3) refers to electrons of one spin orientation only. In zero magnetic field the Pauli exclusion principle allows two electrons with opposing spins to occupy each state.

At the absolute zero of temperature, $T = 0$ K, the function $f(E) \equiv f^0(E)$ is unity for $E < \eta(0)$ and zero for $E > \eta(0)$. For this case, $\eta = \eta(0)$ is given the special name of Fermi-energy, which we shall denote by E_F,

$$\eta(0) = E_F \tag{8.9.5}$$

and the Fermi–Dirac distribution, $f^0(E)$, may be represented by the unit step function relation

$$f^0(E) = 1 - u(E - E_F) \tag{8.9.6}$$

For actual metals E_F is about 5 eV whilst at room temperature kT is only 0·026 eV. In most practical applications, therefore, $kT \ll E_F$ and the distribution function referring to 0 K provides an adequate representation.

Consider the paramagnetic susceptibility of this ideal electron gas. Take the direction of the applied magnetic field to be the z-axis for the system. The energy of an electron with spin parallel to $\hat{\mathbf{k}}H_0$ will be,

$$E_+ = E_k + \beta H_0 \tag{8.9.7}$$

and for an electron with spin antiparallel to $\hat{\mathbf{k}}H_0$,

$$E_- = E_k - \beta H_0 \tag{8.9.8}$$

In these equations E_k represents the kinetic energy of the electron in the absence of a magnetic field. The distribution of electrons over the energy states described by equations (8.9.7) and (8.9.8) will be given by two separate Fermi–Dirac functions, one for each spin orientation. It is evident that, at 0 K, both of these functions will have the same value for $\eta_H(0)$, which in first order remains equal to E_F. This is because $\eta_H(0)$ will not depend upon the direction of H_0 in the electron gas. However, changing the sign of H_0 is equivalent

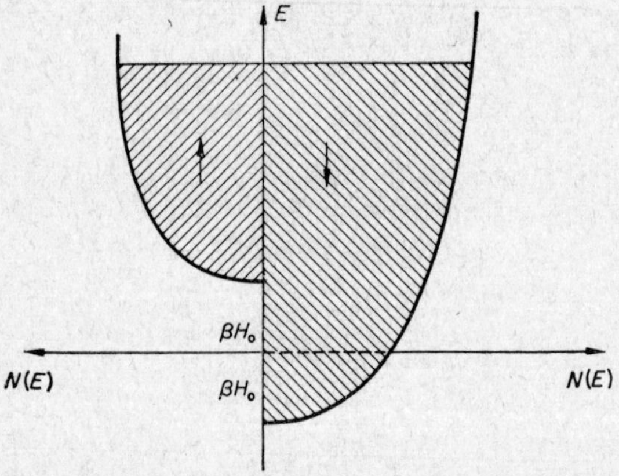

FIG. 8.14 The electron distributions for parallel and antiparallel spins at 0 K.

to reversing its direction and so $\eta_H(0)$ cannot contain a term linear in H_0. Hence to first order $\eta_H(0) = E_F$. The effect of the magnetic field is therefore simply to displace the energy distribution for the electrons with spins parallel to $\hat{\mathbf{k}}H_0$ by an amount $2\beta H_0$ measured with respect to the electrons with spins antiparallel to $\hat{\mathbf{k}}H_0$. The highest occupied energy states have the same energy E_F for both distributions, as shown in Fig. 8.14.

The spin magnetization of the electron gas arises from the difference in the numbers of parallel and antiparallel spins.

$$MV = \beta(N_- - N_+) \tag{8.9.9}$$

$$= \beta \int_0^\infty g(E_k)[f_-^0(E) - f_+^0(E)]\, dE_k \tag{8.9.10}$$

But, from equations (8.9.6), (8.9.7) and (8.9.8),

$$f_-^0(E) = 1 - u(E_k - \beta H_0 - E_F) \approx f^0(E) + \beta H_0\left[\frac{\partial u}{\partial(\beta H_0)}\right]_{H=0}$$

$$f_+^0(E) = 1 - u(E_k + \beta H_0 - E_F) \approx f^0(E) - \beta H_0\left[\frac{\partial u}{\partial(\beta H_0)}\right]_{H=0}$$

and since the differential of the unit step function is the Dirac δ-function,

$$MV = \beta\int_0^\infty g(E_k)\, 2\beta \mathrm{H}_0\, \delta(E_k - E_F)\, \mathrm{d}E_k \tag{8.9.11}$$

$$M = \frac{2\beta^2 H_0}{V} g(E_F) \tag{8.9.12}$$

The susceptibility per unit volume is

$$\chi^p = \frac{2\beta^2}{V} g(E_F) \tag{8.9.13}$$

or, upon using equations (8.9.1) and (8.9.4)

$$\chi^p = \frac{3}{2}\frac{n\beta^2}{E_F} \tag{8.9.14}$$

where $n = \left(\frac{N_{\text{total}}}{V}\right)$ is the total number of electrons per unit volume.

The magnetization of the electron gas described by equation (8.9.12) is usually referred to as the Pauli paramagnetic contribution. There is also an additional diamagnetic term which arises from the translational motion of the electrons in the magnetic field (see Section 9.4.5). This is the Landau diamagnetic contribution which for the ideal electron gas amounts to $-\frac{1}{3}$ of the spin paramagnetic contribution. The net susceptibility of the ideal electron gas is therefore

$$\chi_{\text{ net}} = \frac{n\beta^2}{E_F} \tag{8.9.15}$$

Although this expression has been derived using the distribution function characteristic of 0 K, we may expect it to be reasonably correct over the wide temperature range for which $kT \ll E_F$. The susceptibility of the itinerant electron distribution is not, therefore, expected to vary significantly with temperature over the range usually available for experiment.

A criterion which will determine whether the electron gas will exhibit spontaneous magnetization may be derived in the molecular field approximation. The magnetized state will be stable if the decrease in magnetic energy arising from spontaneous magnetization exceeds the increase in kinetic energy which results from the transfer of electrons between the different spin distributions. If the spontaneous magnetization is $\mathbf{M}$, the molecular field is $\lambda\mathbf{M}$ and the resulting decrease in energy per unit volume is

$$E_m = -\tfrac{1}{2}\lambda M^2 = -\tfrac{1}{2}\lambda\left(\frac{N_- - N_+}{V}\right)^2\beta^2 \tag{8.9.16}$$

This degree of magnetization is achieved by transferring $(N_- - N_+)/2$ electrons from the top of the N_+ distribution to the top of the N_- distribution. The accompanying increase in kinetic energy may be estimated by considering the magnetization process at the point where a fraction α of the final number $(N_- - N_+)/2$ have been transferred. The energy of an electron at the top of the N_+ distribution is, in first order,

$$E_+ = E_F - \alpha\left(\frac{N_- - N_+}{2}\right)\frac{1}{N(E_F)} \tag{8.9.17}$$

whilst the energy of a vacant level at the top of the N_- distribution is

$$E_- = E_F + \alpha\left(\frac{N_- - N_+}{2}\right)\frac{1}{N(E_F)} \tag{8.9.18}$$

Thus the gain in kinetic energy upon transferring an additional $\{(N_- - N_+)/2\}\mathrm{d}\alpha$ electrons at this point is

$$\mathrm{d}E_k = (E_- - E_+)\left(\frac{N_- - N_+}{2}\right)\mathrm{d}\alpha \tag{8.9.19}$$

$$= \tfrac{1}{2}(N_- - N_+)^2\frac{1}{N(E_F)}\alpha\,\mathrm{d}\alpha \tag{8.9.20}$$

The total increase in kinetic energy per unit volume, which arises when the magnetization **M** of equation (8.9.16) is created, is given by

$$\Delta E_k^1 = \frac{1}{V}\int \mathrm{d}E_k = \frac{1}{2}\frac{(N_- - N_+)^2}{V}\frac{1}{N(E_F)}\int_0^1 \alpha\,\mathrm{d}\alpha \tag{8.9.21}$$

$$\Delta E_k^1 = \frac{1}{4}\frac{(N_- - N_+)^2}{V}\frac{1}{N(E_F)} \tag{8.9.22}$$

Spontaneous magnetization will occur if,

$$|E_m| > \Delta E_k^1 \tag{8.9.23}$$

that is, if

$$\frac{2\lambda\beta^2 N(E_F)}{V} > 1 \tag{8.9.24}$$

Spontaneous magnetization is therefore most likely to occur in those metals for which the electrons at the top of the Fermi distribution occupy levels which have a high density of states. Iron, cobalt and nickel are ferromagnetic metals which satisfy this requirement. For these materials, the 3*d* bands have a narrow width in energy and occur at the top of the Fermi distribution, so giving rise to a large value for $N(E_F)$.

The process of spontaneous magnetization will continue until the density of states factor $N(E)$ reaches a sufficiently small value for further electron transfer to be energetically unfavourable. The number of electrons which will have been transferred, when the process is completed, will be determined by the properties of the electron gas and need not be integrally related to the number of atoms in the material. Consequently the magnetic moment per atom need not be an integral number of Bohr magnetons. This conclusion is a significant success for the itinerant electron model. The measured magnetic moment per atom in the ferromagnetic state of iron is $2{\cdot}22\,\beta$, in cobalt the moment is $1{\cdot}72\,\beta$ and in nickel the moment is $0{\cdot}6\,\beta$. The itinerant electron model provides a natural explanation for these non-integral values. They are more difficult to account for in terms of localized spins at particular lattice sites. For this model it is necessary to visualize the ground state of each atom as a combination of different valence configurations, for example a particular nickel atom might be regarded as having a combination of $3d^9$ and $3d^{10}$ configurations. On the other hand the microscopic theory of the exchange interaction can be treated more easily with the model system of localized spins. Exchange interactions in the itinerant electron model are effectively restricted to the molecular field approximation. For many purposes, however, it is unnecessary to distinguish between these two complementary models. The macroscopic properties of a ferromagnetic material can usually be written in terms of an overall angular momentum or energy density. A molecular field treatment is adequate in these circumstances and can be regarded as a formal representation of either the itinerant or the localized electron models.

8.10 Domains in Ferromagnetic Materials

The description of magnetic materials which has been given in the preceding sections of this chapter may suggest that ferromagnetic materials will automatically exhibit a resultant moment in the temperature range below the Curie point. Such a conclusion is not in accordance with experience. For example the usual equilibrium state of a common ferromagnetic metal such as iron or nickel does not display a net magnetic moment. A large moment may, however, be easily produced by the application of a small magnetic field. The presence of a spontaneous magnetization but yet the absence of a net magnetic moment may be readily explained if the ferromagnetic material is composed of many regions. Each individual region will possess a net magnetic moment but the relative orientations of the moments will be such as to produce zero moment over all. The equilibrium configuration will be determined, in practice, by a complicated interplay of effects arising from the exchange interaction, the crystal anisotropy and magnetostrictive strains together with additional factors which arise from impurities, dislocations and atomic vacancies in the material. The present discussion will be limited to some very simple features which are sufficient to indicate that a domain structure can occur.

The development of a domain structure may be inferred from a consideration of Fig. 8.15. Here, Fig. 8.15(*a*) represents a simple, single crystal block of ferromagnetic material uniformly magnetized in the 'upward' direction, which is taken to be an easy axis of magnetization. The block constitutes a single domain and in conventional language there will be extensive magnetic pole distributions with a large external field and a correspondingly large magnetostatic energy. The magnetostatic energy can be reduced by creating two regions with oppositely directed magnetic moments as shown in

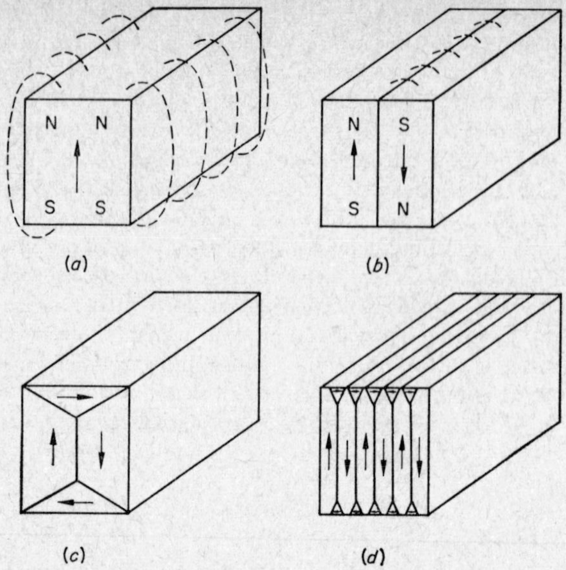

FIG. 8.15 The development of domain structures in a ferromagnetic material.

Fig. 8.15(*b*). In creating these two regions the energy has increased because of the exchange interactions within the boundary layer separating the regions but provided that the size of the block is not too small (linear dimensions not less than about 500 Å) the total energy is reduced by this process. The energy may be further reduced by creating the domains of closure shown in Fig. 8.15(*c*). Here the sample has no surface poles and no overall magnetic moment. However, there will be additional energy due to crystal anisotropy and magnetostrictive effects. The crystal anisotropy arises because, in general, the magnetization within a domain of closure will no longer lie along the easy direction. The magnetostrictive energy arises because the domains of closure are elongated along the direction of magnetization and no longer fit into the volumes assigned to them without introducing strain. These energy contributions can be reduced by increasing the number of domains as shown in Fig. 8.15(*d*). The process will continue until the total energy reaches a minimum.

8.10.1 Domain Width

To estimate the equilibrium size of a domain we require to minimize the energy arising from the crystal anisotropy, the magnetostrictive and the boundary layer (called the Bloch wall) terms. The crystal anisotropy energy and the magnetostrictive energy will be proportional to the volume, V_c, of the domains of closure. The energy arising from the creation of the Bloch walls is proportional to the total area of the walls, A_W. We may therefore write the domain contributions to the energy as

$$E = (E_A + E_L)V_c + E_W A_W \tag{8.10.1}$$

where E_A and E_L are energies per unit volume whose order of magnitude is known from the crystal anisotropy constant and the magnetoelastic coefficient. The energy per unit area of the Bloch wall, E_W, is determined by the interplay of the exchange interaction and the crystal anisotropy. Consider a block of material which has a total volume of 1 cm³ and is composed of domains, each of width D cm, as shown in Fig. 8.16.

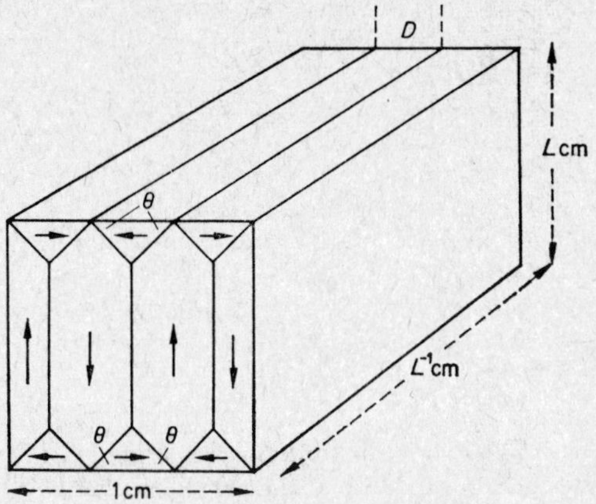

FIG. 8.16. A model domain structure.

Take the angle θ to be 45°. Then, the volume, v_c, of a prismatic domain of closure is,

$$v_c = \frac{1}{2} D \left(\frac{D}{2}\right) L^{-1} \tag{8.10.2}$$

There are $2/D$ such domains of closure in the block, and so,

$$V_c = \frac{2v_c}{D} = \frac{D}{2L} \tag{8.10.3}$$

The total area of the Bloch walls is,

$$A_W = L^{-1}(L - D)\frac{1}{D} + L^{-1}\frac{D}{\sqrt{2}}\left(\frac{4}{D}\right) \tag{8.10.4}$$

or

$$A_W = \frac{1}{D} + \frac{\{2\sqrt{(2)} - 1\}}{L} \tag{8.10.5}$$

From equation (8.10.1), the domain contribution to the energy is,

$$E = (E_A + E_L)\frac{D}{2L} + E_W\left[\frac{1}{D} + \frac{\{2\sqrt{(2)} - 1\}}{L}\right] \tag{8.10.6}$$

This is a minimum when,

$$\frac{\partial E}{\partial D} = 0$$

and

$$D = \sqrt{\left(\frac{2LE_W}{E_A + E_L}\right)} \tag{8.10.7}$$

The domain wall energy E_W is typically 1 erg cm^{-2} in the common ferromagnetic metals. For silicon-iron, which has a cubic crystal structure, the block may be cut so that all the domains are magnetized along equivalent easy directions. In this case the term E_A does not contribute to equation (8.10.1) and the domain width is determined by the magnetostrictive term E_L. Taking $E_L \sim 350$ erg cm^{-3} for silicon-iron, and $L = 1$ cm we obtain

$$D_{SiFe} \sim 8.10^{-2} \text{ cm}$$

Cobalt, on the other hand, is highly anisotropic with predominantly uniaxial symmetry and so, for this material, the domain width is determined by the anisotropy term E_A. Taking $E_A \sim 10^5$ ergs cm^{-3} and $L = 1$ cm for cobalt, we obtain

$$D_{Co} \sim 4.10^{-3} \text{ cm}$$

8.10.2 Domain Wall

The boundary layer which separates two domains is usually referred to as a domain wall or a Bloch wall. Within such a region the magnetization is changing direction continuously and the distance over which the changes take place defines the width of the wall. This width is determined by the interplay of the exchange interaction and the forces arising from crystal

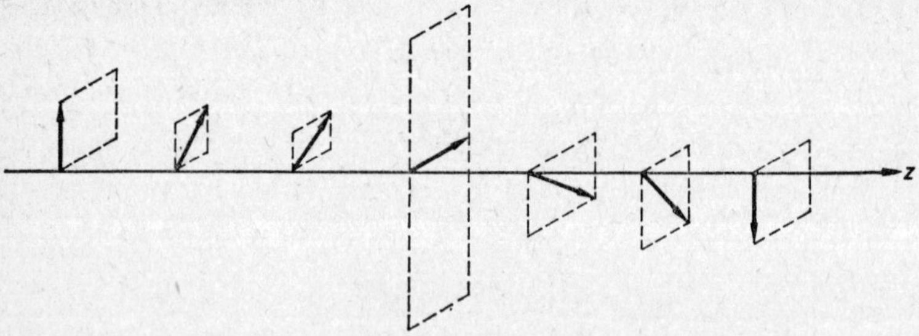

Fig. 8.17 A one-dimensional model for a domain wall.

anisotropy. The exchange interaction restrains the magnetization vectors to be as parallel as possible and so to distribute the changes over a large distance in the crystal. On the other hand, the crystal anisotropy forces act to direct the magnetization quickly into an easy direction and tend to reduce the width of the wall. The competition between these two competing processes leads to a width for the wall determined by a minimum energy configuration.

Consider a simple one-dimensional model for a domain wall in which the magnetization rotates through 180 degrees in traversing the wall. This model is illustrated in Fig. 8.17.

The magnetization vector **M** is to be regarded as constant in magnitude†

† Hence $\mathbf{M}.\mathbf{M} = M^2 =$ constant.

but varying in direction as a continuous function of z. Assume for simplicity that the vector **M** lies in a plane at right angles to the z-axis. From equation (8.7.13) the effective exchange field may be written

$$\mathbf{H}_{\text{ex}} = \lambda(\mathbf{M} + \alpha'\nabla^2\mathbf{M}) \tag{8.10.8}$$

and so the exchange energy density, which is derived from $-\mathbf{M}.\mathbf{H}_{\text{ex}}$, will have the form

$$E = -\{A + B(\mathbf{M}.\nabla^2\mathbf{M})\} \tag{8.10.9}$$

In this equation A is a constant term, derived from the factor **M.M**, which may be ignored for the present discussion where we are calculating the increase in exchange energy density due to the creation of the domain wall. Since **M.M** is constant,

$$\nabla^2(\mathbf{M}.\mathbf{M}) = 0 \tag{8.10.10}$$

and so

$$\{(\nabla M_x.\nabla M_x) + (\nabla M_y.\nabla M_y) + (\nabla M_z.\nabla M_z)\} = -\mathbf{M}.\nabla^2\mathbf{M} \tag{8.10.11}$$

Hence the exchange energy density within the domain wall, may be written as

$$E = B\{(\nabla M_x.\nabla M_x) + (\nabla M_y.\nabla M_y) + (\nabla M_z.\nabla M_z)\} \tag{8.10.12}$$

Take polar angles for the magnetization with respect to the z-axis,

$$\begin{aligned} M_x &= M\sin\theta\cos\phi = M\cos\phi \\ M_y &= M\sin\theta\sin\phi = M\sin\phi \\ M_z &= 0 \end{aligned} \tag{8.10.13}$$

When the wall has a width l, the exchange energy associated with unit area of the wall is,

$$E_{\text{ex}} = \int_0^l E\,\mathrm{d}z = \int_0^l B\left\{\left(\frac{\partial M_x}{\partial z}\right)^2 + \left(\frac{\partial M_y}{\partial z}\right)^2\right\}\mathrm{d}z \tag{8.10.14}$$

$$= BM^2\int_0^l \left(\frac{\partial\phi}{\partial z}\right)^2 \mathrm{d}z \tag{8.10.15}$$

If the magnetization varies in direction uniformly with the distance z, $(\partial\phi/\partial z)$ is constant and equal to π/l. In this case

$$E_{\text{ex}} = \frac{BM^2\pi^2}{l} \tag{8.10.16}$$

The crystal anisotropy energy per unit volume may be denoted by E_A, in accordance with equation (8.10.1), and the anisotropy energy per unit area of the domain wall will be

$$E' = E_A l \tag{8.10.17}$$

Adding together equations (8.10.16) and (8.10.17) the energy per unit area of the wall will be

$$E_{\text{W}} = E_A l + \frac{BM^2\pi^2}{l} \tag{8.10.18}$$

This energy is a minimum for

$$l = M\pi\sqrt{\left(\frac{B}{E_A}\right)} \tag{8.10.19}$$

in which case

$$E_W = \frac{2BM^2\pi^2}{l} = 2M\pi\sqrt{(E_A B)} \qquad (8.10.20)$$

The product BM^2 has a value of about 10^{-6} erg cm^{-1} (on an atomic model note that $BM^2 = (JS^2/a)$ see, for example, equation (8.4.1)) and so, if $E_A \sim 10^5$ erg cm^{-3}, the width of a domain wall is,

$$l \sim 1000 \text{ Å}$$

For a wall of this width,

$$E_W \sim 2 \text{ erg cm}^{-2}.$$

Bibliography

BLEANEY, B. I. and BLEANEY, B., *Electricity and Magnetism*, 2nd edn, O.U.P., 1965
BROWN, W. F., JR., *Magnetostatic Principles in Ferromagnetism*, North-Holland 1962.
EYRING, H., WALTER, J. and KIMBALL, G. E., *Quantum Chemistry*, Wiley, 1949
MARTIN, D. H., *Magnetism in Solids*, Iliffe, 1967
MATTIS, D. C., *The Theory of Magnetism*, Harper and Row, 1965
MORRISH, A. H., *The Physical Principles of Magnetism*, Wiley, 1965
PEIERLS, R. E., *Quantum Theory of Solids*, O.U.P., 1955
VAN VLECK, J. H., *Electric and Magnetic Susceptibilities*, O.U.P., 1932

MICROWAVE RESONANCE

ABRAGAM, A. and BLEANEY, B., *Electron Paramagnetic Resonance of Transition Ions*, O.U.P., 1970
INGRAM, D. J. E., *Spectroscopy at Radio and Microwave Frequencies*, Butterworths, 1955
LAX, B., and BUTTON, K. J., *Microwave Ferrites and Ferrimagnetics*, McGraw-Hill, 1962
PAKE, G. E., *Paramagnetic Resonance*, Benjamin, 1962
SLICHTER, C. P., *Principles of Magnetic Resonance*, Harper and Row, 1963

EXPERIMENTAL METHODS

ARAJS, S. and DUNMYRE, G. R., *Phys. Stat. Sol.* **21**, 191, 1967
BATES, L. F., *Modern Magnetism*, C.U.P., 1948
BOZORTH, R. M., *Ferromagnetism*, Van Nostrand, 1951
High Magnetic Fields, Eds H. Kolm, B. Lax, F. Bitter and R. Mills, M.I.T. Press, 1962

Chapter 9

Electrical Conduction in Solids

9.1 Introduction

In this chapter we shall examine some of the essential features of electrical conduction in crystalline solids. The discussion will be limited to the class of normal electrical conductors and we shall not attempt to account for the phenomenon of superconductivity. We shall base the microscopic theory of electrical conduction on the dynamical motion of electrons in the electric potential field which characterizes the solid. On this model the electronic motion is separable from that of the nuclei which are the basic elements of the crystal lattice. In general we shall regard the nuclei as being at rest and the electrons as moving in an electrostatic potential field. The nuclear motion which actually occurs (lattice waves) will be regarded as a small perturbation which leads to electrical resistance and Joule heating. Superconductivity is therefore immediately excluded from our discussion. This phenomenon arises when the ground state of the electronic system is directly determined by interactions involving the lattice waves.

We shall avoid giving a microscopic theory of the electron–electron interactions. On the present model two classes of electrons are to be distinguished. The first class is composed of 'core' electrons which are localized at particular nuclei and described by characteristic atomic wave functions having the electron–electron interactions already included in principle. These core electrons do not require an explicit representation since they, along with the nuclei, are formally described by the potential field of the crystal lattice. The second class of electrons are the mobile particles which are responsible for the electrical conductivity. Electron–electron interactions are included within this class by assuming, where necessary, that these electrons occupy states in accordance with Fermi–Dirac statistics. The dynamical motion of these mobile electrons, which is our immediate concern, may therefore be derived from the states of one electron moving in the potential field of the crystal.

The potential field will reflect the translation symmetry of the crystal lattice. This periodicity of the lattice may be described by means of three primitive vectors **a**, **b**, **c**. If an arbitrary point is taken as origin within the crystal, then exactly equivalent points are identified by a vector $\mathbf{R}_i$, which is given by the expression,

$$\mathbf{R}_i = l_i\mathbf{a} + m_i\mathbf{b} + n_i\mathbf{c} \tag{9.1.1}$$

where l_i, m_i, n_i are integers or zero. The space distribution described by this equation is called the Bravais lattice of the crystal† and the vectors $\mathbf{R}_i$

† Fourteen different space lattices, each obeying the general relation of equation (9.1.1), are available for actual crystals. The differences arise from the relative orientations and lengths of the vectors **a**, **b**, **c**.

are called lattice vectors. If $V(\mathbf{r})$ represents the potential energy of an electron at the point $\mathbf{r}$ in the lattice, then the point $\mathbf{r} + \mathbf{R}_i$ is an equivalent point and translational symmetry requires that

$$V(\mathbf{r} + \mathbf{R}_i) = V(\mathbf{r}) \tag{9.1.2}$$

$V(\mathbf{r})$ is therefore a periodic potential function.

9.2 Electrons in a Periodic Potential

The Hamiltonian operator for one electron with potential energy $V(\mathbf{r})$ is

$$\mathscr{H} = -\frac{\hbar^2}{2m}\nabla^2 + V(\mathbf{r}) \tag{9.2.1}$$

If $V(\mathbf{r})$ is the periodic potential function from equation (9.1.2) it is evident that the one electron wave functions must reflect the symmetry of the crystal lattice. We introduce, therefore, a set of translational operators $O(\mathbf{R}_i)$ which change the coordinates $\mathbf{r}$ of a spatial function $g(\mathbf{r})$ into $(\mathbf{r} + \mathbf{R}_i)$. Thus,

$$O(\mathbf{R}_i).g(\mathbf{r}) = g(\mathbf{r} + \mathbf{R}_i) \tag{9.2.2}$$

Because $V(\mathbf{r})$ satisfies equation (9.1.2), all the translation operators, $O(\mathbf{R}_i)$, commute with the Hamiltonian given in equation (9.2.1),

$$[\mathscr{H}, O(\mathbf{R}_i)] = 0 \tag{9.2.3}$$

It is possible therefore to choose wave functions for the one electron states which are simultaneous eigenfunctions of the Hamiltonian operator, $\mathscr{H}$, and all the translation operators $O(\mathbf{R}_i)$. Let $\psi(\mathbf{r})$ be one such eigenfunction. Then,

$$O(\mathbf{R}_i).\psi(\mathbf{r}) = \psi(\mathbf{r} + \mathbf{R}_i) = \xi_i\,\psi(\mathbf{r}) \tag{9.2.4}$$

Taking the complex conjugate of this equation, we also have,

$$\psi^*(\mathbf{r} + \mathbf{R}_i) = \xi_i^*\,\psi^*(\mathbf{r}) \tag{9.2.5}$$

and multiplying together equations (9.2.4) and (9.2.5), we have

$$|\psi(\mathbf{r} + \mathbf{R}_i)|^2 = |\xi_i|^2\,|\psi(\mathbf{r})|^2 \tag{9.2.6}$$

The electron density distribution is described by $|\psi|^2$ and this quantity must also have the periodicity of the lattice. Hence,

$$|\psi(\mathbf{r} + \mathbf{R}_i)|^2 = |\psi(\mathbf{r})|^2 \tag{9.2.7}$$

and so, from equation (9.2.6),

$$|\xi_i|^2 = 1 \tag{9.2.8}$$

The eigenvalue ξ_i is one of the complex roots of unity therefore, and we may write

$$\xi_i = e^{j\Gamma_i} \tag{9.2.9}$$

with Γ_i a dimensionless, real quantity.

Consider two successive translations $\mathbf{R}_i$ and $\mathbf{R}_m$. Evidently

$$O(\mathbf{R}_i)\,O(\mathbf{R}_m).\psi(\mathbf{r}) = O(\mathbf{R}_i + \mathbf{R}_m)\,\psi(\mathbf{r}) \tag{9.2.10}$$

and so

$$\xi_i\,\xi_m = \xi_{i+m} \tag{9.2.11}$$

This relation is satisfied if we write

$$\Gamma_i = \mathbf{k}.\mathbf{R}_i \tag{9.2.12}$$

where **k** is an arbitrary vector which has the same value for all translations.† We may therefore write equation (9.2.4) in the form,

$$\psi_{\mathbf{k}}(\mathbf{r}+\mathbf{R}_i) = e^{j\mathbf{k}.\mathbf{R}_i}\,\psi_{\mathbf{k}}(\mathbf{r}) \tag{9.2.13}$$

Here we have written in the subscript **k** to indicate that this vector is to be associated with the eigenvalue of the translation operator. Equation (9.2.13) is usually called Bloch's theorem.

The wave function $\psi_{\mathbf{k}}(\mathbf{r})$ may be written in more explicit forms which have advantages for particular applications. A very useful representation may be constructed from a subsidiary function $u_{\mathbf{k}}(\mathbf{r})$ by writing

$$\psi_{\mathbf{k}}(\mathbf{r}) = e^{j\mathbf{k}.\mathbf{r}}\,u_{\mathbf{k}}(\mathbf{r}) \tag{9.2.14}$$

In this representation $u_{\mathbf{k}}(\mathbf{r})$ has the periodicity of the crystal lattice. We have

$$\begin{aligned}\psi_{\mathbf{k}}(\mathbf{r}+\mathbf{R}_i) &= \Big[\exp\{j\mathbf{k}.(\mathbf{r}+\mathbf{R}_i)\}\Big]\,u_{\mathbf{k}}(\mathbf{r}+\mathbf{R}_i)\\ &= \exp(j\mathbf{k}.\mathbf{R}_i)\Big\{e^{\,j\mathbf{k}.\mathbf{r}}\,u_{\mathbf{k}}(\mathbf{r}+\mathbf{R}_i)\Big\}\end{aligned} \tag{9.2.15}$$

But from Bloch's theorem, equation (9.2.13),

$$\begin{aligned}\psi_{\mathbf{k}}(\mathbf{r}+\mathbf{R}_i) &= \Big\{\exp(j\mathbf{k}.\mathbf{R}_i)\Big\}\,\psi_{\mathbf{k}}(\mathbf{r})\\ &= \exp(j\mathbf{k}.\mathbf{R}_i)\Big\{e^{\,j\mathbf{k}.\mathbf{r}}\,u_{\mathbf{k}}(\mathbf{r})\Big\}\end{aligned} \tag{9.2.16}$$

We therefore require

$$u_{\mathbf{k}}(\mathbf{r}+\mathbf{R}_i) = u_{\mathbf{k}}(\mathbf{r}) \tag{9.2.17}$$

and so $u_{\mathbf{k}}(\mathbf{r})$ must have the periodicity of the crystal lattice.

The function $e^{\,j\mathbf{k}.\mathbf{r}}\,u_{\mathbf{k}}(\mathbf{r})$ is called a Bloch function. Evidently this function reduces to the simple plane wave solution $\psi_{\mathbf{k}}(\mathbf{r}) = u_0\,e^{j\mathbf{k}.\mathbf{r}}$ as the potential energy becomes uniform and $u_{\mathbf{k}}(\mathbf{r})$ tends to a constant value u_0. More generally the periodic function $u_{\mathbf{k}}(\mathbf{r})$ may be expanded as a sum of plane waves. Consider one such wave which we write as

$$f(\mathbf{r}) = \exp(j\mathbf{K}_\mu.\mathbf{r}) \tag{9.2.18}$$

where $\mathbf{K}_\mu$ is a *reciprocal lattice vector* satisfying the relation

$$\mathbf{K}_\mu.\mathbf{R}_i = 2\pi n \tag{9.2.19}$$

with n equal to an integer or zero. The vector $\mathbf{K}_\mu$ is derived from the primitive vectors $\mathbf{a}_r$, $\mathbf{b}_r$, $\mathbf{c}_r$ for the *reciprocal lattice* by a relation analogous to equation (9.1.1)

$$\mathbf{K}_\mu = l_\mu\mathbf{a}_r + m_\mu\mathbf{b}_r + n_\mu\mathbf{c}_r \tag{9.2.20}$$

with l_μ, m_μ, n_μ, integers or zero. The primitive vectors $\mathbf{a}_r$, $\mathbf{b}_r$, $\mathbf{c}_r$ for the reciprocal lattice are defined from the corresponding vectors **a**, **b**, **c**, for the direct lattice. The relations are

$$\begin{matrix}\mathbf{a}_r.\mathbf{a}=2\pi & \mathbf{a}_r.\mathbf{b}=0 & \mathbf{a}_r.\mathbf{c}=0\\ \mathbf{b}_r.\mathbf{a}=0 & \mathbf{b}_r.\mathbf{b}=2\pi & \mathbf{b}_r.\mathbf{c}=0\\ \mathbf{c}_r.\mathbf{a}=0 & \mathbf{c}_r.\mathbf{b}=0 & \mathbf{c}_r.\mathbf{c}=2\pi\end{matrix} \tag{9.2.21}$$

from which equation (9.2.19) follows immediately.

† Note that $\mathbf{k}.\mathbf{R}_i$ is dimensionless and so **k** is an arbitrary vector in 'reciprocal space'.

The function $f(\mathbf{r})$ in equation (9.2.18) has the translation symmetry of the crystal lattice. We have

$$f(\mathbf{r}+\mathbf{R}_i) = \exp\{j\mathbf{K}_\mu.(\mathbf{r}+\mathbf{R}_i)\} = e^{\,j2\pi n}\exp(j\mathbf{K}_\mu.\mathbf{r}) = f(\mathbf{r}) \qquad (9.2.22)$$

The function $u_\mathbf{k}(\mathbf{r})$ may therefore be written as a sum of these plane wave functions with appropriate coefficients,

$$u_\mathbf{k}(\mathbf{r}) = \sum_{\mathbf{K}_\mu} b_{\mathbf{k},\mathbf{K}_\mu}\exp(j\,\mathbf{K}_\mu.\mathbf{r}) \qquad (9.2.23)$$

and the corresponding Bloch function is

$$\psi_\mathbf{k}(\mathbf{r}) = e^{j\mathbf{k}.\mathbf{r}}\,u_\mathbf{k}(\mathbf{r}) = \sum_{\mathbf{K}_\mu} b_{\mathbf{k},\mathbf{K}_\mu}\exp\{j\,(\mathbf{k}+\mathbf{K}_\mu).\mathbf{r}\} \qquad (9.2.24)$$

Note that this summation is over the reciprocal lattice vectors $\mathbf{K}_\mu$.

Equation (9.2.24) is of course, only one of the possible representations which may be used to construct a wave function $\psi_\mathbf{k}(\mathbf{r})$ satisfying the Bloch theorem of equation (9.2.13). Another representation may be constructed, for example, from atomic wave functions located at particular lattice sites. Write an atomic function centred on lattice site $\mathbf{R}_\lambda$ as $\phi_a(\mathbf{r}-\mathbf{R}_\lambda)$. We may obtain a function which obeys the Bloch theorem of equation (9.2.13) by taking a combination of the atomic functions in the form

$$\psi_\mathbf{k}(\mathbf{r}) = \frac{1}{\sqrt{N}}\sum_\lambda \{\exp(j\mathbf{k}.\mathbf{R}_\lambda)\}\,\phi_a(\mathbf{r}-\mathbf{R}_\lambda) \qquad (9.2.25)$$

The atomic function $\phi_a(\mathbf{r}-\mathbf{R}_\lambda)$ refers to the atom in the crystal lattice and so will not in general be a simple atomic orbital. However, the representation of equation (9.2.25) is particularly useful, when $\phi_a(\mathbf{r}-\mathbf{R}_\lambda)$ can be written as a linear combination of a small number of free atom orbitals, ϕ^0, for example, as

$$\phi_a(\mathbf{r}-\mathbf{R}_\lambda) = \sum_\mu c_\mu\,\phi^0_\mu(\mathbf{r}-\mathbf{R}_\lambda) \qquad (9.2.26)$$

with c_μ a constant multiplier. Equation (9.2.25) becomes

$$\psi_\mathbf{k}(\mathbf{r}) = \frac{1}{\sqrt{N}}\sum_{\lambda,\mu} \{\exp(j\mathbf{k}.\mathbf{R}_\lambda)\}\,c_\mu\,\phi^0_\mu(\mathbf{r}-\mathbf{R}_\lambda) \qquad (9.2.27)$$

This type of wave function provides a representation which is closely related to the visual model in which we regard the mobile electron as spending considerable time near an atomic nucleus and 'hopping' from one lattice site to another. The functions are usually called 'tight binding' wave functions.

9.2.1 The Brillouin Zone

Equations (9.2.20) and (9.2.21) define a reciprocal space corresponding to a given crystal lattice. It is possible to construct an elementary cell in reciprocal space such that all the one electron states determined by the periodic potential of the direct crystal lattice are characterized by **k** vectors lying in this cell. This elementary cell is called the Brillouin Zone.†

To construct the Brillouin zone choose a particular lattice point in the reciprocal lattice as origin. Draw in vectors connecting the origin to other points of the reciprocal lattice and construct the planes which are the perpendicular bisectors of these vectors. The smallest cell containing the origin which is bounded by these planes is the elementary cell of the reciprocal lattice and is also the required Brillouin zone. The solid figure which corresponds to this elementary cell is usually relatively simple and the surface is determined by a small number of bisecting planes. The vector equation for a plane bounding the Brillouin zone is evidently given by‡

$$\mathbf{k}.(\tfrac{1}{2}\mathbf{K}_\mu) = \tfrac{1}{4}\,|\mathbf{K}_\mu|^2 \tag{9.2.28a}$$

or

$$2\mathbf{k}.\mathbf{K}_\mu = |\mathbf{K}_\mu|^2 \tag{9.2.28b}$$

where $\mathbf{K}_\mu$ is one of the particular reciprocal lattice vectors used for constructing the zone.

An eigenfunction $\psi_{\mathbf{k}}(\mathbf{r})$, derived from the periodic potential, is characterized by the wave vector **k**. If **k** corresponds to a point in reciprocal space which is outside the Brillouin zone it will be related to a vector $\mathbf{k}_0$ inside the zone through a reciprocal lattice vector $\mathbf{K}_\lambda$. We may write in general

$$\mathbf{k} = \mathbf{k}_0 + \mathbf{K}_\lambda \tag{9.2.29}$$

From equation (9.2.19),

$$\exp(j\mathbf{k}.\mathbf{R}_i) = \exp(j\mathbf{k}_0.\mathbf{R}_i) \tag{9.2.30}$$

and so the Bloch theorem of equation (9.2.13) is satisfied with the same eigenvalue ξ_i for both **k** and $\mathbf{k}_0$. We may therefore regard the vector indices **k**, $\mathbf{k}_0$ as equivalent from this point of view and restrict the values of **k** to lie within the Brillouin zone. The energy values and eigenfunctions for the different values of **k** which correspond to the same value of $\mathbf{k}_0$ will not be the same, however, and so, if **k** is restricted to lie within the Brillouin zone a given value of $\mathbf{k}_0$ will correspond to a number of possible energy states. For this reason it is customary to write the wave functions as $\psi_{n,\mathbf{k}_0}(\mathbf{r})$ where the subscript n, called the 'band index', identifies the energy and the eigenfunction for the state.

As an example of this general discussion consider a crystal lattice which has the body-centred cubic structure.§ Taking rectangular coordinate axes with respect to a corner of the unit cell, as shown in Fig. 9.1, it appears

† More correctly we should call this zone the first Brillouin zone. Higher order Brillouin zones may be defined but these are not relevant to our discussion.

‡ Recall that **k** is an arbitrary position vector in reciprocal space.

§ Metals which have this type of crystal structure are the alkali metals with one mobile electron per atom, Ba with two mobile electrons per atom, and the more complex metals V, Cr, Mo and W.

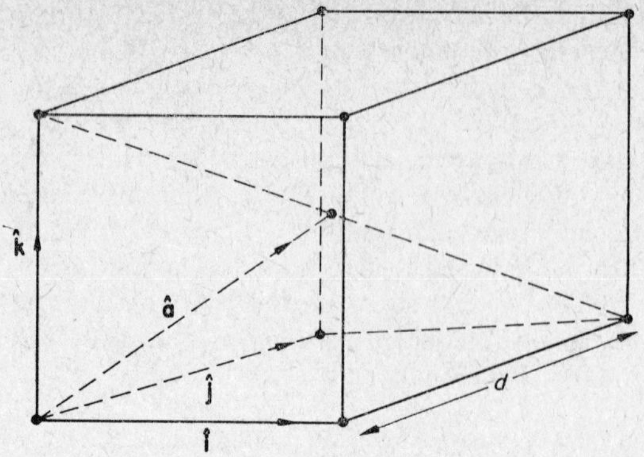

FIG. 9.1 A body-centred cubic crystal lattice primitive cell.

that this crystal structure is formed from two interpenetrating simple cubic lattices, generated by the relations

$$\mathbf{R} = l\hat{\mathbf{i}}d + m\hat{\mathbf{j}}d + n\hat{\mathbf{k}}d$$
$$\mathbf{R}' = (l + \tfrac{1}{2})\hat{\mathbf{i}}d + (m + \tfrac{1}{2})\hat{\mathbf{j}}d + (n + \tfrac{1}{2})\hat{\mathbf{k}}d \qquad (9.2.31)$$

where l, m, n are integers and d is the length of a cube edge. If, however, we choose as primitive vectors the set

$$\mathbf{a} = \left(\frac{d}{2}\right)(\hat{\mathbf{i}} + \hat{\mathbf{j}} + \hat{\mathbf{k}})$$
$$\mathbf{b} = \left(\frac{d}{2}\right)(\hat{\mathbf{i}} + \hat{\mathbf{j}} - \hat{\mathbf{k}}) \qquad (9.2.32)$$
$$\mathbf{c} = \left(\frac{d}{2}\right)(\hat{\mathbf{i}} - \hat{\mathbf{j}} + \hat{\mathbf{k}})$$

all points in the crystal are described by the one equation,

$$\mathbf{R}_i = l_i\mathbf{a} + m_i\mathbf{b} + \mathbf{n}_i\mathbf{c} \qquad (9.2.33)$$

with l_i, m_i, n_i integers. Equations (9.2.32) and (9.2.33) therefore define a Bravais lattice for the body-centred cubic crystal.

The primitive vectors for the reciprocal lattice are constructed to satisfy equations (9.2.21).† A suitable set is,

$$\mathbf{a}_r = \left(\frac{2\pi}{d}\right)(\hat{\mathbf{j}} + \hat{\mathbf{k}})$$
$$\mathbf{b}_r = \left(\frac{2\pi}{d}\right)(\hat{\mathbf{i}} - \hat{\mathbf{k}}) \qquad (9.2.34)$$
$$\mathbf{c}_r = \left(\frac{2\pi}{d}\right)(\hat{\mathbf{i}} - \hat{\mathbf{j}})$$

† The vector solutions for equations (9.2.21) are

$$\mathbf{a}_r = \frac{2\pi(\mathbf{b}\wedge\mathbf{c})}{\mathbf{a}.(\mathbf{b}\wedge\mathbf{c})},\ \mathbf{b}_r = \frac{2\pi(\mathbf{c}\wedge\mathbf{a})}{\mathbf{a}.(\mathbf{b}\wedge\mathbf{c})},\ \mathbf{c}_r = \frac{2\pi(\mathbf{a}\wedge\mathbf{b})}{\mathbf{a}.(\mathbf{b}\wedge\mathbf{c})}$$

Note that these vector relations also define a set of primitive vectors $\mathbf{a}_r$, $\mathbf{b}_r$, $\mathbf{c}_r$, for the Bravais lattice corresponding to a face-centred cubic crystal. The reciprocal lattice for a body-centred cubic crystal is therefore a face-centred cubic network. The Brillouin zone based on equations (9.2.34) is sketched in Fig. 9.2. In this diagram the reference axes are the vector directions $\hat{\mathbf{i}}$, $\hat{\mathbf{j}}$ and $\hat{\mathbf{k}}$. In terms of these vectors,

$$\mathbf{K}_\lambda = \left(\frac{2\pi}{d}\right)\{(m_\lambda + n_\lambda)\hat{\mathbf{i}} + (l_\lambda - n_\lambda)\hat{\mathbf{j}} + (l_\lambda - m_\lambda)\hat{\mathbf{k}}\} \qquad (9.2.35)$$

It is customary to label the zone centre by the symbol Γ and the (100), (111) and (110) axes by Δ, Λ and Σ respectively. The solid figure corresponding to the Brillouin zone is a rhombic dodecahedron and the principal symmetry points, which are frequently used in band structure calculations, are assigned the symbols Γ, H, P and N.

In order to illustrate the one electron energy states for the body-centred cubic crystal consider the limiting case where the periodic potential becomes

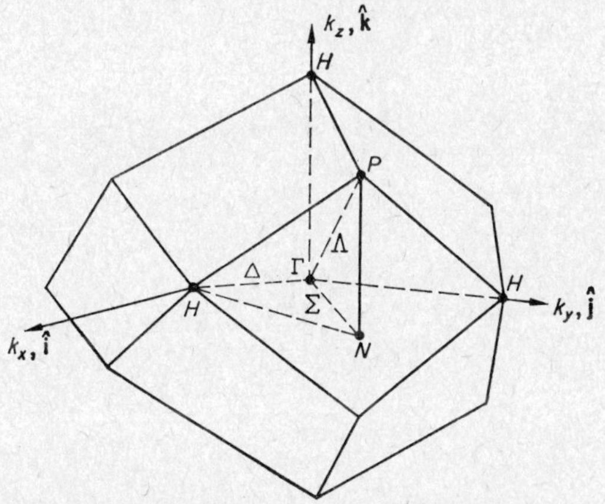

FIG. 9.2 The Brillouin zone for the body-centred cubic lattice; symmetry points and axes are indicated by their appropriate symbols.

very weak but the symmetry properties are preserved in the formulation o the wave functions. We shall have a free electron wave function

$$\psi_{\mathbf{k}}(\mathbf{r}) \sim e^{j\mathbf{k}.\mathbf{r}}$$

with $\mathbf{k}$ spanning the whole of reciprocal space.

In this representation the energy is a simple quadratic function of k, we have not yet emphasized the translation symmetry of the crystal lattice. From equation (9.2.1),

$$\mathscr{H}\,\psi_{\mathbf{k}}(\mathbf{r}) = -\frac{\hbar^2}{2m}\nabla^2\psi_{\mathbf{k}}(\mathbf{r}) = \frac{\hbar^2}{2m}|\mathbf{k}|^2\,\psi_{\mathbf{k}}(\mathbf{r}) \qquad (9.2.36a)$$

$$E(\mathbf{k}) = \frac{\hbar^2}{2m}|\mathbf{k}|^2 \qquad (9.2.36b)$$

We now wish to take account of the symmetry of the body-centred cubic lattice and to restrict the representation to wave vectors lying inside the Brillouin zone. If $\mathbf{k}_0$ is a vector inside the zone, we may write in general, from equation (9.2.29)

$$\mathbf{k} = \mathbf{k}_0 + \mathbf{K}_\lambda$$

and $\psi_{\mathbf{k}}(\mathbf{r})$ takes the form of a Bloch function,

$$\begin{aligned}\psi_{n,\mathbf{k}}(\mathbf{r}) &= \exp(\mathrm{j}\mathbf{k}_0.\mathbf{r}) \exp(\mathrm{j}\,\mathbf{K}_\lambda.\mathbf{r}) \\ &= \left\{\exp(\mathrm{j}\mathbf{k}_0.\mathbf{r})\right\} u_{n,\mathbf{k}_0}(\mathbf{r})\end{aligned} \tag{9.2.37}$$

Here $u_{n,\mathbf{k}_0}(\mathbf{r})$ obviously has the translation symmetry of the lattice and the band index n is determined by the particular choice of reciprocal lattice vector $\mathbf{K}_\lambda$. The energy is obtained from equation (9.2.36a),

$$E_n(\mathbf{k}_0) = \frac{\hbar^2}{2m} |\mathbf{k}_0 + \mathbf{K}_\lambda|^2 \tag{9.2.38}$$

which we may write in terms of the vector components as

$$E_n(\mathbf{k}_0) = \frac{\hbar^2}{2m} \left\{(k_{0x} - K_{\lambda x})^2 + (k_{0y} - K_{\lambda y})^2 + (k_{0z} - K_{\lambda z})^2\right\} \tag{9.2.39}$$

with the x-, y-, z-axes oriented along the vector directions $\hat{\mathbf{i}}, \hat{\mathbf{j}}, \hat{\mathbf{k}}$.

Consider the energy as a function of $\mathbf{k}_0$ along the (100) direction in the Brillouin zone which was identified by the symbol Δ in Fig. 9.2. For this direction $\mathbf{k}_0 = \hat{\mathbf{i}}k_{0x}$ and $k_{0y} = k_{0z} = 0$. The lowest band is characterized by $\mathbf{K}_\lambda = 0$, and we have

$$E_1(\mathbf{k}_0) = \frac{\hbar^2}{2m} k_{0x}^2 \tag{9.2.40}$$

This band terminates at the H point of the Brillouin zone, where $k_{0x} = (2\pi/d)$ $k_{0y} = k_{0z} = 0$. Hence, along the symmetry direction Δ, $E_1(\mathbf{k}_0)$ is a quadratic function of k_{0x}, increasing continuously from zero at the zone centre to the value

$$E_1\left(\frac{2\pi}{d}\right) = \frac{\hbar^2}{2m}\left(\frac{2\pi}{d}\right)^2$$

at the H point. The next set of bands is generated by taking reciprocal lattice vectors of the type $(2\pi/d)(1, 1, 0)$ in the $\hat{\mathbf{i}}, \hat{\mathbf{j}}, \hat{\mathbf{k}}$, representation. For these bands we obtain from equation (9.2.39)

$$\begin{aligned}E_2(\mathbf{k}_0) &= \frac{\hbar^2}{2m}\left\{\left(k_{0x} - \frac{2\pi}{d}\right)^2 + \left(\frac{2\pi}{d}\right)^2\right\} \\ E_3(\mathbf{k}_0) &= \frac{\hbar^2}{2m}\left\{k_{0x}^2 + \left(\frac{2\pi}{d}\right)^2 + \left(\frac{2\pi}{d}\right)^2\right\} \\ E_4(k_0) &= \frac{\hbar^2}{2m}\left\{\left(k_{0x} + \frac{2\pi}{d}\right)^2 + \left(\frac{2\pi}{d}\right)^2\right\}\end{aligned} \tag{9.2.41}$$

There are four possible reciprocal lattice vectors for each of these equations and so each band is fourfold degenerate. Note that the band $E_2(\mathbf{k}_0)$ decreases in energy from the value

$$E_2(0) = 2\,\frac{\hbar^2}{2m}\left(\frac{2\pi}{d}\right)^2$$

at the zone centre to meet the band $E_1(\mathbf{k}_0)$ at the H point.

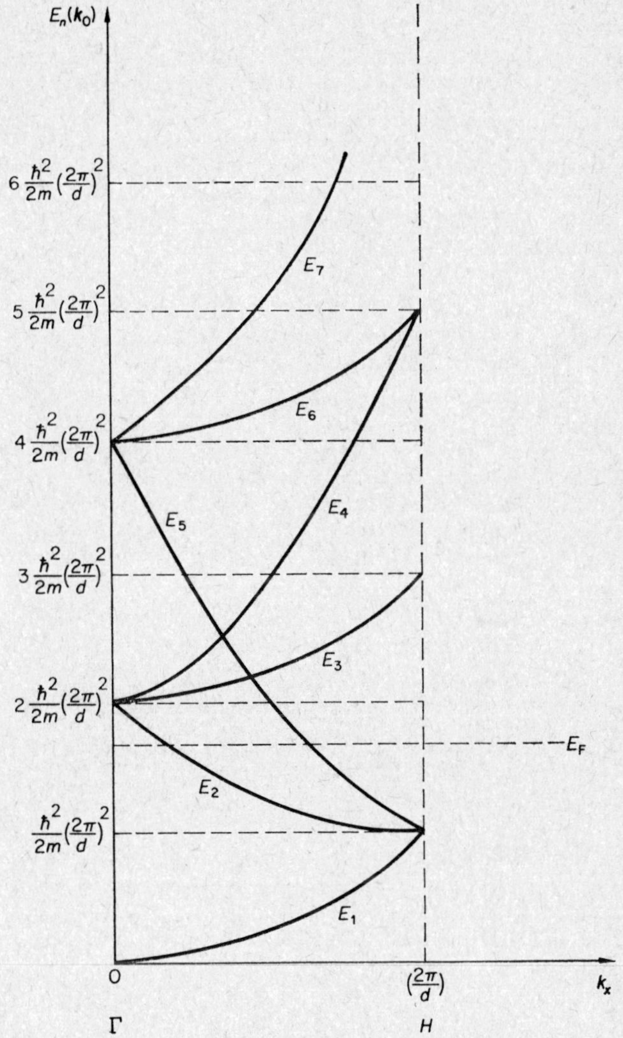

FIG. 9.3 Energy bands for a free electron in a body-centred cubic crystal. The energy is plotted as a function of wave vector along a (1, 0, 0) direction in the Brillouin zone.

Another set of bands may be generated by taking reciprocal lattice vectors of the type $(2\pi/d)$ (2, 0, 0).

$$E_5(\mathbf{k}_0) = \frac{\hbar^2}{2m}\left(k_{0x} - \frac{4\pi}{d}\right)^2$$

$$E_6(\mathbf{k}_0) = \frac{\hbar^2}{2m}\left\{k_{0x}^2 + \left(\frac{4\pi}{d}\right)^2\right\} \tag{9.2.42}$$

$$E_7(\mathbf{k}_0) = \frac{\hbar^2}{2m}\left(k_{0x} + \frac{4\pi}{d}\right)^2$$

For these energy bands, $E_5(\mathbf{k}_0)$ and $E_7(\mathbf{k}_0)$ are non-degenerate whereas $E_6(\mathbf{k}_0)$ is fourfold degenerate. The complete set of bands described by equations (9.2.40), (9.2.41) and (9.2.42) are sketched in Fig. 9.3.

A similar calculation may be carried through for the face-centred cubic crystal structure.† The primitive vectors for the reciprocal lattice (which is a body-centred cubic network), are given by

$$\mathbf{a}_\mathrm{r} = \left(\frac{2\pi}{d}\right)(\hat{\mathbf{i}} + \hat{\mathbf{j}} + \hat{\mathbf{k}})$$

$$\mathbf{b}_\mathrm{r} = \left(\frac{2\pi}{d}\right)(\hat{\mathbf{i}} + \hat{\mathbf{j}} - \hat{\mathbf{k}}) \tag{9.2.43}$$

$$\mathbf{c}_\mathrm{r} = \left(\frac{2\pi}{d}\right)(\hat{\mathbf{i}} - \hat{\mathbf{j}} + \hat{\mathbf{k}})$$

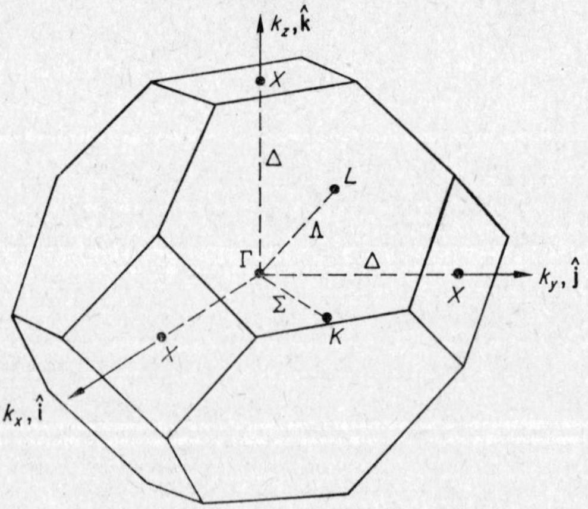

FIG. 9.4 The Brillouin zone for the body-centred cubic lattice; symmetry points and axes are indicated by their appropriate symbols.

The Brillouin zone for this structure is sketched in Fig. 9.4. The simplest reciprocal lattice vectors are now of the type $(2\pi/d)(1, 1, 1)$ and $(2\pi/d)(2, 0, 0)$ when referred to the vector axes $\hat{\mathbf{i}}$, $\hat{\mathbf{j}}$, $\hat{\mathbf{k}}$. The coordinates of the symmetry point X are therefore like $(2\pi/d, 0, 0)$. The energy bands for a (1, 0, 0) direction in the Brillouin zone are shown in Fig. 9.5.

† Metals which have this type of crystal structure are Cu, Ag, Au, Al, Ni, Pd and Pt.

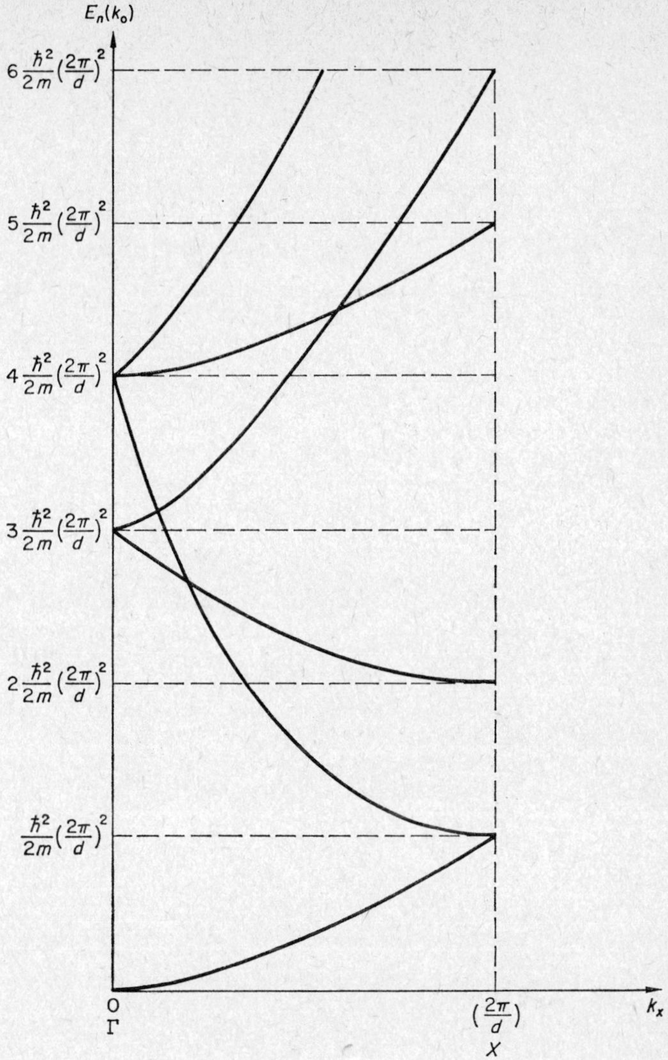

FIG. 9.5 Energy bands for a free electron in a face-centred cubic crystal. The energy is plotted as a function of wave vector along a (1, 0, 0) direction in the Brillouin zone.

The present discussion has been based on a simple free electron model with a vanishingly small periodic potential. The principal effect of including a finite periodic potential in the calculation of the energy bands is the removal of many of the degeneracies. In addition, the slopes of the energy bands are usually zero at the centre and the boundaries of the Brillouin zone. There may now be ranges of energy in which no electron states are available. These are the 'energy gaps' which play a particularly important role in the discussion of the electrical properties of semiconductors and also provide the basis for distinguishing metallic conductors from insulating solids. The band structure of silicon which is sketched in Fig. 9.6 illustrates some of these features.

Finally we note that, in principle, reciprocal space is filled with a set of Brillouin zones, one centred on each lattice point. In this complete representation a given energy band $E_n(\mathbf{k})$ is periodic in $\mathbf{k}$ repeating its values in each zone as $\mathbf{k}$ spans the whole of reciprocal space. The restriction of $\mathbf{k}$ to

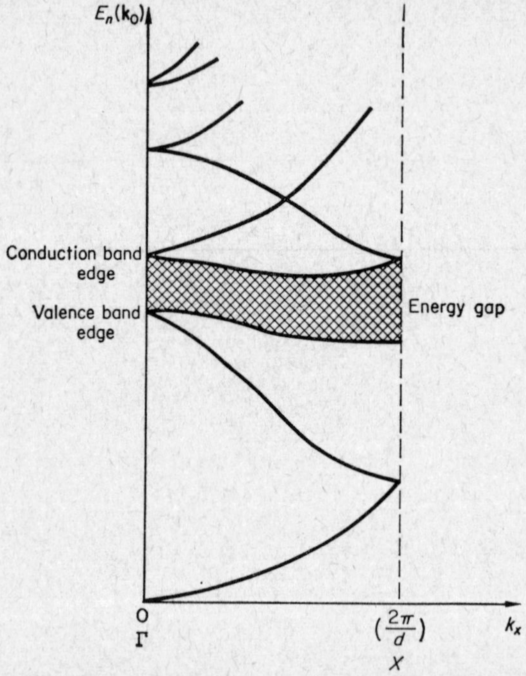

FIG. 9.6 The energy band structure of silicon (after L. Kleinman and J. C. Phillips, *Phys. Rev.* **118,** 1963 (1960)). The shaded area indicates the energy gap between occupied and unoccupied states in the pure crystal.

values $\mathbf{k}_0$ inside the Brillouin zone is therefore a convention, but for our purposes it is a useful and practical simplification of the visual model for the energy bands which we shall require.

9.2.2 The Fermi Surface

The electrons responsible for the electrical conductivity of a solid may usually be regarded as moving along trajectories which have essentially constant energy. The surfaces of constant energy in wave vector space are therefore particularly important for describing the motion of these electrons. Of these surfaces the most important is the Fermi surface which is well defined at the absolute zero of temperature and bounds all those states which are occupied (see Section 8.9, page 442). When we use the representation for $\psi_{\mathbf{k}}(\mathbf{r})$ in which the wave vector spans the whole of reciprocal space the energy is a single valued function of $\mathbf{k}$ and the Fermi surface is straightforwardly the boundary surface which encloses the occupied states. The Fermi surface of this unrestricted representation may, however, be mapped into the Brillouin zone by translations which involve a reciprocal lattice vector. Within the Brillouin zone the energy is a multivalued function of $\mathbf{k}_0$, consequently the reduced Fermi surface will have a complicated topology and it is generally important to assign the different segments of the surface to their appropriate energy bands.

We may illustrate the characteristic features of the Fermi surface by considering the limiting case of the free electron model which was discussed in the previous section (see pages 457–461). Imagine that we gradually add

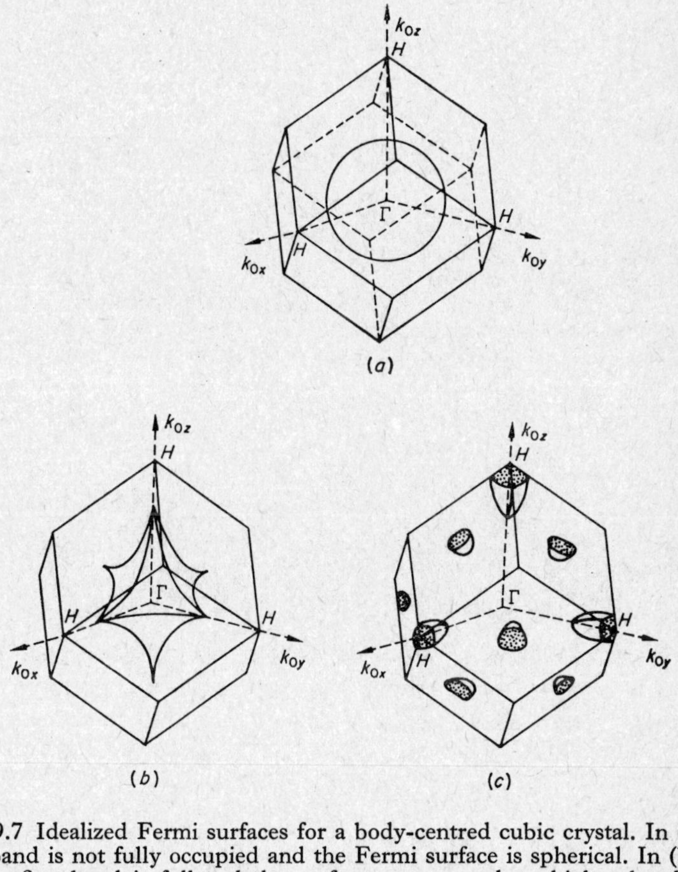

FIG. 9.7 Idealized Fermi surfaces for a body-centred cubic crystal. In (a) the first band is not fully occupied and the Fermi surface is spherical. In (b) and (c) the first band is full and the surfaces correspond to higher bands. For example, (b) might represent the band $E_2(\mathbf{k}_0)$ and (c) the bands $E_3(\mathbf{k}_0)$, $E_5(\mathbf{k}_0)$. Regions bounded by convex surfaces are occupied.

electrons to a lattice which is initially 'empty'. The Fermi surface is at first spherical, with radius

$$k_{0\mathrm{F}} = \left(\frac{2mE_{\mathrm{F}}}{\hbar^2}\right)^{1/2}$$

The electrons conform to the Fermi–Dirac distribution, and so, as their number increases the Fermi sphere will expand until eventually it touches the surface of the Brillouin zone. For a body-centred cubic crystal this will occur first of all at the N point which has coordinates $(\pi/d)(1,1,0)$. The energy bands for a (110) direction in the Brillouin zone are somewhat similar to those sketched in Fig. 9.3 but of course with a different scale (Fig. 9.3 refers

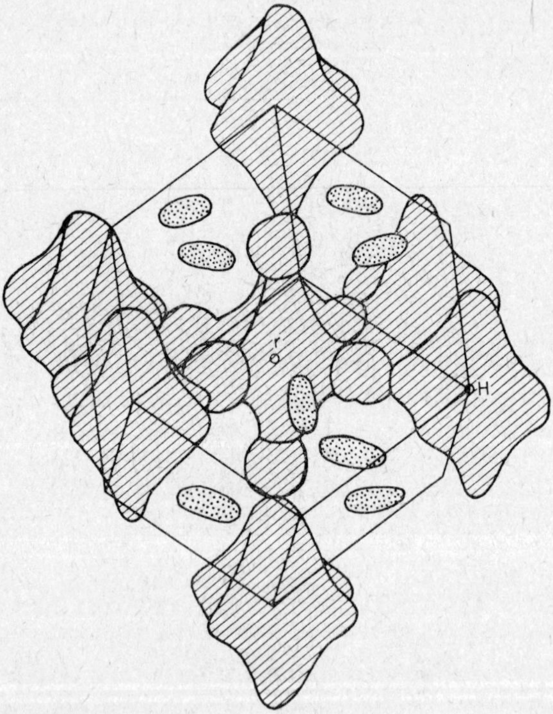

FIG. 9.8 The Fermi surfaces for a body-centred cubic metal such as Cr, Mo or W, as drawn by L. F. Mattheiss, *Phys. Rev.* **139,** A1893(1965), from the calculations of N. Lomer, *Proc. Phys. Soc.* **84,** 327, (1964). In this figure the central 'jack' bounds states occupied by electrons whilst the octahedral regions at H and the lenses at N bound regions of bands unoccupied by electrons.

to a (100) direction in the Brillouin zone). Consequently, as more electrons are added some will continue to go into the first band filling states with wave vectors differing from N and others will occupy states in the next energy band. The Fermi surface for the electrons in the second band will recede from the boundary of the Brillouin zone, a process which is sometimes described as 'reflection' of the Fermi surface from the zone boundary. When, for example, the Fermi energy occurs at the value shown in Fig. 9.3, it is evident that, for electrons with wave vectors corresponding to points on the (100) axis, the first band $E_1(\mathbf{k}_0)$ is completely filled and bands $E_2(\mathbf{k}_0)$, $E_5(\mathbf{k}_0)$, are partially occupied. Segments of the Fermi surface may therefore be drawn within the Brillouin zone which map out this constant energy surface for the different bands. The topology of these surfaces is now quite complicated. Typical examples are sketched in Figs. 9.7 and 9.8.

9.2.3 Electron Velocity, Acceleration and Effective Mass

The velocity of an electron in a state with wave vector $\mathbf{k} = \mathbf{k}'$ may be derived by constructing a wave packet from a group of Bloch functions with wave vectors centred on $\mathbf{k}'$. The group velocity of this wave packet is identified with the particle velocity of the electron. We shall need to write the wave function as a travelling wave and so we must include the time dependence explicitly. An electron in the band, labelled n, in the Brillouin zone representation will have energy $E_n(\mathbf{k})$ and a time-dependent wave function

$$\Psi_{n,\mathbf{k}}(\mathbf{r}, t) = \psi_{n,\mathbf{k}}(\mathbf{r}) \exp\left\{-\mathrm{j}\frac{E_n(\mathbf{k})}{\hbar}t\right\} \tag{9.2.44}$$

$$= u_{n,\mathbf{k}}(\mathbf{r}) \exp\left[\mathrm{j}\left\{\mathbf{k}.\mathbf{r} - \frac{E_n(\mathbf{k})}{\hbar}t\right\}\right] \tag{9.2.45}$$

The corresponding wave packet is a combination of these waves, similar to the representation used previously for a group of electromagnetic waves (see Section 2.8). We write

$$W_{n,\mathbf{k}'}(\mathbf{r}, t) = \int a_n(\mathbf{k})\, u_{n,\mathbf{k}}(\mathbf{r}) \exp\left[\mathrm{j}\left\{\mathbf{k}.\mathbf{r} - \frac{E_n(\mathbf{k})}{\hbar}t\right\}\right] \mathrm{d}^3\mathbf{k} \tag{9.2.46}$$

The distribution function $a_n(\mathbf{k})$ is assumed to be sharply peaked about the chosen value of $\mathbf{k} = \mathbf{k}'$. If $u_{n,\mathbf{k}}(\mathbf{r})$ varies slowly with $\mathbf{k}$ this function may be taken outside the integral sign, to give

$$W_{n,\mathbf{k}'}(\mathbf{r}, t) = u_{n,\mathbf{k}'}(\mathbf{r}) \int_{\Delta k} a_n(\mathbf{k}) \exp\left[\mathrm{j}\left\{\mathbf{k}.\mathbf{r} - \frac{E_n(\mathbf{k})}{\hbar}t\right\}\right] \mathrm{d}^3\mathbf{k} \tag{9.2.47}$$

Expand $E_n(\mathbf{k})$ to first order about $\mathbf{k} = \mathbf{k}'$,

$$E_n(\mathbf{k}) = E_n(\mathbf{k}') + \{\nabla_{\mathbf{k}} E_n(\mathbf{k})\}_{\mathbf{k}'}.(\mathbf{k} - \mathbf{k}') \tag{9.2.48}$$

Equation (9.2.47) may now be written as

$$W_{n,\mathbf{k}'}(\mathbf{r}, t) = u_{n,\mathbf{k}'}(\mathbf{r}) \exp\left[\mathrm{j}\left\{\mathbf{k}'.\mathbf{r} - \frac{E_n(\mathbf{k}')}{\hbar}t\right\}\right]$$
$$\times \int_{\Delta\mathbf{k}} a_n(\mathbf{k}) \exp\left[\!\left[\mathrm{j}\left[\mathbf{r} - \frac{\{\nabla_{\mathbf{k}} E_n(\mathbf{k})\}_{\mathbf{k}'}}{\hbar}t\right].(\mathbf{k} - \mathbf{k}')\right]\!\right] \mathrm{d}^3\mathbf{k} \tag{9.2.49}$$

Evidently the wave packet which we have constructed has the form of a fundamental wave function $\Psi_{n,\mathbf{k}'}(\mathbf{r}, t)$ modulated by an amplitude function $A_n(\mathbf{k}')$,

$$A_n(\mathbf{k}') = \int_{\Delta\mathbf{k}} a_n(\mathbf{k}) \exp\left[\!\left[\mathrm{j}\left[\mathbf{r} - \frac{\{\nabla_{\mathbf{k}} E_n(\mathbf{k})\}_{\mathbf{k}'}}{\hbar}t\right]\right]\!\right].(\mathbf{k} - \mathbf{k}')\, \mathrm{d}^3\mathbf{k} \tag{9.2.50}$$

This amplitude function has the same value for all points which satisfy the equation

$$\mathbf{r} - \frac{\{\nabla_{\mathbf{k}} E_n(\mathbf{k})\}_{\mathbf{k}'}}{\hbar} t = \text{constant} \tag{9.2.51}$$

These points move with the constant velocity

$$\mathbf{V}(\mathbf{k}') = \frac{1}{\hbar} \{\nabla_{\mathbf{k}} E_n(\mathbf{k})\}_{\mathbf{k}'} \tag{9.2.52}$$

which defines the group velocity for the wave packet $W_{n,\mathbf{k}'}(\mathbf{r}, t)$ and the particle velocity for the electron state with wave vector $\mathbf{k} = \mathbf{k}'$.

Note that for free electrons,

$$E_n(\mathbf{k}) = \frac{\hbar^2 |\mathbf{k}^2|}{2m} = \frac{\hbar^2}{2m}(k_x^2 + k_y^2 + k_z^2)$$

and so

$$|\mathbf{V}(\mathbf{k})| = \sqrt{\left\{\frac{2E(\mathbf{k})}{m}\right\}} \tag{9.2.53}$$

which coincides with the classical result for a particle of mass m. This is a special case, however, and equation (9.2.52) also shows that in a solid the electron velocity does not necessarily increase with increasing energy. For example, near the top of a band the slope of $E_n(\mathbf{k})$ may decrease and the velocity will also decrease, falling to zero if $\nabla_{\mathbf{k}} E_n(\mathbf{k})$ vanishes at the boundary of the Brillouin zone.

The equation of motion for an electron under the action of a perturbing force $\mathbf{F}$ may be derived from the work done in unit time. The relation between the work done and the increase in energy is

$$\frac{\mathrm{d}}{\mathrm{d}t} E_n(\mathbf{k}) = \mathbf{F}.\mathbf{V}(\mathbf{k}) \tag{9.2.54}$$

In addition,

$$\frac{\mathrm{d}}{\mathrm{d}t} E_n(\mathbf{k}) = \nabla_{\mathbf{k}} E_n(\mathbf{k}).\frac{\mathrm{d}\mathbf{k}}{\mathrm{d}t} \tag{9.2.55}$$

Assuming that the electron remains within the same band labelled, n, and that the perturbing force does not destroy the description in terms of Bloch functions, we obtain from equations (9.2.52), (9.2.54) and (9.2.55),

$$\hbar \frac{\mathrm{d}\mathbf{k}}{\mathrm{d}t} = \mathbf{F} \tag{9.2.56}$$

We regard this relation as the equation of motion for an electron which is described by a wave function with wave vector $\mathbf{k}$.

Comparison of equation (9.2.56) with the corresponding classical equation of motion, involving the momentum of the particle, leads to the concept of effective mass, m^*. We seek to write an equation of the form,

$$\hbar \frac{\mathrm{d}\mathbf{k}}{\mathrm{d}t} = \mathbf{F} = m^* \frac{\mathrm{d}\mathbf{V}}{\mathrm{d}t} \tag{9.2.57}$$

and an acceleration,

$$\frac{\mathrm{d}\mathbf{V}}{\mathrm{d}t} = \frac{\hbar}{m^*}\frac{\mathrm{d}\mathbf{k}}{\mathrm{d}t} = \frac{1}{m^*}\mathbf{F} \tag{9.2.58}$$

However, for an electron in a crystal, we should no longer restrict m^* to be a simple isotropic, scalar quantity. We may expect m^*, now, to reflect the symmetry of the crystal lattice. From equation (9.2.52), write

$$\frac{\mathrm{d}}{\mathrm{d}t}\mathbf{V}(\mathbf{k}) = \frac{1}{\hbar}\frac{\mathrm{d}}{\mathrm{d}t}\{\nabla_{\mathbf{k}}\, E_n(\mathbf{k})\} \tag{9.2.59}$$

Consider the x component of this equation,

$$\frac{\mathrm{d}V_x(\mathbf{k})}{\mathrm{d}t} = \frac{1}{\hbar}\left[\frac{\partial^2 E_n}{\partial k_x^2}\frac{\mathrm{d}k_x}{\mathrm{d}t} + \frac{\partial^2 E_n}{\partial k_y\,\partial k_x}\frac{\mathrm{d}k_y}{\mathrm{d}t} + \frac{\partial^2 E_n}{\partial k_z\,\partial k_x}\frac{\mathrm{d}k_z}{\mathrm{d}t}\right] \tag{9.2.60}$$

Taking into account similar equations for the components $\mathrm{d}V_y(\mathbf{k})/\mathrm{d}t$, $\mathrm{d}V_z(\mathbf{k})/\mathrm{d}t$, the complete relation between $\mathrm{d}\mathbf{V}(\mathbf{k})/\mathrm{d}t$ and $\mathrm{d}\mathbf{k}/\mathrm{d}t$ may be written in matrix form as,

$$\frac{\mathrm{d}}{\mathrm{d}t}\begin{bmatrix} V_x(\mathbf{k}) \\ V_y(\mathbf{k}) \\ V_z(\mathbf{k}) \end{bmatrix} = \frac{1}{\hbar}\begin{bmatrix} \dfrac{\partial^2 E_n}{\partial k_x^2} & \dfrac{\partial^2 E_n}{\partial k_y\,\partial k_x} & \dfrac{\partial^2 E_n}{\partial k_z\,\partial k_x} \\ \dfrac{\partial^2 E_n}{\partial k_x\,\partial k_y} & \dfrac{\partial^2 E_n}{\partial k_y{}^2} & \dfrac{\partial^2 E_n}{\partial k_z\,\partial k_y} \\ \dfrac{\partial^2 E_n}{\partial k_x\,\partial k_z} & \dfrac{\partial^2 E_n}{\partial k_y\,\partial k_z} & \dfrac{\partial^2 E_n}{\partial k_z^2} \end{bmatrix}\begin{bmatrix} \dfrac{\mathrm{d}k_x}{\mathrm{d}t} \\ \dfrac{\mathrm{d}k_y}{\mathrm{d}t} \\ \dfrac{\mathrm{d}k_z}{\mathrm{d}t} \end{bmatrix} \tag{9.2.61}$$

Comparing this equation with equation (9.2.58) it is evident that the motion of the electron in the crystal may be described in terms of an equivalent classical acceleration formula if we define an effective mass tensor by the relation

$$\left[\frac{1}{m^*}\right] = \frac{1}{\hbar^2}\begin{bmatrix} \dfrac{\partial^2 E_n}{\partial k_x^2} & \dfrac{\partial^2 E_n}{\partial k_y\,\partial k_x} & \dfrac{\partial^2 E_n}{\partial k_z\,\partial k_x} \\ \dfrac{\partial^2 E_n}{\partial k_x\,\partial k_y} & \dfrac{\partial^2 E_n}{\partial k_y^2} & \dfrac{\partial^2 E_n}{\partial k_z\,\partial k_y} \\ \dfrac{\partial^2 E_n}{\partial k_x\,\partial k_z} & \dfrac{\partial^2 E_n}{\partial k_y\,\partial k_z} & \dfrac{\partial^2 E_n}{\partial k_z^2} \end{bmatrix} \tag{9.2.62}$$

Equation (9.2.58) may now be written as

$$\frac{\mathrm{d}\mathbf{V}}{\mathrm{d}t} = \left[\frac{1}{m^*}\right].\mathbf{F} \tag{9.2.63}$$

with the tensor $[1/m^*]$ playing the part of the reciprocal of the inertial mass in classical dynamics. For example, if the energy function has a simple quadratic dependence upon $|\mathbf{k}|$,

$$E_n(\mathbf{k}) = A\,|\mathbf{k}|^2 = A\,(k_x^2 + k_y^2 + k_z^2) \tag{9.2.64}$$

the effective mass is an isotropic scalar quantity, given by

$$\frac{1}{m^*} = \frac{1}{\hbar^2}\frac{\partial^2 E_n}{\partial k_x^2} = \frac{1}{\hbar^2}\frac{\partial^2 E_n}{\partial k_y^2} = \frac{1}{\hbar^2}\frac{\partial^2 E_n}{\partial k_z^2} = \frac{2A}{\hbar^2} \tag{9.2.65}$$

We may therefore write equation (9.2.64) in the familiar form

$$E_n(\mathbf{k}) = \frac{\hbar^2 \, |\mathbf{k}|^2}{2m^*} \tag{9.2.66}$$

which corresponds directly with the classical energy equation for a particle if we regard $\hbar\mathbf{k}$ as the momentum and m^* as the inertial mass. The analogy between m^* and the classical inertial mass is not so direct, however, for more complicated energy functions. In general we should no longer expect the effective mass to be isotropic nor should we expect it to be a necessarily positive quantity. At the top of a band we may expect $(\partial^2 E_n/\partial k_i \, \partial k_j)$ to be negative and give rise to negative components of the effective mass. This is not to be regarded as unrealistic. The effective mass representation is simply one particular way of interpreting the form of the energy function. Thus in classical mechanics we may write an energy-momentum relation for a free particle,

$$E(\mathbf{p}) = A' \, |\mathbf{p}|^2 = \frac{1}{2m} \, |\mathbf{p}|^2 \tag{9.2.67}$$

and the inertial mass is recognized as $m = 1/2A'$. On band theory the energy of an electron moving in a periodic potential is a more complicated function of the wave vector,

$$E_n(\mathbf{k}) = f(\mathbf{k}) \tag{9.2.68}$$

and the constants appearing in this equation when it is written as a power series in $\mathbf{k}$ need not necessarily be interpreted in terms of a *straightforward* inertial mass.

It is sometimes useful to describe the motion of the assembly of electrons in a nearly full band by the complementary motion of the 'holes' which represent the unoccupied states. These holes are associated with a positive electron charge and a positive effective mass. Consider a band which is filled except for one state $\psi_{n,\mathbf{k}'}(\mathbf{r})$. The motion of all the electrons in the band will determine the motion of the 'hole' in the state $\psi_{n,\mathbf{k}'}(\mathbf{r})$. A full band of electrons will carry no net current since the electron velocities will cancel in pairs. Hence we may write

$$-|e| \sum_{\mathbf{k} \neq \mathbf{k}'} \mathbf{V}(\mathbf{k}) - |e| \, \mathbf{V}(\mathbf{k}') = 0 \tag{9.2.69}$$

or

$$-|e| \sum_{\mathbf{k} \neq \mathbf{k}'} \mathbf{V}(\mathbf{k}) = +|e| \, \mathbf{V}(\mathbf{k}') \tag{9.2.70}$$

But the left-hand side of this equation is the net current carried by the set of electrons in the band omitting the one state $\psi_{n,\mathbf{k}'}(\mathbf{r})$. Evidently this is also the current arising from a single positive electron charge moving with the velocity $\mathbf{V}(\mathbf{k}')$ which is characteristic of the state $\psi_{n,\mathbf{k}'}(\mathbf{r})$. The acceleration of an electron in the state $\psi_{n,\mathbf{k}'}(\mathbf{r})$ is given by equation (9.2.63). This will also be the acceleration associated with the hole in state $\psi_{n,\mathbf{k}'}(\mathbf{r})$ since its motion is determined by the wave packet constructed from the filled electron states in the vicinity of $\mathbf{k}'$. In particular, if we consider the action of a uniform electric field $\mathbf{E}$,

$$\frac{\mathrm{d}}{\mathrm{d}t}\mathbf{V}(\mathbf{k}') = \left[\frac{1}{m^*}\right]_{\mathbf{k}'}\cdot(-|e|\,\mathbf{E}) \tag{9.2.71}$$

which may be written

$$\frac{\mathrm{d}}{\mathrm{d}t}\mathbf{V}(\mathbf{k}') = -\left[\frac{1}{m^*}\right]\cdot(+|e|\,\mathbf{E}) \tag{9.2.72}$$

The acceleration may therefore be interpreted as arising from a positive electron charge in the state $\psi_{n,\mathbf{k}'}(\mathbf{r})$ having a reciprocal mass tensor which is the negative of that given by equation (9.2.62). For states near the top of the band this will generally lead to an effective mass tensor with positive components. Consequently the electrical properties of a band which is nearly filled may often be conveniently interpreted as arising from positively charged holes with positive effective mass.

9.2.4 k.p Perturbation Theory

A very useful perturbation technique has been developed to describe the energy surfaces in the vicinity of a symmetry point in the Brillouin zone. The method has found wide application in the case of semiconducting materials and leads directly to a first order estimate of the effective mass for the carriers.

Consider the operation of the momentum operator $\mathbf{p} = -\mathrm{j}\hbar\nabla$ upon a Bloch function,

$$\mathbf{p}\Big[\{\exp(\mathrm{j}\mathbf{k}_0.\mathbf{r})\}\,u_{n,\mathbf{k}_0}(\mathbf{r})\Big] = \exp(\mathrm{j}\mathbf{k}_0.\mathbf{r})\Big\{(\hbar\mathbf{k}_0 + \mathbf{p})\,u_{n,\mathbf{k}_0}(\mathbf{r})\Big\} \tag{9.2.73}$$

and

$$\begin{aligned} p^2\Big[\{\exp(\mathrm{j}\mathbf{k}_0.\mathbf{r})\}\,u_{n,\mathbf{k}_0}(\mathbf{r})\Big] &= \mathbf{p}.\mathbf{p}\Big[\{\exp(\mathrm{j}\mathbf{k}_0.\mathbf{r})\}\,u_{n,\mathbf{k}_0}(\mathbf{r})\Big] \\ &= \exp(\mathrm{j}\mathbf{k}_0.\mathbf{r})\Big\{(p^2 + 2\hbar\mathbf{k}_0.\mathbf{p} + \hbar^2k_0^2)\,u_{n,\mathbf{k}_0}(\mathbf{r})\Big\} \end{aligned} \tag{9.2.74}$$

The Hamiltonian operator for an electron in the periodic potential of the crystal is given by equation (9.2.1). Hence the function $u_{n,\mathbf{k}_0}(\mathbf{r})$ and the energy $E_n(\mathbf{k}_0)$ are solutions of the equation

$$\left\{\frac{p^2}{2m} + \frac{\hbar}{m}\mathbf{k}_0.\mathbf{p} + \frac{\hbar^2k_0^2}{2m} + v(\mathbf{r})\right\}u_{n,\mathbf{k}_0}(\mathbf{r}) = E_n(\mathbf{k}_0)\,u_{n,\mathbf{k}_0}(\mathbf{r}) \tag{9.2.75}$$

The terms $(\hbar/m)\mathbf{k}_0.\mathbf{p}$ and $(\hbar^2/2m)k_0^2$ are treated by perturbation theory in the vicinity of a symmetry point in the Brillouin zone where the form of the functions $u_{n,\mathbf{k}_0}(\mathbf{r})$ and $E_n(\mathbf{k}_0)$ may already be known. For example, at the centre of the Brillouin zone the unperturbed equation is,

$$\mathscr{H}_0\,u_{n,0}^0(\mathbf{r}) = \left\{\frac{p^2}{2m} + v(\mathbf{r})\right\}u_{n,0}^0(\mathbf{r}) = E_n^0\,u_{n,0}^0(\mathbf{r}) \tag{9.2.76}$$

Equations (9.2.75) and (9.2.76) are immediately applicable to a number of semiconductors having the diamond or zinc blende type of crystal structures. For these materials the bottom of the conduction band and the top of the valence band each occur at $\mathbf{k}_0 = 0$ to first order. The wave functions for an electron in the conduction band at $\mathbf{k}_0 = 0$ has the symmetry of an atomic s function whilst at the top of the valence band the wave function has the symmetry of an atomic p function. Hence we may write for equation (9.2.76)

$$\begin{aligned} (\mathscr{H}_0 - E_c^0)u_{c,0}(\mathbf{r}) &= 0 \\ (\mathscr{H}_0 - E_v^0)u_{v,0}(\mathbf{r}) &= 0 \end{aligned} \tag{9.2.77}$$

Here $E_c^0 - E_v^0 \equiv E_g^0$ is the energy gap at $\mathbf{k}_0 = 0$, and there are three valence band functions corresponding to the atomic functions p_x, p_y and p_z. Assuming that the energy separations to other bands are sufficiently large the conduction and valence band eigenfunctions form an adequate set for the perturbation calculation. We construct the secular determinant therefore from the eigenfunctions $u_{c,0}(\mathbf{r})$, $u_{v,0}(\mathbf{r})$, using the Hamiltonian operator from equation (9.2.75).

$$\begin{vmatrix} E_c^0 + \dfrac{\hbar^2 k_0^2}{2m} - E & k_x P & k_y P & k_z P \\ k_x P & E_v^0 + \dfrac{\hbar^2 k_0^2}{2m} - E & 0 & 0 \\ k_y P & 0 & E_v^0 + \dfrac{\hbar^2 k_0^2}{2m} - E & 0 \\ k_z P & 0 & 0 & E_v^0 + \dfrac{\hbar^2 k_0^2}{2m} - E \end{vmatrix} = 0 \tag{9.2.78}$$

In this determinant the element P is given by

$$\begin{aligned} P &= \frac{\hbar}{m}\left\langle u_{c,0}(\mathbf{r})\right| -j\hbar\frac{\partial}{\partial x}\left|u_{v,0}(X)\right\rangle \\ &= \frac{\hbar}{m}\left\langle u_{c,0}(\mathbf{r})\right| -j\hbar\frac{\partial}{\partial y}\left|u_{v,0}(Y)\right\rangle \\ &= \frac{\hbar}{m}\left\langle u_{c,0}(\mathbf{r})\right| -j\hbar\frac{\partial}{\partial z}\left|u_{v,0}(Z)\right\rangle \end{aligned} \tag{9.2.79}$$

where the symbols X, Y, Z, identify the three valence band functions. A number of elements in the determinant are zero because of the symmetries of the basis functions $u_{c,0}(\mathbf{r})$, $u_{v,0}(\mathbf{r})$. Expand the determinant in equation (9.2.78), to obtain

$$\left\{E_v^0 + \frac{\hbar^2 k_0^2}{2m} - E\right\}^2\left\{\left(E_c^0 + \frac{\hbar^2 k_0^2}{2m} - E\right)\left(E_v^0 + \frac{\hbar^2 k_0^2}{2m} - E\right) - k_0^2 P^2\right\} = 0 \tag{9.2.80}$$

From which

$$E = E_v^0 + \frac{\hbar^2 k_0^2}{2m} \quad \text{(twice)} \tag{9.2.81}$$

$$E = \tfrac{1}{2}(E_c^0 + E_v^0) + \frac{\hbar^2 k_0^2}{2m} \pm \{\tfrac{1}{4}(E_c^0 - E_v^0)^2 + k_0^2 P^2\}^{1/2} \tag{9.2.82}$$

For small values of k_0^2 equation (9.2.82) reduces to

$$E = E_c^0 + \frac{\hbar^2 k_0^2}{2m} + \frac{k_0^2 P^2}{E_g} \tag{9.2.83}$$

$$E = E_v^0 + \frac{\hbar^2 k_0^2}{2m} - \frac{k_0^2 P^2}{E_g} \tag{9.2.84}$$

and so the effective masses obtained from equation (9.2.62) are given by

$$\frac{1}{m_c^*} = \frac{1}{m} + \frac{2P^2}{\hbar^2 E_g} \tag{9.2.85}$$

$$\frac{1}{m_{v1}^*} = \frac{1}{m} - \frac{2P^2}{\hbar^2 E_g} \tag{9.2.86}$$

$$\frac{1}{m_{v2}^*} = \frac{1}{m} \quad \text{(from equation (9.2.81)} \tag{9.2.87}$$

In equations (9.2.85) and (9.2.86) the term in $1/m$ is small compared with $|2P^2/\hbar^2 E_g|$, so that we have

$$m_c^* \approx |m_{v1}^*| \approx \frac{\hbar^2 E_g}{2P^2} \ll m \tag{9.2.88}$$

$$m_{v2}^* = m \tag{9.2.89}$$

These results illustrate two characteristic features of the simple **k.p** perturbation theory. The conduction electron mass and the light hole mass are equal to one another and each is directly proportional to the energy gap. Since it is found empirically that the matrix element P does not vary greatly among different semiconductors belonging to the same crystal structure class, the electron and light hole masses vary in proportion to the energy gaps for the semiconductors. Typical materials to which the theory may be usefully applied are given in Table 9.1.

TABLE 9.1

Material	*Conduction electron mass*† (*units of m*)	*Hole masses*† (*units of m*)	*Energy gap* E_g (0 K) (*eV*)
InSb	0·016	0·016 0·4	0·24
InAs	0·024	0·026 0·41	0·42
GaSb	0·042	0·052 —	0·81
GaAs	0·067	0·12 0·68	1·52

† m is the free electron mass.

This method of evaluating the energy surfaces and effective masses may be further refined by including a larger number of bands in the calculation and by taking into account the spin-orbit coupling interaction. The principle of the calculation is unchanged but the mathematics is more complicated and the problem is usually treated by second order perturbation theory rather than by seeking the complete solution from the secular determinant corresponding to equation (9.2.78). With these additional refinements the degeneracy of the two bands given by equation (9.2.81) is removed and the two heavy hole masses are no longer equal to the free electron mass.

9.3 Electrical Conductivity

In Chapter 1 (Section 1.5.1) we developed a simple theory of linear electrical conduction (Ohm's Law). A finite electrical conductivity or electrical resistance was the result of collision processes which were characterized by a collision time parameter $\langle t \rangle$. This parameter measured the average time between collisions for the electron charge carriers which were regarded as classical particles. The first order effects of the collision processes were the destruction of the continuous accelerations arising from the applied electric field and the redistribution of the electron velocities. The electrons, therefore, acquired a net drift velocity rather than an acceleration and so Ohm's Law could be explained. The energy losses at each individual collision, which gave rise to the Joule heating in the material, were taken to be small and consequently the collisions could be regarded as essentially elastic processes. In the present section we shall attempt to give a more detailed description of the collision processes leading to a finite electrical conductivity, but the underlying assumptions will not differ greatly from those of Chapter 1.

A perfect, rigid, crystal lattice gives rise to the ideal periodic potential field which we have discussed in Section 9.2. The electron motion may be described in terms of Bloch wave functions, each function being an eigenfunction of the ideal Hamiltonian and representing a stationary state of the system with no dissipation. The perfect crystal lattice therefore has zero electrical resistance. If, however, the ideal lattice is perturbed either by static imperfections or by thermal vibrations, the simple Bloch functions will no longer be satisfactory eigenfunctions. Provided that the perturbations are small, the new wave functions can be usefully expanded as a linear superposition of the Bloch functions referring to the ideal lattice, and we may interpret this representation as describing the scattering of an electron from an initial state $\psi_{n,\mathbf{k}}(\mathbf{r})$ to a new state $\psi_{n,\mathbf{k}'}(\mathbf{r})$.

We shall assume that the energy of an electron does not change upon being scattered from the state $\psi_{n,\mathbf{k}}(\mathbf{r})$ into the state $\psi_{n,\mathbf{k}'}(\mathbf{r})$. The scattering process simply redistributes electrons over a constant energy surface. This is evidently a reasonable assumption for scattering by a static imperfection such as an impurity atom, a vacancy or a dislocation. For this case the electron is scattered from a centre which is bound to the (rigid) lattice by strong forces. The scattering centre has therefore effectively infinite mass compared with that of the electron and the energy exchange is negligibly small. To estimate the energy exchange upon scattering by a thermal vibration, we describe the electron energy by

$$E(\mathbf{k}) = \frac{\hbar^2 |\mathbf{k}|^2}{2m^*} = \frac{\hbar^2 k^2}{2m^*} \tag{9.3.1}$$

and assume that the interaction creates a phonon particle having energy

$$E(\mathbf{q}) = \hbar\omega(\mathbf{q}) \tag{9.3.2}$$

with a wave vector $\mathbf{q}$. The classical equations for conservation of energy and momentum give

$$\frac{\hbar^2 k^2}{2m^*} = \hbar\omega(\mathbf{q}) + \frac{\hbar^2 k'^2}{2m^*} \tag{9.3.3}$$

$$\mathbf{k} = \mathbf{q} + \mathbf{k}' \tag{9.3.4}$$

The fractional change in energy for the electron is

$$\frac{\Delta E}{E} = \frac{\hbar\omega(\mathbf{q})}{\hbar^2 k^2/2m^*} = \frac{\hbar c_s q}{\hbar^2 k^2/2m^*} = \frac{2c_s}{v(\mathbf{k})} \cdot \frac{q}{k} \tag{9.3.5}$$

where c_s is the velocity of sound in the material and $v(\mathbf{k})$ is the velocity of the electron. It is reasonable to take $q \sim k$ when estimating a value for $\Delta E/E$ since, from equation (9.3.4)

$$\mathbf{k}'.\mathbf{k}' = k'^2 = k^2 + q^2 - 2\mathbf{k}.\mathbf{q} \tag{9.3.6}$$

Substituting for k'^2 in equation (9.3.3) gives

$$\frac{\hbar k}{m^*} = \frac{c_s + (\hbar q/2m^*)}{\cos\theta} \tag{9.3.7}$$

$$k \geqslant \tfrac{1}{2} q \tag{9.3.8}$$

Hence, from equation (9.3.5)

$$\frac{\Delta E}{E} \approx \frac{2c_s}{v(\mathbf{k})} \tag{9.3.9}$$

For a metal, we take $v(\mathbf{k}) \sim 5.10^8$ cm s^{-1} at the Fermi surface and $c_s \sim 5.10^5$ cm s^{-1}. The fractional change in energy for the electron is therefore very small and the collisions are closely elastic. For a semiconductor the electrons will have a thermal distribution of velocities. At room temperature it is reasonable to take $v(\mathbf{k}) \sim 10^7$ cm s^{-1}, for carriers having a free electron mass, and c_s again equal to 5.10^5 cm s^{-1}. For this case an electron may change its energy by about 10% in a collision. In many semiconductors the change in energy will be considerably smaller than this value because the effective mass of a charge carrier is frequently much less than the free electron mass. It is therefore still quite a good approximation to regard the collision process as elastic.

9.3.1 Current Density

Consider a group of electrons with wave vectors in the range $\mathbf{k}$ to $\mathbf{k} + d\mathbf{k}$ and velocities $v(\mathbf{k})$. If there are $n_\mathbf{k}$ of these electrons per unit volume of the material, the total current density is given by equations (1.1.10) and (1.5.5) of Chapter 1,

$$\mathbf{J} = -\frac{1}{c}\sum_{\mathbf{k}} n_{\mathbf{k}}\,|e|\,\mathbf{V}(\mathbf{k}) \tag{9.3.10}$$

Describe the distribution of electrons over the states $\mathbf{k}$ by a distribution function $f(\mathbf{k})$, such that $f(\mathbf{k})\,g(\mathbf{k})\,d^3\mathbf{k} \equiv n_{\mathbf{k}}$ is the number of electrons per unit volume with wave vectors between $\mathbf{k}$ and $\mathbf{k} + d\mathbf{k}$, whilst the function $g(\mathbf{k})$ is just the density of electron states in $\mathbf{k}$ space. The current density may therefore be written as

$$\mathbf{J} = -\frac{|e|}{c}\int \mathbf{V}(\mathbf{k})\,f(\mathbf{k})\,g(\mathbf{k})\,d^3\mathbf{k} \tag{9.3.11}$$

The density of states in $\mathbf{k}$ space is†

$$g(\mathbf{k}) = \frac{1}{4\pi^3} \tag{9.3.12}$$

where we have taken account of the two spin orientations. Hence

$$\mathbf{J} = -\frac{|e|}{4\pi^3 c}\int \mathbf{V}(\mathbf{k}) f(\mathbf{k})\,d^3\mathbf{k} \tag{9.3.13}$$

Note that, if $f^0(\mathbf{k})$ is the distribution function with no applied electric field, there will be no net current density and so,

$$0 = -\frac{|e|}{4\pi^3 c}\int \mathbf{V}(\mathbf{k}) f^0(\mathbf{k})\,d^3\mathbf{k} \tag{9.3.14}$$

Suppose that the electric field is applied along the z-direction, $E = \hat{\mathbf{k}}E_z$. The equation of motion for an electron in this field is given by equation (9.2.56). If we neglect collision processes for the moment,

$$\hbar\frac{dk_z}{dt} = -|e|\,E_z \tag{9.3.15}$$

Evidently each representative point acquires a velocity in $\mathbf{k}$ space,

$$\frac{dk_z}{dt} = -\frac{|e|\,E_z}{\hbar} \tag{9.3.16}$$

If, therefore, we consider an elementary volume in $\mathbf{k}$ space,

$$d^3\mathbf{k} \equiv \Delta k_x\,\Delta k_y\,\Delta k_z,$$

bounded by planes at k_x and $(k_x + \Delta k_x)$, k_y and $(k_y + \Delta k_y)$, k_z and $(k_z + \Delta k_z)$, the electric field will cause electrons (representative points) to drift into and out of this volume through the planes at k_z and $(k_z + \Delta k_z)$. The number of electrons which enter in unit time through the plane at k_z is

† The density of states in *momentum space* is given by

$$G(\mathbf{p})d^3\mathbf{p} = \frac{1}{h^3}dp_x\,dp_y\,dp_z$$

per unit volume.

$\frac{1}{4\pi^3} f(k_x, k_y, k_z)(\mathrm{d}k_z/\mathrm{d}t)\,\Delta k_x\,\Delta k_y$, at an instant when the distribution function has the value $f(\mathbf{k})$. The number leaving in unit time at the same instant is $\frac{1}{4\pi^3} f(k_x, k_y, k_z + \Delta k_z)(\mathrm{d}k_z/\mathrm{d}t)\,\Delta k_x\,\Delta k_y$. The net rate of increase in the number of electrons in this elementary volume of **k** space, due to the action of the electric field, is therefore

$$\left[\frac{\mathrm{d}N(\mathbf{k})}{\mathrm{d}t}\right]_{\text{field}} = \frac{1}{4\pi^3}\Big[f(k_x, k_y, k_z) - f(k_x, k_y, k_z + \Delta k_z)\Big]\frac{\mathrm{d}k_z}{\mathrm{d}t}\Delta k_x\,\Delta k_y \quad (9.3.17)$$

that is,

$$\left[\frac{\mathrm{d}N(\mathbf{k})}{\mathrm{d}t}\right]_{\text{field}} = \frac{1}{4\pi^3}\left(-\frac{\partial f}{\partial k_z}\right)\frac{\mathrm{d}k_z}{\mathrm{d}t}\,\Delta k_x\,\Delta k_y\,\Delta k_z \quad (9.3.18)$$

Electrons will also enter the elementary volume of **k** space as a result of scattering processes. We represent the scattering by an exponential relaxation process which describes the way in which an initial arbitrary distribution $f_{\mathrm{i}}(\mathbf{k})$ will approach the equilibrium configuration $f^0(\mathbf{k})$ under the action of scattering processes alone. We write

$$f(\mathbf{k}) - f^0(\mathbf{k}) = [f_{\mathrm{i}}(\mathbf{k}) - f^0(\mathbf{k})]\mathrm{e}^{-t/\tau(\mathbf{k})} \quad (9.3.19)$$

This equation identifies a characteristic time $\tau(\mathbf{k})$ for the process. Evidently,

$$\left[\frac{\mathrm{d}f(\mathbf{k})}{\mathrm{d}t}\right]_{\text{scattering}} = -\frac{f(\mathbf{k}) - f^0(\mathbf{k})}{\tau(\mathbf{k})} \quad (9.3.20)$$

and the number of electrons which enter the elementary volume of **k** space in unit time as a result of scattering processes is

$$\left[\frac{\mathrm{d}N(\mathbf{k})}{\mathrm{d}t}\right]_{\text{scattering}} = \frac{1}{4\pi^3}\left[\frac{\mathrm{d}f(\mathbf{k})}{\mathrm{d}t}\right]_{\text{scattering}}\Delta k_x\,\Delta k_y\,\Delta k_z \quad (9.3.21)$$

When the electron system has reached a steady state under the action of both the applied electric field and the scattering processes the number of electrons in a given elementary volume of **k** space will remain constant. For this steady state therefore

$$\left[\frac{\mathrm{d}N(\mathbf{k})}{\mathrm{d}t}\right]_{\text{field}} + \left[\frac{\mathrm{d}N(\mathbf{k})}{\mathrm{d}t}\right]_{\text{scattering}} = 0 \quad (9.3.22)$$

From equations (9.3.18), (9.3.20) and (9.3.21),

$$\left(-\frac{\partial f}{\partial k_z}\right)\frac{\mathrm{d}k_z}{\mathrm{d}t} - \left\{\frac{f(\mathbf{k}) - f^0(\mathbf{k})}{\tau(\mathbf{k})}\right\} = 0 \quad (9.3.23)$$

Since we are at present only considering the action of an electric field on the system of electrons, we may to sufficient approximation write†

$$\partial f/\partial k_z = \partial f^0/\partial k_z,$$

† When a magnetic field is present it is necessary to retain $\partial f/\partial k_z$ since the term in $\partial f^0/\partial k_z$ which multiplies $(\mathbf{V} \wedge \mathbf{H})$ vanishes. See page 496.

and, using equation (9.3.16), obtain

$$f(\mathbf{k}) = f^0(\mathbf{k}) + \frac{|e|\, E_z\, \tau(\mathbf{k})}{\hbar}\left(\frac{\partial f^0}{\partial k_z}\right) \tag{9.3.24}$$

Introducing this expression into equation (9.3.13), we obtain the current density resulting from the action of both the applied field and the scattering processes.

$$\mathbf{J} = -\frac{e^2 E_z}{4\pi^3 c\hbar}\int \mathbf{V}(\mathbf{k})\,\tau(\mathbf{k})\left(\frac{\partial f^0}{\partial k_z}\right) \mathrm{d}^3\mathbf{k} \tag{9.3.25a}$$

$$= -\frac{e^2 E_z}{4\pi^3 c\hbar}\int \left(\frac{\partial E}{\partial k_z}\right)\mathbf{V}(\mathbf{k})\,\tau(\mathbf{k})\left(\frac{\partial f^0}{\partial E}\right) \mathrm{d}^3\mathbf{k} \tag{9.3.25b}$$

The elementary volume in **k** space, $\mathrm{d}^3\mathbf{k}$, is now, for convenience, expressed in terms of an element of area $\mathrm{d}S_{\mathbf{k}}$ on a constant energy surface times the normal distance between two such surfaces with energies E and $E + \mathrm{d}E$,

$$\mathrm{d}^3\mathbf{k} = \frac{\mathrm{d}S_{\mathbf{k}}\,\mathrm{d}E}{|\nabla_{\mathbf{k}}\, E(\mathbf{k})|} \tag{9.3.26}$$

Equation (9.3.25a) may therefore be written, using also equation (9.2.52),

$$\mathbf{J} = \frac{e^2 E_z}{4\pi^3 c\hbar^2}\int \tau(\mathbf{k})\,\frac{\nabla_{\mathbf{k}}\, E(\mathbf{k})}{|\nabla_{\mathbf{k}}\, E(\mathbf{k})|}\left(\frac{\partial E}{\partial k_z}\right) \mathrm{d}S_{\mathbf{k}}\left(-\frac{\partial f^0}{\partial E}\right) \mathrm{d}E \tag{9.3.27}$$

For a metal $f^0(\mathbf{k})$ is the Fermi–Dirac distribution function and so, as on page 443, $(-\,\partial f^0/\partial E)$ is the Dirac δ-function, $\delta(E - E_{\mathrm{F}})$. For this case, integration over the energy, E, gives

$$\mathbf{J} = \frac{e^2 E_z}{4\pi^3 c\hbar^2}\int_{\text{Fermi surface}} \tau(\mathbf{k}_{\mathrm{F}})\,\frac{\nabla_{\mathbf{k}}\, E(\mathbf{k})}{|\nabla_{\mathbf{k}}\, E(\mathbf{k})|}\left(\frac{\partial E}{\partial k_z}\right) \mathrm{d}S_{\mathbf{k}} \tag{9.3.28}$$

Use equation (9.3.1) to evaluate $\nabla_{\mathbf{k}}\, E(\mathbf{k})$, $(\partial E/\partial k_z)$, and take spherical coordinates with respect to the k_z direction. The components J_x, J_y, vanish upon integrating over the variable ϕ, and we obtain†

$$J_z = \frac{e^2 E_z}{4\pi^3 c\hbar^2}\int_0^{\pi} \tau(\mathbf{k}_{\mathrm{F}})\frac{\hbar^2}{m^*}\, k_{\mathrm{F}}^3\, 2\pi \cos^2\theta \sin\theta\, \mathrm{d}\theta \tag{9.3.29}$$

Assuming, now, that the collision time parameter $\tau(\mathbf{k}_{\mathrm{F}})$ is independent of θ, we write $\tau(\mathbf{k}_{\mathrm{F}}) \equiv \langle t\rangle$ and take this term outside the integral,

$$J_z = \frac{e^2 E_z}{3\pi^2 c m^*}\langle t\rangle k_{\mathrm{F}}^3 \tag{9.3.30}$$

For the simple spherical Fermi surface which we have considered, the Fermi–Dirac distribution function gives

$$N = \int_0^{k_{\mathrm{F}}} \frac{1}{4\pi^3}\, \mathrm{d}^3\mathbf{k} = \frac{1}{4\pi^3}\left(\frac{4}{3}\pi k_{\mathrm{F}}^3\right)$$

$$= \frac{1}{3\pi^2}\, k_{\mathrm{F}}^3 \tag{9.3.31}$$

† We assume here that $\tau(\mathbf{k}_{\mathrm{F}})$ does not depend on ϕ.

where N is the number of electrons per unit volume. Hence the current density is

$$J_z = \frac{Ne^2E_z}{m^*c}\langle t\rangle \tag{9.3.32}$$

and the electrical conductivity is

$$\sigma = \frac{Ne^2\langle t\rangle}{m^*} \tag{9.3.33}$$

in agreement with the result obtained in Chapter 1.†

For a semiconductor we use Boltzmann statistics to describe the carrier distribution function. Assume that the material is n-type and that the holes make no significant contribution to the electrical conduction. We have

$$f^0(E) = A\,\mathrm{e}^{-E/kT} \tag{9.3.34}$$

If there are N electrons per unit volume

$$\begin{aligned} N &= \int f^0(E)\frac{1}{4\pi^3}\,\mathrm{d}^3\,\mathbf{k} \\ &= A\frac{2^{1/2}(m^*)^{3/2}}{\pi^2\hbar^3}\int E^{1/2}\,\mathrm{e}^{-E/kT}\,\mathrm{d}E \end{aligned} \tag{9.3.35}$$

Here we have used equation (9.3.1) to evaluate $\mathrm{d}^3\mathbf{k}$. The contribution to the current density from a particular energy surface is obtained by first carrying through the integration over $\mathrm{d}S_{\mathbf{k}}$ in equation (9.3.27). Using equation (9.3.1) and again assuming that $\tau(\mathbf{k})$ does not depend upon the angle variable we may perform this integration to obtain for the total current density,

$$J_z = \frac{e^2E_z}{3\pi^2cm^*}\int \tau(E)\,k^3\left(-\frac{\partial f^0}{\partial E}\right)\mathrm{d}E \tag{9.3.36}$$

$$= \frac{Ne^2E_z}{m^*}\left(\frac{2}{3kT}\right)\frac{\int \tau(E)\,E^{3/2}\,\mathrm{e}^{-E/kT}\,\mathrm{d}E}{\int E^{1/2}\,\mathrm{e}^{-E/kT}\,\mathrm{d}E} \tag{9.3.37}$$

From which σ again takes the form given in Chapter 1,‡

$$\sigma = \frac{Ne^2}{m^*}\langle t\rangle \tag{9.3.38}$$

with

$$\langle t\rangle = \left(\frac{2}{3kT}\right)\frac{\int \tau(E)\,E^{3/2}\,\mathrm{e}^{-E/kT}\,\mathrm{d}E}{\int E^{1/2}\,\mathrm{e}^{-E/kT}\,\mathrm{d}E} \tag{9.3.39}$$

† Measured values of σ lead to $\langle t\rangle \approx 10^{-14}$ s. for a metal at room temperature.
‡ For a typical high purity semiconductor at room temperature $t \sim 10^{-13}$ s.

9.3.2 Relaxation Time

The characteristic time $\tau(\mathbf{k})$ introduced in equation (9.3.19) also describes the decay of the current density $\mathbf{J}$. Suppose there is initially a current density $\mathbf{J}_i$ due to an arbitrary distribution $f_i(\mathbf{k})$ and we examine how this current density will decay towards zero under the influence of the scattering processes. The distribution function will return to its equilibrium value in accordance with equation (9.3.19). At any instant the current density is therefore given by equation (9.3.13),

$$\mathbf{J} = -\frac{|e|}{4\pi^3 c}\int \mathbf{V}(\mathbf{k})\, f(\mathbf{k})\, d^3\mathbf{k}$$

with

$$f(\mathbf{k}) = f^0(\mathbf{k}) + \{f_i(\mathbf{k}) - f^0(\mathbf{k})\}\, e^{-t/\tau(\mathbf{k})}$$

Using equation (9.3.14), we obtain

$$\mathbf{J} = -\frac{|e|}{4\pi^3 c}\int \mathbf{V}(\mathbf{k})[f_i(\mathbf{k}) - f^0(\mathbf{k})]\, e^{-t/\tau(\mathbf{k})}\, d^3\mathbf{k} \tag{9.3.40}$$

Assuming that $\tau(\mathbf{k}) = \tau$, a constant independent of $\mathbf{k}$,

$$\mathbf{J} = -\frac{|e|}{4\pi^3 c}\, e^{-t/\tau}\int \mathbf{V}(\mathbf{k})\, f_i(\mathbf{k})\, d^3\mathbf{k} \tag{9.3.41}$$

or

$$\mathbf{J} = \mathbf{J}_i\, e^{-t/\tau} \tag{9.3.42}$$

and the current density decays exponentially with the characteristic time τ.

We now seek to relate the characteristic time $\tau(\mathbf{k})$ to the microscopic transition probability which arises from the scattering processes. If an electron is initially in the state $\psi_{\mathbf{k}}(\mathbf{r})$ at time t, then we shall write $P(\mathbf{k}, \mathbf{k}')\,\delta t$ for the probability of finding this electron in the state $\psi_{\mathbf{k}'}(\mathbf{r})$ at time $t + \delta t$. Usually there is no change in the spin state for the electron when scattered by thermal vibrations or lattice imperfections and so the appropriate density of states in $\mathbf{k}$ space is $g(\mathbf{k}) = 1/8\pi^3$. In addition we shall again assume that the energy-wave vector relation takes the simple form

$$E(\mathbf{k}) = \frac{\hbar^2\,|\mathbf{k}|^2}{2m^*} = \frac{\hbar^2 k^2}{2m^*} \tag{9.3.43}$$

The number of transitions from $\mathbf{k}$ to $\mathbf{k}'$ in the time δt, is obtained by multiplying the probability $P(\mathbf{k}, \mathbf{k}')\,\delta t$, by the number of electrons with a given spin orientation in the element of $\mathbf{k}$ space $d^3\mathbf{k}$ centred on $\mathbf{k}$ and by the number of vacancies in the element $d^3\mathbf{k}'$. Thus,

$$N_{\mathbf{k}\to\mathbf{k}'} = P(\mathbf{k}, \mathbf{k}')\,\delta t\frac{1}{8\pi^3}f(\mathbf{k})\, d^3\mathbf{k}\frac{1}{8\pi^3}\{1 - f(\mathbf{k}')\}\, d^3\mathbf{k}' \tag{9.3.44}$$

Similarly, the number of transitions into $\mathbf{k}$ from $\mathbf{k}'$ in time δt, is

$$N_{\mathbf{k}'\to\mathbf{k}} = P(\mathbf{k}', \mathbf{k})\,\delta t\frac{1}{8\pi^3}\, f(\mathbf{k}')\, d^3\mathbf{k}'\frac{1}{8\pi^3}\{(1 - f(\mathbf{k})\}d^3\mathbf{k} \tag{9.3.45}$$

The net transition rate into **k** from all states **k**′ is therefore,

$$+\left[\frac{\mathrm{d}N(\mathbf{k})}{\mathrm{d}t}\right]_{\text{scattering}} = \sum_{\mathbf{k}'} \frac{N_{\mathbf{k}'\to\mathbf{k}} - N_{\mathbf{k}\to\mathbf{k}'}}{\delta t} \tag{9.3.46}$$

and, by comparison with equation (9.3.21),

$$\left[\frac{\mathrm{d}N(\mathbf{k})}{\mathrm{d}t}\right] = \sum_{\mathbf{k}'} \frac{N_{\mathbf{k}'\to\mathbf{k}} - N_{\mathbf{k}\to\mathbf{k}'}}{\delta t} = \frac{1}{8\pi^3}\left[\frac{\mathrm{d}f(\mathbf{k})}{\mathrm{d}t}\right]_{\text{scattering}} \mathrm{d}^3\mathbf{k} \tag{9.3.47}$$

Equations (9.3.44) and (9.3.45), now lead to

$$\left[\frac{\mathrm{d}f(\mathbf{k})}{\mathrm{d}t}\right]_{\text{scattering}} = \frac{1}{8\pi^3}\int [P(\mathbf{k}', \mathbf{k}) f(\mathbf{k}')\{1 - f(\mathbf{k})\} - P(\mathbf{k}, \mathbf{k}') f(\mathbf{k})\{1 - f(\mathbf{k}')\}]\, \mathrm{d}^3\mathbf{k}' \tag{9.3.48}$$

For the equilibrium distribution functions $f^0(\mathbf{k})$, $f^0(\mathbf{k}')$, the principle of detailed balancing requires that each scattering process and its inverse should occur with equal frequency. Hence

$$P(\mathbf{k}',\mathbf{k}) f^0(\mathbf{k}')\{1 - f^0(\mathbf{k})\} = P(\mathbf{k}, \mathbf{k}') f^0(\mathbf{k})\{1 - f^0(\mathbf{k}')\} \tag{9.3.49}$$

Since we are considering essentially elastic processes with no energy change, these transitions take place over a constant energy surface described by equation (9.3.43) with $|\mathbf{k}|^2 = |\mathbf{k}'|^2$. Consequently,

$$f^0(\mathbf{k}) = f^0(\mathbf{k}') = f^0(E) \tag{9.3.50}$$

with $f^0(E)$ the Fermi–Dirac distribution function corresponding to the energy E. Equation (9.3.49) now gives,

$$P(\mathbf{k}', \mathbf{k}) = P(\mathbf{k}, \mathbf{k}') \tag{9.3.51}$$

and equation (9.3.48) gives

$$\left[\frac{\mathrm{d}f(\mathbf{k})}{\mathrm{d}t}\right]_{\text{scattering}} = \frac{1}{8\pi^3}\int P(\mathbf{k}, \mathbf{k}')[f(\mathbf{k}') - f(\mathbf{k})]\, \mathrm{d}^3\mathbf{k}' \tag{9.3.52}$$

The integration in this equation is over the variable **k**′, and so the characteristic time $\tau(\mathbf{k})$ may be obtained immediately from equation (9.3.20), as

$$\frac{1}{\tau(\mathbf{k})} = -\frac{1}{8\pi^3}\int P(\mathbf{k}, \mathbf{k}')\left[\frac{f(\mathbf{k}') - f(\mathbf{k})}{f(\mathbf{k}) - f^0(\mathbf{k})}\right] \mathrm{d}^3\mathbf{k}' \tag{9.3.53}$$

or

$$\frac{1}{\tau(\mathbf{k})} = \frac{1}{8\pi^3}\int P(\mathbf{k}, \mathbf{k}')\left[1 - \frac{f(\mathbf{k}') - f^0(\mathbf{k}')}{f(\mathbf{k}) - f^0(\mathbf{k})}\right] \mathrm{d}^3\mathbf{k}' \tag{9.3.54}$$

For our problem in electrical conductivity the functions $f(\mathbf{k}), f(\mathbf{k}')$ represent the perturbed distributions arising from the combined action of the applied electric field $\mathbf{E} = \hat{\mathbf{k}}E_z$ and the scattering processes. We may therefore use equation (9.3.24) to write

$$f(\mathbf{k}) - f^0(\mathbf{k}) = \frac{|e|\, E_z}{\hbar}\,\tau(\mathbf{k})\left(\frac{\partial f^0}{\partial k_z}\right)$$

$$f(\mathbf{k}') - f^0(\mathbf{k}') = \frac{|e|\, E_z}{\hbar}\,\tau(\mathbf{k}')\left(\frac{\partial f^0}{\partial k_z'}\right) \tag{9.3.55}$$

Assuming that $\tau(\mathbf{k}) = \tau(\mathbf{k}') = \tau(E)$, and so is independent of direction over the constant energy surface, these equations lead to,

$$\frac{f(\mathbf{k}') - f^0(\mathbf{k}')}{f(\mathbf{k}) - f^0(\mathbf{k})} = \frac{(\partial f^0/\partial k_z')}{(\partial f^0/\partial k_z)} = \frac{(\partial f^0/\partial E)(\partial E/\partial k_z')}{(\partial f^0/\partial E)(\partial E/\partial k_z)} = \frac{k_z'}{k_z} \tag{9.3.56}$$

and equation (9.3.54) may be written,

$$\frac{1}{\tau(E)} = \frac{1}{8\pi^3} \int P(\mathbf{k}, \mathbf{k}') \left[1 - \frac{k_z'}{k_z}\right] \mathrm{d}^3\mathbf{k}' \tag{9.3.57}$$

This integral may be simplified further by choosing a coordinate system oriented with respect to the direction $\mathbf{k}$. If $\mathbf{a}_1$, $\mathbf{a}_2$, $\mathbf{a}_3$ are the unit vectors of this new rectangular coordinate system having $\mathbf{a}_3$ along $\mathbf{k}$, then the representation for $\mathbf{k}$ and the applied field $\mathbf{E}$ becomes,

$$\mathbf{k} = \mathbf{a}_3 \,|\mathbf{k}| \tag{9.3.58}$$

$$\mathbf{E} = (\mathbf{a}_1 \cos\alpha + \mathbf{a}_2 \cos\beta + \mathbf{a}_3 \cos\gamma)E_z \tag{9.3.59}$$

Here, $\cos\alpha$, $\cos\beta$, $\cos\gamma$, are the direction cosines for $\mathbf{E}$ with respect to $\mathbf{a}_1$, $\mathbf{a}_2$, $\mathbf{a}_3$. The vector $\mathbf{k}'$ is most conveniently written in terms of spherical polar coordinates with respect to the new axes. Taking $\theta = 0$ along $\mathbf{a}_3$,

$$\mathbf{k}' = (\mathbf{a}_1 \sin\theta\cos\phi + \mathbf{a}_2 \sin\theta\sin\phi + \mathbf{a}_3 \cos\theta)\,|\mathbf{k}| \tag{9.3.60}$$

$$\mathrm{d}^3\mathbf{k}' = k'^2\, \mathrm{d}k' \sin\theta \,\mathrm{d}\theta\, \mathrm{d}\phi \tag{9.3.61}$$

and,

$$\frac{k_z'}{k_z} = \frac{\mathbf{k}'.\mathbf{E}}{\mathbf{k}.\mathbf{E}} = \frac{\cos\alpha\sin\theta\cos\phi + \cos\beta\sin\theta\sin\phi}{\cos\gamma} + \cos\theta \tag{9.3.62}$$

The transition probability $P(\mathbf{k}, \mathbf{k}')$ is not expected to depend upon the azimuthal angle ϕ. Consequently we may integrate equation (9.3.57) over the angle ϕ, using equations (9.3.61) and (9.3.62), to obtain

$$\frac{1}{\tau(E)} = \frac{1}{4\pi^2} \int P(\mathbf{k}, \mathbf{k}')(1 - \cos\theta)\, k'^2\, \mathrm{d}k' \sin\theta\, \mathrm{d}\theta \tag{9.3.63}$$

To derive a more explicit relation for the transition probability $P(\mathbf{k}, \mathbf{k}')$ we consider a model for the actual crystal lattice in which the deviations from the ideal structure are suddenly created and then frozen in. Let $V(\mathbf{r})$ be the periodic potential energy for an electron in the ideal lattice. Suppose that at a particular instant in time the ideal lattice is suddenly distorted giving rise to a new non-periodic potential function $V_1(\mathbf{r})$ which represents the effect of thermal vibrations, impurities and defects. The difference between $V_1(\mathbf{r})$ and $V(\mathbf{r})$ may be regarded as a perturbation which gives rise to transitions between the Bloch states $\psi_{\mathbf{k}}(\mathbf{r})$ and $\psi_{\mathbf{k}'}(\mathbf{r})$. We make use of the time-dependent perturbation theory outlined in Section 7.4. Write a wave function in the distorted lattice as a linear sum of Bloch functions appropriate to the perfect crystal,

$$\Psi(\mathbf{r}, t) = \sum_{\mathbf{k}} a_{\mathbf{k}}(t)\, \psi_{\mathbf{k}}(\mathbf{r}) \exp\left\{-\mathrm{j}\frac{E(\mathbf{k})}{\hbar}t\right\} \tag{9.3.64}$$

Initially the lattice is undistorted and the electron is in the state described by $\psi_{\mathbf{k}}(\mathbf{r})$. Hence $a_{\mathbf{k}}(0) = 1$ and all other $a_{\mathbf{k}'}(0)$ are zero. Let

$$U(\mathbf{r}) = V_1(\mathbf{r}) - V(\mathbf{r}) \tag{9.3.65}$$

be the perturbation which is constant in time except for being switched on at $t = 0$. At any subsequent time the coefficients $a_{\mathbf{k}'}(t)$ are obtained from equation (7.4.11).

$$a_{\mathbf{k}'}(t) = \frac{1}{\mathrm{j}\hbar}\int_0^t \Big[\exp{(\mathrm{j}/\hbar)\{E(\mathbf{k}') - E(\mathbf{k})\}t}\Big]\langle\mathbf{k}'|\, U\,|\mathbf{k}\rangle\,\mathrm{d}t \tag{9.3.66}$$

$$= \frac{\langle\mathbf{k}'|\, U\,|\mathbf{k}\rangle\Big[\Big[1 - \exp[(\mathrm{j}/\hbar)\{E(\mathbf{k}') - E(\mathbf{k})\}t]\Big]\Big]}{E(\mathbf{k}') - E(\mathbf{k})} \tag{9.3.67}$$

and

$$|a_{\mathbf{k}'}(t)|^2 = 4|\,\langle\mathbf{k}'|\, U\,|\mathbf{k}\rangle\,|^2\,\frac{\sin^2\Big[\{E(\mathbf{k}') - E(\mathbf{k})\}(t/2\hbar)\Big]}{E(\mathbf{k}') - E(\mathbf{k})} \tag{9.3.68}$$

The probability of finding the electron in the state $\psi_{\mathbf{k}'}(\mathbf{r})$ after the time t is measured by $|a_{\mathbf{k}'}(t)|^2$ and so the time rate of increase of this quantity is the probability function $P(\mathbf{k}, \mathbf{k}')$

$$P(\mathbf{k}, \mathbf{k}') = \frac{\mathrm{d}}{\mathrm{d}t}\;|a_{\mathbf{k}'}(t)|^2 = \frac{2}{\hbar}|\langle\mathbf{k}'|\, U\,|\mathbf{k}\rangle|^2\,\frac{\sin\Big[\{E(\mathbf{k}') - E(\mathbf{k})\}(t/\hbar)\Big]}{E(\mathbf{k}') - E(\mathbf{k})} \tag{9.3.69}$$

For any reasonable time interval we may use the δ-function representation (compare equation (4.1.16)), and write

$$P(\mathbf{k}, \mathbf{k}') = \frac{2\pi}{\hbar}|\,\langle\mathbf{k}'|\, U\,|\mathbf{k}\rangle\,|^2\,\delta\{E(\mathbf{k}') - E(\mathbf{k})\} \tag{9.3.70}$$

$$= \frac{2\pi}{\hbar}|\,\langle\mathbf{k}'|\, U\,|\mathbf{k}\rangle\,|^2\,\delta(E' - E)$$

Using the relation $k'\,\mathrm{d}k' = (m^*/\hbar^2)\,\mathrm{d}E'$ from equation (9.3.43) and integrating over E', we obtain, from equation (9.3.63),

$$\frac{1}{\tau(E)} = \frac{m^*\sqrt{(2m^*E)}}{2\pi\,\hbar^4}\int|\langle\mathbf{k}'|\, U\,|\mathbf{k}\rangle|^2(1 - \cos\theta)\sin\theta\,\mathrm{d}\theta \tag{9.3.71}$$

The calculation of the relaxation time from first principles depends therefore upon the detailed evaluation of the matrix element $\langle\mathbf{k}'|\, U\,|\mathbf{k}\rangle$. We shall only carry through the calculation for two relatively simple but important processes in order to illustrate the underlying physical features. We shall use the free electron approximation and take as our examples scattering from thermal vibrations of the lattice (lattice scattering) and scattering from ionized impurity centres.

9.3.3 Lattice Scattering—the Deformation Potential

The crystal is regarded as an isotropic elastic continuum and the frozen in thermal vibrations are represented by a superposition of waves in the form

$$\mathbf{R}(\mathbf{r}) = \sum_{\mathbf{q}} \boldsymbol{\alpha}(\mathbf{q})\, e^{j\mathbf{q}.\mathbf{r}} \tag{9.3.72}$$

Here $\mathbf{R}(\mathbf{r})$ is the displacement vector at the position $\mathbf{r}$ and $\mathbf{q}$ is the wave vector of the associated thermal vibration. The perturbing potential $U(\mathbf{r})$ is taken to be a linear function of the deformation of the perfect crystal as measured by the dilatation $D(\mathbf{r})$. The dilatation is the fractional increase in volume produced by the deformation. We write

$$U(\mathbf{r}) = E_1\, D(\mathbf{r}) \tag{9.3.73}$$

but†

$$D(\mathbf{r}) = \nabla . \mathbf{R}(\mathbf{r}) \tag{9.3.74}$$

and so equation (9.3.72) gives,

$$U(\mathbf{r}) = E_1 \sum_{\mathbf{q}} j\mathbf{q}.\boldsymbol{\alpha}(\mathbf{q})\, e^{j\mathbf{q}.\mathbf{r}} \tag{9.3.75}$$

The constant E_1 is called the deformation potential constant, it may be determined experimentally from the pressure dependence of the electrical conductivity.

We may evaluate the matrix element $\langle \mathbf{k}'|\, U\, |\mathbf{k}\rangle$ in the free electron approximation using

$$\psi_{\mathbf{k}}(\mathbf{r}) = e^{j\mathbf{k}.\mathbf{r}} \tag{9.3.76}$$

These wave functions are normalized to unit volume.

$$\begin{aligned}\langle \mathbf{k}'|\, U\, |\mathbf{k}\rangle &= \int e^{-j\mathbf{k}'.\mathbf{r}} \left\{ E_1 \sum_{\mathbf{q}} j\mathbf{q}.\boldsymbol{\alpha}(\mathbf{q})\, e^{j\mathbf{q}.\mathbf{r}} \right\} e^{j\mathbf{k}.\mathbf{r}}\, d^3\mathbf{r} \\ &= E_1 \sum_{\mathbf{q}} j\mathbf{q}.\boldsymbol{\alpha}(\mathbf{q}) \int e^{j(\mathbf{k}'-\mathbf{k}+\mathbf{q}).\mathbf{r}}\, d^3\mathbf{r} \end{aligned} \tag{9.3.77}$$

Using the δ-function representation that was given in Chapter 4, namely, $(2\pi)^3\, \delta(\mathbf{a}) = \int e^{j\mathbf{a}.\mathbf{r}} d^3\mathbf{r}$, and writing the summation over $\mathbf{q}$ as an integral, we obtain

$$\langle \mathbf{k}'|\, U\, |\mathbf{k}\rangle = \frac{E_1}{8\pi^3} \int j\mathbf{q}.\boldsymbol{\alpha}(\mathbf{q})\, \delta(\mathbf{k}' - \mathbf{k} + \mathbf{q})\, d^3\mathbf{q} \tag{9.3.78}$$

$$= j\frac{E_1}{8\pi^3}\{(\mathbf{k}' - \mathbf{k}).\boldsymbol{\alpha}(\mathbf{q})\} \tag{9.3.79}$$

with $\mathbf{q} = (\mathbf{k}' - \mathbf{k})$.

In order to derive the temperature variation of the relaxation time, we shall estimate a value for the term $|(\mathbf{k}' - \mathbf{k}).\boldsymbol{\alpha}(\mathbf{q})|^2$ using the law of equipartition of energy. First averaging the function $|(\mathbf{k}' - \mathbf{k}).\boldsymbol{\alpha}(\mathbf{q})|^2$ over all values of the squared cosine of the angle between $(\mathbf{k}' - \mathbf{k})$ and $\boldsymbol{\alpha}(\mathbf{q})$, a numerical factor of $\frac{1}{3}$ is obtained,

$$|(\mathbf{k}' - \mathbf{k}).\boldsymbol{\alpha}(\mathbf{q})|^2 = \tfrac{1}{3}|\mathbf{k}' - \mathbf{k}|^2 \langle|\boldsymbol{\alpha}(\mathbf{q})|^2\rangle \tag{9.3.80}$$

† See, for example, Kittel, C., *Introduction to Solid State Physics*, Wiley, 1953, p. 45.

with $\langle|\boldsymbol{\alpha}(\mathbf{q})|^2\rangle$ a mean value for $|\boldsymbol{\alpha}(\mathbf{q})|^2$. The energy of an elastic wave having amplitude $\boldsymbol{\alpha}(\mathbf{q})$ and frequency $\omega_{\mathbf{q}}$ is proportional to $\omega_{\mathbf{q}}^2\,|\boldsymbol{\alpha}(\mathbf{q})|^2$. We therefore use the law of equipartition to write

$$\omega_{\mathbf{q}}^2\langle|\boldsymbol{\alpha}(\mathbf{q})|^2\rangle \propto \tfrac{3}{2}kT \tag{9.3.81}$$

where we have written $\frac{3}{2}kT$ explicitly to take account of the three polarizations for an elastic wave. Consequently equation (9.3.79) leads to

$$|\langle\mathbf{k}'|\,U\,|\mathbf{k}\rangle|^2 \propto E_1^2\,|\mathbf{k}'-\mathbf{k}|^2\,\frac{T}{\omega_{\mathbf{q}}^2} \tag{9.3.82}$$

$$\propto \frac{E_1^2 T}{c_s^2} \tag{9.3.83}$$

where c_s is the velocity of sound in the material. This relation is frequently written in terms of the Debye temperature θ_D for the crystal. Using the relation

$$k\theta_D = \hbar q_D c_s \tag{9.3.84}$$

we obtain

$$|\langle\mathbf{k}'|\,U\,|\mathbf{k}\rangle|^2 \propto \frac{q_D^2 E_1^2}{\theta_D^2}T \tag{9.3.85}$$

At high temperatures lattice vibrations are available to scatter the electrons over a wide range of angles determined by the relation $\mathbf{q} = (\mathbf{k}' - \mathbf{k})$. The integral over the variable θ in equation (9.3.71) is not therefore expected to be very sensitive to temperature, and so

$$\frac{1}{\tau(E)} \propto \frac{(m^*)^{3/2}E^{1/2}q_D^2E_1^2}{\theta_D^2}T \tag{9.3.86}$$

For a metal, equation (9.3.28) indicates that $\tau(E)$ is to be evaluated at the Fermi surface. Hence

$$\frac{1}{\langle t\rangle} \propto \frac{1}{\tau(E_F)} \propto \frac{(m^*)^{3/2}E_F^{1/2}q_D^2E_1^2}{\theta_D^2}T \tag{9.3.87}$$

$$\propto T \tag{9.3.88}$$

The high temperature electrical conductivity of a metal is proportional to T^{-1}; the electrical resistance is proportional to T. If $\rho(T)$ is the resistivity of a metal at temperature T K and $\rho(\theta_D)$ is the corresponding quantity at the Debye temperature, it is evident from equation (9.3.87) that

$$\frac{\rho(T)}{\rho(\theta_D)} = \frac{T}{\theta_D} \tag{9.3.89}$$

Thus the graph of $\rho(T)/\rho(\theta_D)$ against the reduced temperature T/θ_D should give the same straight line for all metals at high temperatures ($T > \theta_D$). Experimental results for the metals Al, Cu, Na, Au and Ni do in fact indicate that this relation is quite well satisfied.

For a semiconductor, equations (9.3.39), (9.3.71) and (9.3.83) give†

$$\langle t\rangle \propto \frac{E_1^{-2}}{kT}\frac{\int ET^{-1}\,\mathrm{e}^{-E/kT}\,\mathrm{d}E}{\int E^{1/2}\,\mathrm{e}^{-E/kT}\,\mathrm{d}E}$$

$$\propto E_1^{-2}T^{-3/2} \tag{9.3.90}$$

which leads at once to the $T^{-3/2}$ power law for the carrier mobility μ,

$$\mu = \frac{e\langle t\rangle}{m^*}$$

when scattering from lattice vibrations is the dominant relaxation process.

At low temperatures it is necessary to examine more closely the part played by the integration over the angle variable θ in equation (9.3.71). This is an important consideration in the case of a metal but is not usually significant for a semiconductor. For a metal at low temperatures the scattering will be limited to small angles. Consider a spherical Fermi surface. The relation between q and the electron wave vector k_F is

$$q = 2k_F \sin(\theta/2) \tag{9.3.91}$$

At low temperatures q/k_F will be a small quantity. We may estimate an upper bound from

$$(\hbar\omega_q)_{max} = \hbar c_s q_{max} \approx kT \tag{9.3.92}$$

which leads to a maximum angle for scattering,

$$\theta_{max} = \frac{q_{max}}{k_F} \approx \frac{kT}{\hbar c_s k_F} \tag{9.3.93}$$

We again use the law of equipartition to assign an energy $\frac{3}{2}kT$ to the thermal vibrations with $q \leqslant q_{max}$ and take account of the limited thermal excitation by restricting the scattering to angles $\theta \leqslant \theta_{max}$. Equations (9.3.71) and (9.3.85) give

$$\frac{1}{\langle t\rangle} = \frac{1}{\tau(E_F)} \propto \int_0^{\theta_{max}} \frac{q_D^2 E_1^2 T}{\theta_D^2}(1-\cos\theta)\sin\theta\,\mathrm{d}\theta \tag{9.3.94}$$

$$\propto \frac{q_D^2 E_1^2 T}{\theta_D^2}\int_0^{\theta_{max}} \theta^3\,\mathrm{d}\theta$$

$$\propto \frac{T^5}{\theta_D^6} \tag{9.3.95}$$

The electrical resistance of a metal is therefore expected to be proportional to T^5 at low temperatures. This relation is usually assumed to be valid for an ideal metal but the experimental verification is incomplete because in practice the low temperature resistivity is usually dominated by scattering from static imperfections.

† $\int \exp(-\alpha u^2)\, u^3\, \mathrm{d}u = \frac{1}{2\alpha^2}$; $\int \exp(-\alpha u^2)\, u^2\, \mathrm{d}u = \frac{1}{4}\sqrt{\left(\frac{\pi}{\alpha^3}\right)}$

For semiconductors, the relaxation time due to scattering by thermal vibrations is described reasonably well by equation (9.3.90) even at low temperatures. In these materials the number of carriers having energies greater than kT is negligibly small and so all permissible final states are accessible as a result of scattering by lattice vibrations. The integration over the angle θ in equation (9.3.71) is, therefore, essentially independent of temperature and the $T^{-3/2}$ law is in principle satisfactory down to very low temperatures. In practice, the usefulness of equation (9.3.90) is limited at low temperatures because the dominant scattering mechanism in this temperature range arises from ionized impurities and other static imperfections. At high temperatures the present discussion is also inadequate because optical modes of the lattice vibrations may be thermally excited. For the semiconductors germanium and silicon, in the temperature range where acoustic lattice scattering is dominant (roughly 20–100 K), the electron mobility varies as $T^{-1\cdot66}$ and $T^{-1\cdot5}$, respectively.† These power laws are in good agreement with the prediction from the simple theory leading to equation (9.3.90).

9.3.4 Ionized Impurity Scattering

Impurities which differ in valency from the host lattice will give rise to electrostatic fields which may be regarded as perturbations on the periodic potential field for the perfect crystal. Since the mobile electrons, which are responsible for electrical conduction within the material, can redistribute themselves around the impurity in such a way as to cancel the field at large distances, the perturbing potential does not usually take the simple Coulomb form. Instead it is common to use a screened Coulomb potential for the perturbation

$$U(\mathbf{r}) = \frac{A}{|\mathbf{r}|} \mathrm{e}^{-\lambda|\mathbf{r}|} \tag{9.3.96}$$

In its simplest form, this equation is written explicitly as

$$U(\mathbf{r}) = \frac{(\Delta Z)e^2}{\varepsilon\,|\mathbf{r}|} \mathrm{e}^{-\lambda|\mathbf{r}|} \tag{9.3.97}$$

where ΔZ is the valency difference for the impurity atom and ε is the macroscopic dielectric constant of the material (in a metal ε is taken equal to unity).

We shall evaluate the matrix element $\langle\mathbf{k}'|\,U\,|\mathbf{k}\rangle$ in the free electron approximation, again taking wave functions

$$\psi_{\mathbf{k}}(\mathbf{r}) = \mathrm{e}^{\mathrm{j}\mathbf{k}.\mathbf{r}} \tag{9.3.98}$$

which are normalized to unit volume. Using equation (9.3.96) for $U(\mathbf{r})$,

$$\langle\mathbf{k}'|\,U\,|\mathbf{k}\rangle = A\int \frac{1}{|\mathbf{r}|}\,\mathrm{e}^{-\lambda|\mathbf{r}|}\,\mathrm{e}^{\mathrm{j}(\mathbf{k}-\mathbf{k}').\mathbf{r}}\,\mathrm{d}^3\mathbf{r} \tag{9.3.99}$$

Write $(\mathbf{k} - \mathbf{k}') = \boldsymbol{\kappa}$ and choose spherical polar coordinates with the axis defined by $\theta = 0$ along the direction $\boldsymbol{\kappa}$. Then

$$\begin{aligned}(\mathbf{k} - \mathbf{k}').\mathbf{r} &= |\boldsymbol{\kappa}|\,|\mathbf{r}|\cos\theta = \kappa r\cos\theta \\ \mathrm{d}^3\mathbf{r} &= 2\pi r^2\,\mathrm{d}r\,\sin\theta\,\mathrm{d}\theta\end{aligned} \tag{9.3.100}$$

† Logan, R. A. and Peters, A. J., *J. appl. Phys.* **31**, 122, 1960.

and
$$\langle \mathbf{k}'|\,U\,|\mathbf{k}\rangle = 2\pi A \int \frac{1}{r}\,e^{-\lambda r}\,e^{j\kappa r\cos\theta}\,r^2\,dr\,\sin\theta\,d\theta \qquad (9.3.101)$$
$$= \frac{2\pi A}{j\kappa}\int e^{-\lambda r}\,(e^{+j\kappa r} - e^{-j\kappa r})\,dr$$
$$= \frac{2\pi A}{\kappa^2 + \lambda^2} \qquad (9.3.102)$$

The relaxation time is given by equation (9.3.71)
$$\frac{1}{\tau(E)} = \frac{2\pi A^2 m^*\sqrt{(2m^*E)}}{\hbar^4}\int \frac{(1-\cos\theta)\sin\theta}{(\kappa^2+\lambda^2)^2}\,d\theta \qquad (9.3.103)$$

For scattering over a simple spherical constant energy surface,
$$E(\mathbf{k}) = \frac{\hbar^2\,|\mathbf{k}|^2}{2m^*} = \frac{\hbar^2 k^2}{2m^*}$$
and
$$\kappa = 2k\sin(\theta/2) \qquad (9.3.104)$$

Hence,
$$\frac{1}{\tau(E)} = \frac{2\pi A^2 m^*\sqrt{(2m^*E)}}{\hbar^4\lambda^4}\int_0^1 \frac{8y^3\,dy}{[1+by^2]^2} \qquad (9.3.105)$$

with $y = \sin(\theta/2)$ and $b = (2k/\lambda)^2$.
Integrate this equation by parts,
$$\frac{1}{\tau(E)} = \frac{2\pi A^2 m^*\sqrt{(2m^*E)}}{\hbar^4\lambda^4}\cdot\frac{4}{b^2}\left\{\log_e(1+b) - \frac{b}{(1+b)}\right\} \qquad (9.3.106)$$

but
$$b = \left(\frac{2k}{\lambda}\right)^2 = \frac{8m^*E}{\hbar^2\lambda^2}$$

so that
$$\frac{1}{\tau(E)} = \frac{\pi A^2}{4}\cdot\frac{1}{(2m^*)^{1/2}}\cdot\frac{1}{E^{3/2}}\left\{\log_e(1+b) - \frac{b}{1+b}\right\} \qquad (9.3.107)$$

It is not very rewarding to attempt to interpret this equation in detail for a metallic conductor. Note, however, that the simple relation for the constant A given in equation (9.3.97) indicates that the resistivity will vary as the square of the valency difference for the impurity atom. This feature appears to be verified by experiment and is known as Linde's rule. Equation (9.3.107) finds a wider application to semiconductors. The term in brackets is taken to be constant over the energy range which is significant for the thermal average of $\tau(E)$. Consequently equation (9.3.39) leads to†

$$\langle t\rangle \propto \frac{1}{T}\,\frac{\int (m^*)^{1/2}E^3\,e^{-E/kT}\,dE}{\int E^{1/2}\,e^{-E/kT}\,dE} \qquad (9.3.108)$$

$$\langle t\rangle \propto (m^*)^{1/2}T^{3/2} \qquad (9.3.109)$$

† $\int e^{-\alpha u^2}\,u^7\,du = \frac{3}{\alpha^4}$; $\int e^{-\alpha u^2}\,u^2\,du = \frac{1}{4}\sqrt{\left(\frac{\pi}{\alpha^3}\right)}$

and the carrier mobility is

$$\mu \propto (m^*)^{-1/2} T^{3/2} \tag{9.3.110}$$

When ionized impurity scattering is the dominant relaxation process the carrier mobility should increase with increasing temperature. This conclusion seems to be verified by experiment and a temperature dependence lying between T and $T^{3/2}$ appears to be characteristic of the ionized impurity mobility for electrons in germanium.

At moderately low temperatures the ionized impurity scattering and the lattice vibration scattering may each make a significant contribution to the total relaxation process. In these circumstances the transition probability will be the sum of the individual probabilities (which are assumed to describe independent processes) and so equation (9.3.53), for example, leads to

$$\left[\frac{1}{\tau(E)}\right]_{\text{total}} = \left[\frac{1}{\tau(E)}\right]_{\text{lattice}} + \left[\frac{1}{\tau(E)}\right]_{\text{ionized}} \tag{9.3.111}$$

More generally, it is customary to write

$$\left[\frac{1}{\tau(E)}\right]_{\text{total}} = \sum_m \frac{1}{\tau_m(E)} \tag{9.3.112}$$

where the $\tau_m(E)$ refer to the different independent scattering processes arising from thermal vibrations, ionized impurities, neutral impurities, dislocations, strains, etc.

9.4 Electrical Conduction in a Magnetic Field

A magnetic field has a significant influence upon the motion of the charge carriers which are responsible for electrical conduction in a material. New effects such as magneto-resistance, cyclotron resonance and the Hall effect occur. We shall emphasize the physical features underlying these phenomena by first examining a simple model using a classical equation of motion. In a subsequent section we shall undertake a more complete discussion involving the Boltzmann transport equation.

Consider a single charge carrier system with spherical energy surfaces in **k** space which are described, for example, by equation (9.3.1),

$$E(\mathbf{k}) = \frac{\hbar^2 |\mathbf{k}|^2}{2m^*} = \frac{\hbar^2 k^2}{2m^*} \tag{9.4.1}$$

Neglecting collision processes for the moment, the motion of a charge is described by equation (1.1.7) of Chapter 1 and equation (9.2.56)

$$\hbar \frac{d\mathbf{k}}{dt} = \mathbf{F} = q\left[\mathbf{E} + \frac{1}{c}(\mathbf{v} \wedge \mathbf{B})\right] \tag{9.4.2}$$

Assume that the material is non-magnetic in the sense that $\mathbf{B} = \mathbf{H}$, and note that equation (9.4.1) implies that $\mathbf{v}(\mathbf{k}) = (\hbar\mathbf{k}/m^*)$. The effect of the electric field alone is to give each representative point in **k** space a velocity

$$\dot{\mathbf{k}} = \frac{q\mathbf{E}}{\hbar} \tag{9.4.3}$$

whereas the effect of a steady magnetic field alone is to circulate the representative points around a contour on a constant energy surface in **k** space. There is no energy change for the carrier under the action of the magnetic field **H** since,

$$\mathbf{F}.\mathbf{v}(\mathbf{k}) = \frac{q}{c}[\mathbf{v}(\mathbf{k}) \wedge \mathbf{H}].\mathbf{v}(\mathbf{k}) = 0 \qquad (9.4.4)$$

Consider therefore a plane perpendicular to the direction of **H.** Equation (9.4.2) shows that when $\mathbf{E} = 0$, $d\mathbf{k}/dt$ lies in this plane. The representative point initially identified by **k** will move around the contour determined by the intersection of this plane with the constant energy surface appropriate to **k,** as shown in Fig. 9.9.

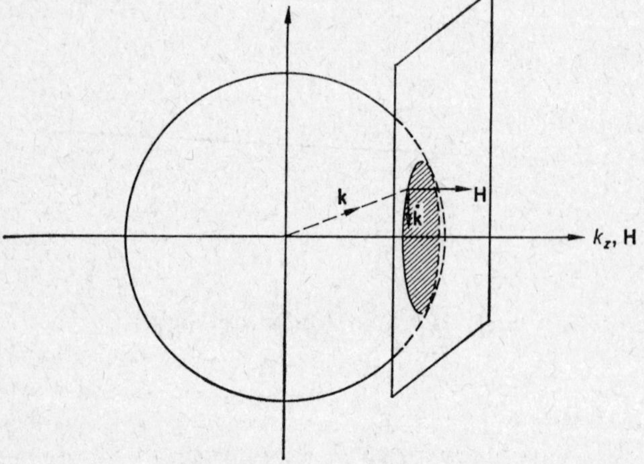

FIG. 9.9 Motion of a representative point in **k** space under the action of a steady magnetic field.

For most experimental measurements the force arising from the magnetic field is greater than (and often considerably greater than) the force arising from the electric field. Consider the electric force acting on an electron, $\mathbf{F}_e = -|e|\,\mathbf{E} = -1{\cdot}6 \times 10^{-12}\mathbf{E}$ dynes when **E** is measured in volts cm^{-1}. For a metal with conductivity $\sigma \sim 0{\cdot}5 \times 10^6$ (ohm cm)$^{-1}$, and a current density 1 amp cm^{-2}, $|\mathbf{E}| \sim 2 \times 10^{-6}$ volt cm^{-1} and $|\mathbf{F}_e| \sim 3 \times 10^{-18}$ dynes. For a semiconductor we take $\sigma \sim 10^{-2}$ (ohm cm)$^{-1}$, and the corresponding value for $|\mathbf{E}|$ is 10^2 volt cm^{-1}, with $|\mathbf{F}_e| \sim 1{\cdot}6 \times 10^{-10}$ dynes. The magnetic force is $|\mathbf{F}_h| \sim 1{\cdot}6 \times 10^{-20}vH$ dynes, with v measured in cm s^{-1} and H in gauss. For a metal, $v \sim 5 \times 10^{-8}$ cm s^{-1} at the Fermi surface and so, with $H = 10^3$ gauss, $|\mathbf{F}_h| \sim 10^{-8}$ dynes. This force is therefore considerably greater than F_e which was estimated above. Even in the case of a semiconductor, for which we take $v \sim 10^7$ cm s^{-1} at room temperature, to give $|\mathbf{F}_h| \sim 1{\cdot}6 \times 10^{-10}$ dynes the electric and magnetic forces have the same order of magnitude.

9.4.1 Magnetoresistance and the Hall Effect

If a magnetic field **H** is applied to a conductor in a direction perpendicular to the current flow **I**, an e.m.f. is generated across the conductor in a direction perpendicular to both **H** and **I**. This is the Hall effect. The electrical resistance of a conductor is also modified by the presence of a magnetic field. There are two configurations which are commonly used for studying the magnetoresistance, (a) the current flow **I** is along the direction **H** and the experiment measures the longitudinal effect, (b) the current flow **I** is perpendicular to the direction of **H** and the experiment measures the transverse effect. Experiments involving the Hall effect and the magnetoresistance are usually carried out under isothermal conditions in order to avoid any complications which may arise from thermoelectric phenomena.

Include the effect of scattering processes in the equation of motion for a charge carrier by writing equation (9.4.2) as

$$\hbar\frac{d\mathbf{k}}{dt}+\frac{\hbar}{\tau}\mathbf{k}=q\left[\mathbf{E}+\frac{1}{c}(\mathbf{v}\wedge\mathbf{H})\right] \tag{9.4.5}$$

Evidently τ is a relaxation time. For spherical energy surfaces described by equation (9.4.1), we may write,

$$m^*\frac{d\mathbf{v}}{dt}+\frac{m^*}{\tau}\mathbf{v}=q\left[\mathbf{E}+\frac{1}{c}(\mathbf{v}\wedge\mathbf{H})\right] \tag{9.4.6}$$

and for $\mathbf{H}=\hat{\mathbf{k}}H_z$,

$$m^*\frac{dv_x}{dt}+\frac{m^*}{\tau}v_x=q\left[E_x+\frac{1}{c}v_yH_z\right] \tag{9.4.7a}$$

$$m^*\frac{dv_y}{dt}+\frac{m^*}{\tau}v_y=q\left[E_y-\frac{1}{c}v_xH_z\right] \tag{9.4.7b}$$

$$m^*\frac{dv_z}{dt}+\frac{m^*}{\tau}v_z=qE_z \tag{9.4.8}$$

Note that equation (9.4.8) does not depend upon the magnetic field. There is no longitudinal magnetoresistance effect for this simple model system. Thus, if E_z is taken to be constant, the velocity v_z after the elapse of a time interval long compared with τ, is $v_z(\infty)=(q\tau/m^*)E_z$, and the electrical conductivity is given by the standard formula $\sigma_z=(Nq^2\tau)/m^*$. The magnetic field dependent effects arise from equations (9.4.7a) and (9.4.7b). An elegant technique for solving these equations makes use of the complex summation $(v_x+jv_y)\equiv v_+$. Evidently,

$$m^*\frac{dv_+}{dt}+\left(\frac{m^*}{\tau}+\frac{jqH_z}{c}\right)v_+=q(E_x+jE_y) \tag{9.4.9}$$

which may be written as,

$$\frac{dv_+}{dt}+\left(\frac{1}{\tau}+j\omega_0\right)v_+=\frac{q}{m^*}(E_x+jE_y) \tag{9.4.10}$$

with

$$\omega_0=\frac{qH_z}{m^*c} \tag{9.4.11}$$

Take $E_x + jE_y = (E_x^0 + jE_x^0)\, e^{j\omega t}$, then, writing for brevity

$$\left(\frac{1}{\tau} + j\omega_0\right) = \alpha,$$

$$v_+ = v_0\, e^{-\alpha t} + \frac{q\,(E_x^0 + jE_y^0)}{m^*(j\omega + \alpha)}[e^{j\omega t} - e^{-\alpha t}] \tag{9.4.12}$$

and the steady state solution is

$$v_+ = \frac{q}{m^*}\frac{(E_x^0 + jE_y^0)\, e^{j\omega t}}{\left[\dfrac{1}{\tau} + j(\omega + \omega_0)\right]} \tag{9.4.13}$$

Consequently v_+ will exhibit some kind of resonance response when $\omega = -\omega_0$. This is the cyclotron resonance phenomenon which, for a simple electron system, occurs when $\omega = (|e|\, H_z)/(m^*c)$.

It is customary to measure the Hall effect and the magnetoresistance with time independent fields. Putting $\omega = 0$ in equation (9.4.13),

$$v_+(0) = \frac{q}{m^*}\frac{(E_x^0 + jE_y^0)}{\left(\dfrac{1}{\tau} + j\omega_0\right)} \tag{9.4.14}$$

For time independent fields, v_x and v_y are real quantities which can be obtained immediately from the real and imaginary parts of $v_+(0)$. The current density will be given by

$$J_x = \frac{Nq}{c}\langle v_x\rangle; \qquad J_y = \frac{Nq}{c}\langle v_y\rangle \tag{9.4.15}$$

where $\langle v_x\rangle$ and $\langle v_y\rangle$ are averages taken over the distribution function (compare equation (9.3.11)). Separating the real and imaginary parts, we obtain,

$$J_x = \frac{Nq^2}{m^*c}\left(E_x^0\left\langle\frac{\tau}{1 + \omega_0^2\tau^2}\right\rangle + E_y^0\omega_0\left\langle\frac{\tau^2}{1 + \omega_0^2\tau^2}\right\rangle\right) \tag{9.4.16}$$

$$J_y = \frac{Nq^2}{m^*c}\left(E_y^0\left\langle\frac{\tau}{1 + \omega_0^2\tau^2}\right\rangle - E_x^0\omega_0\left\langle\frac{\tau^2}{1 + \omega_0^2\tau^2}\right\rangle\right) \tag{9.4.17}$$

The Hall probes may be identified with the y-axis of coordinates and measure E_y^0. The probes are effectively open circuit and so we take $J_y = 0$. Hence

$$\frac{E_y^0}{E_x^0} = \omega_0\frac{\langle\tau^2(1 + \omega_0^2\tau^2)^{-1}\rangle}{\langle\tau(1 + \omega_0^2\tau^2)^{-1}\rangle} \tag{9.4.18}$$

$$J_x = \frac{Nq^2}{m^*c}E_x^0\left[\langle\tau(1 + \omega_0^2\tau^2)^{-1}\rangle + \frac{\omega_0^2\{\langle\tau^2(1 + \omega_0^2\tau^2)^{-1}\rangle\}^2}{\langle\tau(1 + \omega_0^2\tau^2)^{-1}\rangle}\right] \tag{9.4.19}$$

If τ is independent of energy or, as in the case of a metal, is to be evaluated at the Fermi energy, the averages are unnecessary and,

$$J_x = \frac{Nq^2\tau}{m^*c}E^0 \tag{9.4.20}$$

$$E_y^0 = \omega_0\tau E_x^0 = \frac{1}{Nq}J_xH_z \tag{9.4.21}$$

In these circumstances there is no transverse magnetoresistance ($\sigma_x \equiv cJ_x/E_x$ does not depend upon H_z) and the Hall coefficient, R, which is defined by the relation

$$R = \frac{E_y^0}{J_x H_z} \tag{9.4.22}$$

measures directly the sign of the charge and the number density of the carriers.†

It appears that a simple metal with a spherical Fermi surface will exhibit neither a longitudinal nor a transverse magnetoresistance. This conclusion remains unchanged after carrying through a more rigorous analysis based on the Boltzmann transport equation. In practice, however, all metals are found to show these effects and generally the longitudinal and transverse magnetoresistances are of the same order of magnitude. The empirical relation for the change in resistivity, ρ, as a function of magnetic field is,

$$\frac{\Delta\rho}{\rho_0} = AH_z^2 \tag{9.4.23}$$

where ρ_0 is the resistivity in zero field.‡ In the polyvalent metals the magnetoresistance may be explained immediately by the complex nature of the Fermi surface and the existence of more than one type of carrier (for example, there may be holes and electrons or the effective mass may vary over the Fermi surface). For the simple alkali metals, the existence of a magnetoresistance effect probably indicates that the relaxation time τ varies over the Fermi surface.

In the case of a semiconductor, it is necessary to take account of the energy dependence of τ. The experimental conditions frequently satisfy the relation $\omega_0\tau \ll 1$. We may therefore expand the denominators in equations (9.4.18) and (9.4.19), to obtain,

$$J_x = \frac{Nq^2}{m^*c} E_x^0 \left(\langle\tau\rangle - \omega_0^2\langle\tau^3\rangle + \omega_0^2 \frac{\langle\tau^2\rangle^2}{\langle\tau\rangle - \omega_0^2\langle\tau^3\rangle} \right) \tag{9.4.24}$$

$$= \frac{Nq^2}{m^*} E_x^0 \left(\frac{\langle\tau\rangle^2 - 2\omega_0\langle\tau^3\rangle\langle\tau\rangle + \omega_0^2\langle\tau^2\rangle^2}{\langle\tau\rangle - \omega_0^2\langle\tau^3\rangle} \right) \tag{9.4.25}$$

and so

$$E_x^0 = c\rho_0 J_x \left(1 + \omega_0^2 \frac{\langle\tau^3\rangle\langle\tau\rangle - \langle\tau^2\rangle^2}{\langle\tau\rangle^2} \right) \tag{9.4.26}$$

with

$$\rho_0 = \frac{1}{\sigma_0} = \frac{m^*}{Nq^2} \cdot \frac{1}{\langle\tau\rangle} \tag{9.4.27}$$

† For a simple metal $|R| \sim 10^4$ cm³ coulomb⁻¹ corresponding to approximately one electron per atom. Note the relation $R = 10^8\, V_y z / I_x H_z$ gives R in cm^3 coulomb^{-1} when V_y is measured in volts, z in centimetres, I_x in amperes and H_z in gauss.

‡ For a pure metal at liquid helium temperatures the order of magnitude of A is 10^{-10} gauss^{-2}.

Hence the transverse magnetoresistance is given by

$$\frac{\Delta\rho}{\rho_0} = \frac{\rho(H) - \rho_0}{\rho_0} = \frac{1}{\rho_0}\left(\frac{E_x^0}{cJ_x}\right) - 1 \tag{9.4.28}$$

$$= \left(\frac{q}{m^*c}\right)^2\left(\frac{\langle\tau^3\rangle\langle\tau\rangle - \langle\tau^2\rangle^2}{\langle\tau\rangle^2}\right)H_z^2 \tag{9.4.29}$$

which may be written as†

$$\frac{\Delta\rho}{\rho_0} = \xi_\rho\left(\frac{q}{m^*c}\right)^2\langle\tau^2\rangle\, H_z^2 \tag{9.4.30}$$

with ξ_ρ the dimensionless factor,

$$\xi_\rho = \frac{\langle\tau^2\rangle}{\langle\tau\rangle^2}\left[\frac{\langle\tau^3\rangle\langle\tau\rangle}{\langle\tau^2\rangle^2} - 1\right] \tag{9.4.31}$$

When τ is determined by lattice wave scattering, $\tau(E) \propto E^{-1/2}T^{-1}$ and $\xi_\rho = 0{\cdot}32$. Alternatively, when ionized impurity scattering is dominant $\tau(E) \propto E^{3/2}$ and $\xi_\rho = 1{\cdot}08$.

The Hall coefficient is obtained from equations (9.4.18) and (9.4.26),

$$\frac{E_y^0}{E_x^0} = \omega_0\frac{\langle\tau^2\rangle}{\langle\tau\rangle}; \qquad E_x^0 = \rho_0 cJ_x \tag{9.4.32}$$

Evidently

$$E^0 = \frac{1}{Nq}\frac{\langle\tau^2\rangle}{\langle\tau\rangle^2}J_xH_z \tag{9.4.33}$$

and‡

$$R = \frac{1}{Nq}\frac{\langle\tau^2\rangle}{\langle\tau\rangle^2} = \xi_R\frac{1}{Nq} \tag{9.4.34}$$

When τ is determined by lattice wave scattering $\xi_R = 3\pi/8 = 1{\cdot}18$, whereas, when ionized impurity scattering is dominant $\xi_R = 315\pi/512 = 1{\cdot}93$.

It is evident that the simple model which we have considered, although adequate for establishing the physical principles underlying the Hall effect and the magnetoresistance, is unlikely to yield quantitative results. The sensitivity of the low field phenomena to the correct averages for the relaxation times presents a serious problem in the interpretation of the experimental data. Moreover, the magnitudes of these magnetic effects are very dependent upon the detailed shape of the constant energy surfaces and in an actual semiconductor it will generally be necessary to take account of the simultaneous presence of electrons and holes. Some simplification in the Hall effect may be achieved by working with sufficiently high magnetic fields, for which $\omega_0\tau \gg 1$. In order to satisfy this condition with presently available laboratory fields it is usually necessary to carry out the experiments at low temperatures. Provided that quantum effects can be neglected, the con-

† For a semiconductor $\Delta\bar{\rho}/\rho H^2 \sim 10^{-10}$ gauss^{-2}.

‡ For high purity germanium and silicon $R \sim 10^5$ cm^3 coulomb^{-1} at room temperature.

tributions from the different types of carriers can be summed to give, in place of equations (9.4.16) and (9.4.17),

$$J_x = \sum_r \frac{n_r q_r^2}{m_r^* c}\left(E_x^0\left\langle\frac{\tau_r}{1+\omega_{0r}^2\tau_r^2}\right\rangle + E_y^0\omega_{0r}\left\langle\frac{\tau_r^2}{1+\omega_{0r}^2\tau_r^2}\right\rangle\right) \quad (9.4.35)$$

$$J_y = \sum_r \frac{n_r q_r^2}{m_r^* c}\left\{E_y^0\left\langle\frac{\tau_r}{1+\omega_{0r}^2\tau_r^2}\right\rangle - E_x^0\omega_{0r}\left\langle\frac{\tau_r^2}{1+\omega_{0r}^2\tau_r^2}\right\rangle\right\} \quad (9.4.36)$$

Taking $J_y = 0$, $\omega_{0r}\tau_r \gg 1$,

$$E_x^0 = \frac{1}{\omega_{0r}}\left\langle\frac{1}{\tau_r}\right\rangle E_y^0 \ll E_y^0 \quad (9.4.37)$$

and

$$J_x = \sum_r \frac{n_r q_r^2}{m_r^* c}\cdot\frac{1}{\omega_{0r}}E_y^0 = \left(\sum_r \pm n_r\right)\frac{|e|\,E_y^0}{H_z} \quad (9.4.38)$$

where the positive and negative signs refer to holes and electrons, respectively. The Hall coefficient at sufficiently high magnetic fields is therefore

$$R_\infty = \frac{1}{|e|}\cdot\frac{1}{\left(\sum_r \pm n_r\right)} \quad (9.4.39)$$

and is independent of the averages which so seriously affect the coefficient when measured in low magnetic fields.

We shall conclude this discussion of magnetic effects for the simple charge system by deriving an explicit relation for the power absorption arising from cyclotron resonance. Let the electric field components E_x, E_y, retain the exponential time dependence† $e^{j\omega t}$. We shall need to evaluate v_x in order to derive the mean power loss $\langle P\rangle$ using equation (1.5.44) of Chapter 1.

$$\langle P\rangle = \tfrac{1}{2}\,\mathrm{Re}\,[c\mathbf{J}.\mathbf{E}^*] \quad (9.4.40)$$

It is not possible to obtain v_x immediately upon taking the real part of v_+ in equation (9.4.12) because v_x and v_y are now, themselves, complex quantities. Use equations (9.4.7a) and (9.4.7b) to derive the differential equation for $v_- \equiv v_x - jv_y$,

$$\frac{dv_-}{dt} + \left(\frac{1}{\tau} - j\omega_0\right)v_- = \frac{q}{m^*}(E_x - jE_y) \quad (9.4.41)$$

The steady state solution corresponding to equation (9.4.13) is

$$v_- = \left(\frac{q}{m^*}\right)\frac{(E_x^0 - jE_y^0)\,e^{j\omega t}}{\{(1/\tau) + j(\omega - \omega_0)\}} \quad (9.4.42)$$

Assume that an oscillating electric field is applied along the x-direction of coordinates so that $E_y = 0$. Evaluate $v_x = \frac{1}{2}(v_+ + v_-)$.

$$v_x = \frac{1}{2}\left(\frac{q}{m^*}\right)\frac{2\tau(1 + j\omega\tau)}{(1 + j\omega\tau)^2 + \omega_0^2\tau^2}E_x^0\,e^{j\omega t} \quad (9.4.43)$$

† We neglect propagation effects, see footnote, page 496, and page 502.

Neglect the mathematical complexity introduced by taking averages, and write

$$J_x = \frac{1}{c} N q v_x$$

so that,

$$\begin{aligned}\langle P \rangle &= \tfrac{1}{2}\,\mathrm{Re}\,[cJ_x E_x^*] \\ &= \tfrac{1}{2}\frac{Nq^2\tau}{m^*}(E_x^0)^2\,\mathrm{Re}\left[\frac{(1+\mathrm{j}\omega\tau)}{1+(\omega_0^2-\omega^2)\tau^2+2\mathrm{j}\omega\tau}\right] \qquad (9.4.44)\\ &= \tfrac{1}{2}\sigma_0(E_x^0)^2\left[\frac{1+(\omega_0^2+\omega^2)\tau^2}{\{1+(\omega_0^2-\omega^2)\tau^2\}^2+4\omega^2\tau^2}\right] \qquad (9.4.45)\end{aligned}$$

with $\sigma_0 = Nq^2\tau/m^*$. Evidently there is a maximum in the power absorption near $\omega = \pm\omega_0$. This is the resonance response referred to previously on page 490, which provides a direct method of measuring the effective mass for the charge carriers. At resonance, where $H_z = (m^*c/q)\omega$, the power absorption is greater than that in zero magnetic field by a factor $(\omega^2\tau^2)/2$ provided that $\omega\tau \gg 1$. This is a large effect, for example $\omega^2\tau^2/2 = 50$ if $\omega\tau = 10$.

9.4.2 Boltzmann Transport Equation

The discussion of the distribution function in Section 9.3 was confined to the particular case of a uniform spatial distribution of carriers under the action of an electric field. The use of the distribution function is not necessarily so restricted. We may extend the previous discussion to include magnetic fields and a non-uniform spatial distribution of carriers such as, for example, arises when temperature gradients are present in the material. Assume that there exists a distribution function $f(\mathbf{k}, \mathbf{r}, t)$ which measures the number of carriers in the state $\psi_{\mathbf{k}}$ within the elementary volume space at the point $\mathbf{r}$ at time t. Written explicitly, the number of carriers in the volume element $\mathrm{d}^3\mathbf{r}$ having wave vectors in the range $\mathrm{d}^3\mathbf{k}$ at time t is

$$\mathrm{d}N(\mathbf{k}, \mathbf{r}, t) = f(\mathbf{k}, \mathbf{r}, t)\, g(\mathbf{k})\, \mathrm{d}^3\mathbf{k}\, \mathrm{d}^3\mathbf{r} \qquad (9.4.46)$$

where $g(\mathbf{k})$ is again the density of states in $\mathbf{k}$ space. The number of carriers in this six-dimensional elementary unit of $\mathbf{k}$ and volume space may change because of the action of external fields, because of spatial diffusion and because of relaxation processes. An external field will cause representative points to enter and leave the elementary unit $\mathrm{d}^3\mathbf{k}$ and, if the field is non-uniform, will modify the carrier distribution as a function of $\mathbf{r}$. In $\mathbf{k}$ space, the net effect is given by extending equation (9.3.18) to an arbitrarily directed force $\mathbf{F}$ which will give rise to a velocity for points in $\mathbf{k}$ space according to the relation $\mathbf{F} = \hbar(\mathrm{d}\mathbf{k}/\mathrm{d}t)$. Representative points will enter and leave the elementary unit $\mathrm{d}^3\mathbf{k}$ through each face of the cube $\mathrm{d}k_x\,\mathrm{d}k_y\,\mathrm{d}k_z$, so that equation (9.3.18) becomes,

$$\begin{aligned}\left(\frac{\mathrm{d}N}{\mathrm{d}t}\right)_{\text{field}} &= -\left(\frac{\partial f}{\partial k_x}\frac{\mathrm{d}k_x}{\mathrm{d}t} + \frac{\partial f}{\partial k_y}\frac{\mathrm{d}k_y}{\mathrm{d}t} + \frac{\partial f}{\partial k_z}\frac{\mathrm{d}k_z}{\mathrm{d}t}\right) g(\mathbf{k})\,\mathrm{d}^3\mathbf{k}\,\mathrm{d}^3\mathbf{r} \qquad (9.4.47)\\ &= -\left[\frac{\mathrm{d}\mathbf{k}}{\mathrm{d}t}\cdot\nabla_{\mathbf{k}} f\right] g(\mathbf{k})\,\mathrm{d}^3\mathbf{k}\,\mathrm{d}^3\mathbf{r} \qquad (9.4.48)\end{aligned}$$

Changes in the spatial distribution of the carriers, whether arising from non-uniform fields or from temperature gradients, may be evaluated collectively and the originating processes need not be separately identified at present. If the velocity of a carrier is $\mathbf{v}(\mathbf{k}, \mathbf{r}, t)$, then the net accumulation of carriers within the elementary unit $d^3\mathbf{r}$, in unit time, is obtained by considering the motion across each face of the cube $dx\,dy\,dz$. Thus for the faces at x and $x + dx$, there will be a contribution

$$[f(\mathbf{k}, x, y, z, t) - f(\mathbf{k}, x + dx, y, z, t)]v_x(\mathbf{k}, x, y, z, t)\,dy\,dz\,g(\mathbf{k})\,d^3\mathbf{k} \quad (9.4.49)$$

$$= -\frac{\partial f}{\partial x} v_x(\mathbf{k}, \mathbf{r}, t)\,d^3\mathbf{r}\,g(\mathbf{k})\,d^3\mathbf{k} \quad (9.4.50)$$

and the total rate of accumulation in the element $d^3\mathbf{r}$, is

$$\left[\frac{dN}{dt}\right]_{\text{space}} = -\left[\mathbf{v}\,.\,\nabla_{\mathbf{r}} f\right] g(\mathbf{k})\,d^3\mathbf{k}\,d^3\mathbf{r} \quad (9.4.51)$$

Relaxation processes will give rise to changes in wave vector $\mathbf{k}$ within a given volume element $d^3\mathbf{r}$. These will be described by equations (9.3.20) and (9.3.21) which were derived previously,

$$\left[\frac{dN}{dt}\right]_{\text{scattering}} = \left[\frac{df}{dt}\right]_{\text{scattering}} g(\mathbf{k})\,d^3\mathbf{k}\,d^3\mathbf{r} \quad (9.4.52)$$

$$= -\frac{f(\mathbf{k}, \mathbf{r}, t) - f^0(\mathbf{k}, \mathbf{r}, t)}{\tau(\mathbf{k})} g(\mathbf{k})\,d^3\mathbf{k}\,d^3\mathbf{r} \quad (9.4.53)$$

Consequently the rate of change of the number of carriers in the elementary unit $d^3\mathbf{k}\,d^3\mathbf{r}$ due to all causes is given by

$$\left[\frac{dN}{dt}\right]_{\text{total}} = \left[\frac{dN}{dt}\right]_{\text{fields}} + \left[\frac{dN}{dt}\right]_{\text{space}} + \left[\frac{dN}{dt}\right]_{\text{scattering}} \quad (9.4.54)$$

and hence we may write

$$\frac{df}{dt} = -\frac{d\mathbf{k}}{dt}\cdot\nabla_{\mathbf{k}} f - \mathbf{v}\,.\,\nabla_{\mathbf{r}} f - \frac{f - f^0}{\tau(\mathbf{k})} \quad (9.4.55)$$

This is the Boltzmann transport equation.

If the distribution function is uniform in space $\nabla_{\mathbf{r}} f = 0$. The steady state corresponds to $df/dt = 0$ when a time invariant electric field is applied to the material. The current density is therefore derived from

$$\mathbf{J} = \frac{q}{c}\int \mathbf{v}(\mathbf{k})\,f(\mathbf{k})\,g(\mathbf{k})\,d^3\mathbf{k} \quad (9.4.56)$$

with $f(\mathbf{k})$ the solution of the differential equation

$$f(\mathbf{k}) + \tau(\mathbf{k})\frac{d\mathbf{k}}{dt}\,.\,\nabla_{\mathbf{k}} f(\mathbf{k}) - f^0(\mathbf{k}) = 0 \quad (9.4.57)$$

Writing $\hbar \dfrac{d\mathbf{k}}{dt} = q\mathbf{E}$, we obtain

$$f(\mathbf{k}) = f^0(\mathbf{k}) - \frac{q\tau(\mathbf{k})}{\hbar}[\mathbf{E}\,.\,\nabla_{\mathbf{k}} f(\mathbf{k})] \quad (9.4.58)$$

which, to sufficient approximation, may be written†

$$f(\mathbf{k}) = f^0(\mathbf{k}) - q\,\tau(\mathbf{k})[\mathbf{E}.\mathbf{v}(\mathbf{k})]\,\frac{\partial f^0}{\partial E} \tag{9.4.59}$$

Hence

$$\mathbf{J} = \frac{q^2}{c}\int \tau(\mathbf{k})\,\mathbf{v}(\mathbf{k})[\mathbf{E}.\mathbf{v}(\mathbf{k})]\left(-\frac{\partial f^0}{\partial E}\right) g(\mathbf{k})\,\mathrm{d}^3\mathbf{k} \tag{9.4.60}$$

which is a slightly more general form of equation (9.3.25b).

9.4.3 Cyclotron Resonance

As an example of the technique for solving the Boltzmann equation when a time-dependent electric field is present together with a constant magnetic field, we consider the description of cyclotron resonance. When time-dependent fields are present $\mathrm{d}f/\mathrm{d}t$ will no longer be zero in the steady state. Assuming that the distribution is uniform in space, equation (9.4.55) becomes

$$\frac{\mathrm{d}f}{\mathrm{d}t} = -\frac{\mathrm{d}\mathbf{k}}{\mathrm{d}t}.\nabla_{\mathbf{k}}f - \frac{f - f^0}{\tau(\mathbf{k})} \tag{9.4.61}$$

Equation (9.4.59) suggests that we may seek a solution in the form

$$f(\mathbf{k}, t) = f^0(\mathbf{k}) - G(\mathbf{k}, t)\,\frac{\partial f^0}{\partial E} \tag{9.4.62}$$

For cyclotron resonance an electric field $\mathbf{E}$ with harmonic time dependence is used together with a steady field $\mathbf{H}$.‡

$$\mathbf{E} = \mathbf{E}_0\,\mathrm{e}^{\mathrm{j}\omega t}; \qquad \hbar\frac{\mathrm{d}\mathbf{k}}{\mathrm{d}t} = q\left[\mathbf{E} + \frac{1}{c}\mathbf{v}\wedge\mathbf{H}\right] \tag{9.4.63}$$

It seems reasonable to suppose that $f(\mathbf{k}, t)$ and $G(\mathbf{k}, t)$ will also be harmonic time functions. Assuming $G(\mathbf{k}, t) = G_0(\mathbf{k})\,\mathrm{e}^{\mathrm{j}\omega t}$, equation (9.4.62), leads to,

$$\frac{\mathrm{d}f(\mathbf{k}, t)}{\mathrm{d}t} = -\mathrm{j}\omega\, G(\mathbf{k}, t)\,\frac{\partial f^0}{\partial E} \tag{9.4.64}$$

and, to first order,

$$\nabla_{\mathbf{k}}f = \hbar\,\mathbf{v}(\mathbf{k})\,\frac{\partial f^0}{\partial E} - \nabla_{\mathbf{k}}\,G(\mathbf{k}, t)\,\frac{\partial f^0}{\partial E} \tag{9.4.65}$$

Hence we obtain from equation (9.4.61)

$$-\mathrm{j}\omega\, G(\mathbf{k}, t) = -\frac{q}{\hbar}\left[\mathbf{E} + \frac{1}{c}\mathbf{v}\wedge\mathbf{H}\right].[\hbar\mathbf{v} - \nabla_{\mathbf{k}}\,G(\mathbf{k}, t)] + \frac{1}{\tau}\,G(\mathbf{k}, t) \tag{9.4.66}$$

† $\nabla_{\mathbf{k}}f = \left(\hat{\mathbf{i}}\frac{\partial f}{\partial k_x} + \hat{\mathbf{j}}\frac{\partial f}{\partial k_y} + \hat{\mathbf{k}}\frac{\partial f}{\partial k_z}\right) = \left(\hat{\mathbf{i}}\frac{\partial f}{\partial E}\frac{\partial E}{\partial k_x} + \hat{\mathbf{j}}\frac{\partial f}{\partial E}\frac{\partial E}{\partial k_y} + \hat{\mathbf{k}}\frac{\partial f}{\partial E}\frac{\partial E}{\partial k}\right)$

$= \left(\hat{\mathbf{i}}\frac{\partial E}{\partial k_x} + \hat{\mathbf{j}}\frac{\partial E}{\partial k_y} + \hat{\mathbf{k}}\frac{\partial E}{\partial k_z}\right)\frac{\partial f}{\partial E} = (\nabla_{\mathbf{k}}E)\,\frac{\partial f}{\partial E} \approx (\nabla_{\mathbf{k}}E)\,\frac{\partial f^0}{\partial E}$

‡ We neglect propagation effects. The present discussion therefore applies most directly to semiconducting materials. The underlying physical features, however, also apply to metals, where, with an appropriate field configuration, 'Azbel-Kaner cyclotron resonance' is observed.

and the equation for $G_0(\mathbf{k})$ is in first order,

$$q\mathbf{E}_0 . \mathbf{v} - \frac{q}{\hbar c}(\mathbf{v} \wedge \mathbf{H}) . \nabla_{\mathbf{k}} G_0 - \left(\frac{1}{\tau} + j\omega\right) G_0 = 0 \qquad (9.4.67)$$

since $(\mathbf{v} \wedge \mathbf{H}) . \mathbf{v} = 0$.

The solution of this equation may be carried through by first defining a new variable t_H from the relation

$$\hbar \frac{d\mathbf{k}}{dt_H} = \frac{q}{c}(\mathbf{v} \wedge \mathbf{H}) \qquad (9.4.68)$$

Evidently t_H is the time variable which would be appropriate to the motion of a representative point in **k** space when no electric fields or relaxation processes are present. In terms of this variable, equation (9.4.68) shows that a representative point in **k** space will describe a contour which is given by the intersection of a plane perpendicular to **H** and a constant energy surface.† Such a contour is shown in Fig. 9.10, where the magnetic field is directed

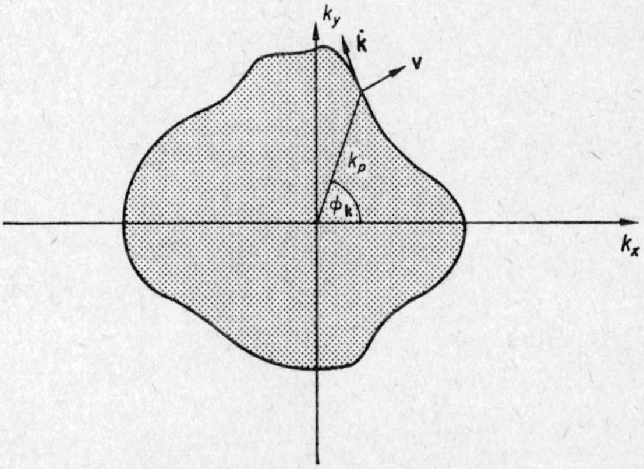

FIG. 9.10 The motion of a representative point along a constant energy contour in **k** space. The magnetic field is normal to the plane of the paper.

into the plane of the paper and the energy increases away from the origin. The motion of a representative point in **k** space is a periodic function of t_H. To determine the period T_H which characterizes the circulation around the contour take cylindrical coordinates denoted by k_ρ, ϕ_k, k_z with the axis of k_z along the direction of **H**. Evidently

$$k_x = k_\rho \cos \phi_k, \quad k_y = k \sin \phi_k, \quad k_z = k_z \qquad (9.4.69a)$$

and

$$\mathbf{v}(\mathbf{k}) = \frac{1}{\hbar} \nabla_{\mathbf{k}} \mathbf{E} = \frac{1}{\hbar}\left[\mathbf{a}_\rho \frac{\partial E}{\partial k_\rho} + \mathbf{a}_\phi \frac{1}{k_\rho} \frac{\partial E}{\partial \phi_{\mathbf{k}}} + \mathbf{a}_z \frac{\partial E}{\partial k_z}\right] \qquad (9.4.69b)$$

† Compare the discussion on page 488.

with $\mathbf{a}_\rho$, $\mathbf{a}_\phi$, $\mathbf{a}_z$ the mutually orthogonal unit vectors for the cylindrical coordinate system. Consider the component of equation (9.4.68) along $\mathbf{a}_\phi$.

$$k_\rho \frac{\mathrm{d}\phi_k}{\mathrm{d}t} = -\frac{q}{\hbar^2 c} H_z \left(\frac{\partial E}{\partial k_\rho}\right) \tag{9.4.70}$$

and so

$$\oint \mathrm{d}t = T_H = -\frac{\hbar^2 c}{qH_z} \oint_0^{2\pi} \frac{k_\rho \, \mathrm{d}\phi_k}{(\partial E/\partial k_\rho)} \equiv \frac{2\pi}{\omega_c} \tag{9.4.71}$$

In general the value of T_H will depend upon the particular coordinate k_z at which the plane perpendicular to **H** is drawn. Only when the constant energy surface is a sphere or an ellipsoid will it be possible for T_H to be independent of k_z.

Note that equation (9.4.71) defines a cyclotron effective mass for the carriers. Define m^*_c by the relation

$$\omega_c = \frac{qH_z}{m_c^* c} \tag{9.4.72}$$

then

$$m_c^* = -\frac{\hbar^2}{2\pi} \oint_0^{2\pi} \frac{k_\rho \, \mathrm{d}\phi_k}{(\partial E/\partial k_\rho)} \tag{9.4.73}$$

In order to evaluate the power absorption due to cyclotron resonance, use equation (9.4.68) to write equation (9.4.67) as,

$$q\mathbf{E}_0 . \mathbf{v} - \frac{\mathrm{d}\mathbf{k}}{\mathrm{d}t_H} \cdot \nabla_\mathbf{k} G_0 - \left(\frac{1}{\tau} + \mathrm{j}\omega\right) G_0 = 0 \tag{9.4.74}$$

Hence

$$\frac{\mathrm{d}G_0}{\mathrm{d}t_H} + \left(\frac{1}{\tau} + \mathrm{j}\omega\right) G_0 = q\mathbf{E}_0 . \mathbf{v} \tag{9.4.75}$$

This is a first order linear differential equation with the steady state solution

$$G_0 = \exp\left\{-\left(\frac{1}{\tau} + \mathrm{j}\omega\right) t_H\right\} \int q\mathbf{E}_0 . \mathbf{v} \exp\left\{\left(\frac{1}{\tau} + \mathrm{j}\omega\right) t_H\right\} \mathrm{d}t_H \tag{9.4.76}$$

We have already noted that the motion of a representative point in **k** space is periodic in the variable t_H. Hence in the steady state, $\mathbf{v}(\mathbf{k})$ is also periodic in t_H and we may expand $\mathbf{v}(\mathbf{k})$ as a Fourier series,

$$\mathbf{v}(\mathbf{k}) = \sum_{-\infty}^{\infty} \mathbf{v}_n(\mathbf{k}) \exp(\mathrm{j}n\omega_c t_H) \tag{9.4.77}$$

This equation expresses $\mathbf{v}(\mathbf{k})$ in terms of the ideal motion described by equation (9.4.68). Consequently the components $v_x(\mathbf{k})$, $v_y(\mathbf{k})$, will have no term with $n = 0$ since their representative points simply circulate about the direction of **H**. On the other hand, $v_z(\mathbf{k})$ will in first order be unaffected by **H** and so will be dominated by the term with $n = 0$. From equation (9.4.76),

$$G_0 = \exp\left\{-\left(\frac{1}{\tau} + j\omega\right)t_H\right\} \int q\mathbf{E}_0 . \sum \mathbf{v}_n(\mathbf{k}) \exp\left[\left\{\frac{1}{\tau} + j(\omega + n\omega_c)\right\}t_H\right] dt_H \tag{9.4.78}$$

$$= q \sum_{-\infty}^{\infty} \frac{\mathbf{E}_0 . \mathbf{v}_n(\mathbf{k}) \exp(jn\omega_c t_H)}{\left[\frac{1}{\tau} + j(\omega + n\omega_c)\right]} \tag{9.4.79}$$

The current density is

$$\mathbf{J} = \frac{q}{c} \int \mathbf{v}(\mathbf{k}) f(\mathbf{k})\, g(\mathbf{k})\, d^3\mathbf{k}$$

$$= \frac{q}{c} e^{j\omega t} \int \mathbf{v}(\mathbf{k})\, G_0 \left(-\frac{\partial f^0}{\partial E}\right) g(\mathbf{k})\, d^3\mathbf{k} \tag{9.4.80}$$

that is

$$\mathbf{J} = \frac{q^2}{c} e^{j\omega t} \int \left\{\sum_{-\infty}^{\infty} \mathbf{v}_{n'} \exp(jn'\omega_c t_H)\right\}$$

$$\times \left\{\sum_{-\infty}^{\infty} \frac{\mathbf{E}_0 . \mathbf{v}_n \exp(jn\omega_c t_H)}{\left[\frac{1}{\tau} + j(\omega + n\omega_c)\right]}\right\}\left(-\frac{\partial f^0}{\partial E}\right) g(\mathbf{k})\, d^3\mathbf{k} \tag{9.4.81}$$

We select from this expression the terms which are independent of t_H, and so to this approximation, $\mathbf{J}$ is simply a harmonic time function,

$$\mathbf{J} = \mathbf{J}_0 e^{j\omega t}$$

with

$$\mathbf{J}_0 \propto \mathbf{E}_0 \frac{q^2\tau}{c} \sum \left[\frac{1}{\{1 + j(\omega + n\omega_c)\tau\}} + \frac{1}{\{1 + j(\omega - n\omega_c)\tau\}}\right]$$

$$\propto \mathbf{E}_0 \frac{q^2\tau}{c} \sum_{n=1}^{\infty} \frac{(1 + j\omega\tau)}{(1 + j\omega)^2 + n^2\omega_c^2\tau^2} \tag{9.4.82}$$

with τ an average value for the relaxation time. The summation has been restricted to the lower limit $n = 1$ in order to emphasize the usual experimental conditions for cyclotron resonance where $\mathbf{E}_0$ is perpendicular to $\mathbf{H}$ and so only the transverse components of $\mathbf{v}$ are relevant.

The power absorption due to cyclotron resonance is again obtained from

$$\langle P \rangle = \tfrac{1}{2} \operatorname{Re}[c\mathbf{J}.\mathbf{E}^*]$$

If the constant of proportionality is written as P_0,

$$\langle P \rangle = P_0 E_0^2 \left[\frac{1 + (n^2\omega_c^2 + \omega^2)\tau^2}{\{1 + (n^2\omega_c^2 - \omega^2)\tau^2\}^2 + 4\omega^2\tau^2}\right] \tag{9.4.83}$$

which is similar in form to equation (9.4.45). There will, however, be additional maxima in the power absorption at harmonics of the fundamental cyclotron frequency. These will arise when the constant energy surfaces are of low symmetry and the Fourier expansion for $\mathbf{v}(\mathbf{k})$ involves a number of terms. Recall also that ω_c in equation (9.4.83) is a function of k_z and so different parts of an energy surface may give rise to different cyclotron frequencies.

Figure 9.11 shows an experimental absorption curve which illustrates this discussion. The curve represents the differential of the power absorption as a function of magnetic field for a sample of *p*-type germanium. The differential function $\partial\langle P\rangle/\partial H$ rather than $\langle P\rangle$ was measured simply for experimental convenience. The zeros of the differential curve correspond to maxima in the power absorption arising from cyclotron resonance of the hole carriers in the germanium. It can be seen that a number of absorption lines are present. The most intense lines arise from carriers with $k_z = 0$ and refer to two different energy bands, one having an effective mass of about $0{\cdot}08\ m_e$ and known as the light hole band, the other having an effective mass of about $0{\cdot}36\ m_e$ and known as the heavy hole band. In addition to these intense

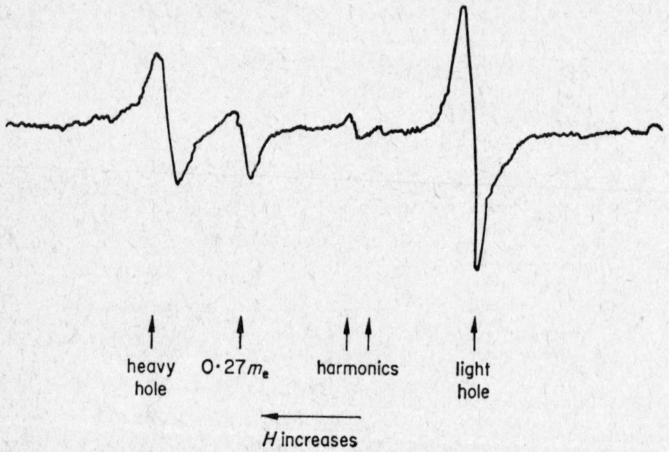

FIG. 9.11 Pen recorder trace of the differential absorption lines in high-purity *p*-germanium at 4 K and 8·8 mm wavelength with the magnetic field in the ⟨110⟩ direction.

absorption lines, the heavy hole carriers give rise to subsidiary lines at harmonics of the fundamental and to a line with $m^* \sim 0{\cdot}27\ m_e$ which is due to carriers with a non zero value for k_z. The effective masses which are measured in these experiments are directly related to the form of the constant energy surface through the term $(\partial E/\partial k_\rho)$ in equation (9.4.71). Cyclotron resonance experiments in semiconductors provide, therefore, detailed information about the energy function $E(\mathbf{k})$. In addition the widths of the absorption lines measure the relaxation time τ.

9.4.4 Anomalous Skin Effect

As an example of the technique for solving the Boltzmann transport equation when the electromagnetic field arises from a propagating wave, we shall consider the conditions which lead to the anomalous skin effect. This discussion will also indicate the conditions under which it is reasonable to take $\nabla_\mathbf{r} f = 0$ in equation (9.4.55).

We summarize the discussion of the normal skin effect given previously in Chapter 2. In the harmonic approximation the electromagnetic field equations may be written

$$\mathbf{H} = -\frac{c}{\mathrm{j}\omega\mu}\nabla\wedge\mathbf{E}; \quad \mathbf{E} = \frac{c}{\mathrm{j}\omega\varepsilon_c}\nabla\wedge\mathbf{H} \tag{9.4.84}$$

$$\nabla.\mathbf{E} = 0; \quad \nabla.\mathbf{H} = 0 \tag{9.4.85}$$

$$\varepsilon_c = \varepsilon - \mathrm{j}\frac{4\pi\sigma}{\omega} \tag{9.4.86}$$

Consider the propagation of a plane wave along the z-axis, into a semi-infinite block of metal bounded by the x, y plane. Take the x-, y-directions such that

$$\begin{aligned} \mathbf{E} &= \mathbf{n}_x E_{0x} \exp\{\mathrm{j}(\omega t - \mathbf{n}_z.\mathbf{r})\} \\ \mathbf{H} &= \mathbf{n}_y H_{0y} \exp\{(\mathrm{j}\omega t - \tilde{k}\mathbf{n}_z.\mathbf{r})\} \end{aligned} \tag{9.4.87}$$

with $\mathbf{n}_x$, $\mathbf{n}_y$, $\mathbf{n}_z$ unit vectors.
Equations (9.4.84) give,

$$\begin{aligned} \tilde{k}E_{0x}(\mathbf{n}_z \wedge \mathbf{n}_x) &= \frac{\omega\mu}{c}H_{0y}\mathbf{n}_y \\ \tilde{k}H_{0y}(\mathbf{n}_z \wedge \mathbf{n}_y) &= -\frac{\omega\varepsilon_c}{c}E_{0x}\mathbf{n}_x \end{aligned} \tag{9.4.88}$$

Hence

$$\begin{aligned} -\frac{\omega\varepsilon_c}{c}E_{0x}\mathbf{n}_z &= \tilde{k}(\mathbf{n}_z \wedge H_{0y}\mathbf{n}_y) \\ &= \tilde{k}\mathbf{n}_z \wedge (\mathbf{n}_z \wedge \mathbf{n}_x)\frac{c\tilde{k}}{\omega\mu}E_{0x} \\ &= -\frac{c\tilde{k}^2}{\omega\mu}E_{0x}\mathbf{n}_x \end{aligned} \tag{9.4.89}$$

and so

$$\tilde{k}^2 = \frac{\omega^2\varepsilon_c\mu}{c^2} \tag{9.4.90}$$

Assuming

$$\frac{4\pi\sigma}{\omega} \gg 1, \quad \varepsilon = \mu = 1$$

$$\tilde{k}^2 = -\mathrm{j}\frac{4\pi\sigma\omega}{c^2} \tag{9.4.91}$$

$$\tilde{k} = (1-\mathrm{j})\sqrt{\left(\frac{2\pi\sigma\omega}{c^2}\right)} = (1-\mathrm{j})\frac{1}{\delta} \tag{9.4.92}$$

where δ is the penetration depth. In this equation the electrical conductivity σ is a function of frequency and is to be evaluated using the Boltzmann transport equation. We seek a solution for equation (9.4.55),

$$\frac{\mathrm{d}f}{\mathrm{d}t} + \frac{\mathrm{d}\mathbf{k}}{\mathrm{d}t}\cdot\nabla_{\mathbf{k}}f + \mathbf{v}.\nabla_r f + \frac{f - f^0}{\tau} = 0$$

Since we have a wave propagating along the z-axis, we extend the relation used in equation (9.4.62) and write

$$f(\mathbf{k}, z, t) = f^0(\mathbf{k}) - G(\mathbf{k}, z, t)\frac{\partial f^0}{\partial E} \tag{9.4.93}$$

$$= f^0(\mathbf{k}) - G_0(\mathbf{k}) \exp \{\mathrm{j}(\omega t - \tilde{k}\mathbf{n}_z . \mathbf{r})\}\frac{\partial f^0}{\partial E} \tag{9.4.94}$$

In addition, we have

$$\hbar\frac{\mathrm{d}\mathbf{k}}{\mathrm{d}t} = q\mathbf{E} = \mathbf{n}_x qE_{0x} \exp \{\mathrm{j}(\omega t - \tilde{k}\mathbf{n}_z . \mathbf{r})\} \tag{9.4.95}$$

$$\nabla_{\mathbf{k}} f = \hbar\mathbf{v}\frac{\partial f^0}{\partial E} \qquad \text{in first order} \tag{9.4.96}$$

and so the Boltzmann equation may be written,

$$\left(\mathrm{j}\omega + \frac{1}{\tau}\right)G + \mathrm{j}v_z\tilde{k}G - q\mathbf{E} . \mathbf{v}(\mathbf{k}) = 0 \tag{9.4.97}$$

Cancelling throughout the exponential factor, we have

$$G_0 = \frac{q\tau\mathbf{E}_0 . \mathbf{v}(\mathbf{k})}{1 + \mathrm{j}\omega\tau + \mathrm{j}v_z\tilde{k}\tau} \tag{9.4.98}$$

Neglecting the spatial dependence of $f(\mathbf{k}, z, t)$, implies $\nabla_{\mathbf{r}} f = 0$, and is equivalent to neglecting the term $\mathrm{j}v_z\tilde{k}\tau$ in this equation. For this case

$$G_0 = \frac{q\tau\mathbf{E}_0 . \mathbf{v}(\mathbf{k})}{1 + \mathrm{j}\omega\tau} \tag{9.4.99}$$

and the current density is,

$$\mathbf{J} = \frac{q}{c}\int \mathbf{v}(\mathbf{k})\, f(\mathbf{k})\, g(\mathbf{k})\, \mathrm{d}^3\mathbf{k}$$

$$\mathbf{J} = \frac{q^2}{c} \exp \{\mathrm{j}(\omega t - \tilde{k}\mathbf{n}_z . \mathbf{r})\} \int \frac{\tau\mathbf{v}(\mathbf{k})[\mathbf{E}_0 . \mathbf{v}(\mathbf{k})]}{1 + \mathrm{j}\omega\tau}\left(-\frac{\partial f^0}{\partial E}\right) g(\mathbf{k})\, \mathrm{d}^3\mathbf{k} \tag{9.4.100}$$

Comparing this equation with equations (9.4.60) and (9.3.25b), we can see that the electrical conductivity for a simple metal with a spherical Fermi surface and an isotropic relaxation time is given by

$$\sigma(\omega) = \frac{Nq^2}{m^*}\left(\frac{\tau}{1 + \mathrm{j}\omega\tau}\right) = \frac{\sigma_0}{1 + \mathrm{j}\omega\tau} \tag{9.4.101}$$

Here σ_0 is the electrical conductivity for a time-independent electric field.

The wave vector for the electromagnetic wave is obtained from equation (9.4.92),

$$\tilde{k} = (1 - j)\frac{1}{\delta_0\sqrt{(1 + \mathrm{j}\omega\tau)}} \tag{9.4.102}$$

where $\delta_0 = \sqrt{(c^2/2\pi\sigma_0\omega)}$ is the low frequency penetration depth. This will be a satisfactory value for the wave vector provided that equation (9.4.99) is adequate, or, from equation (9.4.98), provided that

$$\mathrm{j}v_z\tilde{k}\tau \ll (1 + \mathrm{j}\omega\tau) \tag{9.4.103}$$

Write $v_z\tau = l$, so that $|l|$ is a measure of the mean free path for the carriers. Using the relation $AA^* = |A|^2$, equation (9.4.103) requires that

$$\frac{|l|}{\delta_0} \ll \frac{1}{\sqrt{2}}(1 + \omega^2\tau^2)^{3/4} \tag{9.4.104}$$

Evidently this inequality should be satisfied at low frequencies where δ_0 is large and at high frequencies where $\omega\tau$ is large.† There may, however, be an intermediate frequency range where equation (9.4.104) is not satisfied. Thus for a pure metal at very low temperatures a typical value of τ may be 10^{-11} s, with $|l| \sim 10^{-3}$ cm and $\delta_0 \sim 10^{-5}$ cm at $\omega \sim 10^{10}$ c/s. In these circumstances $|l|/\delta_0 \sim 10^2$ whilst $\omega\tau \sim 10^{-1}$. The wave vector $\tilde{k}$ will no longer be given

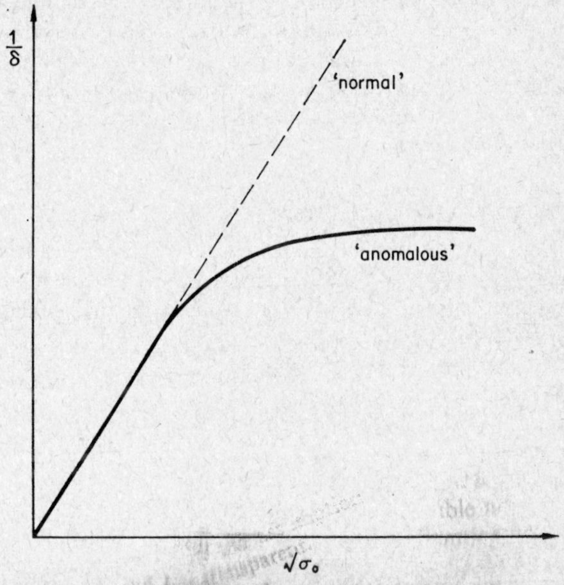

FIG. 9.12 The relationship between the reciprocal penetration depth and the square root of the d.c. conductivity for the 'normal' and 'anomalous' skin effect regimes.

by the 'normal' relation, equation (9.4.102), and correspondingly equation (9.4.92) indicates that the penetration depth will be anomalous. The metal is said to exhibit the anomalous skin effect.

The general solution of the Boltzmann transport equation for conditions applicable to the anomalous skin effect presents considerable mathematical difficulties. It is no longer valid to assume that the distribution function is determined simply by the instantaneous local value of the electric field (this assumption is implicit in equation (9.4.94) and in all our previous discussion), but instead the function should reflect the influence of the fields experienced by the carriers at previous times and previous spatial positions. The theory

† Some care is necessary at very high frequencies where the assumption that $4\pi\sigma/\omega \gg 1$ ceases to be valid; here the metal becomes transparent.

indicates that the reciprocal penetration depth, δ^{-1}, regarded as a function of the d.c. conductivity σ_0 should reach a limiting value rather than tending to infinity continuously in proportion to $\sqrt{\sigma_0}$. Thus if σ_0 is increased, by lowering the temperature at a particular frequency, the reciprocal penetration depth should approach a constant value. Figure 9.12 shows the general form of the relation between the reciprocal of the penetration depth and the square root of the d.c. conductivity. In the 'normal' regime there is a linear relation which is indicated by the broken straight line. The experimental results are generally in agreement with the theoretical predictions, but it is usually not possible to carry through a detailed numerical comparison. The quantitative predictions are dependent on the particular manner in which the carriers are reflected from the boundary surface of the metal (for example whether the reflection is specular or diffuse) and in any case refer only to the most simple spherical energy surface. In addition there are experimental difficulties. The measurements must be carried out with extremely pure, strain free, specimens otherwise the residual resistance arising from the static imperfections will mask the anomalous effect.

9.4.5 Elementary Quantum Theory

We shall illustrate the fundamental features of the quantum theory for a carrier moving in a magnetic field by considering a single charge with spherical energy surfaces. In the absence of a magnetic field, the energy surface is given for example by

$$E(\mathbf{k}) = \frac{\hbar^2|\mathbf{k}|^2}{2m^*} = \frac{\hbar^2 k^2}{2m^*} \tag{9.4.105}$$

The Hamiltonian operator in the presence of a magnetic field $\mathbf{H}$, is obtained from equation (7.4.18),†

$$\mathscr{H} = \frac{1}{2m^*}\left(\mathbf{p} - \frac{q\mathbf{A}}{c}\right)^2 \tag{9.4.106}$$

with $\mathbf{p}$ the linear momentum operator $\mathbf{p} = -\mathrm{j}\hbar\nabla$, and $\mathbf{A}$ the magnetic vector potential. Take $\mathbf{H}$ along the z coordinate axis, $\mathbf{H} = \hat{\mathbf{k}}H_z$, so that the vector potential $\mathbf{A}$ may be written with components $(-\frac{1}{2}yH_z, \frac{1}{2}xH_z, 0)$. Define, for convenience, a new operator $\mathbf{P}$ such that

$$\mathbf{P} = \left(\mathbf{p} - \frac{q\mathbf{A}}{c}\right) \tag{9.4.107}$$

In terms of this new operator the Hamiltonian is simply

$$\mathscr{H} = \frac{\mathbf{P}.\mathbf{P}}{2m^*} \tag{9.4.108}$$

† We shall neglect spin throughout this discussion. This is consistent with the previous discussion of the classical theory and is justified in the quantum theory because, in first order, the spin direction does not change in a transition between two energy states.

The commutation rules for **P** are derived from the definition of equation (9.4.107). We know that

$$[p_x p_y - p_y p_x] = [p_x, p_y] = 0 = [p_x, p_z] = [p_y, p_z]$$
$$[x, p_x] = \mathrm{j}\hbar, \quad [x, p_y] = 0, \text{ etc.} \tag{9.4.109}$$

hence,

$$[P_x, P_y] = \mathrm{j}\hbar\frac{qH_z}{c}; \quad [P_x, P_z] = [P_y, P_z] = 0 \tag{9.4.110}$$

The energy eigenvalues may be obtained by writing the Hamiltonian operator from equation (9.4.108) as,

$$\mathscr{H} = \frac{1}{2m^*}(P_x^2 + P_y^2 + P_z^2)$$
$$= \frac{1}{2m^*}(P_x^2 + P_y^2) + \frac{1}{2m^*}P_z^2 \tag{9.4.111}$$
$$= \mathscr{H}_{\mathrm{t}} + \mathscr{H}_z \tag{9.4.112}$$

From the definition of **P** the 'transverse' Hamiltonian operator, $\mathscr{H}_{\mathrm{t}}$, is a function only of x, y, whereas the 'longitudinal' Hamiltonian operator, $\mathscr{H}_z$, is a function only of z. We may therefore seek eigenfunctions in the form

$$\psi(\mathbf{r}) = \psi(x, y)\,\psi(z) \tag{9.4.113}$$

and so

$$\psi(z)\,\mathscr{H}_{\mathrm{t}}\,\psi(x, y) + \psi(x, y)\,\mathscr{H}_z\,\psi(z) = E\,\psi(x, y)\,\psi(z)$$

with eigenvalues

$$E = E_{\mathrm{t}} + E_z \tag{9.4.114}$$

Evidently,

$$-\frac{\hbar^2}{2m^*}\frac{\partial^2\psi(z)}{\partial z^2} = E_z\,\psi(z)$$

$$\psi(z) \sim \exp(\mathrm{j}k_z z); \qquad E_z = \frac{\hbar^2 k_z^2}{2m^*} \tag{9.4.115}$$

The motion in the z-direction is therefore the same as that for a free particle. It is not influenced by the presence of the magnetic field and there is a continuous set of eigenvalues. To determine the energy arising from the motion in the plane perpendicular to **H**, combine the operators P_x, P_y, to form,

$$P_+ = P_x + \mathrm{j}P_y; \quad P_- = P_x - \mathrm{j}P_y \tag{9.4.116}$$

The commutation rule is obtained from equations (9.4.110),

$$[P_+, P_-] = 2m^*\hbar\omega_0 \tag{9.4.117}$$

with $\omega_0 = qH_z/m^*c$. The Hamiltonian operator $\mathscr{H}_{\mathrm{t}}$, may now be written,

$$\mathscr{H}_{\mathrm{t}} = \frac{1}{2m^*}(P_x^2 + P_y^2) = \frac{1}{2m^*}\{P_+P_- - \mathrm{j}[P_x, P_y]\}$$

$$\mathscr{H}_{\mathrm{t}} = \frac{1}{2m^*}P_+P_- + \tfrac{1}{2}\hbar\omega_0 \tag{9.4.118}$$

This Hamiltonian closely resembles that discussed previously in connection with spin waves (compare Section 8.5). We therefore define a ground state wave function $|0\rangle$ by the relation

$$P_- |0\rangle = 0 \tag{9.4.119}$$

and obtain a set of eigenfunctions describing the excited states

$$\psi_t(n) = (P_+)^n |0\rangle \tag{9.4.120}$$

Evidently

$$\begin{aligned} P_+ P_- \{P_+ |0\rangle\} &= P_+ \{P_- P_+ |0\rangle\} \\ &= 2m^* \hbar\omega_0 P_+ |0\rangle \end{aligned} \tag{9.4.121}$$

consequently

$$\mathscr{H}_t \{P_+ |0\rangle\} = (1 + \tfrac{1}{2})\hbar\omega_0 P_+ |0\rangle \tag{9.4.122}$$

and

$$\mathscr{H}_t \{(P_+)^n |0\rangle\} = (n + \tfrac{1}{2})\hbar\omega_0 (P_+)^n |0\rangle \tag{9.4.123}$$

The transverse motion is therefore characterized by a set of discrete energy eigenvalues,

$$E_t = (n + \tfrac{1}{2})\hbar\omega_0 \tag{9.4.124}$$

The total energy is obtained from equations (9.4.114), (9.4.115),

$$E = (n + \tfrac{1}{2})\hbar\omega_0 + \frac{\hbar^2 k_z^2}{2m^*} \tag{9.4.125}$$

When considering the interaction of a charge carrier with an electromagnetic field, it is usually assumed that an induced energy transition takes place with no change in k_z. This is because the momentum (h/λ) associated with the electromagnetic field is negligible in comparison with that of the carrier for most experimental conditions. For this case, $\Delta k_z = 0$, and the transitions between adjacent energy levels correspond to

$$\Delta E(n, n + 1) = \hbar\omega_0 \tag{9.4.126}$$

These transitions give rise to the cyclotron resonance absorption discussed previously.

The discrete energy eigenvalues associated with the transverse Hamiltonian give rise to quantized orbital motion in a plane perpendicular to the magnetic field. Consider the time rate of change of the dynamical variable corresponding to the operator **P**. This quantity may be derived from relations of the form†

$$\frac{dP_x}{dt} = \frac{j}{\hbar}[\mathscr{H} P_x - P_x \mathscr{H}] \tag{9.4.127}$$

where dP_x/dt is written for the expectation value $\langle n| dP_x/dt |n\rangle$ and similarly $[\mathscr{H} P_x - P_x \mathscr{H}]$ refers to $\langle n| [\mathscr{H} P_x - P_x \mathscr{H} |n\rangle$. We obtain

$$dP_x/dt = \omega_0 P_y; \quad dP_y/dt = -\omega_0 P_x; \quad dP_z/dt = 0 \tag{9.4.128}$$

The corresponding equations for the spatial coordinates are,

$$\frac{d}{dt}x = \frac{1}{m^*}P_x; \quad \frac{d}{dt}y = \frac{1}{m^*}P_y; \quad \frac{d}{dt}z = \frac{1}{m^*}P_z \tag{9.4.129}$$

† See, for example, SCHIFF, L. I., *Quantum Mechanics*, McGraw-Hill, 1949.

Combine equations (9.4.128) and (9.4.129),

$$\frac{d}{dt}\left(x + \frac{1}{m^*\omega_0}P_y\right) = 0$$
$$\frac{d}{dt}\left(y - \frac{1}{m^*\omega_0}P_x\right) = 0 \qquad (9.4.130)$$

and so we may write,

$$x + \frac{1}{m^*\omega_0}P_y = \text{constant} = x_0$$
$$y - \frac{1}{m^*\omega_0}P_x = \text{constant} = y_0 \qquad (9.4.131)$$

If we identify the coordinates for the centre of the orbit with x_0, y_0, the radius of an orbit is given by

$$r_n^2 = (x - x_0)^2 + (y - y_0)^2 \qquad (9.4.132)$$
$$= \frac{1}{(m^*\omega_0)^2}\{P_x^2 + P_y^2\}$$
$$= \frac{2E_t}{m^*\omega_0^2} \qquad (9.4.133)$$

This equation is exactly the same as that which is obtained for a circular orbit on classical theory. The energy E_t is quantized, however, and so,

$$r_n^2 = (n + \tfrac{1}{2})\frac{2\hbar}{m^*\omega_0} \qquad (9.4.134)$$

The orbits in momentum space are also quantized. Consider first a particular spherical energy surface in the absence of a magnetic field,

$$E(\mathbf{k}) = \frac{\hbar^2}{2m^*}(k_x^2 + k_y^2) + \frac{\hbar^2}{2m^*}k_z^2$$
$$= E_t + \frac{\hbar^2}{2m^*}k_z^2 \qquad (9.4.135)$$

The allowed states are distributed continuously over the whole of the energy surface. All values of k_z up to the maximum value $(k_z)_{\max} = \sqrt{\{(2m^*E)/\hbar^2\}}$ are allowed since E_t is a continous function. In the presence of a magnetic field, however, the transverse energy E_t is quantized. Consider now a spherical constant energy surface in **P** space.

$$E(\mathbf{P}) = \frac{1}{2m^*}(P_x^2 + P_y^2) + \frac{1}{2m^*}P_z^2$$
$$= (n + \tfrac{1}{2})\hbar\omega_0 + \frac{1}{2m^*}P_z^2 \qquad (9.4.136)$$

There will only be allowed states of those values of P_z (note $P_z = \hbar k_z$) on the constant energy surface for which $E_t = (n + \tfrac{1}{2})\hbar\omega_0$. Thus the allowed orbits in **P** space occur at values of P_z for which

$$\frac{1}{2m^*}P_z^2 = E - (n + \tfrac{1}{2})\hbar\omega_0 \qquad (9.4.137)$$

In particular, there will be an allowed orbit at $P_z = 0$, only if

$$E = (n + \tfrac{1}{2})\hbar\omega_0 \tag{9.4.138}$$

This relation has an important application to metals and degenerate semiconductors. For a simple material there will be an orbit at $P_z = 0$ if equation (9.4.138) is satisfied with $E = E_F$. As the magnetic field increases we can imagine an orbit of given quantum number n moving outwards at $P_z = 0$ and passing through the surface of the Fermi sphere when

$$E_F = (n + \tfrac{1}{2})\frac{\hbar q H_z}{m^* c} \tag{9.4.139}$$

Successive orbits will pass through the Fermi surface at $P_z = 0$ with a periodicity in H_z^{-1} given by

$$\Delta\left(\frac{1}{H_z}\right) = \frac{\hbar q}{m^* c}\left(\frac{1}{E_F}\right) \tag{9.4.140}$$

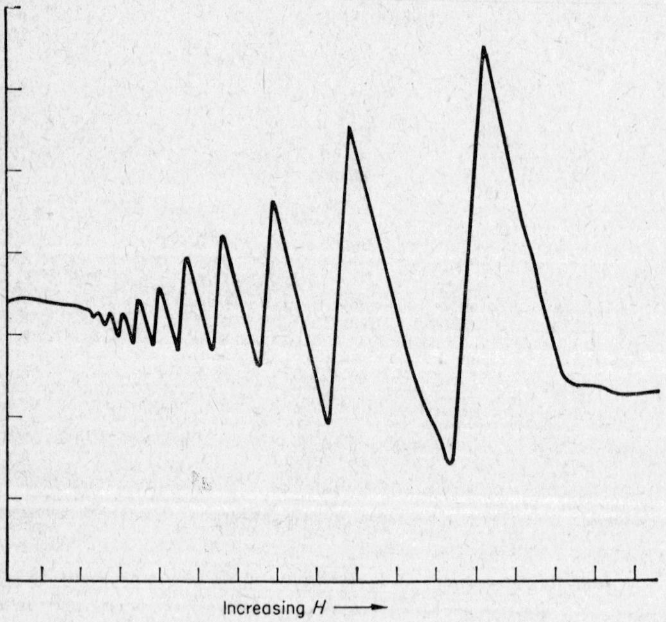

FIG. 9.13 De Haas–Shubnikov oscillations in bismuth. This record was taken by undergraduates, as part of the practical course in Physics at Oxford University, using apparatus designed by Dr R. A. Stradling.

This gives rise to a periodic modulation of the magnetic susceptibility and also the electrical conductivity for the material. In the case of the magnetic susceptibility, it is known as the de Haas-van Alphen effect, and in the case of the electrical conductivity, it is known as the De Haas-Shubnikov effect (see Fig. 9.13).

Bibliography

BLATT, F. J., 'Theory of Mobility of Electrons in Solids', in *Solid State Physics* Vol. 4, Eds. Seitz and Turnbull, Academic Press, 1957

BROOKS, H., 'Theory of the Electrical Properties of Germanium and Silicon', in *Advances in Electronics and Electron Physics*, Ed. L. Marton, Academic Press, 1955

CALLAWAY, J., *Energy Band Theory*, Academic Press, 1964

Handbuch der Physik, Vol. 19, Ed. S. Flugge, Springer-Verlag, 1956

LONG, D., *Energy Bands in Semiconductors*, Wiley, 1968

SHOCKLEY, W., *Electrons and Holes in Semiconductors*, Van Nostrand, 1956

ZIMAN, J. M., *Principles of the Theory of Solids*, C.U.P., 1969

Index